Advances in Intelligent Systems and Computing

Volume 236

Series editor

Janusz Kacprzyk, Warsaw, Poland

For further volumes:
http://www.springer.com/series/11156

About this Series

The series "Advances in Intelligent Systems and Computing" contains publications on theory, applications, and design methods of Intelligent Systems and Intelligent Computing. Virtually all disciplines such as engineering, natural sciences, computer and information science, ICT, economics, business, e-commerce, environment, healthcare, life science are covered. The list of topics spans all the areas of modern intelligent systems and computing.

The publications within "Advances in Intelligent Systems and Computing" are primarily textbooks and proceedings of important conferences, symposia and congresses. They cover significant recent developments in the field, both of a foundational and applicable character. An important characteristic feature of the series is the short publication time and world-wide distribution. This permits a rapid and broad dissemination of research results.

B. V. Babu · Atulya Nagar
Kusum Deep · Millie Pant
Jagdish Chand Bansal
Kanad Ray · Umesh Gupta
Editors

Proceedings of the Second International Conference on Soft Computing for Problem Solving (SocProS 2012), December 28–30, 2012

Volume I

Springer

Editors
B. V. Babu
Institute of Engineering and Technology
JK Lakshmipat University
Jaipur
Rajasthan
India

Atulya Nagar
Department of Computer Science
Liverpool Hope University
Liverpool
UK

Kusum Deep
Department of Mathematics
Indian Institute of Technology Roorkee
Roorkee
Uttaranchal
India

Millie Pant
Department of Paper Technology
Indian Institute of Technology Roorkee
Roorkee
India

Jagdish Chand Bansal
Department of Applied Mathematics
South Asian University
New Delhi
India

Kanad Ray
Umesh Gupta
Institute of Engineering and Technology
JK Lakshmipat University
Jaipur
Rajasthan
India

ISSN 2194-5357 ISSN 2194-5365 (electronic)
ISBN 978-81-322-1601-8 ISBN 978-81-322-1602-5 (eBook)
DOI 10.1007/978-81-322-1602-5
Springer New Delhi Heidelberg New York Dordrecht London

Library of Congress Control Number: 2013955048

Printed on acid-free paper

Springer is part of Springer Science+Business Media (www.springer.com)

Foreword

I am delighted that the Second International Conference on “Soft Computing for Problem Solving (SocProS 2012)” was organized by the Institute of Engineering and Technology of our University from December 28–30, 2012.

l. Zadeh, founder of soft computing had stated that soft computing differs from conventional (hard) computing in that, unlike hard computing, it is tolerant of imprecision, uncertainty, partial truth and approximation. In effect, the role model for soft computing is the human mind. The guiding principle of soft computing being: Exploit the tolerance for imprecision, uncertainty, partial truth and approximation to achieve tractability, robustness and low solution cost. Every single one of these methods can be called soft computing method.

The soft computing models are based on human reasoning and are closer to human thinking. Soft computing is not a melange; it can be viewed as a founding component for the emerging field of conceptual intelligence. It is the fusion of methodologies assigned to model and enable solutions to real-world problems which are not modelled or too difficult to model mathematically. The core of soft computing comprises neural networks, fuzzy logic and genetic computing. In the coming years, soft computing will play a more important role in many areas, including software engineering. Soft computing is still growing. It is still somewhat in the formative stage and the definite components that comprise soft computing have not yet been designed. More new sciences are still merging into soft computing. SocProS 2012 was aimed to reach out to such development.

I am happy to know that Springer is publishing the Proceedings of this conference as AISC book series. I am confident that the readers will be benefited by the research inputs provided by delegates during the conference. This book will certainly provide an enriching and rewarding experience to all for academic networking.

I congratulate the editorial team and Springer for meticulously working on all the details of the book. I also congratulate all the authors of research papers for their contribution to SorProS 2012.

Jaipur, September 27, 2013

Upinder Dhar

About JK Lakshmipat University

Set up in the pink city of Jaipur, India and spread over a sprawling 30 acre campus, JK Lakshmipat University (JKLU) offers state-of-the-art academic infrastructure and world class faculty for conducting Graduate, Post Graduate and Ph.D. programmes in the mainstream disciplines of Management and Engineering. Promoted by JK Organisation, one of India's leading industrial conglomerates with a rich heritage of over 100 years, the University offers students an open, green and high-tech learning environment, combining the serene settings of the Gurukuls of yesteryears with the technological advancements of the new age.

The Institute of Management offers MBA (Full-Time Residential), MBA (Family Business and Entrepreneurship), MEA (Master of Educational Administration), a 5-year integrated dual degree (BBA + MBA) and Ph.D. programme in Management, besides PG Diploma in Tourism Administration and PG Diploma in Family Business & Entrepreneurship. Specializations include Finance and Accounting, Marketing, HRM, International Business and Information Technology. In order to balance conceptual framework with the industry practices, the delivery of each course is done in close consultation with the corporate world.

The University has also set up a 'Management Development Centre' for practicing Managers aiming to build capabilities through short-term programmes. It also organizes faculty development programmes for the benefit of academia.

The Institute of Engineering and Technology offers 4 year B.Tech, 2 year M.Tech, 5 year integrated dual degree (B.Tech + MBA) and (B.Tech + M.Tech), and Ph.D. programme in Chemical Engineering, Civil Engineering, Computer Science Engineering, Electrical Engineering, Electronics and Communications Engineering, Information Technology and Mechanical Engineering.

The curriculum is designed to enrich the students with the knowledge and relevant skills to prepare them not only to face the contemporary world but also to make them future ready to effectively perform in leadership roles assigned to them. The curriculum is updated to integrate changes that are taking place in the business environment.

JKLU is visualized to emerge as a premier institution of higher learning with global standards; it has tied up with a few renowned universities, such as University of Houston (USA), Hanyang University (South Korea), St. Cloud State University (USA), University of Wales (UK) and Szechenyi Istvan University (Hungary) for cooperation in the field of Faculty Development, Students Exchange and Research. The University has also signed an MoU with IBM India Ltd. for establishing a 'Centre of Technology Excellence' for undertaking technology development projects involving faculty members and students.

JKLU has been set up under Rajasthan Private Universities Act by 'Lakshmipat Singhania Foundation for Higher Learning'.

Preface

Unlike hard computing, the guiding principle of soft computing is exploiting the tolerance for imprecision, uncertainty, partial truth, and approximation to achieve tractability, robustness, and low solution cost, the role model being the human mind. The areas that come under the purview of Soft Computing include Fuzzy Logic, Neural Computing, Evolutionary Computation, Machine Learning and Probabilistic Reasoning, with the latter subsuming belief networks, chaos theory and parts of learning theory. The successful applications of soft computing and its rapid growth suggest that the impact of soft computing will be felt increasingly in the coming years. Soft computing is likely to play a very important role in science and engineering, but eventually its influence may extend much farther.

After the successful completion of the First International Conference on Soft Computing for Problem Solving in 2011 (SocProS 2011) at IIT Roorkee, the 2nd of the series, Second International Conference on Soft Computing for Problem Solving (SocProS 2012), was held in JK Lakshmipat University (JKLU), Jaipur, during December 28–30, 2012. It had been a matter of great privilege and pleasure to organize SocProS 2012 at JKLU, Jaipur. The response had been overwhelming and heartwarming. We had received 312 paper submissions of which 200, i.e., 64 % had been accepted for presentation. The review process included double review of submitted papers by an International team of reviewers using Easy Chair as online conference management software. There were 20 Technical Sessions including two Special Sessions, besides five Keynote Addresses by eminent Academicians and Scientists from different parts of the world that made SocProS 2012 an enriching experience to all the participants. In addition, there was one Workshop on Game Programming and one Special Tutorial on Brain like computing as a part of this conference. There were 144 Programme Committee members and 20 International Advisory Committee members from across the world associated with SocProS 2012.

We would like to express our sincere gratitude to Dr. Madhukar Gupta, Divisional Commissioner, Government of Rajasthan for gracing the occasion as the Chief Guest for the Inaugural Session.

We would also like to extend our heartfelt gratitude to all Programme Committee and International Advisory Committee members. We sincerely thank the Keynote Speakers Prof. Dipankar Dasgupta from University of Memphis—USA,

Prof. Anirban Bandyopadhyay from National Institute of Material Sciences (NIMS)—Japan, Prof. Ravindra Gudi from IIT Bombay—Mumbai, Prof. D. Nagesh Kumar from IISc—Bangalore and Prof. Ajoy Ray from Bengal Engineering and Science University (BESU)—Howrah.

We are grateful to Prof. Anirban Bandyopadhyay for conducting a Special Tutorial also in addition to delivering Keynote Address. We thank Mr. Arijit Bhattacharyya from Virtual Infocom—Kolkata for conducting the Workshop.

We thank the invited guests for accepting our invitation and also for chairing the technical sessions. We thank Dr. Kannan Govindan and Dr. P. C. Jha for delivering the Invited Talks. We express sincere thanks to the entire national and local organizing committee members for their continuous support and relentless cooperation right from the conception to execution in making SocProS 2012 a memorable event. We thank all the participants who had presented their research papers and attended the conference. A special mention of thanks is due to our student volunteers for the spirit and enthusiasm they had shown throughout the duration of the event, without which it would had been difficult for us to organize such a successful event.

Thanks are due to Springer for Publishing the Conference Proceedings. We hope that the Proceedings will prove helpful towards understanding about Soft Computing in teaching as well as in their research and will inspire more and more researchers to work in this interesting, challenging and ever growing field of Soft Computing.

We are thankful to the JKLU family for the support given in making this mega event successful. We sincerely hope that the delegates had certainly taken home several pleasant memories of SocProS 2012 with them.

We are honoured and it has been a proud privilege to be associated with this Second International Conference on SocProS 2012 as its General Chairs.

Dr. B. V. Babu
Dr. Atulya Nagar
Dr. Kusum Deep
Dr. Millie Pant
Dr. Jagdish Chand Bansal
Dr. Kanad Ray
Dr. Umesh Gupta

Committees

Organizing Committee

Patron	Dr. Upinder Dhar, JK Lakshmipat University, Jaipur, India
General Chairs	Dr. B. V. Babu, JK Lakshmipat University, Jaipur, India Dr. Atulya Nagar, Liverpool Hope University (LHU), Liverpool, UK
	Dr. Kusum Deep, IIT Roorkee, Roorkee, India
Programme Committee Chairs	Dr. Millie Pant, IIT Roorkee, Roorkee, India
	Dr. Jagdish C. Bansal, South Asian University, New Delhi, India
Local Organizing Chairs	Dr. Kanad Ray, JK Lakshmipat University, Jaipur, India
	Dr. Umesh Gupta, JK Lakshmipat University, Jaipur, India
Special Session Chairs	Dr. Millie Pant, IIT Roorkee, Roorkee, India
	Dr. Jagdish C. Bansal, ABV-IIITM, Gwalior, India
	Mr. Alok Agarwal, JK Lakshmipat University, Jaipur, India
Publicity Chairs	Dr. Sandeep Kumar Tomar, JK Lakshmipat University, Jaipur, India
	Dr. Sonal Jain, JK Lakshmipat University, Jaipur, India
Best Paper Chairs	Dr. Ravidra Gudi, IIT Bombay, Mumbai, India
	Dr. D. Nagesh Kumar, IISc, Banglore, India
	Dr. P. C. Jha, Delhi University, New Delhi, India
Steering Committee	Dr. Atulya Nagar
	Dr. B. V. Babu
	Dr. Kusum Deep
	Dr. Millie Pant
	Dr. Jagdish C. Bansal

International Advisory Committee

Prof. Zbigniew Michalewicz	University of Adelaide, Adelaide, Australia
Dr. Gurvinder S. Baicher	University of Wales, UK
Prof. Lipo Wang	Nanyang Technological University, Singapore
Prof. Patrick Siarry,	Université de Paris, France
Prof. Michael N. Vrahatis	University of Patras, Greece
Prof. Helio J. C. Barbosa	Federal University of Juiz de Fora, Brazil
Prof. S. K. Singh	CSIR—Central Building Research Institute, Roorkee, India
Prof. Wei-Chiang Samuelson Hong	Oriental Institute of Technology, Taiwan
Prof. P. K. Kapur	Amity University, Noida, India
Prof. Samrat Sabat	University of Hyderabad, India
Prof. S. S. Rao	University of Miami, Florida, USA
Prof. Pramod Kumar Singh	ABV-IIITM, Gwalior, India
Prof. Montaz Ali	Witwatersrand University, Johannesburg, South Africa
Prof. Andries P. Engelbrecht	University of Pretoria, South Africa
Prof. Suresh Chandra	Indian Institute of Technology New Delhi, New Delhi, India
Prof. Kalyanmoy Deb	Indian Institute of Technology Kanpur, Kanpur, India
Prof. D. Nagesh Kumar	Indian Institute of Science, Bangalore, India
Prof. Nirupam Chakraborti	Indian Institute of Technology Kharagpur, Kharagpur, India
Prof. K. C. Tan	National University of Singapore, Singapore
Prof. Roman R. Poznanski	Universiti Tunku Abdul Rahman (UTAR), Malaysia

Programme Committee

Dr. Abhijit Sanyal
Dr. Abhay Jha
Dr. Abhijit Sarkar
Dr. Abhinay Pandya
Dr. Adel Aljumaily
Dr. K. K. Shukla
Dr. K. P. Singh
Dr. Kusum Deep
Dr. Lalit Awasthi
Dr. Laxman Tawade

Dr. Aitorrodriguez Alsina
Dr. A. J. Umbarkar
Dr. Akila Muthuramalingam
Dr. Amit Pandit
Dr. Andres Muñoz
Dr. Anil Parihar
Dr. Antonio Jara
Dr. Anupam Singh
Dr. Anuradha Fukane
Dr. Aram Soroushian
Dr. Arnab Nandi
Dr. Arshin Rezazadeh
Dr. Ashish Gujarathi
Dr. Ashraf Darwish
Dr. Ashwani Kush
Dr. AsokeNath
Dr. Atulya Nagar
Dr. Ayoub Khan
Dr. B. V. Babu
Dr. Balaji Venkatraman
Dr. Balakrishna Maddali
Dr. Banani Basu
Dr. Bharanidharan Shanmugam
Dr. Bratin Ghosh
Dr. Carlos Fernandezllatas
Dr. Chu-Hsing Lin
Dr. Ciprian Dobre
Dr. D. Nagesh Kumar
Dr. Dante Tapia
Dr. Dipika Joshi
Dr. Dipti Gupta
Dr. Eduard Babulak
Dr. Farhad Nematy
Dr. Francesco Marcelloni
Dr. G. Shivaprasad
Dr. G. R. S. Murthy
Dr. Gauri S. Mittal
Dr. Gendelman Oleg
Dr. Gurvinder Singh-Baicher
Dr. Leonard Barolli
Dr. Manjaree Pandit
Dr. Manoj Saxena
Dr. Manoj Thakur
Dr. Mansaf Alam
Dr. Manu Augustine
Dr. Mario Koeppen
Dr. Mehul Raval
Dr. Millie Pant
Dr. Mohammad Ahoque
Dr. Mohammad Reza Nouri Rad
Dr. Mohammed Abdulqadeer
Dr. Mohammed Rokibul Alam Kotwal
Dr. Mohdabdul Hameed
Dr. Mourad Abbas
Dr. Mrutyunjaya Panda
Dr. Munawar A. Shaik
Dr. Musrrat Ali
Dr. Ninansajeeth Philip
Dr. Nitin Merh
Dr. O. P. Vyas
Dr. Philip Moore
Dr. Poonam Sharma
Dr. Pramod Kumar Singh
Dr. Punam Bedi
Dr. Pushpinder Patheja
Dr. Radha Thangaraj
Dr. Rajesh Sanghvi
Dr. Ram Ratan
Dr. Ramesh Babu
Dr. K. C. Raveendranathan
Dr. Ravi Sankar Vadali
Dr. Razibhayat Khan
Dr. Ritu Agarwal
Dr. Rodger Carroll
Dr. Sami Habib
Dr. Sandeep Kumar Tomar
Dr. Sanjeev Singh
Dr. Shaojing Fu

Dr. Hadushmebrahtu Adane
Dr. Hariganesh Shanmugam
Dr. Harris Michail
Dr. Hemant Mehta
Dr. Hitesh Shah
Dr. Hugo Proença
Dr. Ivica Boticki
Dr. Jagdish Chand Bansal
Dr. Javier Bajo
Dr. Jayant M. Modak
Dr. Jerlang Hong
Dr. Joanna Kolodziej
Dr. Jose Molina
Dr. J. T. Pal Singh
Dr. Juan Mauricio
Dr. K. Mustafa
Dr. Kadian Davis
Dr. Kamal Kant
Dr. Kanad Ray
Dr. Kannammal Sampath Kumar
Dr. Karthik Sindhya
Dr. Katheej Parveen
Dr. Kavita Burse
Dr. Kazumi Nakamatsu
Dr. Kedarnath Das
Dr. Khaled Abdullah
Dr. Khushboo Hemnani
Dr. Kishanrao Kalitkar
Dr. Shashi Barak
Dr. Shashibhushan Kotwal
Dr. Shirshu Varma
Dr. Shyam Lal
Dr. S. M. Sameer
Dr. Sonal Jain
Dr. Sotirios Ziavras
Dr. Soumyabrata Saha
Dr. Sucheta Datt
Dr. Sudeepto Bhattacharya
Dr. Sudhir Warier
Dr. Sumithra Devika
Dr. Sunilkumar Jha
Dr. Suparna Dasgupta
Dr. Suresh Sankaranarayanan
Dr. Surya Prakash
Dr. Sushil Kulkarni
Dr. Thanga Raj
Dr. Trilochan Panigrahi
Dr. Tzungpei Hong
Dr. Umesh Chandra Pati
Dr. Umesh Gupta
Dr. Uzay Kaymak
Dr. Vidya Dhamdhere
Dr. Vivek Tiwari
Dr. Wei-Chiang Samuelson Hong
Dr. Yoseba Penya
Dr. Yusuke Nojima

Contents

Part I
Genetic Algorithm for Problem Solving (GAPS)

Insulin Chart Prediction for Diabetic Patients Using Hidden Markov Model (HMM) and Simulated Annealing Method

Ravindra Nath and Renu Jain

Abstract Most of the diabetic patients need to take insulin before every meal. The doctors have to decide insulin doses for every patient according to the patient's previous records of doses and sugar levels measured at regular intervals. This paper proposes a hidden Markov model to predict the insulin chart for a patient and uses simulated annealing search algorithm to efficiently implement the model. The one-month chart maintained by the patient has been used to train the model, and the prediction for next fifteen days is done on the basis of the trained data. We discussed the results with the university medical doctor; he was very pleased to see to the result obtained.

Keywords Hidden Markov model (HMM) · Randomized algorithm (RA) · Simulated annealing (SA) · Diabetic patient chart prediction (DPCP)

1 Introduction

Hidden Markov model has various applications in the area of speech recognition, bioinformatics (DNA sequences and gene recognitions) [1–4], climatology, acoustics [1, 5], etc. In addition to this, HMM has been applied for prediction problems like stock marketing [6–8] and forecasting. Mostly researches have to work with the third problem of HMM for training and prediction of the data. Out of many probabilistic models, HMM is most popular due to its mathematical foundation of model.

R. Nath · R. Jain (✉)
Department of Computer Science and Engineering, University Institute of Engineering and Technology, Chattrapati Shahuji Maharaj University, Kanpur, Uttar Pradesh 208024, India
e-mail: rnkatiyar@gmail.com

R. Jain
e-mail: jainrenu@gmail.com

B. V. Babu et al. (eds.), *Proceedings of the Second International Conference on Soft Computing for Problem Solving (SocProS 2012), December 28–30, 2012*, Advances in Intelligent Systems and Computing 236, DOI: 10.1007/978-81-322-1602-5_1,

In this paper, we have taken the application of medical science, i.e., preparation of insulin chart for diabetic patients. We have used HMM to predict the insulin chart taking the data of a diabetic patient. We have used simulated annealing as a randomized algorithm. Training is done taking one-month chart maintained by the patient.

Simulated annealing [9–11] is a randomized search method that can perform global search within the defined searching space giving local maxima or global maxima. In our previous paper [12, 13], we modeled HMM learning problem as a discrete search problem and solved that discrete problem using different randomized search algorithms. In this paper, insulin chart prediction is modeled as HMM, and HMM learning problem is solved using simulated annealing algorithm. Experimental results show that SA evaluates HMM parameters quite fast and accurately on the basis of previous data giving good prediction results.

The organization of the paper is as follows. Section 2 briefly describes the data set taken. Section 3 explains HMM, and Sect. 4 explains in detail the HMM used and results obtained. Section 5 is about results and discussions.

2 Data Set Information

The data set used in this paper was taken from http://archive.ics.uci.edu/ml/ called UCI Repository. Diabetics patient records can be obtained from two sources: an automatic electronic recording device and paper records. The automatic device has an internal clock to timestamp events, whereas paper records provide 'logical time' slots (breakfast, lunch, dinner, bedtime). Diabetic files consist of four fields per record: (1) date in MM-DD-YYYY format, (2) time in XX:YY format, (3) code, and (4) value.

The code field is deciphered as follows: 33 = Regular insulin dose, 34 = NPH insulin dose, 35 = Ultralente insulin dose, 48 = Unspecified blood glucose measurement, 57 = Unspecified blood glucose measurement, etc.

We have taken two-month data (insulin chart) for code 33, i.e., regular insulin dose of a patient. Assuming first-month data as training data, next one-month chart is predicted and compared with the actual data.

3 The Hidden Markov Model

HMM is a probabilistic model useful for finite-state stochastic sequence structures. Stochastic sequences are called observation sequences, i.e., $O = O_1\ O_2 \ldots, O_T$, where T is the length of the observed sequence. HMM with N states ($S_1, S_2, \ldots S_N$) can be characterized by a set of parameters (A, B, π) is called the model of HMM $\lambda = (A, B, \pi)$

In order to characterize an HMM completely, following elements are needed [8–10, 14, 15]

N: The number of states in the model

M: The number of distinct observation symbols M per state

A: The state transition probability distribution

$$A = \{a_{ij}\}, a_{ij} = p(q_t = s_j | q_{t-1} = S_i)$$

$$\sum_{i=1}^{j} a_{ij} = 1 \text{ where } a_{ij} \geq 0, 1 \leq i \leq N \text{ and } 1 \leq j \leq N$$

B: The observation symbol probability distribution in state j

$$B = b_j(k) = P(V_k \; at \; t | q_t = S_j)$$

$$\sum_{k=1}^{M} b_j(k) = 1 \quad b_j(k) \geq 0$$

π: The initial-state distribution $\pi_i = P(q_1 = S_i)$

The three main problems of HMM are as follows: evaluation problem, decoding, problem, and learning problem.

(1) HMM evaluation problem: Compute $P(O|\lambda)$, the probability of the observation sequence $O = O_1 \, O_2 \, O_3 \ldots O_T$, given the model $\lambda = (A, \; B, \; \pi)$.
(2) HMM decoding problem: Uncover the hidden part of the model, i.e., find the optimal state sequence, for the given observation sequences $O = O_1 O_2 \, O_3 \ldots O_T$, given the model $\lambda = (A, \; B, \; \pi)$.
(3) HMM learning problem: Model parameters (A, B, π) are adjusted such that $P(O|\lambda)$ is maximized.

In this paper, we have considered the third problem of HMM, i.e., the learning problem or training problem and tried to solve it.

3.1 Hidden Markov Model for a Diabetic Patient

To completely define a problem of HMM, we need to define three probabilities: state transition probabilities, observation symbols probabilities, and initial-state probabilities. It was observed that a patient takes the medicine at breakfast (08:00), lunch

Table 1 Actual input of doses for one month

Date	Code	Breakfast	Lunch	Dinner	Bedtime
21-Apr-91	33	9	0	7	0
22-Apr-91	33	10	2	7	0
23-Apr-91	33	11	0	7	0
24-Apr-91	33	10	4	0	5
25-Apr-91	33	9	4	7	2
26-Apr-91	33	9	5	7	0
27-Apr-91	33	10	0	8	0
28-Apr-91	33	10	0	7	0
29-Apr-91	33	9	5	8	0
30-Apr-91	33	10	4	7	0
1-May-91	33	10	4	7	0
2-May-91	33	10	5	7	0
3-May-91	33	10	5	7	0
4-May-91	33	10	5	7	0
5-May-91	33	10	5	7	2
6-May-91	33	10	5	7	0
7-May-91	33	10	5	7	0
8-May-91	33	10	5	7	0
9-May-91	33	9	4	7	0
.	.	.	.	.	.
.	.	.	.	.	.

(12:00), dinner (18:00), and bedtime (22:00) or morning, noon, evening, and night, and the amount of insulin (code=33) varies from value 2 to 11. The data set contains 4 values per day corresponding to 4 slots, i.e., breakfast, lunch, dinner, and bedtime. We have taken these four slots as four states where S_1 corresponds to breakfast, S_2 corresponds to lunch, S_3 corresponds to dinner, and S_4 corresponds to bedtime. In addition to this, we have taken 10 observation symbols on the basis of insulin dose values given to the patient. Table 1 shows the actual amount of insulin give to the patient for thirty days.

Further, whole one-month data were divided into slots of three days and training is done taking three-day data repeatedly. For training, simulated annealing algorithm is used, and at the end of the process, a model λ is obtained. Hence, our model λ has four states (S_1, S_2, S_3, and S_4), and we assume that the patient starts his doses from morning, i.e., there is a very high probability that the patient will be in state 1. So, we take initial probability as:

$$\pi = [0.925\ 0.025\ 0.025\ 0.025]$$

After examining the patient's previous data, we roughly initialize the state transition probabilities and symbol emitting probabilities generating initial $A = a_{ij}$ and $B = B_j(k)$ as follows:

$$A = a_{ij} = \begin{matrix} [0.025\ 0.325\ 0.625\ 0.025 \\ 0.025\ 0.025\ 0.925\ 0.025 \\ 0.925\ 0.025\ 0.025\ 0.025 \\ 0.250\ 0.250\ 0.250\ 0.250] \end{matrix}$$

$$B = b_j(k) =$$
$$\begin{matrix} [.0090\ .0091\ .0091\ .0091\ .0091\ .0091\ .0091\ .0091\ .3391\ .3391\ .2491 \\ .6090\ .3091\ .0091\ .0091\ .0091\ .0091\ .0091\ .0091\ .0091\ .0091\ .0091 \\ .0090\ .0091\ .0091\ .0091\ .0091\ .0091\ .9000\ .0091\ .0091\ .0091\ .0091 \\ .9090\ .0091\ .0091\ .0091\ .0091\ .0091\ .0091\ .0091\ .0091\ .0091\ .0091] \end{matrix}$$

The observation sequence for first three days will be:

$$\text{O} = [O_1\ O_2\ O_3\ O_4\ O_5\ O_6\ O_7\ O_8\ O_9\ O_{10}\ O_{11}\ O_{12}]$$

Taking the initial model, we keep training using simulated annealing [11, 16, 17] method for every three days and get a model λ, and then, it is again trained taking all the previous observation sequences getting a λ_{Final}. Taking λ_{Final} as model, observation sequence is predicted by matching $P(O|\lambda)$ values with previous model values, and these processes are shown as follows.

3.2 HMM Training by Simulated Annealing Algorithm

The HMM training is performing in the following steps.

(a) Take an initial values A, B, and π.
(b) Using simulated annealing method and adjusting the values of A and B, a set of values of A and B is found having maximum values say λ_1.
(c) for the next sequence, initial values of A and B are taken for the model λ_1, and then, step 'b' is repeated to get a new model λ_2. This way we continue and $\lambda_1, \lambda_{12} \ldots, \lambda_{10}$ models are created as shown in the Table 2.

(d) To further refine the model, we took the model λ_{10}as a starting model and using simulated annealing, got refined new model $\lambda_{\text{new1}}, \lambda_{\text{new2}}, \ldots, \lambda_{\text{new10}}$ as shown in Table 3 and corresponding values of A_{10}, B_{10}, and π_{10} are given in the Table 3.

Table 2 Observation sequence and corresponding $p(O|\lambda)$ values

S.no.	Training sequence	HMM	Log10 ($p(O|\lambda_i)$)
1	9 1 7 1 10 2 7 1 11 1 7 1	**λ1**	−9.6952
2	10 4 1 5 9 4 7 2 9 5 7 1	**λ2**	−10.797
3	10 1 8 1 10 1 7 1 9 5 8 1	**λ3**	−8.7954
4	10 4 7 1 10 4 7 1 10 5 7 1	**λ4**	−7.3351
5	10 5 7 1 10 5 7 1 10 5 7 2	**λ5**	−3.0548
6	10 5 7 1 10 5 7 1 10 5 7 1	**λ6**	−6.4811
7	9 4 7 1 10 4 8 1 10 4 7 1	**λ7**	−8.9516
8	9 4 7 1 10 4 8 1 10 4 7 1	**λ8**	−9.8487
9	10 6 7 1 10 1 7 1 9 5 7 1	**λ9**	−10.7967
10	10 5 7 1 10 4 7 1 9 5 7 1	**λ10**	−8.1579

Table 3 Training sequence and corresponding $p(O|\lambda)$ values (second iteration values)

S.no.	Training sequence	HMM	log10($p(O|\lambda\max)$)
1	9 1 7 1 10 2 7 1 11 1 7 1	$\lambda_{\mathbf{new1}}$	−12.5923
2	10 4 1 5 9 4 7 2 9 5 7 1	$\lambda_{\mathbf{new2}}$	−13.5613
3	10 1 8 1 10 1 7 1 9 5 8 1	$\lambda_{\mathbf{new3}}$	−8.9966
4	10 4 7 1 10 4 7 1 10 5 7 1	$\lambda_{\mathbf{new4}}$	−6.6894
5	10 5 7 1 10 5 7 1 10 5 7 2	$\lambda_{\mathbf{new5}}$	−3.1633
6	10 5 7 1 10 5 7 1 10 5 7 1	$\lambda_{\mathbf{new6}}$	−2.8713
7	9 4 7 1 10 4 8 1 10 4 7 1	$\lambda_{\mathbf{new7}}$	−10.6096
8	9 4 7 1 10 4 8 1 10 4 7 1	$\lambda_{\mathbf{new8}}$	−6.7935
9	10 6 7 1 10 1 7 1 9 5 7 1	$\lambda_{\mathbf{new9}}$	−8.5143
10	10 5 7 1 10 4 7 1 9 5 7 1	$\lambda_{\mathbf{new10}}$	−6.619

$$\mathrm{Pi}_{10} = [0.9250\ 0.0250\ 0.0250\ 0.0250]$$

$$B_{10} = \begin{matrix} [0.0130\ 0.0091\ 0.0091\ 0.0091\ 0.0091\ 0.0091\ 0.0091\ 0.0091\ 0.0131\ 0.0091\ 0.0091 \\ 0.0090\ 0.0091\ 0.0091\ 0.0111\ 0.9031\ 0.0131\ 0.0091\ 0.0091\ 0.0091\ 0.0091\ 0.0091 \\ 0.0090\ 0.0091\ 0.0091\ 0.0091\ 0.0091\ 0.0091\ 0.0091\ 0.6071\ 0.0111\ 0.0091\ 0.0091 \\ 0.6051\ 0.3031\ 0.0091\ 0.0131\ 0.0091\ 0.0091\ 0.0091\ 0.0091\ 0.0091\ 0.0151\ 0.0091 \end{matrix}$$

$$A_{10} = \begin{matrix} 0.0270\ 0.9170\ 0.0290\ 0.0270 \\ 0.0250\ 0.0330\ 0.9170\ 0.0250 \\ 0.6170\ 0.0330\ 0.0350\ 0.3150 \\ 0.9110\ 0.0270\ 0.0330\ 0.0290 \end{matrix}$$

(e) $\lambda_{\text{new}10}$ is assumed as λ_{final}, which we used for prediction of sequence.

4 Prediction

After training the HMM [18], the procedure can be described as for the predicting the observation sequence. For predicting $O_i + 1$, we use $P(O_i|\lambda_{\text{final}})$ (O_i is ith the observation sequence) to find those events which have closest $P(O_i|\lambda_{\text{final}})$ value. If we assume there are two closest values O_{k1} and O_{k2} ; then, we evaluate two possible predicting values $O_{p1} = O_i + (O_{k1+1} - O_{k1})$ and $O_{p2} = O_i + (O_{k2+1} - O_{k2})$. For example, for predicting O_{11}, we use the following steps:

(a) For predicting O_{11}, we found $P(O_4/\lambda_{\text{final}})$ and $P(O_8/\lambda_{\text{final}})$, which are the closest observation sequences. (b) We find the differences of $O_5 - O_4$ and $O_9 - O_8$, and then, these differences are added to O_{10} giving two predicted observation sequences O_{p1} and O_{p2}.

Predicted Observation

($O_{p1} = 10\ 7\ 7\ 0\ 10\ 0\ 7\ 0\ 9\ 5\ 7\ 2$) and ($O_{p2} = 10\ 8\ 7\ 0\ 10\ 5\ 6\ 0\ 9\ 5\ 7\ 0$)

(c) We again evaluate the $P(O_{p1}|\lambda_{\text{Final}})$ and $P(O_{p2}|\lambda_{\text{Final}})$ values, and O_{p2} chosen because it is observed that $P(O_{p2}|\lambda_{\text{Final}})$ is higher value than $P(O_{p1}|\lambda_{\text{Final}})$. Therefore, O_{p2} observation sequence because the predicted sequence.

Similarly using steps a, b, and c, we predict O_{P11}, O_{P12}, O_{P13}, O_{P14} and O_{P15} as follows.

$$O_{P11} = 10\ 6\ 7\ 0\ 10\ 5\ 6\ 0\ 9\ 5\ 7\ 0 \quad O_{P12} = 9\ 3\ 8\ 0\ 10\ 7\ 8\ 0\ 9\ 4\ 7\ 0 \quad O_{P13} = 10\ 2\ 7\ 0\ 10\ 6\ 6\ 0\ 9\ 5\ 7\ 0$$

$$O_{P14} = 10\ 6\ 7\ 0\ 10\ 9\ 6\ 0\ 9\ 5\ 7\ 0 \quad O_{P15} = 9\ 3\ 7\ 0\ 10\ 7\ 8\ 0\ 9\ 4\ 7\ 0$$

Actual observation sequences are O_{a11}, O_{a12}, O_{a13}, O_{a14} and O_{a15} as follows (Table 4):

Table 4 Comparison between actual data and predicted data

Date	Code	Actual data				Predictable data			
		Breakfast	Lunch	Dinner	Bedtime	Breakfast	Lunch	Dinner	Bedtime
21-May-91	33	9	2	7	0	10	6	7	0
22-May-91	33	9	4	8	0	10	5	6	0
23-May-91	33	10	0	7	0	9	5	7	0
24-May-91	33	10	3	8	0	9	3	8	0
25-May-91	33	10	7	2	0	10	7	8	0
26-May-91	33	11	0	7	2	9	4	7	0
27-May-91	33	11	2	7	0	10	2	7	0
28-May-91	33	9	4	7	0	10	6	6	0
29-May-91	33	10	4	7	0	9	5	7	0
30-May-91	33	9	3	7	0	10	6	7	0
31-May-91	33	10	3	7	0	10	9	6	0
1-Jun-91	33	11	3	7	0	9	5	7	0
2-Jun-91	3	9	0	7	0	9	3	7	0
3-Jun-91	33	9	4	7	0	10	7	8	0
4-Jun-91	33	10	4	7	0	9	4	7	0

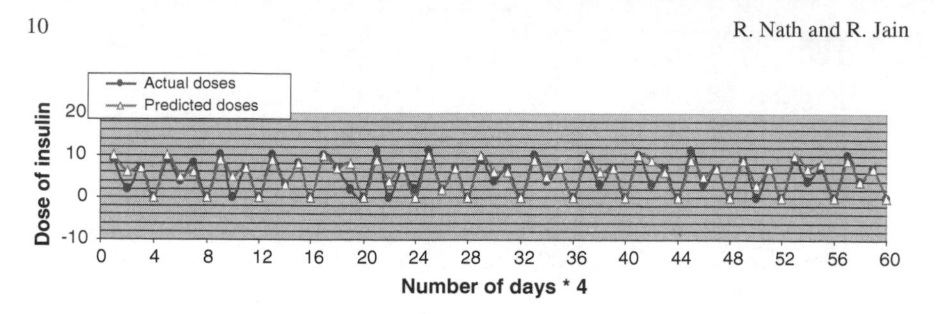

Fig. 1 Graph between actual doses and predicted doses for fifteen days

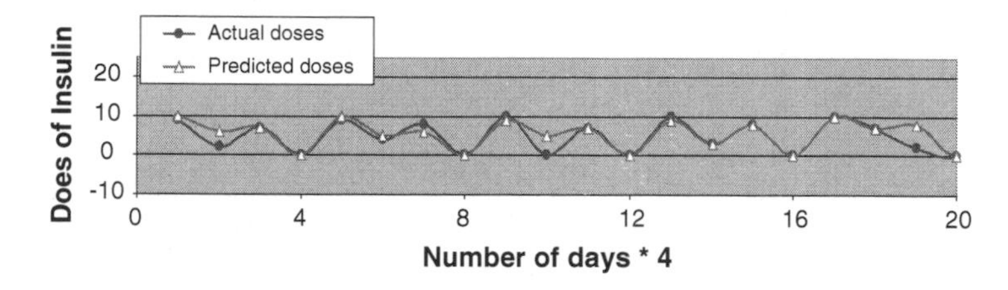

Fig. 2 Graph between actual doses and predicted doses for five days

$$O_{a11} = 9\ 2\ 7\ 0\ 9\ 4\ 8\ 0\ 10\ 0\ 7\ 0 \quad O_{a12} = 10\ 3\ 8\ 0\ 10\ 7\ 2\ 0\ 11\ 0\ 7\ 2$$
$$O_{a13} = 11\ 2\ 7\ 0\ 9\ 4\ 7\ 0\ 10\ 4\ 7\ 0$$
$$O_{a14} = 9\ 3\ 7\ 0\ 10\ 3\ 7\ 0\ 11\ 3\ 7\ 0 \quad O_{a15} = 9\ 0\ 7\ 0\ 9\ 4\ 7\ 0\ 10\ 4\ 7\ 0$$

5 Results and Discussions

In this study, we have taken the problem of predicting insulin chart for diabetic patients. Figures 1 and 2 show the comparison between predicted data and actual data, and it can be observed that results are very encouraging. We have discussed our results with our university medical doctor Dr. Chaman Kumar; he was very enthusiastic to see the results. To start with, the results obtained can be very useful to guide the junior doctors who prepare the complete chart for the patients. However, we need to make it user friendly before it can be tested in actual practice by the doctors. The solution of third problem, i.e., the learning problem has exponential complexity (N^T) and HMM learning problem is solved using randomized search algorithm in polynomial time. We would like to compare the results by implementing other randomized algorithms.

Acknowledgments The authors of the paper are highly grateful to Dr. Chaman Kumar (MBBS, MD), CSJM University, Kanpur, for discussing the results with us and encouraging us to do more work in this direction.

Appendix A

http://archive.ics.uci.edu/ml/ called UCI Repository.

References

1. Durbin, R., Eddy, R., Krogh, A.: Graeme Mitchison. Cambridge University Press, Biological Sequence Analysis (2005)
2. Dan, E. Krane., Michael, L. Raymer.: Fundamental Concepts of Bioinformatics. Pearson Education first edition (2000)
3. Antequera, F., Bird, A.: CpG islands as the genomic footprints of promoters that are associated with replication origins. Curr Biol. **9**(17), 661–667 (1999)
4. Larsen, F., Gundersen, G., Lopez, R., Prydz, H.: CpG islands as gene markers in the human genome. Genomics **13**(4), 1095–1107 (1992)
5. Eren, Akdemir., Tolga.: The use of articulator motion information in automatic speech segmentation., Science Direct Received 5 October 2007; received in revised form 7 March 2008; accepted 17 April (2008)
6. Md. Rafiul, Hassan., Baikunth, Nath.: Stock market forecasting using HMM: a new approach. In: Proceeding of international conference on intelligent systems design and applications(ISDA) (2005)
7. Behrooz, Nobakht., Cart-Edward, Joseph., Babak, Loni.: Stock market analysis and prediction Using HMM LIACS
8. Sundararajan, R.: Stock market trend and predection using Markov models. RTCSP 09, Department of ECE, Amrita Vishwa vidyapeeth, Coimbatore
9. Kwong, S., Chau, C.W., Man, K.F., Tang, K.S.: Optimization of HMM topology and its model parameters by genetic algorithm. Pattern Recognitions **34**, 509–522 (2001)
10. The Metropolis Algorithm, Statistical systems and simulated annealing
11. Mantawy, A.H., Abdul Mazid, L., Selim, Z.: Integrating genetic algorithms. Tabu search and Simulated Annealing for the unit commitment problem. IEEE Trans Power Syst **14**(3) August (1999)
12. Ravindra, Nath., Renu, Jain.: Estimating HMM learning parameters using genetic algorithm. In: International conference on computational intelligence applications (2010)
13. Ravindra, Nath., Renu, Jain.: Using randomized search algorithms to estimate HMM learning parameters. In: IEEE international advanced computing conference (IACC) (2009)
14. Rabiner, L.R.: A tutorial on HMM and selected applications in speech recognition. Proc IEEE **77**(2), 267–296 (1977)
15. Liu, Y., Lin, Y., Chen Z.: Using hidden markov model for information extraction based on multiple templates. In: Porch of the international conference on natural language processing and knowledge engineering pp. 394–399, (2003)
16. Coleman, C. M.: Investigation of simulated annealing, Ant-colony optimization, and genetic algorithms for self-structuring antennas. **52**(4), April (2004)
17. Pirlot, Mark: General local search method. European J. Opera. Res. **92**, 493–511 (1996)
18. Wegener, I.: Randomized search heuristics as an alternative to exact optimization. University of Dortmund, Department of the Computer Science, Technical report February 2004

A Single Curve Piecewise Fitting Method for Detecting Valve Stiction and Quantification in Oscillating Control Loops

S. Kalaivani, T. Aravind and D. Yuvaraj

Abstract Stiction is one of the most common problems in the spring-diaphragm type control valves, which are widely used in the process industry. In this paper, a procedure for single curve piecewise fitting stiction detection method and quantifying valve stiction in control loops based on ant colony optimization has been proposed. The single curve piecewise fitting method of detecting valve stiction is based on the qualitative analysis of the control signals. The basic idea of this method is to fit two different functions, triangular wave and sinusoidal wave, to the controller output data. The calculation of stiction index (SI) is introduced based on the proposed method to facilitate the automatic detection of stiction. A better fit to a triangular wave indicates valve stiction, while a better fit to a sinusoidal wave indicates non-stiction. This method is time saving and easiest method for detecting the stiction. Ant colony optimization (ACO), an intelligent swarm algorithm, proves effective in various fields. The ACO algorithm is inspired from the natural trail following behaviour of ants. The parameters of the Stenman model estimated using ant colony optimization, from the input–output data by minimizing the error between the actual stiction model output and the simulated stiction model output. Using ant colony optimization, Stenman model with known nonlinear structure and unknown parameters can be estimated.

S. Kalaivani (✉)
Electronics and Instrumentation Engineering, Muthayammal Engineering College, Namakkal, India
e-mail: instrokalai@gmail.com

T. Aravind
Computer Science Engineering, Muthayammal Engineering College, Namakkal, India
e-mail: aravindnkl@gmail.com

D. Yuvaraj
Instrumentation and Control Engineering, Tamilnadu College of Engineering, Coimbatore, India
e-mail: pciyuvaraj@gmail.com

B. V. Babu et al. (eds.), *Proceedings of the Second International Conference on Soft Computing for Problem Solving (SocProS 2012), December 28–30, 2012*, Advances in Intelligent Systems and Computing 236, DOI: 10.1007/978-81-322-1602-5_2, © Springer India 2014

Keywords Control valve stiction · Stenman model · Single curve piecewise fitting · Ant colony optimization

1 Introduction

Large-scale, highly integrated processing plants include some hundreds or even thousands of control loops. The aim of each control loop is to maintain the process at the desired operating conditions, safely and efficiently. A poorly performing control loop can result in disrupted process operation, degraded product quality, higher material or energy consumption and thus decreased plant profitability.

Nonlinearities such as stiction, backlash and dead band cause oscillations in the process output. They may be present in the process itself or in the control valves. Among the many types of nonlinearities in control valves, stiction is the most common problem in the control valves, which are widely used in the process industry. It hinders the proper movement of the valve stem and consequently affects control loop performance. As the presence of oscillation in a control-loop increases the variability of the process variables, thus causing inferior quality products and larger rejection rates, it is important to detect and quantify stiction. The single curve piecewise fitting method involves fitting the single curve of OP to both triangular and sinusoidal waves using least square estimation (LSE). A better fit to a triangular wave indicates valve stiction, while a better fit to a sinusoidal wave indicates nonstiction. This method is time saving and easiest method for detect the stiction. All valves are sticky to some extent, it is important to quantify stiction. The quantification is implemented by an ant colony optimization procedure. The ant colony optimization procedure involves certain steps to estimate the parameter values. The parameter estimation is done by minimizing the objective function. The error (e) is the difference between actual stiction model output (y) and the dynamic stiction model output (y_m). It is used as the criterion to correct the model parameters, so as to identify the parameters of the actual process.

2 Valve Stiction Model

The present work focuses on pneumatic control valves, which are widely used in the process industry. The general structure of pneumatic control valve is shown in Fig. 1.

Stiction is a portmanteau word formed from the two words static friction. Stiction is the static friction that prevents an object from moving and when the external force overcomes the static friction the object starts moving [1]. The presence of stiction impairs proper valve movement, i.e. the valve stem may not move in response to the output signal from the controller or the valve positioner. To check the behaviour

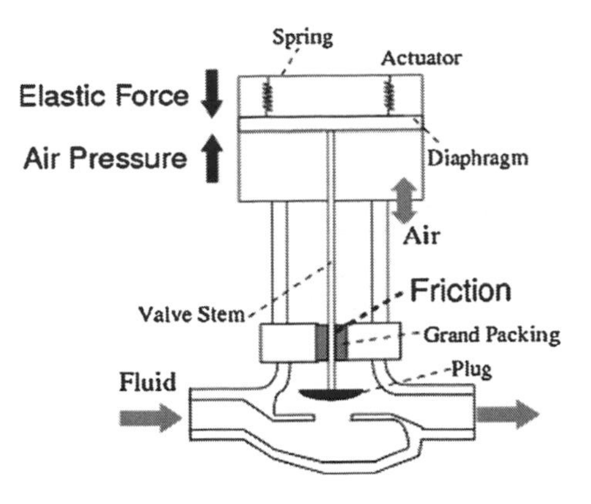

Fig. 1 Structure of pneumatic control valve

of valve moment by modelling, the stiction detection made by physical model and Stenman model used to quantification process.

2.1 Physical Model of Valve Friction

The purpose of this section is to understand the physics of valve friction and reproduce the behaviour seen in real plant data. For a pneumatic sliding stem valve, the force balance equation based on Newton's second law can be written as,

$$M\frac{d2x}{dt2} = \sum Force = Fa + Fr + Ff + Fp + Fi \tag{1}$$

where,

M = Mass of the moving parts, x = Relative stem position.
$F_a = Au$ = Force applied by pneumatic actuator (A = Area of the diaphragm, u = Actuator air pressure or the valve input signal).
$F_r = -kx$ = Spring force (k = Spring constant).
$F_p = -\alpha \Delta P$ = Force due to fluid pressure drop (α = plug unbalance area, ΔP = Fluid pressure drop across the valve).
F_i = Extra force required to force the valve to be into the seat.
F_f = Friction force includes static and moving friction.

where

$$F_f = \begin{cases} F(v) & \text{if } v \neq 0 \\ -(F_a + F_r) & \text{if } v = 0 \text{ and } |F_a + F_r| \leq F_s \\ -\text{Fs sign } (F_a + F_r) & \text{if } v = 0 \text{ and } |F_a + F_r| > F_s \end{cases} \tag{2}$$

Table 1 Values of F_s and F_c for different levels of stiction

Magnitude of stiction	F_s(lbf)	F_c (lbf)
Weak stiction	384	320
Strong stiction	600	500

$$F(v) = -F_c sgn(v) - vF_v - (F_s - F_c)\exp(-v/vs)2sgn(v) \tag{3}$$

The expression for the moving friction is in the first line of equation and comprises a velocity independent term F_c known as Coulomb friction and a viscous friction term vF_v that depends linearly upon velocity. Both act in opposition to the velocity, as shown by the negative signs.

The second line in equation is the case when the valve is stuck. F_s is the maximum static friction. The velocity of the stuck valve is zero and not changing; therefore, the acceleration is zero also. Thus, the right-hand side of Newton's law is zero, so $F_f = -(F_a + F_r)$.

The third line of the model represents the situation at the instant of breakaway. At that instant, the sum of forces is $(F_a + F_r) - F_s sgn(F_a + F_r)$, which is not zero if $|F_a + F_r| > F_s$. Therefore, the acceleration becomes nonzero and the valve starts to move. Here, F_i and F_p assumed to be zero because of their negligible contribution in the model Table 1.

3 Proposed Single Curve Piecewise Fitting Detection Method

Single curve piecewise fitting method of detecting valve stiction is based on the qualitative analysis of the control signals [2]. The basic idea of this method is to fit two different functions, triangular wave and sinusoidal wave, to the controller output (OP) data. The response of physical model is considered as the valve stiction, the stiction detection method is based on the single curve piecewise fitting results of the output signal of first integrating processes, and finally, according the calculation of stiction index, it is introduced based on the proposed method to facilitate the automatic detection of stiction [3].

3.1 Method Description

The single curve piecewise fitting based identification algorithm can be summarized in the following steps:

Step 1: Simulated the closed-loop stiction model in pneumatic control valves.
Step 2: M output data points are generated from the system to be identified [4].

Step 3: In the case of stiction-induced oscillations, the valve position switches back and forth periodically, which results in a rectangular wave.

Step 4: The first integrator after the valve in the control loop converts it into a triangular wave.

Step 5: A sinusoidal external disturbance results in sinusoidal controller output (OP) and process variable (PV), as the integration of a sine wave results in a sinusoidal wave with phase shift.

Step 6: A marginally stable control loop also results in smooth sinusoidal-shaped controller output (OP) and process variable (PV) for the same reason as for a sinusoidal external disturbance.

Step 7: Random initial values for parameters of the nonlinearities in the appropriate range are generated. Choose the any single curve from the controller output [5].

Step 8: Generate the single sine wave and triangular wave and fit with controller output, the objective function for each particle in the initial population is evaluated.

Step 9: Judge end of the iteration and output the best solution, while a better fit to a sinusoidal wave indicates nonstiction. A better fit to a triangular wave indicates valve stiction.

Step 10: According to the stiction index, value used to find the magnitude of stiction.

3.2 Stiction Index

The stiction index (SI) is defined as the ratio of the MSE of the sinusoidal fit to the summation of the MSEs of both sinusoidal and triangular fits: SI is bounded to the interval [0, 1]. The mathematical expression for SI can be written as

$$\mathrm{SI} = \frac{\mathrm{MSEsin}}{\mathrm{MSEsin} + \mathrm{MSEtri}} \tag{4}$$

$\mathrm{SI} = 0$ indicates nonstiction, where S(t) fits a sinusoidal wave perfectly ($\mathrm{MSEsin} = 0$), $\mathrm{SI} = 1$ indicates stiction, where S(t) fits a triangular wave perfectly ($\mathrm{MSEtri} = 0$). For real process data, an SI close to 0 would indicate nonstiction, while an SI close to 1 would indicate stiction. SI is around 0.5, which means MSEsin = MSEtri, and it is undetermined.

Based on the experience, the following rules are recommended:

$$\begin{aligned} &\mathrm{SI} \le 0.4 = \text{ no stiction} \\ &0.4 < \mathrm{SI} < 0.6 = \text{undetermined} \\ &\mathrm{SI} \ge 0.6 = \text{stiction} \end{aligned} \tag{5}$$

4 Data-Driven Stiction Model

Stenman et al.[1] with reference to a private communication with Hagglund reported a one-parameter data-driven stiction model. Stenman proposed a single-parameter data-driven stiction model based on d. Since physical model has certain disadvantages, a single-parameter data-driven model is used for quantifying stiction.

The model is described as follows:

$$x(t) = \begin{cases} x(t-1) & \text{if } |u(t) - x(t-1) \leq d| \\ u(t) & \text{otherwise} \end{cases} \quad (6)$$

where u and x are the valve input and output, $x(t-1)$ and $x(t)$ represent past and present stem positions, $u(t)$ is the actual controller output, and d is the valve stiction band. The model compares the difference between the current input $(u(t))$ to the valve and the previous output $(x(t-1))$ of the valve with the dead band. A real valve can stick anywhere whenever the input reverses direction.

5 Ant Colony Optimization

Ant colony algorithm was first introduced by E. Bonabeau and M. Dorigo in 1991, and the algorithm is a simulation-based evolution process of the real ant seeking food. In 1992, ant colony optimization (ACO) takes inspiration from the foraging behaviour of some ant species. A foraging ant deposits a chemical (pheromone) on the ground which increases the probability that the other ant will follow the same path. This type of communication is also known as stigmergy [6].

The basic procedure of ACO involves certain steps to estimate the unknown parameters of the system. The flowchart of the basic ant colony optimization is shown in Fig. 2.

The main principle of ACO is to minimize the objective function which is also represented as fitness function. If this objective function does not reach the minimum value, the next iteration starts by updating the pheromones. The pheromone is updated till the objective function reaches the minimum value [7].

6 Principle and Implement of Parameter Estimation Using ACO

The framework of ACO-based parameter estimation of the Stenman stiction model is illustrated in Fig. 3. The quantification of process nonlinearity can help decide whether to implement a nonlinear controller or not. It is important to measure the degree of nonlinearity of a process under various input excitation signals or operating

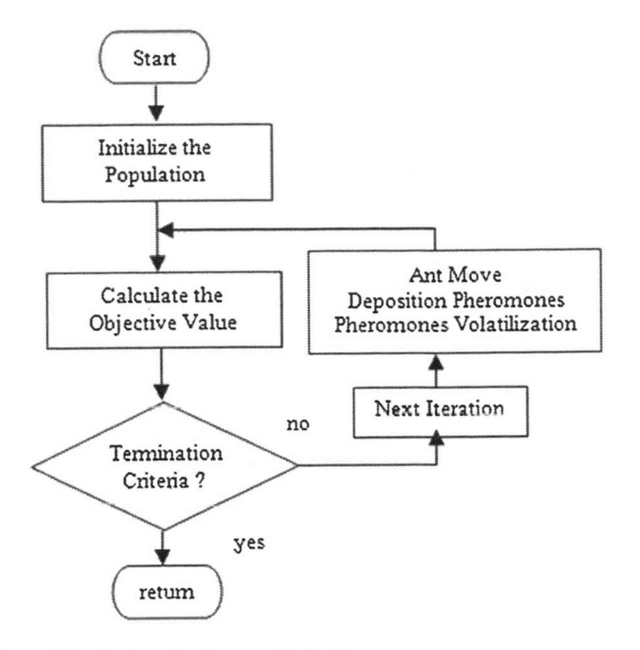

Fig. 2 Flowchart of the basic ant colony optimization

conditions [6]. The quantification is implemented by an ant colony optimization procedure. The open-loop response is obtained for Stenman stiction model.

Since, Stenman model is a single-parameter model, the valve stiction band (d) is to be estimated by obtaining the difference between the actual stiction model output $y(t)$ and simulated stiction model output $y_m(t)$, $u(t)$ is the system input signal that can be used in common to both the actual stiction model and simulated stiction model

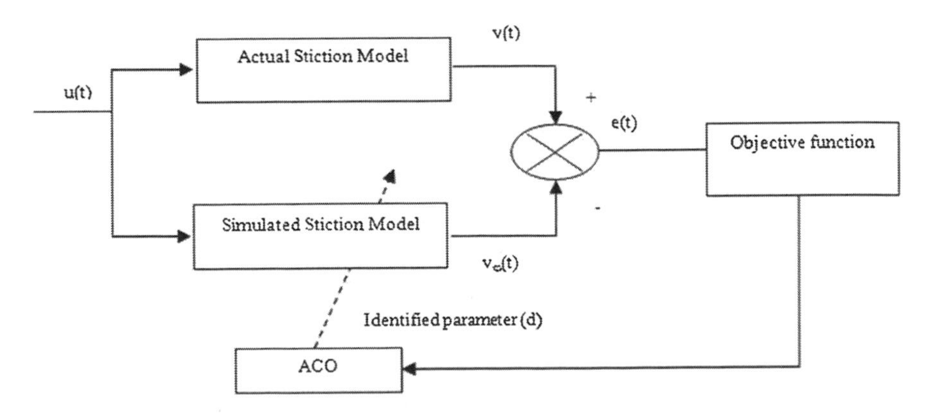

Fig. 3 ACO-based parameter estimation procedure

The following objective function (fitness function) can be defined so as to determine how well the estimates fit the system [8].

$$F = \sum_{t=1}^{M} (y(t) - y_m(t))^2 \tag{7}$$

The ant colony optimization automatically adjusts the parameters of the simulated stiction model. The ACO procedure is used to minimize the objective function which is the difference between the actual stiction model output $y(t)$ and the simulated stiction model output $y_m(t)$.

6.1 Algorithm for Parameter Estimation

6.1.1 Initialize the Pheromone

For constructing a solution, an ant chooses at each construction step $t = 1, \ldots m$, a value for decision variable x_i in m dimensional problem. While termination condition not met, do [10].

```
Procedure ACO
begin
        Initialize the pheromone
        while (stopping criterion not satisfied) do
          Position each ant in a starting point
          while (stopping when every ant has
                    build a solution) do
for each ant do
                    Chose position for next task by
                    pheromone trail intensity
          end for
          end while
update the pheromone
end while
end
```

6.1.2 Ant Solution Construction

The tour length for the kth ant, L_k, the quantity of pheromone added to each edge belonging to the completed tour is given by the following equation

$$\Delta\tau_{ij}^{k}(t) \begin{cases} \frac{Q}{L_k} & \text{where edge } (i, j) \in T_k(t) \\ 0 & \text{if edge } (i, j) \notin T_k(t) \end{cases} \tag{8}$$

where τ_{ij} is the trail intensity which indicates the intensity of the pheromone on the trail segment (ij), and Q represents the pheromone quantity [10].

6.1.3 Pheromone Update

After performing local searching, the pheromone table is updated by using the former ants. The pheromone decay in each edge of a tour is given by

$$\tau_{ij}\,(t+1) = (1-\rho)\,\tau_{ij}\,(t) + \Delta\tau_{ij}\,(t) \tag{9}$$

where $\rho \in (0, 1)$ is the trail persistence or evaporation rate. The greater the value of ρ is, the less the impact of past solution is. When the ant completes its tour, the local pheromone updating is done. The value of $\Delta\tau_j$ is defined as follows:

$$\Delta\tau_j = \frac{1}{T_{ik}} \tag{10}$$

where T_{ik} is the shortest path length that searched by kth ant at ith iteration. When the ant completes its tour, if it finds the current optimal solution, it can lay a larger intensity of the pheromone on its tour, and the global pheromone updating is applied and the value of $\Delta\tau_j$ is given by

$$\Delta\tau_j = \frac{D}{T_{\mathrm{op}}} \tag{11}$$

where T_{op} is the current optimal solution, and D is the encouragement coefficient.

7 Results and Discussion

In this section, several simulations are performed for detecting pneumatic control valve stiction in the closed loop of a physical model using MATLAB/Simulink software. The control valve stiction is detected by obtaining the closed-loop response of valve stiction. To study the effect of stiction, a first-order process with a time-delay was simulated using a pneumatic control valve modelled using Newton's second law.

$$G(s) = \frac{1.54e^{-1.07s}}{5.93s+1} \tag{12}$$

The model parameters used in the simulation are given below. The values of fs and fc are as per the Table 3.1. $A = 1{,}000\ \mathrm{in}^2$, $k = 300\,\mathrm{lbf.in}^{-1}$, $M = 3\,\mathrm{lb}$, $F_v = 3.5\,\mathrm{lbf.s.in}^{-1}$, vs $= 0.01$. Figures 4a and 4b show the variations of controller output, plant output and valve output in the presence of weak stiction and strong stiction, respectively.

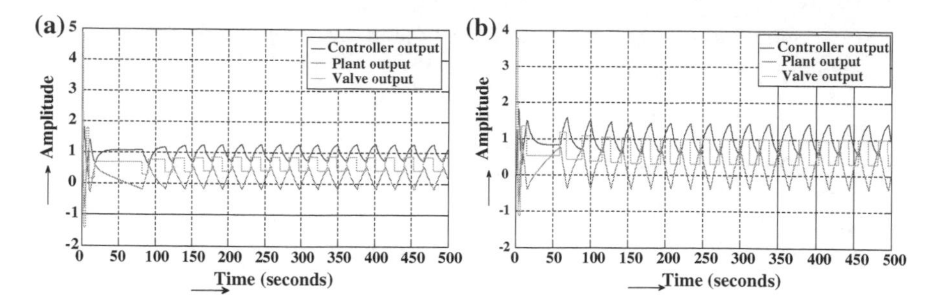

Fig. 4 **a** Closed-loop response of physical model in the case of weak stiction, **b** closed-loop response of physical model in the case of strong stiction

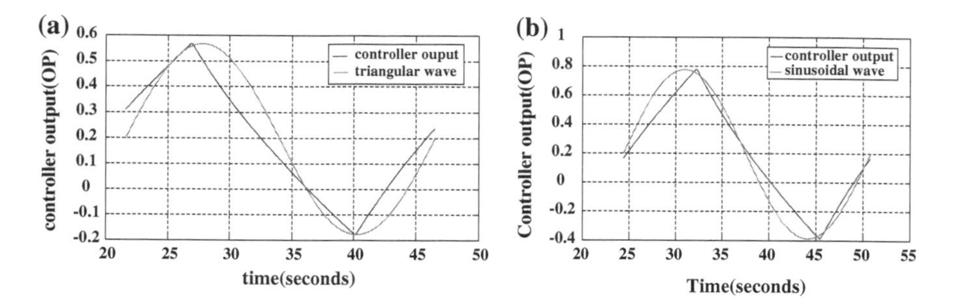

Fig. 5 **a** Response of curve fitting method for weak stiction, **b** response of curve fitting method for strong stiction

Table 2 Stiction index for different levels of stiction for curve fitting method

Magnitude of stiction	Stiction index
Weak stiction	0.5342
Strong stiction	0.8008

The above figure shows the process output and plant output produces triangle wave and the stiction valve produce the rectangular-shaped output. The above figs. 6a and 6b show the variations of controller output for weak stiction and strong stiction using sinusoidal fitting.

The single curve controller output is fitting with single curve triangular wave form, and the mean square error (MSE) value was calculated in weak stiction and strong stiction, the mean square error value was calculated in weak and strong stiction. The figs. 7(a) and 7(b) show the variations of controller output for weak stiction and strong stiction using triangular fitting. Stenman model is a single parameter model, the valve stiction band (d) is to be estimated by obtaining the difference between the actual stiction model output $y(t)$ and simulated stiction model output $y_m(t)$, and $u(t)$ is the system input signal that can be used in common to both the actual stiction model and simulated stiction model. The trajectories of estimated parameters for weak stiction and strong stiction shown in figs. 7(a) and 7(b)

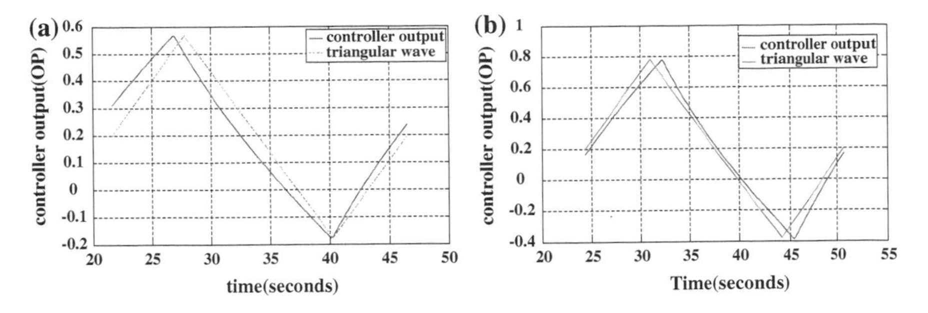

Fig. 6 **a** Response of curve fitting method for weak stiction, **b** response of curve fitting method for strong stiction

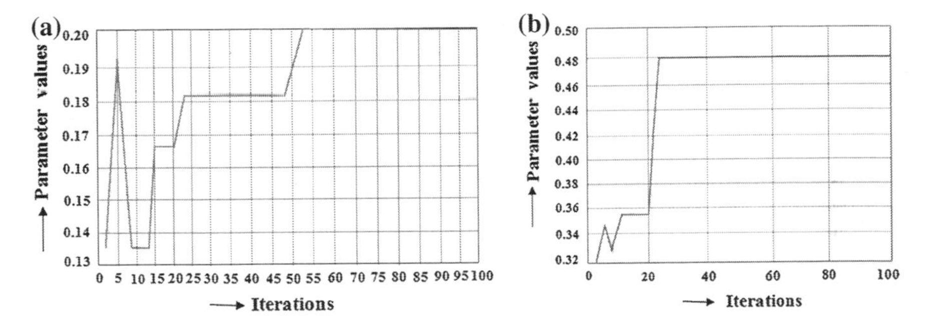

Fig. 7 **a** Trajectories of estimated parameters for weak stiction ($d = 0.2$), **b** trajectories of estimated parameters for strong stiction ($d = 0.5$)

The population and iteration values are 20 and 100, respectively. The parameter 'd' is initialized from 0.01 and is increased up to 10. The evaporation rate ρ is 0.2, and the parameter Q is 100. A control valve stiction model with weak stiction ($d = 0.2$), and strong stiction ($d = 0.5$) cases are simulated in the control loop.

Due to the presence of stiction the quantification of stiction is essential. The quantification of stiction is done by using ant colony optimization procedure [9]. By using ACO algorithm, the stiction parameters are estimated, when the objective function reaches the minimum value and the process is repeated for 100 iterations Table 3.

Table 3 ACO-based optimization

Magnitude of stiction	Actual stiction model parameter (d) value	Simulated stiction model parameter (d) value
Weak stiction	0.2	0.2
Strong stiction	0.5	0.48

It shows the parameter estimation is done by minimizing the objective function. The error, e(t), is the difference between actual strong stiction model output y(t) and the simulated strong stiction model output ym(t). It is used as the criterion to correct the model parameters, so as to estimate the parameters of the actual process. Here, 50 seconds time is taken to find the optimal value. The estimates of the recovered stiction model are very close to the true values.

8 Conclusion

In this paper, the dynamics of the stiction phenomenon found in the pneumatic control valve is understood by the physical model. The stiction found in the pneumatic control valve is modelled using first principles and implemented using MATLAB/Simulink software environment. The physical model involves several parameters, but the magnitude of stiction is based on the two parameters such as maximum static friction (fs) and Coulomb friction (fc). The closed-loop response for the physical model is obtained and the various detection methods such as the single curve piecewise fitting method implemented in the controller output. Due to the presence of stiction, the quantification of stiction is essential. The quantification of stiction is done by using ant colony optimization procedure. The ant colony optimization procedure involves certain steps to estimate the parameter values. The parameter estimation is done by minimizing the objective function. The error (e) is the difference between actual stiction model output (y) and the dynamic stiction model output (ym). It is used as the criterion to correct the model parameters, so as to estimate the parameters of the actual process.

References

1. Choudhury, M.A.A.S., Thornhill, N.F., Shah, S.L.: Modelling valve stiction. Control Eng. Pract. **13**, 641–658 (2004)
2. Rossi, M., Scali, C.: A comparison of techniques for automatic detection of stiction: simulation and application to industrial data. J. proc. control **15**, 505–514 (2005)
3. Horch, A.: A simple method for detection of stiction in control valves. Control Eng. Pract. **7**, 1221–1231 (1999)
4. He, Q.p., Pottmann.: Detection of valve stiction using curve fitting. Internal Report, Process Dynamics and Control. Dupont Engineering, (2003).
5. He, Q.P., Wang, J., Pottmann, M., Qin, S.J.: A curve fitting method for detecting valve stiction in oscillating control loops. Ind. Eng. Chem. Res. **46**, 4549–4560 (2007)
6. Dorigo, M., Maniezzo, V., Colorni, A.: Ant system: optimization by a colony of cooperating agents. IEEE Trans. Syst. Man Cybern. B **26**(1), 29–41 (1996)
7. Dorigo, M., Stützle, T.: Ant Colony Optimization. MIT Press, Cambridge (2004)
8. Toliyat, H.A., Levi, E., Raina, M.: A review of RFO induction motor parameter estimation techniques. IEEE Trans. Ene. Conv. **18**, 271–283 (June 2003)
9. Sivagamasundari, S., Sivakumar, D.: Estimation of valve stiction using particle swarm optimization. J. Sens. Transducers **129**, 149–162 (2011)
10. Kim, J.-W., Kim, S.W.: Parameter identification of induction motors using dynamic encoding algorithm for searchs (DEAS). IEEE Trans. Ene. Conv. **20**, 16–24 (2005)

Feed Point Optimization of Fractal Antenna Using GRNN-GA Hybrid Algorithm

Balwinder Singh Dhaliwal and Shyam S. Pattnaik

Abstract The design of miniaturized and efficient patch antennas has been a main topic of research in the past two decades. The fractal patch antennas have provided a good solution to this problem. But, in fractal antennas, finding the location of optimum feed point is a very difficult task. In this chapter, a novel method of using GRNN-GA hybrid model is presented to find the optimum feed location. The results of this hybrid model are compared with the simulation results of IE3D which are in good agreement.

Keywords Fractal antenna · Generalized regression neural networks · Genetic algorithm · Hybrid algorithm

1 Introduction

A fractal antenna is one that has been shaped in a fractal fashion, either through bending or shaping a volume, or through introducing holes. They are based on fractal shapes such as the Sierpinski triangle, Mandelbrot tree, Koch curve, and Koch island etc. [1]. There has been a considerable amount of recent interest in the possibility of developing new types of antennas that employ fractal rather than Euclidean geometric concepts in their design. Sierpinski gasket is an example of mostly explored fractal antenna. Other geometries used for fractal antennas include Sierpinski carpet, Hilbert, Koch, and Crown square antenna [2].

B. S. Dhaliwal (✉)
Guru Nanak Dev Engineering College, Ludhiana, Punjab, India
e-mail: bs_dhaliwal@gndec.ac.in

S. S. Pattnaik
National Institute of Technical Teachers Training and Research, Chandigarh, India

B. V. Babu et al. (eds.), *Proceedings of the Second International Conference on Soft Computing for Problem Solving (SocProS 2012), December 28–30, 2012*, Advances in Intelligent Systems and Computing 236, DOI: 10.1007/978-81-322-1602-5_3,

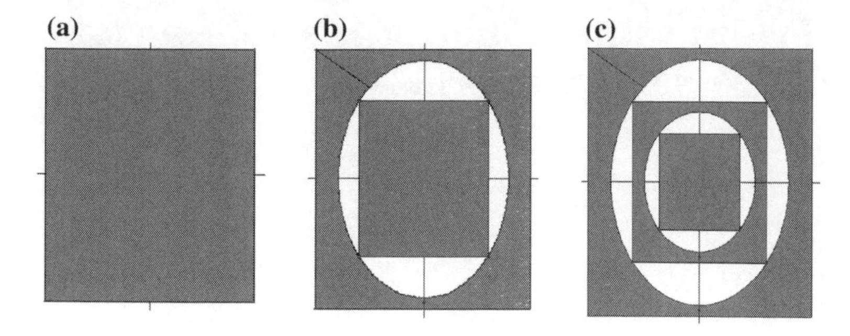

Fig. 1 **a** Base geometry (zeroth iteration), **b** first iteration geometry, **c** second iteration geometry

A fractal antenna based on rectangular base shape is proposed in this presented work. The base geometry or zeroth iteration is a rectangle with side lengths of 39.3 and 48.4 mm as shown in Fig. 1a. To obtain the first iteration geometry shown in Fig.1b, an ellipse is cut from the base shape of Fig. 1a. The primary axis radius of the ellipse is 16 mm, and secondary axis radius of ellipse is 22.62 mm. Then, a rectangle is inserted in the area from where the ellipse is cut, such that all four corners of inserted rectangle are connected with the boundary of ellipse cut. The side lengths of the inserted rectangle are 24.18 and 29.64 mm. The dimensions of the ellipse and rectangle are selected so that the inserted rectangle is 60 % of size of the base rectangle. To obtain the second iteration geometry shown in Fig.1c, the same procedure is applied on the first iteration geometry i.e., an ellipse is cut, and then a rectangle is inserted. The primary axis radius of the ellipse cut is 10 mm, and secondary axis radius is 12.92 mm. The side lengths of the inserted rectangle are 14.5 and 17.78 mm. The dimensions of the ellipse and rectangle are selected so that the inserted rectangle is 60 % of size of the first iteration rectangle. The height of substrate used is 3.175 mm with dielectric constant and loss tangent of 2.2 and 0.0009, respectively.

In antenna design, feeding techniques are very important as they ensure that antenna structure operates at full power of transmission. Especially at high frequencies, designing of feeding techniques becomes a more difficult process. One of the most common techniques used for feeding antennas is coaxial probe feed technique. The location of feed (i.e., feed point) is very important in antenna performance. The feed point must be located at that point on the patch, where the input impedance is 50 ohms for the resonant frequency. But, it is not an easy task to achieve specially for small-size antennas. This problem is further complex in case of fractal antennas because of the complex geometry of different iterations.

In this chapter, a novel method of finding feed point using hybrid GRNN-GA model is proposed. The following section describes GRNN, GA, and proposed hybrid algorithm. The results are given in Sect. 3, and present work is concluded in Sect. 4.

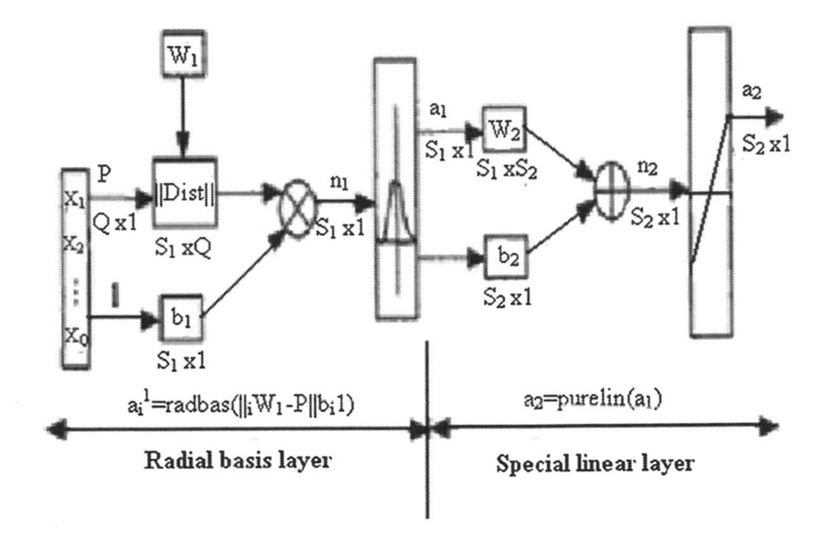

Fig. 2 Generalized regression neural network model

2 Hybrid GRNN-GA Model for Feed Point Calculation

2.1 Generalized Regression Neural Networks

Back-propagation neural networks (BPNN) is a widely used model of the neural network paradigm and has been applied successfully in applications in a broad range of areas. However, BPNN, in general, has slow convergence speed, and there is no guarantee at all that the absolute minima can be achieved. This disadvantage can be overcome by using generalized regression neural networks (GRNN).

The GRNN were first proposed by Sprecht in 1991, which are feed-forward neural network model based on nonlinear regression theory. The network structure is shown in Fig. 2. GRNN consists of a radial basis function network layer and a linear network layer. The transfer function of hidden layer is radial basis function. The basis function of hidden layer nodes in network adopts Gaussian function, which is a non-negative and nonlinear function of local distribution and radial symmetry attenuation for central point, and generates the responses to input signal locally. GRNN employs the smoothing factor as a parameter in learning phase. The single smoothing factor is selected to optimize the transfer function for all nodes. To reduce computational time, GRNN performs one-pass training through the network [3].

2.2 Genetic Algorithms

Genetic algorithm belongs to a class of probabilistic methods called "Evolutionary Algorithms" based on the principles of selection and mutation. GA was introduced by J. Holland, and it is based on natural evolution theory of Darwin. It is a population-based algorithm, and they find their application in various engineering problems. Usually, a simple GA consists of the following operations: selection, crossover, mutation, and replacement. First, an initial population composed of a group of chromosomes is generated randomly. These chromosomes represent the problem's variables. The fitness values of the all chromosomes are evaluated by calculating the objective function in a decoded form. A particular group of chromosomes is selected from the population to generate the offspring by the defined genetic operations such as crossover and mutation. The fitness of the offspring is evaluated in a similar fashion to their parents. The chromosomes in the current population are then replaced by their offspring, based on a certain replacement strategy. Such a GA cycle is repeated until a desired termination criterion is reached (for example, a predefined number of generations are produced). If all goes well throughout this process of simulated evolution, the best chromosomes in the final population can become a highly evolved solution to the problem. To overcome the possibility of being trapped in local minima, in GA, the mutation operation in the chromosomes is employed. GA has been applied in a large number of optimization problems in several domains, telecommunication, routing, and scheduling, and it proves its efficiency to obtain a good solution. It has also been extensively used for a variety of problems in antenna design during the last decade [4–6].

2.3 Proposed Hybrid Algorithm

The genetic algorithm (GA) technique uses the objective function for the optimization and without which the optimization technique has no meaning. But in case of fractal antennas, closed-form mathematical formulation for finding the optimum feed location is not available. Thus, a novel method of objective function formulation has been presented, in which generalized regression neural networks are used as the fitness function. This technique can be used everywhere, particularly in those cases where the objective function formulation is difficult, or the objective function is improper. The procedure adopted to find the optimum feed location of the proposed fractal antenna is given below.

- Data Set Generation: Data set for the training of GRNN has been prepared by using IE3D software. The antenna is simulated for different feed locations (x_i, y_i), and the return loss for the corresponding feed location points has been taken as output. The center of antenna is considered at (0, 0) all cases.

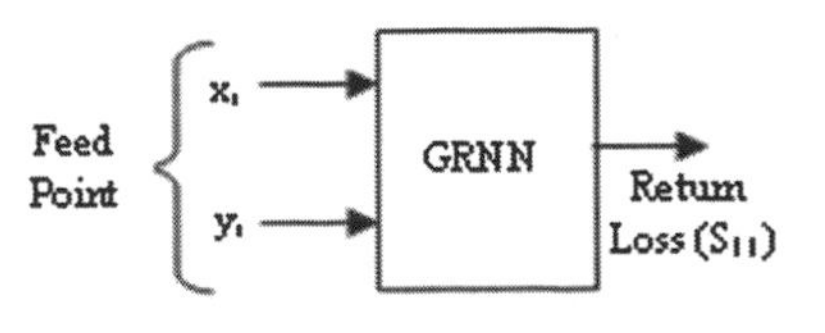

Fig. 3 Training model of GRNN

- Training the ANN: The above data set has been used to train the GRNN in MATLAB. Sufficient number of training samples has been used to train the network. Fig. 3 shows the model for the training of GRNN network.
- The genetic algorithm optimization technique has been implemented using MATLAB, and the trained GRNN network has been used as objective function for the GA algorithm. The GA minimizes the objective function, and it gives the feed location where return loss is minimum, as output.
- The optimum feed points given by GA have been simulated using IE3D software, and return loss (S_{11}) is found. The simulation values are compared with the hybrid model results in order to check the accuracy of the results.

3 Results and Discussion

Three different models have been trained, one for each geometry. For the training of GRNN, the spread constant is required to be set, and it is taken as 0.1 for all the 3 models. These trained GRNN models have been then used as objective functions for the genetic algorithm. The parameters of GA are as follows: The population size is 50 and the crossover probability is 0.8. GA is run for 500 iterations for each model. The optimum feed locations provided by this hybrid GRNN-GA model along with the minimized return loss are given in Table 1.

The feed locations obtained from hybrid algorithm are simulated using IE3D software. The return loss (S_{11}) for all geometries is found. The simulated values of return loss are almost same as given by hybrid algorithm for base geometry and second iteration geometry, and it is better than hybrid model value for first iteration geometry. The comparison of hybrid model results and simulation results is given in Table 2 which shows a reasonable match.

Table 1 Results of hybrid GRNN-GA model

Geometry	Minimized return loss (S_{11}) in dB	Optimized feed location	
		x_i	y_i
Base geometry	−34.370	8.152	13.326
First iteration	−36.142	8.348	9.005
Second iteration	−40.349	−9.432	12.466

Table 2 Comparison of hybrid model and simulation results

Geometry	S_{11} (dB) Hybrid model value	S_{11} (dB) Simulation value	Resonant frequency f_r (GHz)
Base geometry	−34.370	−33.816	2.34
First iteration	−36.142	−43.535	1.85
Second iteration	−40.349	−40.351	1.82

4 Conclusion

The optimum feed points of the three geometries are different. As the iterations of fractal antenna increases, the optimum feed location changes, thus affecting the performance of the antenna. A novel approach based on GRNN-GA hybrid algorithm has been implemented successfully to locate the optimum feed point for each generation of fractal geometry. The results obtained using GRNN-GA hybrid algorithm are in good agreement with the simulation results obtained using IE3D software. This approach can be used for any other geometry.

References

1. Puente, C., Romeu, J., Pous, R., Cardama, A.: On the behavior of the sierpinski multiband fractal antenna. IEEE Trans. Antenna Propagation **46**(4), 517–524 (1998)
2. Dehkhoda, P., Tavakoli, A.: A crown square microstrip fractal antenna. IEEE Antennas and Propagation Society International Symposium **3**, 2396–2399 (2004)
3. Jeatrakul, P., Wong, K.W.: Comparing the performance of different neural networks for binary classification problems. International Symposium on Natural Language Processing. 111–115 (2009)
4. Cengiz, Y., Tokat, H.: Linear antenna array design with use of genetic, memetic and tabu search optimization algorithms. Prog Electromagnetics Res C. **1**, 63–72 (2008)
5. Pattnaik, S.S., Khuntia, B., Panda, D.C., Neog, D.K., Devi, S.: Calculation of optimized parameters of rectangular microstrip patch antenna using genetic algorithm. Microwave Optical Technol Lett **23**(4), 431–433 (2003)
6. Johnson, J.M., Rahmat, S.Y.: Genetic algorithms and method of moments (GA/MOM) for the design of integrated antennas. IEEE Trans. Antennas Propagation **47**(10), 1606–1614 (1999)

Diversity Maintenance Perspective: An Analysis of Exploratory Power and Function Optimization in the Context of Adaptive Genetic Algorithms

Sunanda Gupta and M. L. Garg

Abstract In order to increase the probability of finding optimal solution, GAs must maintain a balance between the exploration and exploitation. Maintaining population diversity not only prevents premature convergence but also provides a better coverage of the search space. Diversity measures are traditionally used to analyze evolutionary algorithms rather than guiding them. This chapter discusses the applicability of updation phase of binary trie coding scheme [BTCS] in introducing as well as maintaining population diversity. Here, the robustness of BTCS is compared with informed hybrid adaptive genetic algorithm (IHAGA), which works by adaptively changing the probabilities of crossover and mutation based on the fitness results of the respective offsprings in the next generation.

Keywords Genetic algorithm · Multidimensional knapsack problem · Diversity maintenance

1 Introduction

Genetic algorithms (GAs) have been successfully applied to various optimization problems where one intends to find an optimum or approximate solution to a problem that has a huge size of solution space. However, one of the major concerns in using evolutionary algorithms to search a complex state space is the problem of premature convergence, especially for combinatorial optimization problems like multidimensional knapsack problem (MKP), where the landscape is multipeaked; the probability of search sticking to local optima is all the more high. Because genetic

S. Gupta (✉) · M. L. Garg
School of Computer Science and Engineering, S.M.V.D.U, Katra, India
e-mail: sunanda.gupta@smvdu.ac.in; gupta.sunanda@gmail.com

M. L. Garg
e-mail: garg.ml@smvdu.ac.in

B. V. Babu et al. (eds.), *Proceedings of the Second International Conference on Soft Computing for Problem Solving (SocProS 2012), December 28–30, 2012*, Advances in Intelligent Systems and Computing 236, DOI: 10.1007/978-81-322-1602-5_4,

programming is highly stochastic, we do not expect to obtain clear rules about exact levels of diversity. We aim to draw general conclusions and "rules of thumb" from the investigation of evolving populations with different measures of diversity. Given a specific landscape structure—defined by the search space, objective function, then relying on problem-specific knowledge for navigating this structure in order to extract helpful information from the search space, would make the optimization faster and more effective. This paper investigates the characteristic issues of BTCS [1] (which incorporates this strategy) and compares it with IHAGA [2] for solving test instances of combinatorial optimization problem on MKP. The simulation results show that the proposed strategy significantly improves the computational efficiency of GAs. The rest of the chapter is organized as follows. In the section that follows, a brief review of the BTCS scheme and the research work going on in the field of using adaptive crossover and mutation operators for achieving diversity is provided. Section 3 describes the BTCS bucket updation phase vis-a-vis population diversity. Experimental results are presented in Sect. 4. Section 5 summarizes the main contributions of the chapter.

2 Related Work

2.1 Binary Trie Coding Scheme

Binary trie coding scheme [BTCS] creates and maintains a diverse population of highly fit individuals capable of adapting quickly to fitness landscape change [1]. BTCS provides three major contributions related to duplicate elimination and premature convergence in a steady-state GA. The first contribution of BTCS is the virtually compressed binary trie structure (VCBT). VCBT when integrated with GA proves to be beneficial in determining duplicates among all the generations and replacing them with unique individuals [1, 3]. The second contribution is to demonstrate that preventing duplicates results in improved performance. It effectively avoids what is usually a difficult trade-off between achieving fast search and sustaining diversity and thereby provides means to avoid premature convergence. The third contribution of BTCS is that it relies on problem-specific knowledge in fragmenting the search space into feasible and infeasible regions and then pruning the infeasible regions. This chapter discusses as to how bucket updation phase of BTCS incorporates the effective measures pertaining to population diversity without using adaptive crossover and mutation operators.

2.2 Adaptive Crossover and Mutation Rate

The significance of crossover operator in controlling GA performance has long been acknowledged in GA research which can be dated back to 1980s [4]. A number of guidelines exist in the literature for setting the values of crossover probability [5]. Some studies particularly focused on finding optimal crossover rates [6]. These heralded the need for self-adaption of the crossover or mutation rates. In [7], an adaptive genetic algorithm was proposed, in which crossover and mutation probabilities were varied according to the fitness values of the solutions. There were also works on devising adaptive crossover operators instead of varying the crossover rates [8]. Several operators were employed, and the probabilities of applying each operator were adapted according to the performance of the offsprings generated by the operator. Since then, several similar works have also been done [9].

The choice of mutation rate is also critical to GA's performance [10]. Various researchers have come up with novel approaches to implement the adaptive mutation into a GA. Some approaches to adaptive mutation control employ parent fitness in determining mutation probability [11]. If selected, highly fit individuals undergo low levels of mutation (minimal disruption), while low-fitness individuals are subjected to large rates of disruptive mutation. A measure of population diversity is employed by [12] and [13] in adapting mutation probabilities. In a similar vein, Zhang et al. [14] adapt crossover and mutation according to parameters extracted from a K-means clustering algorithm. Thus, many researchers have emphasized on using adaptive mutation so as to improve GA's performance as it facilitates the finding of global optimum more efficiently [15].

Although the adaptive crossover and mutation rates are hot spots in the study of genetic search, the BTCS scheme proposed by us [1] does not explicitly employ any scheme to adaptively mutate or crossover. For analyzing the robustness of BTCS, we compare it with informed hybrid adaptive genetic algorithm (IHAGA). This scheme works by adjusting its cross-adaptive rate and mutation rate according to the situation surrounding the fitness of the individual [2]. In the course of crossover and mutation, the probabilities of crossover and mutation are adjusted adaptively according to the following formulas:

$$Pc = \begin{cases} \dfrac{Pc_{\max} - Pc_{\min}}{1 + \exp\left(Ax\left(\dfrac{2\left(f' - f_{\text{avg}}\right)}{f_{\max} - f_{\text{avg}}} - 1\right)\right)} + Pc_{\min} & f' \geq f_{\text{avg}} \\ Pc_{\min} & f' \leq f_{\text{avg}} \end{cases}$$

$$Pm = \begin{cases} \dfrac{Pm_{\max} - Pm_{\min}}{1 + \exp\left(Ax\left(\dfrac{2\left(f' - f_{\text{avg}}\right)}{f_{\max} - f_{\text{avg}}} - 1\right)\right)} + Pm_{\min} & f' \geq f_{\text{avg}} \\ Pm_{\min} & f' \leq f_{\text{avg}} \end{cases}$$

where $Pc_{\max}$ and $Pc_{\min}$ denote the lower limit and the upper limit of probability of crossover, respectively. $f_{\max}$ and f_{avg} denote the maximal fitness and the average fitness of population, respectively, f' denotes the higher fitness of two crossover individuals, f denotes the fitness of the individuals, and $A = 9.903438$ [2].

3 BTCS Bucket Updation Phase

3.1 Buckets and Their Significance

The buckets correspond to the leaf nodes in a VCBT [1, 3]. The aim of buckets is to maintain a continuous presence on as many peaks as possible. Population's spatial information is obtained with computationally inexpensive buckets. It provides important information in addition to the address of the trie structure existing under it. This information is used to identify potentially local convergence. Buckets are significant in dividing the population into an exploration section and exploitation section. It monitors and measures diversity at synchronized time intervals and accordingly attempts to control or promote diversity during the evolution. Identifying such measures allows better prediction for run performance and improved understanding of the population and enables the design of efficient operators.

3.2 Bucket Updation

The contribution of updation phase in the BTCS scheme is twofold [1, 3]. The first contribution of updation phase is to manage the size of VCBT structure. The size of VCBT structure can be kept small by pruning fully explored regions of the search space. The second contribution is to monitor convergence and introduce diversity so as to avoid local entrapment.

3.2.1 Guided Crossover Operator

The proposed procedure works by randomly selecting buckets with criterion value 1, and then exchanges information by copying their best strings [1, 3]. The copying of best feasible boundary solution of one to another is done only if (*new bucket_sum* + *old bucket_best_sum*) are feasible and *new bucket_sum* is greater than *old bucket_sum*. Doing this restricts the copying of strings between any two selected buckets randomly. Bucket_sum and bucket_best_sum are two variables that are unique to each bucket. They are used for storing the sum of included objects from root to bucket_position and sum of bucket + 1 position till n (the number of objects), respectively. GA with the proposed method distributes the individuals more widely compared to simple GA, where the individuals represent the local optima for that region. The aim is to identify feasible regions in the landscape that could replace less fit individuals by more promising samples from the unexplored sections of the search space. This prevents entrapment in local optima by including new individuals from the unexplored regions of the search space.

Table 1 Average execution time of BTCS in comparison with IHAGA

n	m	α	Simple GA		BTCS_IMO		IHAGA	
			A.B.S.T	A.E.T	A.B.S.T	A.E.T	A.B.S.T	A.E.T
100	5	0.25	9.6	345.9	10.85	109.47	8.14	**31.13**
		0.5	23.5	347.3	26.32	120.23	19.92	**76.41**
		0.75	26.9	361.7	33.36	123.02	23.19	**90.43**
	10	0.25	97.5	384.1	104.20	**115.12**	83.33	192.05
		0.5	97.3	418.9	111.90	143.6	84.61	**129.86**
		0.75	16.8	462.6	19.15	159.56	14.36	**129.53**
	30	0.25	177.4	604.5	198.69	202.68	150.00	**199.49**
		0.5	118	782.1	130.98	247.74	113.90	**218.99**
		0.75	90	904.2	80.10	315.34	80.10	**253.18**
250	5	0.25	50.7	682	34.19	**216.03**	34.48	265.98
		0.5	276.7	709.4	191.59	**257.54**	185.39	333.42
		0.75	195.9	763.3	127.12	**241.78**	137.13	534.31
	10	0.25	359	870.9	258.16	**290.65**	290.43	566.09
		0.5	342.2	931.5	249.91	**295.06**	281.63	596.16
		0.75	129.1	1011.2	95.91	**320.3**	104.44	455.04
	30	0.25	582.9	1499.5	332.51	**493.45**	472.73	794.74
		0.5	901.5	1980	601.20	**643.14**	720.30	1207.80
		0.75	1059.3	2441.4	754.22	**815.23**	840.02	1440.43
500	5	0.5	291.3	1345.9	142.10	**416.09**	236.83	969.05
		0.75	386.2	1412.6	188.40	**499.32**	317.84	974.69
	10	0.5	562.2	1728.8	274.20	**615.35**	490.24	1383.04
		0.75	937.6	1931.7	457.40	**715.34**	792.27	1564.68
	30	0.5	1121.6	3198.9	547.10	**1334.56**	923.08	2600.71
		0.75	1903.3	3888.2	928.40	**1231.49**	1545.48	3110.56

3.2.2 Adaptive Selection Parameter Control

This takes place when there is some form of feedback from the search that serves as input to the mechanism used to determine the change in the strategy parameters. During this phase, the *avg* corresponding to the worst and best individuals within that bucket is checked. It is computed as the average of all the individuals within that bucket. The *new avg* value will drop if more boundary solutions between the worst and average interval are generated and would increase if more boundary solutions between average and best interval are generated. During this phase, that bucket is selected, whose *new avg* has increased and at least approximately more than 60 % of the region within that bucket has already been explored. The aim of phase 2 in bucket updation is to prevent the unnecessary delay caused in exploring those regions of search space where the probability of best solution to exist is very limited. The phase 2 describes the buckets' solution space diversity from a fitness perspective, i.e., a measure of diversity of healthy individuals. *BDS Bucket Updation* employs ASPC

Table 2 Percentage gaps for BTCS and IHAGA

n	m	α	GA	BTCS_IMO	IHAGA
100	5	0.5	0.4564	0.4613	0.46200
		0.75	0.3212	0.2884	0.32119
	10	0.5	0.7982	0.7774	0.79838
		0.75	0.4813	0.4697	0.48100
	30	0.5	1.3457	1.3145	1.36953
		0.75	0.8321	0.8296	0.81546
250	5	0.5	0.1253	0.1183	0.12525
		0.75	0.0811	0.0752	0.08759
	10	0.5	0.2543	0.2362	0.25429
		0.75	0.1572	0.1513	0.15710
	30	0.5	0.5321	0.5267	0.54877
		0.75	0.3112	0.2972	0.32431
500	5	0.5	0.0441	0.0443	0.04631
		0.75	0.0378	0.0429	0.04271
	10	0.5	0.1134	0.0946	0.11907
		0.75	0.0712	0.0501	0.07903
	30	0.5	0.2635	0.2387	0.26908
		0.75	0.1747	0.1738	0.17905

to regulate selection pressure. ASPC's objective is to create a diversity of health in the population, i.e., the diversity of high-fitness individuals.

4 Experimental Results

Tables 1 and 2 illustrate the comparison of results of BTCS with those of IHAGA [2] for solving the MKP. The results of BTCS and IHAGA are based on our own computations. Table 1 provides the average best solution time (ABST) and the average execution time (AET) for both BTCS and IHAGA. The bold highlights in Table 1 show the optimal average execution time among the two for varying n and m values, where n is the number of objects and m is the number of constraints. It is clear from Table 1 that IHAGA outperforms BTCS computationally, for smaller values of n. The cost of constructing VCBT results in an increase in the average execution time. However, for larger instances, despite the time utilized in the construction of VCBT structure, BTCS is effective in reaching the optimal solution in comparison with IHAGA. The ability to work with unique boundary individuals facilitates faster convergence. The probability of recurrence of individuals in the subsequent generations results in deviation from path, leading to optimality, for larger set of individuals in the case of IHAGA despite its ability to guide the search. Table 2 further provides the average percentage gaps for the two approaches. For both, the BTCS and IHAGA, the percentage gaps (100 $\times$ (optimum LP – optimum GA)/optimum LP) relative to the solution of LP relaxation were computed. Here GA refers to special cases of GA, i.e., BTCS and IHAGA. It can be inferred from the results of Table 2 that BTCS outperforms IHAGA for all test instances under consideration. BTCS has provided better coverage of the search space and has been found to be successful in providing solutions of better quality in comparison with IHAGA.

5 Conclusion

The aim of updation phase in the BTCS scheme has been the exploring of promising regions while concentrating the search on hyperplanes that are likely to contain good solutions. The GCO and ASPC focus on extracting information about the selected buckets before deciding on the introduction of diversity. Its advantage is that at the point of near convergence, late in a GA run, such diversion reduces the probability of GA to entrap in local convergence and thus provides better solutions.

In our approach, we have not used a mutation parameter, which should be adapted explicitly. Instead, it is the principle of working with unique chromosomes (or individuals) in the VCBT structure, which guarantees automatic mutation. Furthermore, our approach still concentrates on using one-point crossover operator with a fixed probability of 0.70. This is attributed to the deeper nature of BTCS scheme, which permits only good optimal solutions from the search space to participate in the process of evolution. Working with unique boundary solutions assists in maintaining an optimum level of diversity among the individuals.

References

1. Sunanda, Garg M.L.: Binary trie coding scheme—An intelligent genetic algorithm avoiding premature convergence. Int. J. Comput. Math. Taylor & Francis. (2012). doi:10.1080/00207160.2012.742514.
2. Yanqin, M., Jianchen W.: Improved hybrid adaptive genetic algorithm for solving knapsack problem. In: Proceedings of the 2nd International Conference and Information Processing, pp. 644–647 IEEE, (2011)
3. Sunanda, Garg, M.L.: GA implementation of the multi dimensional knapsack problem using compressed binary tries. Advances in computational research (ISSN: 0975–3273), 43–46 (2009)
4. Goldberg, D.: Genetic Algorithms in Search, Optimization and Machine Learning. Addison Wesley Publishing Company Inc, Massachusetts (1989)
5. Schaffer, J.D., Carvana, R.A., Eshelman, L.J.: R. A study of control parameters affecting online performance of genetic algorithms for function optimization. In: Proceedings of the Third International Conference on Genetic Algorithms, Das (1989)
6. Ochoa, G., Harvey, I., Buxton, H.: On recombination and optimal mutation rates. In: Proceedings of Genetic and Evolutionary Computation Conference, pp. 488–495 (1999)
7. Srinivas, M., Parnaik, L.: Adaptive probabilities of crossover and mutation in genetic algorithms. IEEE Trans. Syst. Man. Cybern. **3**, 1841–1844 (2003)

8. Vekaria, K., Clark, C.: Biases introduced by adaptive recombination operators. In: Proceedings of Genetic and Evolutionary Computation Conference, 670–677 (1999)
9. Ono, I., Kita, H., Kobayashi, S.: A robust real coded genetic algorithm using unimodal normal distribution crossover augmented by uniform crossover: Effects of self-adaptation of crossover probabilities. In: Proceedings of Genetic and Evolutionary Computation Conference, pp. 496–503 (1999)
10. Cervantes, J., Stephens, C.R.: Limitations of existing mutation rate heuristics and how a rank GA overcomes them. IEEE Trans. evol. comput. **13**(2), 369–397 (2009)
11. Liu, D., Feng, S.: A novel adaptive genetic algorithms. Proc. Int. Conf. Mach. Learn. Cybern. **1**, 414–416 (2004)
12. Zhu, K.: A diversity controlling adaptive genetic algorithm for the vehicle routing Problem with time windows. In: Proceedings. 15th IEEE International Conference on Tools with Artificial Intelligence, pp. 176–183 (2003)
13. Hagras, H., Pounds-Cornish, A., Cooley, M., Callaghan, V., Clarke, G.: Evolving spiking neural network controllers for autonomous robots. In: Proceedings IEEE International Conference on Robotics and Automation, vol. 5, pp. 4620–4626 (2004)
14. Zhang, J., Chung, H., Lo, W., et al.: Clustering based adaptive crossover and mutation probabilities for genetic algorithms. IEEE Trans. Evol. Comput. **11**(3), 326–335 (Jun. 2007)
15. Lobo, F. G., Lima, C. F., Michalewicz, Z (eds.): Parameter setting in evolutionary algorithms. volume 54 of Studies in Computational Intelligence Springer, (2007)

Use of Ant Colony System in Solving Vehicle Routing Problem with Time Window Constraints

Sandhya Bansal, Rajeev Goel and C. Mohan

Abstract Vehicle routing problem with time window constraints (VRPTW) is an extension of the original vehicle routing problem (VRP). It is a well-known NP-hard problem which has several real-life applications. Meta-heuristics have been often tried to solve VRTPW problem. In this paper, an attempt has been made to develop a suitable version of Ant colony optimization heuristic to efficiently solve VRPTW problem. Experimentation with the developed version of Ant colony optimization has shown that it succeeds in general in obtaining results obtained with earlier version and often even better than the results that are better than the corresponding results available in literature which have been obtained using even previously developed hybridized versions of ACO. In many cases, the obtained results are comparable with the best results available in literature.

Keywords Vehicle routing problem · Ant colony optimization · Heuristics · Optimization

1 Introduction

Vehicle routing problem (VRP) [21, 22] lies at the heart of logistics and distribution management that is presently being used by the companies engaged in delivery and collection of goods. Since the conditions and constraints vary from one situation to another, several variants [3, 11, 18] of basic problem have been proposed in

S. Bansal (✉)
Maharishi Markandeswar University, Mullana, India
e-mail: Sandhya12bansal@gmail.com

R. Goel · C. Mohan
Ambala College of Engineering and Applied Research, Ambala, India
e-mail: rajiv_2709@rediffmail.com

B. V. Babu et al. (eds.), *Proceedings of the Second International Conference on Soft Computing for Problem Solving (SocProS 2012), December 28–30, 2012*, Advances in Intelligent Systems and Computing 236, DOI: 10.1007/978-81-322-1602-5_5,

literature. This paper addresses itself to developing an efficient Ant-colony-system-based heuristic for solving VRPTW.

VRPTW [23, 26] problem consists of finding the minimum set of routes for identical capacity vehicles originating and terminating at a central depot such that each customer is served once and only once, given that the exact number of customers and their demands are known. There are also constraints of time windows in that each customer must be served in a specified slot of time. Objective of VRPTW is to find the minimum of total distance travelled and/or the minimum number of vehicles required which can accomplish this job. It is a NP-hard problem where the number of feasible solutions grows exponentially as the number of customer's increases. The work on this problem available in literature can be divided into two classes: exact optimization techniques and heuristic-based (approximate) algorithms. In the first category, the works by [21, 24, 27] can be cited. The methods developed in these papers have been able to efficiently solve some of the Solomon benchmark problems [5, 20, 31]. In the second category, a very large number of heuristic approaches such as tabu search, genetic algorithms, ant colony algorithms, simulated annealing, large neighborhood search, variable neighborhood based algorithms, and multi-phase approaches have been tried [6, 7, 20, 26, 29].

Among heuristic-based optimization techniques, ACO is a more recent optimization heuristic proposed by Dorigo et al. [4, 12–16]. ACO imitates real ant behavior to search for optimal solutions. ACO-based optimization techniques tried thus far for solving VRPTW problem generally use distance and pheromone as search guide parameters. This paper proposes a version of ACO which incorporates besides these two parameters waiting time, urgency to serve and the bias factor also as search guide parameters. Our objective has been to see whether inclusion of these additional parameters can further improve the performance of ACO algorithm for solving VRPTW.

The rest of the paper is organized as follows. Section 2 presents mathematical model of VRPTW problem. Conventional use of ACO heuristic in solving VRPTW problem is presented in Sect. 3. Proposed modifications in this algorithm are presented in Sect. 3.2. Computational experimentation using the modified ant colony system (ACS) algorithm is presented in Sect. 4. Comparison of the obtained results with those earlier available in literature obtained through is done Sect. 5 and certain conclusions drawn.

2 Mathematical Description of VRPTW

In this section, we briefly describe the mathematical model of VRPTW.

The VRP is a complicated combinatorial optimization problem. It has received considerable attention in the past decades because of its practical importance in the fields of transportation, distribution, and logistics. VRPTW is a generalization of the classical VRP with the additional restriction of time window constraints. The

VRPTW can be modeled in mathematical terms through a complete weighted digraph as follows.

Let $G = (V, A)$ where $V = \{v_0, v_1, v_2 \ldots v_n\}$ be a set of nodes, where v_0 represents the depot that holds a fleet of vehicles and $v_1, v_2 \ldots v_n$ denote a set of n customers which are to be served by these vehicles. Each customer has an associated demand q_i, service time s_i, a service time window $[e_i, l_i]$. Also $A = [\{v_i, v_j\}(i, j = 0, 1, 2, \ldots n, i \neq j)]$ is the set of arcs connecting various nodes, having distance d_{ij} as weights. If a vehicle reaches a customer v_i before specified time e_i, it needs to wait until e_i in order to service the customer. The service has to be provided before close time of window at l_i. The depot has also time window $[e_0, l_0]$. No vehicle is to leave the depot before e_0 and all should come back before l_0. The load-carrying capacity of all vehicles is same and all travel with identical constant speed. The objective of the VRPTW is to service all the customers without violating vehicle capacity constraints and time window constraints using minimum number of vehicles that travel minimum possible total distance.

Mathematically, the problem is usually expressed as:

Minimize

$$F = \sum_{i=0}^{N} \sum_{j=0}^{N} \sum_{K=1}^{V} c_{ij} x_{ij}^{v} \tag{1}$$

Subject to:

$$\sum_{v=1}^{V} \sum_{j=1}^{N} x_{ij}^{v} \leq V \quad \text{for } i = 0 \tag{2}$$

$$\sum_{v=1}^{V} x_{ij}^{v} = \sum_{j=1}^{N} x_{ji}^{v} \leq 1 \quad \text{for } i = 0 \text{ and } v \in \{1, \ldots, V\} \tag{3}$$

$$\sum_{v=1}^{V} \sum_{j=0}^{N} x_{ij}^{v} = 1 \quad \text{for } i \in \{1, \ldots, N\} \tag{4}$$

$$\sum_{v=1}^{V} \sum_{i=0}^{N} x_{ij}^{v} = 1 \quad \text{for } j \in \{1, \ldots, N\} \tag{5}$$

$$\sum_{i=0}^{N} c_i \sum_{j=0}^{N} x_{\text{ij}}^{v} \leq q_v \quad \text{for } v \in \{1, \ldots, V\} \tag{6}$$

$$\sum_{v=1}^{V} \sum_{i=0}^{N} x_{\text{ij}}^{k} \left(t_i + t_{\text{ij}} + s_i + w_i\right) = t_j \quad \text{for } j \in \{1, \ldots, N\} \tag{7}$$

$$e_i \leq (t_i + w_i \leq l_i) \quad \text{for } i \in \{1, \ldots, N\} \tag{8}$$

$$t_0 = w_0 = s_0 = 0 \tag{9}$$

$x_{\text{ij}} = 1$ if vehicle k travels from customer i to customer j, and 0 otherwise ($i \neq j; i, j = 0, 1, \ldots, N$).

Here,

V denotes total number of vehicles,
N total number of customers,
c_i customer $i (i = 1, 2, \ldots, N)$ and c_0 delivery depot,
d_{ij} traveling distance between customer i and customer j,
t_{ij} travel time between customer i and customer j,
q_i demand of customer i,
q_v loading capacity of each vehicle, (loading capacity of all vehicles are identical).
e_i earliest permitted arrival time of vehicle at customer i;
l_i latest permitted arrival time of vehicle at customer i;
s_i service time of customer i;

This is an optimization problem in which, (1) is the objective function of the problem which is to be minimized subject to constraints (2)–(9).

In this optimization model, decision variables are as follows:

V total number of vehicles required;
t_i arrival time of vehicle V at customer i;
w_i waiting time of vehicle at customer i before service can be started;

Objective function (1) ensures that total distance travelled by all the vehicles is minimized. The first constraint (2) specifies that there are at the most V vehicles going out of the depot. The set of constraint (3) ensures that every vehicle starts from and ends at the delivery depot. The next two sets of constraints (4) and (5) restrict the assignment of each customer to exactly one vehicle route. The next set of constraints (6) ensures that the loading capacity of no vehicle is exceeded. The constraints of set (7) are the maximum travel time constraint. Remaining constraints (8) guarantee schedule feasibility with respect to time windows.

In formulating the above mathematical model, the following assumptions have been made:

- Identical vehicles with known capacities Q are used.
- All vehicles travel with identical constant velocity.
- Every vehicle leaves the depot and returns to the depot within specified time window $[l_0, s_0]$.
- Demand of each customer is q_i is known.
- Each customer is serviced by one and only one vehicle.
- The total demand of any customer is not more than the capacity of the vehicle.

3 Use of ACS in Solving VRPTW

VRPTW being an NP-hard problem, its exact solution is not known in general. Therefore, large numbers of alternative algorithms have been proposed for solving it. In this section, we first present conventional ACS-based approach for solving VRPTW problem and then present our proposed modification in it. The basic philosophy of ACS approach is to use a suitable positive feedback mechanism to reinforce those arcs of the graph that belong to a good solution. This mechanism is implemented associating pheromone levels with each arc which are then updated in proportion to the goodness of solutions found.

While presenting ACS-based algorithm, for solving VRPTW, we shall use the term 'tour' to denote a set of routes followed by all ants (vehicles) which are able to serve all the nodes of the graph as per their requirements under specified conditions. Our problem is to determine an optimal tour. The ACS algorithm commonly used for solving VRPTW is given below.

3.1 ACS Algorithm for Solving VRPTW:

Step 1. Construction of an initial feasible route:

(a) Each ANT starts from the depot and the set of customers included in its route is empty.

(b) The ant selects the next customer to visit from the list of feasible customers based upon the probabilistic formula (10).

(c) After serving the customer storage capacity and the time used thus far of the Ant is updated and the process continued. Ant returns to the depot when either of the capacity constraint or time window constraint of the depot is satisfied.

(d) We next check whether all the customers have been served or not. If all the customers have been served, stop else send a new ant (vehicle) to visit the remaining destinations.
(e) Continue till all customers served.
(f) Calculate total distance travelled and the number of vehicles used and compute the objective function value for the complete route using (1) (which gives total distance travelled by all used vehicles).

Step 2. Construct a specified number of feasible tours as in step 1.
Step 3. From among these constructed feasible tours, choose the tour which uses minimum number of ants (vehicles). In case of a tie, choose the tour in which total travelled distance is minimum (or vice-versa depending upon whether greater priority is to be given to minimize the number of vehicles used or total distance travelled).

3.1.1 Selection of Next Customer

In setting up of a feasible tour, each ant constructs a path that visits certain customer before returning back to depot. In the previous studies using ACO/ACS for solving VRPTW, the ant (vehicle), currently located at node i, selects the next node j to move to using the transition rule,

$$j = \begin{cases} \arg\max_{j \in \phi i} \left\{ \tau_{\mathrm{ij}}^{\alpha} \eta_{\mathrm{ij}}^{\beta} \right\} & \text{if } q \leq q_0 \\ J & \text{otherwise} \end{cases} \tag{10}$$

where

$$\eta_{\mathrm{ij}} = 1/d_{\mathrm{ij}} \text{ is a heuristic-based parameter.} \tag{11}$$

and $J \in \phi_i$ is randomly chosen according to the probability

$$p_{\mathrm{iJ}}^{k} = \frac{\tau_{\mathrm{iJ}}^{\alpha} \eta_{\mathrm{iJ}}^{\beta}}{\sum_{u \in \phi_i} \tau_{\mathrm{iu}}^{\alpha} \eta_{\mathrm{iu}}^{\beta}} \tag{12}$$

Here, set ϕ_i contains the cities not visited so far.

In (10), $q \sim U\,(0, 1)$, and $q_o \in [0, 1]$ is a user-specified value of parameter q.

In (11), d_{ij} is Euclidian distance between i and j and τ_{ij} is the amount of pheromone on the path between current location i and next possible location j. Also α, β are the positive constants that determine the importance of η verses τ.

The transition rule (10) creates a bias toward customers connected by short distances and having large amount of pheromone. The parameter q_o balances exploration and exploitation. if $q \leq q_o$, the algorithm exploits (favoring the best nearest customer). Otherwise if $q > q_o$, the algorithm explores selecting node $j \in \phi_i$ randomly.

3.1.2 Pheromone Update for New Tour

After construction of a complete feasible route, the pheromones are laid for the next path depending upon the total traveled distance (L) of that route. For each arc $v_i \rightarrow v_j$ that was used in the previous feasible route, the pheromone trail is increased by $\Delta\tau_{\text{ij}}$. Furthermore, part of existing pheromone is also allowed to evaporate [4, 11, 14, 17, 34]. Thus in the next route, pheromones are updated according to the following

$$\tau_{\text{ij}} = \tau_{\text{ij}}\,(1-\rho) + \Delta\tau_{\text{ij}} \tag{13}$$

$$\text{where } \Delta\tau_{\text{ij}} = Q/L \tag{14}$$

Here, ρ is parameter that controls rate of evaporation of pheromone.

3.2 Proposed Modifications in ACS Algorithm

Following modifications have been introduced in the conventional algorithms for solving VRPTW problem using ACS heuristics.

1. Whereas earlier approaches using ant colony technique for solving VRPTW problem have primarily given importance to distance and pheromone only to guide the heuristic [3, 32, 34], we in our present study have used besides these two parameters such as urgency to serve, waiting time and bias parameter also for this purpose. In (11), heuristic-based parameter η_{ij} only gives importance to the distance in determining the heuristic parameter. However, it was observed on experimentation that in addition to the distance waiting time, urgency to serve, and biasing should also be given importance in deciding the choice of next customer [11, 32].
   As a result in our present study choice of η_{ij} defined in (11) has been modifies as:

$$\eta_{\text{ij}} = \frac{1}{(d_{\text{ij}} + w_j)^{\lambda}} * \frac{1}{(l_j - a_j)^{\gamma}} * \frac{1}{\left(I_{\max} - I_j\right)\delta} \tag{15}$$

   In (13), a_j is arrival time of vehicle at customer j and w_j defined as

$$w_j = \begin{cases} e_j - a_j & \text{if } e_j > a_j \\ 0 & \text{otherwise} \end{cases} \tag{16}$$

is the waiting time at customer j before service can be started. Also $l_j - a_j$, $a_j < l_j$, is the difference between the latest arrival time l_j and actual arrival time a_j at customer j. It is a measure of urgency of customer j to be served, emphasizing that those customers whose time window is going expire soon be given priority. Also $I_{\max} - I_j$, (where I_j is the number of consecutive times the customer j who could be next visited from the present customer has not been visited and $I_{\max}$ is a user-specified maximum permissible value of $I_j(I_j < I_{\max})$ is a measure of bias factor).

2. In order to prevent the slow convergence of the algorithm when specified number of initial tours have been generated, we update the pheromone for the next tour using the best solution among the solution provided by m feasible routes [17]. In order to prevent local optimization and increase the probability of obtaining higher quality of solution upper and lower values of pheromone have been specified as $1/\sum 2d_{0i}$ and $1/\min(d_{\mathrm{ij}})$ respectively, where d_{0i} is distance from the depot to the customer.
3. In conventional studies, total travelled distance has been minimized irrespective of vehicles needed. However keeping in view the fact that cost of obtaining a vehicle and its maintenance is generally much more than fuel cost we have tried to minimize total number of vehicles required as a first priority and total travelled distance as a second priority.

4 Implementation of the Modified ACS Algorithm

In this section, we summarize our computational experience of using the modified ACS algorithm for solving some of the Solomon benchmark [5] problems.

Solomon generated a set of 56 problems which have been frequently used in literature to check the performance of the developed algorithm. This set is divided into three categories namely C, R, and RC. In C category problems, customers are clustered either geographically or according to the time windows. R types of problems have uniform distribution of customers. Category RC is hybrid of problems of R and C set. In our present study, we have chosen 15 problems of 25 customers, 10 problems of 50 customers, and 10 problems of 100 customers from all these three sets. To solve these problems, the proposed algorithm was coded in MATLAB 7.0 at Intel Core 2 Duo 2.0 GHz. After experimentation, it was observed that the following values of parameters proved most suited for solving these problems. The number of initial feasible tours $= 10, \alpha = 1, \beta = 1, \lambda = 5.5, \gamma = 3.5, \delta = 1, Q = 250$ and $q_0 = 0.9$. All problems were run for maximum of 2,500 iterations, Tables 1, 2 and 3 present a summary of our results and their comparison with the best-known routing solutions compiled from different heuristics available in literature as per our information.

It may be noticed that whereas in our study, we have used only ACS, most of the earlier studies with ACO/ACS usually are hybrid in the sense that after completion of search with ACO/ACS, search is further carried out with some other optimization

Table 1 Comparison of best-known results with the results generated by proposed algorithms for Solomon's 25 customers set problem

Problem	Best	Worst	Using conventional ACO's	Best known [Ref.]
C201	215.54/2	215.54/2	222.53/2	214.7/2 [10]
R101	618.33/8	619.17/8	625.23/8	617.1/8 [20]
R102	563.35/7	573.15/7	605.45/7	547.1/7 [20]
R105	556.72/5	556.72/5	600.13/5	530.5/6 [20]
R109	442.63/5	448.54/5	510.31/5	441.3/5 [20]
RC101	462.15/4	462.15/4	507.87/4	461.4/4 [20]
RC105	416.16/4	416.88/4	435.97/4	411.3/4 [20]
RC106	346.51/3	346.51/3	402.11/3	345.5/3 [20]
RC201	432.30/2	432.30/2	412.34/3	360.2/3 [10]
RC202	376.61/2	381.75/2	400.72/2	338.0/3 [18]
RC203	433.94/1	433.94/1	454.78/2	326.9/3 [19]
RC204	331.29/1	333.36/1	370.56/1	299.7/3 [8]
RC205	386.15/2	386.15/2	413.37/3	338.0/3 [23]
RC207	358.92/2	367.92/2	387.16/2	298.3/3 [19]
RC208	309.85/1	309.85/1	313.76/1	269.1/2 [8]

Table 2 Comparison of best-known results with the results generated by proposed algorithms for Solomon's 50 customers set problem

Problem	Best	Worst	Using conventional ACO's	Best known [Ref.]
C101	363.25/5	363.25/5	363.25/5	362.5/5 [20]
C201	444.96/2	444.96/2	402.43/3	360.2/3 [20]
C205	444.57/2	444.57/2	407.58/2	360.2/3 [20]
R101	1053.04/12	1054.84/12	1107.18/12	1044/12 [20]
R201	882.32/2	893.56/3	900.72/3	791.9/6 [20]
R202	869.42/2	870.06/2	898.68/3	791.9/6 [20]
R203	741.3/2	764.3/2	612.32/5	605.3/5 [20]
R206	711.6/2	711.6/2	645.56/4	632.4/4 [20]
R209	722.24/2	735.20/2	619.23/4	600.6/4 [20]
RC101	951.07/8	962.80/8	987.97/8	944/8 [20]

heuristic also (such as local search, genetic algorithm [7, 21, 27]). In order to compare our present results with performance of earlier versions of ACO only (without use of any hybrid), we repeated our experimentation with those versions without using any other add on optimization heuristic. The results of this study are also presented in the Table 1 (column 4) for comparison.

The proposed algorithm has produced some improved results with lesser number of vehicles used (however, with some increase in routing length compared to the best available in literature. This is due to the priority that we assigned to minimize the total number of vehicles used visa vis total distance travelled).

Table 3 Comparison of best-known results with the results generated by proposed algorithms for Solomon's 100 customers set problem

Problem	Best	Worst	Using conventional ACO's	Best known [Ref.]
C101	828.94/10	828.94/10	830.37/10 (828.94/10)* [9]	828.94/10 [30]
C102	874.20/10	875.36/10	917.53/10 (828.94/10)* [34]	828.94/10 [30]
C105	828.94/10	828.94/10	830.37/10 (828.94/10)* [9]	828.94/10 [30]
C106	856.18/10	857.91/10	875.71/10 (828.94/10)* [9]	828.94/10 [30]
C107	830.60/10	838.42/10	842.67/10 (828.94/10)* [34]	828.94/10 [30]
C201	591.56/3	591.56/3	594.23/3 (591.56/3)* [34]	591.56/3 [20]
C205	591.5/3	595.33/3	598.28/3 (588.88/3)* [25]	588.88/3 [20]
R101	1714.26/19	1725.65/19	1845.12/19 (1670.66/19)* [25]	1645.79/19 [30]
R102	1558.19/17	1575.69/17	1613.34/18 (1535.52/17)* [25]	1486.12/17 [30]
R105	1519.55/14	1544.86/14	1853.45/18 (1365.23/15)* [25]	1377.11/14 [30]

Note * indicates results available in literature using hybrid versions of ACO's

5 Conclusions

In this paper, a modified version of ACS is proposed.

In our proposed algorithm, we have modified the heuristic-based parameter and pheromone updation rules, used in conventional ACO for solving VRPTW. An extensive computational study on a set of benchmark test problems has been conducted. The experimental results show that the proposed algorithm even when used by itself is competitive with the earlier versions of ACO even when these are hybridized with certain other heuristics. We have obtained certain results in which lesser number of vehicles are needed than those reported in literature. However, in most of such cases, total distance travelled is slightly greater. Lesser number of vehicles means less initial investment in purchase of vehicles and less maintenance cost. (However, there is slight increase in fuel cost if total distance travelled is more).

The results are encouraging and we propose to direct further study toward use of proposed algorithm for solving the dynamic VRPTW

References

1. Andres, F.M.: An iterative route construction and improvement algorithm for the vehicle routing problem with soft time windows. Transp. Res. Part B **43**, 438–447 (2010)
2. Azi, N., Gendreau, M., Potvin, J.Y.: An exact algorithm for a vehicle routing problem with time windows and multiple use of vehicles. Eur. J. Oper. Res. **202**(3), 756–763 (2010)
3. Balseiro, S.R., Loiseau, I., Ramone, J.: An ant colony algorithm hybridized with insertion heuristics for the time dependent vehicle routing problem with time windows. Comput. Oper. Res. **38**, 954–966 (2011)
4. Bell, J.E., McMullen, P.R.: Ant colony optimization techniques for the vehicle routing problem. Adv. Eng. Inform. **18**, 41–48 (2004)
5. Best solutions to Solomon problems identified by heuristics. (SINTEF VRP page): http://www.sintef.no/static/am/opti/projects/top/vrp/bknown.html

6. Bianchessi, N., Righini, G.: Heuristic algorithms for the vehicle routing problem with simultaneous pick-up and delivery. Comput. Oper. Res. **34**, 578–594 (2007)
7. Bräysy, O.: Local search and variable neighborhood search algorithms for the vehicle routing problem with time windows. University of Vaasa, Finland, Doctoral Dissertation (2001)
8. Chabrier, A.: Vehicle routing problem with elementary shortest path based column generation. Comput. Oper. Res. **33**(10), 2972–2990 (2006)
9. Chen, C.-H., et al.: A hybrid ant colony system for vehicle routing problem with time windows. J. East. Asia Soc. Transp. Stud. **6**, 2822–2836 (2005)
10. Cook, W., Rich, J.L.: A parallel cutting plane algorithm for the vehicle routing problem with time windows. Working Paper of Computational and Applied Mathematics, Rice University, Houston, TX (1999)
11. Donati, A.V., et al.: Time dependent vehicle routing problem. Eur. J. Oper. Res. (2006). doi:10.1016/j.ejor.06.047
12. Dorigo, M., Birattari, M., Stützle, T.: Ant colony optimization: Artificial ants as a computational intelligence technique. IEEE Comput. Intell. Mag. **1**, 28–39 (2006)
13. Dorigo, M., Blum, C.: Ant colony optimization theory: A survey. Theor. Comput. Sci. **334**, 243–278 (2005)
14. Dorigo, M., Gambardella, L.M.: Ant colony system: A cooperative learning approach to the traveling salesman problem. IEEE Trans. Evol. Comput. **1**, 53–66 (1997)
15. Dorigo, M., Maniezzo, V., Colorni, A.: Ant system: Optimization by a colony of cooperating agents. IEEE Trans. Syst. Man Cybern. Part B. Cybern. **26**, 29–41 (1996)
16. Gambardella, L.M., Taillard, E., Agazzi, G.: MACS-VRPTW: A Multiple Ant Colony System for Vehicle Routing Problems with Time Windows New Ideas in Optimization. McGraw-Hill, London (1999)
17. Garcia-Najera, A., Bullinaria, J.A.: An improved multi-objective evolutionary algorithm for the vehicle routing problem with time windows. Comput. Oper. Res. **38**(1), 287–300 (2011)
18. Irnich, S., Villeneuv, D.: The shortest path problem with k-cycle elimination (k > 3): Improving a branch-and-price algorithm for the VRPTW INFORMS. J. Comput. **18**(3), 391–406 (2003)
19. Kallehauge, B., Larsen, J., Madsen, O.B.G.: Lagrangean duality and non-differentiable optimization applied on routing with time windows. Comput. Oper. Res. **33**(5), 1464–1487 (2006)
20. Kohl, N., Desrosiers, J., Madsen, O.B.G., Solomon, M.M., Soumis, F.: 2-path cuts for the vehicle routing problem with time windows. Transp. Sci. **33**(1), 101–116 (1999)
21. Laporate, G.: The vehicle routing problem: An overview of exact and approximate algorithms. Eur. J. Oper. Res. **59**, 345–358 (1997)
22. Laporte, G.: Fifty years of vehicle routing. Transp. Sci. **43**, 408–416 (2011)
23. Larsen J.: Parallelization of the vehicle routing problem with time windows. Ph.D. thesis IMM-PHD-1999-62, Department of Mathematical Modelling, Technical University of Enmark, Lyngby, Denmark (1999)
24. Lau, H.C., et al.: Vehicle routing problem with time windows and a limited number of vehicles. Eur. J. Oper. Res. **148**, 559–569 (2003)
25. Ma, X.: Vehicle routing problem with time windows based on improved ant colony algorithm. In Proceedings of IEEE Computer Society Second International Conference on Information Technology and Computer Science, Washington, DC, USA, pp. 94–97 (2010)
26. Masrom, S., Nasir, A.M., Abidin, S.Z., Rahman, A.S.A.: Software framework for vehicle routing problem with hybrid metaheuristic algorithms. In Proceedings of the Applied Computing Conference, Angers, pp. 55–61 (2011)
27. Mester, D.: An evolutionary strategies algorithm for large scale vehicle routing problem with capacitate and time windows restrictions. In Proceedings of the Conference on Mathematical and Population Genetics, University of Haifa, Israel (2002)
28. Qi, C., Sun, Y.: An improved ant colony algorithm for VRPTW. In IEEE International Conference on Computer Science and, Software Engineering (2008)
29. Rizzoli, A.E., Montemanni, R., Lucibello, E., Gambardella, L.M.: Ant colony optimization for real-world vehicle routing problems. Swarm Intell **1**(2), 135–151 (2007)

30. Rochat, Y., Taillard, E.: Probabilistic diversification and intensification in local search for vehicle routing. J. Heuristics **1**, 147–167 (1995)
31. Solomon, M.M.: Algorithms for vehicle routing and scheduling problems with time window constraints. Oper. Res. **35**, 254–265 (1987)
32. Yu, B., Yang, Z.Z., Yao, B.: An improved ant colony optimization for vehicle routing problem. Eur. J. Oper. Res. **196**, 171–176 (2009)
33. Yuichi, N., Olli, B.: A powerful route minimization heuristic for the vehicle routing problem with time windows. Oper. Res. Lett. **37**, 333–338 (2009)
34. Zhang, X., Tang, L.: A new hybrid ant colony optimization algorithm for the vehicle routing problem. Pattern Recogn. Lett. **30**, 848–855 (2009)

Energy Saving Model for Sensor Network Using Ant Colony Optimization Algorithm

Doreswamy and S. Narasegouda

Abstract In this paper, we propose an energy saving model for sensor network by finding the optimal path for data transmission using ant colony optimization (ACO) algorithm. The proposed model involves (1) developing a relational model based on the correlation among sensors both in spatial and in temporal dimensions using DBSCAN clustering, (2) identifying a set of sensors which represents the network state, and (3) finding the best path for transmission of data using ACO algorithm. Experimental results show that the proposed model reduces the energy consumption by reducing the amount of data acquiring and query processing using the representative sensors and ensures that the transmission is done on the best path which minimizes the probability of retransmission of data.

Keywords Ant colony optimization · DBSCAN · Data mining · Sensor network

1 Introduction

Revolution in technology has made sensors an integral part of our life. Sensors are used in various applications to make human life safer, comfortable, and profitable. In many areas such as agriculture, smart parking, structural health, traffic control, fire detection, air and/or water pollution monitoring, environmental monitoring, surveillance, sensor nodes are deployed to collect and process the data to aid users in decision making. These tiny sensor nodes are equipped with limited battery supply. And the performance, life span of the network, depends on the energy of the

S. Narasegouda (✉)
Department of P.G. Studies and Research in Computer Science,
Mangalore University, Mangalagangothri, Mangalore, India
e-mail: srinivasnpatil@gmail.com

Doreswamy
e-mail: doreswamyh@yahoo.com

B. V. Babu et al. (eds.), *Proceedings of the Second International Conference on Soft Computing for Problem Solving (SocProS 2012), December 28–30, 2012*, Advances in Intelligent Systems and Computing 236, DOI: 10.1007/978-81-322-1602-5_6,

sensor nodes. Due to the limited battery power, sensor networks' life span rapidly decreases resulting in the failure of the applications. Researchers have observed that data processing and data transmission are the main causes for the decline in sensor node energy. Hence, various methodologies have been proposed in order to reduce query processing and data transmission.

In a densely deployed sensor network, correlated data are frequently generated by different sensors. In the past, data mining technique such as clustering has been exploited [1–3] to analyze the correlation among sensors both in spatial and in temporal dimensions to reduce the amount of data acquiring and query processing. Other methodologies such as data aggregation, clustering, efficient routing protocols have also been applied to reduce the energy consumption in sensor networks.

The rest of the paper is organized as follows. An overview of the literature survey is given in Sect. 2. Section 3 presents the proposed model. Experimental results are discussed in Sect. 4, and conclusion is given in Sect. 5.

2 Literature Review

Energy-efficient routing strategy using nonlinear min–max programming problem with convex product was proposed in [4]. The selection of sensor representatives to reduce query processing and save energy was first proposed by [5], where selection was made by exchanging the messages between the neighboring nodes. In [1], clustering techniques were exploited to select a set of representative sensor nodes for query processing, resulting in the reduction in the data collection and energy consumption. Energy-efficient data gathering algorithm Energy-efficient Routing Algorithm to Prolong Lifetime (ERAPL) was proposed by [6]. SeReNe framework was proposed by [2] to develop energy saving model for wireless sensor networks. SeReNe framework exploited the clustering technique to select the set of sensor representatives. Using traveling salesman problem concept, an efficient transmission path is estimated. In [3], a technique was developed for the selection of cluster heads to reduce energy consumption called Cluster-based Routing for Top k Querying (CRTQ). Chong et al. [7] proposed a rule-based framework called Context Awareness in Sensing Environments (CASE) to save energy in sensor networks. In [8], an energy-efficient routing protocol for acquiring correlated data in sensor network by considering the issues such as energy of the sensor node, multi-hop data aggregation was developed. To reduce the number of message exchange between source and destination, data aggregation technique was designed by [9].

Another approach to save energy in sensor network is to select the best transmission path which minimizes the retransmission rate. But selecting the optimal path from a finite set sensor node is a combinatorial optimization problem. Hence, in this paper, we have proposed to develop energy saving model for sensor network by finding an optimal path for transmission by applying ant colony optimization (ACO) algorithm on a subset of sensor nodes, which best represents the network state.The ant colony optimization was proposed in [10, 11] and is used in the past for finding an optimal solution in many combinatorial problems such as traveling salesman problem, assignment problems, scheduling problems. [12].

3 Proposed Model

The proposed model works on the phenomena of finding the optimized data transmission path using ACO algorithm on the relational model of sensor network to reduce query processing, data retransmission, and energy consumption.

In relational model, clustering techniques are exploited to find correlation among the sensor observations in both spatial and temporal dimensions. Among clustering techniques, partitioning clustering algorithms such as k-means are only capable of finding circle-shaped clusters, whereas density-based clustering algorithms such as DBSCAN is capable of identifying non-spherical-shaped clusters and DBSCAN is very less sensitive in the presence of outliers. Hence, DBSCAN is used in our model. By applying the DBSCAN algorithm [13], different clusters are formed. In each cluster, representative sensors are selected based on the measurement tendency strategy as explained in [1, 2]. These set of representative sensors symbolize the entire sensor network. Hence, query processing is reduced by querying only the set of representative sensors instead of querying all sensor nodes in the network.

In any sensor network, sensor nodes consume more energy to transmit a packet than to collect the data. Due to lack of connectivity strength, packet may not reach the destination. In such cases, retransmission of lost packet results in excessive energy consumption. In order to minimize the loss of packets, we propose to find the optimized transmission path among the representative sensors by applying ACO algorithm. To estimate the best transmission path, we have considered the probability of packets reaching from source to destination as a parameter. In [14], ACO algorithm is explained in detail. The pseudo-code for ACO algorithm is given below.

```
Set parameters: alpha,beta,rho,Q,ant_num,Max_time.
Initialize: Sensor_Distance_Graph,
Ant_to_random_trail,
Determine initial best path and its length
Initialize pheromone trails
While Max_time
 Update Ants
 Update pheromones
 Construct Ant Solutions
 If current solution is better than previous
  Initialize: Best solution=current solution
End While
```

The proposed model is summarized as follows:

1. Calculate spatiotemporal measures using DBSCAN.
2. Select representative sensors.
3. Calculate optimized transmission path using ACO algorithm.

4 Experimental Results

The proposed model is implemented using Visual Studio 2010 with C#. Experiment has been conducted on publicly available [15] dataset. It contains 2.3 million observations and three tables, namely *location table* containing information about x and y coordinates of sensors, *aggregate connectivity table* containing information about probability of a packet reaching from source to destination, and *data table* containing information about features such as date, time, epoch, sensor id, temperature, humidity, light, voltage. By applying the data mining preprocessing techniques such as data cleaning, data smoothing using binning, dataset has been reduced to 6.5 lakhs.

In our experiment, before applying clustering technique, entire dataset has been divided into three parts. Each part contains readings taken for 12 days. DBSCAN is applied on each part separately. And DBSCAN parameters are set as, epsilon value eps $=$ 1.75 and minimum point to form cluster minPts $=$ 25. In our experiment, we found that cluster 1 contains measurements from all sensors and most of the other clusters formed were containing faulty data; hence, they were discarded. After clusters are formed, representative sensors and set of sensors it representing are obtained using measuring tendency. Twenty-three representative sensors have been identified, which can be used to represent the network state. In Fig 1, purple rectangle

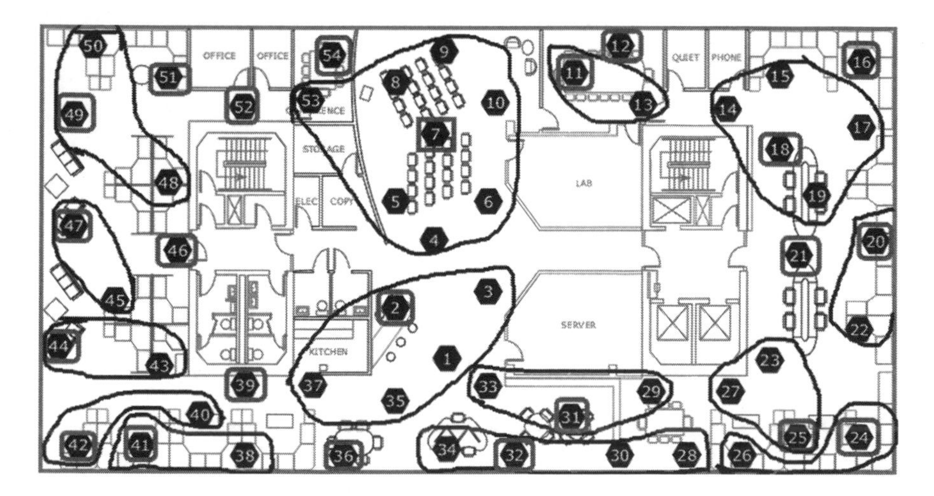

Fig. 1 Intel Berkeley research lab

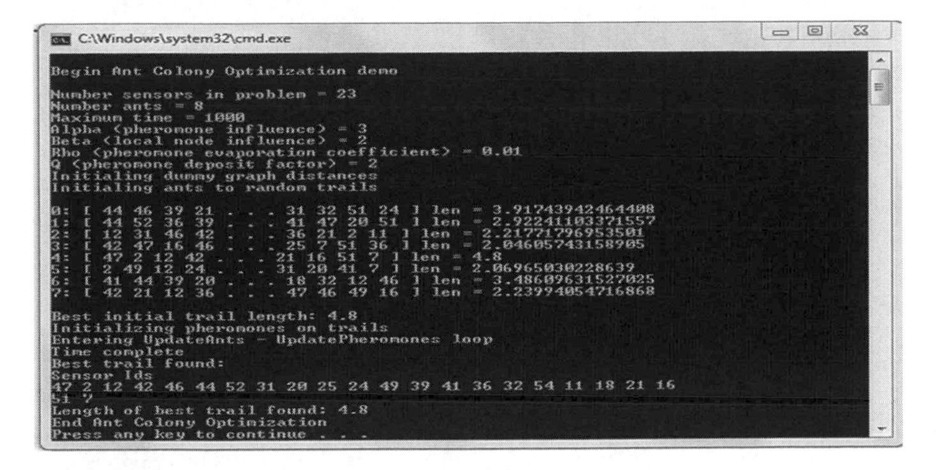

Fig. 2 Result of ACO

Table 1 Default values of ACO

Parameter	Default value
Influence of pheromone on direction (*alpha*)	3
Influence of adjacent node distance (*beta*)	2
Pheromone decrease factor (*rho*)	0.01
Pheromone increase factor (*Q*)	2.0
Number of ants (*Ant_num*)	8
Maximum time (*Max_time*)	1000

represents the representative sensors, and a boundary is drawn to show the set of correlated sensors they represent and representative sensors without any boundary indicates that they represent only themselves.

The optimal transmission path among the representative sensors is calculated using ACO algorithm where the best solution is defined as the transmission path whose connectivity strength is higher than other transmission paths. The default parameter set for ACO algorithm is given in Table 1. The optimal transmission path found by the ACO algorithm is **{47, 2, 12, 42, 46, 44, 52, 31, 20, 25, 24, 49, 39, 41, 36, 32, 54, 11, 18, 21, 16, 51, 7 }**, and the strength of the best trail found is **4.8** (result is shown in Fig. 2).

From experimental results, we observed that (1) out of 54 only 23 sensor nodes need to be queried to model the network state, i.e., energy required to collect the data can be reduced since query processing is reduced by 57.40 %. (2) Optimized path for data transmission is obtained.

5 Conclusion

In any sensor network, the main reason for energy consumption is data acquiring, query processing, and data transmission. Furthermore, an excessive energy is consumed when data are retransmitted due to failure in reaching the destination. Data acquiring and query processing can be reduced by analyzing the correlation among the sensors, and data retransmission can be reduced by selecting the best transmission path which increases the possibility of packet being delivered to destination successfully.

A relational model is developed using DBSCAN to analyze the correlation among sensors both in spatial and in temporal dimensions. Using relational model, a set of representative sensors are selected, which best represents the network's state. Instead of using all sensor nodes, only representative sensors are used for querying the sensor network, resulting in the reduction in amount of data acquiring and query processing. In order to reduce the retransmission of data, we applied ACO algorithm to find the best transmission path which increases the probability of data being delivered from source to sink successfully.

Experimental results show that the proposed model reduces the energy consumption by reducing the amount of data collection and query processing using the representative sensors and ensures that the transmission is done on the best path which minimizes the need for retransmission of data. In future, we wish to address the other issues such as estimation of missing data, prediction of data, developing a visualization model for sensor networks.

References

1. Baralis, E., Cerquitelli, T., D'Elia, V.: Modeling a sensor network by means of clustering. In: Proceedings of the 18th International Workshop on Database and Expert Systems Applications, pp. 177–181 (2007)
2. Apiletti, D., Baralis, E., Cerquitelli, T.: Energy saving models for wireless sensor networks. Knowl. Inf. Syst. **28**, 615–644 (2010)
3. Mo, S., Chen, H., Li, Y.: Clustering based routing for top k query in wireless sensor networks. EURASIP J. Wireless Commun. Netw. 1–13 (2011)
4. Shiou, C.W., Lin, Y.S., Cheng, H.C., Wen, Y.F.: Optimal energy efficient routing for wireless sensor networks. In: Proceedings of 19th International Conference on Advanced Information Networking and Applications, pp. 325–330 (2005)
5. Kotidis, Y.: Snapshot queries towards data centric sensor networks. In: IEEE Proceedings of 21st International Conference on Data, Engineering. pp. 131–142 (2005)
6. Zhu, Y., Wu, W., Pan, J., Tang, Y., et al.: An energy efficient data gathering algorithm to prolong lifetime of wireless sensor. Networks **33**, 639–647 (2010)
7. Chong, S.K., Gaber, M.M., Krishnaswamy, S., Loke, S.W., et al.: Energy conservation in wireless sensor networks a rule based approach. Knowl. Inf. Syst. **28**(3), 579–614 (2011)
8. Zeydan, E., Tureli, D.K., Comaniciu, C., Tureli, U.: Energy efficient routing for correlated data in wireless sensor networks. Ad Hoc Netw. **10**, 962–975 (2011)
9. Hung, C.C., Peng, W.C.: Optimizing in-network aggregate queries in wireless sensor networks for energy saving. Data Knowl. Eng. **70**, 617–641 (2011)

10. Dorigo, M., Maniezzo, V., Colorni, A.: Positive feedback as a search strategy. Technical Report, Dipartimento di Elettronica, Politecnico di Milano, Italy, pp. 91–016 (1991)
11. Dorigo, M., Maniezzo, V., Colorni, A.: Ant system optimization by a colony of cooperating agents. IEEE Trans. Syst. Man Cybern. B Cybern. **26**, 29–41 (1996)
12. Dorigo, M., Blum, C.: Ant colony optimization theory a survey. Theor. Comput. Sci. **344**, 243–278 (2005)
13. Han, J., Kamber, M.: Data Mining concepts and techniques. Morgan Kaufmann Publishers, San Francisco (2006)
14. Dorigo, M., Birattari, M., Stutzle, T.: Ant colony optimization. IEEE Comput. Intell. Mag. 28–39 (2006)
15. Intel Berkeley Research Lab.: http://db.csail.mit.edu/labdata/labdata.html

Multi-Objective Optimization of PID Controller for Coupled-Tank Liquid-Level Control System Using Genetic Algorithm

Sanjay Kr. Singh, Nitish Katal and S. G. Modani

Abstract The main aim of this chapter is to obtain optimal gains for a PID controller using multi-objective genetic algorithm used in a coupled-tank liquid-level control system. Liquid level control system is a nonlinear system and finds a wide application in petrochemical, food processing, and water treatment industries, and the quality of control directly affects the quality of products and safety. This chapter employs the use of multi-objective genetic algorithm for the optimization of the PID gains for better plant operations in contrast to conventional tuning methods and GA. The simulations indicate that better performance is obtained in case of multi-objective genetic algorithm-optimized PID controller.

Keywords PID controller · Multi-objective genetic algorithm · PID optimization · Liquid level control

1 Introduction

Coupled-tank liquid-level control is the center to many diverse industrial applications ranging from petrochemical, food processing to nuclear power generation [1]. The main objective of this system is to control the flow of liquid between tanks so that optimum levels are maintained in both the tanks [2].

S. K. Singh
Department of ECE, Anand International College of Engineering, Jaipur, Rajasthan, India
e-mail: sksingh.mnit@gmail.com

N. Katal(✉)
Department of ECE, ASET, Amity University, Jaipur, Rajasthan, India
e-mail: nitishkatal@gmail.com

S. G. Modani
Malaviya National Institute of Technology, Jaipur, Rajasthan, India

B. V. Babu et al. (eds.), *Proceedings of the Second International Conference on Soft Computing for Problem Solving (SocProS 2012), December 28–30, 2012*, Advances in Intelligent Systems and Computing 236, DOI: 10.1007/978-81-322-1602-5_7, © Springer India 2014

In this chapter, coupled-tank liquid-level system has been considered, and the PID controller is implemented for either maintaining the liquid level at a desired set point, disturbance rejection or to be used for moving the liquid set point. For designing the PID controller, classical method of Ziegler Nichols has been used, followed by the optimization using multi-objective genetic algorithm. The gain parameters have been tuned with respect to the objective function, stated as "Sum of integral of the squared error and the sum of integral of absolute error". According to the results obtained, considerably better results have been obtained in case of multi-objective genetic algorithm-optimized PID controllers when compared to Ziegler-Nichols method in their respective step response on the system.

2 Mathematical Modeling of Coupled-Tank Liquid-Level System

Considering the coupled-tank system, is in Fig. 1. The dynamic equations of the system, by considering the flow balances for each tank, the equations for rate of change of fluid volume in tanks are as [3, 4]:

$$\text{For Tank 1} : Q_i - Q_1 = A\frac{\mathrm{d}H_1}{\mathrm{d}t} \tag{1}$$

$$\text{For Tank 2} : Q_1 - Q_0 = A\frac{\mathrm{d}H_2}{\mathrm{d}t} \tag{2}$$

where

H_1, H_2 Height of tank 1 and 2
A Cross sectional area of tank 1 and 2
Q_1, Q_2 Flow rate of the fluid
Q_i Pump flow rate

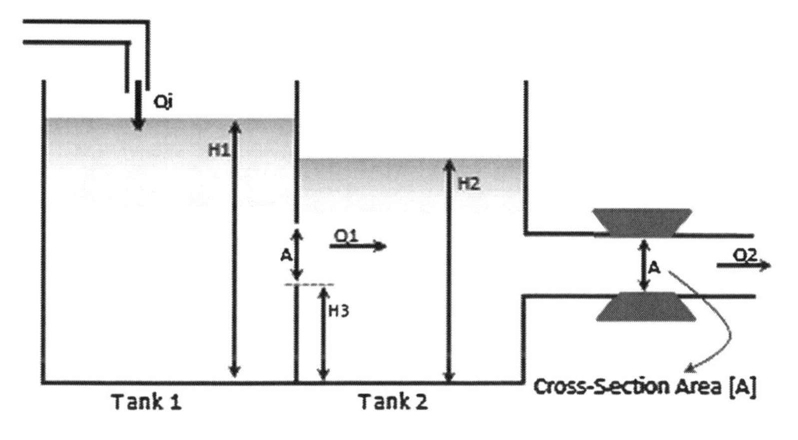

Fig. 1 Schematic representation of the coupled-tank system

The steady-state representation of the coupled-tank system can be given as follows:

$$\begin{bmatrix} \dot{h}_1 \\ \dot{h}_2 \end{bmatrix} = \begin{pmatrix} -k_1/A & k_1/A \\ k_1/A & -\frac{(k_1+k_2)}{A} \end{pmatrix} \begin{bmatrix} h_1 \\ h_2 \end{bmatrix} + \begin{bmatrix} 1/A \\ 0 \end{bmatrix} q_i \tag{3}$$

Taking the Laplace transformation of Eq. 3, the transfer function is obtained in Eq. 4.

$$G(s) = \frac{1/k_2}{\left(\frac{A^2}{k_1 k_2}\right).s^2 + \left(\frac{A(2k_1+k_2)}{k_1 k_2}\right).s + 1} = \frac{1/k_2}{(sT_1+1)(sT_2+1)}$$

where

$$T_1 T_2 = A^2/k_1 k_2$$
$$T_1 + T_2 = \frac{A(2k_1+k_2)}{k_1 k_2}$$
$$k_1 = \frac{\alpha}{2\sqrt{H_1 - H_2}}$$
$$\text{and } k_2 = \frac{\alpha}{2\sqrt{H_2 - H_3}}$$

Using; $H_1 = 18$ cm, $H_2 = 14$ cm, $H_3 = 6$ cm, $\alpha = 9.5$ (constant for coefficient of discharge), $H = 32$; the transfer function can be obtained in Eq. 4.

$$G(s) = \frac{0.002318}{s^2 + 0.201.s + 0.00389} \tag{4}$$

3 Designing and Optimization of PID Controllers

PID controllers are the most widely used controllers in the industrial control processes [5], and 90 % of the controllers today used in industry are alone PIDs. The general equation for a PID controller can be given by Eq. 5.

$$C(s) = K_p.R(s) + K_i \int R(s)\mathrm{d}t + K_d \frac{\mathrm{d}R(s)}{\mathrm{d}t} \tag{5}$$

where K_p, K_i and K_d are the controller gains, $C(s)$ is output signal, and $R(s)$ is the difference between the desired output and output obtained [6].

3.1 PID Tuning Using Ziegler Nichols

Ziegler Nichols is the most operative method for tuning the PID controllers. But, this method is limited for application till ratio of 4:1 for the first two peaks in closed-loop response, leading to an oscillatory response [7]. Initially, unit-step response is derived (Fig. 2) followed by the computation of the PID gains as suggested by Ziegler-Nichols as in Table 1.

3.2 PID Optimization Using Genetic Algorithm

Genetic algorithms have vanguard advantage of wider adaptability to any constraints and hence are considered as one of the most robust optimization algorithms [8]. Optimization of the PID controllers with genetic algorithms focuses on obtaining the best possible solution for the three PID gains $[K_p, K_i, K_d]$ by minimizing the objective function. For the optimal tuning of the controller, the minimization of the integral square error (ISE) has been carried out.

$$ISE = \int_0^{T_s} e^2(t)\, dt$$

The optimization has been carried out using Global Optimization Toolbox and Simulink [9] with a population size of 20, scattered crossover, both-side migration and roulette-wheel-based selection. The PID gains obtained by optimal tuning using GA are represented in Table 2, and Fig. 3 shows the closed-loop response of the GA-optimized controllers. Figure 4 represents the plot for best and mean fitness

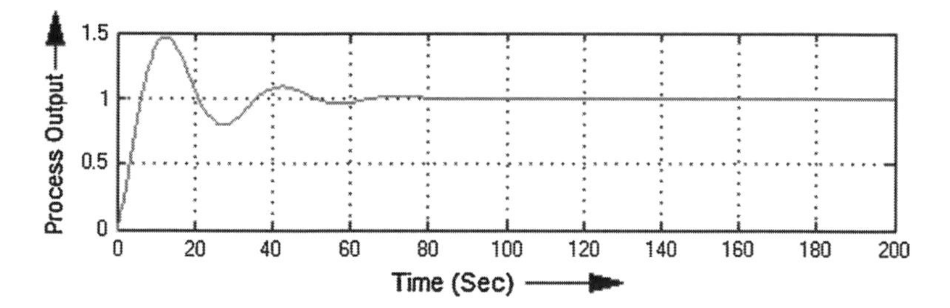

Fig. 2 Closed-loop step response of the system with ZN-PID controller

Table 1 PID parameters estimated by Ziegler-Nichols

PID gains	Value
K_p	28.214
K_i	4.155
K_d	47.89

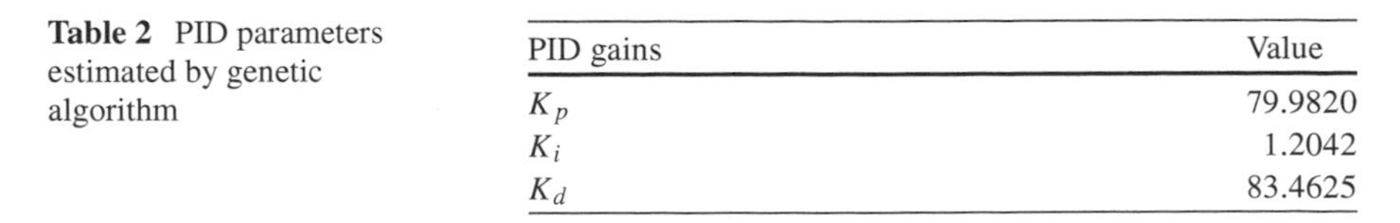

Table 2 PID parameters estimated by genetic algorithm

PID gains	Value
K_p	79.9820
K_i	1.2042
K_d	83.4625

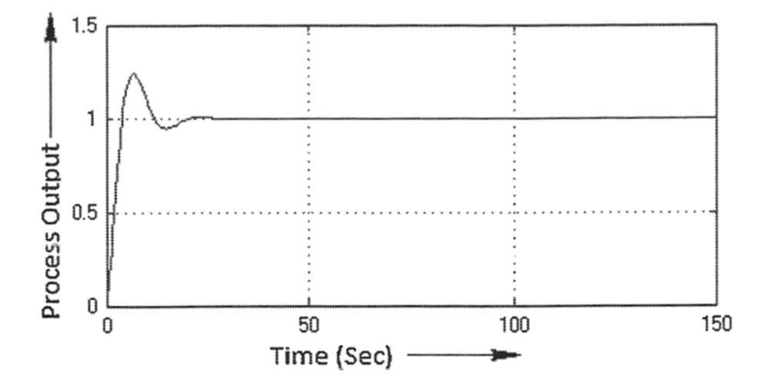

Fig. 3 Closed-loop step response of the system with GA-PID controller

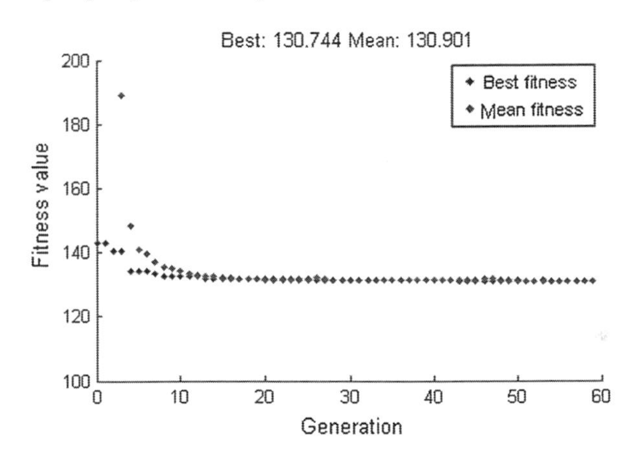

Fig. 4 Plot for the best and average fitness values of the genetic algorithm optimization

values across various generations obtained while optimizing the PID controller using Genetic Algorithm.

3.3 PID Optimization Using Multi-Objective Genetic Algorithm

Since the Ziegler-Nichols tuned PID controllers give an oscillatory response, they are not optimum for implementation for plant. PID optimization using multi-objective genetic algorithm aims at obtaining an optimal Pareto solution, simultaneously

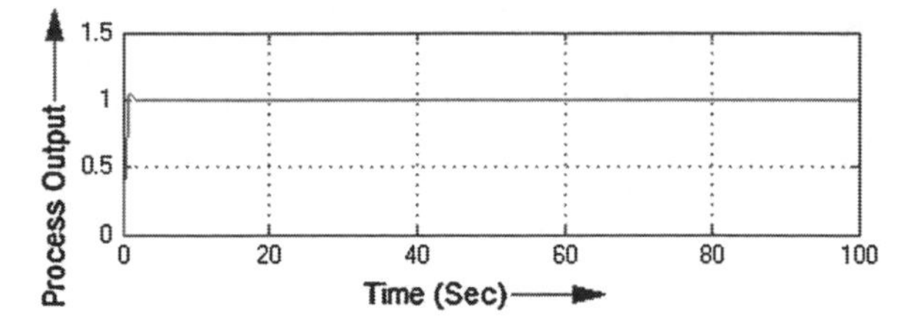

Fig. 5 Closed-loop response using Mobj-GA-optimized PID controllers

Table 3 PID parameters estimated by multi-objective genetic algorithm

PID gains	Value
K_p	255.1
K_i	5.5
K_d	1249.96

improving the objective function of both the objectives O_1 and O_2, given as follows:

First objective function is integral square error (ISE) which discards the large amplitudes, and second objective function is integral absolute error (IAE) which gives the measure of the systems performance. ISE tends to suppress the larger errors, while ISE tends to suppress the smaller errors [10]. The algorithm used here is NSGA-II, which using the controlled elitist genetic algorithm boosts obtaining the better fitness value of the individuals; and if the value is less, it still favors increasing the diversity of the population [11, 12]. Diversity of the populations/gains is controlled by the elite members of the population, while elitism is controlled by Pareto fraction and at Pareto Front also bound the number of individuals.

$$ISE = O_1 = \int_0^{T_s} e^2(t)\,\mathrm{d}t \quad and \quad IAE = \int_0^{T_s} |u(t)|\,\mathrm{d}t$$

The system implementation and optimization have been carried out in MATLAB and Simulink [9] environment using Global Optimization Toolbox. The population

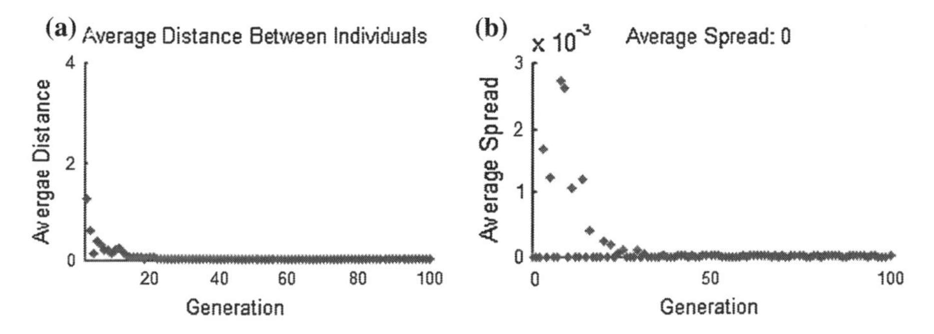

Fig. 6 Plot for the **a** Average distance **b** Average spread between individuals

size of 45 has been considered with adaptive feasible mutation function, heuristic crossover, and the selection of individuals on the basis of tournament with a tournament size of 2. A hybrid function of Fitness Goal Attain (*fgoalattain*) is used, which further minimizes the function after GA terminates. Figure 5 shows the closed-loop response, and the optimized PID parameters are shown in Table 3. In Fig. 6a, distance between members of each generation is shown, and Fig. 6b gives the plot for average Pareto spread, which is the change in distance measure with respect to the previous generations.

4 Results and Discussion

In this chapter, the implementation and simulations of the system has been carried out in Simulink. Initially, the gains of the PID have been estimated using Ziegler Nichols rules [13] which give an oscillatory response, followed by the optimization by genetic algorithm and multi-objective genetic algorithm. The computed parameters are implemented for obtaining the closed-loop response of the system. Figure 7 shows the compared closed-loop step response graph, clearly indicating that better results are obtained in case of multi-objective genetic algorithm-optimized PID controller with decreased overshoot percentage and rise and settling time values. Table 4 represents the numerical data of the results obtained.

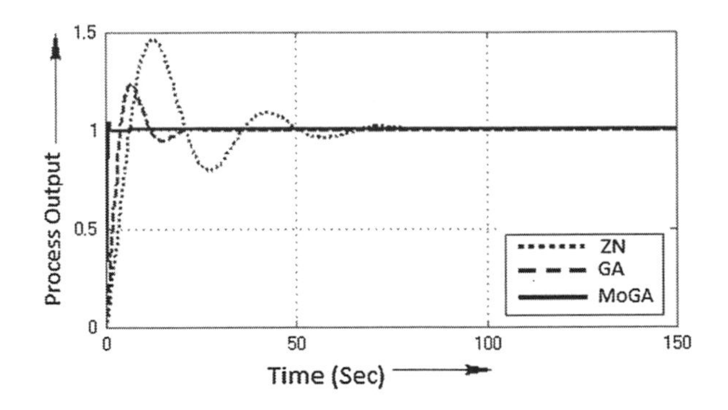

Fig. 7 Comparative closed-loop response of the ZN, GA, and MoGA-optimized PID controllers

Table 4 Comparison of the results

Method of design	Overshoot percentage	Rise time (s)	Settling time (s)
Ziegler-Nichols	46.4	4.83	62.4
Genetic algorithm	23.7	2.93	18.5
Multi-objective GA	4.47	0.504	1.41

5 Conclusion

The use of multi-objective genetic algorithm for the optimization of PID controller offers better results in terms of decreased overshoot percentage and rise and settling times as compared to Ziegler Nichols and genetic algorithm-tuned PIDs, thus offering better operation for the coupled-tank liquid-level control and better plant safety and performance.

References

1. Bhuvaneswari, N. S., G. Uma, and T. R. Rangaswamy. "Adaptive and optimal control of a non-linear process using intelligent controllers." Applied Soft Computing 9.1 (2009): 182–190. Elsevier Ltd.
2. Capón-García, Elisabet, Espuña, Antonio, Puigjaner, Luis: Statistical and simulation tools for designing an optimal blanketing system of a multiple-tank facility. Chemical Engineering Journal **152**(1), 122–132 (2009)
3. B. Seth, D.S. J, "Liquid Level Control", Control System Laboratory (ME413), IIT Bombay (2006–07).
4. Elke Laubwald, "Coupled Tanks Systems 1", www.control-systems-principles.co.uk.
5. Åström, K. J., Albertos, P. and Quevedo, J. 2001. PID Control. Control Engineering Practice, 9,159–1161.J.
6. Norman S. Nise, 2003, Control System Engineering, 4th Edition.
7. Goodwin, G.C., Graebe, S.F., Salgado, M.E.: Control System Design. Prentice Hall Inc., New Jersey (2001)
8. Larbes, C., Aït Cheikh, S. M., Obeidi, T., & Zerguerras, A. (2009). Genetic algorithms optimized fuzzy logic control for the maximum power point tracking in photovoltaic system. Renewable Energy, Elsevier Ltd. 34(10), 2093–2100.
9. MATLAB and SIMULINK Documentation.
10. Corriou, Jean-Pierre. Process Control: Theory and Applications. Springer. 2004. Page132-133.
11. Grefenstette, J.J.: Optimization of Control Parameters for Genetic Algorithms. IEEE Trans. Systems, Man, and, Cybernetics **SMC–16**(1), 122–128 (1986)
12. Konak, Abdullah, Coit, David W., Smith, Alice E., et al.: Multi-objective optimization using genetic algorithm. Reliability Engineering and Safety System **91**, 992–1007 (2006). Elsevier Ltd
13. Ziegler, J.G., Nichols, N.B.: Optimum settings for automatic controllers. Transactions of the ASME. **64**, 759–768 (1942)

Comparative Performance Analysis of Particle Swarm Optimization and Interval Type-2 Fuzzy Logic-Based TCSC Controller Design

Manoj Kumar Panda, G. N. Pillai and Vijay Kumar

Abstract In this paper, an interval type-2 fuzzy logic controller (IT2FLC) is proposed for thyristor-controlled series capacitor (TCSC) to improve power system damping. It has been tested on the single-machine infinite-bus (SMIB) system. The proposed controller performance is compared with particle swarm optimization (PSO) and type-1 fuzzy logic controller (T1FLC)-based TCSC. In this problem, the PSO algorithm is applied to find out the optimal values of parameters of lead-lag compensator-based TCSC controller. The comparative performance is analyzed based on the simulation results obtained for rotor speed deviation and power angle deviation plot, and it has been found that for damping oscillations of SMIB system, the proposed IT2FLC is quite effective. The proposed controller is also robust subjected to different operating conditions and parameter variation of the power system.

Keywords Particle swarm optimization · Type-2 fuzzy system · TCSC · Fuzzy logic controller

1 Introduction

The particle swarm optimization (PSO) algorithm is a population-based, stochastic and multi-agent parallel global search technique [1]. The PSO algorithm is based on the mathematical modeling of various collective behaviors of the living creatures that

M. K. Panda (✉) · V. Kumar
Department of Electronics and Computer Engineering, I. I. T Roorkee, Roorkee, India
e-mail: pandadec@iitr.ernet.in

V. Kumar
e-mail: vijecfec@iitr.ernet.in

G. N. Pillai
Department of Electrical Engineering, I. I. T Roorkee, Roorkee, India
e-mail: gnathfee@iitr.ernet.in

B. V. Babu et al. (eds.), *Proceedings of the Second International Conference on Soft Computing for Problem Solving (SocProS 2012), December 28–30, 2012*, Advances in Intelligent Systems and Computing 236, DOI: 10.1007/978-81-322-1602-5_8,

display complex social behaviors. In the PSO algorithm, while a particle is developing a new situation, both the cognitive component of the relative particle and the social component generated by the swarms are used. This situation enables the PSO algorithm to effectively develop the local situations into global optimum solutions [2]. PSO is a computational intelligence-based technique that is not largely affected by the size and nonlinearity of the problem and can converge to the optimal solution in many problems where most analytical methods fail to converge. It can therefore be effectively applied to different optimization problems in power systems [1]. The PSO is successfully applied in almost all areas of power system engineering like reactive power and voltage control, economic dispatch, power system reliability and security [1]. Very few applications of type-2 fuzzy logic to power system problems were reported in literature [3–6].

Thyristor-controlled series capacitor (TCSC) is one of the important members of flexible AC transmission systems (FACTS) family for damping the power oscillations also to enhance the transient stability [7, 8]. Over the years, artificial intelligence techniques [9–11] being used in developing TCSC models.

In this paper, a comparison has been made between the performance of three types of TCSC controller, i.e., PSO optimized lead lag compensator based TCSC (PSOLLC) in which the time constants and gain of LLC are optimized, a fuzzy logic control (FLC)-based TCSC controller, and the proposed IT2FL-based TCSC controller. Simulation is carried out for single-machine infinite-bus (SMIB) system. The effectiveness of the proposed controller is also tested at all loading conditions with transmission line reactance variation.

Section 2 of this paper describes about basic theory of PSO and its application in TCSC controller design. Type-2 fuzzy logic controller (IT2FLC)-based TCSC is presented in Sect. 3. Results are given and discussed in Sect. 4 and finally conclusion is presented in Sect. 5.

2 Particle Swarm Optimized Lead-Lag Compensator-based TCSC

In this paper, the PSO technique is applied to find out the optimal values of TCSC controller gain K_T and time constants T_{1T} and T_{3T} as shown in Fig. 1.

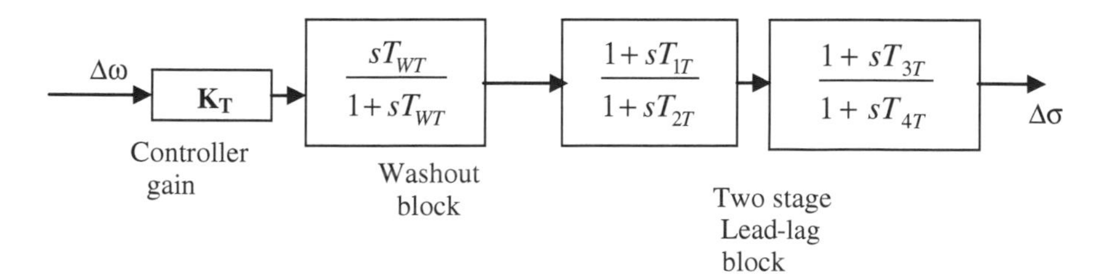

Fig. 1 TCSC controller block diagram

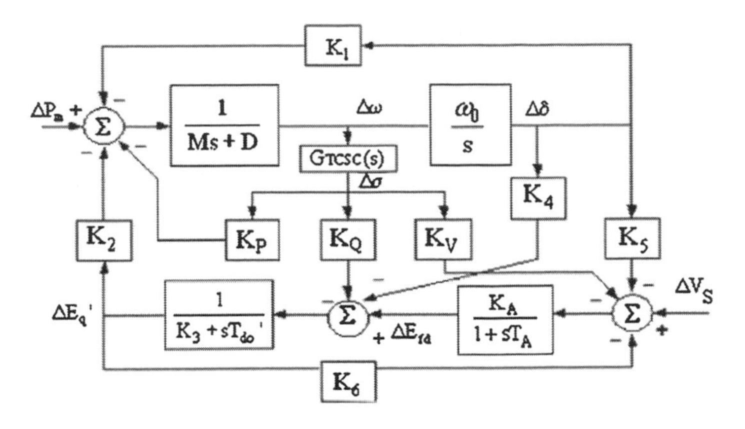

Fig. 2 Modified Phillips–Heffron model of SMIB system using TCSC [12]

2.1 TCSC Controller Description

The TCSC controller consists of a gain block having gain K_T, a washout block, which is a high-pass filter to allow signals associated with oscillations in input signal to pass unchanged and a two-stage phase-compensation block to compensate for the phase lag between the input and output signal.

The transfer function of the TCSC controller is

$$y = K_T \left(\frac{sT_{wT}}{1 + sT_{wT}} \right) \left(\frac{1 + sT_{1T}}{1 + sT_{2T}} \right) \left(\frac{1 + sT_{3T}}{1 + sT_{4T}} \right) x \tag{1}$$

where y is the output signal and x is the input signal of the TCSC controller, respectively. The TCSC controller is connected in the SMIB system model as shown in the Fig. 2. The objective of the TCSC controller is to minimize the power system oscillations after a disturbance to improve the stability by contributing a damping torque [10].

2.2 Application of PSO for Computing Optimum TCSC Controller Parameter

The problem is formulated as an optimization problem for the TCSC controller (as shown in Fig. 1). In this case, the washout time constant T_{WT} and the time constant of the two-stage lead-lag block T_{2T} and T_{4T} are prespecified. The controller gain K_T and time constant T_{1T} and T_{3T} of lead-lag compensator are to be determined applying PSO. As mentioned earlier, the aim of the TCSC-based controller is to minimize the power system oscillations after a disturbance to improve the stability. These oscillations are reflected in the deviation in the generator rotor speed ($\Delta\omega$).

The objective function considered here is an integral time absolute error of the speed deviations, i.e.,

$$J = \int_{t=0}^{t=t_1} |\Delta\omega| \cdot t \cdot \mathrm{d}t \tag{2}$$

The aim is to minimize this objective function to improve the system response in terms of the settling time and overshoot. In this optimization problem, the different parameters chosen are as follows:

Swarm size = 20, maximum number of generations = 100, $C_1 = C_2 = 2.0$, $w_{\text{start}} = 0.9$ and $w_{\text{end}} = 0.4$ [15]. The optimized values of TCSC-based controller parameters obtained by PSO are as follows:

$K_T = 62.9343$, $T_{1T} = 0.1245$ and $T_{3T} = 0.1154$.

3 Interval Type-2 Fuzzy Logic-based TCSC Controller

It is a well-known fact that the conventional fuzzy logic controller (type-1 FLC) has the limitations that, it cannot handle or accommodate the linguistic and numerical uncertainties associated with dynamical systems because its membership grade is crisp in nature. Type-2 fuzzy logic systems outperformed the type-1 fuzzy logic systems because of the membership functions of an IT2FLS are fuzzy and also contain a footprint of uncertainty (Fig. 3). IT2FLC design is based on the concept of interval type-2 fuzzy logic system. The structure of IT2FLC is same as the conventional fuzzy logic controller structure except, one type reducer block is introduced between the inference engine and defuzzifier block because the output of the inference engine is a type-2 output fuzzy set and before applying it to the defuzzifier for getting the crisp input, it has to be converted to a type-1 fuzzy set.

The block diagram of an IT2FLC is shown in Fig. 3b which contains five interconnected blocks, i.e., fuzzifier, rules, inference, type reducer, and defuzzifier. There is a mapping exist between crisp inputs to crisp outputs of the IT2FLS and is expressed as $Y = f(X)$. The principle of working of the IT2FLC is very much similar to the type-1 fuzzy logic controller (T1FLC). It is important to note that increasing the type of fuzzy system only enhances the degree of fuzziness of the system and all other principles of conventional fuzzy logic like inferencing procedure, defuzzification techniques holds good for both type [6].

In this problem, the conventional lead-lag compensator-based TCSC is used in the modified Phillips–Heffron model block diagram, and GTCSC(s) is replaced by an IT2FLC. First, the SMIB system model is simulated using a conventional T1FLC and then with IT2FLC. The rule base is same for both FLC and IT2FLC. The inputs considered here are speed ($\Delta\omega$) and its derivative ($\Delta\omega'$). The output is the change in conduction angle ($\Delta\sigma$). Triangular and gaussian type membership functions have been used for the mamdani-type FLC and IT2FLC, respectively. Centroid-type

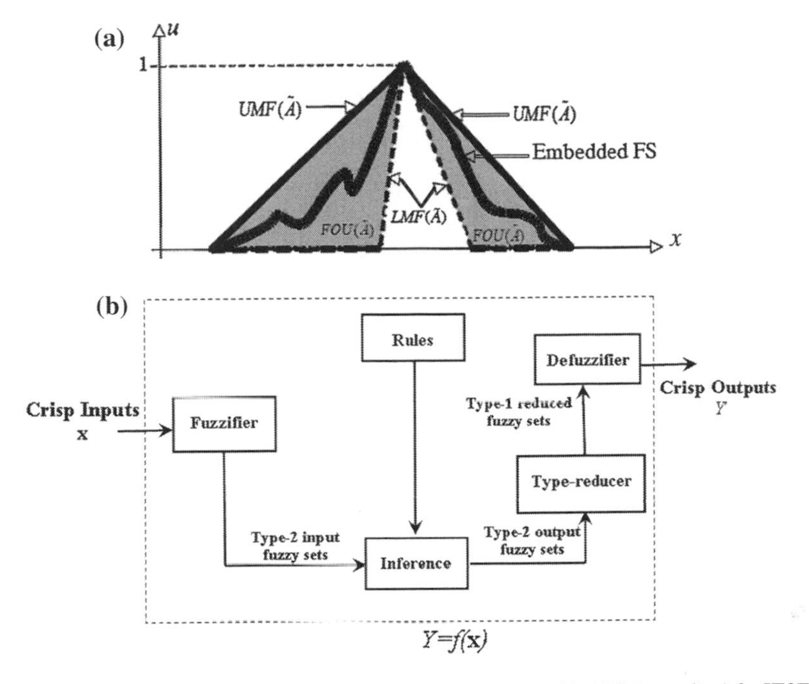

Fig. 3 **a** FOU (*shaded*), LMF (*dashed*), UMF (*solid*) and an embedded FS (*wavy line*) for IT2FSÃ. **b** Block diagram of IT2FLC [6]

Table 1 Rule base table for both FLC and IT2FLC

$\Delta\sigma$		$\Delta\omega \rightarrow$				
		NB	NS	ZO	PS	PB
$\Delta\omega'$	NB	NB	NB	NB	NM	NS
$\downarrow$	NS	NM	NS	NS	ZE	PS
	ZO	NM	NS	ZE	PS	PM
	PS	NS	ZE	PS	PS	PM
	PB	PS	PM	PB	PB	PB

Where NB-Negative Big, NS-Negative Small, ZO-Zero Error, PS-Positive Small and PB-Positive Big are the name of the membership functions

defuzzification method is used for the FLC design. The performance of the SMIB system is analyzed.

The performance of the controller is studied and is also validated at different operating conditions.

4 Results and Discussion

First, the SMIB power system model is simulated using the lead-lag compensator-based TCSC. The PSO algorithm was used to find the optimal values $K_T = 62.9343$, $T_{1T} = 0.1245$ and $T_{3T} = 0.1154$. Washout time constant T_{WT} and the time

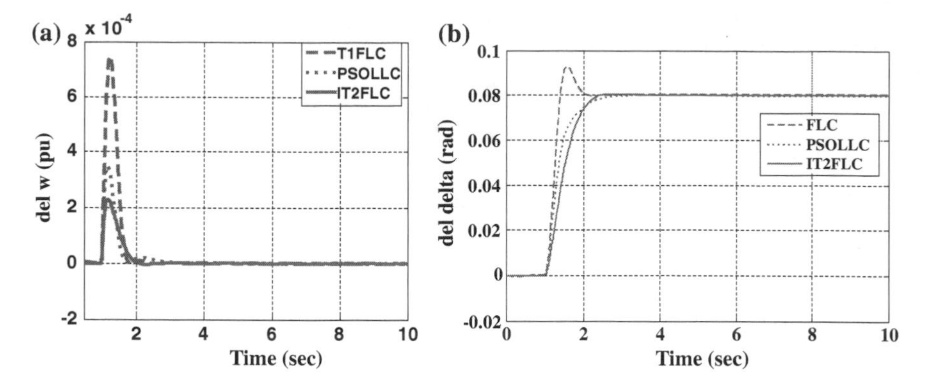

Fig. 4 **a** Rotor speed deviation. **b** Power angle deviation plot for 5 % step increase in mechanical power at nominal loading condition

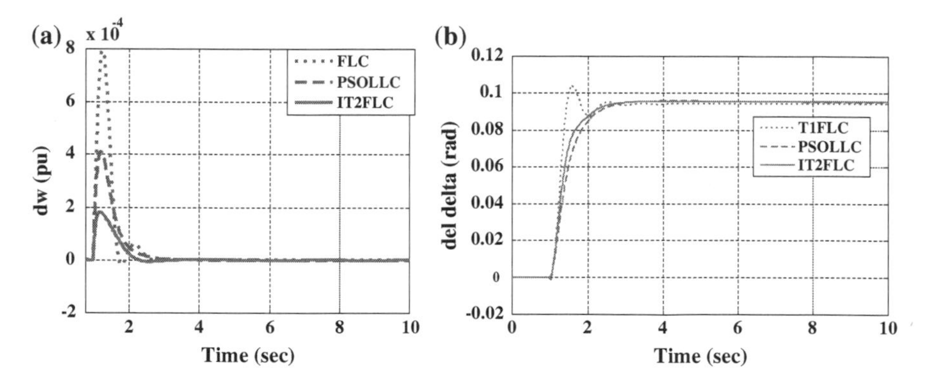

Fig. 5 **a** Rotor speed deviation. **b** Power angle deviation plot for 5 % step increase in mechanical power input with 50 % increase in line reactance at nominal loading

constant of the two-stage lead-lag block T_{2T} and T_{4T} are prespecified. These values are considered in the TCSC structure for simulation. Three loading conditions are taken, i.e., nominal, light, and heavy loading. The real (P) and reactive power (Q) values for the two loading conditions are as follows.

(1) Nominal loading (pu) $\rightarrow$ P = 0.9 and Q = 0.469 (2) Heavy loading (pu) $\rightarrow$ P = 1.02 and Q = 0.5941.

Second, the GTCSC(S) block of SMIB model is replaced by the conventional fuzzy logic controller designed with the principle as discussed in the Sect. 3. Third, the GTCSC(S) block of SMIB system is replaced by the IT2FLC designed with the procedure as depicted in previous section. The rule base as shown in Table 1 is designed based on the generalized performance of power system oscillations employing TCSC. The effectiveness and robustness of the controllers are also evaluated at (1) different loading conditions (2) disturbance of 5 % step increase in reference mechanical power input (3) variation of transmission line reactance.

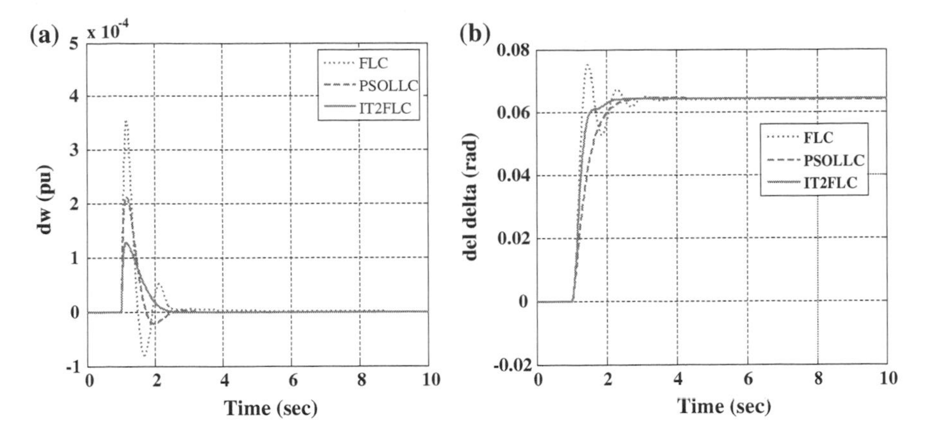

Fig. 6 **a** Rotor speed deviation. **b** Power angle deviation plot for 5 % step increase in mechanical power input at heavy loading with 10 % decrease in line reactance

Figures 4, 5 and 6 shows the rotor speed deviation and power angle deviation plots at disturbance of 5 % step increase in mechanical power input at different loading conditions and varying the transmission line reactance. It is analyzed from all the responses that the magnitude of overshoot and the settling time in all the speed deviation plots is less in case of IT2FLC compared to both PSOLLC and FLC. For the same condition, it is observed from the power angle deviation plots that there is no overshoot contributed by the PSOLLC and IT2FLC, but the IT2FLC response is faster compared to other two. FLC contributed some overshoot, but all are settling approximately at the same time.

It is found from all three results that the proposed controller is an effective and robust one compared to PSOLLC and FLC-based TCSC in providing good damping of low-frequency oscillations and to improve the system voltage profile. Although PSO-based algorithms are simple concept, easy to implement and computationally efficient, but there is possibility of trapped in local minima when handling more constrained problems due to the limited searching capacity.

5 Conclusion

In this paper, the performance of IT2FLC and FLC-based TCSC controllers are compared with a PSO optimized lead-lag compensator-based TCSC controller. The IT2FL-based TCSC controller surpasses the FL- and PSO-tuned TCSC controller performance at different loading conditions.

Appendix

All data are in per unit (pu) unless specified. Generator: $M = 9.26\,\text{s}$, $D = 0$, $X_d = 0.973$, $X_q = 0.55$, $X'_d = 0.19$, $T'_{\text{do}} = 7.76, f = 60\,\text{Hz}$, $X = 0.997$, Exciter: $K_A = 50$, $T_A = 0.05\,\text{s}$, TCSC: $X_{\text{TCSC0}} = 0.2169$, $X_C = 0.2X$, $X_P = 0.25X_C$.

References

1. Valle, Y.D., Venayagamoorthy, G.K., Mohagheghi, S., Hernandez, J.C., Harley R.G.: Particle swarm optimization: basic concepts, variants and applications in power systems. In: IEEE transaction on evolutionary computation. vol. 12(2), pp. 171–195 (2008)
2. Civicioglu, P., Besdok, E.: A conceptual comparison of the Cuckoo search, particle swarm optimization, differential evolution and artificial bee colony algorithm. Artif Intell Rev. (2011). doi:10.1007/s10462-011-9276-0
3. Aguero, J.R., Vargas, A.: Calculating functions of interval type-2 fuzzy numbers for fault current analysis. IEEE Trans. Fuzzy Syst. **15**(1), 31–40 (2007). February
4. Lin, P.Z., Lin, C.M., Hsu, C.F., Lee, T.T.: Type-2 fuzzy controller design using a sliding-mode approach for application to DC-DC converters. IEE Proc. Electr. Power Appl. **152**(6), 1482–1488 (2005). Nov
5. Tripathy, M., Mishra, S.: Interval type-2 based thyristor controlled series capacitor to improve power system stability. IET Gener. Transm. Distrib. **5**(2), 209–222 (2011)
6. Panda, M.K., Pillai, G.N., Kumar, V.: Design of an Interval Type-2 Fuzzy Logic Controller for Automatic Voltage Regulator System. Electric Power Compon. Syst. **40**(2), 219–235 (2012)
7. Hingorani, N.G., Gyugyi, L.: Understanding FACTS: Concepts and Technology of Flexible ac Transmission Systems. IEEE Press, New York (2000)
8. Mathur, R.M., Verma, R.K.: Thyristor-based FACTS Controllers for Electrical Transmission Systems. IEEE Press, Piscataway (2002)
9. Panda S.,, Padhy, N.P., Patel, R.N.: Genetically optimized TCSC controller for transient stability improvement. Int. J. Comput. Inf. Syst. Sci. Eng. 1(1) 2008
10. Panda, S., Padhy, N.P.: Comparison of particle swarm optimization and genetic algorithm for FACTS-based controller design. Electric Power Syst. Res. (2008)
11. Hameed, S., Das, B., Pant, V.: A self tuning PI controller for TCSC to improve power system stability. Electric Power Syst. Res. **78**, 1726–1735 (2008)
12. Padiyar,K.R.: Power System Dynamics: Stability and Control. Wiley, New York (1996)

Improved Parallelization of an Image Segmentation Bio-Inspired Algorithm

Javier Carnero, Hepzibah A. Christinal, Daniel Díaz-Pernil, Rául Reina-Molina and M. S. P. Subathra

Abstract In this paper, we give a solution for the segmentation problem using membrane computing techniques. There is an important difference with respect to the solution presented in Christinal et al. [6], we use multiple membranes. Hence, the parallel behavior of the algorithm with respect to the previous works has been improved.

Keywords Image segmentation · Digital topology · Membrane computing · Tissue-like P systems · Parallel computing

1 Introduction

Membrane systems [8] are distributed and parallel computing devices processing multisets of objects in compartments delimited by membranes. Computation is carried out by applying given rules to every membrane content, in a maximally parallel non-deterministic way, although other semantics are being explored.

J. Carnero · D. Díaz-Pernil · R. Reina-Molina
Research Group on Computational Topology and Applied Mathematics, Department of Applied Mathematics, University of Sevilla, Sevilla, Spain
e-mail: javier@carnero.net

D. Díaz-Pernil
e-mail: sbdani@us.es

R. Reina-Molina
e-mail: raureimol@alum.us.es

H. A. Christinal (✉) · M. S. P. Subathra
Karunya University, Coimbatore, India
e-mail: hepzia@yahoo.com

M. S. P. Subathra
e-mail: sumiolivia@gmail.com

B. V. Babu et al. (eds.), *Proceedings of the Second International Conference on Soft Computing for Problem Solving (SocProS 2012), December 28–30, 2012*, Advances in Intelligent Systems and Computing 236, DOI: 10.1007/978-81-322-1602-5_9, © Springer India 2014

Segmentation [11] is the process of splitting a digital image into sets of pixels in order to make it simpler and easier to analyze. Segmentation is typically used to locate region of interest (ROI) in medical images or in satellite image by finding the frontiers among regions. Segmentation has shown its utility in bordering tumors and other pathologies, computer-guided surgery or the study of anatomical structure, but also in techniques which are not thought to produce images, but it produces positional information as electroencephalography (EEG), or electrocardiography (ECG). Locating a ROI is a hard task even for the expert human eye, mainly due to problems such as noise and the degradation of colors. Technically, the process consists of assigning a label to each pixel, in such way that pixels with the same labels form a meaningful region.

Here, a solution is given for the segmentation problem using a membrane computing device: tissue-like P systems. Initially, systems with only one working cell were used (see Carnero et al. [2], Christinal et al. [6]). Formally, membrane computing was used, but the key of these models was not considered: the membranes. In order to address this, return the membranes to their previous role and to improve the inherent parallelism of these models (see Christinal et al. [6]) using multiple membranes, increase the scalability for future parallel software implementations of the algorithm. A brief proof has been added for the ideas which are shown in Reina-Molina et al. [9]. Also a software tool implemented in Python showing the use of this algorithm is presented.

In the literature, one can find several attempts for bridging problems from digital imagery with membrane computing. The following are a few examples: the works done by Subramanian et al. [3, 4] and a few more problems from digital imagery have been solved in the framework of membrane computing (see Christinal et al. [5, 6]).

The paper is organized as follows: First, a family of tissue-like P systems is designed to obtain a segmentation of a 2D digital image using multiple cells. Moreover, an overview of the computation of this algorithm and a complexity study are presented in the following section. Next, an implementation of the algorithm using Python and some examples are shown. Finally, future work is presented.

2 Image Processing: Segmentation Problem

The *m-D Segmentation Problem with k auxiliary cells (mDSP-kC)*can be described as follows: Given a m-D Digital Image I, of size n^m, to determine the edge pixels of this image using k auxiliary cells.

There are two usual problems when we work with real images, the noise and degradation of colors. The former arises when we process or analyze an image whose color has no relation with those in its environment. The later takes place when we look at a digital image and we can see different colors connecting two adjacent regions, blurring the common edge of them. Our aim is to define the boundaries of these regions. These boundaries are considered as small regions where the colors of pixels gradually change from one side to another side of each region.

The usual definition of edge pixel presents problems from a practical point of view, when noise and degradation of colors are taken into account, because a lot of border pixels that are not edge pixels are considered. Hence, some cleaning processes must be applied to our image in order to obtain better results in edge pixels detection.

Next, we will show that *mDSP-kC* can be solved in constant time (with respect to the number of pixels of the image) by a family of tlP systems.

Formally, a *tissue-like P system* (*tlP System*) of degree $q \geq 1$ with input is a tuple of the form $\Pi = (\Gamma, \Sigma, \xi, w_1, \ldots, w_q, R, i_{\Pi,} o_{\Pi})$ where

(a) Γ is a finite *alphabet*, whose symbols will be called *objects*, $\Sigma \subset \Gamma$ is the input alphabet and $\xi \subseteq \Gamma$ is the alphabet of objects in the environment. (b) $w1, \ldots, wq$ are strings over Γ representing the multisets of objects associated with the cells at the initial configuration. (c) R is a finite set of communication rules of the following form: $(i, u/v, j)$, for $i, j \in \{0, 1, 2, \ldots, q\}, i \neq j, u, v \in \Gamma^*$ and (d) $i_\Pi \in \{1, 2, \ldots, q\}$ is the input cell and $o_\Pi \in \{0, 1, 2, \ldots, q\}$ is the output cell.

A tlP system of degree $q > 1$ can be seen as a set of q cells labeled by $1, 2, \ldots, q$. We will use 0 to refer to the label of the environment, i_Π and o_Π denote the input region and the output region (which can be the region inside a cell or the environment), respectively.

Let us construct a family $\pi = \{p(n, m, k) \colon n, m, k \in N\}$ where each system of the family will process every instance u of the problem given by a m-D image I with n^m pixels and using k auxiliary cells. More formally, we define $s(u) = \langle n, m, k \rangle = \langle n, \langle m, k \rangle \rangle$, where $\langle x, y \rangle = (x + y)(x + y + 1)/2 + x$ is the Gödel mapping. In order to provide a suitable encoding of this instances into the systems, we will use the objects $a_{i_1}, \ldots, i_m$ with $1 \leq i_1, \ldots, i_m \leq n$, to represent the pixels of the image, and we will provide $cod(u)$ as the initial multiset for the system, where $cod(u)$ is the multiset $a'_{i_1}, \ldots, i_m$.

Then, given an instance u of the *mDSP-kC* problem, the system $\pi(s(u))$ with input $cod(u)$ gives a solution to this problem, implemented in the following stages:

- Cleaning noise.
- Homogenizing colors using a general thresholding in color space.
- Segmenting image process.

The family $\pi = \{\pi(n, 2, k) \colon n, k \in N\}$ of tlP systems of degree $k + 1$ is defined as follows: for each $n, k \in N$, $\pi(n, 2, k) = (\Gamma, \Sigma, \xi, w_1, \ldots, w_{k+1}, R, i_\Pi, o_\Pi)$, defined as follows:

- $\Gamma = \Sigma \cup \left\{a_{ij}a''_{ij}, \overline{a}_{ij}, A_{ij}, A'_{ij}, A''_{ij}, \overline{A}_{ij} \colon 1 \leq i, j \leq n, a \in C\right\}$,
  $\Sigma = \left\{a'_{ij} \colon 1 \leq i, j \leq n, a \in C\right\}, \xi = \Gamma - \Sigma$,
- $w_1 = *_{ij}; *_{ji}$ with $i = 0, n + 1, 0 \leq j \leq n + 1, w_2 = \cdots = w_{k+2} = T^{[n^2/k]}$,
- R is the following set of communication rules:
  - $\left(1, a'_{ij}/a^8_{ij}A_{ij}, 0\right)$ for $0 \leq i, j \leq n + 1$ and $a \in C \cup \{*\}$

- $\left(1, \begin{matrix} c_{i-1j-1} & d_{i-1j} & e_{i-1j+1} \\ b_{ij-1} & A_{ij} & f_{ij+1} \\ o_{i+1j-1} & h_{i+1j} & g_{i+1J+1} \end{matrix} \Big/ T, t\right)$

  for $1 \leq i, j \leq n$, $a, b, c, d, e, f, g, h, o \in C \cup \{*\}$ and $2 \leq t \leq k+1$ indicating an auxiliary working cell.

- $\left(t, \begin{matrix} c_{i-1j-1} & d_{i-1j} & e_{i-1j+1} \\ b_{ij-1} & A_{ij} & f_{ij+1} \\ o_{i+1j-1} & h_{i+1j} & g_{i+1j+1} \end{matrix} \Big/ z'_{ij}, 0\right)$

  for $1 \leq i, j \leq n$, $a, b, c, d, e, f, g, h, o \varepsilon C \cup \{*\}$. We take μ as the number of pixels adjacent to the *ij* position with colors in C and $* = 0$. Then, $av = (b + c + d + e + f + g + j + o)/\mu$ and $z = \max\{s \in C: s \leq av\}$ and $|z - av| > \rho_1$, where $\rho_1 \in (, +\infty)$.

- $\left(t, \begin{matrix} c_{i-1j-1} & d_{i-1j} & e_{i-1j+1} \\ b_{ij-1} & A_{ij} & f_{ij+1} \\ o_{i+1j-1} & h_{i+1j} & g_{i+1j+1} \end{matrix} \Big/ a'_{ij}, 0\right)$

  for $1 \leq$i,j $\leq n, a, b, c, d, e, f, g, h, o \in C \cup \{*\}$. We take μ as the number of pixels adjacent to the ij position with colors in C and $* = 0$. $av = (b + c + d + e + f + g + j + o)/\mu$ and $z = \max\ \{s \in C: s \leq av\}$ and $|a - av| \leq \rho_1$, where $\rho_1 \in (, +\infty)$.

- $(t, b'_{\mathrm{ij}}/A'_{\mathrm{ij}}, 0)$ for $1 \leq i, j \leq n, \nu = (|C|/\rho_2), l = 0, 1, 2, \ldots, \rho_2$. If $b \in C$ then a $\in C$

  $$(a < b \leq a + (\nu - 1) \text{ and } a = \nu \cdot l) \text{ or } (b = a = \nu \cdot l) \text{ and, if } b = * \text{ then } A = *.$$

- $(t, A'_{ij}/T, 1)$ for a $\in C, 0 \leq i, j \leq n+1$ and $2 \leq t \leq k+1$.
- $(t, A'_{ij}/A''_{ij}\overline{a}^{8}_{ij}, 0)$ for a $\in C, \cup\{*\}$ and $0 \leq i, j \leq n+1$.
- $\left(0, \begin{matrix} \bar{c}_{i-1j-1} & \bar{d}_{i-1j} & \bar{e}_{i-1j+1} \\ \bar{b}_{ij-1} & A''_{ij} & \bar{f}_{ij+1} \\ i_{i+1j-1} & \bar{h}_{i+1j} & \bar{g}_{i+1j+1} \end{matrix} \Big/ T, t\right)$

  For $1 \leq i, j \leq n$ and $a, b, c, d, e, f, g, h, i \in C \cup \{*\}$.
- $(t, A''_{ij}\overline{b}_{kl}/\overline{A}_{ij}, 0)$ for $1 \leq i, j, k, l \leq n$, $(i, j), (k, l)$ adjacent pixels; i.e. $(i,j) \in \Delta$ and $a, b \in$ C and $a < b$.
- $(t, \overline{A}_{ij}/T, 1)$ for $a \in C$ and $1 \leq i, j \leq n$.
- $i_\Pi = o_\Pi = 1$.

2.1 Overview

The input data of the system are given by the set $\{a'_{ij}: a' \in C, 1 \leq i, j \leq n\}$, which consists of the pixels of an image.

The computation starts in cell 1 which contains the initial image encoded by the objects a'_{ij}, along the edge pixels $*_{ij}$. Cells $2, 3, \ldots, k+1$ contain enough copies of objects T called workflow markers. The unique rule that can be applied to this initial configuration is the first one, which is designed to mark non-edge pixels (a'_{ij}, $a \in C$) with A_{ij} and to create enough copies of objects a'_{ij} and $*_{ij}$, which will be used later in workflow distribution. This step will be called *copies creation.*

Then, the only rule that can be applied is the second one, which sends one object A_{ij} and its neighborhood represented by adjacent objects a_{kl} to one working cell (cells $2, 3, \ldots, k+1$) to be processed later. This step will be called *object distribution.*

After these two initial steps, for the third configuration of the system, only rules of the third and fourth type can be applied. These rules transform one object A'_{ij} into one object z''_{ij} , where z is the nearest color in C to the average color in the neighborhood of object A_{ij} and a is considered as noise, or color a itself in other cases.

Next, only the fifth rule can be applied, which transforms objects b''_{ij} into objects A'_{ij} , where a is a color in a subset of C defining a general thresholding with respect to the colors. Subsequently, the sixth rule is the one available, for obtaining the sixth configuration of the system, with the original image preprocessed (noise cleaned and thresholded) in order to improve the segmentation process.

Once all the preprocessing work is done, the image to be segmented is represented as objects A'_{ij} where a is either a color in C or $*$. The next two steps consist in applying seventh and eighth rules, respectively, which send to the working cells an object A''_{ij} and its neighborhood represented by adjacent pixels $\bar{a}_{kl}$. From these steps, we get the eighth configuration.

The next configuration of the system is obtained by applying the rules in the ninth scheme, marking with $\bar{A}_{ij}$ the edge pixels A''_{ij} with respect to an adjacent pixel $\bar{b}_{kl}$ with greater color. The last configuration is got by sending the edge pixels $\bar{A}_{ij}$ back to the cell 1.

2.2 Complexity Aspects

A little study of the complexity aspects of this solution is given by Fig. 1 showing this is an efficient algorithm from a theoretical point of view (Table 1).

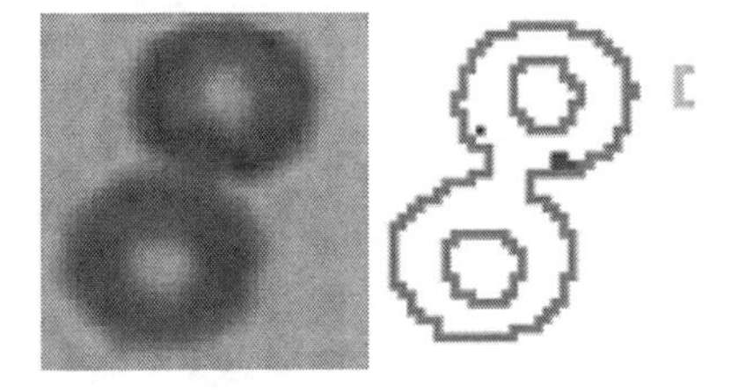

Fig. 1 Test image (on the *left*) and its segmented one (on the *right*). Next, we will show that *mD*

3 Implementation of the Algorithm

The algorithm proposed in this paper is implemented in Python Rossum et al. [10], an interpreted multiparadigm programming language which is powerful and easy to learn, read, and write. We have focused our implementation in the use of matrices as every alphabet object is represented by a matrix. This approach allows us to use the Numpy library (Ascher et al. [1]) for working effectively with large arrays of numbers. This way of facing the implementation of membrane objects seems to make available an easy adaptation of the algorithm to parallel architectures as GPU or multicores CPU.

The segmentation algorithm is implemented through the definition of the following functions:

- *BioSeg*. This function achieves the full segmentation process by calling other auxiliary functions described below. First of all, it creates a dictionary for storing the initial state of the computation. Next, a new computation is performed by distributing the work flow through the auxiliary cells. Then, cleaning and thresholding stages are performed into each auxiliary cell, following which a new computation is carried out by integrating all the information processed by the auxiliary cells into the first one. Next, the image cleaned and thresholded is distributed among the auxiliary cells to be segmented. Finally, the processed information is returned to the initial cell to be returned.
- *createInitialState*. This function creates the initial structure for storing the initial state of the P system.
- *distributionWork*. This function distributes every column and its adjacent ones to the corresponding auxiliary cell. In the P system design, this task is nondeterministically carried out. However, in the current implementation, the auxiliary cell for every pixel and its adjacent is deterministically chosen.
- *integrationWork*. This function integrates the processed pixels back into the first cell, forgetting the surrounding ones.
- *cleaningStep*. This function performs the image cleaning in every auxiliary cell.
- *thresholdingStep*. This function achieves the color thresholding in every auxiliary cell.

Table 1 Complexity aspects, where the size of the input data is $O(n^2)$, $|C| = h$ is the number of colors of the image, and k is the number of working cells

mDSP-kC Problem	
Complexity	
Number of steps of computation	9
Resources needed	
Size of the alphabet	$8n^2 + 4n + 5$
Initial number of cells	$k + 1$
Initial number of objects	$(n + 2)^2$
Number of rules	$O(n^2 \cdot h^9 \cdot k)$
Upper bound for the length of the rules	10

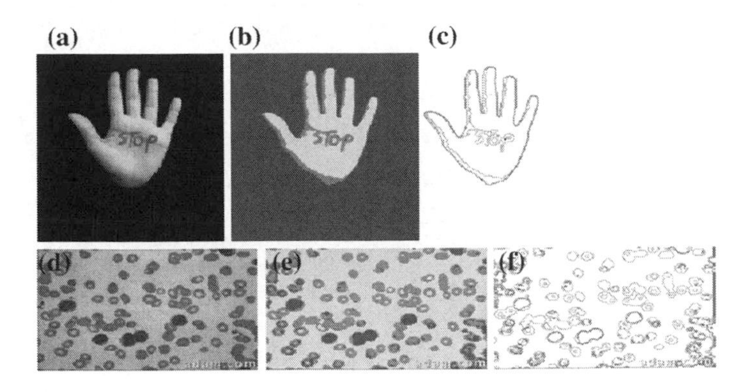

Fig. 2 Examples of images at the beginning of the algorithm (**a**) and (**d**), after cleaning and thresholding steps (**b**) and (**e**) and after segmentation step (**c**) and (**f**)

- *segmentationStep*. This function makes the segmentation process in every auxiliary cell.

The algorithm design implemented allows us to rewrite the code in order to take advantage of massively parallel devices as GPUs. However, in the current development stage, we have not implemented it yet and the code executes sequentially.

On the other hand, both cleaning, thresholding and segmentation steps can be developed using different approaches than those selected by us. Hence, the segmentation algorithm can evolve from the easier one in this work to more complex one, also designed to be executed in parallel. These changes do not alter the behavior of the complete algorithm, but they can improve the final result.

3.1 Examples

We present in this section several examples of the application of the implementation proposed in this paper.

The process is illustrated in Fig. 2, where two images are shown in the initial state, after cleaning, thresholding and segmentation. Both processes have been carried out using 1 as cleaning threshold, 3 as color threshold, and 5 as segmentation threshold.

4 Future Work

Our next step in this research line will be the real parallel implementation of the algorithm using the massive parallelism present in current GPUs. It is also planned to study other noise cleaning algorithms and also more elaborated segmentation algorithms, aiming for better results.

References

1. Ascher, D., Dubois, P.F., Hinsen, K., Hugunin, J., Oliphant, T.: Numerical Python, Lawrence Livermore National Laboratory, Livermore, California, USA, 2001. Available at http://numpy.scipy.org/
2. Carnero, J., Diaz-Pernil, D., Molina-Abril, H., Real, P.: Image segmentation inspired by cellular models using hardware programming. Image-A **2**(4), 25–28 (2010)
3. Ceterchi, R., Gramatovici, R., Jonoska, N., Subramanian, K.G.: Tissue-like P systems with active membranes for picture generation. Fundam. Inf. **56**(4), 311–328 (2003)
4. Ceterchi, R., Mutyam, M., Paun, G., Subramanian, K.G.: Array-rewriting P systems. Nat. Comput. **2**(3), 229–249 (2003)
5. Christinal, H.A., Diaz-Pernil, D., Real, P.: P systems and computational algebraic topology. J. Math. Comput. Model. **52**(11–12), 1982–1996 (2010)
6. Christinal, H.A., Diaz-Pernil, D., Real, P.: Region-based segmentation of 2D and 3D images with tissue-like P systems. Patt. Recogn. Lett. **32**(16), 2206–2212 (2011)
7. Diaz-Pernil, D., Gutierrez-Naranjo, M.A., Molina-Abril, H., Real, P.:A bio-inspired software for segmenting digital images. In: Nagar, A.K., Thamburaj, R., Li, K., Tang, Z., Li, R. (eds.) Proceedings of the 2010 IEEE Fifth International Conference on Bio-Inspired Computing: Theories and Applications BIC-TA, IEEE Computer. Society, vol. 2, pp. 1377–1381 (2010)
8. Paun, G.: Computing with membranes. Tech. Rep. 208, Turku Centre for Computer Science, Turku, Finland (November 1998).
9. Reina-Molina, R., Carnero, J., Diaz-Pernil, D.: Image segmentation using tissue-like P systems with multiple auxiliary cells. Image-A **1**(3), 143–150 (2010)
10. Rossum, G. Van and Drake, F.L. (eds.) Python Reference Manual, PythonLabs, Virginia, USA, 2001. Available at http://www.python.org
11. Shapiro, L.G., Stockman, G.C.: Computer Vision Upper Saddle River. Prentice Hall PTR, NJ (2001)

A Novel Hardware/Software Partitioning Technique for System-on-Chip in Dynamic Partial Reconfiguration Using Genetic Algorithm

Janakiraman N. and Nirmal Kumar P.

Abstract Hardware/software partitioning is a common method used to reduce the design complexity of a reconfigurable system. Also, it is a major critical issue in hardware/software co-design flow and high influence on the system performance. This paper presents a novel method to solve the hardware/software partitioning problems in dynamic partial reconfiguration of system-on-chip (SoC) and observes the common traits of the superior contributions using genetic algorithm (GA). This method is stochastic in nature and has been successfully applied to solve many non-trivial polynomial hard problems. It is based on the appropriate formulation of a general system model, being therefore independent of either the particular co-design problem or the specific partitioning procedure. These algorithms can perform decomposition and scheduling of the target application among available computational resources at runtime. The former have been entirely proposed by the authors in previous works, while the later have been properly extended to deal with system-level issues. The performance of all approaches is compared using benchmark data provided by MCNC standard cell placement benchmark netlists. This paper has shown the solution methodology in the basis of quality and convergence rate. Consequently, it is extremely important to choose the most suitable technique for the particular co-design problem that is being confronted.

Keywords Hardware/software partitioning · Genetic algorithm · Dynamic partial reconfiguration · System-on-chip

N. Janakiraman (✉)
Anna University, Chennai, India
e-mail: janakiramanforu@yahoo.com

P. N. Kumar
Anna University, Chennai, India
e-mail: nirmal.p@annauniv.edu

B. V. Babu et al. (eds.), *Proceedings of the Second International Conference on Soft Computing for Problem Solving (SocProS 2012), December 28–30, 2012*, Advances in Intelligent Systems and Computing 236, DOI: 10.1007/978-81-322-1602-5_10,

1 Introduction

Hardware/software partitioning is a method of dividing a complex heterogeneous system into hardware co-processor functions and its compatible software programs. It is a prominent practice that can realize results greater than the software-only or hardware-only solutions in system-on-chip (SoC) design. This technique can improve the system performance [1] and reduce the total energy consumption [2]. The proposed partial dynamic reconfiguration method does not depend on any tool. It uses a set of algorithms to detect crucial code regions, compilation/synthesize of hardware/software modules, and updating of communication logic. Hence, it could tune up the system to give full efficiency without disruption of other SoC-related operations. Here, the genetic algorithm (GA) is used for optimization process. This is essential in system-level design, since decision-making process affects the total performance of system. This paper presents a novel system partitioning technique with in-depth analysis. The paper is organized as follows. Section 2 briefs about the previous works in this field. Section 3 presents the proposed system model for partitioning problem. Section 4 gives the results and its analysis. Section 5 concludes the paper and discusses about the future work. Last section provides the list of references.

2 Related Works

When compared to dynamic partitioning using standard software, the run-time (or) partial dynamic reconfigurable systems had attained superior performance with manually specified predetermined hardware regions. Multiple choices of preplanned reconfigurations were rapidly executed in a run-time reconfigurable system using PipeRench architecture [3] and dynamically programmable gate arrays (DPGA) [4]. The binary-level partitioning technique [5] was provided a good solution compared to source-level partitioning methods due to the functionality of any high-level language and software compiler. Since the satisfaction of performance was not considered for the cost function of this system, it may be failed to find out local minima. A mapping technique for nodes and hardware/software components was developed in [6] called GCLP algorithm. The hardware cost was minimized by the incorporation of hill-climbing heuristic algorithm with the hardware/software partitioning algorithm [7].

3 System Model for Partitioning

The problem resolution requires the system model definition to represent the important issues in the hardware/software co-design for a specific problem [8]. The system partitioning problem model is represented by the task graph (TG) flow diagram. TG is a model of directed and acyclic graph (DAG) flow with weight vectors. Formally,

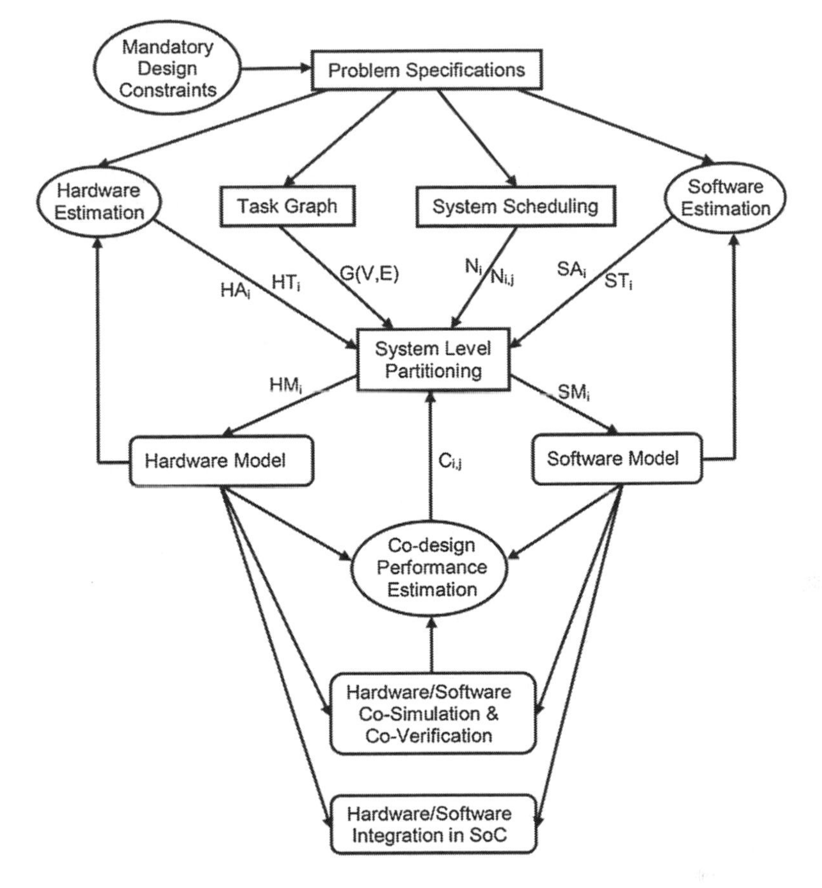

Fig. 1 System model for partitioning

it is defined as $G = (V, E)$, where 'V' represents the nodes and 'E' represents the edges. The flow direction is represented by each edge. Due to reducing the complexity of TG, it can be modified as one starting node and one ending node. Figure 1 represents the overview of the partitioning procedure. Design constraints and design specifications are given as the input to the partitioning process as a high-level specification language. The nodes can act as giant pieces of information like tasks and processes of coarse granularity or tiny types like instructions and operations of fine granularity approach.

After the system space estimation, every node is tagged with some attributes. Giant pieces of data for a node ($V_{i,j}$) are represented by 5 attributes as follows:

(1) Hardware area ($HA_{i,j}$).
(2) Hardware implementation time ($HT_{i,j}$).
(3) Software memory size ($SS_{i,j}$).
(4) Software execution time ($ST_{i,j}$).

(5) The average execution time in numbers $(N_{\mathrm{i,j}})$.

Shortly,

Hardware module $\left(\mathrm{HM}_{\mathrm{i,j}}\right) = \left(\mathrm{HA}_{\mathrm{i,j}}\right) + \left(\mathrm{HT}_{\mathrm{i,j}}\right) + (N_{\mathrm{i,j}})$
Software module $\left(\mathrm{SM}_{\mathrm{i,j}}\right) = \left(\mathrm{SS}_{\mathrm{i,j}}\right) + \left(\mathrm{ST}_{\mathrm{i,j}}\right) + (N_{\mathrm{i,j}})$

Communication values $(C_{\mathrm{i,j}})$ of every node are represented by three components as follows:

(1) Transfer time $(\mathrm{TT}_{\mathrm{i,j}})$
(2) Synchronization time $(\mathrm{SynT}_{\mathrm{i,j}})$
(3) The average communication time in numbers $(M_{\mathrm{i,j}})$

Shortly,
Communication value of node $\left(C_{\mathrm{i,j}}\right) = \left(\mathrm{TT}_{\mathrm{i,j}}\right) + \left(\mathrm{SynT}_{\mathrm{i,j}}\right) + (M_{\mathrm{i,j}})$

$$C_{\mathrm{i,j}} = \frac{(N_i * \Delta\mathrm{TT}_i) + \left(N_j * \Delta\mathrm{TT}_j\right) + (\mathrm{SynT}_{\mathrm{i,j}})}{(\mathrm{HT}_i) + (\mathrm{HT}_j)}$$

where $(\Delta\mathrm{TT}_i) = (\mathrm{ST}_i) - (\mathrm{HT}_i)$ and $(\Delta\mathrm{TT}_j) = \left(\mathrm{ST}_j\right) - \left(\mathrm{HT}_j\right)$.

Efficiency of the hardware/software system partitioning process is based on the target architecture and its mapping technique. Hence, this work considers the 'Dynamically Reconfigurable Architecture for Mobile Systems' (DReAM) as target architecture. Execution of hardware and software processes should be concurrently in the standard processor and the application-specific co-processor. This partitioning process concludes the assignment of modules to implement the hardware and software stages, implementation schedule (timing), and the communication interface between software and hardware modules. In general, this partitioning solution can be validated by the measurement of eminent attributes like performance and cost parameters. Hence, this paper used as three quality attributes related to design elements as follows:

(1) The estimated hardware area is A_E, and the maximum available area is A.
(2) The estimated design latency is T_E, and the maximum allowed latency is T.
(3) The estimated software (or) memory space is M_E, and the maximum available space is M.

Static-list scheduling method is used for the scheduling process [9]. It is a subtype of resource-constrained scheduling algorithm. This scheduler considers the timing estimation of every vertex and its interconnections. This scheduler unit provides the design latency (T_E) and the cost of communication for hardware–software co-design. Based on the hardware and software implementations, another four parameters are considered for co-design realization.

When the entire system is implemented in hardware,

(1) The minimum design latency is MinT.
(2) The maximum hardware area is MaxA.

When the entire system is implemented in software,

(1) The maximum design latency is MaxT.
(2) The maximum memory space is MaxM.

These parameters are used to create the bounding constraints for the design space. $0 \leq A \leq \text{MaxA}$; $0 \leq M \leq \text{MaxM}$; $\text{MinT} \leq T \leq \text{MaxT}$.

3.1 System Operations

The design specifications are given in the format of ISPD98 benchmark suite [10] circuit netlist. This partitioning process has three stages.

In first stage, the processing of design specifications is divided into three subtasks. The first subtask is the separation of hardware (HA_i and HT_i) and software (SS_i and ST_i) estimations from the design specifications. The second subtask is to translate the design specifications into a hypergraph-based control data flow graph (CDFG) representation $G = (V, E)$. The third subtask is scheduling (N_i and $N_{\text{i,j}}$) of each operations in the CDFG with satisfaction of the design constraints and the priority of operations.

In second stage, the outputs of these three tasks are given into the system-level partitioning module through the registers. It has three functionalities. The operational-level analysis is the first process, used to classify the tasks whether it is suitable for hardware realization or software execution. Next, the allocation process is used to allocate the required supporting entities like functional units, interconnections, and storage elements for the scheduled hardware and software systems. This allocation is based on the speed constraint (i.e., parallel processing) and the area constraint (i.e., dynamic partial reconfiguration). Finally, an absolute data path is generated by integrating components in the basis of hardware and software partitions. Then, the partitioning data are given to the specific hardware (HM_i) and software (SM_i) models.

In third stage, the hardware and software models are executed separately and the outcomes are compared with their estimated values (i.e., first stage). If any controversy arises, the feedbacks are given to the second-stage process. This looping process is continued till the satisfaction of all criterions.

Next, the performance ($C_{\text{i,j}}$) of hardware–software co-design is estimated and compared with target performance metrics. If any misalignment arises, the feedback is indicated to the system-level partitioning stage. Then, the entire second and third stages are recompiled, till the achievement of target performance measures. Finally, the hardware/software co-simulation and co-verification is performed, and then, the SoC is realized.

3.2 Hardware/Software Estimation

The CDFG file is given to the input of both hardware and software estimations with the settings of target technology files and processor specifications. The hardware execution is a parallel process since the specifications are modeled in VHDL library. The software execution is a sequential process since the specifications are modeled in C code. The GA technique is used to optimize these parallel and sequential processes.

Hardware estimation is based on the high-level synthesizable components, to share the control and data path between hardware and software processes. GA is used to optimize this resource sharing process [11]. The quality measures are closely associated with performance metrics like execution, implementation, transfer, and synchronization times commonly called reaction time. This reaction time is associated with each node in each execution of local DFG. For convenient, the CDFG is split into several small DFGs called local DFGs.

The response times for
Routine statements, $T_{\mathrm{RS}} = T_{\mathrm{DFG}}$
Conditional statements, $T_{\mathrm{CS}} = \sum_{n} P_n T_{\mathrm{DFGn}}$;

n—Number of iterations
P_n—Probabilities of iterations of outcomes

Looping statements, $T_{\mathrm{LS}} = nT_{\mathrm{DFG}}$;

$$T_{\mathrm{CDFG}} = F(T_{\mathrm{DFG1}}, F_{\mathrm{DFG1}}, \ldots, T_{\mathrm{DFGi}}, F_{\mathrm{DFGi}}) + F(T_{\mathrm{DFG1}}, F_{\mathrm{DFG1}}, \ldots, T_{\mathrm{DFGj}}, F_{\mathrm{DFGj}})$$

$$\mathrm{MinT} = \alpha[(\mathrm{MaxA} * C_{\mathrm{i,j}}) + \sum_{i} T_i N_{\mathrm{i,j}}]$$

T_i—Time delay for each node
α—Co-estimation factor

$$\mathrm{MaxT} = \mathrm{MinT} + \beta \sum_{i} [T_i \sum_{j=1}^{R_i} N_{\mathrm{i,j}}]$$

R_i—Required components of each node 'i'
β—Constant, since MaxT is a higher-order term
F_i—Number of fixed components for each node 'i'

$$T_{\mathrm{CDFG}} = \mathrm{MinT} + \beta \sum_{i} [\frac{T_i}{F_i} \sum_{j=F_i+1}^{R_i} N_{\mathrm{i,j}}]$$

Register Estimation: [12]
Many input multiplexers = (i*MUXs)

State machine-based control logic is used to control lines, $\log_2 i$

ROM size, $(\mathrm{STA}^*[(1+\log_2 i)\left(\mathrm{REG}+\sum_i F_i\right)+\log_2 S])$bits

STA—Number of states
REG—Number of registers

Software estimation is based on the calculation of memory space occupied by instruction set and user-defined data types and data structures. The average queuing time for each memory access can be modeled as T_q, and the number of access is represented by N_{mem}. This calculation is necessary to estimate $(\mathrm{TT}_{\mathrm{i,j}})$ and $\left(\mathrm{SynT}_{\mathrm{i,j}}\right)$.

Hardware estimation $(T_{\mathrm{HM}}) = (T_{(\mathrm{CDFG,HM})}) + \alpha T_q(N_{\mathrm{mem,HM}})$

Software estimation $(T_{\mathrm{SM}}) = (T_{(\mathrm{CDFG,SM})}) + T_q(N_{(\mathrm{mem,SM})})$

Co-estimation $(T_{\mathrm{HM/SM}}) = \sigma(T_q) + \varphi(\frac{N_{\mathrm{mem}}}{T_q})$; where σ and φ are complex structures.

4 Analyses of Results

All the hardware/software partitioning algorithms have been experimented in a set of benchmark suites provided by ISPD'98, whose characterization is shown in Table 1. Size and values of the system graph should bound within the design space. All these examples are illustrated in the form of directed and acyclic graphs to specify the certain coarse–grain tasks. Every example has been tested in different constraints, but it always within the specified boundary conditions. The results are summarized in Table 2. These results will be analyzed from both qualitative and quantitative perspectives. The qualitative aspects will be mainly represented by the resulting cost of the solutions obtained from each method, under different constraints. The quantitative issues will be shown by means of the computation time resulting from each technique.

Table 1 Design characteristics for ISPD'98 benchmark suite

Circuit	# Cells	# Pads	# Modules	# Nets	# Pins
ibm01	12,506	246	12,752	14,111	50,566
ibm02	19,342	259	19,601	19,584	81,199
ibm03	22,853	283	23,136	27,401	93,573
ibm04	27,220	287	27,507	31,970	105,859
Ibm05	28,146	1,201	29,347	28,446	126,308

Table 2 Results acquired with the ISPD'98 examples

Example	Constraints			Genetic algorithm			
	Area (CLBs)	Time (ns)	Memory (Bytes)	A_E	T_E	M_E	**Fitness**
ibm01	121,800	10,200	52,670	118,146	9,384	46,350	0.9233
	103,080	8,670	44,770	101,637	8,020	41,189	0.9437
ibm02	154,700	12,600	55,980	140,170	11,230	49,823	1.0000
	193,375	15,750	48,980	172,104	15,435	51,429	0.9733
ibm03	171,200	14,200	48,090	154,896	12,040	38,953	1.0000
	111,280	9,230	57,708	103,521	8,769	54,823	1.0000
ibm04	182,200	15,900	56,460	173,090	14,469	62,106	0.9866
	258,724	19,239	50,814	234,597	16,546	46,749	0.9600
ibm05	198,300	16,800	62,210	180,453	13,776	58,478	0.8900
	97,167	12,432	81,495	92,309	10,940	84,755	0.9566

5 Conclusion and Future Work

In this paper, the commonly used biologically inspired optimization algorithm, which addresses the hardware/software partitioning problem for SOC designs, is implemented using clustering approach as well as their performance is evaluated. This evaluation process does not have any constraints on the cluster size and the number of clusters. Hence, this evaluation approach is quiet suitable to be used in reducing the design complexity of systems. This paper had shown how this problem can be solved by means of very different partitioning techniques at runtime of the system (dynamic partial reconfiguration). The problem resolution has been based on the definition of a common system model that allows the comparison of different procedures. These extensions have improved previous implementations, because they include some issues previously not considered. The constraints of these algorithms have been integrated into the cost function in a general and efficient way. This genetic algorithm-based dynamic partitioning technique has produced an average of 16.19 % accuracy in hardware/software partitioning compared to [13] and [14].

A future study could extend the system model to encompass other quality attributes, like power consumption, influence of communications, and the degree of parallelism. Also, the hybrid algorithms of these biologically inspired algorithms and their compilation are currently under study.

Acknowledgments This work was supported in part by All India Council for Technical Education —Quality Improvement Programme scheme 2010. Access to research and computing facilities was provided by the Anna University and K.L.N. College of Engineering.

References

1. Gajski, D.D., Vahid, F., Narayan, S., Gong, J.: SpecSyn—an environment supporting the specify-explore-refine paradigm for Hardware/Software system design. IEEE Trans. VLSI Syst. **6**(1), 84–100 (1998)
2. Henkel, J.: A low power Hardware/Software partitioning approach for core-based embedded systems. In: Proceedings of the 36th ACM/IEEE Conference on Design Automation, pp. 122–127 (1999)
3. Goldstein, S.C., Schmit, H., Budiu, M., Moe, M., Taylor, R.R.: PipeRench—a reconfigurable architecture and compiler. IEEE Computer **33**, 70–77 (2000)
4. DeHon, A.: DPGA-coupled microprocessors-commodity ICs for the early 21st century. In: Proceedings of FCCM (1994)
5. Stitt, G., Vahid, F.: Hardware/Software partitioning of software binaries. In: IEEE/ACM International Conference on Computer Aided Design, pp. 164–170 (2002)
6. Karypis, G., Aggarwal, R., Kumar, V., Shekhar, S.: Multilevel hypergraph partitioning—application in VLSI domain. IEEE Trans. VLSI Syst. **20**(1) (1999)
7. Alpert, C. J.: The ISPD98 circuit benchmark suite. In: Proceedings of the 1998 International Symposium on Physical Design, pp. 80–85 (1998)
8. Jiang, Y., Zhang, H., Jiao, X., Song, X., Hung, W.N.N., Gu, M., Sun, J.: Uncertain model and algorithm for Hardware/Software partitioning. IEEE Comp. Soc. Annu. Symp. VLSI 243–248 (2012)
9. Al-Wattar, A., Areibi, S., Saffih, F.: Efficient on-line Hardware/Software task scheduling for dynamic run-time reconfigurable systems. In: 26th International Parallel and Distributed Processing Symposium Workshops & PhD, Forum, pp. 401–406 (2012)
10. Goldberg, D.E.: Genetic Algorithms in Search, Optimization and Machine Learning. Pearson Education (2004)
11. Sheng, W., He, W., Jiang, J., Mao, Z.: Pareto optimal temporal partition methodology for reconfigurable architectures based on multi-objective genetic algorithm. In: 26th International Parallel and Distributed Processing Symposium Workshops and PhD, Forum, pp. 425–430 (2012)
12. Mazumder, P., Rudnik, E.M.: Genetic Algorithms for VLSI Design, Layout and Test Automation. Pearson Education (2003)
13. Luo, L., He, H., Dou, Q., Xu, W.: Hardware/Software partitioning for heterogeneous multicore SoC using genetic algorithm. In: Second International Conference on Intelligent System Design and Engineering Application, pp. 1267–1270 (2011)
14. Su, L., Zhang, X.: Research on an SOC Software/Hardware partition algorithm based on undirected graphs theory. In: IEEE International Conference on Computer Science and Automation Engineering, pp. 274–278 (2012)

Solving School Bus Routing Problem Using Hybrid Genetic Algorithm: A Case Study

Bhawna Minocha and Saswati Tripathi

Abstract School bus routing involves transporting students from predefined locations to school using a fleet of buses with varying capacity. This paper describes a real-life problem of the school bus routing. The overall goal of this study is to develop a route plan for the school bus service so that it is able to serve the students in an efficient and economical manner with maximum utilization of the capacity of buses using hybrid genetic algorithm.

Keywords School bus routing problem · Genetic algorithm · Vehicle routing problem with time windows

1 Introduction

Approximately 23 million public school students travel through 400,000 school buses twice daily for going to school and back from the school. It has been estimated that out of these one to two million students travel in school buses (National Association of State Directors of Pupil Transportation Services 1998).

School bus routing and scheduling is a transportation area which needs an in-depth study for improving the service quality and reducing operating costs. The school bus routing problem consists of a set of students dispersed in a region who have to travel to and back from their schools every day. The main goal of any school transportation system is to provide safe efficient and reliable transportation for its students. It is therefore necessary to efficiently assign students to designated bus

B. Minocha (✉)
Amity School of Computer Sciences, Noida, India
e-mail: bminocha@amity.edu

S. Tripathi
Indian Institute of Foreign Trade, Kolkata, India
e-mail: saswati@iift.ac.in

B. V. Babu et al. (eds.), *Proceedings of the Second International Conference on Soft Computing for Problem Solving (SocProS 2012), December 28–30, 2012*, Advances in Intelligent Systems and Computing 236, DOI: 10.1007/978-81-322-1602-5_11, © Springer India 2014

stops and determine the appropriate locations of different bus stops and routes and the bus schedules to minimize the total operating cost while satisfying requirements of the school and the children.

According to Swersey et al. [8], "In school bus transportation the two most visible problems are routing and scheduling. In the routing problem every student is assigned to a bus stop and those particular stops are sum up to form routes. In the morning a bus follows these routes, from one stop to another, picking up the students and carrying them to school. In the scheduling problem, particular buses are assigned to particular routes."

According to Spasovic et al. [7], "School Bus Routing is a version of the traveling salesman problem, normally referred to the group of vehicle routing problems (VRP), also with or with no time window constraints. Three factors that make School Bus Routing unique are: (1) Efficiency (the cost to run a school bus) (2) Effectiveness (how well the demand for service is fulfilled) (3) Equity (fairness of the school bus for each student)."

Ke et al. [3] present a variety of model formulations of school bus routing problem. Li et al. [4] solved a case study treating the problem as multi-objective combinatorial optimization problem. Thangiah et al. [9] discuss the routing of school buses in rural areas. Recently, a complete review of school bus routing problems has been provided by Park et al. [6].

Bus routing has gained the attention of many researchers in different fields. Whereas some researchers are focusing on the designing new algorithms, others are advancing existing algorithms and applying existing algorithms to the real-world problems. These have been discussed by Toth et al. [10] and Golden et al. [1] in their respective books.

The aim of school bus routing is to transport students in the safest, most economical, and convenient manner such that the following objectives and constraints are met:
Objectives:

- Minimize the total number of buses required.
- Minimize the total distance traveled by the buses.

Constraints:

- The number of students must not exceed the number of seats on each bus.
- The time taken by buses must not exceed the specified limit.
- Each bus stop is to be allocated to only one bus.
- Each bus must pick all the students allocated on that route within specified time windows and must reach the school before it starts.

Thus, school bus routing is essentially vehicle routing problem with time window (VRPTW). In VRPTW, a set of vehicles with fixed and identical capacity is to be routed from a central depot to a set of geographically scattered locations (cities, stores, schools, customers, warehouses, etc.), which have varying demands and predefined time windows. The vehicle can visit the location in this specified time window only.

The objective of VRPTW is to service all the customers as per their requirement while minimizing the number of vehicles required as well as the total travel distance by all the vehicles used without violating capacity constraints of the vehicles and the location's time window requirement such that each location is visited once and only once by one of the vehicles. All the routes are to start and ultimately end at the depot. In the present study, we deal with routing the buses to appropriate bus stops.

The genetic algorithm (GA) approach was proposed by Holland [2] in 1975. It is an adaptive heuristic search method that mimics evolution through natural selection. It works by combining selection, crossover, and mutation operations of genes. The selection procedure drives the population toward a better solution, while crossover uses genes of selected parents to produce new offsprings that form the next generation. The genetic algorithm approach has now become popular as it helps in finding reasonably good solutions for complex mathematical problems and NP-hard problems like vehicle routing problem.

2 Optimal Allocation of School Buses in Vehicle Routing Planning of Blooming Dales School: A Case Study

The case study considered by us is the school bus routing problem of Blooming Dales School situated in Ganga Nagar, Rajasthan, India. The school has primary wing and high school wing in which presently students from nursery grade to grade X study. Students from all over the town come there for schooling as it is very famous school in the city imparting quality education to the students. An important problem before the school management is to provide transport the students to and from the school which is safe, economical, and convenient to the students who do not have their own means of travel to the school.

Students travel to the school by different modes of transports. However, a large number of them depend upon school buses. Presently, the school has a fleet of six buses each with a seating capacity of 60 seats. The school starts at 8:00 am, so all the buses have to reach the school campus by 7:50 am. The students go to their allocated bus stop from their respective homes and board the school buses from there. The students have been clustered into groups. Each group is associated with one of 50 bus stops. The buses are required to depart and return to the school within an allotted fifty minutes time period. Table 1 shows the data provided by the school.

The travel plan currently being followed by the school management is shown in Table 2. The currently running routes have been established intuitively and providing the bus facility to 286 students. Every single route is serviced by a single vehicle. All routes start and end at 0 indicate that all buses start their journey from school and ends at school. Each bus stop is represented by a unique number, and all students at a particular bus stop are serviced by a single bus.

Table 1 Distance and demands of case study

Bus stop	No. of students boarding at bus stop	Distance of bus stop from school (in km)	Scheduled arrival time of bus at bus stop
0	0	0	7:00
1	12	5.1	7:33
2	4	7.9	7:22
3	9	9.5	7:17
4	4	5.7	7:33
5	10	7.2	7:28
6	6	6.7	7:24
7	4	4.2	7:34
8	8	4.7	7:32
9	6	9.7	7:19
10	4	7.9	7:26
11	6	6.3	7:30
12	3	8.9	7:20
13	6	8.3	7:17
14	4	8.9	7:21
15	3	4.9	7:36
16	9	8.4	7:21
17	3	5.9	7:33
18	6	7.4	7:25
19	8	6.2	7:26
20	7	9.4	7:20
21	6	6.4	7:31
22	4	6.9	7:29
23	3	7.8	7:24
24	3	7.9	7:26
25	8	6.7	7:28
26	8	6.9	7:27
27	4	9	7:16
28	6	8.9	7:23
29	2	10.1	7:19
30	4	5.2	7:31
31	3	4.4	7:38
32	4	3.9	7:40
33	7	4.1	7:38
34	12	5.9	7:31
35	2	8.5	7:24
36	5	5.4	7:35
37	8	7.2	7:22
38	4	5.2	7:34
39	7	4.9	7:35
40	8	7.4	7:25
41	4	5.7	7:28
42	2	9.3	7:22
43	4	7.5	7:27

(continued)

Table 1 (continued)

Bus stop	No. of students boarding at bus stop	Distance of bus stop from school (in km)	Scheduled arrival time of bus at bus stop
44	16	8.4	7:18
45	9	8.5	7:23
46	8	6.1	7:30
47	3	8.8	7:16
48	2	9.9	7:18
49	6	7.7	7:00
50	2	4.1	7:37

Table 2 Current bus route plan

Bus	Route	Start time (am)	Return time (am)
1	0–29–28–35–10–5–21–17–15–31–0	7:00	7:46
2	0–48–20–14–24–43–22–4–39–33–0	7:00	7:46
3	0–3–12–16–23–40–25–34–38–0	7:00	7:45
4	0–47–13–49–37–6–19–41–8–7–0	7:00	7:43
5	0–27–44–2–18–26–46–1–0	7:00	7:45
6	0–9–42–45–11–30–36–50–32–0	7:00	7:44

2.1 Problem Identification

We analyzed the data provided by the school management and calculated the capacity utilization of each bus, which is shown in Table 3. The table shows details for each bus route which includes the starting time from the school and time at which bus return to the school, the number of stops it covers and total number of students it served during the entire trip. The capacity utilization (students/capacity) has also been computed and listed in the Table 3. This shows that some of the buses are under utilized.

2.2 Design of More Efficient Bus Routes

The goal of this study is to develop a route plan for the school bus service so that it is able to serve all the students efficiently with maximum utilization of the buses. For this, we have first transformed the problem into the format of a VRPTW. In order to convert the given school bus routing problem into VRPTW problem, we need the latest arrival time of bus and required service time at each stop. We have added four minutes to the scheduled arrival time of bus at each stop to determine the latest acceptable arrival time of bus at a designated stop. Service time is actually the

Table 3 Current bus routes with their capacity utilization

Bus	Route	Start time (am)	Return time (am)	No. of stops	No. of students	Vehicle capacity	Capacity utilization (%)
1	0–29–28–35–10–5–21–17–15–31–0	7:00	7:46	9	39	60	65
2	0–48–20–14–24–43–22–4–39–33–0	7:00	7:46	9	42	60	70
3	0–3–12–16–23–40–25–34–38–0	7:00	7:45	8	56	60	93.33
4	0–47–13–49–37–6–19–41–8–7–0	7:00	7:43	9	53	60	88.33
5	0–27–44–2–18–26–46–1–0	7:00	7:45	7	58	60	96.66
6	0–9–42–45–11–30–36–50–32–0	7:00	7:44	8	38	60	63.33
Total				50	286	360	79.44

stoppage time of bus at each stop. This is equal to total time required by all students to board the bus. This has been taken to be the number of students boarding the bus from that stop multiplied by ten. (We are assuming here that each student needs about 10 s to board the bus.)

The buses depart from the school to the farthest bus stop points and start picking up students from various bus stops in their return journey to the school. The travel time t_{0i} is computed as d_{oi}/v, where d is the distance of bus stop i from school, and v is the average running speed of the bus (we have assumed the average speed of the bus to be 30 km/h). The present problem in reality is a VRPTW problem with some constraints. Buses cannot travel randomly from one stop to another. They have to follow the roads available. For instance, there is no direct path between many bus stops such as 45 and 35, 3 and 29, 42 and 48, 47 and 27, 18 and 19, 16 and 37. Similarly, one can reach bus stop 42 only from bus stop nine.

3 Use of Genetic Algorithm to Solve School Bus Routing Problem

Keeping all points in view, we developed a VRPTW-based model for this problem. A genetic algorithm is developed to solve it. After building the initial population, all individuals are evaluated according to the fitness criteria. The evolution continues with a three-way tournament selection in which good individuals are selected for reproduction. In each generation, two best individuals are preserved for the next generation without being subjected to genetic operations. The problem-specific route-exchange crossover and mutation operations have been designed and applied to modify the selected individuals to form new feasible individuals for the population. Details of the algorithm are available in Minocha et al. [5]. The proposed algorithm has been coded in C++.

The travel plan obtained by genetic algorithm is shown in Table 4. All the bus departs and arrives at the school within the time window. Now, bus no. 1 and bus no. 2 cover more bus stops, but bus No. 6 covers lesser bus stops than the original route plan. The overall capacity utilization of the buses though remains same. But if we replace the bus no. 6 with a bus with half the capacity, it will increase the overall capacity utilization to 86.67 %.

4 Use of Hybrid Genetic Algorithm to Solve School Bus Routing Problem

Although genetic algorithms can rapidly locate the region in which the global optimum exists, they take a relatively long time to locate the exact local optimum in the region of convergence. On contrary, local searches (valid in a small region of search

Table 4 New bus routes by genetic algorithm

Bus	Route	Start time (am)	Return time (am)	No. of stops	No. of students	Vehicle capacity	Capacity utilization (%)
1	0–29–9–42–28–35–10–5–21–17–15–31–0	7:00	7:48	11	47	60	78.34
2	0–48–20–14–24–43–22–11–4–39–33–0	7:00	7:47	10	42	60	80
3	0–3–12–16–23–40–25–34–38–0	7:00	7:45	8	56	60	93.33
4	0–47–13–49–37–6–19–41–8–7–0	7:00	7:43	9	53	60	88.33
5	0–27–44–2–18–26–46–1–0	7:00	7:45	7	58	60	96.66
6	0–45–30–36–50–32–0	7:00	7:44	5	24	60	40
Total				50	286	360	79.44

Table 5 New bus routes by hybrid genetic algorithm

Bus	Route	Start time (am)	Return time (am)	No. of stops	No. of students	Vehicle capacity	Capacity utilization (%)
1	0–29–9–42–28–35–10–5–21–17–36–15–31–32–0	7:00	7:49	13	56	60	93.33
2	0–48–20–14–45–24–43–22–11–4–39–33–0	7:00	7:48	11	57	60	95
3	0–3–12–16–23–40–25–34–38–50–0	7:00	7:46	9	58	60	96.66
4	0–47–13–49–37–6–19–41–30–8–7–0	7:00	7:44	10	57	60	95
5	0–27–44–2–18–26–46–1–0	7:00	7:45	7	58	60	96.66
Total				50	286	300	95.33

space) are quick in finding an optimal solution. A combination of a genetic algorithm and a local search method can speed up the search to locate the exact or near-exact global optimum. These are now properly known as hybridized GAs. In such an algorithm, applying a local search to the solutions that are guided by a genetic algorithm to the most promising region can accelerate convergence to the global optimum.

In this study, we incorporate two new local search heuristics to search for better routing solutions of VRPTW. These searches are as follows:

1. *Changing next neighbor* (*CNN*): In this case having selected an individual randomly from the population, a node says C_j is randomly chosen from one of its routes. An effort is now made to replace its next neighbor C_{j+1} by some alternative acceptable node, say C_k. If it is possible, then we terminate the selected route at C_{j+1} and all the nodes from C_{j+2} onwards are reinserted in suitable places in other existing routes. The route from which new C_{j+1} was picked up is joined together at the next node of that route. If such an arrangement of routes is not possible, then the same individual is returned.
2. *Reinserting random node* (*RRN*): In this case again after selecting an individual randomly from the population, a customer C_j is randomly selected from one of its routes. An attempt is made to insert it in another existing route at an appropriate place. If that becomes possible we join the route from where C_j was taken at the next node of that route. This creates a new feasible solution. If it is not possible, then the same individual is returned.

The genetic algorithm used in previous section is combined with these local searches to yield hybridized genetic algorithm. Any one of the alternative local search heuristics is applied randomly. The travel plan obtained by hybrid genetic algorithm is shown in Table 5. All the bus departs and arrives at the school within the time window, but covers more bus stops. Hence, the overall requirement of buses is reduced by one, and thus, capacity utilization of buses is increased.

5 Conclusion

In the present study, we have solved a practical real-life situation using hybrid genetic algorithm. New travel plan obtained yields a better solution as compared to the currently used one both in the terms of number of buses required and in terms of the overall capacity utilization of the buses. This has motivated the school management to use the suggested route plan. Less number of buses indicates saving in terms of cost of bus, its maintenance, running expanses (including fuel, salary of driver, conductor). In fact with the increase in fuel prices, transportation schedules should be planned efficiently.

References

1. Golden, B., Raghavan, S., Wasil, E.: The Vehicle Routing Problem: Latest Advances and New Challenges. Springer, Berlin (2008)
2. Holland, J.H.: Adaptation in Natural and Artificial System. The University of Michigan Press, Michigan (1975)
3. Ke, X., Caron, R.J., Aneja, Y.P.: The school bus routing and scheduling problem with homogenous bus capacity: formulations and their solutions. http://www.uwindsor.ca/math/sites/uwindsor.ca.math/files/05-06.pdf (2006)
4. Li, L.Y., Fu, Z.: The school bus routing: a case study. J. Oper. Res. Soc. **53**, 552–558 (2002)
5. Minocha, B. Tripathi S., Mohan, C.: Solving vehicle routing and scheduling problems using hybrid genetic algorithm. In: IEEE Proceedings of 3rd International Conference on Electronics Computer Technology-ICECT 2011, vol. 2, pp. 189–193. IEEE. http://ieeexplore.ieee.org/xpl/freeabs_all.jsp?arnumber=5941682&reason=concurrency
6. Park, J., Kim, B.: The school bus routing problem: a review. Eur. J. Oper. Res. **202**(2), 311–319 (2010)
7. Spasovic, L., Chien, S., Kelnhofer-Feeley, C., Wang, Y., Hu, Q.: A methodology for evaluating of school bus routing-a case study of riverdale, New Jersey. Transportation Research Board 80th Annual Meeting Washington, D.C. Springer, Berlin. http://transportation.njit.edu/nctip/publications/No01-2088.pdf (2001)
8. Swersey, A.J., Wilson, B.: Scheduling school buses. Manag. Sci. **30**(7), 844–853 (1984)
9. Thangiah, S.R., Fergany, A., Wilson, B., Pitluga, A., Mennell, W.: School bus routing in rural school districts. computer-aided systems in public transport. Lecture Notes in Economics and Mathematical Systems, vol. 600(II), pp. 209–232, Springer, Berlin (2008)
10. Toth, P., Vigo. D.: The Vehicle Routing Problem. Monographs on Discrete Mathematics and Applications. SIAM, Philadelphia, (2002)

Taguchi-Based Tuning of Rotation Angles and Population Size in Quantum-Inspired Evolutionary Algorithm for Solving MMDP

Nija Mani, Gursaran, A. K. Sinha and Ashish Mani

Abstract Quantum-inspired evolutionary algorithms (QEAs) have been successfully used for solving search and optimization problems. QEAs employ quantum rotation gates as variation operator. The selection of rotation angles in the quantum gate has been mostly performed intuitively. This paper presents tuning of the parameters by designing experiments using well-known Taguchi's method with massively multimodal deceptive problem as the benchmark.

Keywords Robust design · Multimodal · Deceptive · Optimization

1 Introduction

Quantum-inspired evolutionary algorithms (QEAs) have been successfully applied in solving wide variety of real-life difficult optimization problems, where near-optimal solutions are acceptable and efficient deterministic techniques are not known [1]. QEAs are EAs inspired by the principles of quantum mechanics. They are developed by drawing some ideas from quantum mechanics and integrating them in the current framework of EA. QEAs have performed better than classical evolutionary

N. Mani (✉) · Gursaran · A. K. Sinha
Department of Mathematics, Dayalbagh Educational Institute, Dayalbagh, Agra, India
e-mail: nijam@acm.org

Gursaran
e-mail: gursaran.db@gmail.com

A. K. Sinha
e-mail: arunsinha47@gmail.com

A. Mani
USIC, Dayalbagh Educational Institute, Dayalbagh, Agra, India
e-mail: ashish.mani@ieee.org

B. V. Babu et al. (eds.), *Proceedings of the Second International Conference on Soft Computing for Problem Solving (SocProS 2012), December 28–30, 2012*, Advances in Intelligent Systems and Computing 236, DOI: 10.1007/978-81-322-1602-5_12,

algorithms (EAs) on many complex problems [1] as they provide better balance between exploration and exploitation due to probabilistic representation of solutions.

The canonical QEA proposed by Han and Kim [2] employs Q-bit as the smallest unit of information, which is essentially a probabilistic bit. The Q-bit is modified by using quantum gates, which are implemented as unitary matrix [2]. A quantum gate known as rotation gate has been widely used in many QEA implementations [1]. It acts as the main variation operator that rotates Q-bit strings to obtain good candidate solutions for the next iteration. It takes into account the relative current fitness level of the individual and the attractor and also their binary bit values for selecting the magnitude and direction of rotation. The magnitude of rotation is a tunable parameter and is selected from a set of eight rotation angles. The value of the rotation angles are problem dependent and require tuning.

The other parameters that require tuning in QEA are population size, group size, and migration period. The effect of population size, group size, and migration period on the performance of QEA have been studied in some detail [2]; however, the eight rotation angles have been mostly set by ad hoc experimentation, which is not a best practice as per [3]. It has also been recommended in [3] that proper design of experimentation should be employed for determining the parameter values.

Taguchi had proposed fractional design of experiments which have been very successful in identifying the parameters that have maximum influence on the process. Though Taguchi's method was developed primarily for improving quality of the product by incorporating it in design and manufacturing process, it has been used in some efforts for tuning of evolutionary algorithms [4] as the process of search in EAs is analogous to industrial process affected by set of tunable parameters with the objective function fitness as the quality characteristics [4]. Taguchi's design of experiment are based on a special set of orthogonal arrays that does not guarantee optimality (unlike factorial design of experiments) but has been shown to find better parameters values than ad hoc experimentation [3]. Further, it does not suffer from curse of dimensionality as in case of factorial design of experiments, in which number of experiments quickly become impractical to execute, e.g., in this study, nine parameters with five levels each have been considered, so according to factorial design of experiments, the total number experiments to be conducted is 9^5, which is a hooping 59,049 experiments. It can be argued that even after considering more than 59,000 experiments, the optimality is not guaranteed in principle as it is dependent on choice of levels, which is determined subjectively. Taguchi's fractional design of experiment reduces the number experiments to a manageable 50 experiments only, which is a reduction by a factor of about 1,200.

2 Quantum-Inspired Evolutionary Algorithm

The potential advantages offered by quantum computing [1] have led to the development of approaches that suggest ways to integrate aspects of quantum computing with evolutionary computation [5]. The first attempt was made by Narayan and Moore [6] to use quantum interpretation for designing a quantum-inspired genetic algorithm. A number of other hybridizations have also been proposed, of which the most popular proposal has been made by Han and Kim [2], which primarily hybridizes the superposition and measurement principles of quantum computing in evolutionary framework by implementing qubit as Q-bit.

A qubit is the smallest information element in quantum computer, which is quantum analog of classical bit. The classical bit can be either in state 'zero' or in state 'one,' whereas a quantum bit can be in a superposition of basis states in a quantum system [2]. The Q-bit string acts as genotype of the individual and the binary bit string formed by collapsing Q-bit forms the phenotype of the individual. The process of measuring or collapsing Q-bit is performed by generating a random number between 0 and 1 and comparing it with $|\alpha|^2$. If the random number is less than $|\alpha|^2$, then the Q-bit collapses to 0 or else to 1 and this value is assigned to the corresponding binary bit. Further, the Q-bit is modified by using quantum gates or operators, which are also unitary in nature [7]. The quantum gates are implemented in QEA as unitary matrix, and further details are available in [2].

A quantum gate known as rotation gate has been employed in [2]. It acts as the main variation operator that rotates Q-bit strings to obtain good candidate solutions for the next iteration. It requires an attractor [8] toward, which the Q-bit would be rotated. It further takes into account the relative current fitness level of the individual and the attractor and also their binary bit values for determining the magnitude and direction of rotation. The magnitude of rotation is a tunable parameter and is selected from a set of eight rotation angles viz., $\theta_1, \theta_2, \ldots \theta_8$. The value of the rotation angles is problem dependent and require tuning [2].

The quantum-inspired evolutionary algorithm is as follows [2]:

```
(a) t = 0; Define Group Size;
(b) initialize Q(t);
(c) make P(t) by observing the states of Q(t);
(d) evaluate P(t);
(e) store the best solutions among P(t) into B(t);
(f) while (termination condition is not met) {
(g)         t = t + 1;
(h)         make P(t) by observing the states of Q(t-1);
(i)         evaluate P(t);
(j)         Determine the attractors Atr(t);
(k)         update Q(t) according to P(t) and Atr(t) using Q-gate;
(l)         store the best solutions among B(t-1) and P(t) into B(t);
(m)         store the best solution b among B(t);
(n)         if(migration condition)
(o)         migrate b or b_j^t to B(t) globally or locally, respectively
    }
```

In step (b), the qubit register $Q(t)$ containing Q-bit strings for all the individuals is initialized randomly. In step (c), the binary solutions in $P(0)$ are constructed by observing the states of $Q(0)$. In a quantum computer, the act of observing a quantum state collapses it to a single state. However, collapsing into a single state does not occur in QEA, since QEA runs on a digital computer, not on a quantum computer. In step (d), each binary solution is evaluated to give a measure of its fitness. In step (e), the initial best solutions are then selected among the binary solutions $P(0)$, and stored into $B(0)$. In steps (f) and (g), iteratively, the binary solutions in P(t) are formed by observing the states of $Q(t-1)$ as in step (c), and each binary solution is evaluated for the fitness value. In step (j), the attractors are determined for each individual according to the strategy. In step (k), Q-bit individuals in $Q(t)$ are updated by applying Q-gates by taking into account Atr(t), b and $P(t)$, which is defined as a variation operator of QEA. The variation operator is rotation gate. In steps (l) and (m), the best solutions among $B(t-1)$ and $P(t)$ are selected and stored into $B(t)$, and if the best solution stored in $B(t)$ is better fitted than the stored best solution b, the stored solution is replaced by the new one.

In step (n), a migration condition is checked and if satisfied, the best solution b is migrated to $B(t)$ or the best among some of the solutions in $B(t)$, b_j^t, is migrated to them. The migration condition and local groups are taken to be design parameters and have to be chosen appropriately for the problem at hand.

It is suggested that QEA should have a local migration in every iteration and global migration after every 100 iterations. The group size is taken as five.

3 Design of Experiments

The design of experiment follows guidelines given in [3] and is outlined below:

1. The process objective describing the quality of the algorithm is the objective fitness function value.
2. The design parameters are the eight rotation angle parameters (θ_1 to θ_8) in the Q-gate and the population size. The number of levels has been taken as five for each parameter so that we can investigate them thoroughly. The value of each level for all the parameters is shown in Table 1.
3. There are a total of nine parameters and five levels, so the orthogonal array listed in L50 has been selected [9] for deciding the experiments; hence, a total of fifty experiments have been conducted.
4. Thirty runs of QEA with different sets of random numbers have been conducted for each experiment. The experiments data have been collected for mean and variance of objective fitness value.
5. The primary interest in design of EAs is to have better mean value rather than signal to noise ratio. Therefore, data analysis has been reported in this work for the mean value only.

The experimentation is performed by using massively multimodal deceptive problem as benchmark, with size $k = 40$, which is a large instance [10]. This problem

Table 1 Parameter levels

Parameters	L1	L2	L3	L4	L5
θ_1	0	$-0.0025\,\pi$	$-0.005\,\pi$	$-0.01\,\pi$	$-0.05\,\pi$
θ_2	0	$-0.0025\,\pi$	$-0.005\,\pi$	$-0.01\,\pi$	$-0.05\,\pi$
θ_3	0	$+0.005\,\pi$	$+0.01\,\pi$	$+0.05\,\pi$	$+0.1\,\pi$
θ_4	0	$-0.0025\,\pi$	$-0.005\,\pi$	$-0.01\,\pi$	$-0.05\,\pi$
θ_5	0	$-0.005\,\pi$	$-0.01\,\pi$	$-0.05\,pi$	$-0.1\,\pi$
θ_6	0	$+0.0025\,\pi$	$+0.005\,\pi$	$+0.01\,\pi$	$+0.05\,\pi$
θ_7	0	$+0.0025\,\pi$	$+0.005\,\pi$	$+0.01\,\pi$	$+0.05\,\pi$
θ_8	0	$+0.0025\,\pi$	$+0.005\,\pi$	$+0.01\,\pi$	$+0.05\,\pi$
Population size	10	15	25	35	50

is difficult for EAs as the numbers of local maxima are very large as compared with global maxima. Further, the fitness landscape is such that EA finds it very convenient to reach the local maxima, but searching global maxima is extremely difficult as it is located at the extreme ends of subproblem string.

4 Results and Analysis

The results of the experiments conducted by employing Taguchi's method are summarized in Table 2. The results show that best set of parameters values identified from the experiments are $-0.005\,\pi$ for θ_1, $-0.01\,\pi$ for θ_2, $+0.1\,\pi$ for θ_3, $-0.05\,\pi$ for θ_4, $-0.005\,\pi$ for θ_5, $+0.01\,\pi$ for θ_6, $+0.005\,\pi$ for θ_7, $+0.0025\,\pi$ for θ_8, 25 for population size. Table 3 shows comparison between the performance of QEA with parameters' value tuned with Taguchi's method and the performance of QEA with parameters' value at recommended setting derived by ad hoc experimentation as reported in the literature [1, 2]. The result in Table 3 indicates the superiority of the Taguchi's method over ad hoc experimentation for finding suitable values of parameters.

Table 2 Results of parameter optimization

L	θ_1	θ_2	θ_3	θ_4	θ_5	θ_6	θ_7	θ_8	Pop size
1	37.13	35.38	32.23	35.74	37.43	33.34	36.23	33.83	33.52
2	36.25	36.64	36.02	34.95	**38.63**	37.87	36.28	**37.59**	36.64
3	**37.49**	35.12	37.76	36.00	35.33	35.06	**36.79**	37.42	**37.17**
4	32.74	**37.09**	35.65	35.70	32.88	**37.92**	34.47	35.19	36.98
5	36.49	35.87	**38.77**	**36.29**	35.96	35.64	36.32	35.88	35.68

Table 3 Comparison between Adhoc-tuned and Taguchi-tuned QEAs

	Adhoc-tuned QEA	Taguchi-tuned QEA
Optimal	40	
Best	40	40
Median	39.64058	40
Worst	38.5623	40
Mean	39.4968	40
SD	0.418	0.0
Percentage success runs	26.67	100

The ad hoc-tuned QEA had all θs set to zero except $\theta_3 (= \ 0.01\,\pi)$ and $\theta_5 (= -0.01\,\pi)$ and with population size as 25, and the comparison has been made over 30 independent runs.

5 Conclusion

QEAs are evolutionary algorithms, which provide better balance between exploration and exploitation and have been widely used for solving difficult problems. This paper highlights issues in tuning of parameters and shows the utility of Taguchi-based method for tuning of parameters, especially the eight rotation angles and population size.

Further studies will include more comprehensive study by including other parameters on a suite of difficult problems.

References

1. Zhang, G.: Quantum-inspired evolutionary algorithms: a survey and empirical study. J Heuristics **17**, 303–351 (2011)
2. Han, K.H., Kim, J.H.: Quantum-inspired Evolutionary Algorithm for a Class of Combinatorial Optimization. IEEE Trans. on Evo. Comp. **6**(6), 580–593 (2002)
3. Adenso-Diaz, B., Laguna, M.: Fine-Tuning of Algorithms using Fractional Experimental Designs and Local Search. Op. Res. **54**(1), 99–114 (2006)
4. Hippolyte, J.L., Bloch, C., Chatonnay P., Espanet, C., Chamagne, D., and Wimmer, G.: Tuning an Evolutionary Algorithm with Taguchi Methods and Application to the dimensioning of an Electrical Motor. Proc. CSTST-2008, 265–272 (2008).
5. Sofge, D. A.: Prospective Algorithms for Quantum Evolutionary Computation. Proc. QI-2008, College Publications, UK, 2008.
6. Narayanan, A. and Moore M. : Quantum-inspired genetic algorithms. Proc. IEEE CEC-1996, 61–66, (1996).
7. Nielsen, M.A. and Chuang, I.L.: Quantum Computation and Quantum Information. Cambridge University Press, Cambridge.

8. Platelt, M.D., Schliebs, S., Kasabov, N.: A Verstaile Quantum-inspired Evolutionary Algorithm. Proc. of IEEE CEC **2007**, 423–430 (2007)
9. Design of experiments via Taguchi methods: orthogonal arrays available at https://controls.engin.umich.edu/wiki/index.php/Design_of_experiments_via_taguchi_methods:_orthogonal_arrays
10. Alba, E., Dorronsoro, B.: The exploration / exploitation tradeoff in dynamic cellular genetic algorithms. IEEE Trans. Evo. Co. **8**(2), 126–142 (2005)

Part II
Soft Computing for Mathematics and Optimization (SCMO)

Simultaneous Feature Selection and Extraction Using Fuzzy Rough Sets

Pradipta Maji and Partha Garai

Abstract In this chapter, a novel dimensionality reduction method, based on fuzzy rough sets, is presented, which simultaneously selects attributes and extracts features using the concept of feature significance. The method is based on maximizing both relevance and significance of the reduced feature set, whereby redundancy therein is removed. The chapter also presents classical and neighborhood rough sets for computing relevance and significance of the feature set and compares their performance with that of fuzzy rough sets based on the predictive accuracy of nearest neighbor rule, support vector machine, and decision tree. The effectiveness of the proposed fuzzy rough set-based dimensionality reduction method, along with a comparison with existing attribute selection and feature extraction methods, is demonstrated on real-life data sets.

1 Introduction

Dimensionality reduction is a process of selecting a map by which a sample in an m-dimensional measurement space is transformed into an object in a d-dimensional feature space, where $d < m$. The problem of dimensionality reduction has two aspects, namely formulation of a suitable criterion to evaluate the goodness of a feature set and searching the optimal set in terms of the criterion. The major mathematical measures so far devised for the estimation of feature quality can be broadly classified into two categories, namely feature selection in measurement space and feature selection in a transformed space. The techniques in the first category generally reduce the

P. Maji (✉) · P. Garai
Machine Intelligence Unit, Indian Statistical Institute, 203, BT Road,
Kolkata 700108, India
e-mail: pmaji@isical.ac.in

P. Garai
e-mail: parthagarai_r@isical.ac.in

B. V. Babu et al. (eds.), *Proceedings of the Second International Conference on Soft Computing for Problem Solving (SocProS 2012), December 28–30, 2012*, Advances in Intelligent Systems and Computing 236, DOI: 10.1007/978-81-322-1602-5_13, © Springer India 2014

dimensionality of measurement space by discarding redundant or least information-carrying features. On the other hand, those in second category utilize all information contained in the measurement space to obtain a new transformed space, thereby mapping a higher dimensional pattern to a lower dimensional one. This is referred to as feature extraction [1].

An optimal feature subset selected or extracted by a dimensionality reduction method is always relative to a certain feature evaluation criterion. In general, different criteria may lead to different optimal feature subsets. However, every criterion tries to measure the discriminating ability of a feature or a subset of features to distinguish different class labels. One of the main problems in real-life data analysis is uncertainty. Some of the sources of this uncertainty include incompleteness and vagueness in class definitions. In this background, the rough set theory [2] has gained popularity in modeling and propagating uncertainty. Rough sets can be used to find most informative feature subset of original attributes from a given data with discretized attribute values [3, 4]. However, there are usually real-valued data and fuzzy information in real-world applications. In rough sets, the real-valued features are divided into several discrete partitions, and the dependency or quality of approximation of a feature is calculated. The inherent error that exists in discretization process is of major concern in the computation of the dependency of real-valued features. Combining fuzzy and rough sets provides an important direction in reasoning with uncertainty for real-valued data [5]. They are complementary in some aspects. The generalized theories of rough fuzzy computing have been applied successfully to feature selection of real-valued data set [5–7]. Also, neighborhood rough sets [8] are found to be suitable for both numerical and categorical data sets.

On the other hand, a feature extraction technique such as principal component analysis (PCA), linear discriminant analysis, and independent component analysis [1] generates a new set of features using a mapping function that takes some linear or nonlinear combination of original features. While PCA uses a linear orthogonal transformation to project a sample space containing possibly correlated variables into a different space with uncorrelated variables, independent component analysis decomposes a multidimensional feature vector into statistically independent components to reveal the hidden factors from a set of random variables [1]. In general, a feature extraction technique provides a richer feature subset than that obtained using a feature selection algorithm with a higher cost. Hence, it is very difficult to decide whether to select a feature from original measurement space or to extract a new feature by transforming the existing features for a given data set. A dimensionality reduction algorithm needs to be formulated ,which can simultaneously select or extract features depending upon the criteria, integrating the merits of both feature selection and extraction techniques.

In this regard, a novel dimensionality reduction algorithm is proposed based on fuzzy rough sets, which simultaneously selects and extracts features from a given data set. Using the concept of feature significance, the feature set in each iteration is partitioned into three subsets, namely insignificant, dispensable, and significant feature sets. The insignificant feature set is discarded from the current feature set, while significant feature set is used to select or extract a feature in the next iteration.

Depending on the quality of features present in the dispensable set of current iteration, a new feature is extracted or an existing feature is selected from the dispensable set for reduced feature set. In effect, the final reduced feature set may simultaneously contain some original features of measurement space and extracted new features of transformed space, which are both relevant and significant. The effectiveness of the proposed fuzzy rough dimensionality reduction method, along with a comparison with other methods, is demonstrated on a set of real-life data.

2 Proposed Dimensionality Reduction Method

In this section, a new dimensionality reduction method is presented, integrating fuzzy rough sets and the merits of feature selection and extraction techniques.

2.1 Fuzzy Rough Sets

A crisp equivalence relation induces a crisp partition of the universe and generates a family of crisp equivalence classes. Correspondingly, a fuzzy equivalence relation generates a fuzzy partition of the universe and a series of fuzzy equivalence classes or fuzzy knowledge granules. This means that the decision and condition attributes may all be fuzzy [9]. Let $< \mathbb{U}, \mathbb{A} >$ represents a fuzzy approximation space and X is a fuzzy subset of $\mathbb{U}$. The fuzzy $\mathbb{P}$-lower and $\mathbb{P}$-upper approximations are then defined as follows [9]:

$$\mu_{\underline{\mathbb{P}}X}(F_i) = \inf_x \{\max\{(1 - \mu_{F_i}(x)), \mu_X(x)\}\} \quad \forall i \tag{1}$$

$$\mu_{\overline{\mathbb{P}}X}(F_i) = \sup_x \{\min\{\mu_{F_i}(x), \mu_X(x)\}\} \quad \forall i \tag{2}$$

where F_i represents a fuzzy equivalence class belonging to $\mathbb{U}/\mathbb{P}$, the partition of $\mathbb{U}$ generated by $\mathbb{P} \subseteq \mathbb{A}$, and $\mu_X(x)$ represents the membership of object x in X. These definitions diverge a little from the crisp upper and lower approximations, as the memberships of individual objects to the approximations are not explicitly available. As a result of this, the fuzzy lower and upper approximations can be defined as [5]

$$\mu_{\underline{\mathbb{P}}X}(x) = \sup_{F_i \in \mathbb{U}/\mathbb{P}} \min\{\mu_{F_i}(x), \mu_{\underline{\mathbb{P}}X}(F_i)\} \tag{3}$$

$$\mu_{\overline{\mathbb{P}}X}(x) = \sup_{F_i \in \mathbb{U}/\mathbb{P}} \min\{\mu_{F_i}(x), \mu_{\overline{\mathbb{P}}X}(F_i)\}. \tag{4}$$

The tuple $< \underline{\mathbb{P}}X, \overline{\mathbb{P}}X >$ is called a fuzzy rough set. This definition degenerates to traditional rough sets when all equivalence classes are crisp. The membership of an object $x \in \mathbb{U}$ belonging to the fuzzy positive region is

$$\mu_{\mathrm{POS}_{\mathbb{C}}(\mathbb{D})}(x) = \sup_{X \in \mathbb{U}/\mathbb{D}} \mu_{\underline{\mathbb{C}}X}(x) \tag{5}$$

where $\mathbb{A} = \mathbb{C} \cup \mathbb{D}$. Using the definition of fuzzy positive region, the dependency function can be defined as follows [5]:

$$\gamma_{\mathbb{C}}(\mathbb{D}) = \frac{|\mu_{\mathrm{POS}_{\mathbb{C}}(\mathbb{D})}(x)|}{|\mathbb{U}|} = \frac{1}{|\mathbb{U}|} \sum_{x \in \mathbb{U}} \mu_{\mathrm{POS}_{\mathbb{C}}(\mathbb{D})}(x). \tag{6}$$

2.2 Feature Significance

Let $\mathbb{U} = \{x_1, \cdots, x_i, \cdots, x_n\}$ be the set of n samples and $\mathbb{C} = \{\mathscr{A}_1, \cdots, \mathscr{A}_j, \cdots, \mathscr{A}_m\}$ denotes the set of m features of a given data set. Define $\gamma_{\mathscr{A}_i}(\mathbb{D})$ as the relevance of the feature $\mathscr{A}_i$ with respect to the class label or decision attribute $\mathbb{D}$. The relevance represents the quality of a feature or degree of dependency of decision attribute $\mathbb{D}$ on condition attribute $\mathscr{A}_i$. To what extent a feature is contributing to calculate the joint relevance or dependency can be calculated by the significance of that feature. The change in dependency when a feature is removed from the set of features is a measure of the significance of the feature.

Definition 1 The significance of a feature $\mathscr{A}_j$ with respect to another feature $\mathscr{A}_i$ can be defined as follows:

$$\sigma_{\{\mathscr{A}_i, \mathscr{A}_j\}}(\mathscr{A}_j, \mathbb{D}) = \gamma_{\{\mathscr{A}_i, \mathscr{A}_j\}}(\mathbb{D}) - \gamma_{\mathscr{A}_i}(\mathbb{D}). \tag{7}$$

Hence, the significance of a feature $\mathscr{A}_j$ is the change in dependency when the feature $\mathscr{A}_j$ is removed from the set $\{\mathscr{A}_i, \mathscr{A}_j\}$. The higher the change in dependency, the more significant feature $\mathscr{A}_j$ is. If significance is 0, then feature $\mathscr{A}_j$ is dispensable. The following properties can be stated about the measure:

1. $\sigma_{\{\mathscr{A}_i, \mathscr{A}_j\}}(\mathscr{A}_j, \mathbb{D}) = 0$ if and only if the feature $\mathscr{A}_j$ is dispensable in the set $\{\mathscr{A}_i, \mathscr{A}_j\}$.
2. $\sigma_{\{\mathscr{A}_i, \mathscr{A}_j\}}(\mathscr{A}_j, \mathbb{D}) < 0$ if the feature $\mathscr{A}_i$ is more relevant than the feature set $\{\mathscr{A}_i, \mathscr{A}_j\}$.
3. $\sigma_{\{\mathscr{A}_i, \mathscr{A}_j\}}(\mathscr{A}_j, \mathbb{D}) > 0$ if the feature $\mathscr{A}_j$ is significant with respect to another feature $\mathscr{A}_i$.
4. $\sigma_{\{\mathscr{A}_i, \mathscr{A}_j\}}(\mathscr{A}_j, \mathbb{D}) \neq \sigma_{\{\mathscr{A}_i, \mathscr{A}_j\}}(\mathscr{A}_i, \mathbb{D})$ (asymmetric).

2.3 Simultaneous Feature Selection and Extraction

The high-dimensional real-life data set generally may contain a number of nonrelevant and insignificant features. The presence of such features may lead to a reduction in the useful information. Ideally, the reduced feature set obtained using a dimensionality reduction algorithm should contain features those have high relevance with the classes, while the significance among them would be as high as possible. The relevant and significant features are expected to be able to predict the classes of the samples. Hence, to assess the effectiveness of the features, both relevance and significance need to be measured quantitatively. The proposed dimensionality reduction method addresses the above issues through the following three phases:

1. computation of the relevance of each feature present in original feature set;
2. determination of the insignificant, dispensable, and significant feature sets; and
3. extraction of a relevant feature from the dispensable set.

The fuzzy rough set is used to compute both relevance and significance of features. The insignificant feature set is discarded from the whole feature set, while the significant feature set is used to select or extract significant features for reduced feature set. Let $\gamma_{\mathscr{A}_i}(\mathbb{D})$ represents the relevance of feature $\mathscr{A}_i \in \mathbb{C}$. The proposed algorithm starts with a single feature $\mathscr{A}_i$ that has the highest relevance value. Based on the significance values of all other features, the feature set $\mathbb{C}$ is then partitioned into three subsets, namely insignificant set I_i, dispensable set D_i, and significant set S_i, which are defined as follows:

$$I_i = \{\mathscr{A}_j | \sigma_{\{\mathscr{A}_i, \mathscr{A}_j\}}(\mathscr{A}_j, \mathbb{D}) < -\delta_i; \mathscr{A}_j \neq \mathscr{A}_i \in \mathbb{C}\} \quad (8)$$

$$D_i = \{\mathscr{A}_j | -\delta_i \leq \sigma_{\{\mathscr{A}_i, \mathscr{A}_j\}}(\mathscr{A}_j, \mathbb{D}) \leq \delta_i; \mathscr{A}_j \neq \mathscr{A}_i \in \mathbb{C}\} \quad (9)$$

$$S_i = \{\mathscr{A}_j | \sigma_{\{\mathscr{A}_i, \mathscr{A}_j\}}(\mathscr{A}_j, \mathbb{D}) > \delta_i; \mathscr{A}_j \neq \mathscr{A}_i \in \mathbb{C}\} \quad (10)$$

where $\mathbb{C} = I_i \cup D_i \cup S_i \cup \{\mathscr{A}_i\}$ and δ_i is a predefined threshold value corresponding to the feature $\mathscr{A}_i$.

The insignificant set I_i represents the set of features those are insignificant with respect to the candidate feature $\mathscr{A}_i$ of the current iteration. Hence, the insignificant set I_i should be discarded from the whole feature set $\mathbb{C}$ as the presence of such insignificant features may lead to a reduction in the useful information. If insignificant features are present in the reduced feature set, they may reduce the classification or clustering performance. The significant set S_i consists of a set of features those are significant with respect to the feature $\mathscr{A}_i$. In other words, the set S_i represents the set of features of $\mathbb{C}$ those have the significance values with respect to the feature $\mathscr{A}_i$ greater than the threshold δ_i. This set is considered in the next iteration to select or extract a new feature.

On the other hand, the dispensable set D_i is used for extracting a new feature in the current iteration. As the significance values of the features present in the dispensable set are very low, they form a group of similar features. These features may be considered to generate a new feature. However, the similar features of dispensable set may be in phase or out of phase with respect to each other. Hence, the following definition can be used to extract a new feature $\overline{\mathscr{A}}_i$ from the dispensable set of features D_i:

$$\overline{\mathscr{A}}_i = \frac{\mathscr{A}_i + \sum \lambda_j \mathscr{A}_j}{1 + \sum |\lambda_j|} \text{ where } \lambda_j \in \{-1, 0, 1\} \text{ and } \mathscr{A}_j \in D_i. \quad (11)$$

To find out the value of λ_j for each feature $\mathscr{A}_j \in D_i$, the following greedy algorithm can be used. Let $\mathscr{A}_i$ be the initial representative of the set D_i. The representative of D_i is refined incrementally. By searching among the features of set D_i, the current representative is merged and averaged with other features, both in phase and out of phase, such that the augmented representative $\overline{\mathscr{A}}_i$ increases the relevance value. The merging process is repeated until the relevance value can no longer be improved. If a feature $\mathscr{A}_j \in D_i$ in phase (respectively, out of phase) with the feature $\mathscr{A}_i$ increases the relevance value, then $\lambda_j = 1$ (respectively, $\lambda_j = -1$). On the other hand, the value of $\lambda_j = 0$ if feature $\mathscr{A}_j$ does not increase the relevance value, irrespective of the phases.

After extracting the feature $\overline{\mathscr{A}}_i$ from the dispensable set D_i using (11), the insignificant feature set I_i and used features of D_i are discarded from the whole feature set $\mathbb{C}$. From the remaining features of $\mathbb{C}$, another feature $\mathscr{A}_j$ is selected by maximizing the following condition:

$$\gamma_{\mathscr{A}_j}(\mathbb{D}) + \frac{1}{|\mathbb{S}|} \sum_{\overline{\mathscr{A}}_i \in \mathbb{S}} \sigma_{\{\overline{\mathscr{A}}_i, \mathscr{A}_j\}}(\mathscr{A}_j, \mathbb{D}) \quad (12)$$

where $\mathbb{S}$ is the already selected or extracted feature set. The process is repeated to select or extract more features. In the proposed dimensionality reduction method, both relevance and significance of a set of features are computed using fuzzy rough sets. If $\lambda_j = 0$ for all $\mathscr{A}_j \in D_i$, then the extracted feature $\overline{\mathscr{A}}_i$ at a particular iteration is actually the candidate feature $\mathscr{A}_i$ of the original feature set $\mathbb{C}$. Hence, the proposed dimensionality reduction method generates a reduced feature set $\mathbb{S}$ that may simultaneously contain some selected features of original measurement space and some extracted features of transformed feature space, which are both relevant and significant.

3 Experimental Results

The performance of the proposed fuzzy rough simultaneous attribute selection and feature extraction method is extensively studied and compared with that of some existing feature selection and extraction algorithms, namely maximal-relevance (Max-Relevance) and maximal-relevance maximal-significance (MRMS) [4] frameworks with classical, neighborhood, and fuzzy rough sets, quick reduct (Max-Dependency and rough sets) [3], fuzzy rough quick reduct (Max-Dependency and fuzzy rough sets) [5], neighborhood quick reduct (Max-Dependency and neighborhood rough sets) [8], minimal-redundancy maximal-relevance (mRMR) framework [10], fuzzy rough set-based mRMR framework (fuzzy rough mRMR) [7], and PCA [1].

3.1 Performance of Various Rough Set Models

In dimensionality reduction method, the reduced feature set is always relative to a certain feature evaluation index. In general, different evaluation indices may lead to different reduced feature subsets. To establish the effectiveness of fuzzy rough sets over Pawlak's or classical and neighborhood rough sets, extensive experiments are done on various data sets. Table 1 presents the comparative performance of different rough set models for simultaneous attribute selection and feature extraction tasks. The results and subsequent discussions are presented in this table with respect to the classification accuracy of the K-NN, SVM, and C4.5 on test samples considering the optimum parameter values.

From the results reported in Table 1, it can be seen that the proposed dimensionality reduction method based on fuzzy rough sets attains maximum classification accuracy of the K-NN, SVM, and C4.5 in most of the cases. Out of 12 cases of training–testing, the proposed method with fuzzy rough sets achieves highest classification accuracy in 10 cases, while that with classical or Pawlak's rough sets attains it only in 2 cases. The better performance of the fuzzy rough sets is achieved due to the fact that it can capture uncertainties associated with the data more accurately.

3.2 Performance of Different Algorithms

Finally, Table 2 compares the performance of the proposed fuzzy rough simultaneous feature selection and extraction algorithm with that of different existing feature selection and extraction algorithms on various data sets.

From the results reported in Table 2, it is seen that the proposed dimensionality reduction method achieves highest classification accuracy of SVM, C4.5, and K-NN in 11 cases out of total 12 cases, while the PCA attains highest classification accuracy in only 1 case. The proposed method also provides higher classification accuracy than the max-Relevance, max-Dependency, and MRMS criteria in all cases, irrespective of the classifiers, rough sets, and data sets used.

Hence, all the results reported in Table 2 confirms that the proposed fuzzy rough dimensionality reduction method selects a set of features having highest classification

Table 1 Performance of various rough set models on different data sets

Different data sets	Test accuracy of K-NN			Test accuracy of SVM			Test accuracy of C4.5		
	Classical	Neighbor	Fuzzy	Classical	Neighbor	Fuzzy	Classical	Neighbor	Fuzzy
Satimage	84.65	87.50	87.55	84.45	83.60	87.35	82.40	83.95	87.20
Segmentation	87.61	85.76	86.24	91.38	91.38	92.33	90.90	89.98	90.33
Leukemia II	91.07	91.07	94.64	91.07	90.17	93.75	91.07	90.17	93.75
Breast II	84.21	94.73	94.73	89.47	94.73	94.73	100	100	100

Table 2 Comparative performance analysis of different methods

Methods/ algorithms	Different rough sets	Satimage			Segmentation			Leukemia II			Breast II		
		K-NN	SVM	C4.5	K-NN	SVM	C4.5	K-NN	SVM	C4.5	K-NN	SVM	C4.5
	Classical	69.20	67.25	67.45	63.16	57.89	63.16	81.25	81.25	81.25	67.90	78.86	77.48
Max-relevance	Neighborhood	77.70	74.65	79.60	78.95	73.68	73.68	83.04	80.36	80.36	76.81	78.86	77.48
	Fuzzy	77.90	76.40	80.00	78.95	78.95	78.95	83.04	82.14	82.14	82.10	80.86	78.76
	Classical	70.40	67.30	67.45	63.16	57.89	68.42	82.14	82.14	82.14	67.90	82.48	81.78
Max-dependency	Neighborhood	83.20	81.70	81.45	78.95	73.68	73.68	85.71	83.03	83.03	81.45	82.48	82.80
	Fuzzy	83.20	83.00	82.80	78.95	78.95	78.95	85.71	84.82	84.82	84.19	83.76	82.76
	Classical	74.00	73.85	74.10	72.67	74.10	74.67	84.82	85.71	85.71	84.21	84.21	84.21
MRMS	Neighborhood	83.40	85.00	85.15	80.52	82.57	83.24	87.50	89.29	88.39	84.21	89.47	89.47
	Fuzzy	84.10	84.10	84.10	80.76	83.95	85.14	88.39	90.18	89.29	89.47	94.74	94.74
Classical mRMR		75.45	75.40	75.35	72.81	73.76	74.33	84.82	84.82	84.82	84.21	84.21	89.47
Fuzzy rough mRMR		83.95	84.60	83.70	80.33	84.10	84.71	87.50	89.29	90.18	89.47	89.47	94.74
PCA		82.55	83.95	82.00	78.94	89.47	94.73	80.35	78.59	79.46	77.30	79.50	74.10
Proposed fuzzy rough		87.55	87.35	87.20	86.24	92.33	90.33	94.64	93.75	93.75	94.73	94.73	100

accuracy of the K-NN, SVM, and C4.5 in most of the cases, irrespective of the data sets. Also, the proposed method can potentially yield significantly better results than the existing algorithms. The better performance of the proposed method is achieved due to the fact that it provides an efficient way to simultaneously select and extract features for classification. In effect, a reduced set of features having maximum relevance and significance is being obtained using the proposed method.

4 Conclusion

This chapter presents a novel dimensionality reduction method, integrating judiciously the theory of fuzzy rough sets and merits of both attribute selection and feature extraction. An efficient algorithm is introduced by performing simultaneous feature selection and extraction. It uses the concept of fuzzy rough feature significance for finding significant and relevant features of real-valued data sets. Finally, the effectiveness of the proposed method is presented, along with a comparison with other related algorithms, on a set of real-life data sets.

Acknowledgments This work is partially supported by the Indian National Science Academy, New Delhi (grant no. SP/YSP/68/2012). The work was done when one of the authors, P. Garai, was a DST-INSPIRE Fellow of the Department of Science and Technology, Government of India.

References

1. Duda, R.O., Hart, P.E., Stork, D.G.: Pattern Classification and Scene Analysis. Wiley Inc., New York (2000)
2. Pawlak, Z.: Rough Sets: Theoretical Aspects of Resoning About Data. Kluwer, Dordrecht (1991)
3. Chouchoulas, A., Shen, Q.: Rough set-aided keyword reduction for text categorisation. Appl. Artif. Intell. **15**(9), 843–873 (2001)
4. Maji, P., Paul, S.: Rough set based maximum relevance-maximum significance criterion and gene selection from microarray data. Int. J. Approximate Reasoning **52**(3), 408–426 (2011)
5. Jensen, R., Shen, Q.: Semantics-preserving dimensionality reduction: rough and fuzzy-rough-based approach. IEEE Trans. Knowl. Data Eng. **16**(12), 1457–1471 (2004)
6. Maji, P., Pal, S.K.: Rough-Fuzzy Pattern Recognition: Applications in Bioinformatics and Medical Imaging. Wiley, New Jersey (2012)
7. Maji, P., Pal, S.K.: Feature selection using f-information measures in fuzzy approximation spaces. IEEE Trans. Knowl. Data Eng. **22**(6), 854–867 (2010)
8. Hu, Q., Yu, D., Liu, J., Wu, C.: Neighborhood rough set based heterogeneous feature subset selection. Inf. Sci. **178**, 3577–3594 (2008)
9. Dubois, D., Prade, H.: Rough fuzzy sets and fuzzy rough sets. Int. J. Gen. Syst. **17**, 191–209 (1990)
10. Peng, H., Long, F., Ding, C.: Feature selection based on mutual information: criteria of max-dependency, max-relevance, and min-redundancy. IEEE Trans. Pattern Anal. Mach. Intell. **27**(8), 1226–1238 (2005)

A Fuzzy Programming Approach for Bilevel Stochastic Programming

Nilkanta Modak and Animesh Biswas

Abstract This article presents a fuzzy programming (FP) method for modeling and solving bilevel stochastic decision-making problems involving fuzzy random variables (FRVs) associated with the parameters of the objectives at different hierarchical decision-making units as well as system constraints. In model formulation process, an expectation model is generated first on the basis of the fuzzy random variables involved with the objectives at each level. The problem is then converted into a FP model by considering the fuzzily described chance constraints with the aid of applying chance constrained methodology in a fuzzy context. After that, the model is decomposed on the basis of tolerance ranges of fuzzy numbers associated with the parameters of the problem. To construct the fuzzy goals of the decomposed objectives of both decision-making levels under the extended feasible region defined by the decomposed system constraints, the individual optimal values of each objective at each level are calculated in isolation. Then, the membership functions are formulated to measure the degree of satisfaction of each decomposed objectives in both the levels. In the solution process, the membership functions are converted into membership goals by assigning unity as the aspiration level to each of them. Finally, a fuzzy goal programming model is developed to achieving the highest membership degree to the extent possible by minimizing the under deviational variables of the membership goals of the objectives of the decision makers (DMs) in a hierarchical decision-making environment. To expound the application potentiality of the approach, a numerical example is solved.

Keywords Bilevel programming · Stochastic programming · Fuzzy random variable · Fuzzy goal programming

N. Modak · A. Biswas (✉)
Department of Mathematics, University of Kalyani, Kalyani 741235, India
e-mail: nmodak9@gmail.com

A. Biswas
e-mail: abiswaskln@rediffmail.com

B. V. Babu et al. (eds.), *Proceedings of the Second International Conference on Soft Computing for Problem Solving (SocProS 2012), December 28–30, 2012*, Advances in Intelligent Systems and Computing 236, DOI: 10.1007/978-81-322-1602-5_14,

1 Introduction

Bilevel programming (BLP) is considered as a hierarchical decision-making problem in which decision makers (DMs) locating at two hierarchical levels independently controls a vector of decision variables for optimizing his/her own pay off by taking serious attention to the benefit of the other. In the decision-making situation, although the execution of decision is sequential from higher level to lower level, the decision for optimizing the objective of the upper level DM (leader) is often affected by the reaction of the lower level DM (follower) due to his/her dissatisfaction with the decision. In such cases, the problem for proper distribution of decision power to the DMs is often encountered and decision deadlock arises.

To solve BLP problems (BLPPs), several approaches are developed by some pioneer researchers in the field, viz. Kuhn–Tucker approach [1], kth best solution approach [2], penalty function approach [3], complementary pivot approach [4]. But the use of classical approaches developed so far often leads to the paradox that the decision of the follower dominates the decision of leader and the methods do not always provide satisfactory decision in a highly conflicting hierarchical decision situation.

Considering uncertainties involved with most of the mathematical programming, probabilistic programming and fuzzy programming (FP) advanced to deal with different inexact parameters values and inherent uncertain characteristic of decision parameters. The stochastic programming (SP) is a field of probabilistic programming where the parameters associated with the objectives are random variables. The methodology to solve this kind of problem was introduced by Charnes and Cooper [5] for its potential applications in various real-life planning problems [6]. SP is further classified as chance constrained programming (CCP) if some or more constraints are probabilistic in nature following some sort of probability distribution. Different aspects of CCP were studied [7] in the past.

In model formulation process, it is often found that the parameters involved with the model cannot be defined precisely. Under this situation, FP is appeared as a powerful technique to solve optimization problem, which was developed by Zimmermann [8], Klir and Yuan [9], and others based on fuzzy set theory [10]. The main difficulty with conventional FP approach is that re-evaluation of the problem again and again by redefining membership values of the objectives is involved to reach satisfactory decision.

To avoid such computational difficulties, fuzzy goal programming (FGP) [11] for solving BLPP has been presented by Biswas and Bose [12] in the recent past. Considering fuzziness involved with different CCP or SP problems, Luhandjula and Joubert [13] developed some technique by using fuzzy random variable (FRV) to solve CCP problems in fuzzy environment. A methodology for solving CCP problems through expectation model by using FGP technique has been recently studied by Biswas and Modak [14].

In real-life decision-making context, it is often found that the probability of occurrence of the objective of the DM at higher level is not the same as that of lower level, and also, the parameters involved with the objectives as well as system constraints are not defined precisely. To capture this type of uncertainties, stochastic BLP is considered in this article.

In the present study, FGP approach is adopted to solve fuzzy stochastic BLPP where the parameters associated with the objectives as well as with system constraints are FRVs and fuzzy numbers. In model formulation process, the objectives of the problem is first approximated by using expectation model (E-Model) [14] and the system constraints are converted into fuzzy constraints by applying general CCP methodology. Then, the model is decomposed on the basis of the tolerance ranges of the fuzzy numbers associated with the objectives as well as with the system constraints by the concept of α-cut of fuzzy numbers in order to defuzzify the problem. Individual objective values are obtained to construct the membership functions of each of the decomposed objectives within the extended feasible region. The membership functions are then converted into membership goals by assigning unity as the aspiration level of each of them. Lastly, an FGP model is constructed to achieving the highest membership degree to the extent possible by the DM by minimizing the group regret consisting of under deviational variables in decision-making environment.

2 Formulation of Bilevel Stochastic Programming Problem

In a stochastic bilevel decision-making situation, let F_1 and F_2 be the objective function of leader and follower respectively. Also, let x_1 be the controlling vector of leader and x_2 be the controlling vector of follower. Then, a bilevel SP problem with chance constraints is stated as

Find $X(x_1,\ x_2)$ so as to

$$\underset{x_1}{\text{Max}}\ F_1 = \tilde{c}_{11}x_1 + \tilde{c}_{12}x_2 \text{ and for given } x_1,\ x_2 \text{ solves } \underset{x_2}{\text{Max}}\ F_2 = \tilde{c}_{21}x_1 + \tilde{c}_{22}x_2$$

$$\text{subject to } \Pr(\tilde{a}_{11}x_1 + \tilde{a}_{12}x_2 \le \tilde{b}_1) \ge p_1;\ \Pr(\tilde{a}_{21}x_1 + \tilde{a}_{22}x_2 \le \tilde{b}_2) \ge p_2;\ x_1,\ x_2 \ge 0 \tag{1}$$

where $\tilde{c}_{11}, \tilde{c}_{12}, \tilde{c}_{21}, \tilde{c}_{22}$, and $\tilde{b}_2$ are the vectors of normally distributed independent FRVs with fuzzily defined mean and variance; $\tilde{a}_{11}, \tilde{a}_{12}$ are the matrices of normally distributed independent FRVs; $\tilde{b}_1$ is a vector of right-sided fuzzy number; $\tilde{a}_{21}, \tilde{a}_{22}$ are matrices of triangular fuzzy numbers; p_1, p_2 are vectors of real numbers. With the consideration of the above model, the FP model is derived in the next section.

3 FP Model Construction

Considering the FRVs associated with the objectives at each level DM, an E-model is generated in the following subsection.

3.1 Fuzzy E-model of Objective Function

Since $\tilde{c}_{11}, \tilde{c}_{12}, \tilde{c}_{21}, \tilde{c}_{22}$are the vectors of normally distributed independent FRVs, let $E(\tilde{c}_{11})$, $E(\tilde{c}_{12})$, $E(\tilde{c}_{21})$, and $E(\tilde{c}_{22})$ are the respective mean values of the FRVs $\tilde{c}_{11}, \tilde{c}_{12}, \tilde{c}_{21}, \tilde{c}_{22}$ associated with objective functions of leader and follower. Then, the objectives of equivalent E-model of the given problem (1) are presented as

$$\underset{x_1}{\text{Max}}\ E(F_1) = E(\tilde{c}_{11})x_1 + E(\tilde{c}_{12})x_2 \text{ where for given } x_1, x_2 \text{ solves } \underset{x_2}{\text{Max}}$$
$$E(F_2) = E(\tilde{c}_{21})x_1 + E(\tilde{c}_{22})x_2. \quad (2)$$

Since $E(\tilde{c}_{11})$, $E(\tilde{c}_{12})$, $E(\tilde{c}_{21})$, and $E(\tilde{c}_{22})$ are vectors of fuzzy numbers, introducing $\tilde{M}_{11}, \tilde{M}_{12}, \tilde{M}_{21}$, and $\tilde{M}_{22}$ are vectors of fuzzy numbers with the assigned values $\tilde{M}_{11} = E(\tilde{c}_{11})$, $\tilde{M}_{12} = E(\tilde{c}_{12})$, $\tilde{M}_{21} = E(\tilde{c}_{21})$, and $\tilde{M}_{22} = E(\tilde{c}_{22})$.

In a fuzzy decision-making situation, it is to be assumed that the mean associated with the vectors of FRVs $\tilde{c}_{11}, \tilde{c}_{12}, \tilde{c}_{21}, \tilde{c}_{22}$ are triangular fuzzy numbers.

A triangular fuzzy number $\tilde{a}$ can be represented by a triple of three real numbers as $\tilde{a} = \left(a^L, a, a^R\right)$. The membership function of a triangular fuzzy number is of the form

$$\mu_{\tilde{a}}(x) = \begin{cases} 0 & \text{if} \quad x < a^L \text{ or } x > a^R \\ \left[x - a^L\right] / \left[a - a^L\right] & \text{if} \quad a^L \le x \le a \\ \left[a^R - x\right] / \left[a^R - a\right] & \text{if} \quad a \le x \le a^R \end{cases}$$

where a^L and a^R denote, respectively, the left and right tolerance values of the fuzzy number $\tilde{a}$. Now considering the following triangular fuzzy numbers associated with the mean values of the vectors FRVs of the objectives of the DMs with the form $\tilde{c}_{11}, \tilde{c}_{12}, \tilde{c}_{21}, \tilde{c}_{22}$ as $\tilde{M}_{11} = \left(M_{11}^L,\ M_{11},\ M_{11}^R\right)$, $\tilde{M}_{12} = (M_{12}^L, M_{12}, M_{12}^R)$, $\tilde{M}_{21} = (M_{21}^L, M_{21}, M_{21}^R)$, $\tilde{M}_{22} = (M_{22}^L, M_{22}, M_{22}^R)$, the objective functions in (2) are decomposed as

$$\left.\begin{array}{l} \underset{x_1}{\text{Max}}\ E(F_1)^L = \left\{M_{11}^L + (M_{11} - M_{11}^L)\alpha\right\}x_1 + \left\{M_{12}^L + (M_{12} - M_{12}^L)\alpha\right\}x_2 \\ \underset{x_1}{\text{Max}}\ E(F_1)^R = \left\{M_{11}^R - (M_{11}^R - M_{11})\alpha\right\}x_1 + \left\{M_{12}^R - (M_{12}^R - M_{12})\alpha\right\}x_2 \end{array}\right\} \text{(Leader}'\text{problem)} \quad (3)$$

$$\left.\begin{array}{l} \underset{x_2}{\text{Max}}\ E(F_2)^L = \left\{M_{21}^L + (M_{21} - M_{21}^L)\alpha\right\}x_1 + \left\{M_{22}^L + (M_{22} - M_{22}^L)\alpha\right\}x_2 \\ \underset{x_2}{\text{Max}}\ E(F_2)^R = \left\{M_{21}^R - (M_{21}^R - M_{21})\alpha\right\}x_1 + \left\{M_{22}^R - (M_{22}^R - M_{22})\alpha\right\}x_2 \end{array}\right\} \text{(Follower's problem)} \quad (4)$$

3.2 Conversion of Fuzzy Chance Constraints into Fuzzy Constraints

The parameters involved with the chance constraints are treated here as FRVs following normal distribution. Under this consideration, it is assumed that $E(\tilde{a}_{11})$, $\mathrm{Var}(\tilde{a}_{11})$, $E(\tilde{a}_{12})$, $\mathrm{Var}(\tilde{a}_{12})$ and $E\left(\tilde{b}_2\right)$, $\mathrm{Var}\left(\tilde{b}_2\right)$ be the respective mean and variance of the matrices $\tilde{a}_{11}, \tilde{a}_{12}$, and the vector $\tilde{b}_2$ whose entries are normally distributed FRVs.

Now applying CCP methodology in fuzzy environment, the constraints in (1) is converted into the equivalent fuzzy constraints as

$$\begin{aligned} &E(\tilde{a}_{11})x_1 + E(\tilde{a}_{12})x_2 + \Phi^{-1}(p_1)\sqrt{\mathrm{Var}(\tilde{a}_{11})x_1^2 + \mathrm{Var}(\tilde{a}_{12})x_2^2} \le \tilde{b}_1 \\ &\tilde{a}_{21}x_1 + \tilde{a}_{22}x_2 \le E\left(\tilde{b}_2\right) + \Phi^{-1}(1-p_2)\sqrt{\mathrm{Var}\left(\tilde{b}_2\right)}, x_1, \ x_2 \ge 0 \end{aligned} \tag{5}$$

Here, $\Phi(.)$ represents cumulative distribution function of the standard normal variate.

Since $\tilde{a}_{11}, \tilde{a}_{12}$ are the matrices of normally distributed FRVs and the decision variables $x_1, \ x_2 \ge 0$ are unknowns, then two FRVs, $\tilde{d}_1, \tilde{d}_2$, are introduced by $\tilde{d}_1 = \tilde{a}_{11}x_1$, and $\tilde{d}_2 = \tilde{a}_{12}x_2$ whose respective mean and variance are given by $m_{\tilde{d}_1} = E(\tilde{a}_{11})x_1, m_{\tilde{d}_2} = E(\tilde{a}_{12})x_2, \sigma^2_{\tilde{d}_1} = \mathrm{Var}(\tilde{a}_{11})x_1, \sigma^2_{\tilde{d}_2} = \mathrm{Var}(\tilde{a}_{12})x_2$. Also, let $m_{\tilde{b}_2} = E\left(\tilde{b}_2\right)$ and $\sigma^2_{\tilde{b}_2} = \mathrm{Var}\left(\tilde{b}_2\right)$ be the respective mean and variance of the normally distributed FRV $\tilde{b}_2$.

Now applying mean and variance to the elements of the matrices of FRVs, the above constraints in (5) is converted into the following form

$$\begin{aligned} m_{\tilde{d}_1} + m_{\tilde{d}_2} + \Phi^{-1}(p_1)\left\{\sigma_{\tilde{d}_1} + \sigma_{\tilde{d}_2}\right\} \le \tilde{b}_1, \ \tilde{a}_{21}x_1 + \tilde{a}_{22}x_2 \le m_{\tilde{b}_2} + \Phi^{-1} \\ (1-p_2)\sigma_{\tilde{b}_2}, \ x_1, \ x_2 \ge 0 \end{aligned} \tag{6}$$

In a fuzzy decision-making situation, it is to be assumed that the mean and variance associated with the elements of matrices of the FRVs $\tilde{d}_1, \tilde{d}_2$, and the vectors of FRVs $\tilde{b}_2$ are the matrices of triangular fuzzy numbers. With the form

$$m_{\tilde{d}_1} = \left(m^L_{\tilde{d}_1}, \ m_{\tilde{d}_1}, \ m^R_{\tilde{d}_1}\right); \quad \sigma^2_{\tilde{d}_1} = \left(\sigma^{2^L}_{\tilde{d}_1}, \sigma^2_{\tilde{d}_1}, \ \sigma^{2^R}_{\tilde{d}_1}\right); m_{\tilde{d}_2} = \left(m^L_{\tilde{d}_2}, \ m_{\tilde{d}_2}, \ m^R_{\tilde{d}_2}\right);$$

$$\sigma^2_{\tilde{d}_2} = \left(\sigma^{2^L}_{\tilde{d}_2}, \sigma^2_{\tilde{d}_2}, \ \sigma^{2^R}_{\tilde{d}_2}\right)$$

$$m_{\tilde{b}_2} = \left(m^L_{\tilde{b}_2}, m_{\tilde{b}_2}, m^R_{\tilde{b}_2}\right); \quad \sigma^2_{\tilde{b}_2} = \left(\sigma^{2L}_{\tilde{b}_2}, \sigma^2_{\tilde{b}_2}, \sigma^{2R}_{\tilde{b}_2}\right),$$

$$\tilde{a}_{21} = (a^L_{21}, a_{21}, a^R_{21}), \tilde{a}_{22} = (a^L_{22}, a_{22}, a^R_{22}).$$

Further, the fuzzy numbers $\tilde{b}_1$ are considered as one-sided fuzzy numbers as

$$\mu_{b_1}(x) = \begin{cases} 1 & \text{if} \quad x \leq b_1 \\ [8.5 - x]/0.5 & \text{if} \quad b_1 \leq x \leq b_1 + \delta_1 \\ 0 & \text{if} \quad x \geq 8.5 \end{cases}$$

On the basis of the lower and upper tolerance limits of the triangular fuzzy numbers, the constraints in (6) can be expressed as

$$\begin{aligned}
&\left\{m^L_{\tilde{d}_1} + (m_{\tilde{d}_1} m^L_{\tilde{d}_1})\alpha\right\} + \left\{m^L_{\tilde{d}_2} + (m_{\tilde{d}_2} m^L_{\tilde{d}_2})\alpha\right\} \\
&\qquad + \Phi^{-1}(p_1)\sqrt{\left\{\sigma^{2L}_{\tilde{d}_1} + (\sigma^2_{\tilde{d}_1} - \sigma^{2L}_{\tilde{d}_1})\alpha\right\} + \left\{\sigma^{2L}_{\tilde{d}_2} + (\sigma^2_{\tilde{d}_2} - \sigma^{2L}_{\tilde{d}_2})\alpha\right\}} \\
&\qquad \leq \{E(b_1) + \delta_1 - \alpha\delta_1\} \\
&\left\{m^R_{\tilde{d}_1} - (m^R_{\tilde{d}_1} - m_{\tilde{d}_1})\alpha\right\} + \left\{m^R_{\tilde{d}_2} - (m^R_{\tilde{d}_2} - m_{\tilde{d}_2})\alpha\right\} \\
&\qquad + \Phi^{-1}(p_1)\sqrt{\left\{\sigma^{2R}_{\tilde{d}_1} - (\sigma^{2R}_{\tilde{d}_1} - \sigma^2_{\tilde{d}_1})\alpha\right\} + \left\{\sigma^{2R}_{\tilde{d}_2} - (\sigma^{2R}_{\tilde{d}_2} - \sigma^2_{\tilde{d}_2})\alpha\right\}} \\
&\qquad \leq \{E(b_1) + \delta_1 - \alpha\delta_1\} \\
&\left\{a^L_{21} + (a_{21} - a^L_{21})\alpha\right\} x_1 + \left\{a^L_{22} + (a_{22} - a^L_{22})\alpha\right\} x_2 \leq \left\{m^L_{b_2} + (m_{b_2} - m^L_{b_2})\alpha\right\} \\
&\qquad + \Phi^{-1}(1 - p_2)\left\{\sigma^L_{b_2} + (\sigma_{b_2} - \sigma^L_{b_2})\alpha\right\} \\
&\left\{a^R_{21} - (a^R_{21} - a_{21})\alpha\right\} x_1 + \left\{a^R_{22} - (a^R_{22} - a_{22})\alpha\right\} x_2 \leq \left\{m^R_{b_2} - (m^R_{b_2} - m_{b_2})\alpha\right\} \\
&\qquad + \Phi^{-1}(1 - p_2)\left\{\sigma^R_{b_2} + (\sigma^R_{b_2} - \sigma_{b_2})\alpha\right\} \\
&\qquad 0 \leq \alpha \leq 1
\end{aligned} \tag{7}$$

Hence, the BLPP model under the decomposed set of system constraints is presented as

Find $X(x_1, x_2)$ so as to

$$\left.\begin{aligned}
&\underset{x_1}{\text{Max}}\ E(F_1)^L = \left\{M^L_{11} + (M_{11} - M^L_{11})\alpha\right\} x_1 + \left\{M^L_{12} + (M_{12} - M^L_{12})\alpha\right\} x_2 \\
&\underset{x_1}{\text{Max}}\ E(F_1)^R = \left\{M^R_{11} - (M^R_{11} - M_{11})\alpha\right\} x_1 + \left\{M^R_{12} - (M^R_{12} - M_{12})\alpha\right\} x_2
\end{aligned}\right\} \text{(Leader's problem)}$$

$$\left.\begin{aligned}
&\underset{x_2}{\text{Max}}\ E(F_2)^L = \left\{M^L_{21} + (M_{21} - M^L_{21})\alpha\right\} x_1 + \left\{M^L_{22} + (M_{22} - M^L_{22})\alpha\right\} x_2 \\
&\underset{x_2}{\text{Max}}\ E(F_2)^R = \left\{M^R_{21} - (M^R_{21} - M_{21})\alpha\right\} x_1 + \left\{M^R_{22} - (M^R_{22} - M_{22})\alpha\right\} x_2
\end{aligned}\right\} \text{(Follower's problem)}$$

subject to the constraints given in (7);

$$x_1,\ x_2 \geq 0 \tag{8}$$

3.3 Construction of Membership Functions

In a bilevel decision-making context, the DMs are very much interested to optimize their own payoffs to the extent possible by paying serious attention to the benefit of the others. To assess the fuzzy goals of the objectives of the DMs, the independent optimal solutions are determined first.

Let $E(F_1)_B^L, E(F_1)_B^R$ and $E(F_1)_W^L, E(F_1)_W^R$ be the respective best and worst achieved expected decomposed objective values of the leader when the above model (8) is solved only by considering leader's problem independently.

Similarly, let the independent best and worst achieved expected decomposed objective values of the follower are appeared as $E(F_2)_B^L, E(F_2)_B^R$ and $E(F_2)_W^L, E(F_2)_W^R$.

Hence, the fuzzy goals of the expected values of the decomposed objectives of leader and follower are appeared as

$$E(F_1)^L \gtrsim E(F_1)_B^L,\ E(F_1)^R \gtrsim E(F_1)_B^R,\ E(F_2)^L \gtrsim E(F_2)_B^L,\ E(F_2)^R \gtrsim E(F_2)_B^R$$

On the basis of upper and lower tolerance values of the fuzzy goals, the following membership functions are developed to measure the degree of satisfaction of the each level DMs.

$$\mu_{E(F_1)^L}(x) = \begin{cases} 0 & \text{if } E(F_1)^L \le E(F_1)_W^L \\ \left[E(F_1)^L - E(F_1)_W^L\right] / \left[E(F_1)_B^L - E(F_1)_W^L\right] & \text{if } E(F_1)_W^L \le E(F_1)^L \le E(F_1)_B^L \\ 1 & \text{if } E(F_1)^L \ge E(F_1)_B^L \end{cases}$$

$$\mu_{E(F_1)^R}(x) = \begin{cases} 0 & \text{if } E(F_1)^R \le E(F_1)_W^R \\ \left[E(F_1)^R - E(F_1)_W^R\right] / \left[E(F_1)_B^R - E(F_1)_W^R\right] & \text{if } E(F_1)_W^R \le E(F_1)^R \le E(F_1)_B^R \\ 1 & \text{if } E(F_1)^R \ge E(F_1)_B^R \end{cases}$$

The membership functions corresponding to $E(F_2)^L$ are to be defined similarly.

4 FGP Model Formulation

In FGP, the membership functions are considered as flexible goals by assigning unity as the aspiration level and introducing under- and over-deviational variables to each of them. Then, the FGP model is formulated by considering each membership function of the fuzzy goals of the DMs as follows

Find $X(x_1, x_2)$ so as to

$$\text{Min D} = w_1^- d_1^- + w_2^- d_2^- + w_3^- d_3^- + w_4^- d_4^-$$

and satisfy

$$\frac{E(F_1)^L(x_1, x_2) - E(F_1)^L_W}{E(F_1)^L_B - E(F_1)^L_W} + d_1^- - d_1^+ = 1; \frac{E(F_1)^R(x_1, x_2) - E(F_1)^R_W}{E(F_1)^R_B - E(F_1)^R_W} + d_2^- - d_2^+ = 1$$
$$\frac{E(F_2)^L(x_1, x_2) - E(F_2)^L_W}{E(F_2)^L_B - E(F_2)^L_W} + d_3^- - d_3^+ = 1; \frac{E(F_2)^R(x_1, x_2) - E(F_2)^R_W}{E(F_2)^R_B - E(F_2)^R_W} + d_4^- - d_4^+ = 1$$

subject to the system constraints in (8)

$$x_1, x_2 \geq 0, d_k^-, \ d_k^+ \geq 0, \ \text{with } d_k^- \cdot d_k^+ = 0 \text{ for } k = 1, 2, 3, 4 \tag{9}$$

where d_k^-, d_k^+ represent the under- and over-deviational variables, respectively, and the fuzzy weights w_k, $(k \in N_4)$ are determined as [14]:

$$w_1 = \left[E(F_1)^L_B - E(F_1)^L_W\right]^{-1}, w_2 = \left[E(F_1)^R_B - E(F_1)^R_W\right]^{-1},$$
$$w_3 = \left[E(F_2)^L_B - E(F_2)^L_W\right]^{-1}, w_4 = \left[E(F_2)^R_B - E(F_2)^R_W\right]^{-1}.$$

The *minsum* GP technique is used to solve the problem (9).

5 A Numerical Example

To illustrate the proposed methodology, the following fuzzy stochastic BLPP is solved. The problem is presented as

Find $X(x_1, \ x_2)$ so as to $\underset{x_1}{\text{Max}} \ F_1 = \tilde{c}_{11}x_1 + \tilde{c}_{12}x_2$ and for given $x_1, \ x_2$ solves $\underset{x_2}{\text{Max}} \ F_2 = \tilde{c}_{21}x_2$

$$\text{subject to} \quad \Pr\left(\tilde{a}_{11}x_1 + \tilde{a}_{12}x_2 \leq \tilde{b}_1\right) \geq 0.95; \Pr\left(\tilde{a}_{21}x_1 + \tilde{a}_{22}x_2 \leq \tilde{b}_2\right) \geq 0.10; x_1, \ x_2 \geq 0 \tag{10}$$

Now, the E-model of the above problem with fuzzy constraints is written as

Find $X(x_1, \ x_2)$ so as to

$\underset{x_1}{\text{Max}} \ E(F_1) = E(\tilde{c}_{11})x_1 + E(\tilde{c}_{12})x_2$ and for given $x_1, \ x_2$ solves $\underset{x_2}{Max} \ E(F_2) = E(\tilde{c}_{21})x_2$

$$\text{subject to } E(\tilde{a}_{11})x_1 + E(\tilde{a}_{12})x_2 + 1.645\sqrt{\text{var}(\tilde{a}_{11})x_1^2 + \text{var}(\tilde{a}_{12})x_2^2} \leq \tilde{b}_1,$$
$$\tilde{a}_{21}x_1 + \tilde{a}_{22}x_2 \leq E(\tilde{b}_2) + 1.28\sqrt{\text{var}(\tilde{b}_2)}, x_1, \ x_2 \geq 0 \tag{11}$$

where $\tilde{c}_{11}, \tilde{c}_{12}, \tilde{c}_{21}$ are normally distributed FRVs with respective mean represented by the following triangular fuzzy numbers

$$E(\tilde{c}_{11}) = (6.5, \ 7, \ 7.5); E(\tilde{c}_{12}) = (2.8, \ 3, \ 3.2); E(\tilde{c}_{21}) = (4.6, \ 5, \ 5.6);$$

Also, $\tilde{a}_{11}, \tilde{a}_{12}, \tilde{b}_2$ are normally distributed independent FRVs with the mean and variance of these variables that are considered as $E(\tilde{a}_{11}) = (0.95, 1, 1.05), E(\tilde{a}_{12}) = (2.95, 3, 3.05)$, $Var(\tilde{a}_{11}) = (24.95, 25, 25.05)$, $Var(\tilde{a}_{12}) = (15.95, 16, 16.05)$, $E(\tilde{b}_2) = (6.95, 7, 7.05)$, $Var(\tilde{b}_2) = (8.95, 9, 9.05)$. Again $\tilde{a}_{21}, \tilde{a}_{22}$ are considered as triangular fuzzy numbers with the form $\tilde{a}_{21} = (3.95, 4, 4.05), \tilde{a}_{22} = (2.95, 3, 3.05)$.

Here, $\tilde{b}_1$ is considered as right-sided fuzzy number which is given by

$$\mu_{b_1}(x) = \begin{cases} 1 & \text{if} \quad b_1 \le 8 \\ [8.5 - b_1]/0.5 & \text{if} \quad 8 \le b_1 \le 8.5 \\ 0 & \text{if} \quad b_1 \ge 8.5 \end{cases}$$

Now, the FP model using the above defined fuzzy numbers takes the form as
Find $X(x_1, x_2)$ so as to

$$\begin{aligned} \underset{x_1}{\text{Max}}\ E(F_1)^L &= (6.5 + 0.5\alpha)x_1 + (2.8 + 0.2\alpha)x_2, \underset{x_1}{\text{Max}}\ E(F_1)^R \\ &= (7.5 - 0.5\alpha)x_1 + (3.2 - 0.2\alpha)x_2\ \underset{x_2}{\text{Max}}\ E(F_2)^L = (4.6 + 0.4\alpha)x_2, \\ \underset{x_2}{\text{Max}}\ E(F_2)^R &= (5.6 - 0.6\alpha)x_2 \end{aligned}$$

Subject to

$$\begin{aligned} &(0.95 + 0.05\alpha)\, x_1 + (2.95 + 0.05\alpha)\, x_2 \\ &\qquad + 1.645\sqrt{(24.95 + 0.05\alpha)\, x_1^2 + (15.95 + 0.05\alpha)\, x_2^2} \le (8.5 - 0.5\alpha) \\ &(1.05 - 0.05\alpha)\, x_1 + (3.05 - 0.05\alpha)\, x_2 \\ &\qquad + 1.645\sqrt{(25.05 - 0.05\alpha)\, x_1^2 + (16.05 - 0.05\alpha)\, x_2^2} \le (8.5 - 0.5\alpha) \\ &(3.95 + 0.05\alpha)\, x_1 + (2.95 + 0.05\alpha)x_2 \le (6.95 + 0.05\alpha) + 1.28\sqrt{(8.95 + 0.05\alpha)} \\ &(4.05 - 0.05\alpha)\, x_1 + (3.05 - 0.05\alpha)x_2 \le (7.05 - 0.05\alpha) + 1.28\sqrt{(9.05 - 0.05\alpha)} \\ &x_1\ ,\ x_2\ \ge\ 0,\ 0 \le \alpha \le 1. \end{aligned} \tag{12}$$

The individual best and worst decision of the leader when calculated in isolation are obtained as $E(F_1)_B^L = 6.003$, $E(F_1)_B^R = 6.922$, $E(F_1)_W^L = 0$, $E(F_1)_W^R = 0$

Again the follower's best and worst solution are found as $E(F_2)_B^L = 4.132$, $E(F_2)_B^R = 4.938$, $E(F_2)_W^L = 0$, $E(F_2)_W^R = 0$.

The fuzzy goals of the DMs are appeared as

$$E(F_1)^L \gtrsim 6.003,\ E(F_1)^R \gtrsim 6.922,\ E(F_2)^L \gtrsim 4.132,\ E(F_2)^R \gtrsim 4.938.$$

Now developing the membership functions and assigning highest achievement level (unity) to the membership goals, the *minsum* FGP model is formulated as

Find $X(x_1, x_2)$ so as to

$$\text{Min D} = d_1^-/6.003 + d_2^-/6.922 + d_3^-/4.132 + d_4^-/4.938$$

and satisfy

$$[(6.5 + 0.5\alpha)x_1 + (2.8 + 0.2\alpha)x_2]/6.003 + d_1^- - d_1^+ = 1; [(7.5 - 0.5\alpha)x_1 + (3.2 - 0.2\alpha)x_2]/6.922 + d_2^- - d_2^+ = 1;$$
$$[(4.6 + 0.4\alpha)x_2]/4.132 + d_3^- - d_3^+ = 1; [(5.6 - 0.6\alpha)x_2]/4.938 + d_4^- - d_4^+ = 1$$

subject to same system constraints in (12)

$$x_1, x_2 \geq 0, 0 \leq \alpha \leq 1 d_k^-, d_k^+ \geq 0, \text{ with } d_k^- \cdot d_k^+ = 0 \text{ for } k = 1, 2, 3, 4. \quad (13)$$

The software LINGO (6.0) is used to solve the problem.

The achieved solutions of problem (13) are found as $x_1 = 0.31$, $x_2 = 0.786$ with the expected decomposed objective values $E(F_1)^L = 4.528$, $E(F_1)^R = 4.84$ of the leader and $E(F_2)^L = 3.93$, $E(F_2)^R = 4.402$ for the follower. Hence, it may be concluded that for different tolerance values of the objectives of the fuzzy parameters, the expected objectives value of the leader lies in the interval [4.528, 4.84] and that of the follower lies in the interval [3.93, 4.402]. The achieved solution is most satisfactory from the view point of achieving desired goal levels of the objectives of both the leader and follower in a hierarchical decision-making context.

6 Conclusions

The methodology developed in this paper captures the simultaneous occurrence of probabilistic and imprecise nature of the parameters associated with the model under one roof. The solution approach can be extended to solve multiobjective stochastic BLPP, fuzzy stochastic nonlinear BLPPs and multilevel stochastic linear programming problem, multilevel multiobjective stochastic linear programming problem in some large hierarchical decision-making organization without any computational difficulties. Finally, it is hoped that the approach may open up new look into the way of solving stochastic hierarchical decision-making problems.

Acknowledgments The authors are grateful to the anonymous reviewers for their comments and suggestions.

References

1. Bard, J., Falk, J.: An explicit solution to the multilevel programming problem. Comput. Oper. Res. **9**, 77–100 (1982)
2. Bialas, W., Karwan, M.: Two level linear programming. Manage. Sci. **30**, 1004–1020 (1984)
3. Aiyoshi, E., Shimizu, K.: Hierarchical decentralized systems and its new solution by barrier method. IEEE Trans. Syst. Man Cybern. **11**, 444–449 (1981)
4. Bialas, W., Karwan M., Shaw, J.: A parametric complementary pivot approach for two level linear programming. Operations Research Programme, State University of New York, Buffalo, p. 80 (1980).
5. Charnes, A., Cooper, W.W.: Chance-constrained programming. Manage. Sci. **6**, 73–79 (1959)
6. Feiring, B.R., Sastri, T., Sim, L.S.M.: A stochastic programming model for water resource planning. Math. Comput. Model. **27**(3), 1–7 (1998)
7. Kataoka, S.: A stochastic programming model. Econometric **31**, 181–196 (1963)
8. Zimmermann, H.J.: Fuzzy programming and linear programming with several objective functions. Fuzzy Sets Syst. **1**, 45–55 (1978)
9. Klir, G.J., Yuan, B.: Fuzzy Sets and Fuzzy Logic: Theory and Applications. Prentice-Hall Inc, New Jersey (1995)
10. Zadeh, L.A.: Fuzzy sets. Inf. Control **8**, 338–353 (1965)
11. Biswas, A., Pal, B.B.: Application of fuzzy goal programming technique to land use planning in agricultural system. Omega **33**, 391–398 (2005)
12. Biswas, A., Bose, K.: A fuzzy programming approach for solving quadratic bilevel programming problems with fuzzy resource constraints. Int. J. Oper. Res. **12**, 142–156 (2011)
13. Luhandjula, M.K., Joubert, J.W.: On some optimization models in a fuzzy stochastic environment. Eur. J. Oper. Res. **207**, 1433–1441 (2010)
14. Biswas, A., Modak, N.: A fuzzy goal programming approach for fuzzy multiobjective stochastic programming through expectation model. Commun. Comput. Inf. Sci. **283**, 124–135 (2012)

Implementation of Intelligent Water Drops Algorithm to Solve Graph-Based Travelling Salesman Problem

Roli Bansal, Hina Agrawal, Hifza Afaq and Sanjay Saini

Abstract Thc travelling salesman problem (TSP) is one of the most sought out NP-hard, routing problems in the literature. TSP is important with respect to some real-life applications, especially when tour is generated in real time. The objective of this paper is to apply the intelligent water drops algorithm to solve graph-based TSP (GB-TSP). The intelligent water drops (IWD) algorithm is a new meta-heuristic approach belonging to a class of swarm intelligence-based algorithm. It is inspired from observing natural water drops that flow in rivers. The idea of path finding of rivers is used to find the near-optimal solution of the travelling salesman problem (TSP).

Keywords Intelligent water drops (IWD) · Travelling salesman problem (TSP) · Swarm intelligence · Graph-based TSP (GB-TSP)

1 Introduction

Soft computing is a term applied to a field within computer science which is used to obtain near-optimal solutions to NP-complete problems in polynomial time. Swarm intelligence is a relatively new field of soft computing [1] which is inspired by nature. Swarm intelligence is based on the collective behavior of decentralized, self-

R. Bansal (✉) · H. Agrawal · H. Afaq · S. Saini
Department of Physics and Computer Science, Dayalbagh Educational Institute, Agra, India
e-mail: dei.rolibansal@gmail.com

H. Agrawal
e-mail: hinaagrawal29@gmail.com

H. Afaq
e-mail: hifza.afaq@gmail.com

S. Saini
e-mail: sanjay.s.saini@gmail.com

B. V. Babu et al. (eds.), *Proceedings of the Second International Conference on Soft Computing for Problem Solving (SocProS 2012), December 28–30, 2012*, Advances in Intelligent Systems and Computing 236, DOI: 10.1007/978-81-322-1602-5_15,

organized systems. It refers to algorithms such as ant colony optimization (ACO) [9], particle swarm optimization (PSO) [4], artificial bee colony [3], bat algorithm [11], and many more. Intelligent water drops (IWD) is an upcoming swarm intelligence-based algorithm. IWD was proposed by Hamed Shah-Hosseini [7] in 2007. IWD algorithm is based on the dynamics of river system, action, and reactions that takes place among the water drops in rivers [6]. A water drop prefers the path having low soil to the path having high soil. In this way, it finds best possible path for itself. This behavior of IWD is used to optimize the tour for travelling salesman problem (TSP).

In the TSP [2], a map of cities is given to the salesman and he is required to visit all the cities to complete his tour such that no city is visited twice except the city it starts with, which it has to visit in the end again to complete its tour. In real-life situation, all cities may not be completely interconnected with direct paths and so paths may not exist between some of the cities, so the map of cities is not completely interconnected. The goal in the TSP is to find the tour with the minimum total length, among all possible tours for the given map.

2 Behavior of Intelligent Water Drops

The behavior of the natural water drop is observed with some properties such as

- Velocity (with which a drop moves)
- Soil (which is carried by the drop)

IWD is based on these two properties of water drops. These properties change with time according to the environment of water drop when IWD flows from source to destination. Initially, IWD has zero amount of soil and nonzero velocity. It carries more soil with high velocity and unloads the soil with low velocity of IWD.

A water drop prefers an easier path to a harder path in an obvious way when it has to choose between several paths that exist in the path from source to destination. Each IWD has a number of possible paths to choose from when it goes from one position to another position. It chooses the path with the low soil and maximum probability.

Every IWD flows in finite length steps from one location to another location. The IWD's velocity depends on the soil between two locations. The IWD's velocity increases on less soil path and decreases on high soil path. Thus, IWD's velocity is inversely proportional to the soil between two locations. An IWD removes some amount of soil from the path and carries it while travelling through that path. It removes the soil from the path depending upon the time it takes to cover the distance between two locations. More soil is removed from the path if the time taken by IWD is high and vice versa. Thus, IWD's soil is inversely proportional to time taken in travelling from current location to next location. This time is calculated by the simple laws of physics linear motion.

3 Solving GB-TSP Using IWD

In this section, steps for solving the graph-based TSP (GB-TSP) are discussed. For geographical problems, where location of cities is given by their Cartesian coordinates and path from any node to any other node exists necessarily (which is simply the Euclidean distance between the two nodes), solution using IWD to such TSP has been given by Hamed Shah-Hosseini as MIWD [5]. In our case, a GB-TSP is represented by a graph (N, E), where the node set N denotes the n cities of the TSP, and the edge set E denotes the paths between the two cities. The considered graphs are non-complete, i.e., it is not necessary that a direct edge exists between every pair of nodes. In fact, we consider graphs where a direct edge may not exist between certain two nodes. This formulation of the problem is much more realistic than the earlier problems. The cost associated with the edges represents distance between cities. However, for the sake of simplicity and similarity with earlier problems, in this paper, the location of cities is given by their Cartesian coordinates and the distance between them is their Euclidean distance. For GB-TSP, we start with a graph which is not completely interconnected. For this, we create an adjacency matrix depicting distances between cities, and then, we remove the edges where there is no path between the cities. In this way, we have a subset of edge set which represent a graph which is not completely interconnected. A solution of the GB-TSP having the graph (N, E) is then an ordered set of n distinct cities.

Now, the following IWD strategy for the GB-TSP is used. Each link of the edge set E has an amount of soil. An IWD visits nodes of the graph through the links. The IWD is able to change the amount of soil on the links. An IWD starts its tour from a random node. The IWD changes the soil of each link that it flows on while completing its tour.

Since there are no standard problems available where the graph is not complete, we have created such test graphs for our experiments. For this, we convert a completely connected graph into a non-complete graph. This conversion of complete graph into non-complete graph is done using the X nearest neighbor (XNN) algorithm [8]. In this method, the links, depicting distances, from one node to all other nodes are taken, and then, a certain percentage of those links are dropped. The dropped links are those which are the largest in that set of links. We repeat the same process for all the nodes in our graph. By converting the complete graph into a non-complete graph, the search space of the problem is reduced.

This algorithm takes a complete graph and drops a given percentage of links from it. For our experiments, a few standard problems are considered and 20 % links are dropped from it.

The IWD algorithm that is used for the GB-TSP is as follows:

1. Initialization of static parameters:

    - Set the number of water drops N_{IWD}, the number of cities N_C, and the Cartesian coordinate of each city i such that $c(i) = [x_i, y_i]^T$ to their chosen constant values. The number of cities and their coordinate values depends on the prob-

lem at hand, while the N_{IWD} is set by the user. We choose N_{IWD} to be equal to or greater than the number of cities.
- Set the parameter number of neighbor cities called neighbor_city. For instance, if we have to drop 20 % links from each node, then neighbor_city is 80 % of $N_C - 1$.
- Parameters for velocity updating: $a_v = c_v = 1$ and $b_v = 0.01$.
- Parameters for soil updating: $a_s = c_s = 1$ and $b_s = 0.01$.
- Initial soil on each link is denoted by the constant InitSoil such that the soil of the link between every two cities i and j is set by soil (i, j) = InitSoil. Here, we choose InitSoil = 10,000.
- Initial velocity of IWD is denoted by the constant InitVel. Velocity of each drop with which they start their tour. Here, InitVel = 200.
- The best tour with minimum tour length (Len(T_B)) is denoted by T_B. Initially, it is set as Len(T_B) = infinity.
- The termination condition is met when maximum number of iterations is reached.

2. Initialization of dynamic parameters:

   - For every IWD, we create an empty visited city list $V_c(\text{IWD}) = \{\}$.
   - Initially, each IWD has velocity equal to InitVal and soil equal to zero.

3. Non-complete graph is generated using XNN algorithm [8] along with its adjacency matrix.

   - Initialize the number of neighbors of each city with the constant neighbor_city.
   - Create the adjacency matrix for new city links.

4. For every IWD, select a city randomly and place that IWD on that city.
5. Update the visited city lists of all IWDs to include the cities just visited.
6. Select the next city:

   - For each IWD, choose the next city j to be visited by IWD when it is in city i with the probability $P_i^{\text{IWD}}(j)$ as given in (1).

$$P_i^{\text{IWD}}(j) = \frac{f(\text{soil}(i, j))}{\sum_{k \notin vc} f(\text{soil}(i, k))} \tag{1}$$

such that $f(\text{soil}(i, j)) = 1/\varepsilon_s + g(\text{soil}(i, j))$
where

$$g(\text{soil}(i, j)) = \begin{cases} \text{soil}(i, j) \text{ if } \min_{l \notin vc(\text{IWD})}(\text{soil}(i, j)) \geq 0 \\ \text{soil}(i, j) - \min_{l \notin vc(\text{IWD})}(\text{soil}(i, j)) \text{ else} \end{cases}$$

Here, $\varepsilon_s = 0.01$. Where vc(IWD) is the visited city list of the IWD.

The probability depends on the soil between the path. IWD selects the city with maximum probability.

7. Update the soil and velocity:

- An IWD in city i wants to go to next city j; then, the amount of soil on this path, i.e., soil (i, j) is used to update the velocity as given by (2).

$$\mathrm{vel}^{\mathrm{IWD}}(t+1) = \mathrm{vel}^{\mathrm{IWD}}(t) + \frac{a_v}{b_v + c_v \cdot \mathrm{soil}(i, j)} \tag{2}$$

- Each IWD, carries some amount of the soil, $\Delta\mathrm{soil}(i, j)$, that the current IWD alters in its current path while travelling between the cities i and j is given by (3).

$$\Delta\mathrm{soil}(i, j) = \frac{a_s}{b_s + c_s \cdot \mathrm{time}(i, j; \mathrm{vel}^{\mathrm{IWD}})} \tag{3}$$

where time taken to travel from city i to city j with velocity $\mathrm{vel}^{\mathrm{IWD}}$ is given by $\mathrm{time}(i, j; \mathrm{vel}^{\mathrm{IWD}}) = c(i) - c(j)/\max(\varepsilon_v, \mathrm{vel}^{\mathrm{IWD}})$

- For each IWD, update the soil of the path traversed by that IWD by removing certain soil from the path as in (4)

$$\mathrm{soil}(i, j) = (1 - \rho) \cdot \mathrm{soil}(i, j) - \rho \cdot \Delta\mathrm{soil}(i, j) \tag{4}$$

Here, $\rho = 0.9$.
Update the soil of each IWD by adding soil removed from the path in present soil of the IWD

$$\mathrm{soil}^{\mathrm{IWD}} = \mathrm{soil}^{\mathrm{IWD}} + \Delta\mathrm{soil}(i, j) \tag{5}$$

$\mathrm{soil}(i, j)$ represents the soil of the path between i and j. $\Delta\mathrm{soil}(i, j)$ represents the soil that IWD carries from that path, and $\mathrm{soil}^{\mathrm{IWD}}$ represents total soil carried by the drop.

8. Each IWD completes its tour by using steps 5 to 8 repeatedly. Then, length of the tour ($\mathrm{Tour}^{\mathrm{IWD}}$) traversed by the IWD is calculated. Then, the tour with minimum length among all IWD tours in this iteration is found. Test the correctness of the minimum path from the adjacency matrix.
9. If iteration best tour (current minimum tour) exists then:

- Update the soil of the paths included in the current minimum tour of the IWD denoted by T_M by (6).

$$\mathrm{soil}(i, j) = (1 - \rho) \cdot \mathrm{soil}(i, j) + \rho \cdot \frac{2 \cdot \mathrm{soils}^{\mathrm{IWD}}}{N_c(N_c - 1)} \forall (i, j) \in T_M \tag{6}$$

- If the minimum tour T_M is shorter than the best tour T_B found so far, then T_B is updated by (7).

$$T_B = T_M \text{and Len}\,(T_B) = \text{Len}\,(T_M) \tag{7}$$

10. Otherwise discard the current minimum tour.
11. Go to step 2 unless the maximum number of iterations is reached.
12. If the maximum number of iterations is reached, then the algorithm stops with the best tour T_B with tour length Len(T_B).

4 Experimental Result

In this section, we present computational results obtained. We evaluated the performance of IWD algorithm for some TSP benchmark problems form TSPLIB [10]. We applied IWD algorithm on self-generated network like a pentagon. Firstly, a five node layout is taken as a complete graph, which then is converted to a non-complete graph using XNN algorithm. This network and its optimal path are shown in Fig. 1.

We also apply this on some benchmark problems such as eil51, eil76, st70, and kroA100 after converting them to non-complete graphs.

The experimental result of benchmark problem is shown in the Table 1.

Table 1 Experimental results for benchmark problems (10 run)

Problems	Optimum length	Average length by IWD
Eil51	426	445
Eil76	538	550
St70	675	748
Kroa100	21,282	24,344

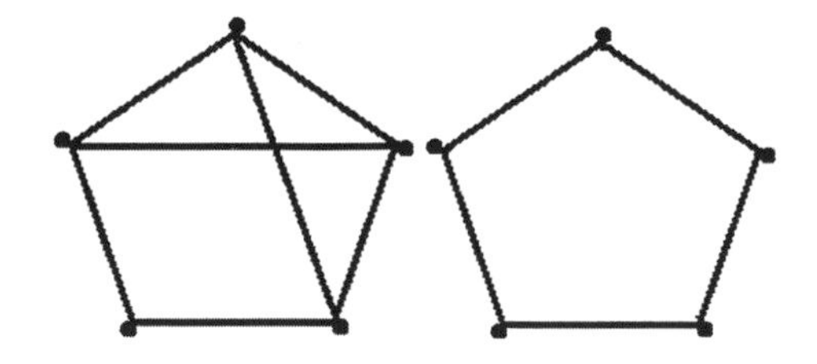

Fig. 1 *Left* the non-complete graph and *right* the optimal path

5 Conclusion

The intelligent water drops algorithm or the IWD algorithm is one of the recent bio-inspired swarm-based optimization algorithms. It gives the optimal solution to various optimization problems. The experimental results show that this algorithm is capable of finding the near-optimal solution. In this paper, we apply IWD on graph-based TSP (non-complete graphs) which gives the near-best optimal solution. We used XNN algorithm to convert the complete TSP graph to non-complete graph. GB-TSP represents the real-life transportation problem.

The IWD algorithm can also be used for solving multiple knapsack problem, n-Queen problem, multilevel thresholding problem, etc.

References

1. Bonabeau, E., Dorigo, M., Theraultz, G.: Swarm Intelligence: From Natural to Artificial Systems. Oxford University Press, New York (1999)
2. Greco, F.: Travelling Salesman Problem. In-Teh is Croatian Branch of I-Tech Education and Publishing KG, Vienna, Austria (2008)
3. Karaboga, D., Akay, B.: A comparative study of artificial bee colony algorithm. Appl. Math. Comput **214**(1), 108–132 (2009)
4. Kennedy, J., Eberhart, R.C.: Particle swarm optimization, pp. 1942–1948. Proceedings of the Fourth IEEE International Conference on Neural Networks, IEEE Service Center, Perth, Australia (1995)
5. Shah-Hosseini, H.: Optimization with the Nature-Inspired Intelligent Water Drops Algorithm. In: Sentos E.W (ed.) Evolutionary Computation, p. 572. I-Tech, Vienna, Austria (2009).
6. Shah-Hosseini, H.: Problem solving by intelligent water drops. Proceedings on IEEE Congress on Evolutionary Computation, pp. 3226–3231. Singapore (2007).
7. Shah-hosseini, H.: The intelligent water drops algorithm : a nature-inspired swarm-based optimization algorithm. Compu. Eng. **1**(1), 71–79 (2009)
8. Taba, M.S.: Solving Traveling Salesman Problem With a Non-complete Graph. Waterloo, Ontario, Canada (2009)
9. Toksari, M.D.: Ant colony optimization for finding the global minimum. Appl. Math. Comput. **176**(1), 308–316 (2006)
10. TSPLIB. (n.d.). Retrieved 08 23, 2012, from http://comopt.ifi.uni-heidelberg.de/software/TSPLIB95/
11. Yang, X.-S.: A new metaheuristic bat-inspired algorithm. In: Nature Inspired Cooperative Strategies for Optimization (NICSO 2010) 284, 65–74 (2010).

Optimization of Complex Mathematical Functions Using a Novel Implementation of Intelligent Water Drops Algorithm

Maneet Singh and Sanjay Saini

Abstract The intelligent water drops (IWD) algorithm was introduced by Hamed Shah Hosseini in 2007, which has been used to solve some of the discrete optimization problems. This chapter introduces a simplified version of the basic IWD algorithm and uses it to find the global minimum of some of the complex mathematical functions. The parameter settings have a very important role in the efficiency of the algorithm. The results demonstrated that the simplified IWD algorithm gave better results as compared to other techniques in less number of iterations.

Keywords Complex mathematical functions · Global minimum · Intelligent water drops

1 Introduction

The optimization problems can be classified into two broad categories namely discrete optimization problems and continuous optimization problem. Discrete optimization problems are those that involve finding values for discrete variables such that constructing an optimal solution for the given objective function. Traveling salesman problem, multi-knapsack problem, and n-queen problem are some of the examples of combinatorial or discrete optimization problems. Continuous optimization problems are those that involve finding real values of the parameters of a given problem. Function minimization and maximization are the examples of continuous optimization problems.

M. Singh (✉) · S. Saini
Department of Physics and Computer Science, Dayalbagh Educational Institute, Agra, India
e-mail: maneetsingh88@gmail.com

S. Saini
e-mail: sanjay.s.saini@gmail.com

B. V. Babu et al. (eds.), *Proceedings of the Second International Conference on Soft Computing for Problem Solving (SocProS 2012), December 28–30, 2012*, Advances in Intelligent Systems and Computing 236, DOI: 10.1007/978-81-322-1602-5_16,

2 Global Minimum

Each function has two kinds of extremes, commonly known as maxima and minima. The maxima corresponds to the largest value of the function, whereas minima corresponds to the smallest value of the function. Global minimum is the term which is used in place of minima to refer to the point where the function takes on the smallest value. The point where the function has the smallest value with respect to its neighborhood and not to the whole function is known as local minima.

3 Recent Work

Intelligent water drops (IWD) algorithm follows the principle of swarm intelligence. Using Swarm Intelligence for optimization problems as specified by Blum et al. [1], has become a very popular and widely used technique. IWD algorithm was proposed by Hosseini [2, 3]. Various problems have been solved using this technique. Kamkar et al. [4] used IWD to solve an NP_hard combinatorial optimization problem i.e., vehicle routing problem. Rayapudi [5] solved economic dispatch problem using IWD. Ochoa et al. [6] used IWD algorithm along with the cultural algorithm (hybrid approach) to improve a shoal in FishVille. Msallam et al. [7] improved the IWD algorithm using adaptive schema to prevent it from premature convergence. Hosseini [8] reviewed various optimization problems (traveling salesman problem, n-queen puzzle, multidimensional knapsack problem and automatic multilevel thresholding) that can be solved using IWD.

4 IWD Algorithm

IWD algorithm is a nature-inspired optimization algorithm proposed by Hamed Shah Hoseini. Shah Hoseini simulated the behavior of water drops flowing in the natural river. The moving water drops all together in the form of a group creates path for the natural river. The water drops affect the environment in which they move, and the environment also effects the movement of water drops. Artificial water drops retain two important properties of natural water drops—velocity and soil. The velocity with which an IWD moves and the soil an IWD carries may change as an IWD flows in the environment. The path containing less soil is considered to be the easiest path to traverse. An IWD may load or unload soil from the path it traverses.

5 Complex Mathematical Functions Taken for Finding Global Minimum

The following mathematical functions are considered for applying IWD to find the global minimum:

1. Rastrigin's Function: This function is defined as:

$$f(x) = 10n + \sum_{i=1}^{n} \left(x_i^2 - 10\cos(2\pi x_i) \right)$$

   In this function, there are two number of variables, several local mimima, and the global minima is $x = (0, 0, \ldots 0)$, with $f(x) = 0$.
2. Rosenbrock's Function: This function is defined as:

$$f(x) = \sum_{i-1}^{n-1} \left[100 \left(x_{i+1}^2 - x_i^2 \right)^2 + \left(1 - x_i^2 \right)^2 \right]$$

   In this function, there are two number of variables, several local mimima, and the global minima is $x = (1, 1, \ldots 1)$, with $f(x) = 0$.
3. Griewank's Function: This function is defined as:

$$f(x) = \sum_{i=1}^{n} \frac{x_i^2}{4,000} - \prod_{i=1}^{n} \cos\left(\frac{x_i}{\sqrt{i}} \right) + 1$$

   In this, the global minima is $x = (0, 0, \ldots 0)$, with $f(x) = 0$.
4. Sphere Function: This function is defined as:

$$f(x) = \sum_{i=1}^{n} x_i^2$$

In this function, the global minima is $x = (0, 0, \ldots 0)$, with $f(x) = 0$.

6 Simplified Intelligent Water Drops Algorithm for Finding Global Minimum

The basic IWD algorithm is being modified and implemented on some of the complex mathematical functions. The simplified IWD algorithm is as follows:

1. Initializing static parameters: n_{iwd} for number of water drops, G_B represents the quality of the total best solution—initially set to worse, *max_iter* for maximum number of iterations, *count_iter* for counting the number of iterations—initially set to zero.
2. Initializing dynamic parameters : velocity of each IWD is initially set to *start_vel*, but as the iteration proceeds, the velocity decreases, soil of each IWD is set to the function value at the point in the space where IWD lies.
3. Every IWD is randomly placed in the two-dimensional space.

4. Each IWD randomly selects some theta values. If the lowest function value for all the directions is less than the current soil of the IWD, then IWD is moved in that particular direction with the distance equal to the velocity of the IWD; otherwise, the IWD will not move for the current iteration.
5. The velocity is reduced by some fixed ratio each time.
6. Compute the iteration best solution I_B from the given solution by all IWD_B, where $\mathrm{IWD}_B = \mathrm{soil}_{\mathrm{iwd}}$.
7. Update the total best solution as the current best solution if the current best solution is better than total best solution; otherwise, the total best solution will remain unchanged.
8. The *count_iter* is incremented by one, and if *count_iter* is less than *max_iter*, then go to step 2 (initializing dynamic parameters).
9. The algorithm gets terminated with G_B as the optimal solution.

7 Results

The results that were obtained after applying the simplified IWD algorithm on complex mathematical functions are compared with the performance of GA and PSO as specified by Valdez et al. [9] and are shown in the Table 1. From the table, we can say that the performance of the simplified IWD algorithm is better than GA and PSO. The performance of the proposed algorithm on Rosenbrock function was compared with the ARSET and ACO algorithms as specified by Toksari [10] (Table 2).

8 Conclusion

In this chapter, a simplified version of IWD algorithm is proposed. The proposed algorithm is tested on some of the benchmark problems, the results are compared with the results of genetic algorithm and particle swarm optimization algorithm, and it has been observed that the performance of the proposed algorithm is better than the GA and PSO. The efficiency of the simplified IWD algorithm was tested in terms of epoch number by comparing with ARSET and ACO. The results clearly depicts

Table 1 Comparison of the performance of simplified IWD with GA and PSO

Mathematical functions	Minima by GA	Minima by PSO	Minima by simplified IWD	Global minimum
Rastrigin	7.36E-07	3.48E-05	0	0
Rosenbrock	2.35E-05	2.46E-03	0	0
Sphere	1.62E-04	8.26E-11	1.5831e-046	0
Griewank	2.552E-05	2.56E-02	0	0

Table 2 Comparison of the performance of simplified IWD with ARSET and ACO for Rosenbrock function

Algorithms	X	y	F(x, y)	Epoch number
ARSET	0.99401	0.997	3.58E-005	10,000
ACO	1.00021	1.00004	1.73E-006	
Simplified IWD	1	1	0	
ARSET	1.0001	1.0001	2.03E-008	30,000
ACO	1	1	5.68E-12	
Simplified IWD	1	1	0	
ARSET	1	1	4.02E-16	50,000
ACO	1	1	0	
Simplified IWD	1	1	0	

that the proposed algorithm is much better also in the case of number of iterations required to produce good results.

References

1. Blum, C., Li, X.: Swarm intelligence in optimization. In: Blum, C., Merkle, D. (eds.) Swarm Intelligence — Introduction and Applications, pp. 43–85. Springer (2008)
2. Hosseini, H.S.: Problem solving by intelligent water drops. In: Proceeding of IEEE Congress on Evolutionary Computation (2007)
3. Hosseini, H.S.: The intelligent water drops algorithm: a nature inspired swarm based optimization algorithm. Int. J. Bio-Insp. Comput. **1**(1/2) (2009)
4. Kamkar, I., Akbarzadeh-T, M.-R., Yaghoobi, M.: Intelligent water drops a new optimization algorithm for solving the vehicle routing problem. In: Proceedings of IEEE International Conference on Systems Man and Cybernetics (2010)
5. Rayapudi, S.R.: An intelligent water drop algorithm for solving economic load dispatch problem. Int. J. Electric. Electron. Eng. **5**, 1 (2011)
6. Ochoa, A., Hernández, A., Bustillos, S., Zamarrón, A.: Water on the water : A hybrid approach using to intelligent water drops algorithm and cultural algorithm to improve a shoal in FishVille. In: Proceedings of 3rd Workshop of Hybrid Intell. Syst. at MICIA (2010)
7. Msallam, M.M., Hamdan, M.: Improved intelligent water drops algorithm using adaptive schema. Int. J. Bio-Inspir. Comput. **3**, 103–111 (2011)
8. Hosseini, H.S.: Optimization with the nature-inspired intelligent water drops algorithm. Evol. Comput. (2009)
9. Valdez, F., Melin, P.: Comparative study of particle swarm optimization and genetic algorithms for complex mathematical functions. J. Autom. Mob. Robot. Intell. Syst. **2**, 43–51 (2008)
10. Toksari, M.D.: Ant colony optimization for finding the global minimum. Appl. Math. Comput. 308–316 (2006)

A New Centroid Method of Ranking for Intuitionistic Fuzzy Numbers

Anil Kumar Nishad, Shailendra Kumar Bharati and S. R. Singh

Abstract In this paper, we proposed a new ranking method for intuitionistic fuzzy numbers (IFNs) by using centroid and circumcenter of membership function and non-membership function of the intuitionistic fuzzy number. The method utilizes the midpoint of the circumcenter of membership and non-membership function of intuitionistic fuzzy number to define the ranking function for IFN satisfying the general axioms of ranking functions. The developed method has been illustrated by some examples and is compared with some existing ranking method to show its suitability.

Keywords Intuitionistic fuzzy sets (IFS) · Trapezoidal intuitionistic fuzzy number · Triangular intuitionistic fuzzy number (TIFN) · Membership function · Non-membership function · Ranking function

1 Introduction

Decision-making problems need the processing of information for getting an optimal solution of a problem in a specific situation. But in general the information available is often imprecise and vague and many times contains uncertainty, and thus, such situation demands its handling by non-traditional methods. The fuzzy set theory developed by Zadeh [14] immerged as a potential tool in theory of optimization to deal with imprecision and vagueness in parameters. Further dealing with sociometric

A.K. Nishad (✉) · S. K. Bharati · S. R. Singh
Department of Mathematics, Banaras Hindu University, Varanasi 221005, India
e-mail: anil.k.nishad16@gmail.com

S. K. Bharati
e-mail: skmaths.bhu@gmail.com

S. R. Singh
e-mail: singh_shivaraj@rediffmail.com

B. V. Babu et al. (eds.), *Proceedings of the Second International Conference on Soft Computing for Problem Solving (SocProS 2012), December 28–30, 2012*, Advances in Intelligent Systems and Computing 236, DOI: 10.1007/978-81-322-1602-5_17,

problems of decision-making one is encountered by a situation where information available also contains hesitation factor in addition to belonging and non-belonging. Intuitionistic fuzzy set theory was developed by Atanassov [1, 2] to deal with such situations in an effective way. Such optimization problems in general involve the intuitionist fuzzy numbers. Here, while defining arithmetic operations on intuitionistic fuzzy number (IFN), its ranking is needed to have comparison among IFNs. Unlike to ranking of fuzzy numbers, one has also to take the cognition of non-membership functions in ranking of IFNs. Thus, this ranking becomes an interesting property of intuitionistic fuzzy number, and various workers have proposed several methods of ranking. In order to develop a standard for raking methods, Wang and Kerre [13] proposed six axioms which a reasonable ranking method is desired to satisfy. Further, Grzegorzewski [5] studied distances and ordering in a family of IFNs. A significant work on ranking method was carried out by Michell [7] for fuzzy numbers. Theory of ranking methods for IFN was further enriched by Su [12], Nayagam [10], Abbasbandy [3], and Nan and Li [8]. Recently, Rao and Shanker [11], Dubey and Mehra [4] , Nagoorgani and Ponnalagu [9] proposed some ranking methods for normal IFNs. In the present study, we have studied the various aspects of some ranking methods on six standard axioms and have proposed a general method for ranking of IFNs satisfying the six standard axioms for its application to decision-making problems.

2 Preliminaries

Definition 1 *(Fuzzy Set)* If X is a collection of objects denoted generically by x, then a fuzzy set $\tilde{A}$ in X is a set of ordered pairs: $\tilde{A} = \{(x, \mu_{\tilde{A}}(x)) \,|x \in X\}$, where $\mu_{\tilde{A}}(x)$ is called the membership function or grade of membership of x in $\tilde{A}$ that maps X to the membership space M [0, 1].

Definition 2 *(Intuitionistic Fuzzy Set)* If X is a collection of objects, then an intuitionistic fuzzy set $\tilde{A}$ in X is defined as : $\tilde{A} = \{(x, \mu_{\tilde{A}}(x), \nu_A(x)) \,|x \in X\}$, where $\mu_{\tilde{A}}(x), \nu_A(x)$ is called the membership and non-membership function of x in $\tilde{A}$ respectively.

Definition 3 *[Trapezoidal Intuitionistic Fuzzy Number (TFN)]* An intuitionistic fuzzy set (IFS) $\tilde{A} = \{(x, \mu_{\tilde{A}}(x), \nu_A(x)) \,|x \in X\}$ of R is said to be an intuitionistic fuzzy number if μ_A, and ν_A are fuzzy numbers with $\nu_A \leq \mu^c_{A,}$ where $\mu^c_{A,}$ denotes the complement of μ_A. A TFN with parameters $a' \leq a \leq b \leq c \leq d \leq d'$ denoted by $\tilde{A} = \langle (a, b, c, d, \mu_A,), (a', b, c, d', \nu_A) \rangle$ is a IFS on real line $\mathbb{R}$ whose membership function and non-membership function are defined as follows:

$$\mu_{\tilde{A}}(x) = \begin{cases} \frac{(x-a)w}{(b-a)} & \text{if } a \leq x < b \\ w & \text{if } b \leq x < c \\ \frac{(d-x)w}{(d-c)} & \text{if } c \leq x < d \end{cases}$$

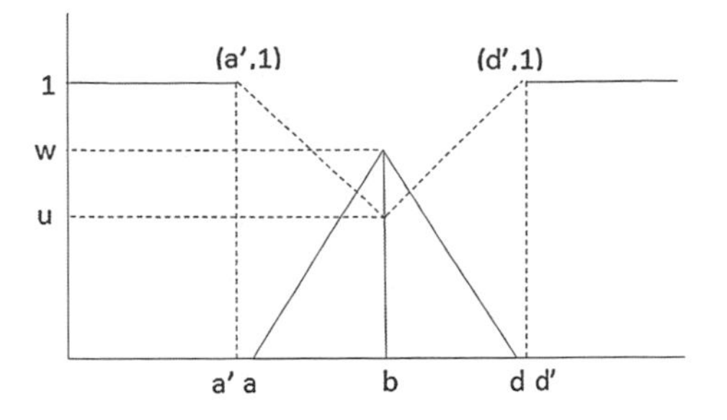

Fig. 1 Membership and non-membership function of triangular intuitionistic fuzzy number

and

$$\nu_A(x) = \begin{cases} \frac{(x-a')u+(b-x)}{(b-a')} & \text{if } a' \leq x < b \\ u & \text{if } b \leq x < c \\ \frac{(x-d')u+(c-x)}{(c-d')} & \text{if } c \leq x < d' \end{cases}$$

respectively. The values w and u represent the maximum degree of membership and the minimum degree of non-membership function, respectively, such that $\mu_A : X \to [0, 1]$ and $\nu_A : X \to [0, 1]$ and $0 \leq w + u \leq 1$.

Definition 4 *(Triangular Intuitionistic Fuzzy Number (TIFN))* If in a trapezoidal IFN we take (b=c), then it becomes a triangular IFN with the parameters $a' \leq a \leq b \leq d \leq d'$ and denoted by $\tilde{A} = \langle (a, b, d, \mu_A,), (a', b, d', \nu_A) \rangle$ Fig. 1.

3 Ranking of Intuitionistic Fuzzy Numbers

Ranking of fuzzy numbers is an important arithmetic property of IFNs. It provides comparison between two IFNs. This property of ranking is used in decision-making problems in uncertain environment. In view of belonging, non-belonging and hesitations factor of an IFN several ranking methods have been developed. Thus, in order to standardize these methods, Wang and Kerre [13] proposed six axioms . Let M be an ordering method and S the set of fuzzy quantities for which the method M can be applied and A be finite subset of S.

A_1. For any arbitrary finite subset $\tilde{A}$ of S and $\tilde{a} \in \tilde{A} \tilde{a} \gtrsim \tilde{a}$ by M on $\tilde{A}$.

A_2. For an arbitrary finite subset $\tilde{A}$ of S and $(\tilde{a}; \tilde{b}) \in \tilde{A}^2, \tilde{a} \gtrsim \tilde{b}$, and $\tilde{b} \gtrsim \tilde{a}$ by M on $\tilde{A}$, we should have $\tilde{a} \sim \tilde{b}$ by M on $\tilde{A}$.

A_3. For an arbitrary finite subset $\tilde{A}$ of S and $(\tilde{a}; \tilde{b}; \tilde{c}) \in \tilde{A}^3, \tilde{a} \gtrsim \tilde{b}$ and $\tilde{b} \gtrsim \tilde{c}$ by M on $\tilde{A}$ we should have $\tilde{a} \gtrsim \tilde{c}$ by M on $\tilde{A}$

A_4. For an arbitrary finite subset $\tilde{A}$ of S and $(\tilde{a}, \tilde{b}) \in \tilde{A}^2$ inf sup $(\tilde{a}) \geq$ sup supp $(\tilde{b})$, we should have $\tilde{a} \geq \tilde{b}$ by M on $\tilde{A}$

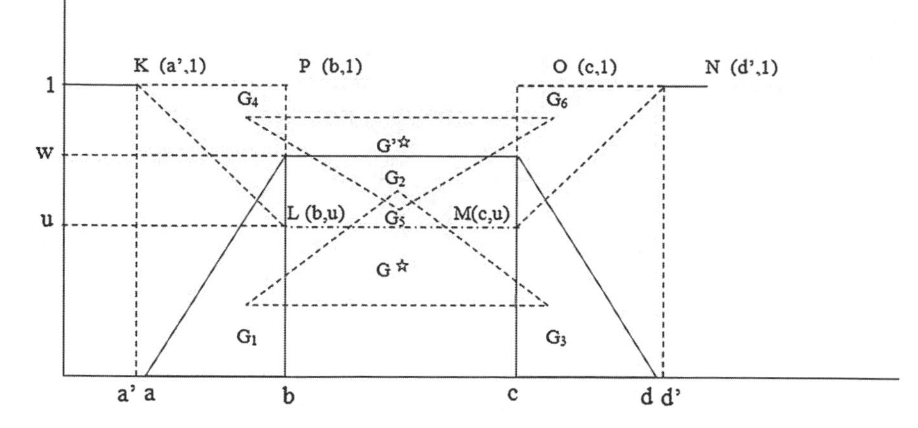

Fig. 2 Centroids of membership function and non-membership function

A_5. Let S and S' be two arbitrary finite sets of fuzzy quantities in which M can be applied and $\tilde{a}$ and $\tilde{b}$ are in $s \cap s'$. We obtain the ranking order $\tilde{a} > \tilde{b}$ by M on S' iff $\tilde{a} > \tilde{b}$ by M on S.
A_6. Let $\tilde{a}, \tilde{b}, \tilde{a}+\tilde{c}, \tilde{b}+\tilde{c}$ be element of S, If $\tilde{a} \gtrsim \tilde{b}$ by M on $\{\tilde{a}, \tilde{b}\}$, then $\tilde{a}+\tilde{c} \gtrsim \tilde{b}+\tilde{c}$ by M on $\{\tilde{a}+\tilde{c}, \tilde{b}+\tilde{c}\}$.

3.1 Proposed Ranking Method

Centroid of TIFN is considered as the balancing point of trapezoidal IFN for membership function and non-membership function are shown Fig. 2. Midpoints of circumcenters of centroid of membership function and non-membership function is taken as point of references to define the ranking of generalized trapezoidal IFN, and these points (circumcenters) are considered as balancing point of each individual membership function and non-membership function.

Consider a generalized trapezoidal IFN, $\tilde{A} = \langle (a, b, c, d, \mu_A,), (a', b, c, d', \nu_A) \rangle$ as shown in Fig. 2.

In which centroid of the three plain figures of membership function are $G_1 = (\frac{(a+2b)}{3}, \frac{w}{3})$, $G_2 = (\frac{(b+c)}{2}, \frac{w}{2})$, and $G_3 = (\frac{(2c+d)}{3}, \frac{w}{3})$, respectively. These centroids are non-collinear and they form a triangle whose circumcenter is $C_{\tilde{A}}(x_0, y_0)$ with vertices G_1, G_2 and G_3 of membership function of generalized trapezoidal IFN. Let $\tilde{A} = \langle (a, b, c, d, w), (a', b, c, d', u) \rangle$ is a intuitionistic fuzzy number whose circumcenter is given as:

$$C_{\tilde{A}}(x_0, y_0) = (\frac{(a+2b+2c+d)}{6}, \frac{(2a+b-3c)(c+2d-3b)+5w^2}{12w})$$

And centroid points of non-membership function are $G_4 = (\frac{(a\prime+2b)}{3}, \frac{u+2}{3})$, $G_5 = (\frac{(b+c)}{2}, \frac{1+u}{2})$, and $G_6 = (\frac{(2c+d\prime)}{3}, \frac{2+u}{3})$, respectively forms a triangle G_4, G_5, G_6 whose circumcenters is given as:

$$C'_{\tilde{A}}(x_0, y_0) = \left(\frac{(a'+2b+2c+d')}{6}, \frac{(-2a\prime-b+3c)\left(c+2d'-3b\right)+7-2u-5u^2}{12(1-u)}\right) \quad (1)$$

Thus, we have two circumcenters $C_{\tilde{A}}(x_0, y_0)$ and $C'_{\tilde{A}}(x_0, y_0)$, now take the mid-point of these circumcenters of generalized trapezoidal IFNs $S_{\tilde{A}}(\bar{x}_0, \bar{y}_0)$ which is given as :

$$S_{\tilde{A}}(\bar{x}_0, \bar{y}_0) = (\frac{(a'+a+4b+4c+d'+d)}{12}, \frac{(1-u)\left[(2a+b-3c)(c+2d-3b)+5w^2\right)+w[(-2a'-b+3c)(c+2d\prime-3b)+(7-2u-5u^2)}{24w(1-u)})$$

Definition 5 For a generalized trapezoidal IFN $\tilde{A} = \langle (a, b, c, d, w), (a', b, c, d', u) \rangle$ with midpoint of circumcenters of membership and non-membership function $S_{\tilde{A}}(\bar{x}_0, \bar{y}_0)$, we define index associated with the ranking as $I_\alpha = \alpha\bar{y}_0 + (1-\alpha)\bar{x}_0$I, where $\alpha \in [0, 1]$ is the index of optimism which represent the degree of optimism of decision maker. If $\alpha = 0$, we have a pessimistic decision maker, and if $\alpha = 1$, then we have optimistic decision maker or a neutral with $\alpha = .5$. The ranking function of trapezoidal IFN $\tilde{A} = \langle (a, b, c, d, w), (a', b, c, d', u) \rangle$ is a function which maps the set of all IFNs to a set of real number and is defined as $\mathrm{R}(\tilde{A}) = \sqrt{\bar{x}_0^2 + \bar{y}_0^2}$, which is the Euclidean distance of midpoint of circumcenters of centroids of membership function and non-membership function of trapezoidal IFN. Now using above definition, we define ranking between IFNs as follows :

Let $\tilde{A}$ and $\tilde{B}$ be two IFNs then

1. $\tilde{A} > \tilde{B}$, if $(R(\tilde{A}) > R(\tilde{B}))$
2. $\tilde{A} < \tilde{B}$, if $(R(\tilde{A}) < R(\tilde{B}))$

But if $R(\tilde{A}) = R(\tilde{B})$, then this definition of ranking fails.

For such situation, we use second ranking function defined as $I_{(\alpha,\beta)}(\tilde{A}) = \beta(\frac{\bar{x}_0+\bar{y}_0}{2}) + (1-\beta)I_\alpha(\tilde{A})$ where $\beta \in [0, 1]$ and represent the weight of central value and $(1-\beta)$ is the weight associated with the extreme values of $\bar{x}_0 and \bar{y}_0$. This ranking function of trapezoidal IFN also holds for the triangular IFN since triangular IFNs are special case of trapezoidal IFN, i.e., let $\tilde{A} = \langle (a, b, c, d, w), (a', b, c, d', u) \rangle$ be a trapezoidal IFN and if take $b = c$ it becomes triangular IFN, the midpoint of circumcenters of membership and non-membership function of triangular IFN is given by

$$S_{\tilde{A}}(\bar{x}_0, \bar{y}_0) = (\frac{(a' + a + 8b + d' + d)}{12}, \frac{(1-u)\,[4\,(a-b)\,(d-b) + 5w^2) + w[4\left(b-a'\right)(d\prime - b) + (7 - 2u - 5u^2)}{24w(1-u)})$$

and holds for ranking function defined above.

Example 1 Let $\tilde{A} = \langle(0.4, 0.5, 0.7; 0.8), (0.3, 0.5, 0.8; 0.1)\rangle$, $\tilde{B} = \langle(0.4, 0.6, 0.7; 0.6), (0.5, 0.6, 0.9; 0.3)\rangle$, $\tilde{C} = \langle(0.2, 0.6, 0.7; 0.9), (0.1, 0.6, 0.8; 0.3)\rangle$, are three IFNs which are to be compared

Here, using the above ranking method, we get

$S_{\tilde{A}}(\bar{x}_0, \bar{y}_0) = (0.516, 0.486)$, $S_{\tilde{B}}(\bar{x}_0, \bar{y}_0) = (0.608, 0.475)$, $S_{\tilde{C}}(\bar{x}_0, \bar{y}_0) = (0.55, 0.511)$.

This on further computation gives

$$R(\tilde{A}) = 0.7088 \quad R(\tilde{B}) = 0.771 \quad R(\tilde{C}) = 0.750 \Rightarrow \tilde{B} > \tilde{C} > \tilde{A}$$

Example 2 Let $\tilde{A} = \langle(0.1, 0.3, 0.5; 1), (0.1, 0.3, 0.5; 0)\rangle$ and $\tilde{B} = \langle(-0.5, -0.3, -0.1; 1), (-0.5, -0.3, -0.1; 0)\rangle$ are two IFNs.

The above method gives $S_{\tilde{A}}(\bar{x}_0, \bar{y}_0) = (0.3, 0.5)$, $S_{\tilde{B}}(\bar{x}_0, \bar{y}_0) = (-0.3, 0.5)$ and hence $\mathrm{R}(\tilde{A}) = \mathrm{R}(\tilde{B}) = \frac{\sqrt{34}}{10}$

Thus, we use second definition for its ranking

1. For a pessimistic decision maker, i.e., $\alpha = 0$

$$I_{(0.\beta)}(\tilde{A}) = \beta(0.4) + (1-\beta)0.3$$
$$I_{(0.\beta)}(\tilde{\beta}) = \beta(0.1) - (1-\beta)0.3 \Rightarrow \tilde{A} > \tilde{B}$$

2. For a optimistic decision maker, i.e., $\alpha = 1$

$$I_{(1.\beta)}(\tilde{A}) = \beta(0.4) + (1-\beta)0.5$$
$$I_{(1.\beta)}(\tilde{B}) = \beta(0.1) + (1-\beta)0.5 \Rightarrow \tilde{A} > \tilde{B}$$

3. For a neutral decision maker, i.e., $\alpha = 0$.

$$I_{(0.5.\beta)}(\tilde{A}) = \beta(0.4) + (1-\beta)0.4$$
$$I_{(0.5.\beta)}(\tilde{B}) = \beta(0.1) + (1-\beta)0.4 \Rightarrow \tilde{A} > \tilde{B}$$

$\Rightarrow \tilde{A} > \tilde{B}$ Thus, we see that the ranking order of IFN is same for all cases

Example 3 Let $\tilde{A} = \langle(0.1, 0.3, 0.5; 1), (0.1, 0.3, 0.5; 0)\rangle$, $\tilde{B} = \langle(-0.5, -0.3, -0.1; 1), (-0.5, -0.3, -0.1; 0)\rangle$ are two IFNs.

Then, $-\tilde{A} = \langle(-0.5, -0.3, -0.1; 1), (-0.5, -0.3, -0.1; 0)\rangle$ and

$-\tilde{B} = \langle(0.1, 0.3, 0.5; 1), (0.1, 0.3, 0.5; 0)\rangle, and S_{\tilde{A}}(\bar{x}_0, \bar{y}_0) = (-0.3, 0.5)$, $S_{\tilde{B}}(\bar{x}_0, \bar{y}_0) = (0.3, 0.5)$,

Thus, $R(\tilde{A}) = R(\tilde{B}) = \frac{\sqrt{34}}{10}$ and discrimination of triangular IFN is not possible. Thus, for this case, we use the second definition of ranking function and get the following,

1. For a pessimistic decision maker, i.e., $\alpha = 0$

$$I_{(0.\beta)}(-\tilde{A}) = \beta(0.1) - (1 - \beta)0.3$$
$$I_{(0.\beta)}(-\tilde{B}) = \beta(0.1) - (1 - \beta)0.3 \Rightarrow -\tilde{A} < -\tilde{B}$$

2. For an optimistic decision maker, i.e., $\alpha = 1$

$$I_{(1.\beta)}(-\tilde{A}) = \beta(0.1) + (1 - \beta)0.5$$
$$I_{(1.\beta)}(-\tilde{B}) = \beta(0.4) + (1 - \beta)0.5 \Rightarrow -\tilde{A} < -\tilde{B}$$

3. For a neutral decision maker, i.e., $\alpha = 0.5$

$$I_{(0.5.\beta)}(-\tilde{A}) = \beta(0.1) + (1 - \beta)0.4$$
$$I_{(0.5.\beta)}(-\tilde{B}) = \beta(0.4) + (1 - \beta)0.4 \Rightarrow -\tilde{A} < \tilde{B}$$

Now from the above Example 2 and Example 3, we see that $\tilde{A} > \tilde{B} \Rightarrow -\tilde{A} < -\tilde{B}$

3.1.1 Comparison with Dubey and Mehra Method

Example 4 Let $\tilde{A} = \langle(9, 10, 20; 1), (9, 10, 20; 0)\rangle$, $\tilde{B} = \langle(8.7, 8.8, 8.9; 1), (8.7, 8.8, 8.9; 0)\rangle$ are two IFNs, then Dipty ranking method fails to satisfy axiom A_4 (For an arbitrary finite subset $\tilde{A}$ of S and $(\tilde{a}, \tilde{b}) \in \tilde{A}^2$ inf sup $(\tilde{a}) \geq$ sup supp $(\tilde{b})$, we should have $\tilde{a} \geq \tilde{b}$ by M on $\tilde{A}$) and rank as $\tilde{A} < \tilde{B}$.

But proposed ranking method satisfy axiom A_4 and rank this IFN, i.e., by getting

$$S_{\tilde{A}}(\bar{x}_0, \bar{y}_0) = (18.66, 0.5), S_{\tilde{B}}(\bar{x}_0, \bar{y}_0) = (14.66, 0.5), R(\tilde{A}) = 18.17, R(\tilde{B}) = 14.67 \Rightarrow \tilde{A} > \tilde{B}$$

which is correct according to axiom A_4.

3.1.2 Comparison with Hassan Method

Let $\tilde{A} = \langle(a_1, a_2, a_3, a_4), (b_1, b_2, b_3, b_4,)\rangle$ be a triangular IFN, then characteristic values of membership and non-membership for IFN $\tilde{A}$ with parameter (k) denoted by $C^k_\mu(\tilde{A}), C^k_\nu(\tilde{B})$, respectively, are defined by

$$C^k_\mu(\tilde{A}) = \frac{a_1 + a_3}{2} + \frac{(a_1 - a_2) + (a_4 - a_3)}{2(k+2)}, C^k_\nu(\tilde{B}) = \frac{b_1 + b_4}{2} + \frac{(b_2 - b_1) + (b_3 - b_4)}{2(k+2)}$$

and according to the two IFNs, $\tilde{A}$ and $\tilde{B}$ are compared as follows :

1. For a given k, compare ordering of $\tilde{A}$ and $\tilde{B}$ according to relative position of $C^k_\mu(\tilde{A})$,and $C^k_\mu(\tilde{B})$,
2. If $C^k_\mu(\tilde{A})$ and $C^k_\mu(\tilde{B})$ are equals, then conclude that $\tilde{A}$ and $\tilde{B}$ are equals. Otherwise, rank $\tilde{A}$ and $\tilde{B}$ according to relative position of $-C^k_\nu(\tilde{A})$ and $-C^k_\nu(\tilde{B})$

Example 5 Let $\tilde{A} = \langle(0.4, 0.5, 0.7; 0.8), (0.3, 0.5, 0.8; 0.1)\rangle$ and $(\tilde{B}) = \langle(0.4, 0.5, 0.7; 0.7), (0.3, 0.5, 0.8; 0.1)\rangle$ are two IFNs, then by Hassan as well as Nagoorgani methods it is clear that these two IFNs are equal. But using the proposed method, we have $S_{\tilde{A}}(\bar{x}_0, \bar{y}_0) = (0.516, 0.486), S_{\tilde{B}}(\bar{x}_0, \bar{y}_0) = (0.516, 0.464)$,

$$R(\tilde{A}) = 0.7088, R(\tilde{B}) = 0.6939 \Rightarrow R(\tilde{A}) > R(\tilde{B}), \Rightarrow \tilde{A} > \tilde{B}.$$

4 Conclusion

Thus, we proposed a more general method for ranking of IFNs to provide an appealing and logically interpretation of comparison in IFNs in view of ambiguity. The proposed method satisfy the standard axioms given by Wang and Kerre [13] which is commonly used standard for ranking method of IFNs. We have clearly illustrated with suitable examples that even those specific cases where Dubey and Mehra [4], Hassan [6], and Nagoorgani [9] methods for ranking do not give clear conclusion, the proposed method provides a better comparison to such IFN. Thus, the proposed method of ranking may be suitably used in optimization problems in intuitionistic fuzzy environment for better understanding under ambiguity.

Acknowledgments The authors are thankful to University Grants Commission (UGC), Government of India, and DST-CIMS, Banaras Hindu University, for financial support to carry out this research work.

References

1. Atanassov, K.T.: Intuitionistic fuzzy sets. Fuzzy Sets Syst. **20**, 87–96 (1986)
2. Atanassov, K.T.: interval valued intuitionistic fuzzy sets. Fuzzy Sets Syst. **31**, 343–349 (1989)
3. Abbasbandy S, Ranking of fuzzy numbers, some recent and new formulas, in IFSA-EUSFLAT (2009), pp. 642–646
4. Dipti, Dubey, Aparna, Mehra: Fuzzy linear programming with triangular intuitionistic fuzzy number, EUSFLAT-LFA 2011. Advances in intelligent System Research, Atlantis Press, vol. 1, issue 1, pp. 563–569 (2011)
5. Grzegorzewski P, Distances and orderings in a family of intuitionistic fuzzy numbers, in Proceedings of the Third Conference on Fuzzy Logic and Technology (Eusflat03), pp. 223–227, (2003)
6. Hassan, M.N.: A new Ranking Method for Intuitionistic Fuzzy Numbers. Int. J. Fuzzy Syst. **12**(1), 80–86 (2010)
7. Mitchell H.B., Ranking Intuitionistic fuzzy numbers, Int. J. Uncertainty Fuzziness Knowl. Based Syst. **12**(3), 377–386 (2004)
8. Nan, J.X., Li, D.-F.: A lexicographic method for matrix games with payoffs of triangular intuitionistic fuzzy numbers. Int. J. Comput. Intell. Syst. **3**, 280–289 (2010)
9. Nagoorgani, A.: Ponnalagu K. A New Approach on Solving Intuitionistic Fuzzy Linear Programming Problem, Appl. Math. Sci. **6**(70), 3467–3474 (2012)
10. Nayagam V.L., Vankateshwari G, and SivaramanG, Ranking of intuitionistic fuzzy numbers. IEEE International Conference on Fuzzy Systems, pp. 1971–1974 (2008)
11. Rao P.H.B., Shankar N.R., Ranking fuzzy numbers with a distance method using circumcenter of centroids and an index of modality. Adv. Fuzzy Syst. Article ID 178308, (2011). doi:10.1155/2011/178308
12. Su, J.S.: Fuzzy programming based on interval valued fuzzy numbers and ranking. Int. J. Contemp. Math. Sci. **2**, 393–410 (2007)
13. Wang, X.: Kerre E.E., Reasonable properties for the ordering of fuzzy quantities (I). Fuzzy Sets Syst. **118**, 375–385 (2001)
14. Zadeh, L.: A. Fuzzy sets. Inf. Control **8**, 338–353 (1965)

Solution of Multi-Objective Linear Programming Problems in Intuitionistic Fuzzy Environment

S. K. Bharati, A. K. Nishad and S.R. Singh

Abstract In the paper, we give a new method for solution of multi-objective linear programming problem in intuitionistic fuzzy environment. The method uses computation of the upper bound of a non-membership function in such way that the upper bound of the non-membership function is always less than the upper bound of the membership function of intuitionistic fuzzy number. Further, we also construct membership and non-membership function to maximize membership function and minimize non-membership function so that we can get a more efficient solution of a probabilistic problem by intuitionistic fuzzy approach. The developed method has been illustrated on a problem, and the result has been compared with existing solutions to show its superiority.

Keywords Multi-objective programming · Positive ideal solution · Intuitionistic fuzzy sets · Intuitionistic fuzzy optimization

1 Introduction

Atanassov [1] generalized the fuzzy sets to intuitionistic fuzzy sets to deal with imprecision, vagueness, and uncertainty for a class of problems in a better way. In fuzzy sets, we consider only belonging of an element to a set, whereas in intuitionistic fuzzy set theory, we consider both the belonging and the non-belonging as membership and non-membership functions. Intuitionistic fuzzy set, with this

S. R.Singh · S. K. Bharati (✉) · A. K. Nishad
Department of Mathematics, Banaras Hindu University, Varanasi 221005, India
e-mail: srsingh_mathbhu@rediffmail.com

S. K. Bharati
e-mail: skmaths.bhu@gmail.com

A. K. Nishad
e-mail: anil.k.nishad16@gmail.com

B. V. Babu et al. (eds.), *Proceedings of the Second International Conference on Soft Computing for Problem Solving (SocProS 2012), December 28–30, 2012*, Advances in Intelligent Systems and Computing 236, DOI: 10.1007/978-81-322-1602-5_18,

property, emerged as more powerful tool in dealing with vagueness and uncertainty than fuzzy set. Angelov [3] proposed a method for solving multi-objective programming problems in intuitionistic fuzzy environment. Further, Atanassov and Gargov in [2] generalized intuitionistic fuzzy sets and proposed several new properties to intuitionistic fuzzy sets which made IFS suitable to deal with problems of optimizations. De et al. [4], Mondal and Samanta [12] proposed some properties of intuitionistic fuzzy sets to make it more suitable for various applications. For dealing with multi-objective programming, goal programming emerged as more powerful to provide its solutions, and Mohamed [11] studied relationship in goal programming and fuzzy programming. Etoh et al. [8] considered a probabilistic problem in fuzzy environment for its solution, and this problem was further studied by Garg and Singh [7] for suitability of fuzzy solution of a probabilistic problem. Jana and Roy [9] considered a multi-objective intuitionistic fuzzy linear programming approach for solution of transportation problem, and Mahapatra et al. [13] studied intuitionistic fuzzy mathematical programming on reliability optimization model. Another direction in optimization under interval-valued intuitionistic fuzzy emerged with the work of Li [10] who considered linear programming method for MADM with interval-valued intuitionistic fuzzy sets. Dubey et al. [5, 6] considered the linear programming problems with triangular intuitionistic fuzzy number interval uncertainty in intuitionistic fuzzy set (IFS). Recently Nachammai and Thangaraj [14], Nagoorgani, Ponnalagu and Shahrokhi et al. [16] have also studied the solutions of linear programming problems in intuitionistic fuzzy environment. Here, we construct the membership and non-membership functions and have applied the developed algorithm for solution of an probabilistic problem by intuitionistic fuzzy approach.

2 Preliminaries

Since Zadeh [17] generalized the set theory as fuzzy set theory to deal with information available in imprecise form, many new properties have been developed for fuzzy set and numerous applications have been developed. It was Zimmermann [18] who considered a fuzzy programming with several objectives. As Atanassov [1, 2] theories are considered the generalization of fuzzy set to intuitionistic fuzzy set, it is needed to study the basics of intuitionistic fuzzy to develop an application of this IFS. Thus, here we reproduce some of its fundamentals to make the study self-sufficient.

Definition 1 An intuitionistic fuzzy set $\tilde{A}$ assigns to each element x of the universe X a membership degree $\mu_{\tilde{A}}(x) \in [0, 1]$ and non-membership degree $\nu_{\tilde{A}}(x) \in [0, 1]$ such that $\mu_{\tilde{A}}(x) + \nu_{\tilde{A}}(x) \leq 1$. A IFS is mathematically represented as $\{\langle x, \mu_{\tilde{A}}(x), \nu_{\tilde{A}}(x)\rangle | x \in X\}$ where $1 - \mu_{\tilde{A}}(x) - \nu_{\tilde{A}}(x)$ is called hesitancy margin.

Example Let A be set of countries with elected government, and let x be a member of A. Let $M(x)$ be the percentage of the electorate that voted for the government, $N(x)$ the percentage that voted against. If we take $\mu_{\tilde{A}}(x) = \frac{M(x)}{100}$, $\nu_{\tilde{A}}(x) = \frac{N(x)}{100}$

then $\mu_{\tilde{A}}(x)$ gives the degree of support, $\nu_{\tilde{A}}(x)$ the degree of opposition and $h_{\tilde{A}}(x) = 1 - \mu_{\tilde{A}}(x) - \nu_{\tilde{A}}(x)$ stand for indeterminacy which is the portion that cast bad votes: invalid votes, abstinent.

2.1 Intuitionistic Fuzzy Number

An IFS $\tilde{A} = (\mu_{\tilde{A}}, \nu_{\tilde{A}})$ of real numbers is said to be an intuitionistic fuzzy number if $\mu_{\tilde{A}}$ and $\nu_{\tilde{A}}$ are fuzzy numbers. Hence, $A = (\mu_{\tilde{A}}, \nu_{\tilde{A}})$ denotes an intuitionistic fuzzy number if $\mu_{\tilde{A}}$, and $\nu_{\tilde{A}}$ are fuzzy numbers with $\nu_{\tilde{A}} \leq \mu_{\tilde{A}}^C$, where $\mu_{\tilde{A}}^C$ denotes the complement of $\mu_{\tilde{A}}$.

Some operations on intuitionistic fuzzy sets are as follows:

$$\tilde{A} \cap \tilde{B} = \{\langle x, \min(\mu_{\tilde{A}}(x), \mu_{\tilde{B}}(x)), \max(\nu_{\tilde{A}}(x), \nu_{\tilde{B}}(x))\rangle \,|x \in X\}$$
$$\tilde{A} \cup \tilde{B} = \{\langle x, \max(\mu_{\tilde{A}}(x), \mu_{\tilde{B}}(x)), \min(\nu_{\tilde{A}}(x), \nu_{\tilde{B}}(x))\rangle \,|x \in X\}$$

3 Optimization in Intuitionistic Fuzzy Set

Various studies of optimization problems in fuzzy environment showed the suitability of considering optimization problems in fuzzy environment. The reason for the success was quite obvious that a small violation in constraints leads to more efficient solution. Further studies revealed that fuzzy optimization formulations are more flexible and allow better range of solutions especially when boundaries are not sharp. As a matter of fact in case of multi-objective programming problem, we search an optimal compromise solution rather than optimal solution. This idea of getting compromise solution in intuitionistic fuzzy environment needs to maximize the degree of acceptance to objective functions and constraints and to minimize the rejection of objective functions and constraints.

Consider the intuitionistic fuzzy optimization problem as generalization of fuzzy optimization problem under taken by Angelov [3] and is given as

$$\min \; f_i(x), \quad i = 1, 2, \ldots m$$

Such that

$$g_j(x) \leq 0, \quad j = 1, 2, \ldots n$$

where x is decision variables, $f_i(x)$ denotes objective functions, and $g_j(x)$ denotes the constraint functions. m and n denote the number of objective(s) and constraints, respectively.

Theorem 1 *For objective function of maximization problem, the upper bound for non-membership function is always less than that of the upper bound of membership function.*

Proof From definition of IFS, sum of the degree of rejection and acceptance is less than unity.

If, U_k^{μ} and L_k^{μ} are upper and lower bound, respectively, for the membership function and similarly U_k^{ν} and L_k^{ν} are upper and lower bound, respectively, for the non-membership function, then

$$\mu_k(f_k(x)) + \nu_k(f_k(x)) < 1 \quad \text{for all} \quad k = 1, 2, \ldots, K$$

or

$$\frac{f_i(x) - L_k^{\mu}}{U_k^{\mu} - L_k^{\mu}} + \frac{U_k^{\nu} - f_k(x)}{U_k^{\nu} - L_k^{\nu}} < 1$$

Case 1 If possible, let $U_k^{\nu} = U_k^{\mu}$, then we have

$$\frac{f_k(x) - L_k^{\mu}}{U_k^{\mu} - L_k^{\mu}} + \frac{U_k^{\nu} - f_k(x)}{U_k^{\mu} - L_k^{\nu}} < 1$$

this gives $L_k^{\nu} < L_k^{\mu}$ which is contradicting the fact that lower bound of the membership and non-membership is equal; hence, $U_k^{\nu} \neq U_k^{\mu}$.

Case 2 Let us consider $L_k^{\nu} = L_k^{\mu}$, then we have

$$\frac{f_k(x) - L_k^{\mu}}{U_k^{\mu} - L_k^{\mu}} + \frac{U_k^{\nu} - f_k(x)}{U_k^{\nu} - L_k^{\mu}} < 1$$

Which imply that $U_k^{\nu} < U_k^{\mu}$.

Case 3 Let us consider $L_k^{\nu} = L_k^{\mu} + \varepsilon_k, \varepsilon_k > 0$ for all $k = 1, 2, \ldots, K$.

$$\frac{f_k(x) - L_k^{\mu}}{U_k^{\mu} - L_k^{\mu}} + \frac{U_k^{\nu} - f_k(x)}{U_k^{\nu} - L_k^{\mu}} < 1$$

$$U_k^{\mu} > U_k^{\nu} + \varepsilon_k \frac{U_k^{\nu} - f_k(x)}{f_k(x) - L_k^{\mu} - \varepsilon_k}$$

i.e., $U_k^{\mu} > U_k^{\nu}$ hence $U_k^{\mu} > U_k^{\nu}$.

3.1 Computational Algorithm

Using the above-mentioned theorem and with the method by Anglev [3], we develop the following algorithm for getting solution of a multi-objective programming problem in intuitionistic fuzzy environment:

Step 1: Take one objective function out of given k objectives and solve it as a single objective subject to the given constraints. From obtained solution vectors, find the values of remaining $(k-1)$ objective functions.

Step 2: Continue the step 1 for remaining $(k-1)$ objective functions. If all the solutions are same, then one of them is the optimal compromise solution.

Step 3: Tabulate the solutions thus obtained in step 1 and step 2 to construct the positive ideal solution (PIS) as given in Table 1.

Step 4: From PIS, obtain the lower bounds and upper bounds for each objective functions, where f_k^* and f_k' are the maximum and minimum values, respectively.

Step 5: Set upper and lower bounds for each objective for degree of acceptance and degree of rejection corresponding to the set of solutions obtained in step 4.

For membership functions:

$$U_k^{\mu} = \max(Z_k(X_r)) \quad \text{and} \quad L_k^{\mu} = \min(Z_k(X_r)), \quad 1 \leq r \leq k.$$

For non-membership functions:

$$U_k^{\nu} = U_k^{\mu} - \lambda(U_k^{\mu} - L_k^{\mu}) \quad \text{and} \quad L_k^{\nu} = L_k^{\mu} \quad 0 < \lambda < 1.$$

Step6: Consider the membership function $\mu_k(f_k(x))$ and non-membership function $\nu_k(f_k(x))$ as following linear functions:

$$\mu_k(f_k(x)) = \begin{cases} 0 & \text{if } f_k(x) \leq L_k^{\mu} \\ \dfrac{f_k(x) - L_k^{\mu}}{U_k^{\mu} - L_k^{\mu}} & \text{if } L_k^{\mu} \leq f_k(x) \leq U_k^{\mu} \\ 1 & \text{if } f_k(x) \geq U_k^{\mu} \end{cases}$$

Table 1 Positive ideal solution

	f_1	f_2	f_3	....	f_k	X
max f_1	f_1^*	$f_2(X_1)$	$f_3(X_1)$	...	$f_k(X_1)$	X_1
max f_2	$f_1(X_2)$	f_2^*	$f_3(X_2)$	...	$f_k(X_2)$	X_2
max f_3	$f_1(X_3)$	$f_2(X_3)$	f_3^*	...	$f_k(X_3)$	X_3
⋮	⋮					⋮
⋮	⋮					⋮
max f_k	$f_1(X_k)$	$f_2(X_k)$	$f_3(X_k)$	...	$f_k^*(X_k)$	X_k
	f_1'	f_2'	f_3'	...	f_k'	

$$\nu_k(f_k(x)) = \begin{cases} 0 & \text{if } f_k(x) \geq U_k^{\nu} \\ \dfrac{U_k^{\nu} - f_k(x)}{U_k^{\nu} - L_k^{\nu}} & \text{if } L_k^{\nu} \leq f_k(x) \leq U_k^{\nu} \\ 1 & \text{if } f_k(x) \leq L_k^{\mu} \end{cases}$$

Figure of the membership function and non-membership function for maximization type objective function are shown in Fig. 1.

Step 7: An intuitionistic fuzzy optimization technique for MOLP problem as taken in this section with such membership and non-membership functions can be written as

$$\begin{aligned} \text{Maximize} \quad & \mu_k(f_k(x)) \\ \text{Minimize} \quad & \nu_k(f_k(x)) \\ \text{Subject to} \quad & \mu_k(f_k(x)) + \nu_k(f_k(x)) \leq 1, \\ & \mu_k(f_k(x)) \geq \nu_k(f_k(x)), \\ & \nu_k(f_k(x)) \geq 0, \\ & g_j(x) \leq b_j, \quad x \geq 0, \\ \text{for} \quad & k = 1, 2, \ldots, K; \quad j = 1, 2, \ldots, m. \end{aligned}$$

Now the above problem may be equivalently written in a linear programming problem as

$$\begin{aligned} \text{Maximize} \quad & (\alpha - \beta) \\ \text{Subject to} \quad & \alpha \leq \mu_k(f_k(x)), \\ & \beta \geq \nu_k(f_k(x)), \\ & \alpha + \beta \leq 1, \\ & \alpha \geq \beta, \\ & \beta \geq 0, \\ & g_j(x) \leq b_j, x \geq 0, \quad k = 1, 2, \ldots K; \quad j = 1, 2, \ldots, m. \end{aligned}$$

This linear programming problem can be easily solved by a simple method.

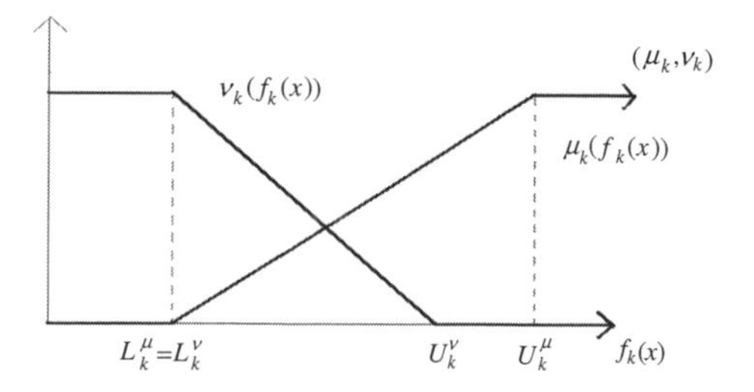

Fig. 1 Membership and non-membership functions

4 Numerical Illustration

In this section, the developed algorithm is implemented by a numerical example. We consider the problem as undertaken by Garg and Singh [7], Itoh [8] in which a farmer has to grow carrot, radish, cabbage, and Chinese cabbage in a season under areas x_1, x_2, x_3, and x_4 (unit 10 acres $= 1000\,\text{m}^2$), respectively. The farmer has a total land of 10 acres and a max labor work time available to him is 260 h. The profit coefficients (unit 10,000 Japanese Yen) and work time for the crops are given in the Table 2 .

The complete mathematical formulation of the above problem is as follows:

$$\begin{aligned}
\text{Maximize } z_1 &= 29.8x_1 + 10.4x_2 + 13.8x_3 + 19.8x_4 \\
\text{Maximize } z_2 &= 23.9x_1 + 21.4x_2 + 49.2x_3 + 32.8x_4 \\
\text{Maximize } z_3 &= 37x_1 + 16x_2 + 3.6x_3 + 9.7x_4 \\
\text{Maximize } z_4 &= 19.3x_1 + 26.6x_2 + 48.4x_3 + 75.6x_4
\end{aligned} \tag{1}$$

Subject to the constraints

$$\begin{aligned}
&x_1 + x_2 + x_3 + x_4 \leq 10 \\
&6.9x_1 + 71x_2 + 2x_3 + 33x_4 \leq 260, \\
&x_1, x_2, x_3, x_4 \geq 0.
\end{aligned}$$

The solution procedure of the above problem involves the following steps:

Step 1: The solution choosing one by one objective as single objective function programming problem
Maximize $z_1 = 29.8x_1 + 10.4x_2 + 13.8x_3 + 19.8x_4$
Subject to the constraints

$$\begin{aligned}
&x_1 + x_2 + x_3 + x_4 \leq 10 \\
&6.9x_1 + 71x_2 + 2x_3 + 33x_4 \leq 260 \\
&x_1, x_2, x_3, x_4 \geq 0.
\end{aligned}$$

Table 2 Values of various parameters available to the problem

Random variable	Carrot	Radish	Cabbage	Chinese cabbage	Probability percentage of profit coefficients
	c_{i1}	c_{i2}	c_{i3}	c_{i4}	
c_1	29.8	10.4	13.8	19.8	10
c_2	23.9	21.4	49.2	32.8	50
c_2	37.0	16.0	3.6	9.7	10
c_2	6.9	26.6	48.4	75.6	30
Work time	6.9	71	2	33	

The optimal solution to this linear programming problem is

$$x_1 = 10, x_2 = 0, x_3 = 0\, x_4 = 0, (z_1)_1 = 298.$$

And with this solution vectors, the value of other objective functions are as

$$(z_2)_1 = 239, \quad (z_3)_1 = 370 \quad \text{and} \quad (z_4)_1 = 193.$$

Step 2: Solve linear programming problem for z_2, z_3, z_4 subject to constraints and find values of remaining objective functions.

Step 3: Tabulate the values as given below to form PIS as given in Table 3.

Step 4: Find lower and upper bounds for each case of max z_1, max z_2, max z_3, max z_4 which are

$$U_1^{\mu} = 298, L_1^{\mu} = 138; U_2^{\mu} = 492, L_2^{\mu} = 239; U_3^{\mu} = 370, L_3^{\mu} = 36;$$
$$U_4^{\mu} = 694.58, \ L_4^{\mu} = 193.$$

Step 5: Set the upper and lower bounds of each objective for degree of rejections as

$$L_k^{\mu} = L_k^{\nu}$$
$$U_k^{\nu} = U_k^{\mu} + \lambda(U_k^{\mu} - L_k^{\mu}) = (1-\lambda)U_k^{\mu} + \lambda L_k^{\mu}, \quad k = 1, 2, 3, 4.$$

Which for $\lambda = 0.6$ becomes

$$U_k^{\nu} = (1-0.6)U_k^{\mu} + 0.6L_k^{\mu} = 0.4U_k^{\mu} + 0.6L_k^{\mu}$$
$$U_1^{\nu} = 202, \ L_1^{\nu} = 138, \ U_2^{\nu} = 340.2, \ L_2^{\nu} = 239, \quad U_3^{\nu} = 169.6, \ L_3^{\nu} = 36,$$
$$U_4^{\nu} = 393.63, \ L_4^{\nu} = 193.$$

Step 6: Construction of membership functions:

Table 3 Positive ideal solution

	z_1	z_2	z_3	z_4	
Max z_1	298	239	370	484	X_1
Max z_2	138	492	36	484	X_2
Max z_3	298	239	370	193	X_3
Max z_4	184.44	365.06	83.21	694.85	X_4

$$\mu_1(x) = \frac{29.8x_1 + 10.4x_2 + 13.8x_3 + 19.8x_4 - 138}{(298 - 138)},$$
$$\mu_2(x) = \frac{23.9x_1 + 21.4x_2 + 49.2x_3 + 32.8x_4 - 239}{(492 - 239)},$$
$$\mu_3(x) = \frac{37x_1 + 16x_2 + 3.6x_3 + 9.7x_4 - 36}{(370 - 36)},$$
$$\mu_4(x) = \frac{19.3x_1 + 26.6x_2 + 48.4x_3 + 75.6x_4 - 139}{(694.58 - 193)}.$$

Construction of non-membership functions:

$$\nu_1(x) = \frac{202 - 29.8x_1 - 10.4x_2 - 13.8x_3 - 19.8x_4}{(202 - 138)},$$
$$\nu_2(x) = \frac{340.2 - 23.9x_1 - 21.4x_2 - 49.2x_3 - 32.8x_4}{(340.2 - 239)},$$
$$\nu_3(x) = \frac{169.6 - 37x_1 - 16x_2 - 3.6x_3 - 9.7x_4}{(169.6 - 36)},$$
$$\nu_4(x) = \frac{393.63 - 19.3x_1 - 26.6x_2 - 48.4x_3 - 75.6x_4}{(393.63 - 193)}.$$

Step 7: The above problem (1) is now equivalently written to a linear programming problem as

$$\begin{aligned}
\text{Maximize} \quad & (\alpha - \beta), \\
\text{Subject to} \quad & 29.8x_1 + 10.4x_2 + 13.8x_3 + 19.8x_4 - 138 \geq 160\alpha, \\
& 23.9x_1 + 21.4x_2 + 49.2x_3 + 32.8x_4 - 239 \geq 253\alpha, \\
& 37x_1 + 16x_2 + 3.6x_3 + 9.7x_4 - 36 \geq 334\alpha, \\
& 19.3x_1 + 26.6x_2 + 48.4x_2 + 75.6x_4 - 193 \geq 501.58\alpha, \\
& 202 - 29.8x_1 - 10.4x_2 - 13.8x_3 - 19.8x_4 \leq 64\beta, \\
& 340.2 - 23.9x_1 - 21.4x_2 - 49.2x_3 - 32.8x_4 \leq 101.2\beta, \\
& 169.6 - 37x_1 - 16x_2 - 3.6x_3 - 9.7x_4 \leq 133.6\beta, \\
& 393.63 - 19.3x_1 - 26.6x_2 - 48.4x_3 - 75.6x_4 \leq 200.63\beta, \qquad (2) \\
& x_1 + x_2 + x_3 + x_4 \leq 10, \\
& 6.9x_1 + 71x_2 + 2x_3 + 33x_4 \leq 260, \\
& \alpha + \beta \leq 1, \\
& \alpha \geq \beta, \quad \beta, x_1, x_2, x_3, x_4 \geq 0
\end{aligned}$$

The above problem (2), a linear programming problem is solved by MATLAB and the solutions obtained are

$$x_1 = 4.14$$

Table 4 Profit in different probabilistic cases

Probability cases for profit coefficients (%)	Profit by proposed intuitionistic fuzzy optimization technique	Profit by Garg and Singh method [7]	Profit by Itoh method [8]
10	216.46	207.37	268.4
50	352.98	348.58	280.4
10	186.80	181.07	303.5
30	419.07	410.54	274.8
Weighted profit	342.53	336.29	280.08

$$x_2 = 0.00$$
$$x_3 = 3.79$$
$$x_4 = 2.06$$
$$\alpha = 0.45$$
$$\beta = 0.00$$

Putting these values in the problem, the profit obtained are as in (Table 4).

5 Conclusion

The objective of this paper was to develop a method to solve a probabilistic programming problem in an intuitionistic fuzzy optimization environment. Here, the developed method first considers the conversion of the probabilistic programming problem into a multi-objective programming problem. This is done by considering the objective function corresponding to probabilistic cases as one objective function of the said multi-objective programming. Thus, such converted multi-objective programming problem is solved with one objective at a time to construct the PIS. Thus, in order to obtain a best compromise solution of the situation, we construct the membership function and non-membership functions for the solutions, and thus, we introduce intuitionistic fuzzy parameters. Using the intuitionistic fuzzy optimization approach, the problem is transformed into an equivalent linear programming problem. The linear programming problem thus obtained has been solved by using MATLAB. The result thus obtained has been compared with the existing solution, and clearly, the proposed method gives a better solution than existing solutions.

Acknowledgments Authors are thankful to University Grants Commission (UGC), Government of India, for financial support to carry out this research work.

References

1. Atanassov, K.T.: Intuitionistic fuzzy sets. Fuzzy Sets and Systems. **20**, 87–96 (1986)
2. Atanassov, K.T., Gargov, G.: interval valued intuitionistic fuzzy sets. Fuzzy Sets and Systems. **31**, 343–349 (1989)
3. Angelov, P.: P Optimization in an intuitionistic fuzzy environment. Fuzzy Sets and Systems. **86**, 299–306 (1997)
4. De, S.K., Biswas, R.A.: Ray R. Some operations on intuitionistic fuzzy sets, Fuzzy Sets and Systems. **114**, 474–487 (2000)
5. Dipti, Dubey, Mehra, Aparna: Linear programming with Triangular Intuitionistic Fuzzy Number, EUSFLAT-LFA 2011. Advances in Intelligent Systems Research **1**(1), 563–569 (2011). Atlantis Press
6. Dipti, Dubey, Suresh, Chandra, Aparna, Mehra: Fuzzy linear programming under interval uncertainty based on IFS representation. Fuzzy Sets and Systems. **188**(1), 68–87 (2012)
7. Garg, A., Singh, S.R.: Optimization under uncertainty in agricultural production planning. iconcept pocket journal: Computational Intelligence for Financial Engineers **1**(1), 1–12 (2010)
8. Itoh, T., Ishii, H., Nanseki, T.: Fuzzy crop planning problem under uncertainty in agriculture management. Int. J. Production Economics. **81–82**, 555–558 (2003)
9. Jana, B., Roy, T.K.: Multiobjective intuitionistic fuzzy linear programming and its application in transportation model **NIFS–13–1–34–51**, 1–18 (2007)
10. Li, D.F.: Linear programming method for MADM with interval valued intuitionistic fuzzy sets. Expert Systems and Applications. **37**, 5939–5945 (2010)
11. Mohamed, R.H.: The relationship between goal programming and fuzzy programming. Fuzzy Sets and Systems **89**, 215–222 (1997)
12. Mondal, T.K., Samanta, S.K.: Generalized intuitionistic fuzzy set. Journal of Fuzzy Math **10**, 839–861 (2002)
13. Mahaparta, G.S., Mitra, M., Roy, T.K.: Intuitionistic fuzzy multiobjective mathematical programming on reliability optimization model. International Journal of Fuzzy Systems. **12**(3), 259–266 (2010)
14. Nachamani, A.L., Thangaraj, P.: Solving Intuitionistic fuzzy linear programming problems by using similarity measures. European Journal of Scientific Research **72**(2), 204–210 (2012)
15. Nagoorgani, A., Ponnalagu, K.: A new approach on solving intuitionistic fuzzy linear programming problem. Applied Mathematical Sciences **6**(70), 3467–3474 (2012)
16. Shahrokhi M, Bernard A, Shidpour H, An integrated method using intuitionistic fuzzy set and linear programming for supplier selection problem, 18th IFAC World congress Milano(Italy)Aug18-Sept2, 2011, 6391–6395.
17. Zadeh, L.: A Fuzzy Sets. Information and Control **8**, 338–353 (1965)
18. Zimmermann, H.: J. Fuzzy programming and linear programming with several objective functions. Fuzzy Sets and Systems **1**, 45–55 (1978)

Analyzing Competitive Priorities for Machine Tool Manufacturing Industry: ANP Based Approach

Deepika Joshi

Abstract In an attempt to study the success factors for the competitiveness of machine tool manufacturing (MTM) industry an in-depth study of 10 manufacturers located in India was carried out. Performance measures, especially which are related to supply chain (SC) activities, are also the part of competitive priorities [16, 17]. It can be seen that systematic identification and prioritization of SC performance indicators would help managers to integrate them into corporate strategy. An ANP approach is used to analyze the dynamic, large, and complex attribute decision. To perform the related computations, a programming platform of MATLAB™ software suite was operated. Research findings unveiled that flexibility and quality dimensions are of foremost significance in the development of sector under study, followed by delivery indicators. However, the companies believed that cost constituents need to be focused to achieve overall SC competitiveness.

Keywords Analytic network process (ANP) · Supply chain performance indicators · Competitiveness · Machine tool manufacturing industry · India

1 Introduction

Machine tool manufacturing (MTM) industry in India is performing a vital role in advancing national competitiveness. In particular, it is instrumental in promoting the output of Indian manufacturing industry and in general, the Indian economy. Typically, growth of any industry depends on the ability to manage its supply chain (SC) activity [28]. SC of MTM industry is composed of large number of firms. Consequently, it necessitates concerted efforts to overcome the overwhelming complexity and associated challenges of this developing sector. It will not only bring structural

---

D. Joshi (✉)
Research/Faculty Associate, Gautam Buddha University, Greater Noida, India
e-mail: joshi.deepikaa@gmail.com

B. V. Babu et al. (eds.), *Proceedings of the Second International Conference on Soft Computing for Problem Solving (SocProS 2012), December 28–30, 2012*, Advances in Intelligent Systems and Computing 236, DOI: 10.1007/978-81-322-1602-5_19,

transformation in business operations but will also compel engineers and managers to inculcate strategic thinking while managing value chain elements.

The spurred industry reforms have radically shifted the significance of competitive priorities in SC management activity. For example, in auto-component manufacturing, cost and flexibility drives the SC performance whereas in retail sector delivery leads to successful business operations [16, 17]. Thus, there are always some sector specific performance indicators leading to overall SC competitiveness. A systematic identification of performance indicators and their priorities is necessary in order to plan and implement suitable strategies for industrial competitiveness. Most of the previous studies on MTM sector were conducted considering the whole gamut of competitive priorities and factors affecting these priorities. Few of such studies, which are exclusively considered competitive priorities, were nation specific like China, US, and Japan [33]. It has been noted, however, that there is hardly any research study which reports the prioritization of competitive priorities specific to Indian MTM industry. On the whole, this perceived gap among existing research studies shapes the main ground of the work presented in this paper.

In Sect. 2 a review of literature is presented to disseminate knowledge regarding SC performance indicators. Section 3 presents a research methodology exclusively designed to attain the research objective. Section 4 portrays the research findings. Section 5 discusses the results and strategies for its managerial implications. Section 6 concludes the paper while providing avenues for future research.

2 Review of Literature

In the present era of globalization and industrialization, competitive priorities like cost, delivery, flexibility, and quality (CDFQ) which are critical to operation's success of the firm [31]. Neely et al., Ho et al. and Singh et al. are few of the numerous proponents who suggested the need of competitive priorities for SC competitiveness building [14, 21, 28]. Identifying competitive priorities for industrial application lead to overall competitiveness.

2.1 Competitive Priorities

Traditional operations management literature considers cost as the simplest measure of competitiveness. Herein, labor cost, raw material cost, R&D cost, manufacturing cost, etc., are found to be important ones [1, 9] and [18]. With the objective of SC responsiveness, inventory control is another major concern for manufacturing based industries [2]. Fuss and Waverman highlighted the impact of variation in inter-country costs such as costs related to labor and raw material, toward cost competitiveness [10]. Due to increased distances the costs related to distribution have become a major concern for managers [2]. Managers control it by focusing on storage facilities at client's location. Flexibility as a determinant of competitiveness has been discussed. It effects the business operations to capture global market opportunities [30]. The various types of flexibilities discussed in literature are—volume flexibility, process flexibility, product-mix, delivery flexibility, distribution flexibility, new product development, and design flexibility [27]. However, the type of flexibility which would best make a business competitive depends on available resources, goals and objectives of a particular firm.

Similarly, researchers mentioned that specific quality norms reduce the defects and enhance the perceived quality level of a product. This also advocates the consistency of process and product design. Durability, performance, conformance, reliability, and design characteristics are various commonly used dimensions of quality [11]. Quality parameters are reflected in higher value of returns on investment, defect-free products, goodwill, strong brand loyalty, and higher chance of repeat purchase. Fulfilling customer demand through on-time delivery leads to competitiveness [13]. Today all major business activities, right from procurement of raw material to distribution of finished goods, mark delivery as a distinctive indicator of their performance. The delivery capability of a firm depends on factors like delivery speed, vehicle speed, delivery date and time [12, 13]. Considering delivery decision as a significant part of SC strategy leads to strategic and operational competitiveness.

Literature review emphasizes the significance in SC performance for competitiveness building. The mentioned performance indicators comprise of internal as well as external performance indicators. Executing these competitive priorities firms can realizes its business goals and objectives. Literature also unveiled the majority of research on SC competitiveness are aimed at competitive priorities [32, 33]. The targeted sectors were auto component firms [16, 17, 20, 22] and grocery and retail [5], PC manufacturing company [24]. But dynamic business environment calls for most up-to-date industry specific Key Performance Indicators [15]. Thus a realistic and easy-to-implement framework is crucial to any SC management activity.

2.2 Multi-Criteria Decision-Making Technique

The survey of literature unveiled a variety of MCDM methods which are universally recognized approaches for MCDM problems. Taking into account both outranking methods and multi attribute utility theory (MAUT), these methods are: analytic hierarchy process (AHP) [25], data envelopment analysis (DEA) [4], GRA [7], rough set approach (RSA) [23], and analytic network process (ANP) [26]. Choice among these to be used would depend on the type of available data, ease of understanding, and nature of the required decision.

Literature review unveiled that in real life business situations, business elements are dependent on each other. Unlike AHP, it is not compulsory to have such dependence in hierarchical form, where the lower level is dependent on the upper one [3, 19, 25]. Similarly, the technique like RSA, which simply classifies the attributes on the basis of 'if-then' decision rules, is least suited, especially for the task of prioritization in modern business environment [29]. Moreover, with information complexity in various business facets, the DEA technique falls short in dealing with fuzzy and imprecise information [8]. MCDM techniques like GRA, DEA, and RSA involves rigorous mathematical calculations which require that managers spend a great deal of time to learn and understand the functioning and implementation of a chosen approach. Thus, a technique which can lead to weights of individual attributes is the need of industrial decision makers.

An apparent technique which can overcome the limitations associated with the above mentioned MCDM techniques was the need of strategists and managers. Such a technique should be clear and precise and most importantly, it should be time bound and easy to implement by managers. To overcome such limitations of AHP, GRA, DEA, and RSA techniques, ANP technique was established. It addresses the issue of prioritization while considering the nature of interdependence.

3 Research Methodology

The current section presents an overview of the overall research approach adopted to discover the fact and accuracy behind the performance of competitive priorities for SC competitiveness.

3.1 Designing of Research Instrument

A well-structured questionnaire was prepared using sixteen performance indicators. These indicators are the dimensions of competitive priorities. For the ease of study symbols were assigned to each criterion. Refer column 3 of Table 2. Questionnaire was divided into two different sections. Section 1 contains a pairwise comparison matrices designed while Sect. 2 is composed of a large number of open-ended questions. Section I was designed on the basis of Satty 9-point scale (2001). The significance of numerical rating is mentioned in Table 1. Section 2 contained questions related to the cost, quality delivery flexibility, business category, and impact of business environment. Unlike few of the research studies, the present research uses open-ended questions to help in gathering the rationale behind the responses of pairwise comparison matrices. During the entire study, the dynamic nature of machine tool industry was discussed in detail with the respondents.

3.2 Profile of the Respondents and Responding Organization

In order to select the responding firms, the directory of India Machine Tool Manufacturers' Association was used. In all, 30 organizations were selected, of which only 10 showed their positive response to participate in research process.

Table 1 Satty 9-point scale [26]

Comparison scale	Verbal scale
1	Equal importance of both elements
3	Moderate importance of one element over another
5	Strong importance of one element over another
7	Very strong importance of one element over another
9	Extreme importance of one element over another
2, 4, 6, 8	Intermediate values

These firms have an annual turnover between 30 million USD to 25 billion USD. In India, these firms are listed in Bombay stock exchange (BSE) and National stock exchange (NSE). Approved firms were considered as the representative section of Indian MTM industry. Managers having minimum experience of 5 years with MTM industry were considered for administering the questionnaire. With the aim of concealing respondent's identity, the firms'/respondents names are not divulged in the paper.

3.3 ANP Technique of Prioritization

As defined by Saaty [26], "ANP is a theory of measurement generally applied to the dominance of influence among several stakeholders, or alternatives with respect to an attribute or a criterion." It is based on the theory of relative measurement. It allows prioritization without making assumptions about the dependence among considered set of elements. Generic steps followed for ANP implementation are: (a) carry out pairwise comparison between the criterions on the basis of the nature of being influenced and influencing, (b) obtain the super-matrix by calculating the weights from pairwise comparison matrices, (c) obtain limit super-matrix by matrix multiplication, and (d) analyze the results based on final assessment. For this research paper, MATLAB™ programming platform was used for ANP computations. Two different programs were written to obtain the final Priority Vector (PV). The first program helped to generate the eigenvectors. The second program was used to obtain the limit super-matrix.

Once the filled pairwise questionnaires were received, these were processed with the first program on the programming platform of MATLAB™ software suite. In all twenty-five eigenvectors were obtained. These eigenvectors were then arranged under the respective control criterion, to obtain the un-weighted super-matrix. The un-weighted super-matrix so obtained was normalized to obtain the weighted super-matrix. This ANP super-matrix reveals the local-priority information of the considered network by representing the overall impact of one criterion on a group of criteria and vice-versa. In this representation, the zeros in the matrix indicate nondependence. Positive numerical values indicate the strength of *being influenced and influencing*

Table 2 Priority vector for competitive priorities

Competitive priorities	Performance indicators	Symbols	Priority vector	Competitive priority contribution
Cost	Labour cost	CPC1	0.04112	0.183197
	Material cost	CPC2	0.04268	
	Manufacturing cost	CPC3	0.03761	
	Inventory cost	CPC4	0.03603	
	Distribution cost	CPC5	0.02575	
Delivery	Delivery dependability	CPD1	0.05312	0.096214
	Delivery speed	CPD2	0.04309	
Flexibility	Product mix	CPF1	0.08818	0.411823
	Volume flexibility	CPF2	0.07407	
	Design flexibility	CPF3	0.12387	
	New product development flexibility	CPF4	0.12571	
Quality	Durability	CPQ1	0.06056	0.308767
	Working condition and safety	CPQ2	0.0608	
	Environmental damage	CPQ3	0.05575	
	Defect rate	CPQ4	0.06274	
	Reliability	CPQ5	0.06892	

(dependence) the criteria by each other. Then, the second program was used to repeatedly multiply the weighted super-matrix by itself, until the entities in the matrix become regularized. The converged matrix so obtained is called a limit super-matrix. The limit super-matrix indicates the final priorities for all the considered elements. The values in any column of the limit super-matrix represent the priority vector for the responding firm.

4 Survey Results

The quantitative results of qualitative variables were obtained by using ANP approach. The priorities so obtained in form of PV elicit the relative weights of each performance indicator in overall SC competitiveness of Indian machine tool industry.

The PV unveiled that:

1. Flexibility (41 % weight) is the most important among all competitive priority. Herein, new product development and design flexibility drives the business growth. Flexibility indicators get affected by almost all cost and quality variables.
2. Quality being the major concern for all manufacturers carries a significant weight of 0.308767. It affects and gets affected by cost and flexibility variables. It gets hold of existing buyers and draws the newer customer demand.

3. Cost falls short to gain the highest contribution in overall SC competitiveness. It is because of the fact that cost advantage associated with labor, manufacturing, and material are higher than other European and American manufacturers.
4. Delivery as competitive priority contributed 10 % weight. Firms' ideal decision regarding both the delivery indicators reduces distribution cost, inventory cost, environmental damage, and enhances the volume flexibility and reliability whereby making Indian machine tool industry a competitive sector.

5 Discussion for Managerial Implications

The questionnaire-based survey of MTM firms of India was carried out. The obtained responses were analyzed to generate priority vectors. Study revealed that all the considered performance indicators are important for SC competitiveness of the sector under study. Based on firm's capability to harness such competitive priorities, the business strategies are planned and implemented [16, 17]. In the following points, the PV and the imperative strategies are discussed for managerial implication.

- Flexibility is found to be the key competitive priority for SC performance of Indian tool manufacturing companies. It is justified by dynamic demand of customers like auto and auto component manufacturers, capital goods industry, consumer durables, and aviation industry. Tool manufacturers are implementing strategies like customization, rapid response, and postponement. It is found that new product development and design flexibility is among the critical dimensions of flexibility. Due to variation in process technology at customers end, the large product mix drives the customer demand.
- The machine tool manufacturers deliver their finished goods to component manufacturers. Their quality in turn determines the production of quality equipment. Presently, Indian machine tool industry is manufacturing its products with ISO certification and international standard of quality/precision and reliability. Companies have entered into joint ventures and alliances with Swiss, German and Chinese counterparts to bring in the state-of-art technology. Computer numerically controlled (CNC) technology is among the most commonly used technique for quality improvement in tool manufacturing. For process improvement companies are executing soft strategies like 5S, Kaizen, and quality circles. Moreover, companies are utilizing third party services for trouble shooting, accurate measurement and adjustment for CNC machines, laser cutter, gas and oil turbine manufacturing, and many more.
- In dynamic business environment, customer responsiveness has become the prime measure of performance. Respondents found delivery as one of the component for fulfilling such responsiveness. Most of the companies were found to be dependable on third-party logistic providers for tools delivery. The advent of information technology had eased the logistic facility, for both national and international

customers. However, respondents mentioned delivery speed as an indicator dependent on other factors like infrastructure, plant location, and vehicle conditions.

- Insignificant variation was observed among the five cost dimensions. Due to major contribution of material cost in overall cost composition, manufacturers are struggling to reduce it. Respondents from studied firms believed that any re-engineering to reduce material cost leads to cost advantage with their immediate competitors. It can also be lowered by reducing the procurement cost and cost of product redesign. Labor cost was found to be the least significant. It is rationalized by the availability of low cost labor with Asian countries especially with India and China. Firms are implementing latest techniques like INVENTORIA to manage the inventory of small-sized tools. However, least effort can be exercised to reduce the inventory cost raised due to storage of heavy machines. It was also found that due to extreme locations of machine tool manufacturers distribution cost needs a greater attention in overall cost structure.

6 Conclusion and Avenues for Future Research

This study highlights the significance of competitive priorities in development of Indian machine tool industry. The findings will guide SC managers in aligning SC strategy with their firm's corporate strategy. It will also help strategy managers in improving their SC competitiveness in global market. Further studies with the aim of considering the impact of other performance indicators like business environment, buyer–supplier relationship, technology, location and infrastructure on competitive priorities are required for more comprehensive insights.

In contrast to the work done by Dangayach and Deshmukh [6] and Joshi et al. [16, 17] on complete Indian manufacturing companies and auto-component sector respectively, the presented work was targeted exclusively for machine tool manufacturing industry. Instead of considering the entire range of manufacturing strategy and related issues, herein the focus was kept on competitive priorities. Similarly, the survey conducted by Zhao [33] was entirely targeted toward the competitiveness of manufacturing enterprises in China. In order to study the SC competitiveness of Indian machine-tool manufacturers, few performance indicators were added. While studying the Indian strategies, it was found that they are rarely different from the ones implemented at China except for greater focus on quality enhancement parameters and cost reduction tactics. With the Indian market strategy of importing technologies from international market like Germany and Switzerland, specially related to CNC will lead to the place of Indian MTM industry in global canvas. Moreover, changing management style at public limited companies and concentration of Department of Heavy Engineering on private firms is boosting the competition. Surely this will lead to the development of Indian manufacturing industry as a whole. It was also found those small and middle scale tool manufacturing firms are still unorganized in their management and technological deals. Prospective research is suggested for the unorganized tool manufacturers in India.

References

1. Balakrishnan, K., Seshadri, S., Sheopuri, A., Iyer, A.: Indian auto-component supply chain at the crossroads. Interfaces **37**(4), 310–323 (2007)
2. Beamon, B.M.: Measuring supply chain performance. Int. J. Oper. Prod. Manage. **19**(3), 275–292 (1999)
3. Buyukyazien, M., Sueu, M.: The analytic hierarchy and analytic network processes. Hacettepe J. Math. Stat. **32**, 65–73 (2003)
4. Charnes, A., Cooper, W.W., Rhodes, E.: Measuring the efficiency of decision making units. Eur. J. Oper. Res. **2**, 429–444 (1978)
5. Chou, C.: Development of comprehensive supply chain performance measurement system: A case study in grocery retail industry. http://dspace.mit.edu/bitstream/handle/1721.1/29520/57308003.pdf (2004)
6. Dangayach, G.S., Deshmukh, S.G.: Evidence of manufacturing strategies in Indian industry: A survey. Int. J. Prod. Econ. **83**(3), 279–298 (2003)
7. Deng, J.L.: The introduction to grey system theory. J. Grey Syst. **1**(1), 1–24 (1989)
8. Dolabi, H., Radfar, R., Nasr, M.: Applied imprecise data envelopment analysis for selecting the efficient method of technology transfer: A case study in automotive parts manufacturing industry corporation. Contemp. Eng. Sci. **4**(1), 13–24 (2011)
9. Edwards, L., Golub, S.: South Africa's international cost competitiveness and exports in manufacturing. World Dev. **32**(8), 1323–1339 (2004)
10. Fuss, M.A., Waverman, L.: Cost and productivity in automobile production: The challenges of Japanese efficiency, Cambridge (2006)
11. Garvin, D.A.: Competing on the eight dimension of quality. Harv. Bus. Rev. **65**(6), 101–109 (1987)
12. Gunasekaran, A., Kobu, B.: Performance measures and metrics in logistics and supply chain management: A review of recent literature (1995–2004) for research and applications. Int. J. Prod. Res. **45**(12), 2819–2840 (2006)
13. Gunasekaran, A., Patel, C., McGaughey, R.E.: A framework for supply chain performance measurement. Int. J. Prod. Econ. **87**(3), 333–347 (2004)
14. Ho, D.C.K., Au, K.F., Newton, E.: Empirical research on supply chain management: A critical review and recommendations. Int. J. Prod. Res. **40**(17), 4415–4430 (2002)
15. Jamil, C.M., Mohamed, R.: Performance measurement system (PMS) in small and medium enterprises (SMES): A practical modified framework. World J. Soc. Sci. **1**(3), 200–212 (2011)
16. Joshi, D., Nepal, B., Rathore, A.P.S., Sharma, D.: On supply chain competitiveness of India automotive component manufacturing industry. Int. J. Prod. Eco. **143**(1), 151–161 (2013)
17. Joshi, D., Rathore, A.P.S., Sharma, D., Nepal, B.: Determinants of competitiveness and their relative importance: A study of Indian auto-component industry. Int. J. Serv. Oper. Manage. **10**(4), 426–448 (2011)
18. Majumdar, S.: How do they plan for growth in auto component business? - A study on small foundries of western India. J. Bus. Venturing **25**(3), 274–289 (2010)
19. Navarro, T.G., Melon, M.G., Martin, D.D., Dutra, S.A.: Evaluation of urban development proposals: An ANP approach. World Acad. Sci. Eng. Tech. **44**, 498–508 (2008)
20. Neely, A., Mills, J., Platts, K., Richard, H., Gregory, M., Bourne, M., Kennerley, M.: Performance measurement system design: Developing and testing a process-based approach. Int. J. Oper. Prod. Manage. **20**(10), 1119–1145 (2000)
21. Neely, A.D., Gregory, M., Platts, K.: Performance measurement system design: A literature review and research agenda. Int. J. Oper. Prod. Manage. **15**(4), 80–116 (1995)
22. Olugu, E.U., Wong, K.Y., Shaharoun, A.M.: A comprehensive approach in assessing the performance of an automobile closed-loop supply chain. Sustainability **2**(4), 871–889 (2010)
23. Pawlak, Z.: Rough sets. Int. J. Comput. Inf. Sci. **11**(5), 341–356 (1982)
24. Ravi, V., Shankar, R., Tiwari, M.K.: Analyzing alternative in reverse logistics for end-of-life computers: ANP and balanced scorecard approach. Comput. Ind. Eng. **48**(2), 327–356 (2005)

25. Saaty, T.L.: The Analytic Hierarchy Process. New York (1980)
26. Saaty, T.L.: Decision Making with Dependence and Feedback: The Analytic Network Process. 2nd edn. Pittsburgh (2001)
27. Sangwan, K.S., Digalwar, A.K.: Evaluation of world-class manufacturing systems: A case of Indian automotive industries. Int. J. Serv. Oper. Manage. **4**(6), 687–708 (2008)
28. Singh, R.K., Garg, S.K., Deshmukh, S.G.: Strategy development for competitiveness: A study on Indian auto component sector. Int. J. Prod. Perform Manage. **56**(4), 285–304 (2007)
29. Tzeng, G.-H., Huang, J.-J.: Multiple Attribute Decision Making: Methods and Applications. CRC Press, USA (2011)
30. Upton, D.: The management of manufacturing flexibility. California Manage. Rev. **36**(2), 72–89 (1994)
31. Van Dierdonck, R., Miller, J.G.: Designing production planning and control systems. J. Oper. Manage. **1**(1), 37–46 (1980)
32. Ward, P.T., Duray, R.: Manufacturing strategy in context: Environment, competitive strategy and manufacturing strategy. J. Oper. Manage. **18**(6), 123–138 (2000)
33. Zhao, X., Yeung, J.H.Y., Zhou, O.: Competitive priorities of enterprises in mainland China. Total Qual. Manage. **13**(3), 285–300 (2002)

A Modified Variant of RSA Algorithm for Gaussian Integers

Sushma Pradhan and Birendra Kumar Sharma

Abstract A Gaussian prime is a Gaussian integer that cannot be expressed in the form of the product of other Gaussian integers. The concept of Gaussian integer was introduced by Gauss [4] who proved its unique factorization domain. In this paper, we propose a modified RSA variant using the domain of Gaussian integers providing more security as compared to the old one.

Keywords RSA public-key cryptosystem · Gaussian integers · Multiprime RSA

1 Introduction

RSA system is the one of most practical public-key password systems. In addition to other domain, it has successfully provided security to the electronic-based commerce. Encryption of plaintext in asymmetric key encryption is based on a public key and a corresponding private key. Document authentication and digital signature are other advantages of RSA public-key cryptosystem. RSA provides security to the plaintext based on factorization problem [5]. There are PKCs other than RSA. Those are Elgamal and Rabin's PKCs. These PKCs provide security-based discrete logarithm problem.

The classical RSA cryptosystem is described in the setting of the ring Zn, the ring of integers modulo a composite integer $n = pq$, where p and q are two distinct odd prime integers. Many aspects of arithmetics over the domain of integers can be carried out to the domain of Gaussian integers $Z[i]$, the set of all complex numbers of the form $a + bi$, where a and b are integers [6]. The RSA cryptosystem was extended

S. Pradhan (✉) · B. K. Sharma
School of Studies in Mathematics, Pt. Ravishankar Shukla University, Raipur, India
e-mail: sushpradhan@gmail.com

B. K. Sharma
e-mail: shramabk07@gmail.com

B. V. Babu et al. (eds.), *Proceedings of the Second International Conference on Soft Computing for Problem Solving (SocProS 2012), December 28–30, 2012*, Advances in Intelligent Systems and Computing 236, DOI: 10.1007/978-81-322-1602-5_20,

domain of Gaussian integers in the papers [2] and [3]. In [2] and [3], the advantages of such extension of RSA were briefly stated in these papers.

Now in this paper, another fast variants of RSA cryptosystems is proposed using arithmetic's modulo of Gaussian integers. Proposed scheme provides more security with same efficiency. Before doing so, in next section, we review the classical RSA PKC. Next, we introduced Gaussian integers and its properties in Sect. 3. In Sect. 4, we present a variant of RSA scheme based on factorization of Gaussian integers with a suitable example. Finally, we conclude with security analysis and comparison with the standard method.

2 Classical RSA Public-Key Cryptosystem

The classical RSA cryptosystem is described as follows: entity A generates the public key by first generating two large random odd prime integers' p and q, each roughly of the same size. Then, entity A computes the modulus $n = pq$ and $\phi(n) = (p-1)(q-1)$, where ϕ is Euler's phi function. Next, entity A selects the encryption exponent e to be any random integer in the interval $(1, \phi(n))$, and which is relatively prime to $\phi(n)$. Using the extended Euclidean algorithm for integers, entity A finds the decryption exponent d, which is the unique inverse of e in Zn. The public key is the pair (n, e) and A's private key is the triplet (p, q, d).

To encrypt a message, entity B first represents the message as an integer m in Zn. Then, entity B obtains A's public-key (n, e), uses it to compute the cipher text $c = m^e (\bmod n)$, and sends c it to entity A. Now, to decrypt c, entity A computes $m = c^d (\bmod n)$ and recovers the original message m.

3 Gaussian Integers

Gaussian integer is a complex number $a + bi$ where both a and b is integers:

$Z[i] = a + bi : a, b \in Z$. Gaussian integers, with ordinary addition and multiplication of complex numbers, form an integral domain, usually written as $Z[i]$. The norm of a Gaussian integer is the natural number defined $|a + bi| = a^2 + b^2$.

Gaussian primes are Gaussian integer's $z = a + bi$ satisfying one of the following properties:

1. If both a and b are nonzero, then $(a + bi)$ is a Gaussian prime iff $(a^2 + b^2)$ is an ordinary prime.
2. If $a = 0$, then bi is a Gaussian prime iff $|b|$ is an ordinary prime and $|b| = 3 \pmod 4$.
3. If $b = 0$, then a is a Gaussian.

J.T. Cross [1] gave a full description for complete residue systems modulo prime powers of Gaussian integers.

4 RSA Algorithms Over the Field of Gaussian Integers

In paper [2], the RSA is extended into the field of Gaussian integers. It is presented as follows:

Key Generation: Generate two large Gaussian primes P and Q. Compute $N = PQ$. Compute ϕ (N) $= (|p| - 1)(|q| - 1)$. Select a random integer e such that $1 < e < \phi(\mathrm{N})$ and gcd (e, $\phi(\mathrm{N}) = 1$). Compute $d = e^{-1} \bmod \phi(N)$. Pair N and e is a public key, and d is the private key.

Encryption: Given a message M (represented as a Gaussian integer), compute

ciphertext

$$C := m^e (\bmod n).$$

Decryption: Compute the original message $M := c^d (\bmod n)$.

Now, we propose the algorithms for the variant of RSA cryptosystem in $Z[i]$ as below:

Key Generation:

Generate b distinct large Gaussian primes α, β, and γ each n/b bits long.
Compute $N = \alpha\beta\gamma$.
Compute ϕ (N) $= (|\alpha| - 1)(|\beta| - 1)(|\gamma| - 1)$.
Select a random integer e such that $1 < e < \phi(N)$ and gcd , (e, ϕ $(N) = 1$)
Compute $d = e^{-1} \bmod \phi(N)$.
Pair (N, e) is a public key, and $(\alpha, \beta, \gamma, d)$ is the private key.

Encryption:

Given a message M (represented as a Gaussian integer) compute cipher text

$$C = m^e (\bmod N)$$

Decryption:

Compute the original message $M = c^d (\bmod N)$.
Following is the example in support of proposed algorithm.

Example

Key generation

Let us select $\alpha = 19$ and $\beta = 5$ and $\gamma = 3$,
Compute the product $N = \alpha\beta\gamma = 285$,

$$\phi(\mathrm{N}) = (19^2 - 1)(5^2 - 1)(3^2 - 1) = 69,120.$$

Choose $e = 3{,}331$.
Then, $d = e^{-1} \bmod \phi(N) = 3331^{-1} \bmod 69{,}120 = 29{,}611$
The public key is $n = 285$, $e = 3{,}331$

Encryption
Let message $M = m_1; m_2 = (555, 444)$

$$\begin{aligned} C = (c_1; c_2) &= m^e (\mathrm{mod}\, N) \\ &= (555, 444)^{3331} \,\mathrm{mod}\, 285 \\ &= (270, 159) \end{aligned}$$

Decryption

$$\begin{aligned} M &= c^d (\mathrm{mod}\, N) \\ &= (270, 159) \,\mathrm{mod}\, 285 \\ &= (555, 444) \end{aligned}$$

5 Security Analyses

The comparison of the classical RSA [7], its Gaussian integer domain in $Z\,[i]$ [2], and our proposed scheme is as follows:

- The generation of primes p, q in classical scheme and Gaussian primes a, b in $Z[i]$ require the same amount of computation. Same in the case with our proposed scheme where an additional prime g in the form of $4k + 3$ would be generated with the same computation.
- The modified Gaussian variant provides more security than the classical method since the number of elements which are chosen to represent the message m is about square of those used in the classical case. scheme would provide security as compared to Gaussian variant, because domain $Z\,[i]$ in our proposed scheme provides a more extension to the range of chosen messages, which make trails more complicated as compared to the Gaussian integer domain [2].
- In [2], Euler's phi function is $\phi = (p^2 - 1)(q^2 - 1)$, whereas in proposed scheme, it is $\phi = (\alpha^2 - 1)(\beta^2 - 1)(\gamma^2 - 1)$. This make the attempt to find the private key d from the public key more complicated as compared to the Gaussian variant [2] in $Z\,[i]$. Thus, our proposed scheme provides more security than the [2]. More so, the computations involved in the Gaussian variant do not require computational procedures different from those of the classical method. Same would be the case with our scheme.
- It is noted that the complexity for programs depends on the complexity of generating the public key. Thus, the classical and proposed algorithms are equivalent since their public-key generation algorithms are identical when restricting the choice of primes to those of the form $4k + 3$. However, our scheme is recommended since it provides a better extension to the message space and the public exponent range as compared to classical one.

6 Conclusion

We modify the computational methods in the domain of Gaussian integers. Lastly, we show how the modified computational methods can be used to extend the RSA algorithm to the domain $Z[i]$. Also, we show that the modified algorithm requires a little additional computational effort than the classical one and accomplishes much greater security.

References

1. Cross, J.T.: The Eulers f -function in the Gaussian integers. Amer. Math. **55**, 518–528 (1995)
2. Elkamchouchi, H., Elshenawy, K., Shaban, H.: Extended RSA cryptosystem and digital signature schemes in the domain of Gaussian integers. The 8th International Conference on Communication Systems, vol. 1 (ICCS'02), pp. 91–95 (2002).
3. El-Kassar, A.N., Haraty, R., Awad, Y., Debnath, N.C.: Modified RSA in the domains of Gaussian integers and polynomials over finite fields. In: Proceedings of international conference computer applications in industry and engineering–CAINE, pp. 298–303 (2005).
4. Gauss, C.F.: Theoria residuorum biquadraticorum. Comm. Soc. Reg. Sci. Gottingen **7**, 1–34 (1832)
5. Menezes, A., Van Oorshot, J., Vanstone, P.C.S.A.: Handbook of Applied Cryptography. CRC Press, Boca Raton, FL (1997)
6. Niven, I., Zukerman, H.S., Montegomery, H.L.: An Introduction to the Theory of Numbers. John Wiley, New York (1991)
7. Rivest, R., Shamir, A., Adleman, L.: A method for obtaining digital signatures and public key cryptosystems. Commun. ACM **21**, 120–126 (1978)

Neural Network and Statistical Modeling of Software Development Effort

Ruchi Shukla, Mukul Shukla and Tshilidzi Marwala

Abstract Many modeling studies that aimed at providing an accurate relationship between the software project effort (or cost) and the involved cost drivers have been conducted for effective management of software projects. However, the derived models are only applicable for a specific project and its variables. In this chapter, we present the use of back-propagation neural network (NN) to model the software development (SD) effort of 18 SD NASA projects based on six cost drivers. The performance of the NN model was also compared with a multi-regression model and other models available in the literature.

Keywords Neural network · Software development · Effort estimation · Regression

R. Shukla (✉)
Department of Electrical and Electronic Engineering Science, University of Johannesburg, Johannesburg, South Africa
e-mail: ruchishuklamtech@gmail.com

M. Shukla
Department of Mechanical Engineering Technology, University of Johannesburg, Johannesburg, South Africa

M. Shukla
Department of Mechanical Engineering, MNNIT, Allahabad, UP, India
e-mail: mukulshukla2k@gmail.com

T. Marwala
Faulty of Engineering and Built Environment, University of Johannesburg, Johannesburg, South Africa
e-mail: tmarwala@uj.ac.z

B. V. Babu et al. (eds.), *Proceedings of the Second International Conference on Soft Computing for Problem Solving (SocProS 2012), December 28–30, 2012*, Advances in Intelligent Systems and Computing 236, DOI: 10.1007/978-81-322-1602-5_21,

1 Introduction

Software companies today are outsourcing a wide variety of their jobs to offshore organizations, for maximizing returns on investments. Estimating the amount of effort, time, and cost required for developing any information system is a critical project management issue. In view of the above, long-term, credible, and optimum forecast of software project estimates in the early stages of a project's life cycle is an almost intractable problem. Often, key information of real-life projects regarding size, complexity, system documentation, vocabulary, annual change traffic, client attitude, multilocation teams, etc. is unavailable. In spite of the availability of more than 100 estimation tools in the market, experience-based reasoning still remains the commonly applied estimation approach owing to some fundamental estimation issues which software developers have struggled with [1].

2 Literature Review

A review of studies on expert estimation of SD effort was presented by [2, 3]. An exploratory analysis of the state of the practice on schedule estimation and software project success prediction is presented in [4]. It was found that the data collection approach, role of respondents, and analysis type had an important impact on software estimation error [5]. Soft-computing- or artificial intelligence (AI)-based approaches are of late being used for more accurate prediction of software effort/cost. Artificial neural networks (ANN) offer a powerful computing architecture capable of learning from experimental data and representing complex, nonlinear, multivariate relationships [6, 7]. Kumar et al. compared the effectiveness of the variants of wavelet neural network (WNN) with many other techniques to forecast the SD effort [8]. Genetic algorithms (GAs) were used for the estimation of COCOMO model parameters of NASA SD projects in [9] while different fuzzy logic-based studies have been conducted [10–12]. Many hybrid schemes (neuro-GA, neuro-fuzzy, grey-GA, fuzzy-grey, etc) have also been investigated [13–15]. Many studies on software prediction have focused on the development of regression models based on historical data [16, 17].

3 Statistical Modeling

This modeling study is based on the SD effort dataset of Bailey and Basili [18] (Table 1—shown partly for brevity reasons). The six input factors are the total lines of code, new lines of code, developed lines of code (DL) (all in kloc), total methodology (ME), cumulative complexity, and cumulative experience, and the output is effort (in man months). Preliminary statistical analysis of the dataset was conducted

Table 1 SD effort dataset [18]

Project no.	Project attributes						Response
	Total lines (kloc)	New lines (kloc)	Developed lines (kloc)	Total methodology	Cumulative complexity	Cumulative experience	Effort (man months)
1	111.9	84.7	90.2	30	21	16	115.8
2	55.2	44	46.2	20	21	14	96
–	–	–	–	–	–	–	–
17	14.8	11.9	12.5	27	23	18	23.9
18	110.3	98.4	100.8	34	33	16	138.3
Correlation Coefficient	0.94	0.97	0.96	0.03	0.65	-0.02	
Covariance	1593.7	1319.3	1456.3	6.98	134.4	−2.82	
Kurtosis	−1.25	−0.38	−0.73	−0.83	−0.27	3.55	−1.26
R-Square	0.88	0.95	0.92	0.00	0.42	0.00	

beforehand including the following: (1) correlation coefficient, (2) covariance, (3) kurtosis, and (4) R-square as presented in Table 1. Initially, from Minitab [19]-based ANOVA, a multivariable linear regression model (Eq. 1) has been fitted. The goodness of this developed model is validated with two other models (Eqs. 2 and 3) given by Sheta and Al-Afeef [15] in Table 2. Based on the high T (or low P) values, the following ranking (in a decreasing order) of the 6 effort drivers has been established: (1) methodology, (2) new LoC, (3) total LoC, (4) cumulative experience, (5) developed LoC, and (6) cumulative complexity. The high R-squared value of 98.3 % and R-Sq(adjusted) values of 97.4 % justify the correctness of the ANOVA.

$$\begin{aligned}\text{Effort} = 41.6 + 0.314\ Tot\_LoC + 0.986\ New\_LoC + 0.116\ Develop\_LoC \\ - 1.57\ Meth - 0.112\ Cum\_Complex + 0.376\ Cum\_Exper\end{aligned} \tag{1}$$

$$E = 1.75992 \times DL - 4.56 \times 10^{-3} \times DL^2 \tag{2}$$

$$E = 2 \times DL - 0.59 \times 10^{-3} ME^2 \times DL \tag{3}$$

The main effect plots for the 6 effort drivers are shown in Fig. 1.

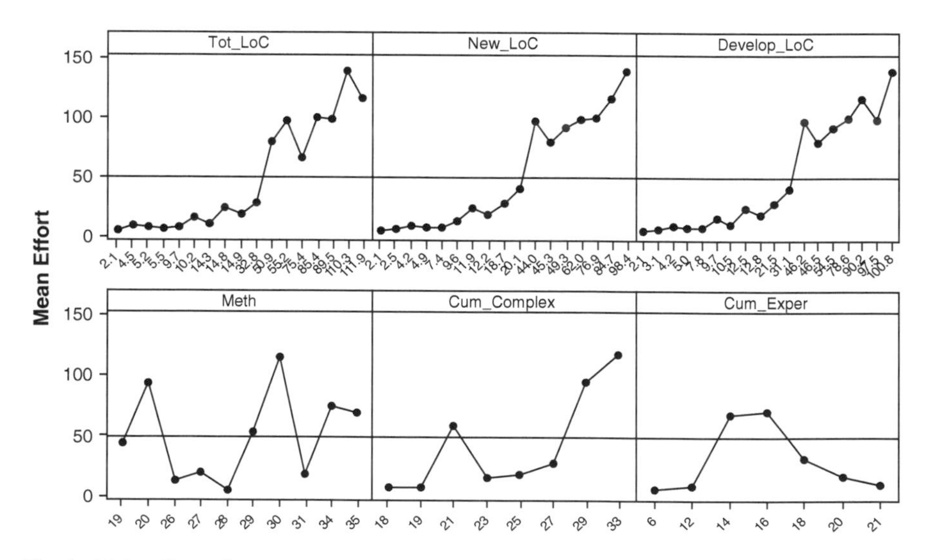

Fig. 1 Main effects plot

4 Neural Network Modeling

Back-propagation (BP) NN modeling for effort estimation has been carried out in this work using the MATLAB (2007b) NN toolbox options. Initially, a simple two-layer BP (6-6-1) NN was employed. The number of hidden nodes in the hidden layer was kept equal to the number of inputs (6 here). The number of hidden neurons was then suitably increased in an orderly hit and trial manner, to decide the final structure of the NN by keeping a check on the convergence rate of training, testing, and validation errors as well as the average percentage error. The learning rate and momentum can also be adjusted for the above purpose (although not varied in the present work).

Before the network is made ready to make estimates, we input the combinations of data inputs and outputs [18] through the network for training (60 %), validation (20 %), and testing (20 %). In our case, the activation functions of both the hidden and output layers were initially chosen to be tan-sigmoid. The same was later changed to the purelin(ear) function in the output layer. We used the two most popular training algorithms i.e., the Levenberg-Marquardt (LM) and the Bayesian regularization (BR) algorithms. The training performance and linear regression analysis (between the network outputs and the corresponding targets) are shown in Figs. 2 and 3. For the LM algorithm, the output tracks the targets reasonably well, and the regression coefficient (R) value is over 0.97 mostly. Similarly, for the BR algorithm-based training with purelin output function, the R values are over 0.99 in nearly all the cases (Fig. 3).

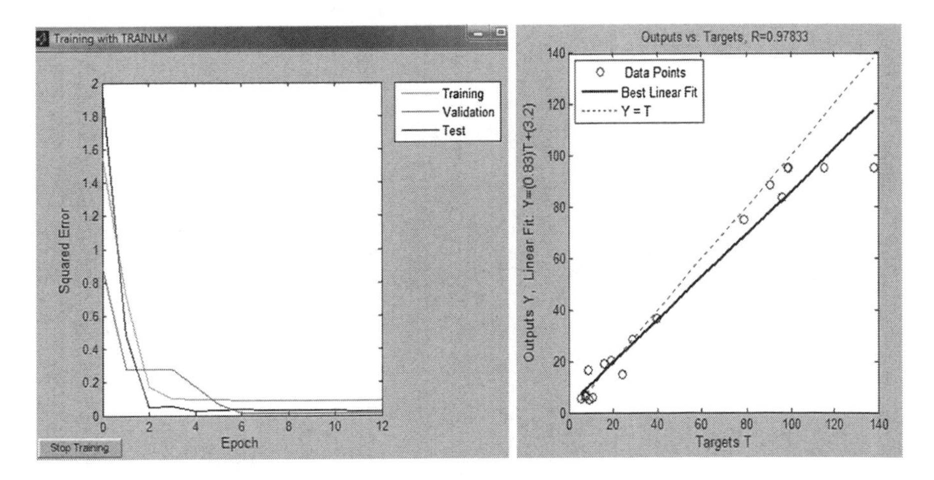

Fig. 2 Levenberg-Marquardt training with tansigmoid function in output layer

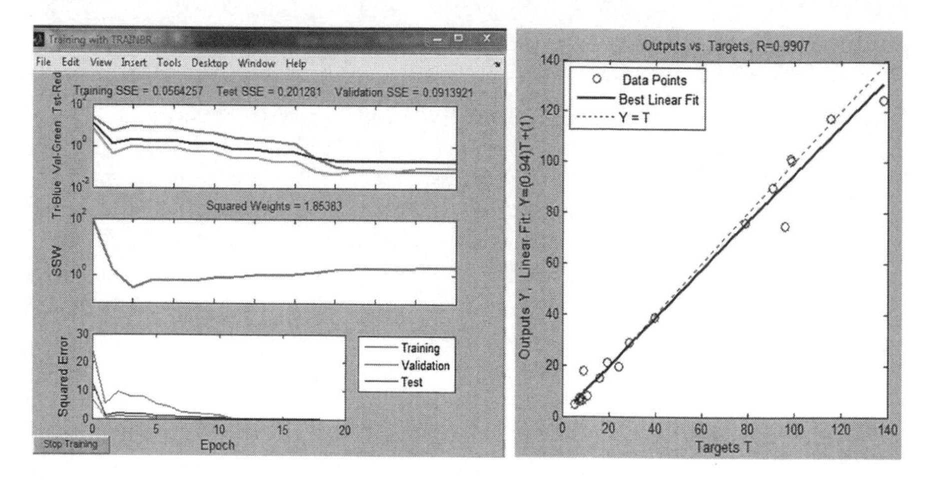

Fig. 3 Bayesian regularization training with purelin(ear) function in output layer

4.1 NN Modeling Tips

Listed underneath are some practical tips for efficient NN modeling.

- NNs are rather sensitive to the number of neurons in the hidden layers. Too few neurons often lead to underfitting, while too many neurons can contribute to overfitting. In this case, inspite of all the training points being well fitted, the fitted curve oscillates largely between these points [20].
- The NN dataset is generally divided in the following ratios: training (50–60), validation (20–25), and testing (20–25).
- Learning rate (alpha) represents how quickly an NN learns ranges from 0 to 1 and is initialized randomly. As with linear networks, a learning rate that is too large leads to unstable learning. Contrarily, a too small learning rate results in much longer training times. Typical values are 0.01–0.05.
- Momentum is a variable, which helps NN to break out of local minima. It may range from 0 to 1. Typical values are around 0.5.
- The threshold function (logsig, tansig, purelin, etc) selection is critical and determines when a node fires propagating a value further through the network. The choice is essentially based on the range and sign of inputs/outputs.
- The BR algorithm which is a modification of the LM algorithm is often used, as it generalizes well and reduces the difficulty of determining the optimum network architecture.
- LM training would normally be used for small- and medium-size networks, if enough memory is available. If memory is a problem, then there are a variety of other fast algorithms available. For large networks, one would probably want to use trainscg (conjugate gradient) or trainrp (resilient BP) algorithms.

- Overfitting is one of the most common problems that occurs in NN training. The training set error becomes a very small value, but the error turns to a large value when new data are presented to the network. An attempt at collecting more data and increasing the size of the training set must be made to prevent the situation of overfitting [20].
- One suggested method to improve network generalization is the use of a just large enough network that provides an adequate fit. Larger is the network used, more complex can be the functions the network can create. A small enough network will not have enough power to overfit the data. The two methods for improving generalization and implemented in MATLAB NN toolbox are regularization and early stopping [20].

5 Results and Discussion

The degree to which a model's estimated effort (MM_{est}) matches the actual or target effort (MM_{act}) is estimated by a percentage relative error. Magnitude of relative error (MRE), which accounts for under and overestimates along with its mean magnitude of relative error (MMRE) is often used in effort estimation analysis.

$$MRE = \left| \frac{MM_{act} - MM_{est}}{MM_{act}} \right| \tag{4}$$

Table 2 (in brief) presents a comparison of the empirical models (Eqs. 1–3) fitted effort and NN effort (for different configurations) with the target effort of [19]. It can be concluded that the present NN framework is able to successfully model the dataset with nearly the following percentage relative error and percentage mean relative error:

1. -10.58 to 8.36 and 4.68 %, respectively, for trainlm with tansig function in output layer and hidden neurons varied from 6 to 20.
2. -12.5 to -9.62 and -10.3 %, respectively, for trainbr with tansig function in output layer and hidden neurons varied from 6 to 20.
3. 0.65 to -3.12 and 0.79 %, respectively, for trainbr with purelin function in output layer and hidden neurons varied from 6 to 20.

The relative error obtained from the developed multi-regression model (Eq. 1) is comparable to other models (Eqs. 2 and 3). A comparison between the mean relative error of the developed NN and regression models and the MMRE of Halstead, Walston-Felix, Bailey-Basili, and Doty models are shown in Table 3 [16].

Table 2 Comparison of empirical model fitted and NN effort with target effort

Project number	Target effort [18]	Predicted effort (Eq. 1)	Percentage error = (1-C3/C2) *100	Predicted effort (Eq. 2)	% error = (1-C5/C2) *100	Percentage error (Predicted effort from Eq. 3)	Percentage error NN output		
							LM	BR tansig	purelin
1	115.8	127.28	−9.91	121.64	−5.05	−14.42	17.8	19.2	−1.1
—									
6	98.4	98.63	−0.23	128.24	−30.33	−49.01	9.0	−54.2	−2.8
—									
18	138.3	133.89	3.19	131.07	5.23	3.94	31.2	31.9	10.0
		Percentage mean error	−5.41		−8.66	−2.76	4.68	−10.3	0.79

Table 3 Comparison of different models

Model name	Model equation	MMRE
Halstead	$E = 5.2(\text{DL})1.50$	0.1479
Walston-Felix	$E = 0.7(\text{DL})0.91$	0.0822
Bailey-Basili	$E = 5.5 + 0.73(\text{DL})1.16$	0.0095
Doty (for DL > 9)	$E = 5.288(\text{DL})1.0$	0.1848
		Mean relative error
Present work—(1) LM-based BPNN	–	0.0468
(2) BR-based BPNN – 1		−0.1030
(3) BR-based BPNN – 2		0.0079
(4) Multilinear regression	Eq. 1	−0.0541

6 Conclusions

Effort estimation is a complex task, and research studies indicate that results in general vary a lot. The market potential for SD and maintenance is huge and constantly growing mainly for financial and online applications. In this work, a twofold approach based on NN and multilinear regression has been carried out for more accurate SD effort estimation

Acknowledgments The authors would like to acknowledge the financial support extended by the Faulty of Engineering and Built Environment, University of Johannesburg.

References

1. Shukla, R. Misra, A.K.: Estimating software maintenance effort - a neural network approach. 1st India, Software Engineering Conference. 107–112 (2008).
2. Jorgensen, M.: A review of studies on expert estimation of software development effort. J Syst. Software. **70**(1–2), 37–60 (2004)
3. Jorgensen, M., Shepperd, M.: A systematic review of software development cost estimation studies. IEEE T. Software Eng. **33**(1), 33–53 (2007)
4. Verner, J.M., Evanco, W.M., Cerpa, N.: State of the practice: an exploratory analysis of schedule estimation and software project success prediction. Inform. Software Tech. **49**(2), 181–193 (2007)
5. Jorgensen, M., Ostvold, K.M.: Reasons for software effort estimation error: impact of respondent role, information collection approach, and data analysis method. IEEE T. Software Eng. **30**(12), 993–1007 (2004)
6. Huang, X., Ho, D., Ren, J., Capretz, L.F.: A soft computing framework for software effort estimation. Soft Comput. **10**, 170–177 (2006)
7. Tronto, I.F.B., Silva, J.D.S., Anna, N.S.: An investigation of artificial neural networks based prediction systems in software project management. J Syst. Software. **81**(3), 356–367 (2008)
8. Kumar, K.V., Ravi, V., Carr, M., Kiran, N.R.: Software development cost estimation using wavelet neural networks. J Syst. Software. **81**(11), 1853–1860 (2008)
9. Sheta, A.: Estimation of the COCOMO model parameters using genetic algorithms for NASA software projects. J. Comput. Sci. **2**(2), 118–123 (2006)

10. Xu, Z.W., Khoshgoftaar, T.M.: Identification of fuzzy models of software cost estimation. Fuzzy Set. Syst. **145**(1), 141–163 (2004)
11. Ahmed, M., Saliu, M.O., Alghamdi, J.: Adaptive fuzzy logic-based framework for software development effort prediction. Inform. Software Tech. **47**(1), 31–48 (2005)
12. Kazemifard, M., Zaeri, A., Ghasem-Aghaee, N., Nematbakhsh, M.A., Mardukhi, F.: Fuzzy emotional COCOMO II software cost estimation (FECSCE) using multi-agent systems. Appl. Soft Comput. **11**(2), 2260–2270 (2011)
13. Shukla, K.K.: Neuro-genetic prediction of software development effort. Inform. Software Tech. **42**, 701–713 (2000)
14. Huang, S.J., Chiu, N.H.: Optimization of analogy weights by genetic algorithm for software effort estimation. Inform. Software Tech. **48**, 1034–1045 (2006)
15. Sheta, A. F., Al-Afeef, A.: A GP effort estimation model utilizing line of code and methodology for NASA software projects. 10th International Conference on Intelligent Systems Design and Applications. 290–295 (2010).
16. Jorgensen, M.: Regression models of software development effort estimation accuracy and bias. Empir. Softw. Eng. **9**, 297–314 (2004)
17. Shukla, R., Misra, A.K.: Software maintenance effort estimation-neural network vs regression modeling approach. Int. J. Comput. Applic. **1**(29), 83–89 (2010)
18. Bailey, J.W., Basili, V.R.: A metamodel for software development resource expenditures. 5th IEEE International Conference on, Software Engineering. 107–116 (1981).
19. www.minitab.com (2012).
20. www.mathworks.com/access/helpdesk/help/pdf_doc/nnet/nnet.pdf (2012).

On α-Convex Multivalent Functions Defined by Generalized Ruscheweyh Derivatives Involving Fractional Differential Operator

Ritu Agarwal and J. Sokol

Abstract In the present investigation, we introduce a class of α-convex multivalent functions defined by generalized Ruscheweyh derivatives introduced by Goyal and Goyal (J. Indian Acad. Math. 27(2):439–456, 2005) which involves a generalized fractional differential operator. The necessary and sufficient condition for functions to belong to this class is obtained. We study properties of this class and derive a theorem about image of a function from this class through generalized Komatu integral operator. Also, the integral representation for the functions of this class has been obtained.

1 Introduction

Let A denote the class of analytic p-valent functions defined on unit disk $U = \{z : |z| < 1\}$ of the form

$$f(z) = z^p + \sum_{k=p+1}^{\infty} a_k z^k. \tag{1}$$

The function $f(z) \in A$ is said to be *p-valent starlike of order* δ if and only if

$$\mathrm{Re}\left(\frac{zf'(z)}{f(z)}\right) > \delta, \ (z \in U;\ 0 \le \delta < p) \tag{2}$$

R. Agarwal (✉)
Malaviya National Institute of Technology, Jaipur, Rajasthan, India
e-mail: ritugoyal.1980@gmail.com

J. Sokol
Institute of Mathematics, University of Rzeszów, ul. Rejtana 16A, 35-310 Rzeszów, Poland
e-mail: jsokol@prz.edu.pl

B. V. Babu et al. (eds.), *Proceedings of the Second International Conference on Soft Computing for Problem Solving (SocProS 2012), December 28–30, 2012*, Advances in Intelligent Systems and Computing 236, DOI: 10.1007/978-81-322-1602-5_22,

The class of starlike functions of order δ is denoted by S^*_δ. On the other hand, a function $f(z) \in A$ is said to be *p-valent convex of order* δ if and only if

$$\operatorname{Re}\left(1 + \frac{zf''(z)}{f'(z)}\right) > \delta, \quad (z \in U; 0 \le \delta < p). \tag{3}$$

The class of convex functions of order δ is denoted by K_δ. It is observed that $f(z) \in K_\delta \Leftrightarrow zf'(z) \in S^*_\delta$.

For $\alpha \in [0, 1]$, let $K_\delta(\alpha)$ denote the family of functions $f(z) \in A$ with $f'(z)f(z)/z \neq 0$ in U such that

$$\operatorname{Re}\left\{(1-\alpha)\frac{zf'(z)}{f(z)} + \alpha\left(\frac{(zf'(z))'}{f'(z)}\right)\right\} > \delta, \quad (z \in U) \tag{4}$$

The functions in class $K_\delta(\alpha)$ are said to be α-convex functions of order δ (see e.g. [1, 2]). We recall the definition of subordination ([3], p. 190). For two functions f and g analytic in U, we say that $f(z)$ is subordinate to $g(z)$ in U and write $f \prec g$ or $f(z) \prec g(z)$, if there exists a Schwarz function $w(z)$, analytic in U with $w(0) = 0$ and $|w(z)| < 1$ such that $f(z) = g(w(z))$, $z \in U$.

The generalized Ruscheweyh derivative operator $\mathscr{J}_p^{\lambda,\mu}$ defined by the author (see e.g. [5, 10]) is given by

$$\mathscr{J}_p^{\lambda,\mu} f(z) = \frac{\Gamma(\mu - \lambda + \nu + 2)}{\Gamma(\nu + 2)\Gamma(\mu + 1)} z^p J_{0,z}^{\lambda,\mu,\nu}(z^{\mu-p} f(z)) \tag{5}$$

where the fractional differential operator $J_{0,z}^{\lambda,\mu,\nu}$ is given by (see e.g. [16])

$$J_{0,z}^{\lambda,\mu,\nu} f(z) = \begin{cases} \frac{1}{\Gamma(1-\lambda)} \frac{d}{dz} \left\{ z^{\lambda-\mu} \int_0^z (z-\zeta)^{-\lambda} \right. \\ \left. \times {}_2F_1\left(\mu-\lambda, -\nu; 1-\frac{\zeta}{z}\right) f(\zeta) d\zeta \right\}, (0 \le \lambda < 1) \\ \frac{d^n}{dz^n} J_{0,z}^{\lambda-n,\mu,\nu} f(z), \ (n \le \lambda < n+1, n \in N) \ (k > \max\{0, \mu-\nu-1\} - 1) \end{cases} \tag{6}$$

provided further that

$$f(z) = O(|z|^k), \ (z \to 0)$$

The generalized Ruscheweyh derivative of $f \in A$, introduced by Goyal and Goyal [5], is defined as

$$\mathscr{J}_p^{\lambda,\mu} f(z) = z^p + \sum_{k=p+1}^{\infty} a_k B_p^{\lambda,\mu}(k) z^k, \tag{7}$$

where

$$B_p^{\lambda,\mu}(k) = \frac{\Gamma(k-p+1+\mu)\Gamma(\nu+2+\mu-\lambda)\Gamma(k+\nu-p+2)}{\Gamma(k-p+1)\Gamma(k+\nu-p+2+\mu-\lambda)\Gamma(\nu+2)\Gamma(1+\mu)} \quad (8)$$

For $\mu = \lambda$, this generalized Ruscheweyh derivative operator reduces to Ruscheweyh derivative operator of order λ (see, e.g. [13]). Further, for $p = 1$, it reduces to ordinary Ruscheweyh derivative of univalent functions [14].

Definition 1 Let q be an univalent function in U with $q(0)=1$ and such that $D = q(U)$ is a convex domain from right half-plane. We define a subclass $\mathcal{M}_\alpha^p(\lambda, \mu, q)$ of α-convex functions f in A for $0 \le \alpha \le 1$ and $\lambda > -1$, satisfying the subordination condition

$$\mathcal{J}(\alpha, \lambda, \mu, f; z) = \left[\frac{(1-\alpha)}{p}\frac{z(\mathcal{J}_p^{\lambda,\mu} f(z))'}{\mathcal{J}_p^{\lambda,\mu} f(z)} + \frac{\alpha}{p}\left(1 + \frac{z(\mathcal{J}_p^{\lambda,\mu} f(z))''}{(\mathcal{J}_p^{\lambda,\mu} f(z))'}\right)\right] \prec q(z) \quad (9)$$

Subclasses of $\mathcal{M}_\alpha^p(\lambda, \mu, q)$ were studied by several authors. To mention a few are:

$$\mathcal{M}_0^1(0,0,q) = S^*(q)$$
$$\mathcal{M}_\alpha^1(0,0,q) = \mathcal{M}_\alpha(q)$$
$$\mathcal{M}_0^1(0,0,q_\gamma) = S^*(\gamma) \text{ where } q_\gamma(z) = \frac{1+(1-2\gamma)z}{1-z}, 0 \le \gamma < 1$$
$$\mathcal{M}_\alpha^1(0,0,q) = \mathcal{M}_\alpha \text{ for } q(z) = \frac{1+z}{1-z}$$
$$\mathcal{M}_\alpha^1(m,m,q) = \mathcal{M}_\alpha(m,0,q), m \in N^*$$

The class $S^*(\gamma)$ is the well-known class of starlike functions of order γ. The class $S^*(q)$ was introduced by Ma and Minda [7], the class $\mathcal{M}_\alpha(q)$ was studied by Ravichandran and Darus [12], $\mathcal{M}_\alpha$ is the class of α-convex functions introduced by Mocanu [9], and $\mathcal{M}_\alpha(m, 0, q)$ makes the object of the papers of Raducanu and Nechita [11]. For $\alpha = 0$, we shall denote the following subclass of functions $f \in A$ as $S_p^*(\lambda, \mu, q)$.

$$S_p^*(\lambda,\mu,q) = \mathcal{M}_0^p(\lambda,\mu,q) = \left\{ f \in A : \mathcal{J}(0,\lambda,\mu,f;z) = \frac{z}{p}\frac{(\mathcal{J}_p^{\lambda,\mu} f(z))'}{\mathcal{J}_p^{\lambda,\mu} f(z)} \prec q(z),\ z \in U \right\} \quad (10)$$

2 Preliminaries

In our present investigation, we shall need the following results concerning Briot–Bouquet differential subordinations.

Theorem 1 [4] *Let $\beta,\ \gamma \in C$, $\beta \neq 0$ and consider the convex function h, such that $Re[\beta h(z) + \gamma] > 0,\ z \in U$. If $p \in \mathcal{H}[h(0), n]$, then*

$$p(z) + \frac{zp'(z)}{\beta p(z) + \gamma} \prec h(z) \Rightarrow p(z) \prec h(z)$$

Theorem 2 [8] *Let q be an univalent function in U and consider θ and ϕ to be analytic functions in a domain $q(U)$. We denote by $Q(z) = zq'(z).\phi[q(z)]$, $h(z) = \theta[q(z)] + Q(z)$ and assume that*

1. *h is convex, or*
2. *Q is starlike.*
   Further suppose that
3. $$Re\frac{zh'(z)}{Q(z)} = Re\left[\frac{\theta'[q(z)]}{\phi[q(z)]} + \frac{zQ'(z)}{Q(z)}\right] > 0$$

If p is an analytic function in U, with $p(0) = q(0)$, $p(U) \subseteq D$ and such that

$$\theta[p(z)] + zp'(z)\phi[p(z)] \prec \theta[q(z)] + zq'(z)\phi[q(z)] = h(z)$$

then $p(z) \prec q(z)$ and q is the best dominant.

3 Main Results

Theorem 3 *Let $\alpha \in [0, 1]$, $\lambda \geq 0$, $\mu > -1$. Then $f \in \mathcal{M}_\alpha^p(\lambda, \mu, q)$ if and only if the function g defined by*

$$g(z) = \mathcal{J}_p^{\lambda,\mu} f(z) \left[\frac{z(\mathcal{J}_p^{\lambda,\mu} f(z))'}{\mathcal{J}_p^{\lambda,\mu} f(z)}\right]^\alpha, \quad z \in U \tag{11}$$

belongs to $S_p^*(\lambda, \mu, q)$. the branch of the power function is chosen such that

$$\left.\left[\frac{z(\mathcal{J}_p^{\lambda,\mu} f(z))'}{\mathcal{J}_p^{\lambda,\mu} f(z)}\right]^\alpha\right|_{z=0} = 1.$$

Proof Calculating the logarithmic derivative of g, we obtain

$$\frac{zg'(z)}{p\,g(z)} = \frac{z(\mathscr{I}_p^{\lambda,\mu} f(z))'}{p\ \mathscr{I}_p^{\lambda,\mu} f(z)} + \frac{\alpha}{p}\left[1 + \frac{z(\mathscr{I}_p^{\lambda,\mu} f(z))''}{(\mathscr{I}_p^{\lambda,\mu} f(z))'} - \frac{z(\mathscr{I}_p^{\lambda,\mu} f(z))'}{\mathscr{I}_p^{\lambda,\mu} f(z)}\right]$$
$$= \mathscr{J}(\alpha, \lambda, \mu, f; z) \prec q(z)$$

The equivalence from the hypothesis follows immediately.

Theorem 4 *If the function* f *belongs to the class* $\mathscr{M}_\alpha^p(\lambda, \mu, q)$, *for a given* $\alpha \in [0, 1]$, $\lambda \geq 0$, $\mu > -1$, *then* $f \in S_p^*(\lambda, \mu, q)$.

Proof Let us denote

$$p(z) = \frac{z\left(\mathscr{I}_p^{\lambda,\mu} f(z)\right)'}{p\left(\mathscr{I}_p^{\lambda,\mu} f(z)\right)}$$

The logarithmic derivative of $p(z)$ is

$$\frac{zp'(z)}{p(z)} = 1 + \frac{z\left(\mathscr{I}_p^{\lambda,\mu} f(z)\right)''}{\left(\mathscr{I}_p^{\lambda,\mu} f(z)\right)'} - \frac{z\left(\mathscr{I}_p^{\lambda,\mu} f(z)\right)'}{\left(\mathscr{I}_p^{\lambda,\mu} f(z)\right)}$$

Since $f \in \mathscr{M}_\alpha^p(\lambda, \mu, q)$,

$$p(z) + \frac{\alpha}{p}\frac{zp'(z)}{p(z)} = \mathscr{J}(\alpha, \lambda, \mu, f; z) \prec q(z)$$

It has been assumed that the function q is convex and that the image $q(U)$ is in the right half-plane. We have $\alpha \in [0, 1]$, therefore,

$$\text{Re}\left[\frac{q(z)}{\alpha}\right] > 0 \ z \in U$$

By applying Theorem 1 for $\beta = \frac{p}{\alpha}$ and $\gamma = 0$, we conclude that $p(z) \prec q(z)$ and hence $f \in S_p^*(\lambda, \mu, q)$.

Theorem 5 *Let* q *be convex function in* U *with* $q(0)=1$ *and* Re $q(z) > 0$. *Also consider* $Q(z) = \alpha\frac{zq'(z)}{p\,q(z)}$ *and* $h(z) = q(z) + Q(z)$, $z \in U$. *If* Q *is a convex function in* U *and* $f \in \mathscr{M}_\alpha^p(\lambda, \mu, h)$ *for an* $\alpha \in [0, 1]$, $\lambda \geq 0$, $\mu > -1$, *then* $f \in S_p^*(\lambda, \mu, q)$.

Proof Define the functions $\theta(w) = w$ and $\phi(w) = \frac{\alpha}{p\,w}$ and notice that the hypothesis of the Theorem 2 are satisfied. It follows that

$$p(z) = \frac{z\left(\mathscr{I}_p^{\lambda,\mu} f(z)\right)'}{p\left(\mathscr{I}_p^{\lambda,\mu} f(z)\right)} \prec q(z)$$

and q is the best dominant. Therefore, $f \in S_p^*(\lambda, \mu, q)$.

The generalized Komatu integral operator $K_{c,p}^{\sigma} : A \to A, c + p > 0, \sigma \geq 0$ introduced by Komatu [6] is defined as

$$K_{c,p}^{\sigma} f(z) = \frac{(c+p)^{\sigma}}{\Gamma \sigma z^{c}} \int_0^z t^{c-1} \log\left(\frac{z}{t}\right)^{\sigma-1} f(t)\mathrm{d}t \tag{12}$$
$$= z^p + \sum_{k=p+1}^{\infty} \left(\frac{c+p}{c+p+k}\right)^{\sigma} a_k z^k$$

Theorem 6 *If $K_{c,p}^{\sigma} f \in \mathscr{M}_{\alpha}^{p}(\lambda, \mu, q)$, then $K_{c,p}^{\sigma+1} f \in S_p^*(\lambda, \mu, q)$, $\lambda \geq 0$, $\mu > -1$.*

Proof Komatu integral operator satisfies the recurrence relation

$$(c+p) K_{c,p}^{\sigma} f(z) = z\left(K_{c,p}^{\sigma+1} f(z)\right)' + c K_{c,p}^{\sigma+1} f(z) \tag{13}$$

Applying generalized Ruscheweyh derivative operator, defined by (5), on both sides of (13), we obtain

$$(c+p)\mathscr{J}_p^{\lambda,\mu}(K_{c,p}^{\sigma} f(z)) = \mathscr{J}_p^{\lambda,\mu}\left(z\, K_{c,p}^{\sigma+1} f(z)\right)' + c\,\mathscr{J}_p^{\lambda,\mu}(K_{c,p}^{\sigma+1} f(z)) \tag{14}$$

Observe that $\mathscr{J}_p^{\lambda,\mu}\left(z\, K_{c,p}^{\sigma+1} f(z)\right)' = z\left(\mathscr{J}_p^{\lambda,\mu}(K_{c,p}^{\sigma+1} f(z))\right)'$. Therefore, for $f \in A$, Eq. (14) result becomes

$$(c+p)\mathscr{J}_p^{\lambda,\mu}(K_{c,p}^{\sigma} f(z)) = z\left(\mathscr{J}_p^{\lambda,\mu}(K_{c,p}^{\sigma+1} f(z))\right)' + c\,\mathscr{J}_p^{\lambda,\mu}(K_{c,p}^{\sigma+1} f(z)) \tag{15}$$

Differentiating (15) w.r.t. z and multiplying by z, we obtain

$$(c+p)z\left[\mathscr{J}_p^{\lambda,\mu}(K_{c,p}^{\sigma} f(z))\right]' \tag{16}$$
$$= (c+1)z\left[\mathscr{J}_p^{\lambda,\mu}(K_{c,p}^{\sigma+1} f(z))\right]' + z^2\left[\mathscr{J}_p^{\lambda,\mu}(K_{c,p}^{\sigma+1} f(z))\right]''$$

Divide (16) by (15) to obtain

$$z\frac{\left(\mathscr{J}_p^{\lambda,\mu}(K_{c,p}^{\sigma} f(z))\right)'}{\mathscr{J}_p^{\lambda,\mu}(K_{c,p}^{\sigma} f(z))} = z\frac{\left(\mathscr{J}_p^{\lambda,\mu}(K_{c,p}^{\sigma+1} f(z))\right)'}{\mathscr{J}_p^{\lambda,\mu}(K_{c,p}^{\sigma+1} f(z))}\left[\frac{c+1+z\dfrac{\left(\mathscr{J}_p^{\lambda,\mu}(K_{c,p}^{\sigma+1} f(z))\right)''}{\left(\mathscr{J}_p^{\lambda,\mu}(K_{c,p}^{\sigma+1} f(z))\right)'}}{c+z\dfrac{\left(\mathscr{J}_p^{\lambda,\mu}(K_{c,p}^{\sigma+1} f(z))\right)'}{\left(\mathscr{J}_p^{\lambda,\mu}(K_{c,p}^{\sigma+1} f(z))\right)}}\right] \tag{17}$$

Using the notations

$$P(z) = \frac{z\left(\mathcal{J}_p^{\lambda,\mu}(K_{c,p}^{\sigma+1}f(z))\right)'}{p\left(\mathcal{J}_p^{\lambda,\mu}(K_{c,p}^{\sigma+1}f(z))\right)} \quad \text{and} \quad p(z) = \frac{z\left(\mathcal{J}_p^{\lambda,\mu}(K_{c,p}^{\sigma}f(z))\right)'}{p\left(\mathcal{J}_p^{\lambda,\mu}(K_{c,p}^{\sigma}f(z))\right)}$$

in (15), we get

$$p(z) = P(z)\frac{c + p\,P(z) + \frac{zP'(z)}{P(z)}}{c + p\,P(z)} = P(z) + \frac{zP'(z)}{c + p\,P(z)}$$

Let $K_{c,p}^{\sigma}f \in \mathcal{M}_{\alpha}^{p}(\lambda, \mu, q)$, then by Theorem 4, $K_{c,p}^{\sigma}f \in S_p^*(\lambda, \mu, q)$ and hence the subordination

$$p(z) \prec q(z) \Rightarrow P(z) + \frac{zP'(z)}{c + p\,P(z)} \prec q(z)$$

holds. As q is a convex function and $\text{Re}[c + p\,P(z)] > 0$, from Theorem 1, with $\beta = p$ and $\gamma = c$, we conclude that $P(z) \prec q(z)$. Thus,

$$P(z) = \frac{z\left(\mathcal{J}_p^{\lambda,\mu}(K_{c,p}^{\sigma+1}f(z))\right)'}{p\left(\mathcal{J}_p^{\lambda,\mu}(K_{c,p}^{\sigma+1}f(z))\right)} \prec q(z),$$

that is, $K_{c,p}^{\sigma+1}f \in S_p^*(\lambda, \mu, q)$.

Put $\sigma = 0$ in the above theorem and denote $K_{c,p}^{1} \equiv L_{c,p}$, which is generalized Libera integral operator, to obtain the important result contained in the following theorem:

Theorem 7 *If $f \in \mathcal{M}_{\alpha}^{p}(\lambda, \mu, q)$, then $L_{c,p}f \in S_p^*(\lambda, \mu, q)$.*

4 A Representation Theorem

Theorem 8 *A function $f \in \mathcal{M}_{\alpha}^{p}(\lambda, \mu, q)$ for an $\alpha \in [0, 1]$, $\lambda \geq 0$, $\mu > -1$ if and only if*

$$\frac{\mathcal{J}_p^{\lambda,\mu}f(z)}{z^p}\left(\frac{z\left(\mathcal{J}_p^{\lambda,\mu}f(z)\right)'}{\left(\mathcal{J}_p^{\lambda,\mu}f(z)\right)}\right)^{\alpha} = \exp\left(p\int_0^z \frac{q(w(z)) - 1}{z}dz\right) \tag{18}$$

where $w(z)$ is analytic in U satisfying $w(0) = 0$ and $|w(z)| \leq 1$.

Proof Since $f \in \mathcal{M}_\alpha^p(\lambda, \mu, q)$, (9) holds, and therefore, there exists a Schwarz function $w(z)$ such that $w(0) = 0$ and $|w(z)| \leq 1$ and

$$\left[\frac{(1-\alpha)}{p}\frac{z(\mathcal{J}_p^{\lambda,\mu} f(z))'}{\mathcal{J}_p^{\lambda,\mu} f(z)} + \frac{\alpha}{p}\left(1 + \frac{z(\mathcal{J}_p^{\lambda,\mu} f(z))''}{(\mathcal{J}_p^{\lambda,\mu} f(z))'}\right)\right] = q(w(z))$$

Rewriting the above equation in the form

$$\left[\frac{(1-\alpha)}{p}\left(\frac{(\mathcal{J}_p^{\lambda,\mu} f(z))'}{\mathcal{J}_p^{\lambda,\mu} f(z)}\right) + \frac{\alpha}{p}\left(\frac{1}{z} + \frac{(\mathcal{J}_p^{\lambda,\mu} f(z))'}{(\mathcal{J}_p^{\lambda,\mu} f(z))'}\right) - \frac{1}{z}\right] = \frac{q(w(z)) - 1}{z}$$

On integrating from 0 to z, we obtain the desired expression upon exponentiation. The converse follows directly by differentiation.

Putting $\mu = \lambda$ in Theorem 8, we obtain the following corollary.

Theorem 9 *A function $f \in \mathcal{M}_\alpha^p(\lambda, q)$ for an $\alpha \in [0, 1]$, $\lambda \geq 0$ if and only if*

$$\frac{D_p^\lambda f(z)}{z^p}\left(\frac{z\left(D_p^\lambda f(z)\right)'}{\left(D_p^\lambda f(z)\right)}\right)^\alpha = \exp\left(p\int_0^z \frac{q(w(z)) - 1}{z} dz\right) \tag{19}$$

where $w(z)$ is analytic in U satisfying $w(0)=0$ and $|w(z)| \leq 1$.

Further, on taking $p = 1$ in the Theorem 9, we can find the corresponding result for univalent functions.

References

1. Acu, M., et al.: On some α-convex functions. Int. J. Open Problems Comput. Sci. Math 1(1), 1–10 (2008).
2. Ali, R.M., Ravichandran, V.: Classes of meromorphic α-convex functions. Taiwanese J. Math. **14**(4), 1479–1490 (2010)
3. Duren, P.L.: Univalent funct. Springer Verlag, New York (1983)
4. Eenigenburg, P.J., Miller, S.S., Mocanu, P.T., Reade, M.O.: On a Briot-Bouquet differential subordination. General inequalities, 3. Int. Series of Numerical Math, pp. 339–348. Birkhauser Verlag, Basel (1983).
5. Goyal, S.P., Goyal, R.: On a class of multivalent functions defined by generalized Ruscheweyh derivatives involving a general fractional derivative operator. J. Indian Acad. Math. **27**(2), 439–456 (2005)
6. Komatu, Y.: On analytic prolongation of a family of operators. Mathematica (Cluj) **32**(55), 141–145 (1990)
7. Ma, W., Minda, D.: A unified treatment of some special classes of univalent functions, proceedings of the conference on complex analysis, Tianjin, China. Int. Press. Conf. Proc. Lect. Notes Anal. 1, 157–169 (1994).

8. Miller, S.S., Mocanu, P.T.: On some classes of first-order differential subordinations. Michig. Math. J. **32**, 185–195 (1985)
9. Mocanu, P.T.: Une propriete de convexite geneealisee dans la theorie de la representation conforme. Mathematica (Cluj) **11**(34), 127–133 (1969)
10. Parihar, H.S., Agarwal, R.: Application of generalized Ruscheweyh derivatives on p-valent functions. J. Math. Appl. **34**, 75–86 (2011)
11. Raducanu, D., Nechita, V.O.: On α-convex functions defined by generalized Ruscheweyh derivatives operator, Studia Univ. "Babes-Bolyai". Mathematica **53**, 109–118 (2008)
12. Ravichandran, V., Darus, M.: On class of α-convex functions. J. Anal. Appl. **2**(1), 17–25 (2004)
13. Ravichandran, V., Seenivasagan, N., Srivastava, H.M.: Some inequalities associated with a linear operator defined for a class of multivalent functions. J. Inequal. Pure and Appl. Math. 4(4), 1–7 (2003). Art. 70.
14. Ruscheweyh, S.T.: New criterion for univalent functions. Proc. Amer. Math. Soc. **49**, 109–115 (1975)
15. Sokol, J.: On a condition for alpha-starlikeness. J. Math. Anal. Appl. **352**, 696–701 (2009)
16. Srivastava, H.M., Saxena, R.K.: Operators of fractional integration and their applications. Appl. Math. Comput. **118**, 1–52 (2001)

A New Expected Value Model for the Fuzzy Shortest Path Problem

Sk. Md. Abu Nayeem

Abstract Here, we consider a network, whose arc lengths are intervals or triangular fuzzy numbers. A new comparison technique based on the expected value of intervals and triangular fuzzy numbers is introduced. These expected values depend on a parameter which reflects the optimism/pessimism level of the decision-maker. Moreover, they can be used for negative intervals or triangular fuzzy numbers.

1 Introduction

Shortest path problem on a network with fuzzy parameters is one of the most studied problems in fuzzy set theory. A wide range of variations in the study of the fuzzy shortest path problem (FSPP) is found in the literature [1–3]. In a recent development, Nayeem and Pal [4] proposed an algorithm to find a fuzzy optimal path to which the decision-maker always satisfies with different grades of satisfaction. Hernandes et al. [5] proposes an iterative algorithm that assumes a generic ranking index for comparing the fuzzy numbers involved in the problem. But, neither of the above works addressed the problem when the arc lengths are imprecise numbers of mixed type. Tajdin et al. [6] gave an algorithm to find a fuzzy shortest path in a network with mixed fuzzy arc lengths using α-cuts. But, this α-cuts are not applicable for intervals, which can be considered as equipossible fuzzy numbers.

Recently, Liu [7, 8] developed a credibility theory including credibility measure, pessimistic value, and expected value as fuzzy ranking methods. Yang and Iwamura [9] introduced the m_λ measure as the linear combination of possibility measure and necessity measure and employed that measure to construct the fuzzy chance-constrained programming models. In this chapter, we have introduced the λ-expected

S. M. A. Nayeem (✉)
Department of Mathematics, Aliah University, DN-41, Sector-V, Salt Lake City, Kolkata 700091, India
e-mail: nayeem.math@aliah.ac.in

B. V. Babu et al. (eds.), *Proceedings of the Second International Conference on Soft Computing for Problem Solving (SocProS 2012), December 28–30, 2012*, Advances in Intelligent Systems and Computing 236, DOI: 10.1007/978-81-322-1602-5_23, © Springer India 2014

value of a fuzzy variable. Using this λ-expected values, Dijkstra's [10] algorithm for classical graphs can be applied to solve the FSPP.

2 Preliminaries and m_λ Measures

In this section, we give the arithmetic and ranking methods of intervals and triangular fuzzy numbers. Also we give a brief introduction to the m_λ measure and define the λ-expected value of fuzzy variables.

An interval number is defined as $A = [a, b] = \{x : a \leq x \leq b\}$, where a and b are real numbers called the left end point and the right end point of the interval A. or, alternatively, $A = \langle m(A), w(A)\rangle$, where $m(A)$ = midpoint of $A = \dfrac{a+b}{2}$, and $w(A)$ = half width of $A = \dfrac{a-b}{2}$. or, an interval $A = [a, b]$ is an equipossible fuzzy variable, whose membership function is given by $\mu_1(x) = \begin{cases} 1, \text{ if } a \leq x \leq b \\ 0, \text{ otherwise.} \end{cases}$

A crisp real number k may be considered as a degenerate interval $[k, k] = \langle k, 0\rangle$.

The sum of two interval numbers $A = [a_1, b_1]$ and $B = [a_2, b_2]$ is given by $A \oplus B = [a_1 + a_2, a_2 + b_2]$. Alternatively, in mean-width notations, if $A = \langle m_1, w_1\rangle$ and $B = \langle m_2, w_2\rangle$, then $A \oplus B = \langle m_1 + m_2, w_1 + w_2\rangle$.

In the following, we give the definition of *acceptability index* in connection with the ranking two intervals, due to Sengupta and Pal [11].

Definition 1 The acceptability index ($\mathscr{A}$-index) of the proposition '$A = \langle m_1, w_1\rangle$ is preferred to $B = \langle m_2, w_2\rangle$' is given by $\mathscr{A}(A \prec B) = \dfrac{m_2 - m_1}{w_1 + w_2}$.

Using this $\mathscr{A}$-index, we may define the following ranking orders.

Definition 2 If $\mathscr{A}(A \prec B) \geq 1$, then A is said to be totally dominating over B in case of minimization, and the case is converse in case of maximization, and this is denoted by $A \prec B$.

Definition 3 If $0 < \mathscr{A}(A \prec B) < 1$, then A is said to be 'partially dominating' over B in the sense of minimization and B is said to be 'partially dominating' over A in the sense of maximization. This is denoted by $A \prec_P B$.

A triangular fuzzy number is given by a triplet $\widetilde{A} = (a, b, c)$ with the membership function $\mu_2(x) = \begin{cases} \dfrac{x-a}{b-a}, \text{ if } a \leq x \leq b \\ \dfrac{x-c}{b-c}, \text{ if } b \leq x \leq c \\ 0, \quad \text{otherwise.} \end{cases}$

i.e., b is the point whose membership value is 1, and $b - a$ and $c - b$ are the left-hand and right-hand spreads, respectively.

In fuzzy optimization theory, the most important fuzzy ranking methods are based on the possibility and necessity measures. The possibility theory was proposed by Zadeh [12] and developed by many researchers such as Dubois and Prade [13]. Let ξ be a fuzzy variable with membership function $\mu_\xi(x)$ and B be an arbitrary subset of R, then the possibility measure of fuzzy event $\{\xi \in B\}$ is defined as Pos $\{\xi \in B\} = \sup_{x \in B} \mu_\xi(x)$. The necessity of this fuzzy event is defined as Nec $\{\xi \in B\} = 1-$ Pos$\{\xi \in B^c\} = 1 - \sup_{x \in B^c} \mu_\xi(x)$.

Liu [7] introduced the credibility measure of a fuzzy event as an average of the possibility and necessity measure as below.

Definition 4 The credibility measure for the chance of a fuzzy event is defined as $\mathrm{Cr}\{\xi \in B\} = \frac{1}{2}(\mathrm{Pos}\{\xi \in B\} + \mathrm{Nec}\{\xi \in B\})$.

Liu and Liu [8] show that the expected value of a fuzzy variable can be defined in terms of the credibility measure as follows.

Definition 5 Let ξ be a fuzzy variable. The expected value of ξ is defined as $E[\xi] = \int_0^{\infty} \mathrm{Cr}\{\xi \geq r\}dr - \int_{-\infty}^{0} \mathrm{Cr}\{\xi \leq r\}dr$, provided that at least one of the two integrals is finite.

But, in reality, most decision-makers are neither absolutely optimistic/ optimistic, nor absolutely neutral. To balance between the optimism and pessimism, a convex combination of the possibility measure, and the necessity measure is introduced by Yang and Iwamura [9]. It gives scope to the decision-maker to set the degree of optimism/ pessimism. Thus, m_λ measure is a generalization of the credibility measure.

Definition 6 Formally, the m_λ measure for the chance of a fuzzy event is defined as $m_\lambda\{\xi \in B\} = \lambda \mathrm{Pos}\{\xi \in B\} + (1 - \lambda)\mathrm{Nec}\{\xi \in B\}$, where the parameter $\lambda \in [0, 1]$ is predetermined by the decision-maker according to the degree of optimism or pessimism.

Clearly, when $\lambda = 0.5$, the m_λ measure reduces to the credibility measure. Based on the m_λ measure, the λ-expected value of a fuzzy variable is defined as follows [14]:

Definition 7 Let ξ be a fuzzy variable. The λ-expected value of ξ is defined as, $E_\lambda[\xi] = \int_0^{\infty} m_\lambda\{\xi \geq r\}dr - \int_{-\infty}^{0} m_\lambda\{\xi \leq r\}dr$, provided that at least one of the two integrals is finite, and the parameter $\lambda \in [0, 1]$ is predetermined by the decision-maker according to the degree of optimism or pessimism.

Example 1 Let ξ be the equipossible fuzzy variable (a, b), i.e., the interval $[a, b]$, then the λ-expected value of ξ is given by $E_\lambda[\xi] = \begin{cases} \lambda b + (1 - \lambda)a, & \text{if } a \geq 0 \\ \lambda(a + b), & \text{if } a < 0 < b \\ \lambda a + (1 - \lambda)b, & \text{if } b \leq 0. \end{cases}$

Example 2 Let η be the triangular fuzzy variable (a, b, c), then the λ-expected value of η is given by $E_\lambda[\eta] = \begin{cases} \frac{a+b}{2} + \frac{\lambda}{2}(c-a), & \text{if } a \geq 0 \\ \frac{b^2+\lambda(bc-ab-ac-a^2)}{2(b-a)}, & \text{if } a < 0 \leq b < c \\ \frac{b^2+\lambda(ab-bc-ac-c^2)}{2(b-c)}, & \text{if } a < b \leq 0 < c \\ \frac{b+c}{2} - \frac{\lambda}{2}(c-a), & \text{if } c \leq 0. \end{cases}$

Liou and Wang [15] considered an ordinance method of fuzzy numbers with integral values. This definition involves the areas relating to the left-hand spread and right-hand spread, and the index is the convex combination of those two.

Remark 1 If ξ is a positive triangular fuzzy number, then the Liou and Wang index of ξ coincides with $E_\lambda[\xi]$.

It can be shown that the λ-expected value operator is linear as like the ordinary expected value operator.

Lemma 1 *Let ξ and η be two independent fuzzy variables, then for any real numbers a and b, we have $E_\lambda[a\xi + b\eta] = aE_\lambda[\xi] + bE_\lambda[\eta]$.*

Using the λ-expected value operator, we can define the following functions to find the fuzzy minimum of two independent fuzzy variables.

Definition 8 Let ξ and η be two independent fuzzy variables, then we say that the fuzzy minimum between ξ and η is ξ if $E_\lambda[\xi] \leq E_\lambda[\eta]$ and denote this by '$\xi \ll \eta$'.

Theorem 1 *Let A and B be two positive intervals, then $A \ll B$ if $A \prec B$.*

Proof Let $A = [a_1, b_1]$ and $B = [a_2, b_2]$, then $A \prec B$ gives $\frac{\frac{a_2+b_2}{2} - \frac{a_1+b_1}{2}}{\frac{b_1-a_1}{2} + \frac{b_2-a_2}{2}} \geq 1$, i.e., $a_2 \geq b_1$.

Now, $E_\lambda[A] = a_1 + (b_1 - a_1)\lambda \leq a_1 + (b_1 - a_1)$ (since, $\lambda \leq 1$)$= b_1 \leq a_2 \leq a_2 + \lambda(b_2 - a_2)$ (since $\lambda \geq 0$)$= E_\lambda[B]$.

Hence the theorem follows. □

Theorem 2 *Let $A = [a_1, b_1]$ and $B = [a_2, b_2]$ be two positive intervals, then $a_1 < a_2$ and $w(A) \leq w(B)$ imply $A \prec_P B$ and $A \ll B$.*

Proof $w(A) \leq w(B)$ gives $b_1 - a_1 < b_2 - a_2$. So, $a_1 < a_2$ and $w(A) \leq w(B)$ together give $b_1 < b_2$. Thus, $m(A) < m(B)$, and hence, $A \prec_P B$.

Again, $E_\lambda[A] = a_1 + \lambda(b_1 - a_1) < a_2 + \lambda(b_1 - a_1) \leq a_2 + \lambda(b_2 - a_2) = E_\lambda[B]$, and hence, $A \ll B$. □

3 Expected Value Model of FSPP

Let us consider a directed network $G = (V, E)$, where V is the set of vertices and E is the set of arcs. Each arc is denoted by an ordered pair (i, j), where $i, j \in V$. We consider that there is only one directed arc (i, j) from i to j. Let the node 1 be the source node and let us assume t as the destination node. We define a path p_{ij} as a sequence $p_{ij} = \{i, (i, i_1), i_1, ..., i_k, (i_k, j), j\}$ of alternating nodes and arcs. The existence of at least one path p_{si} in $G = (V, E)$ is assumed for every node $i \in V - \{s\}$. Let ξ_{ij} be an imprecise number (either an interval or a triangular fuzzy number) associated with the arc (i, j), corresponding to the length necessary to traverse (i, j) from i to j, then the FSPP is formulated as the following linear programming problem:

$$\min f(x) = \sum_{(i,j)\in E} E_\lambda[\xi_{ij}]x_{ij}$$

$$\text{subject to } \sum_j x_{ij} - \sum_j x_{ji} = \begin{cases} 1, & \text{if } i = 1 \\ 0, & \text{if } i \in \{1, 2, \ldots, |V|\} \setminus \{1, t\} \\ -1 & \text{if } i = t, \end{cases}$$

$$x_{ij} = 0 \text{ or } 1, \text{ for } (i, j) \in E.$$

To solve this expected value model of FSPP, we replace each ξ_{ij}, $(i, j) \in E$ with the corresponding λ-expected value $E_\lambda[\xi_{ij}]$ and then we apply the classical shortest path algorithms, like that of Dijkstra [10]. We illustrate this with the help of the following examples.

Example 3 We consider a small-sized network as shown in Fig. 1 having mixed arc lengths with four nodes and five arcs, of which two arcs are of triangular fuzzy length and the other three are of interval length. λ-expected values of the arc lengths for different λ are listed in Table 1. Different fuzzy shortest paths obtained for different λ are shown in Table 2.

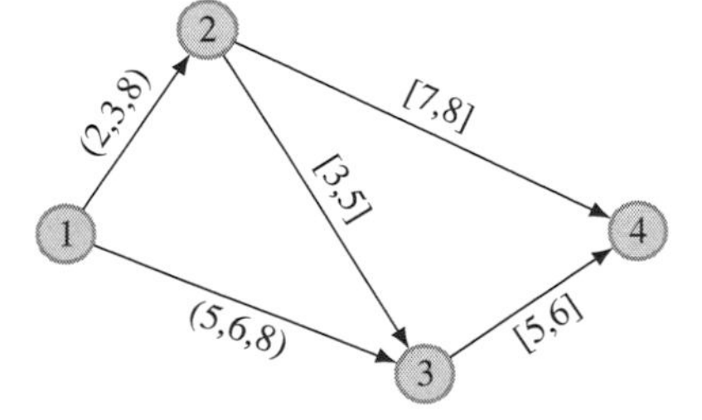

Fig. 1 A small-sized network

Table 1 Arc lengths and expected values

Arc	Length	λ-expected value		
		$\lambda = 0.25$	$\lambda = 0.50$	$\lambda = 0.75$
(1, 2)	(2, 3, 8)	3.25	4.00	4.75
(1, 3)	(5, 6, 8)	5.88	6.25	6.63
(2, 3)	[3, 5]	3.50	4.00	4.50
(2, 4)	[7, 8]	7.25	7.50	7.75
(3, 4)	[5, 6]	5.25	5.50	5.75

Table 2 Different fuzzy shortest paths of the network in Fig 1

λ	Path	Expected length
0.25	$1 \to 2 \to 4$	10.50
0.50	$1 \to 2 \to 4$	11.50
0.75	$1 \to 3 \to 4$	12.38

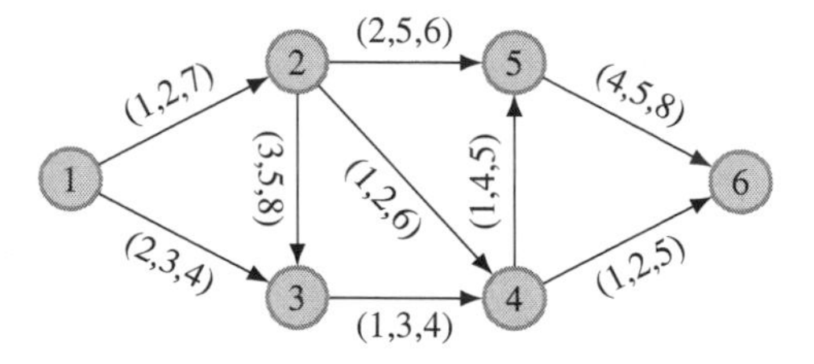

Fig. 2 Another network

Remark 2 It is evident that, as λ increases, the expected length of the FSPP is also increased. So, in a maximization problem, λ would be considered as an index of optimism, whereas in case of minimization, it would be considered as an index of pessimism.

Example 4 For the sake of a comparative study, we consider the network shown in Fig. 2 with same arc lengths as considered by Tajdin et al. [6]. Using our method, we get the FSPP $1 \to 2 \to 4 \to 6$ for $\lambda < 0.5$, and the FSPP $1 \to 3 \to 4 \to 6$ for $\lambda > 0.5$. For $\lambda = 0.5$, there is tie between those two FSPP with the expected length 8.25. The FSPP $1 \to 3 \to 4 \to 6$ has the λ-expected value 5.93 (the distance value obtained by Tajdin et al.) for $\lambda = 0.19$, which clearly reflects a highly optimistic view of the decision-maker.

4 Conclusion

In this chapter, we have considered the fuzzy shortest path problem on a network with mixed arc lengths. Replacing each imprecise arc length by the corresponding λ-expected value, we may apply the Dijkstra's algorithm to solve the problem. If some of the arc lengths are crisp real numbers, then also the method works. It is also worth mentioning that instead of Dijkstra's algorithm, if we use Ford-Moore-Bellman's algorithm [16], then the presence of a negative circuit can be detected, since the λ-expected value of negative intervals or negative triangular fuzzy numbers can also be computed.

References

1. Klein, C.M.: Fuzzy shortest paths. Fuzzy Sets and System **39**, 27–41 (1991)
2. Okada, S., Gen, M.: Order relation between intervals and its application to shortest path problem. Comput. Ind. Eng. **25**, 147–150 (1993)
3. Okada, S., Gen, M.: Fuzzy shortest path problem. Comput. Ind. Eng. **27**, 465–468 (1994)
4. Nayeem, S.M.A., Pal, M.: Shortest path problem on a network with imprecise edge weight. Fuzzy Optim. Decis. Making **4**, 293–312 (2005)
5. Hernandes, F., Lamata, M.T., Verdegay, J.L., Yamakami, A.: The shortest path problem on networks with fuzzy parameters. Fuzzy Sets and Syst. **158**, 1561–1570 (2007)
6. Tajdin, A., Mahdavi, I., Mahdavi-Amiri, N., Sadeghpour-Gildeh, B.: Computing a fuzzy shortest path in a network with mixed fuzzy arc lengths using α-cuts. Comput. Math. Appl. **60**, 989–1002 (2010)
7. Liu, B.: Theory and Practice of Uncertain Programming. Physica-Verlag, Heidelberg (2002)
8. Liu, B.: Uncertainty theory: An introduction to its axiomatic foundations. Springer-Verlag, Berlin (2004)
9. Yang, L., Iwamura, K.: Fuzzy chance-constrained programming with linear combination of possibility measure and necessity measure. Appl. Math. Sci. **2**, 2271–2288 (2008)
10. Dijkstra, E.W.: A note on two problems in connection with graphs. Numerische Mathematik **1**, 269–271 (1959)
11. Sengupta, A., Pal, T.K.: On comparing interval numbers. Eur. J. Oper. Res. **127**, 28–43 (2000)
12. Zadeh, L.A.: Fuzzy sets as a basis for a theory of possibility. Fuzzy Sets and Syst. **1**, 3–28 (1978)
13. Dubois, D., Prade, H.: Possibility theory. Plenum, New York (1988)
14. Nayeem, S.M.A.: An expected value model of quadratic clique problem on a graph with fuzzy parameters. Proceedings of 12th international conference on intelligent systems design and applications (ISDA), Kochi, India, (to appear).
15. Liou, T.-S., Wang, M.-J.: Ranking fuzzy numbers with integral interval. Fuzzy Sets and Syst. **50**, 247–255 (1992)
16. Bellman, R.E.: On a routing problem. Quart. Appl. Math. **16**, 87–90 (1958)

Existence and Uniqueness of Fixed Point in Fuzzy Metric Spaces and its Applications

Vishal Gupta and Naveen Mani

Abstract The main aim of this paper is to prove some fixed point theorems in fuzzy metric spaces through rational inequality. Our results extend and generalize the results of many other authors existing in the literature. Some applications are also given in support of our results.

Keywords Fuzzy metric space · Rational expression · Integral type · Control function

1 Introduction

The foundation of fuzzy mathematics is laid by Zadeh [1] with the introduction of fuzzy sets in 1965. This foundation represents a vagueness in everyday life. Subsequently, several authors have applied various form general topology of fuzzy sets and developed the concept of fuzzy space. In 1975, Kramosil and Michalek [2] introduced concept of fuzzy metric spaces. In 1988, Mariusz Grabiec [3] extended fixed point theorem of Banach and Eldestien to fuzzy metric spaces in the sense of Kramosil and Michalek. In 1994, George et al. [4] modified the notion of fuzzy metric spaces with the help of continuous t-norms. A number of fixed point theorems have been obtained by various authors in fuzzy metric space by using the concept of compatible map, implicit relation, weakly compatible map, R weakly compatible map. (See Section-: [5–13]). Also Saini and Gupta [10, 11] proved some fixed points theorems, on expansion type maps and common coincidence points of

V. Gupta (✉) · N. Mani
Department of Mathematics, Maharishi Markandeshwar University, Mullana
Ambala, Haryana 133001, India
e-mail: vishal.gmn@gmail.com
N. Mani
e-mail: naveenmani81@gmail.com

B. V. Babu et al. (eds.), *Proceedings of the Second International Conference on Soft Computing for Problem Solving (SocProS 2012), December 28–30, 2012*, Advances in Intelligent Systems and Computing 236, DOI: 10.1007/978-81-322-1602-5_24,

R weakly commuting fuzzy maps, in fuzzy metric space. The present paper extends and generalizes the results of Grabeic [3] and also many other authors existing in the literature.

2 Preliminaries

In this section, we define some important definition and results which are used in sequel.

Definition 1 [1] Let X be any set. A fuzzy set A in X is a function with domain X and values in [0, 1].

Definition 2 [14] A binary operation $* : [0, 1] \times [0, 1] \to [0, 1]$ is a continuous t-norms if $([0, 1], *)$ is an abelian topological monoid with the unit 1 such that $a * b \leq c * d$ whenever $a \leq c$ and $b \leq d$ for all $a, b, c, d \in [0, 1]$.

Definition 3 [2] A triplet $(X, M, *)$ is a fuzzy metric space if X is an arbitrary set, $*$ is continuous t-norm and M is a fuzzy set on $X^2 \times (0, \infty)$ satisfying the following conditions, for all $x, y, z \in$ X, such that $t, s \in (0, \infty)$

1. $M(x, y, t) > 0$.
2. $M(x, y, t) = 1$ iff $x = y$.
3. $M(x, y, t) = M(y, x, t)$.
4. $M(x, y, t) * M(y, z, s) \leq M(x, z, t + s)$.
5. $M(x, y, .) : [0, \infty) \to [0, 1]$ is continuous.

Then, M is called a fuzzy metric on X, and $M(x, y, t)$ denotes the degree of nearness between x and y with respect to t.

Definition 4 [3] Let $(X, M, *)$ is a fuzzy metric space then a sequence $\{x_n\} \in X$ is said to be convergent to a point $x \in X$ if $\lim_{n\to\infty} M(x_n, x, t) = 1$ for all $t > 0$.

Definition 5 [3] Let $(X, M, *)$ is a fuzzy metric space then a sequence $\{x_n\} \in X$ is called a Cauchy sequence if $\lim_{n\to\infty} M\left(x_{n+p}, x_n, t\right) = 1$ for all $t > 0$ and $p > 0$.

Definition 6 [3] Let $(X, M, *)$ is a fuzzy metric space then an fuzzy metric space in which every Cauchy sequence is convergent is called complete. It is called compact, if every sequence contains a convergent subsequence.

Lemma 1 [3] For all, $x, y \in X$, $M(x, y, .)$ is non-decreasing.

Lemma 2 [9] If there exist $k \in (0, 1)$ such that $M(x, y, kt) \geq M(x, y, t)$ for all $x, y, \in X$ and $t \in (0, \infty)$, then $x = y$

Now, we prove our main result.

3 Main Results

Theorem 1 *Let $(X, M, *)$ be a complete fuzzy metric space and $f : X \to X$ be a mapping satisfying*

$$M(x, y, t) = 1 \tag{1}$$

and

$$M(fx, fy, kt) \geq \lambda(x, y, t) \tag{2}$$

where

$$\lambda(x, y, t) = min\left\{\frac{M(x, fx, t)\, M(y, fy, t)}{M(x, y, t)}, M(x, y, t)\right\} \tag{3}$$

for all $x, y, \in X$ and $k \in (0, 1)$. Then f has a unique fixed point.

Proof Let us consider $x \in X$ be any arbitrary point in X. Now construct a sequence $\{x_n\} \in X$ such that $fx_n = x_{n+1}$ for all $n \in N$

Claim: $\{x_n\}$ is a Cauchy sequence.

Let us take $x = x_{n-1}$ and $y = x_n$ in (2), we get

$$M(x_n, x_{n+1}, kt) = M(fx_{n-1}, fx_n, kt) \geq \lambda(x_{n-1}, x_n, t) \tag{4}$$

Now

$$\begin{aligned}\lambda(x_{n-1}, x_n, t) &= min\left\{\frac{M(x_{n-1}, x_n, t)\, M(x_n, x_{n+1}, t)}{M(x_{n-1}, x_n, t)}, M(x_{n-1}, x_n, t)\right\}\\ &= min\{M(x_n, x_{n+1}, t), M(x_{n-1}, x_n, t)\}\end{aligned}$$

Now if $M(x_n, x_{n+1}, t) \leq M(x_{n-1}, x_n, t)$, then from Eq. (4)

$$M(x_n, x_{n+1}, kt) \geq M(x_n, x_{n+1}, t)$$

Thus, our claim is immediately follows. Now suppose $M(x_n, x_{n+1}, t) \geq M(x_{n-1}, x_n, t)$ then again from Eq. (4),

$$M(x_n, x_{n+1}, kt) \geq M(x_{n-1}, x_n, t)$$

Now by simple induction, for all n and $t > o$, we get

$$M(x_n, x_{n+1}, kt) \geq M\left(x, x_1, \frac{t}{k^{n-1}}\right) \tag{5}$$

Now for any positive integer q, we have

$$M\left(x_n, x_{n+q}, t\right) \geq M\left(x_n, x_{n+1}, \frac{t}{q}\right) * \ldots.(^q)\ldots. * M\left(x_{n+p-1}, x_{n+p}, \frac{t}{q}\right)$$

By using Eq. (5), we get

$$M\left(x_n, x_{n+q}, t\right) \geq M\left(x, x_1, \frac{t}{qk^n}\right) * \ldots.(^q)\ldots. * M\left(x, x_1, \frac{t}{qk^n}\right)$$

Now taking $\lim_{n\to\infty}$ and using (1), we get

$$\lim_{n\to\infty} M\left(x_n, x_{n+q}, t\right) = 1 \tag{6}$$

This implies, $\{x_n\}$ is a Cauchy sequence. Call the limit z.

Claim: z is a fixed point of f.

Consider

$$M(fz, z, t) \geq M(fz, fx_n, t) * M(x_{n+1}, z_n, t) \geq \lambda\left(z, x_n, \frac{t}{2k}\right) * M(x_{n+1}, z_n, t) \tag{7}$$

Now

$$\lambda\left(z, x_n, \frac{t}{2k}\right) = \min\left\{\frac{M\left(z, fz, \frac{t}{2k}\right) M\left(x_n, fx_n, \frac{t}{2k}\right)}{M\left(z, x_n, \frac{t}{2k}\right)}, M\left(z, x_n, \frac{t}{2k}\right)\right\}$$

Taking $\lim_{n\to\infty}$ in above inequality and using (1), we get

$$\lambda\left(z, z, \frac{t}{2k}\right) = \min\left\{M\left(z, fz, \frac{t}{2k}\right), 1\right\}$$

Now if $M\left(z, fz, \frac{t}{2k}\right) \geq 1$ then $\lambda\left(z, z, \frac{t}{2k}\right) = 1$. Therefore, from (7) and using Definition 3, we get z is a fixed point of f. Now if $M\left(z, fz, \frac{t}{2k}\right) \leq 1$, then $\lambda\left(z, z, \frac{t}{2k}\right) = M\left(z, fz, \frac{t}{2k}\right)$ Hence, from Eq. (7), we get

$$M(fz, z, t) \geq M\left(z, fz, \frac{t}{2k}\right) * M(x_{n+1}, z_n, t) \tag{8}$$

Now taking $\lim_{n\to\infty}$ in (8) and using lemma 2 ,we get $fz = z$. That is z is a fixed point of f.

Uniqueness: Now, we show that z is a unique fixed point of f. Suppose not, then there exist a point $\nu \in X$ such that $fv = v$. Consider

$$1 \geq M(z, v, t) = M(fz, fv, t) \geq \lambda\left(z, v, \frac{t}{k}\right) \geq M\left(fz, fv, \frac{t}{k}\right) \tag{9}$$

Again

$$M\left(fz, fv, \frac{t}{k}\right) \geq M\left(fz, fv, \frac{t}{k^2}\right) \geq \cdots \geq M\left(fz, fv, \frac{t}{k^n}\right) \qquad (10)$$

Taking $\lim_{n\to\infty}$ in (10) and use it in (9), we get $z = v$. Thus, z is unique fixed point of f. This completes the proof of Theorem 1. □

let us define $\Phi = \{\phi/\phi : [0, 1] \to [0, 1]\}$ is a continuous function such that $\phi(1) = 1, \phi(0) = 0, \phi(a) > a$ for each $0 < a < 1$.

Theorem 2 *Let $(X, M, *)$ be a complete fuzzy metric space and $f : X \to X$ be a mapping satisfying*

$$M(x, y, t) = 1 \qquad (11)$$

and

$$M(fx, fy, kt) \geq \phi\{\lambda(x, y, t)\} \qquad (12)$$

where

$$\lambda(x, y, t) = min\left\{\frac{M(x, fx, t)\, M(y, fy, t)}{M(x, y, t)}, M(x, y, t)\right\} \qquad (13)$$

for all $x, y, \in X$, $k \in (0, 1)$, $\phi \in \Phi$. Then, f has a unique fixed point.

Proof Since $\phi \in \Phi$, this implies that $\phi(a) > a$ for each $0 < a < 1$. Thus, from (12)

$$M(fx, fy, kt) \geq \phi\{\lambda(x, y, t)\} \geq \lambda(x, y, t)$$

Now, applying Theorem 1, we obtain the desired result. □

4 Applications

In this section, we give some applications related to our results. Let us define $\Psi : [0, \infty) \to [0, \infty)$, as $\Psi(t) = \int_0^t \varphi(t)\, dt \;\; \forall \; t > 0$, be a non-decreasing and continuous function. Moreover, for each $\epsilon > 0, \varphi(\epsilon) > 0$. Also implies that $\varphi(t) = 0$ iff $t = 0$.

Theorem 3 *Let $(X, M, *)$ be a complete fuzzy metric space and $f : X \to X$ be a mapping satisfying*

$$M(x, y, t) = 1$$

$$\int_0^{M(fx, fy, kt)} \varphi(t)\, dt \geq \int_0^{\lambda(x,y,t)} \varphi(t)\, dt$$

where

$$\lambda(x, y, t) = min\left\{\frac{M(x, fx, t)\, M(y, fy, t)}{M(x, y, t)}, M(x, y, t)\right\}$$

for all $x, y, \in X, \varphi \in \Psi$ *and* $k \in (0, 1)$. *Then,* f *has a unique fixed point.*

Proof By taking $\varphi(t) = 1$ and applying Theorem 1, we obtain the result. □

Theorem 4 *Let* $(X, M, *)$ *be a complete fuzzy metric space and* $f : X \to X$ *be a mapping satisfying*

$$M(x, y, t) = 1$$

$$\int_0^{M(fx, fy, kt)} \varphi(t)\, dt \geq \phi\left\{\int_0^{\lambda(x, y, t)} \varphi(t)\, dt\right\}$$

where

$$\lambda(x, y, t) = min\left\{\frac{M(x, fx, t)\, M(y, fy, t)}{M(x, y, t)}, M(x, y, t)\right\}$$

for all $x, y, \in X$ $\varphi \in \Psi$, $k \in (0, 1)$ *and* $\phi \in \Phi$. *Then,* f *has a unique fixed point. Then,* f *has a unique fixed point.*

Proof Since $\phi(a) > a$ for each $0 < a < 1$, therefore, result follows immediately from Theorem 3. □

Remark 1 Our paper extends and generalizes the result of Grabeic [3] and also many other authors existing in the literature.

Acknowledgments The authors would like to express their sincere appreciation to the referees for their helpful suggestions and many kind comment.

References

1. Zadeh, L.A.: Fuzzy sets. Inform. Control **8**, 338–353 (1965)
2. Kramosil, O., Michalek, J.: Fuzzy metric and statistical metric spaces. Kybornetica **11**, 326–334 (1975)
3. Grabeic, M.: Fixed points in fuzzy metric spaces. Fuzzy Sets Syst. **27**, 385–389 (1988)
4. George, A., Veermani, P.: On some results in fuzzy metric spaces. Fuzzy Sets Syst. **64**, 395–399 (1994)
5. Balasubramaniam, P., Murlisankar, S., Pant, R.P.: Common fixed points of four mappings in a fuzzy metric space. J. Fuzzy Math. **10**(2), 379–384 (2002)
6. Cho, S.H.: On common fixed point theorems in fuzzy metric spaces. Int. Math. Forum **1**(9–12), 471–479 (2006)
7. Cho, Y.J., Sedghi, S., Shobe, N.: Generalized fixed point theorems for compatible mappings with some types in fuzzy metric spaces. Chaos, Solitons and Fractals **39**, 2233–2244 (2009)

8. Gregori, V., Sapena, A.: On fixed point theorems in fuzzy metric spaces. Fuzzy Sets Syst. **125**, 245–252 (2002)
9. Mishra, S.N., Sharma, S.N., Singh, S.L.: Common fixed point of maps on fuzzy metric spaces. Internat. J. Math. Sci **17**, 253–258 (1994)
10. Saini, R.K., Gupta, V.: Common coincidence points of R-Weakly commuting fuzzy maps, thai journal of mathematics, mathematical assoc. of Thailand, 6(1):109–115, (2008) ISSN 1686–0209.
11. Saini, R.K., Gupta, V.: Fuzzy version of some fixed points theorems on expansion type maps in fuzzy metric space. Thai J. Math., Math. Assoc. Thailand **5**(2), 245–252 (2007). ISSN 1686–0209
12. Vasuki, R.: A common fixed point theorem in a fuzzy metric space. Fuzzy Sets and Syst. **97**, 395–397 (1998)
13. Vasuki, R.: Common fixed point for R-weakly commuting maps in fuzzy metric space. Indian J. Pure. Appl. Math. **30**, 419–423 (1999)
14. Schweizer, B., Sklar, A.: Probabilistic metric spaces, North-Holland series in probability and applied mathematics. North-Holland Publishing Co., New York (1983), ISBN: 0-444-00666-4 MR0790314 (86g:54045).

Variable Selection and Fault Detection Using a Hybrid Intelligent Water Drop Algorithm

Manish Kumar, Srikant Jayaraman, Shikha Bhat, Shameek Ghosh and V. K. Jayaraman

Abstract Process fault detection concerns itself with monitoring process variables and identifying when a fault has occurred in the process workflow. Sophisticated learning algorithms may be used to select the relevant process state variables out of a massive search space and can be used to build more efficient and robust fault detection models. In this study, we present a recently proposed swarm intelligence-based hybrid intelligent water drop (IWD) optimization algorithm in combination with support vector machines and an information gain heuristic for selecting a subset of relevant fault indicators. In the process, we demonstrate the successful application and effectiveness of this swarm intelligence-based method to variable selection and fault identification. Moreover, performance testing on standard machine learning benchmark datasets also indicates its viability as a strong candidate for complex classification and prediction tasks.

M. Kumar
Bioinformatics Center, University of Pune, Pune, India
e-mail: rishimanish123@gmail.com

S. Jayaraman · S. Bhat
Centre for Modelling and Simulation, University of Pune, Pune, India
e-mail: srikant@cms.unipune.ac.in

S. Bhat
e-mail: shikha@cms.unipune.ac.in

S. Ghosh · V. K. Jayaraman (✉)
Evolutionary Computing and Image Processing, Centre for Development of Advanced Computing, Pune, India
e-mail: jayaramanv@cdac.in

S. Ghosh
e-mail: shameekg@cdac.in

V. K. Jayaraman
Informatics Centre, Shiv Nadar University, Chithera Village, Dadri District, UP, India

B. V. Babu et al. (eds.), *Proceedings of the Second International Conference on Soft Computing for Problem Solving (SocProS 2012), December 28–30, 2012*, Advances in Intelligent Systems and Computing 236, DOI: 10.1007/978-81-322-1602-5_25,

Keywords Fault detection · Swarm intelligence · Intelligent water drop optimization · Information gain · Support vector machines

1 Introduction

Fault detection and isolation are associated with monitoring a process and identifying when a fault may occur [1, 2]. Generally, a large set of sensor readings may be collected over a long period of time by a process monitoring module. As part of this process, a system can inject faults into the process by random variation of variable measures. A log of all such process data may thus be treated as a training dataset by a learning algorithm for building fault prediction models. However, a limitation of this technique is that it generates a vast amount of complex data. Constructing classification models from such data thus turns out to be very tedious and the resulting model is normally quite inferior due to inclusion of irrelevant and redundant variables in the model. To get around this problem, we can select a small subset of relevant variables from the data. This process is known as variable selection and it helps in reducing the computational load and in increasing the overall classification performance.

Variable selection algorithms may typically be categorized as: wrappers and filters. Wrappers [3–5] use a learning algorithm to estimate the suitability of a subset of variables. In contrast, filters evaluate the capability of a variable considering their inherent characteristics using techniques based on statistical tests and mutual information.

In the following study, we present a recently proposed swarm intelligence-based learning technique known as intelligent water drop (IWD) optimization [6] in conjunction with an entropy-based heuristic ranking and support vector machines as a filter–wrapper algorithm for variable subset selection and simultaneous fault detection.

2 Intelligent Water Drop Algorithm

Natural phenomena are a huge source of inspiration for building swarm intelligence-based techniques. A particular instance of a natural process is found in the optimal selection of path by flowing water sources, while converging to a bigger source of water, say a sea or an ocean. The paths followed by rivers exemplify this nature. At a finer level, the flowing water is basically built up of a swarm of natural water drops. The behaviour of a single water drop thus turns to be of significant importance in the construction and movement of a swarm of water drops. Shah-Hosseini, consequently, extended this concept to introduce the IWD algorithm for the travelling salesman problem (TSP) [6].

The IWD algorithm involves employing a swarm of IWDs characterized by two important properties. These are (1) the IWD's soil content denoted as *soil(IWD)* and (2) the velocity of an IWD denoted by *vel(IWD)*. The *soil(IWD)* and *vel(IWD)* dynamically keep changing based on the path taken by an IWD while flowing through the problem landscape. It is thus pertinent that as an IWD moves, it removes soil from the traversed path and the path soil is also updated dynamically in the process. This joint action may thus contribute to lowering of soil content of certain routes in the problem environment which is associated with the fitness landscape. It has been consequently posited that the paths with lesser soil content are thus the most important for the search of a near optimal solution. The emergent behaviour of a swarm of IWDs therefore governs the construction of an optimal solution for the concerned problem.

Accordingly, when an IWD moves in discrete time steps from its current location i to its next location j, the IWD velocity is increased by a Δvel component which is given by

$$\Delta \mathrm{vel}^{\mathrm{IWD}}(t) = \frac{a_v}{b_v + c_v \times \mathrm{soil}^{2\alpha}(i, j)} \tag{1}$$

Here, a_v, b_v, c_v, and α are algorithm specific parameters. Similarly the soil content of an IWD is also increased by a Δsoil given as

$$\Delta \mathrm{soil}(i, j) = \frac{a_s}{b_s + c_s \times \mathrm{time}^{2\theta}(i, j)} \tag{2}$$

Here, Δsoil indicates the soil content removed by the IWD while moving from location i to $j.time^{2\theta}(i, j)$ denotes the amount of time required for the IWD to move from i to j which is given as

$$\mathrm{time}(i, j) = \frac{\mathrm{HUD}(i, j)}{vel(\mathrm{IWD})} \tag{3}$$

HUD is characterized as a heuristic function which can be used to measure the undesirability of an IWD to select a path from i to j.

Once the IWD properties are updated, the soil content of the concerned solution paths also need to be updated which is possible by following-

$$\mathrm{soil}(i, j) = \rho o \times \mathrm{soil}(i, j) - \rho n \times \Delta \mathrm{soil}(i, j) \tag{4}$$

where ρ_o and ρ_n are between 0 and 1. According to the original IWD algorithm for the TSP, $\rho_o = 1 - \rho_n$.

The most important behavioural characteristic of an IWD lies in its probabilistic selection of a partial solution component where the transition function is given by

$$P(i, j) = \frac{f(\mathrm{soil}(i, j))}{\sum_{k \text{ is unvisited}} f(\mathrm{soil}(i, k))} \tag{5}$$

$$f\left(\mathrm{soil}(i,j)\right) = \frac{1}{\varepsilon + g\left(\mathrm{soil}(i,j)\right)} \tag{6}$$

$$g\left(\mathrm{soil}(i,j)\right) = \begin{cases} \mathrm{soil}(i,j) & \text{if minsoil} \geq 0 \\ \mathrm{soil}(i,j) - \mathrm{minsoil} & \text{if minsoil} < 0 \end{cases} \tag{7}$$

Here, minsoil indicates the least soil available on a path between any location i and j. As illustrated in Eqs. (5–7), the transition probability of an IWD is thus proportional to the soil content available in a path between component i and j. Thus, the lower the soil content of a path, the more the probability of selection of the corresponding solution component.

A discussion of the IWD algorithm for the simultaneous variable selection and fault detection problem is provided in the next section.

3 Hybrid IWD-Based Variable Selection and Fault Detection

The IWD algorithm had earlier been used for solving a variety of discrete combinatorial optimization problems, viz. TSP [6], vehicle routing [7], robot path planning [8]. For the variable selection problem, a solution may be represented as a set of variable indices [4, 5]. For example, if the fault detection dataset is composed of 100 variables/attributes, then a possible solution could be a variable subset comprising of {10, 21, 32, 57, 84} with subset size as 5. Any variable index could thus be a part of the complete solution vector, where the vector size is specified by the user.

Therefore, we initially position each IWD randomly on different variables, from where they commence their flow. Each IWD moves to the next variable by employing the transition probability given by Eq. 5. Once a variable has been visited, a local soil update between variables i and j are performed by Eq. 4 as mentioned before. This process, continues till a complete solution vector is constructed by the IWD, updating the path soil content, the IWD soil and velocity in the process. When a variable subset of the required size is obtained, a corresponding reduced dataset with the given variables is generated. The reduced dataset is thus fed as input to a classifier like SVM [9], which reports back a 10-fold classification cross-validation accuracy (10-fold CVA).The 10-fold CVA is assigned as the fitness function value for the corresponding variable subset. Subsequent IWDs also build up their solution vectors in a similar manner.

At the end of one iteration, the solution vector with the maximum 10-fold CVA is selected as the iteration-best solution (T_{IB}). Subsequently, a certain amount of soil on the edges of the iteration-best solution is decreased based on the quality of the solution. For example, if T_{IB} is given as (5, 11, 18, 21, 76), then the edges to be updated are 5–11, 11–18, 18–21 and 21–27. This can be done according to Eq. 8.

$$\text{soil}\,(i, j) = \rho_s \times \text{soil}\,(i, j) - \rho^{\text{IWD}} \times \frac{1}{(N_{\text{IB}} - 1)} \times \text{soil}_{\text{IB}}^{\text{IWD}} \tag{8}$$

where $\text{soil}_{\text{IB}}^{\text{IWD}}$ represents the soil of the iteration-best IWD. N_{IB} is the number of variables in the solution vector T_{IB}. ρ^{IWD} is the global soil updating parameter, which may be selected from [0, 1]. ρ_s is normally set as $(1+\rho^{\text{IWD}})$.

In addition, we also maintain a global best solution which is given by the maximum of all the iteration-best solutions. The above process is thus repeated till a termination criterion is reached. At this stage, the global best solution is reported as the most optimal solution to the variable selection problem.

3.1 *Information-Gain-Based Heuristic Function*

Owing to a massive search space, it might be advantageous and useful to incorporate additional intelligent information in the form of a variable ranking. We thus employ the information gain (IG) filter as heuristic ranking in the computation of time taken by an IWD (Eq. 3) to move from location i to j as given in (Eq. 9).

$$\text{time}\,(i, j) = \frac{1}{\text{Infogain}\,(j) + \text{vel(IWD)}} \tag{9}$$

IG records the 'information content' of a variable in correlation to the class label for the problem under consideration. So, a higher IG value (corresponding to a relevant variable) aids in minimizing the time taken to traverse from i to j and consequently initiating increased soil removal from the corresponding path given by $\Delta\text{soil}\,(i, j)$ in Eq. (2). IG thus greatly helps in faster convergence by probabilistically favouring soil updates to higher-ranked variables.

4 Results and Discussion

The IWD algorithm has been used to solve the problem of fault detection in the benchmark tennessee eastman process (TEP) [2, 10, 11]. There are 51 variables in the system comprising different pressures, temperatures, etc. monitored over a certain period of time. The data dimensions, variability and dynamics of the process add to the complexity of constructing an efficient classification model. Earlier, several methods have been suggested to solve this problem [2, 10, 11]. We approach this problem typically as a joint subset selection and classification problem by first obtaining an optimal variable subset of a much reduced size and then using SVM to build a binary fault classification model. The faults considered in this case are due to the step in D feed temperature and reactor cooling water inlet temperature as given in [12]. In addition to fault detection, we also considered the ionosphere (34 features) and the

Table 1 Algorithm parameters

Parameters	Values
Number of IWDs	25
Number of iterations	30
ρ_o, ε	0.1, 0.5

Table 2 10-fold cross-validation results

Dataset	SVM	Infogain-SVM	Hybrid IWD
Fault detection	80.89	78.23(15)	92.02(15)
Ionosphere	95.60	95.44(15)	96.86(15)
Wisconsin breast cancer	95.76	94.90(15)	95.95(15)

Wisconsin breast cancer (32 features) datasets for performance benchmarking [13]. The parameters of the hybrid IWD algorithm are as given in Table 1.

We carried out numerous simulations for all cases before arriving at the estimates provided in Table 2. The subset sizes are reported in brackets.

Based on the results, we can infer that the hybrid IWD filter–wrapper technique is clearly more powerful in the selection of important fault detection variables. In contrast, the base SVM without any variable selection and with a filter solely does not help to get the best performances for classification.

5 Conclusion

The IWD algorithm utilizes both filter and wrapper methods to obtain smaller informative subset of variables important for fault detection. The information gain heuristic also provided more possibilities for an effective search space exploration that seems to have helped in the selection of important variables. The algorithm is also simple to implement, flexible and robust since we can adapt it to a given problem and related domain constraints.

Acknowledgments VKJ gratefully acknowledges the Council for Scientific and Industrial Research (CSIR) and Department of Science and Technology (DST), New Delhi, India, for financial support in the form of Emeritus Scientist Grant. The authors also acknowledge the Centre for Modeling and Simulation, University of Pune and C-DAC, Pune, for their kind support.

References

1. Kulkarni, A.J., Jayaraman, V.K., Kulkarni, B.D.: Support vector classification with parameter tuning assisted by agent based technique. Comput. Chem. Eng. **28**, 311–318 (2004)

2. Downs, J.J., Vogel, E.F.: A plant-wide industrial-process control problem. Comput. Chem. Eng. **17**, 245–255 (1993)
3. John, G.H., Kohavi, R., Pfleger, K.: Irrelevant features and the subset selection problem. Proceedings of the Eleventh International Conference on, Machine Learning. 121–129(1994).
4. Gupta A., Jayaraman V. K., Kulkarni. B. D.: Feature selection for cancer classification using ant colony optimization and support vector machines. In Analysis of Biological Data : A Soft Computing Approach. ser. World Scientific, Singapore. 259–280(2006).
5. Nikumbh S., Ghosh S., Jayaraman V. K.: Biogeography-Based Informative Gene Se-lection and Cancer Classification Using SVM and Random Forests. In Proceedings of IEEE World Congress on Computational Intelligence (IEEE WCCI 2012). 187–192(2012).
6. Hosseini, H.S.: The intelligent water drops algorithm: a nature-inspired swarm-based optimization algorithm. International Journal of Bio-Inspired Computation **1**, 71–79 (2009)
7. Kamkar, I.: Intelligent water drops a new optimization algorithm for solving the Vehicle Routing Problem, pp. 4142–4146. In Proceedings of IEEE International Conference on Systems Man and, Cybernetics (2010).
8. Haibin, D., Liu S., Lei, X.: Air robot path planning based on Intelligent Water Drops optimization.In Proceedings of IEEE World Congress on, Computational Intelligence. 1397–1401(2008).
9. Boser, B.E., Guyon, I.M., Vapnik, V. N.: A training algorithm for optimal margin classifiers. In Proceedings of the fifth annual workshop on Computational learning theory, ser. COLT '92. 144–152(1992).
10. Lyman, P.R., Georgakis, C.: Plant-wide control of the Tennessee Eastman problem. Comput. Chem. Eng. **19**, 321–331 (1995)
11. Kulkarni, A., Jayaraman, V.K., Kulkarni, B.D., et al.: Knowledge incorporated support vector machines to detect faults in Tennessee Eastman Process. Comput. Chem. Eng **29**, 2128–2133 (2005)
12. Ricker, N.L.: Tennessee Eastman Challenge Archive http://depts.washington.edu/control/LARRY/TE/download.html
13. UCI Repository http://archive.ics.uci.edu/ml/

Air Conditioning System with Fuzzy Logic and Neuro-Fuzzy Algorithm

Rajani Kumari, Sandeep Kumar and Vivek Kumar Sharma

Abstract Fuzzy logic controls and neuro-fuzzy controls are accustomed to increase the performance of air conditioning system. In this paper, we are trying to provide the new design air conditioning system by exploitation two logics, namely fuzzy logic and neuro-fuzzy management. This paper proposes a set of rule and uses 2 inputs specifically temperature and humidness and 4 outputs specifically compressor speed, fan speed, fin direction and mode of operation. These outputs are rule-based output. At last, compare simulation results of each system exploitation fuzzy logic and neuro-fuzzy management and notice the higher output.

Keywords Fuzzy logic controls · Neuro-fuzzy controls · Air conditioning system · Membership function

1 Introduction

The air conditioning systems are usually found in homes and publicly capsulated areas to make snug surroundings. Air conditioners and air conditioning systems are integral part of nearly each establishment. It includes atmosphere, energy, machinery, physical science and automatic management technology [1, 2].

R. Kumari (✉) · S. Kumar · V. K. Sharma
Jagannath University, Chaksu, Jaipur 303901, India
e-mail: sweetugdd@gmail.com

S. Kumar
e-mail: sandpoonia@gmail.com

V. K. Sharma
e-mail: vivek.kumar@jagannathuniversity.org

B. V. Babu et al. (eds.), *Proceedings of the Second International Conference on Soft Computing for Problem Solving (SocProS 2012), December 28–30, 2012*, Advances in Intelligent Systems and Computing 236, DOI: 10.1007/978-81-322-1602-5_26, © Springer India 2014

1.1 Conventional System

Conventional style strategies need the event of a mathematical model of the control system then use of this model to construct the controller that is represented by the differential equations. The task of dehumidification and temperature decrease goes hand in hand just in case of typical AC. Once target temperature is reached AC seizes to perform sort of a dehumidifier. Within the typical methodology, it is very troublesome to interaction between user preferences, actual temperature and humidness level and it is too nonlinear [3]. Typical AC system controls humidness in its own means while not giving the users any scope for ever changing the point for the targeted humidness. However, this limitation has been overcome by exploitation fuzzy logic management. It is the power to handle nonlinear systems.

1.2 Problem Definition

The optimum limit of temperature that is marked as temperature is 25 °C and saturation point is 11 °C. Standard AC system controls set the target purpose by its own approach. This drawback takes 3 input variables user temperature preference, actual temperature and space saturation point temperature. Fuzzy logic algorithmic program is applied on these variables and finds the ultimate result. User temperature is deducted from actual temperature and then sent it for fuzzification, once this fuzzy arithmetic and criterion is applied on these variables and also the consequence is shipped for defuzzification to urge crisp result.

1.3 Fuzzy Logic Control

Fuzzy logic may be a straightforward however very powerful drawback solving technique with in-depth relevancy. It is presently employed in the fields of business, systems management, physical science and traffic engineering. A fuzzy logic deals with uncertainty in engineering by attaching degrees of certainty to the solution to a logical question. A fuzzy logic system (FLS) will be outlined as the nonlinear mapping of an information set to a scalar output data. Fuzzy logic is employed for management machine and shopper merchandize. Several applications have successfully uses fuzzy logic management, for example environmental management, domestic merchandize and automotive system [4].

The fuzzy sets are quantitatively outlined by membership functions. These functions are generally very straightforward functions that cover a fixed domain of the worth of the system input and output. Fuzzy logic management is primarily rule-based system, and therefore, the performance of it depends on its control rules and membership functions (Fig. 1).

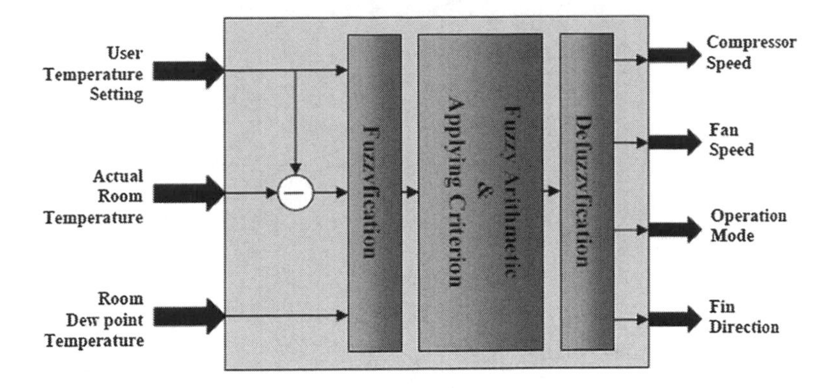

Fig. 1 Block diagram of controller

1.4 Neuro-Fuzzy Logic Control

One of the key issues of the fuzzy logic management is that the problem of selection and style of membership functions for a given downside. Neural networks provide the likelihood of finding the matter of standardization. Neural fuzzy systems will generate formal logic rules and membership functions for advanced systems that a standard fuzzy approach could fail. Hence, combining the adaptive neural networks and formal logic management forms a system known as neuro-fuzzy system. Neuro-fuzzy system is based on the neural network that learned from fuzzy if-then rules. Neural network performance is dependent on the quality and quantity of training samples presented to the network. Neural nets can solve many problems that are either unsolved or inefficiently solved by existing techniques, including fuzzy logic [5, 6].

2 Fuzzy Logic Control Algorithm

Fuzzy logic management primarily based on air conditioning system consists of two inputs that are actual temperature and room temperature dew point (humidity). When measuring actual temperature, the user temperature (Ut) is subtracted from actual temperature for realize the temperature distinction (Td) and sent it for fuzzification. Fuzzy arithmetic and criterion is applied on the input variables, outcome is defuzzified to induce output, and these output signals are distributed to manage the compressor speed. During this case, the range of actual temperature is taken to be 15–50 °C and range of its taken to be 18–30 °C; therefore, the temperature distinction arises between −3 and 32 °C. The input has 2 membership functions. The size over that membership functions for temperature is represented as 0–50 °C and membership functions for humidness is represented as 0–100 %. The output additionally has four

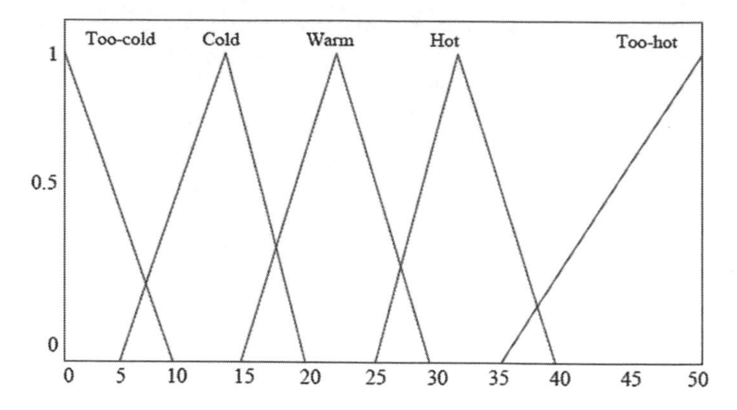

Fig. 2 Temperature membership functions

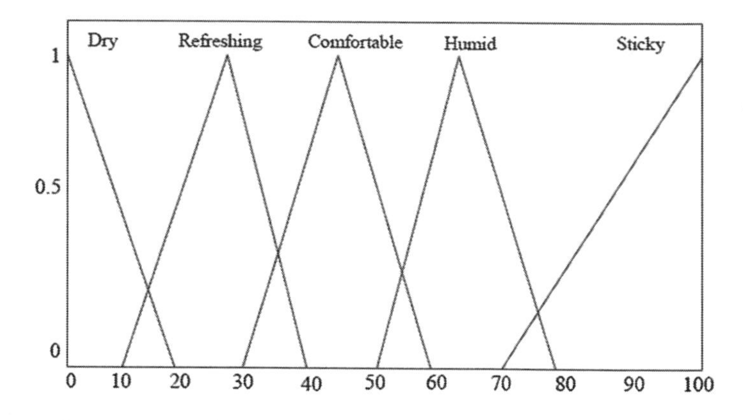

Fig. 3 Humidity membership functions

membership functions particularly compressor speed, fin direction, fan speed and operation mode. The principles base for coming up with is as "IF Temperature is just too cold AND humidness is dry THEN compressor speed is Off, Fin direction is Away, Fan speed is Off and Operation mode is AC" and so on [7, 8] (Table 1).

3 Neuro-Fuzzy Algorithm

Neuro-fuzzy management primarily based on air conditioning system additionally consists of 2 inputs that are actual temperature and space room (humidity). The input, temperature takes the name "input1" (In1) and range is taken to be 0–40 °C for membership function. Similarly, the input, humidness takes the name "input2" (In2) and range is taken to be 5–85 % for membership function. The output, compressor speed amendment the name as "output1" (Out1), fin direction named as "output2"

Table 1 Fuzzy rules for proposed design

Rules	Input		Output			
	Temperature	Humidity	Compressor speed	Fin direction	Fan speed	Operation mode
1	Too cold	Dry	Off	Away	Off	AC
2	Too cold	Refreshing	Off	Away	Off	AC
3	Too cold	Comfortable	Off	Away	Off	AC
4	Too cold	Humid	Off	Away	Very low	AC
5	Too cold	Sticky	Very low	Towards	Low	Dehumidifier
6	Cold	Dry	Off	Away	Off	AC
7	Cold	Refreshing	Off	Away	Off	AC
8	Cold	Comfortable	Very low	Away	Very low	AC
9	Cold	Humid	Very low	Towards	Low	AC
10	Cold	Sticky	Low	Towards	Low	Dehumidifier
11	Warm	Dry	Very low	Away	Very low	AC
12	Warm	Refreshing	Very low	Away	Very low	AC
13	Warm	Comfortable	Low	Away	Low	AC
14	Warm	Humid	Medium	Towards	Medium	Dehumidifier
15	Warm	Sticky	Medium	Towards	Medium	Dehumidifier
16	Hot	Dry	Low	Away	Low	AC
17	Hot	Refreshing	Medium	Away	Medium	AC
18	Hot	Comfortable	Medium	Towards	Medium	AC
19	Hot	Humid	Fast	Towards	Fast	Dehumidifier
20	Hot	Sticky	Fast	Towards	Fast	Dehumidifier
21	Too hot	Dry	Medium	Away	Medium	AC
22	Too hot	Refreshing	Medium	Towards	Medium	AC
23	Too hot	Comfortable	Fast	Towards	Fast	Dehumidifier
24	Too hot	Humid	Fast	Towards	Fast	Dehumidifier
25	Too hot	Sticky	Fast	Towards	Fast	Dehumidifier

(Out2), fan speed named as "output3" (Out3) and operation named as "output4" (Out4). The principles are applied consequently in Table 2.

4 Experimental Results

Result of this experiment is predicated on fuzzy rules and neuro-fuzzy rules. Figures 2 and 3 show input values for fuzzy logic management, and Figs. 4 and 5 show input values for neuro-fuzzy management. Supported these inputs acquire results when simulation of fuzzy logic management is based on air conditioning system that are shown in the following figures. Figure 6 shows the compressor speed memberships of air conditioning system. Compressor speed may be either off or may be varied between 10 and 100 % (Fig. 7).

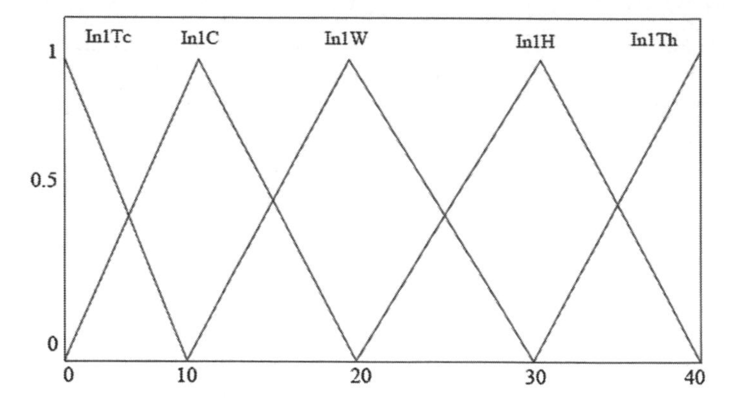

Fig. 4 Input1 membership functions

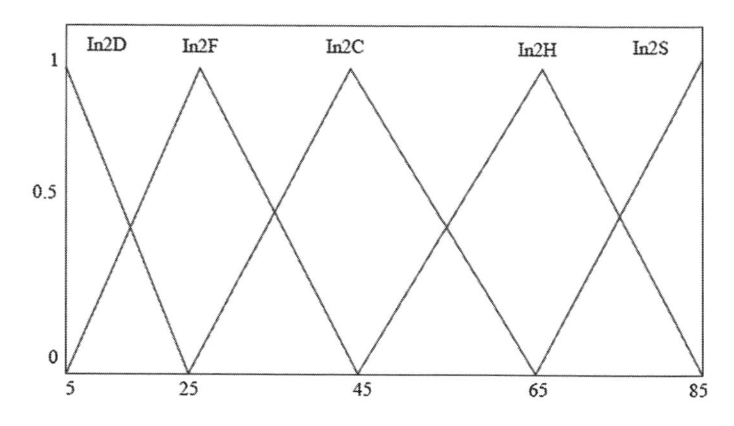

Fig. 5 Input2 membership functions

Figure 8 shows the operation mode memberships of air conditioning system. Mode of operation decides whether AC works like a dehumidifier only or normal. Figure 9 shows the fin direction memberships of air conditioning system. Fin direction directs air from the AC towards or away from occupants.

Figure 10 shows compressor speed with respect to temperature by using fuzzy rules. Figure 11 shows compressor speed with respect to humidity by using fuzzy rules. Figure 12 shows the output1 with respect to input1 by using neuro-fuzzy rules. Figure 13 shows the output1 with respect to input2 by using neuro-fuzzy rules.

Table 2 Neuro-fuzzy rules for proposed design

Rules	Input		Output			
	Temperature (Input1)	Humidity	Compressor speed	Fin direction	Fan speed	Operation mode
1	In1Tc	In2D	Out1Of	Out2A	Out3Of	Out4AC
2	In1Tc	In2R	Out1Of	Out2A	Out3Of	Out4AC
3	In1Tc	In2C	Out1Of	Out2A	Out3Of	Out4AC
4	In1Tc	In2H	Out1Of	Out2A	Out3Vl	Out4AC
5	In1Tc	In2S	Out1Vl	Out2To	Out3L	Out4D
6	In1C	In2D	Out1Of	Out2A	Out3Of	Out4AC
7	In1C	In2R	Out1Of	Out2A	Out3Of	Out4AC
8	In1C	In2C	Out1Vl	Out2A	Out3Vl	Out4AC
9	In1C	In2H	Out1Vl	Out2To	Out3L	Out4AC
10	In1C	In2S	Out1L	Out2To	Out3L	Out4D
11	In1W	In2D	Out1Vl	Out2A	Out3Vl	Out4AC
12	In1W	In2R	Out1Vl	Out2A	Out3Vl	Out4AC
13	In1W	In2C	Out1L	Out2A	Out3L	Out4AC
14	In1W	In2H	Out1M	Out2To	Out3Of	Out4D
15	In1W	In2S	Out1M	Out2To	Out3M	Out4D
16	In1H	In2D	Out1L	Out2A	Out3L	Out4AC
17	In1H	In2R	Out1M	Out2A	Out3M	Out4AC
18	In1H	In2C	Out1M	Out2To	Out3M	Out4AC
19	In1H	In2H	Out1F	Out2To	Out3F	Out4D
20	In1H	In2S	Out1F	Out2To	Out3F	Out4D
21	In1Th	In2D	Out1M	Out2A	Out3M	Out4AC
22	In1Th	In2R	Out1M	Out2To	Out3M	Out4AC
23	In1Th	In2C	Out1F	Out2To	Out3F	Out4D
24	In1Th	In2H	Out1F	Out2To	Out3F	Out4D
25	In1Th	In2S	Out1F	Out2To	Out3F	Out4D

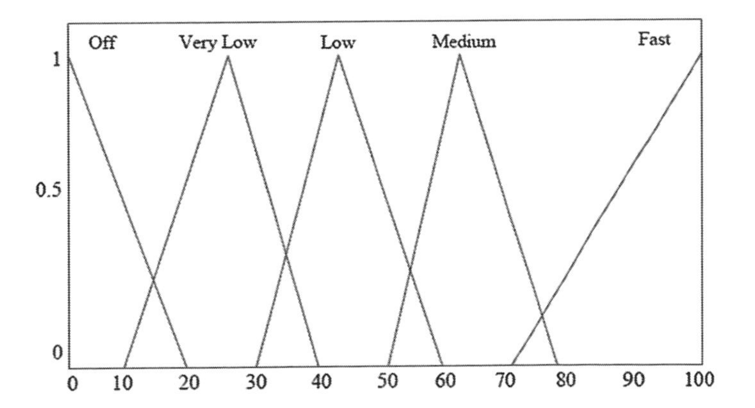

Fig. 6 Compressor speed membership functions

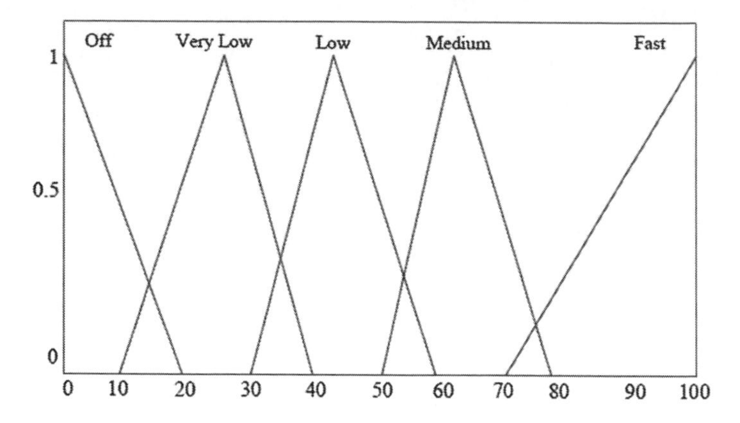

Fig. 7 Fan speed membership functions

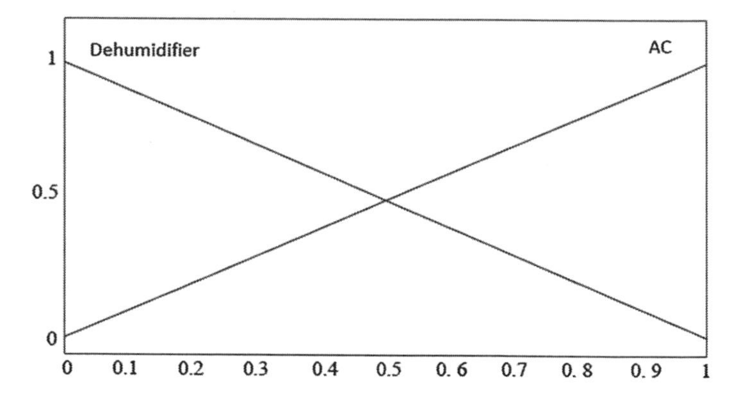

Fig. 8 Operation mode membership functions

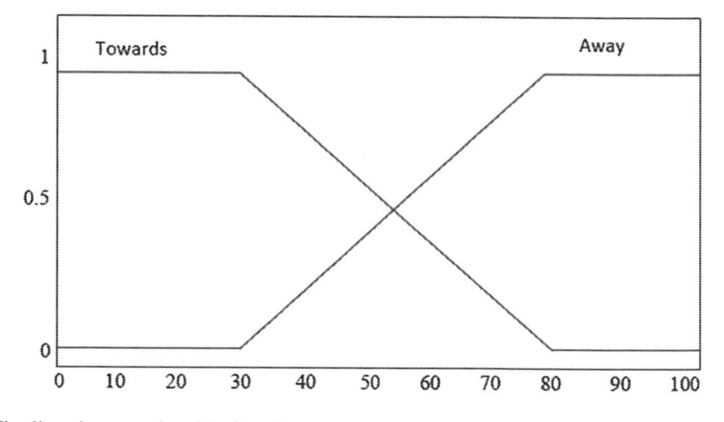

Fig. 9 Fin direction membership function

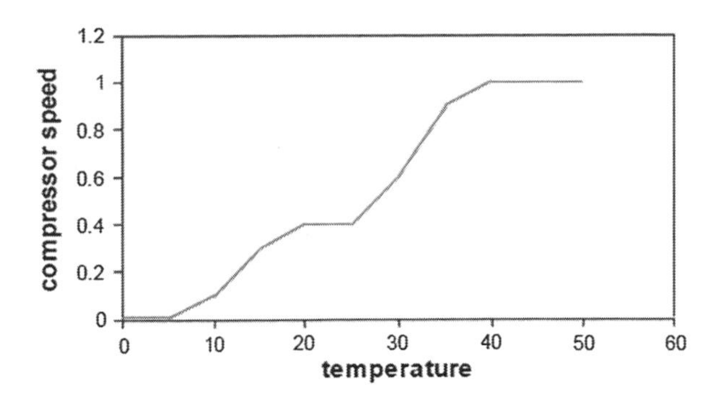

Fig. 10 Compressor speed with temperature

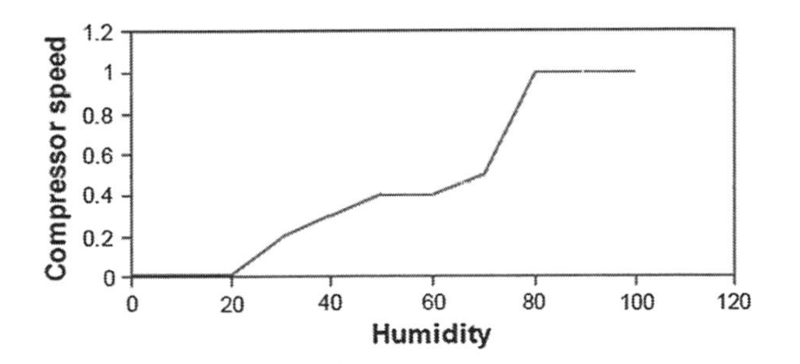

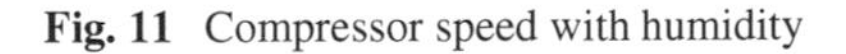

Fig. 11 Compressor speed with humidity

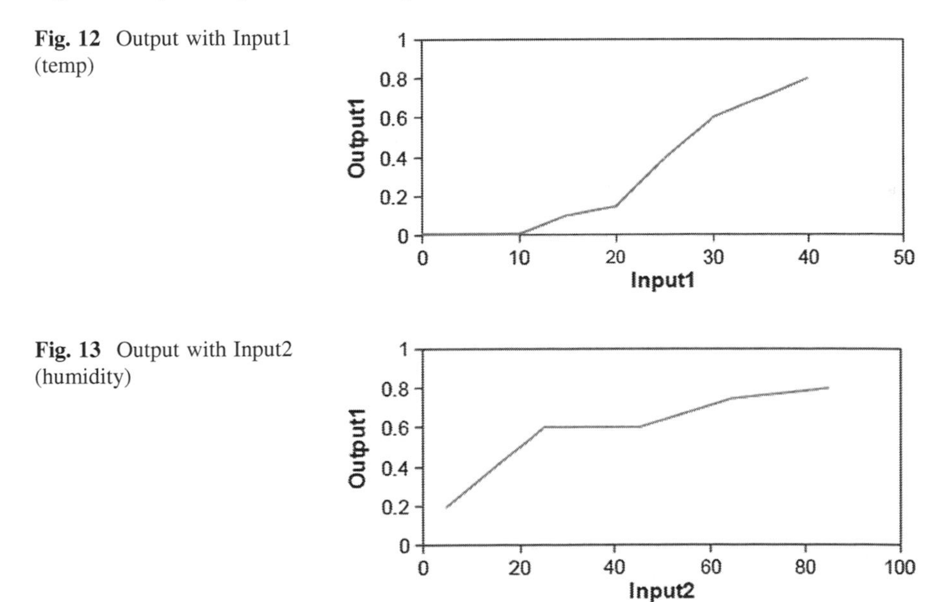

Fig. 12 Output with Input1 (temp)

Fig. 13 Output with Input2 (humidity)

5 Conclusion

Neuro-fuzzy algorithm is better than fuzzy logic algorithm in air conditioning system. Neuro logic algorithm gives a better control than fuzzy logic. In neuro logic algorithm, performance of compressor speed is much better than fuzzy logic algorithm. In fuzzy logic control design, the compressor speed remains constant for temperature range from 35 °C onwards, but in neuro-fuzzy control design, it increases consistently with respect to temperature. By this, it provides proper output and save energy. It controls the room environment and weather.

References

1. Nasution, H., Jamaluddin, H., Syeriff, J.M.: Energy analysis for air conditioning system using fuzzy logic controller. TELKOMNIKA **9**(1), 139–150 (2011)
2. Du, M., Fan, T., Su, W., Li, H.: Design of a new practical expert fuzzy controller in central air conditioning control system. IEEE Pacific-Asia workshop on computational intelligence and industrial application (2008)
3. Passino, K.M., Yurkovich, S.: Fuzzy control, Addison Wesley (1998)
4. Isomursu, P., Rauma, T.: A self-tuning fuzzy logic controller for temperature control of superheated steam. Fuzzy systems, IEEE world congress on computational intelligence, proceedings of the third IEEE conference, vol. 3 (1994)
5. Islam, M.S., Sarker, Z., Ahmed Rafi, K.A., Othman, M.: Development of a fuzzy logic controller algorithm for air conditioning system. ICSE proceedings (2006)
6. Batayneh, W., Al-Araidah, O., Bataineh, K.: Fuzzy logic approach to provide safe and comfortable indoor environment. Int. J. Eng. Sci. Technol. **2**(7) (2010)
7. Abbas, M., Khan, M.S., Zafar, F.: Autonomous room air cooler using fuzzy logic control system. Int. J. Sci. Eng. Res. **2**(5), 74–81 (2011)
8. Hamidi, M., Lachiver, G.: A fuzzy control system based on the human sensation of thermal comfort. Fuzzy systems proceedings, 1998. IEEE world congress on computational intelligence, the IEEE international conference, vol. **1** (1998)

Improving the Performance of the Optimization Technique Using Chaotic Algorithm

R. Arunkumar and V. Jothiprakash

Abstract Optimizing the operations of a multi-reservoir systems are complex because of their larger dimension and convexity of the problem. The advancement of soft computing techniques not only overcomes the drawbacks of conventional techniques but also solves the complex problems in a simple manner. However, if the problem is too complex with hardbound variables, the simple evolutionary algorithm results in slower convergence and sub-optimal solutions. In evolutionary algorithms, the search for global optimum starts from the randomly generated initial population. Thus, initializing the algorithm with a better initial population not only results in faster convergence but also results in global optimal solution. Hence in the present study, chaotic algorithm is used to generate the initial population and coupled with genetic algorithm (GA) to optimize the hydropower production from a multi-reservoir system in India. On comparing the results with simple GA, it is found that the chaotic genetic algorithm (CGA) has produced slightly more hydropower than simple GA in fewer generations and also converged quickly.

Keywords Optimization · Genetic algorithm · Chaotic algorithm · Multi-hydropower system.

1 Introduction

The last decade had witnessed several optimization techniques from conventional linear programming to recently soft computing techniques. Among the soft computing techniques, the genetic algorithm (GA) had been widely used for optimization of

R. Arunkumar (✉) · V. Jothiprakash
Department of Civil Engineering, Indian Institute of Technology Bombay, Mumbai 400076, India
e-mail: arunkumar.r@iitb.ac.in

V. Jothiprakash
e-mail: vprakash@iitb.ac.in

B. V. Babu et al. (eds.), *Proceedings of the Second International Conference on Soft Computing for Problem Solving (SocProS 2012), December 28–30, 2012*, Advances in Intelligent Systems and Computing 236, DOI: 10.1007/978-81-322-1602-5_27,

reservoir operation [1–3]. Application of simple GA technique to solve real-life water resources problems, especially optimizing multi-hydropower system is always cumbersome because of the complex nature, large number of variables and nonlinearity of the problem. Some recent studies reported that simple optimization techniques often succumbs to premature convergence and results in local optimal solution for complex water resources problems [4–6]. Hence, still researches are emerging to introduce new techniques and also to improve the performance of the existing techniques to achieve global optimum and faster convergence.

The search for global optimal solution in evolutionary algorithms begins from the randomly generated initial population. Thus, a better initial population leads to faster convergence and not only saves substantial computational time but also results in global optimal solution. Hence, to generate a good initial population, chaotic algorithm is used, since chaos is highly sensitivity to the initial value, ergodic, and randomness in nature [7]. Yuan et al. [4] proposed a hybrid chaotic genetic algorithm (HCGA) model to prevent premature convergence for a short-term hydropower scheduling problem. Cheng et al. [5] optimized the hydropower reservoir operation using HCGA and reported that the long-term average annual energy production was the best in HCGA and also it converges faster than the standard GA. It was also reported that the combination of chaotic characteristics along with general optimization algorithm would more likely result in global optimal solution. Han and Lu [6] proposed a mutative scale chaos optimization algorithm (MSCOA) for the economic load dispatch problem. Huang et al. [8] optimized the hydropower reservoir with ecological consideration using chaotic genetic algorithm (CGA) approach. These studies show that GA results better when coupled with chaotic algorithm and also escapes premature convergence. Hence in this study, in order to improve the performance of GA, chaotic algorithm is used to generate initial population and tested for a real-life hydropower system. The proposed algorithm is applied to maximize the hydropower production from a multi-reservoir system, namely Koyna hydroelectric project (KHEP).

The KHEP consists of two reservoirs, namely Koyna and Kolkewadi, which has four powerhouses to a total capacity of 1,960 MW [9]. Earlier, Jothiprakash and Arunkumar [10] optimized the operation of Koyna reservoir alone considering it as a single reservoir system. However, in the present study, both the reservoirs are considered and optimized as multi-reservoir system. Apart from hydropower production, it also serves multiple purposes such as irrigation and flood control. In this system, the major powerhouses of the Koyna reservoir and Kolkewadi reservoir are in the western side and the irrigation releases are in the eastern side. The diversion of large quantity of water toward the western side for power production has resulted in disputes from the eastern side stakeholders, and hence, it was limited by Krishna water dispute tribunal [11]. This limiting constraint on the discharge for power production from the western side powerhouses made the system more complex by making the variables hard bound. This hard-binding constraint is considered in this study.

2 Chaotic Genetic Algorithm

GA is a search and optimization algorithm based on the principles of natural genetics [12]. In contrast to traditional optimization technique, GA searches the optimal solution from a randomly generated population within the upper and lower bounds of the variables. Each solution is represented through group of genes (sub-string) called chromosome (string) in the population space (search space). Each gene controls one or more features of the chromosomes. In the present study, the chaotic logistic mapping method is used to generate the initial population. Chaos often exists in nonlinear systems [13] and exhibits many good properties such as ergodicity, stochasticity, and irregularity [7]. May [7] proposed a one-dimensional logistic mapping equation to generate a chaotic sequence. It is given as:

$$Y_{j+1} = \lambda Y_j \left(1 - Y_j\right) \quad j = 1, 2, 3 \ldots \tag{1}$$

where λ is a control parameter and varies between $0 \leq \lambda \leq 4$. The chaotic sequence is produced when λ is equal to 4. The initial random variable (Y_1) is generated in the range between 0 and 1; however, it should not be equal to 0.25, 0.5, and 0.75, since it leads to a deterministic sequence [7]. Thus, the generated sequence using the logistic Eq. (1) is highly irregular and possesses chaotic characteristics. Each variable in the sequence is dependent on initial variable, and a small change in initial value causes a large difference in its long-time behavior, which is the basic characteristic of chaos. This can be correlated with the releases in reservoir operation, such that the releases in the subsequent months depend on previous month releases. Hence in the present study, the initial population is generated as floating-point chaotic values within the upper (UB_j) and lower bounds (LB_j) of the variables using the equation:

$$X_{i,j} = Y_{i,j} \times (\mathrm{UB}_j - \mathrm{LB}_j) + \mathrm{LB}_j \quad i = 1, 2, \ldots, N_p;\ j = 1, 2, \ldots, N_v \tag{2}$$

where $X_{i,j}$ is the 'j^{th}' sub-string of 'i^{th}' string, N_p is the population size, $Y_{i,j}$ is the chaotic variable, and N_v is the number of variables. Once the initial population is generated, the fitness of each string is evaluated using an appropriate fitness function. Based on the fitness value, the strings are selected for crossover and mutation to create a new population for the next generation. The tournament selection is used in the present study, since it provides selective pressure by holding a tournament among the selected individuals [12]. In the tournament selection, the fitness of the randomly selected strings from the population is compared with each other and the string having with higher fitness value will be copied to the mating pool. This process is repeated until the mating pool is filled with strings for generating new offspring for the next generation. The mating pool comprising the winners of the tournament will have higher average fitness value. Then, the strings in the mating pool are made to cross each other to create a new population. The simulated binary crossover (SBX) [14] is applied to create new population. The following steps were performed during SBX operation [12]. In the first step, a random number (u_i) between 0 and 1 is generated.

Then, the spread factor (β_{qi}) is computed using the equation

$$\beta_{\text{qi}} = \begin{cases} (2u_i)^{\frac{1}{(\eta c+1)}}, & \text{if } u_i < 0.5 \\ \left(\frac{1}{2(1-u_i)}\right)^{\frac{1}{(\eta c+1)}}, & \text{otherwise} \end{cases} \tag{3}$$

where η_c is the distribution index for crossover. Deb [12] reported that larger value of η_c produces '*near-parent*' offspring's and vice versa. Then, the off-springs $x_i^{1,t+1}$ and $x_i^{2,t+1}$ are computed from $x_i^{1,t}$ and $x_i^{2,t}$ using the equations,

$$x_i^{1,t+1} = 0.5\left[\left(1+\beta_{\text{qi}}\right)x_i^{1,t} + \left(1-\beta_{\text{qi}}\right)x_i^{2,t}\right] \tag{4}$$

$$x_i^{2,t+1} = 0.5\left[\left(1-\beta_{\text{qi}}\right)x_i^{1,t} + \left(1+\beta_{\text{qi}}\right)x_i^{2,t}\right] \tag{5}$$

where $x_i^{1,t}$ and $x_i^{2,t}$ are the parent string with ith sub-string in the tth generation. If the created offsprings are not within the upper and lower limits, the probability distribution needs to be adjusted accordingly. The new two offsprings are symmetric about the parent to avoid bias toward any particular parent solution in a single crossover operation. After crossover, the strings are subjected to mutation. The mutation operator introduces random changes into the characteristic of the offsprings. Mutation is generally applied at the sub-string level at a very small rate and depends on the length of the string. The mutation reintroduces the genetic diversity into the population and assists the search to escape from the local optima [12]. Then, the fitness of the newly created population is evaluated, and the procedure is continued until the termination criteria are reached.

3 Hydropower Model Development

The objective of the present study is to maximize the power production from all the four powerhouses of the KHEP and is expressed as:

$$\text{Max } Z = \sum_{t=1}^{12}\sum_{n=1}^{4} \text{PH}_{n,t} \tag{6}$$

where $\text{PH}_{n,t}$ is the power production from the powerhouse 'n' during the time period 't' in terms of kWh. The hydropower production from the power plant [15] is given by

$$\text{PH}_{n,t} = K \times R_{n,t} \times \text{HN}_{n,t} \times \eta \tag{7}$$

where K is the constant for converting hydropower production in terms of kilo Watt hour (kWh), $R_{n,t}$ is discharge to the powerhouse 'n' during the time period 't', $\text{HN}_{n,t}$

is the net head available for the powerhouse 'n' during the time period 't', and η is the plant efficiency.

The above objective function is subjected to various constraints. They are:

$$H_{n,t} \geq \text{MDDL}_{n,t} \quad \text{t} = 1, 2 \ldots 12; \ \text{n} = 1, 2, 3, 4 \tag{8}$$

$$\text{PH}_{n,t} \leq P\text{max}_{n,t} \quad \text{t} = 1, 2 \ldots 12; \ \text{n} = 1, 2, 3, 4 \tag{9}$$

$$R_{4,t} \geq \text{ID}_t \quad \text{t} = 1, 2 \ldots 12 \tag{10}$$

$$S_{x,\min} \leq S_{x,t} \leq S_{x,\max} \quad \text{t} = 1, 2 \ldots 12; \ \text{x} = 1, 2 \tag{11}$$

$$S_{1,(t+1)} = S_{1,t} + I_{1,t} - \sum_{n=1}^{1,3,4} R_{n,t} - O_{1,t} - E_{1,t} \quad \text{t} = 1, 2 \ldots 12 \tag{12}$$

$$S_{2,(t+1)} = S_{2,t} + I_{2,t} + R_{1,t} + R_{3,t} - R_{2,t} - O_{2,t} - E_{2,t} \quad \text{t} = 1, 2 \ldots 12 \tag{13}$$

$$O_{x,t} = S_{x,(t+1)} - S_{x,\max} \quad \text{t} = 1, 2 \ldots 12 \tag{14}$$

$$O_{x,t} \geq 0 \quad \text{t} = 1, 2 \ldots 12 \tag{15}$$

where $\text{MDDL}_{n,t}$ is the minimum drawdown level (m) for the powerhouse 'n'; $Pmax_{n,t}$ is the maximum generation capacity (kWh) for the powerhouse 'n'; $R_{4,t}$ is the irrigation release (10^6 m^3); ID_t is the monthly irrigation demand (10^6 m^3); $S_{x,\min}$ is the minimum storage of the reservoir 'x' (10^6 m^3); $S_{x,\max}$ is the maximum storage of the reservoir 'x' (10^6 m^3); $S_{x,t}$ is the storage in the reservoir 'x' (10^6 m^3); $S_{x,(t+1)}$ is the final storage in the reservoir 'x' (10^6 m^3); $I_{x,t}$ is the inflow into the reservoir 'x' (10^6 m^3); $O_{x,t}$ is the overflow from the reservoir 'x' (10^6 m^3); $E_{x,t}$ is the evaporation losses from the reservoir (10^6 m^3); t is the time period.

As already stated, the diversion of water to the western side powerhouses are limited by KWDT [11]. This constraint is given by:

$$\sum_{t=1}^{12} \sum_{n=1,3} R_{n,t} \leq R_{w,\max} \tag{16}$$

$$\sum_{t=1}^{12} R_{4,t} \leq \text{AID}_{\max} \tag{17}$$

where $R_{w,\max}$ is the maximum water that can be diverted to the western side for power production, and $\text{AID}_{\max}$ is the water to be released annually for irrigation to the eastern side. $R_{4,t}$ is the monthly irrigation release on the eastern side of the reservoir. These constraints make the system more complex by limiting the discharge to the powerhouses.

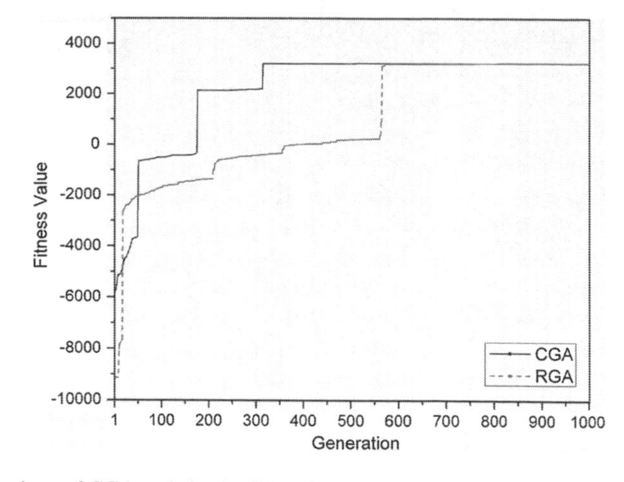

Fig. 1 Comparison of CGA and simple GA technique

4 Results and Discussion

The proposed CGA is applied to a complex multi-powerhouse system, namely KHEP to maximize its power production. The KHEP is one of the largest hydropower projects in India and consists of two reservoirs with various stages of development. The constraint on diverting large quantity of water for power production on the western side powerhouses makes the problem complex. This hard-binding constraint on discharge restricts the operation of the powerhouses on the western side to is full potential. Evaporation is one of the important components of reservoir operation studies, since a considerable amount of water is lost, especially from large reservoirs. Hence in the present study, the regression equation developed by Arunkumar and Jothiprakash [16] for estimating reservoir evaporation is considered and directly incorporated in the continuity equation. Both the CGA and simple GA used in the study employs tournament selection, simulated binary crossover, and random mutation for comparison. The crossover probability of CGA and simple GA is varied from 0.5 to 0.95 with an increment of 0.05 and found that 0.80 resulted better. The mutation probability is fixed as the ratio of the number of variable ($1/n$) [12]. The elitism is also applied to preserve the best strings in the population. The algorithm is evaluated for 1,000 generation for a population size of 250. The constraints of the problems are handled by penalty function approach. Thus, heavy penalties are imposed on fitness function, if the constraints are violated. For each constraint different value of penalties are assumed. The total annual power production shows that CGA has resulted slightly higher power production of 3,225.71 $\times$ 10^6 kWh than simple GA (3,224.23 $\times$ 10^6 kWh). However, the convergence of these techniques over the generation varied largely and is given in Fig. 1. From the figure, it can be observed the convergence to optimal solution by CGA is faster than simple GA. It can also be noted that due to hard-binding constraints on releases, both the techniques have resulted in sub-optimal solution for first few generations. Thus imposing heavy penalty on fitness function leads to negative fitness value and results in sub-optimal solution. However, the CGA has satisfied all the constraints and reaches the optimal solution in lesser generations than simple GA. This also shows that when hard-binding constraints are imposed strictly, the simple GA takes more generations for convergence. The time taken by CGA is 1,790.753 s where as simple GA took 2,418.966 s.

5 Conclusion

In the present study, the chaotic algorithm is combined with genetic algorithm to maximize the power production from a multi-powerhouse system. Based on the performances, it is found that the chaotic genetic algorithm resulted slightly higher power production than simple GA within fewer generations and also converged quickly. This shows that coupling the chaotic algorithm with evolutionary algorithm has enriched the global search of the optimization technique by having better initial population. Thus, it may be concluded that the chaotic algorithm with general optimizer converges quickly to global optimum in lesser generation compared to simple optimization technique.

Acknowledgments The authors gratefully acknowledge the Ministry of Water Resources, Government of India, New Delhi, for sponsoring this research project. The authors also thank Chief Engineer, KHEP, Executive Engineer, Koyna Dam and Executive Engineer, Kolkewadi Dam for providing the necessary data.

References

1. Wardlaw, R., Sharif, M.: Evaluation of genetic algorithms for optimal reservoir system operation. J. Water Resour. Plann. Manage. **125**(1), 25–33 (1999)
2. Sharif, M., Wardlaw, R.: Multireservoir systems optimization using genetic algorithms: case study. J. Comput. Civ. Eng. **14**(4), 255–263 (2000)
3. Jothiprakash, V., Shanthi, G., Arunkumar, R.: Development of operational policy for a multi-reservoir system in India using genetic algorithm. Water Resour. Manage. **25**(10), 2405–2423 (2011)
4. Yuan, X., Yuan, Y., Zhang, Y.: A hybrid chaotic genetic algorithm for short-term hydro system scheduling. Math. Comput. Simul. **59**(4), 319–327 (2002)
5. Cheng, C.T., Wang, W.C., Xu, D.M., Chau, K.: Optimizing hydropower reservoir operation using hybrid genetic algorithm and chaos. Water Resour. Manage. **22**(7), 895–909 (2008)
6. Han, F., Lu, Q.S.: An improved chaos optimization algorithm and its application in the economic load dispatch problem. Int. J. Compt. Math. **85**(6), 969–982 (2008)
7. May, R.M.: Simple mathematical models with very complicated dynamics. Nature **261**(5560), 459–467 (1976)
8. Huang, X., Fang, G., Gao, Y., Dong, Q.: Chaotic optimal operation of hydropower station with ecology consideration. Energy Power Eng. **2**(3), 182–189 (2010)

9. KHEP: Koyna hydro electric project stage-IV. Irrigation Department, Government of Maharashtra, (2005)
10. Jothiprakash, V., Arunkumar, R.: Optimization of hydropower reservoir using evolutionary algorithms coupled with chaos. Water Resour. Manage. (2013). doi:10.1007/s11269-013-0265-8, (Published online)
11. KWDT: Krishna water disputes tribunal: The report of the Krishna water disputes tribunal with the decision. Ministry of water resources, Government of India. New Delhi (2010)
12. Deb, K.: Multi-objective Optimization Using Evolutionary Algorithm. Wiley, New Jersey (2001)
13. Williams, G.P.: Chaos Theory Tamed. Joseph Henry Press, Washington, D.C (1997)
14. Deb, K., Agrawal, R.B.: Simulated binary crossover for continuous search space. Complex Syst. **9**, 115–148 (1995)
15. Loucks, D.P., Stedinger, J.R., Haith, D.A.: Water Resources Systems Planning and Analysis. Prentice Hall Inc, Englewood Cliffs, New Jersey (1981)
16. Arunkumar, R., Jothiprakash, V.: Optimal reservoir operation for hydropower generation using non-linear programming model. J. Inst. Eng. (India) **93**(2), 111–120 (2012)

New Reliable Algorithm for Fractional Harry Dym Equation

Devendra Kumar and Jagdev Singh

Abstract In this paper, a new reliable algorithm based on homotopy perturbation method using Laplace transform, named homotopy perturbation transform method (HPTM), is proposed to solve nonlinear fractional Harry Dym equation. The numerical solutions obtained by the HPTM show that the approach is easy to implement and computationally very attractive.

Keywords Fractional Harry Dym equation · Laplace transform · Homotopy perturbation transform method · He's polynomials · Maple code.

1 Introduction

Fractional differential equations have gained importance and popularity, mainly due to its demonstrated applications in science and engineering. For example, these equations are increasingly used to model problems in fluid mechanics, acoustics, biology, electromagnetism, diffusion, signal processing, and many other physical processes [1–6].

In this paper, we consider the following nonlinear time-fractional Harry Dym equation of the form:

$$D_t^{\alpha} u(x,t) = u^3(x,t) D_x^3 u(x,t), \quad 0 < \alpha \leq 1, \tag{1}$$

D. Kumar (✉)
Department of Mathematics, JaganNath Gupta Institute of Engineering and Technology, Jaipur, Rajasthan 302022, India
e-mail: devendra.maths@gmail.com

J. Singh
Department of Mathematics, Jagan Nath University, Village-Rampura, Tehsil-Chaksu, Jaipur, Rajasthan 303901, India
e-mail: jagdevsinghrathore@gmail.com

B. V. Babu et al. (eds.), *Proceedings of the Second International Conference on Soft Computing for Problem Solving (SocProS 2012), December 28–30, 2012*, Advances in Intelligent Systems and Computing 236, DOI: 10.1007/978-81-322-1602-5_28,

with the initial condition

$$u(x,0) = \left(a - \frac{3\sqrt{b}}{2}x\right)^{2/3}. \tag{2}$$

where $u(x,t)$ is a function of x and t. The derivative is understood in the Caputo sense. In the case of $\alpha = 1$, the fractional Harry Dym equation reduces to the classical nonlinear Harry Dym equation. The exact solution of the Harry Dym equation is given by [7]

$$u(x,t) = \left(a - \frac{3\sqrt{b}}{2}(x+bt)\right)^{2/3}, \tag{3}$$

where a and b are suitable constants. The Harry Dym is an important dynamical equation which finds applications in several physical systems. The Harry Dym equation first appeared in Kruskal and Moser [8] and was discovered by H. Dym in 1973–1974. Harry Dym is a completely integrable nonlinear evolution equation. The Harry Dym equation is very interesting because it obeys an infinite number of conversion laws; it does not possess the Painleve property. It has strong links to the Korteweg-de Vries equation, and applications of this equation were found in the problems of hydrodynamics [9]. Recently, a fractional model of Harry Dym equation was presented by Kumar et al. [10] and approximate analytical solution was obtained by using homotopy perturbation method (HPM).

The HPM was first introduced by He [11]. In a recent paper, a new approach, named homotopy perturbation transform method (HPTM), is introduced by Khan and Wu [12] to handle nonlinear equations. The HPTM is a combination of Laplace transform method, HPM, and He's polynomials.

In this paper, we apply the HPTM to solve the nonlinear time-fractional Harry Dym equation. It provides the solutions in terms of convergent series with easily computable components in a direct way without using linearization, perturbation, or restrictive assumptions.

2 Basic Definitions of Fractional Calculus

In this section, we mention the following basic definitions of fractional calculus.

Definition 2.1 The Riemann–Liouville fractional integral operator of order $\alpha > 0$, of a function $f(t) \in C_\mu, \mu \geq -1$ is defined as [2]

$$J^\alpha f(t) = \frac{1}{\Gamma(\alpha)} \int_0^t (t-\tau)^{\alpha-1} f(\tau) d\tau, (\alpha > 0), \tag{4}$$

$$J^0 f(t) = f(t). \tag{5}$$

For the Riemann–Liouville fractional integral, we have

$$J^{\alpha}t^{\gamma} = \frac{\Gamma(\gamma+1)}{\Gamma(\gamma+\alpha+1)}t^{\alpha+\gamma}. \tag{6}$$

Definition 2.2 The fractional derivative of $f(t)$ in the Caputo sense is defined as [3]

$$D_t^{\alpha} f(t) = J^{m-\alpha} D^n f(t) = \frac{1}{\Gamma(n-\alpha)} \int_0^t (t-\tau)^{m-\alpha-1} f^{(m)}(\tau) d\tau, \tag{7}$$

for $m-1 < \alpha \le m, m \in N, t > 0$.

Definition 2.3 The Laplace transform of the Caputo derivative is given as [6]

$$L[D_t^{\alpha} f(t)] = s^{\alpha} L[f(t)] - \sum_{r=0}^{m-1} s^{\alpha-r-1} f^{(r)}(0+), (m-1 < \alpha \le m). \tag{8}$$

3 Basic Idea of HPTM

We consider a general fractional nonlinear nonhomogeneous partial differential equation with the initial condition of the form:

$$D_t^{\alpha} u(x,t) + Ru(x,t) + Nu(x,t) = g(x,t), \tag{9}$$

$$u(x,0) = f(x), \tag{10}$$

where $D_t^{\alpha} u(x,t)$ is the Caputo fractional derivative of the function $u(x,t)$, R is the linear differential operator, N represents the general nonlinear differential operator, and $g(x,t)$ is the source term.

Applying the Laplace transform (denoted in this paper by L) on both sides of Eq. (9) and using result (8), we have

$$L[u(x,t)] = \frac{f(x)}{s} + \frac{1}{s^{\alpha}} L\,[g(x,t)] - \frac{1}{s^{\alpha}} L\,[R\,u(x,t) + Nu(x,t)]. \tag{11}$$

Operating with the Laplace inverse on both sides of Eq. (11) gives

$$u(x,t) = G(x,t) - L^{-1}\left[\frac{1}{s^{\alpha}} L[R\,u(x,t) + Nu(x,t)]\right], \tag{12}$$

where $G(x,t)$ represents source term and initial conditions separately. Now, we apply the homotopy perturbation method

$$u(x,t)=\sum_{n=0}^{\infty}p^{n}u_{n}(x,t), \tag{13}$$

and the nonlinear term can be decomposed as

$$N\,u(x,t)=\sum_{n=0}^{\infty}p^{n}H_{n}(u), \tag{14}$$

for some He's polynomials $H_n(u)$ [13, 14] that are given by

$$H_{n}(u_{0},u_{1},\ldots,u_{n})=\frac{1}{n!}\frac{\partial^{n}}{\partial p^{n}}\left[N\left(\sum_{i=0}^{\infty}p^{i}u_{i}\right)\right]_{p=0},\ n=0,\ 1,\ 2,\ldots \tag{15}$$

Substituting Eqs. (13) and (14) in Eq. (12), we get

$$\sum_{n=0}^{\infty}p^{n}u_{n}(x,t)=G(x,t)-p\left(L^{-1}\left[\frac{1}{s^{\alpha}}L\left[R\sum_{n=0}^{\infty}p^{n}u_{n}(x,t)+\sum_{n=0}^{\infty}p^{n}H_{n}(u)\right]\right]\right), \tag{16}$$

which is the coupling of the Laplace transform and the homotopy perturbation method using He's polynomials. Comparing the coefficients of like powers of p, the approximations u_0, u_1, u_2, and so on are obtained.

4 Solution of the Problem

In this section, we use the HPTM to solve the time-fractional Harry Dym Eqs. (1)–(2). Applying the Laplace transform on both sides of Eq. (1) and using result (8), we get

$$L[u(x,t)]=\frac{1}{s}\left(a-\frac{3\sqrt{b}}{2}x\right)^{2/3}+\frac{1}{s^{\alpha}}L\left[u^{3}(x,t)D_{x}^{3}u(x,t)\right]. \tag{17}$$

The inverse Laplace transform implies that

$$u(x,t)=\left(a-\frac{3\sqrt{b}}{2}x\right)^{2/3}+L^{-1}\left[\frac{1}{s^{\alpha}}L\left[u^{3}(x,t)D_{x}^{3}u(x,t)\right]\right]. \tag{18}$$

Now applying the homotopy perturbation method, we get

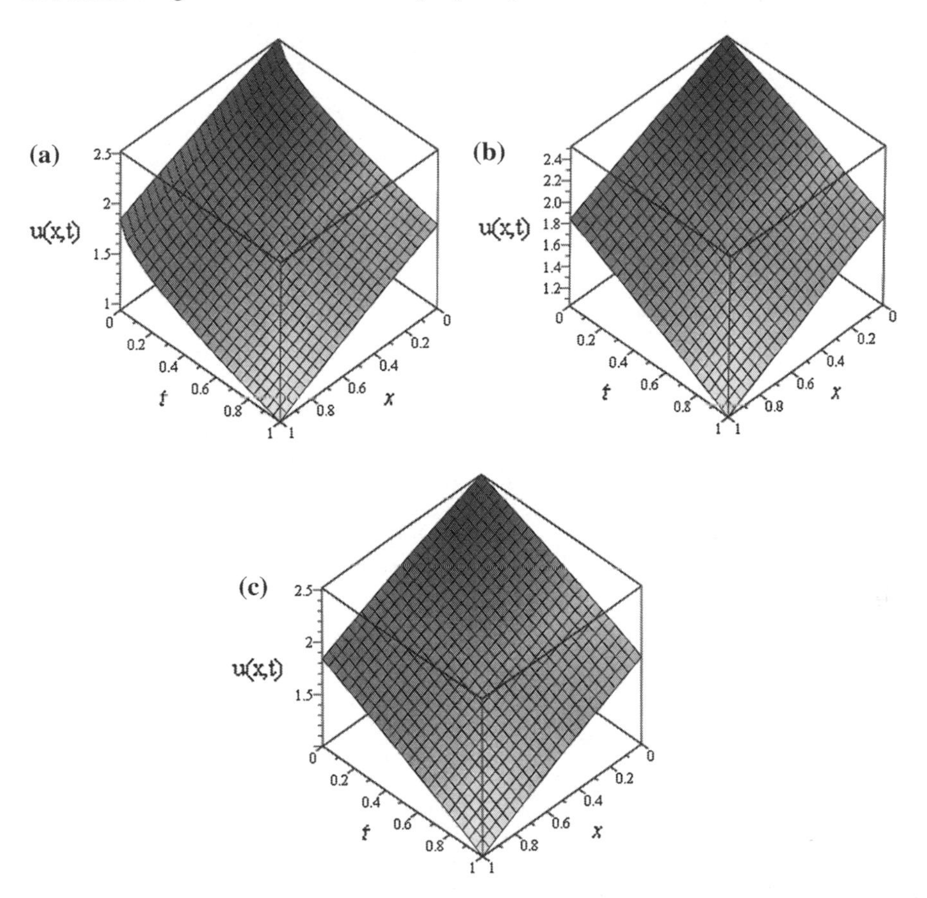

Fig. 1 The behavior of the $u(x,t)$ w.r.t. x and t are obtained, when **a** $\alpha = 1/2$ **b** $\alpha = 1$ **c** exact solution

$$\sum_{n=0}^{\infty} p^n u_n(x,t) = \left(a - \frac{3\sqrt{b}}{2}x\right)^{2/3} + p\left(L^{-1}\left[\frac{1}{s^{\alpha}}L\left[\sum_{n=0}^{\infty} p^n H_n(u)\right]\right]\right), \quad (19)$$

where $H_n(u)$ are He's polynomials [13, 14] that represent the nonlinear terms.

The first few components of He's polynomials are given by

$$\begin{aligned} &H_0(u) = u_0^3 D_x^3 u_0, \\ &H_1(u) = u_0^3 D_x^3 u_1 + 3u_0^2 u_1 D_x^3 u_0, \\ &\vdots \end{aligned} \quad (20)$$

Comparing the coefficients of like powers of p, we have

$$p^0 : u_0(x,t) = \left(a - \frac{3\sqrt{b}}{2}x\right)^{2/3},$$

$$p^1 : u_1(x,t) = L^{-1}\left[\frac{1}{s^\alpha}L\left[H_0(u)\right]\right] = -b^{3/2}\left(a - \frac{3\sqrt{b}}{2}x\right)^{-1/3}\frac{t^\alpha}{\Gamma(\alpha+1)},$$

$$p^2 : u_2(x,t) = L^{-1}\left[\tfrac{1}{s^\alpha}L\left[H_1(u)\right]\right] = -\tfrac{b^3}{2}\left(a - \tfrac{3\sqrt{b}}{2}x\right)^{-4/3}\tfrac{t^{2\alpha}}{\Gamma(2\alpha+1)}, \tag{21}$$

$$\vdots$$

and so on. Thus, the solution $u(x,t)$ of the Eq. (1) is given as

$$u(x,t) = u_0(x,t) + u_1(x,t) + u_2(x,t) + \cdots. \tag{22}$$

The numerical results for the approximate solution obtained by using HPTM and the exact solution given by Mokhtari [7] for constant values of $a = 4$ and $b = 1$ for various values of t, x, and α are shown in Fig. 1a–c. Figure 1b and c clearly shows that when $\alpha = 1$, the approximate solution obtained by the present method is very near to the exact solution.

5 Conclusions

In this paper, the HPTM is successfully applied for solving nonlinear time-fractional Harry Dym equation. It is worth mentioning that the HPTM is capable of reducing the volume of the computational work as compared to the classical methods while still maintaining the high accuracy of the numerical result; the size reduction amounts to an improvement in the performance of the approach. In conclusion, the HPTM may be considered as a nice refinement in existing numerical techniques and might find the wide applications.

References

1. Hilfer, R. (ed.): Applications of Fractional Calculus in Physics, pp. 87–130. World Scientific Publishing Company, Singapore-New Jersey-Hong Kong (2000)
2. Podlubny, I. (ed.): Fractional Differential Equations. Academic Press, New York (1999)
3. Caputo, M.: Elasticita e Dissipazione. Zani-Chelli, Bologna (1969)
4. Miller, K.S., Ross, B.: An Introduction to the fractional Calculus and Fractional Differential Equations. Wiley, New York (1993)
5. Oldham, K.B., Spanier, J.: The Fractional Calculus Theory and Applications of Differentiation and Integration to Arbitrary Order. Academic Press, New York (1974)

6. Kilbas, A.A., Srivastava, H.M., Trujillo, J.J.: Theory and Applications of Fractional Differential Equations. Elsevier, Amsterdam (2006)
7. Mokhtari, R.: Exact solutions of the Harry-Dym equation. Commun. Theor. Phys. **55**, 204–208 (2011)
8. Kruskal, M.D., Moser, J.: Dynamical Systems, Theory and Applications, Lecturer Notes Physics, p. 310. Springer, Berlin (1975)
9. Vosconcelos, G.L., Kaclanoff, L.P.: Stationary solution for the Saffman Taylor problem with surface tension. Phys. Rev. A **44**, 6490–6495 (1991)
10. Kumar, S., Tripathi, M.P., Singh O.P.: A fractional model of Harry Dym equation and its approximate solution. Ain Shams Eng. J. (2012). http://dx.doi.org/10.1016/j.asej.2012.07.001
11. He, J.H.: Homotopy perturbation technique. Comput. Methods Appl. Mech. Eng. **178**, 257–262 (1999)
12. Khan, Y., Wu, Q.: Homotopy perturbation transform method for nonlinear equations using He's polynomials. Comput. Math. Appl. **61**(8), 1963–1967 (2011)
13. Ghorbani, A.: Beyond Adomian's polynomials: He polynomials. Chaos Solitons Fractals **39**, 1486–1492 (2009)
14. Mohyud-Din, S.T., Noor, M.A., Noor, K.I.: Traveling wave solutions of seventh-order generalized KdV equation using He's polynomials. Int. J. Nonlinear Sci. Numer. Simul. **10**, 227–233 (2009)

Floating Point-based Universal Fused Add–Subtract Unit

Ishan A. Patil, Prasanna Palsodkar and Ajay Gurjar

Abstract This paper describes fused floating point add–subtract operations and which is applied to the implementation of fast fourier transform (FFT) processors. The fused operations of an add–subtract unit which can be used both radix-2 and radix-4 butterflies are implemented efficiently with the two fused floating point operations. When placed and routed using a high-performance standard cell technology, the fused FFT butterflies are about may be work fast and gives user-defined facility to modify the butterfly's structure. Also the numerical results of the fused implementations are more accurate, as they use rounding modes is defined as per user requirement.

Keywords Floating point · Fused · Addition · Subtraction · Universal · Different types of rounding

1 Introduction

Traditionally, most DSP applications have used fixed-point arithmetic to reduce delay, chip area, and power consumption. Fixed-point arithmetic has serious problems of overflow, underflow, scaling, etc. Single-precision floating point arithmetic is a potential solution because of no overflow or underflow, automatic scaling [3].

I. A. Patil (✉) · P. Palsodkar
Electronics Engineering Deparment, Yeshwantrao Chavan College of Engineering, Nagpur, Maharashtra, India
e-mail: ishanpatil29@gmail.com

P. Palsodkar
e-mail: palsodkar.prasanna@ieee.org

A. Gurjar
Electronics Engineering Department, Sipna College of Engineering Amravati, Amravati, Maharashtra, India
e-mail: prof_gurjar1928@rediffmail.com

B. V. Babu et al. (eds.), *Proceedings of the Second International Conference on Soft Computing for Problem Solving (SocProS 2012), December 28–30, 2012*, Advances in Intelligent Systems and Computing 236, DOI: 10.1007/978-81-322-1602-5_29, © Springer India 2014

Two new fused floating point element implementation: (1) fused dot product contains multiplies two pair of floating point data, add (or subtract) the product (2) fused add–subtract contains add a pair of floating point data, and simultaneously subtract the same data [1].

It is traditional floating point adder which consist only addition of only valid floating point numbers. They work inefficiently when given data are not a floating point number and also it has some limitation that they can perform only programmer-defined rounding method due this reason if other rounding method we have use then again we have to edit program and change the data for new rounding method, so it can be neglected in proposed floating point adder.

All research is related to Xilinx SPARTAN 6 kit [5, 7] by using compilation software Xilinx version: 13.2. Xilinx is FPGA stimulator to estimate output of from Verilog code in terms of input–output buffers, maximum delay (longest path execution of circuit), area in terms of lookup table (LUTs).This software is used for VHDL and Verilog code implementation, stimulation and generating program for dump on FPGA kit.

2 Basic of Floating Point

The Institute of Electrical and Electronics Engineers (IEEE) Standard for Floating Point Arithmetic (IEEE 754) [4] is a technical standard established by the IEEE. This standard specifies the basic types of representation.

- Half Precision (16-bits or 2-bytes)
- Single Precision (32-bits or 4-bytes)
- Double Precision (64-bits or 8-bytes).

The format of a floating point number comprises 3 types of bits presented in the following Fig. 1.

- **Recall that exponent field is 8 bits for single precision**

- E can be in the range from 0 to 255
- $E = 0$ and $E = 255$ are reserved for special use (discussed later)
- $E = 1$ to 254 are used for normalized floating point numbers
- Bias = 127 (half of 254), val(E) = $E - 127$
- val(E=1) = -126, val(E=127) = 0, val(E=254) = 127.

IEEE 754 standard specifies four modes of rounding

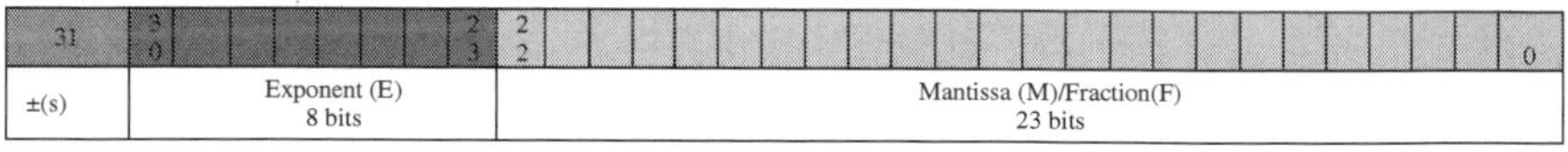

31	30 … 23	22 … 0
±(s)	Exponent (E) 8 bits	Mantissa (M)/Fraction(F) 23 bits

Fig. 1 Single-precision floating point format

1. Round to nearest even: default rounding mode increment result if: rs = "11" or (rs = "10" and fln = '1'). Otherwise, truncate result significant to 1. f1f2...fln
2. Round toward $+\infty$: result is rounded up Increment result if sign is positive and r or s = '1'
3. Round toward $-\infty$: result is rounded down Increment result if sign is negative and r or s = '1'
4. Round toward 0: always truncate result.

3 Normal Addition–Subtraction Rule

$$\begin{array}{cc} +1.1234 & -1.1234 \\ +1.2456 & -1.2456 \\ \hline +2.4690 & -2.4690 \end{array}$$

If we take normal addition or subtraction, consider a two numbers [Greater (G) and Lesser (L)] magnitude wise then only 4 combination are possible which are given in Table 1 [2].

After watching all examples, we came to conclusion that in addition and substation when there same sign number added or substrate then only add or subtract, respectively. Similarly, when two numbers have different signs, then we can use opposite function, i.e., subtract or add.

4 Study of FP Arithmetic Algorithms

Floating Point addition steps

Assume 32 bit binary number and then by applying algorithm for normal addition by calculator and by using FP addition algorithm stepwise.

A = 32′b0__0111_1000__1011_1010_0000_1111_0110_110;
B = 32′b0__0111_0011__0101_0000_0000_0011_1111_111;

Table 1 Signed addition or subtraction rules

Sr. no	Numbers		Operations	Sign used
1	+G	+L	$(+G) + (+L) \Rightarrow G + L$	+
2	−G	−L	$(-G) + (-L) \Rightarrow -G - L \Rightarrow -(G + L)$	+
3	+G	−L	$(+G) + (-L) \Rightarrow G - L$	−
4	−G	+L	$(-G) + (+L) \Rightarrow L - G$ OR $-(G - L)$	−
1	+G	+L	$(+G) - (+L) \Rightarrow G - L$	−
2	−G	−L	$(-G) - (-L) \Rightarrow -G + L \Rightarrow -(G + L)$	−
3	+G	−L	$(+G) - (-L) \Rightarrow G + L$	+
4	−G	+L	$(-G) - (+L) \Rightarrow -G - L$ OR $-(G + L)$	+

A = 1.3490607*E−2, B = 3.2044944*E−4.

4.1 Calculation From Calculator

1st step align decimal point
2nd step add

1.3490607 * E−2
+0.032044944*E−2
+1.381105644*E−2

3rd Normalize result

4.2 Detailed Bitwise Example

$$\begin{array}{lll} S & EXP & Mantissa(M) \\ 0 & 01111000 & 1011\_1010\_0000\_1111\_0110\_110 \\ 0 & 01110011 & 0101\_0000\_0000\_0011\_1111\_111 \end{array} \tag{1}$$

Find Greater no. and lesser no. and assign it [6].

G = 0 01111000 1011_1010_0000_1111_0110_110
L = 0 01110011 0101_0000_0000_0011_1111_111

- **1st step**

  Align radix point by using True exponent value difference
  G_exp_t = 0111_1000(120) − 0111_1111 = 1111_1001(−7)
  L_exp_t = 0111_0011(115) − 0111_1111 = 1111_0100(−12)
  Ed = G_exp-L_exp = 5
  Shift Lesser no. to right by Ed value
  Shift_L_m = 1.0101_0000_0000_0011_1111_111(0)
  Shift_L_m = 0.10101_0000_0000_0011_1111_111(1)
  Shift_L_m = 0.010101_0000_0000_0011_1111_111(2)
  Shift_L_m = 0.0010101_0000_0000_0011_1111_111(3)
  Shift_L_m = 0.00010101_0000_0000_0011_1111111(4)
  Shift_L_m = 0.0000_1010_1000_0000_0001_111_111(5)

- **2nd step** addition of mantissa depend on signs of both no. (Table 1) and store last bits for rounding

  1.1011_1010_0000_1111_0110_110
  +0.0000_1010_1000_0000_0001_111_111
  01.1100_0100_1000_1111_1000_101_111
  Result_m = 01.1100_0100_1000_1111_1000_101_111

- **3rd step** normalize mantissa result

  n_Result_m = 01.1100_0100_1000_1111_1000_101_111
  (For if result is m = 00.0010_0100_1000_1111_1000_101_111
  Normalize now m = 01.0010_0100_0111_1100_0101_111_000)
  (For if result is m = 10.0010_0100_1000_1111_1000_101_111
  Normalize now m = 01.00010_0100_1000_1111_1000_101_111)

- **4th step** Rounding (Round toward 0: always truncate result)

  Round_Result_m = 01.1100_0100_1000_1111_1000_101

- **5th Step** After rounding normalize

  n_Result_m = 01.1100_0100_1000_1111_1000_101
  Final Result sig_G, G_exp, n_Result_m
  Sum = 0 0111_1000 1100_0100_1000_1111_1000_101
  +1.3811056 * E − 2
  +1.381105644 * E − 2 (From actual calculation)

5 Floating Point Adders

This contains original floating point adder and proposed floating point adder. Due three limitations like it will also work on invalid floating point, rounding mode, inefficient swap in greater and lesser number when both number have same sign and same exponent. To remove all limitations, we can see their proposed model satisfied in all three manners.

5.1 Traditional Floating Point Adder

Basic Floating Point Addition Algorithm [1, 3, 8]: The straightforward basic floating point addition algorithm requires the most serial operations. It has the following steps: (Fig. 2)

1. Exponent subtraction: Perform subtraction of the exponents to form the absolute difference $|E_a - E_b| = d$.
2. Alignment: Right shift the significant of the smaller operand by d bits. The larger exponent is denoted Ef .
3. Significant addition: Perform addition or subtraction according to the effective operation. The result is a function of the op-code and the signs of the operands.
4. Conversion: Convert a negative significant result to a sign-magnitude representation. The conversion requires a two's complement operation, including an addition step.

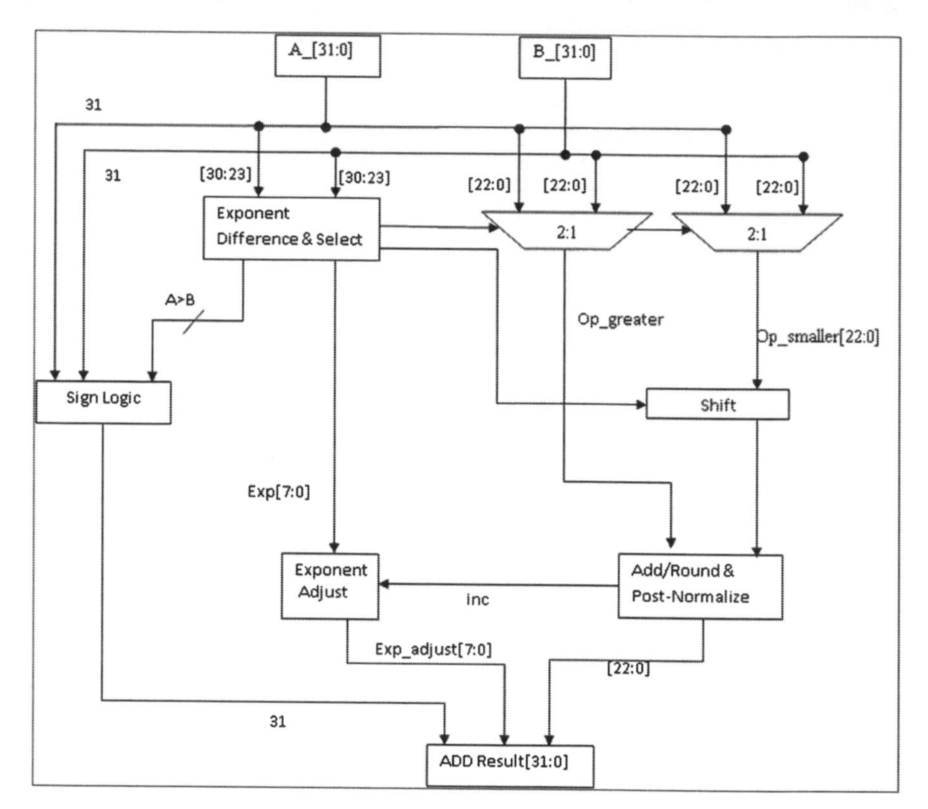

Fig. 2 Traditional floating point adder

5. Leading one detection: Determine the amount of left shift needed in the case of Subtraction yielding cancellation. For addition, determine whether or not a 1-bit right shift is required. Then priority-encode the result to drive the normalizing shifter.
6. Normalization: Normalize the significant and update Exponent appropriately.
7. Rounding: Round the final result by conditionally adding as required by the IEEE standard. If rounding causes an overflow, perform a 1-bit right shift and increment Ef.

5.2 Proposed Floating Point Adder

Proposed Floating Point Addition Algorithm: The straightforward derived floating point addition algorithm requires the most serial operations. It has the following steps: (Fig. 3)

1. Check given number is floating point number or invalid floating point enable other operation or enable not a FP no to show given data is invalid.

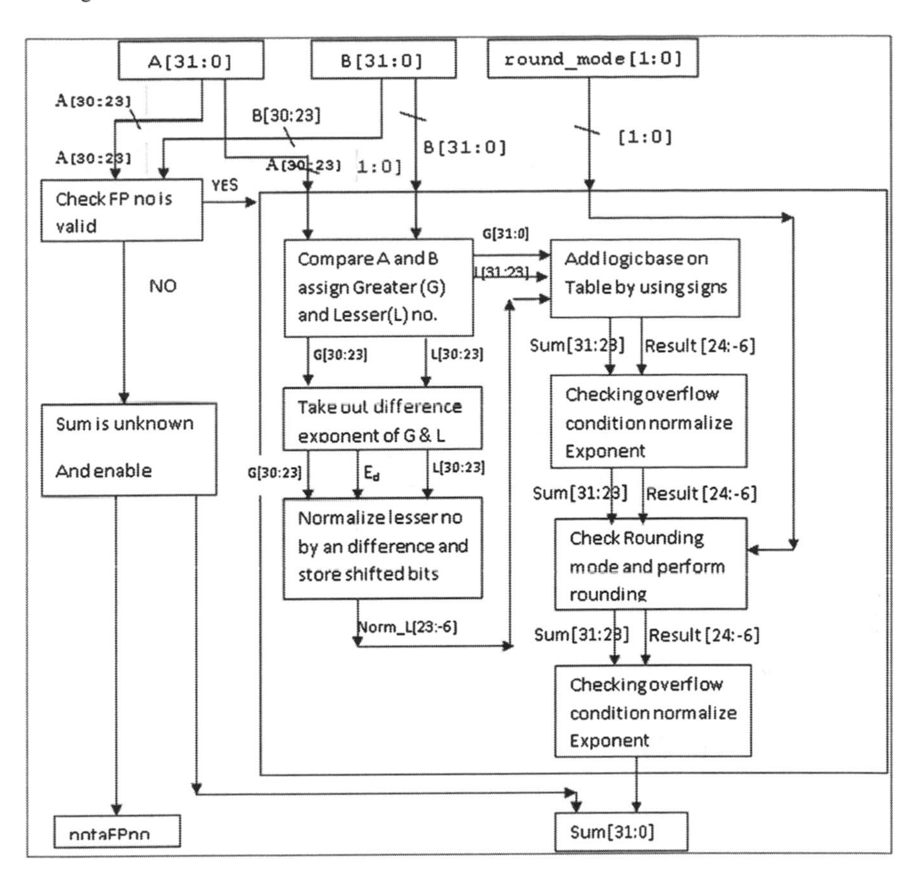

Fig. 3 Proposed floating point adder

2. If given number is valid floating point number then sort out greater and lesser number from data comparing its magnitude.
3. Exponent subtraction: Perform subtraction of greater exponents to lesser exponent $E_G - E_L = d$.
4. Alignment: Right shift the significant of the lesser mantissa by d bits and store last 3 shifted bits.
5. Significant addition: Perform addition according to their signs take decision.
6. Normalize result: check result is overflow or not and then according to condition adjust exponent.
7. Rounding: Check the rounding mode and perform rounding on the result with the help of last 3 stored bits.
8. Normalize result: Check result is overflow or not and then according to condition adjust exponent and Display the result.

To remove all limitation in this, we have use 1st check valid floating point or not then for selective rounding, we have given choice to user mode, i.e., user-defined rounding modes to select round mode by giving signals to round_mode pins, and

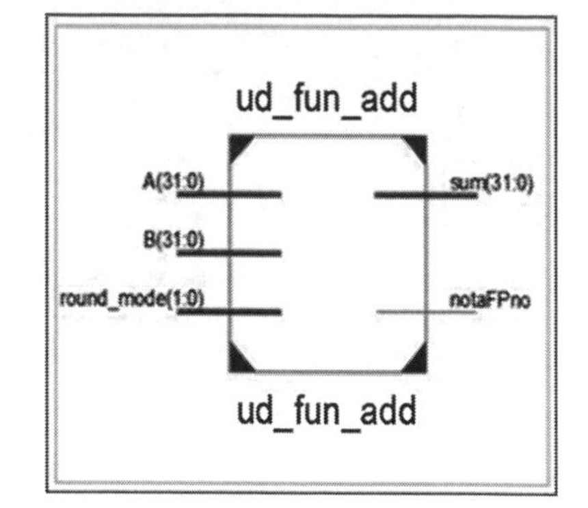

Fig. 4 Adder

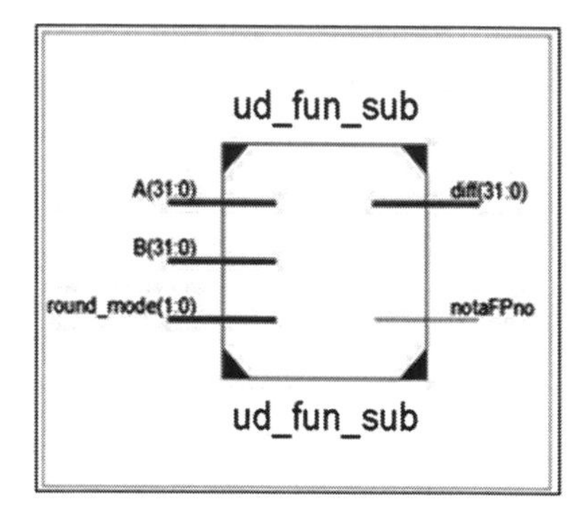

Fig. 5 Subtracter

finally, we have to check exact greater or lesser number by comparing all 31 bits. Similarly, we can create substrater module referring Table 1 and then, we serially and parallel combination of adder and substrater finally we made fuse model of one adder and one substrater and we have advantage that fused required less numbers of LUT's and less delay to get final output as if we use both different models of adder and subtract.

6 Proposed Floating Point Work Result

This shows all main modules implementations in RTL and outputs waveforms with respect to Xilinx SPARTAN 6 kit [7] stimulation on Xilinx software.

6.1 Proposed Models RTL View

Figures 4, 5 and 6.

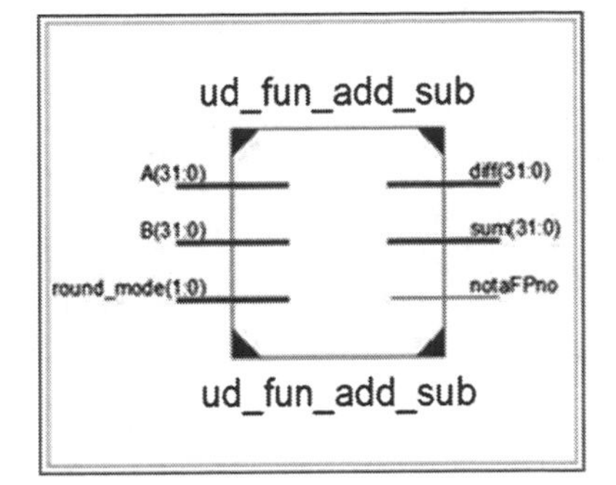

Fig. 6 Add–subtracter

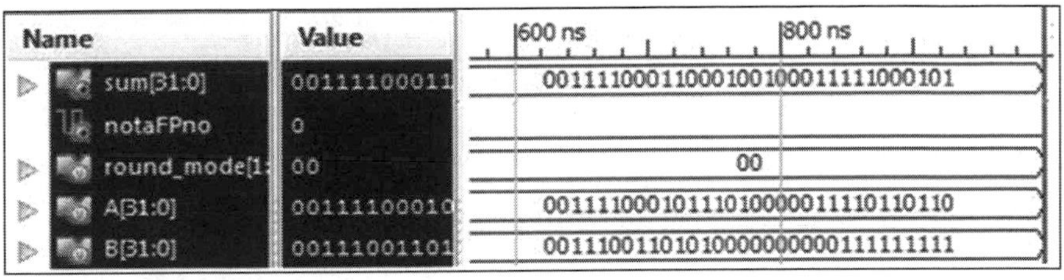

Fig. 7 Adder output waveform

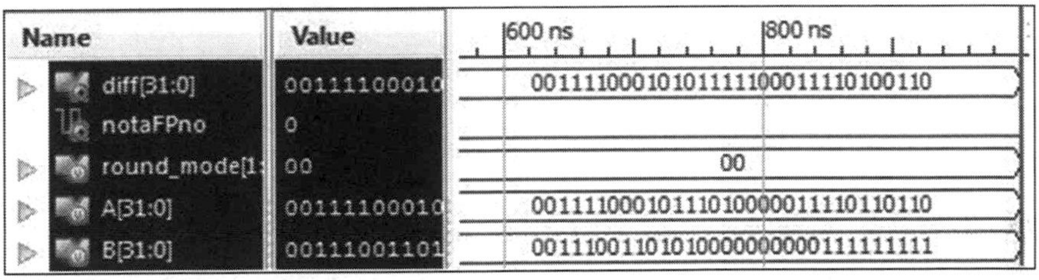

Fig. 8 Subtracter output waveform

6.2 Proposed Models Output Waveform

Figures 7, 8 and 9.

7 Comparisons of all Modules

See Table 2.

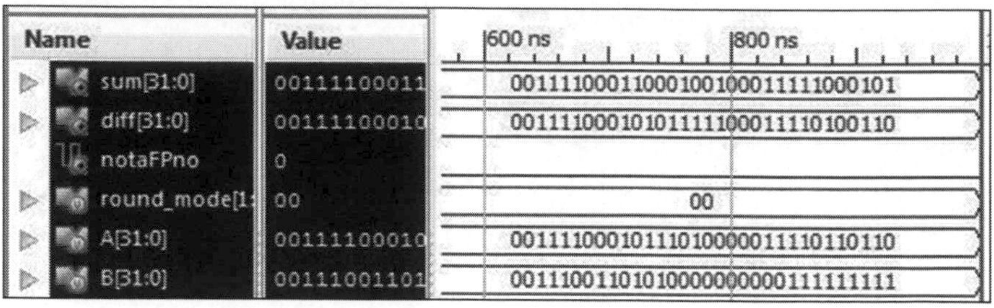

Fig. 9 Add–subtracter output waveform

Table 2 All modules details of implementations

Types	No of LUT used(Area %)	IOBs %	Delay (ns)
FP adder	381 of 9112 (4.18)	99 of 232 (42.67)	35.418
FP subtracter	381 of 9112 (4.18)	99 of 232 (42.67)	35.513
FP Serial AS	762 of 9112 (8.36)	131 of 232 (56.47)	70.518
FP parallel AS	808 of 9112 (8.86)	131 of 232 (56.47)	37.213
Fused FP add–subtract unit	678 of 9112 (7.44)	131 of 232 (56.47)	39.876

Table 3 All modules comparisons in terms of percentage w.r.t. adder

Types	No of LUT used (Area %)	IOBs (%)	Delay (ns)
FP adder	100	100	100
FP subtracter	100	100	100.27
FP serial AS	200	132.32	200
FP parallel AS	212	132.32	105
Fused FP add–subtract unit	177.95	132.32	112.58

Table 4 Basic comparison of programming styles in Verilog on demo floating point adder

Types of programming	Time (ns)	LUTs
Normal adder	26.468	254
Task adder	27.727	308
Function adder	24.533	249

8 Conclusions

In proposed floating point adder have two different functions from traditional floating point adder to provide user-defined adder. First is to check the given data are valid floating point number or not. Second is to give privilege to select rounding modes (i.e., user-defined rounding mode selection and default is truncating). While programming, we used different methods like by using TASK, function, and normal programming. After comparing all types of models of floating point adder, we came to conclusion that we modified traditional adder and able to prove that when we use functions in programming has good result at cost of saving 2 % number of LUTs and 7.8 %

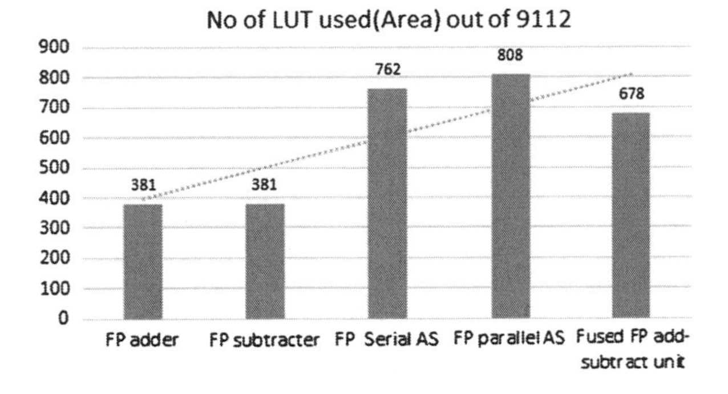

Fig. 10 LUT

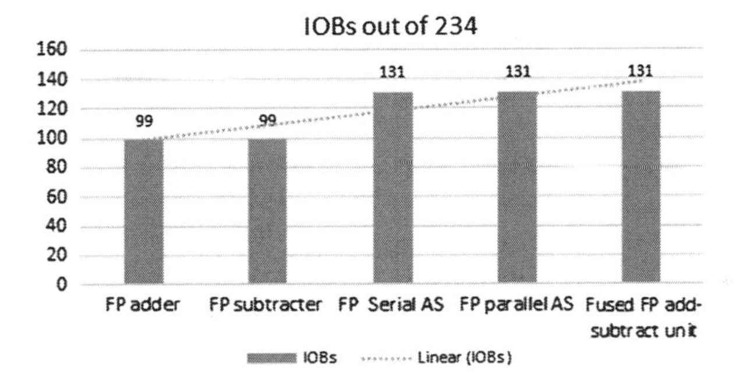

Fig. 11 IOB

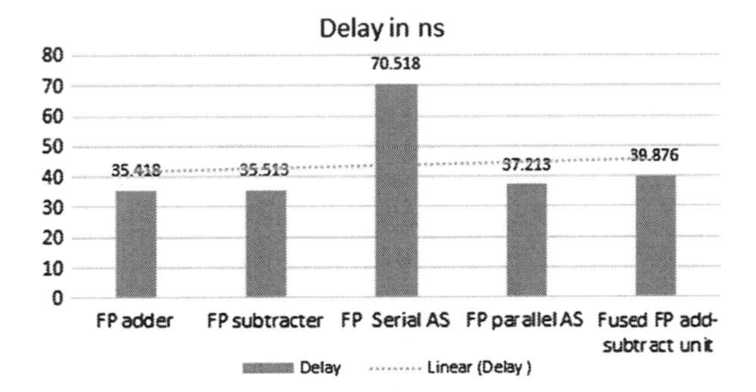

Fig. 12 Delay

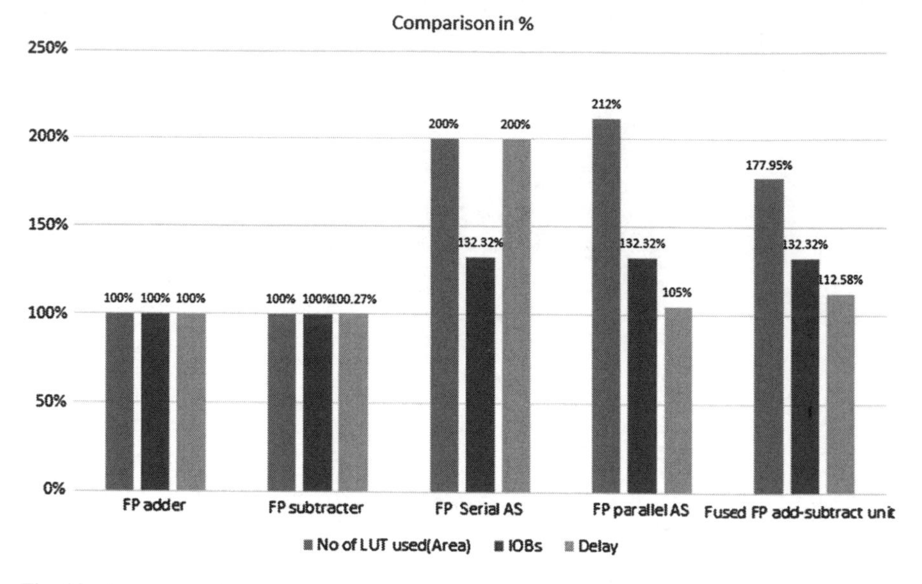

Fig. 13 Comparison in percentage

reduce delay (referring Table 4 Basic comparison of programming styles in Verilog on demo floating point adder). So we finally decided that we are using functions in all Verilog design programs to give best output.

After watching results from Tables 2, 3 and Figs. 10, 11, 12, 13, we came to know that we are saving in proposed floating point add–subtract model is 68 % decrease in IOBs, saving 23 % number of LUTs and 88 % reduce delay as compared to the single adder, single subtracter, serial adder and subtracter, Parallel adder and subtracter.

References

1. Swartzlander Jr, E.E., Saleh, H.H.: FFT Implementation with Fused Foating-Point Operations.IEEE Transactions on Computers, Feb (2012)
2. Jongwook Sohn and Earl E. Swartzlander, Jr., "Improved Architectures for a Fused Floating-Point Add-Subtract Unit" IEEE Transactions on circuits and systems-I, Vol. 59, no. 11, 2012
3. H. Saleh and E.E. Swartzlander, Jr., "A Floating-Point Fused Add-Subtract Unit", Proc. IEEE Midwest Symp. Circuits and Systems (MWSCAS), pp. 519–522, 2008
4. IEEE Standard for Floating-Point Arithmetic, ANSI/IEEE Standard 754–2008, Aug. 2008
5. http://www.xilinx.com/FPGA_series
6. Lecture notes - Chapter 7 - Floating Point Arithmetic: http://pages.cs.wisc.edu/smoler/x86text/lect.notes/arith.flpt.html
7. All o/p are w.r.to XILINX Spartan-6 XC6SLX16 -CSG324C
8. Stuart Franklin Oberman: Design Issues in High Performance Floating Point Arithmetic Units. Stanford University, Ph. D. Dissertation (1996)

New Analytical Approach for Fractional Cubic Nonlinear Schrödinger Equation Via Laplace Transform

Jagdev Singh and Devendra Kumar

Abstract In this paper, a user-friendly algorithm based on new homotopy perturbation transform method (HPTM) is proposed to obtain approximate solution of a time-space fractional cubic nonlinear Schrödinger equation. The numerical solutions obtained by the HPTM indicate that the technique is easy to implement and computationally very attractive.

Keywords Fractional cubic nonlinear Schrödinger equation · Laplace transform · Homotopy perturbation transform method · He's polynomials · Maple code

1 Introduction

Fractional differential equations have found applications in many problems of physics and engineering. For example, these equations are increasingly used to model problems in research areas as diverse as dynamical systems, mechanical systems, control, chaos, chaos synchronization, continuous-time random walks, anomalous diffusive and subdiffusive systems, unification of diffusion and wave propagation phenomenon, and others [1–6]. It is difficult to get the exact analytical solution of fractional order problems especially nonlinear cases. The homotopy perturbation method (HPM) was first introduced by the Chinese mathematician J.H. He in 1998 and was further developed by He [7–9]. This method was applied to handle various problems

J. Singh (✉)
Department of Mathematics, Jagannath University, Village- Rampura, Tehsil-Chaksu, Jaipur, Rajasthan 303901, India
e-mail: jagdevsinghrathore@gmail.com

D. Kumar
Department of Mathematics, Jagannath Gupta Institute of Engineering and Technology, Jaipur, Rajasthan 302022, India
e-mail: devendra.maths@gmail.com

B. V. Babu et al. (eds.), *Proceedings of the Second International Conference on Soft Computing for Problem Solving (SocProS 2012), December 28–30, 2012*, Advances in Intelligent Systems and Computing 236, DOI: 10.1007/978-81-322-1602-5_30,

arising in science and engineering [10–14]. In recent years, many authors have paid attention to study the solutions of linear and nonlinear partial differential equations by using various methods combined with the Laplace transform. Among these are Laplace decomposition method (LDM) [15, 16] and homotopy perturbation transform method (HPTM) [17, 18].

In this article, we consider the following time-space fractional cubic nonlinear Schrödinger equation of the form:

$$i\frac{\partial^{\alpha} u}{\partial t^{\alpha}} + \frac{\partial^{2\beta} u}{\partial x^{2\beta}} + 2|u|^2 u = 0, \quad t > 0,\ 0 < \alpha,\ \beta \leq 1, \tag{1}$$

with the initial condition

$$u(x, 0) = e^{ix}, \quad i = \sqrt{(-1)}. \tag{2}$$

where $u(x, t)$ is a function of x and t. The derivative is considered in the Caputo sense. In the case of $\alpha = 1$ and $\beta = 1$, the fractional cubic nonlinear Schrödinger equation reduces to the classical cubic nonlinear Schrödinger equation. The time-space fractional cubic nonlinear Schrödinger equation has been previously studied by Herzallah and Gepreel [19] and Hemida et al. [20].

In this paper, we apply the HPTM to solve the time-space fractional cubic nonlinear Schrödinger equation. The HPTM is a combination of Laplace transform method, HPM, and He's polynomials. It provides the solutions in terms of convergent series with easily computable components in a direct way without using linearization, perturbation, or restrictive assumptions.

2 Basic Idea of HPTM

To illustrate the basic idea of this method, we consider a general fractional nonlinear nonhomogeneous partial differential equation with the initial condition of the form:

$$D_t^{\alpha} u(x, t) + R\, u(x, t) + N\, u(x, t) = g(x, t), \tag{3}$$

$$u(x, 0) = f(x), \tag{4}$$

where $D_t^{\alpha} u(x, t)$ is the Caputo fractional derivative of the function $u(x, t)$, R is the linear differential operator, N represents the general nonlinear differential operator and $g(x, t)$, is the source term.

Applying the Laplace transform on both sides of Eq. (3), we get

$$L\,[u(x, t)] = \frac{f(x)}{s} + \frac{1}{s^{\alpha}} L\,[g(x, t)] - \frac{1}{s^{\alpha}} L\,[R\, u(x, t) + N\, u(x, t)]. \tag{5}$$

Operating with the Laplace inverse on both sides of Eq. (5) gives

$$u(x,t) = G(x,t) - L^{-1}\left[\frac{1}{s^{\alpha}} L[R\,u(x,t) + N\,u(x,t)]\right], \tag{6}$$

where $G(x,t)$ represents the term arising from the source term and the prescribed initial condition. Now, we apply the HPM

$$u(x,t) = \sum_{n=0}^{\infty} p^n\, u_n(x,t), \tag{7}$$

and the nonlinear term can be decomposed as

$$N\,u(x,t) = \sum_{n=0}^{\infty} p^n H_n(u), \tag{8}$$

for some He's polynomials $H_n(u)$ [21, 22] that are given by

$$H_n(u_0, u_1, \ldots, u_n) = \frac{1}{n!}\frac{\partial^n}{\partial p^n}\left[N\left(\sum_{i=0}^{\infty} p^i u_i\right)\right]_{p=0}, \quad n = 0, 1, 2, \ldots \tag{9}$$

Using (7) and (8) in (6), we get

$$\sum_{n=0}^{\infty} p^n\, u_n(x,t) = G(x,t) - p\left(L^{-1}\left[\frac{1}{s^{\alpha}} L\left[R\sum_{n=0}^{\infty} p^n\, u_n(x,t) + \sum_{n=0}^{\infty} p^n H_n(u)\right]\right]\right), \tag{10}$$

which is the coupling of the Laplace transform and the HPM using He's polynomials. Comparing the coefficients of like powers of p, the components u_0, u_1, u_2 and so on are obtained.

3 Solution of the Problem

In this section, we use the HPTM to solve the time-space fractional cubic nonlinear Schrödinger Eqs. (1), (2). Applying the Laplace transform on both sides of Eq. (1), we get

$$L[u(x,t)] = \frac{e^{ix}}{s} + \frac{i}{s^{\alpha}} L\left[\frac{\partial^{2\beta} u}{\partial x^{2\beta}} + 2u^2\overline{u}\right], \tag{11}$$

where $u^2\overline{u} = |u|^2 u$and $\overline{u}$ is the conjugate of u.

The inverse Laplace transform implies that

$$u(x,t) = e^{ix} + L^{-1}\left[\frac{i}{s^{\alpha}} L\left[\frac{\partial^{2\beta} u}{\partial x^{2\beta}} + 2u^2\overline{u}\right]\right]. \tag{12}$$

Now applying the HPM, we get

$$\sum_{n=0}^{\infty} p^n u_n(x,t) = e^{ix} + p\left(L^{-1}\left[\frac{i}{s^{\alpha}} L\left[\frac{\partial^{2\beta}}{\partial x^{2\beta}}\left(\sum_{n=0}^{\infty} p^n u_n(x,t)\right) + \left(\sum_{n=0}^{\infty} p^n H_n(u)\right)\right]\right]\right), \tag{13}$$

where $H_n(u)$ are He's polynomials that represent the nonlinear terms.

The first few components of He's polynomials are given by

$$\begin{aligned} &H_0(u) = 2u_0^2\bar{u}_0,\\ &H_1(u) = 2(u_0^2\bar{u}_1 + 2u_0u_1\bar{u}_0),\\ &\vdots \end{aligned} \tag{14}$$

Comparing the coefficients of like powers of p, we have

$$\begin{aligned} &p^0 : u_0(x,t) = e^{ix},\\ &p^1 : u_1(x,t) = L^{-1}\left[\tfrac{i}{s^{\alpha}} L\left[\tfrac{\partial^{2\beta}}{\partial x^{2\beta}} u_0 + H_0(u)\right]\right] = c_1 e^{ix}\tfrac{t^{\alpha}}{\Gamma(\alpha+1)},\\ &p^2 : u_2(x,t) = L^{-1}\left[\tfrac{i}{s^{\alpha}} L\left[\tfrac{\partial^{2\beta}}{\partial x^{2\beta}} u_1 + H_1(u)\right]\right] = c_2 e^{ix}\tfrac{t^{2\alpha}}{\Gamma(2\alpha+1)},\\ &p^3 : u_3(x,t) = L^{-1}\left[\tfrac{i}{s^{\alpha}} L\left[\tfrac{\partial^{2\beta}}{\partial x^{2\beta}} u_2 + H_2(u)\right]\right] = c_3 e^{ix}\tfrac{t^{3\alpha}}{\Gamma(3\alpha+1)},\\ &\vdots \end{aligned} \tag{15}$$

and so on. After some calculation, we get

$$\begin{aligned} &c_1 = i(e^{i\pi\beta} + 2),\\ &c_2 = i(e^{i\pi\beta}c_1 + 2\overline{c}_1 + 4c_1),\\ &c_3 = i\left(e^{i\pi\beta}c_2 + 2\left[\overline{c}_2 + 2c_2 + \frac{\Gamma(2\alpha+1)}{(\Gamma(\alpha+1))^2}(2|c_1|^2 + c_1^2)\right]\right),\\ &\vdots \end{aligned} \tag{16}$$

Therefore, the HPTM series solution is

$$u(x,t) = e^{ix}\left(1 + \frac{c_1 t^{\alpha}}{\Gamma(\alpha+1)} + \frac{c_2 t^{2\alpha}}{\Gamma(2\alpha+1)} + \frac{c_3 t^{3\alpha}}{\Gamma(3\alpha+1)} + \cdots\right). \tag{17}$$

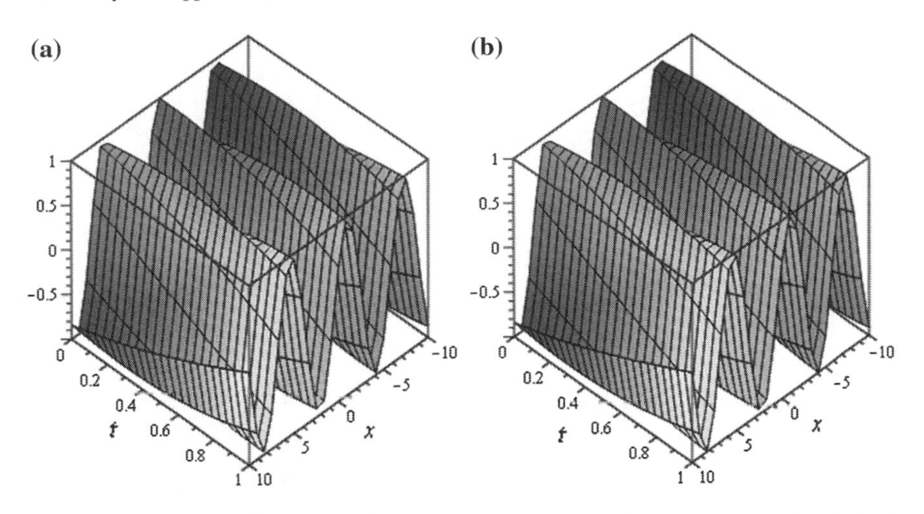

Fig. 1 The behavior of the real part of $u(x, t)$ w.r.t. x and t is obtained, when **a** $\alpha = 1$ and $\beta = 1$ **b** Exact solution

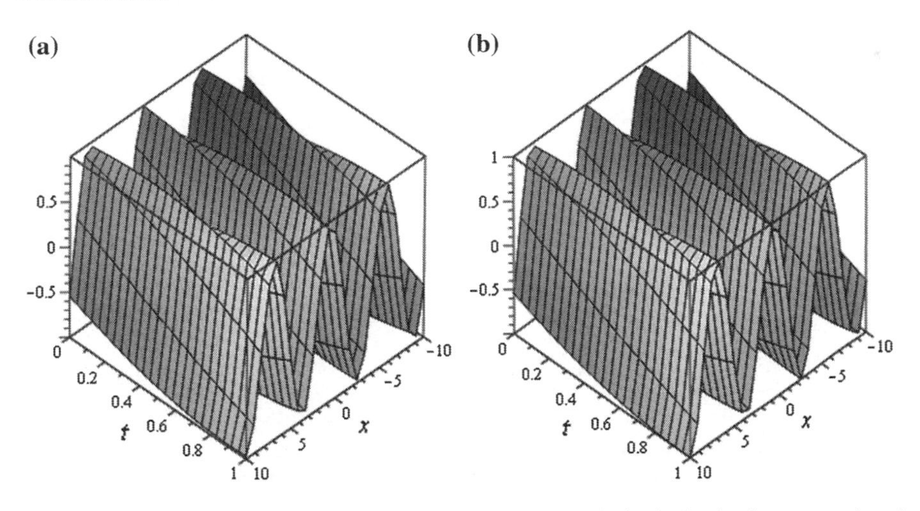

Fig. 2 The behavior of the imaginary part of $u(x, t)$ w.r.t. x and t is obtained, when **a** $\alpha = 1$ and $\beta = 1$ **b** Exact solution

Setting $\alpha = 1$ and $\beta = 1$ in (16) and (17), we reproduce the solution of the problem as follows:

$$u(x, t) = e^{ix}\left(1 + it + \frac{(it)^2}{2!} + \frac{(it)^3}{3!} + \cdots\right). \tag{18}$$

This solution is equivalent to the exact solution in closed form

$$u(x, t) = e^{i(x+t)}. \tag{19}$$

The numerical results for the approximate solution obtained by using HPTM at $\alpha = 1$ and $\beta = 1$ and the exact solution for various values of t and x are shown by Figs. 1 and 2. Figs. 1 and 2 clearly show that, when $\alpha = 1$ and $\beta = 1$, the approximate solution obtained by the HPTM is very near to the exact solution.

4 Conclusions

In this paper, the HPTM is successfully employed for solving time-space fractional cubic nonlinear Schrödinger equation. It is worth mentioning that the HPTM is capable of reducing the volume of the computational work as compared to the classical methods while still maintaining the high accuracy of the numerical result; the size reduction amounts to an improvement of the performance of the approach. Finally, we can conclude the HPTM may be considered as a nice refinement in existing numerical techniques and might find the wide applications.

References

1. Hilfer, R. (ed.): Applications of Fractional Calculus in Physics, pp. 87–130. World Scientific Publishing Company, Singapore, New Jersey, Hong Kong (2000)
2. Podlubny, I.: Fractional Differential Equations. Academic Press, New York (1999)
3. Caputo, M.: Elasticita e Dissipazione. Zani-Chelli, Bologna (1969)
4. Miller, K.S., Ross, B.: An Introduction to the fractional Calculus and Fractional Differential Equations. Wiley, New York (1993)
5. Oldham, K.B., Spanier, J.: The Fractional Calculus Theory and Applications of Differentiation and Integration to Arbitrary Order. Academic Press, New York (1974)
6. Kilbas, A.A., Srivastava, H.M., Trujillo, J.J.: Theory and Applications of Fractional Differential Equations. Elsevier, Amsterdam (2006)
7. He, J.H.: Homotopy perturbation technique. Comput. Methods Appl. Mech. Eng. **178**, 257–262 (1999)
8. He, J.H.: Homotopy perturbation method: a new nonlinear analytical technique. Appl. Math. Comput. **135**, 73–79 (2003)
9. He, J.H.: New interpretation of homotopy perturbation method. Int. J. Mod. Phys. B **20**, 2561–2568 (2006)
10. Abbasbandy, S.: Application of He's homotopy perturbation method to functional integral equations. Chaos, Solitons Fractals **31**, 1243–1247 (2007)
11. Ganji, D.D., Sadighi, A.: Application of He's homotopy perturbation method to nonlinear coupled system of reaction-diffusion equations. Int. J. Nonlinear Sci. Numer. Simul. **7**, 411–418 (2006)
12. Rafei, M., Ganji, D.D.: Explicit solutions of Helmholtz equation and fifth-order KdV equation using homotopy perturbation method. Int. J. Nonlinear Sci. Numer. Simul. **7**, 321–328 (2006)
13. Rafei, M., Ganji, D.D., Daniali, D.: Solution of epidemic model by homotopy perturbation method. Appl. Math. Comput. **187**, 1056–1062 (2007)
14. Ozis, T., Yildirim, A.: Travelling wave solution of KdV equation using He' homotopy perturbation method. Int. J. Nonlinear Sci. Numer. Simul. **8**, 239–242 (2007)

15. Khuri, S.A.: A Laplace decomposition algorithm applied to a class of nonlinear differential equations. J. Appl. Math. **1**, 141–155 (2001)
16. Khan, Y.: An effective modification of the Laplace decomposition method for nonlinear equations. Int. J. Nonlinear Sci. Numer. Simul. **10**, 1373–1376 (2009)
17. Khan, Y., Wu, Q.: Homotopy perturbation transform method for nonlinear equations using He's polynomials. Comput. Math. Appl. **61**(8), 1963–1967 (2011)
18. Singh, J., Kumar, D., Rathore, S.: Application of homotopy perturbation transform method for solving linear and nonlinear Klein-Gordon equations. J. Inf. Comput. Sci. **7**(2), 131–139 (2012)
19. Herzallah Mohamed, A.E., Gepreel, K.A.: Approximate solution to time-space fractional cubic nonlinear Schrödinger equation. Appl. Math. Model. **36**(11), 56–78 (2012)
20. Hemida, K.M., Gerpreel, K.A., Mohamed, M.S.: Analytical approximate solution to the time-space nonlinear partial fractional differential equations. Int. J. Pure Appl. Math. **78**(2), 233–243 (2012)
21. Ghorbani, A.: Beyond adomian's polynomials: He polynomials. Chaos, Solitons Fractals **39**, 1486–1492 (2009)
22. Mohyud-Din, S.T., Noor, M.A., Noor, K.I.: Traveling wave solutions of seventh-order generalized KdV equation using He's polynomials. Int. J. Nonlinear Sci. Numer. Simul. **10**, 227–233 (2009)

An Investigation on the Structure of Super Strongly Perfect Graphs on Trees

R. Mary Jeya Jothi and A. Amutha

Abstract A graph G is super strongly perfect graph if every induced subgraph H of G possesses a minimal dominating set that meets all the maximal cliques of H. In this paper, we have characterized the super strongly perfect graphs on trees. We have presented the results on trees in terms of domination and codomination numbers γ and $\overline{\gamma}$. Also, we have given the relationship between diameter, domination, and codomination numbers in trees.

Keywords Super strongly perfect graph · Minimal dominating set · Domination and codomination numbers and tree

1 Introduction

Graph theory is a growing area in mathematical research and has a large specialized vocabulary. It is rapidly moving into the mainstream of mathematics mainly because of its applications in diverse fields which include biochemistry (genomics), electrical engineering (communications networks and coding theory), computer science (algorithms and computations), and operations research (scheduling) [1].

A tree is a mathematical structure that can be viewed either as a graph or as a data structure. The two views are equivalent, since a tree data structure contains not only a set of elements, but also connections between elements, giving a tree graph. Trees were first studied by Cayley [2, 3]. A tree is a special kind of graph and follows a particular set of rules. It is a simple, undirected, connected, acyclic graph. A tree does not have a specific direction. Depending on how it is to be used, the tree may branch outward while going [4].

R. Mary Jeya Jothi (✉)
Research Scholar, Department of Mathematics, Sathyabama University, Chennai 600119, India
e-mail: jeyajothi31@gmail.com

A. Amutha
Assistant Professor, Department of Mathematics, Sathyabama University, Chennai 600119, India
e-mail: amudhajo@gmail.com

B. V. Babu et al. (eds.), *Proceedings of the Second International Conference on Soft Computing for Problem Solving (SocProS 2012), December 28–30, 2012*, Advances in Intelligent Systems and Computing 236, DOI: 10.1007/978-81-322-1602-5_31, © Springer India 2014

2 Basic Concepts

In this paper, graphs are finite and simple, that is, they have no loops or multiple edges. Let $G = (V, E)$ be a graph. A *clique* in G is a set $X \subseteq V(G)$ of pairwise adjacent vertices. A subset D of $V(G)$ is called a *dominating set* if every vertex in $V–D$ is adjacent to at least one vertex in D. A subset S of V is said to be a *minimal dominating set* if $S–\{u\}$ is not a dominating set for any $u \in S$. The domination number $\gamma(G)$ of G is the smallest size of a dominating set of G. The domination number of its complement $\overline{G}$ is called the codomination number of G and is denoted by $\gamma(\overline{G})$ or simply $\overline{\gamma}$ [5]. A shortest $u–v$ path of a connected graph G is often called a geodesic. The diameter denoted by diam (G) is the length of any longest geodesic. A vertex of degree zero in G is called an isolated vertex of G. The minimum degree of a graph is denoted by $\delta(G)$.

3 Our Results on Super Strongly Perfect Graph

In this paper, we have characterized the super strongly perfect graphs on trees. We have presented the results on trees in terms of domination and codomination numbers γ and $\overline{\gamma}$. Also, we have found the relationship between diameter, domination, and codomination numbers of super strongly perfect graph in trees.

3.1 Super Strongly Perfect Graph (SSP)

A Graph $G = (V, E)$ is super strongly perfect if every induced subgraph H of G possesses a minimal dominating set that meets all the maximal cliques of H. Figures 1 and 2 illustrates the structures of SSP and Non-SSP

Example 1 $\{v_1, v_2\}$ is a minimal dominating set which meets all maximal cliques of G.

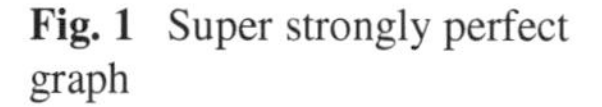
Fig. 1 Super strongly perfect graph

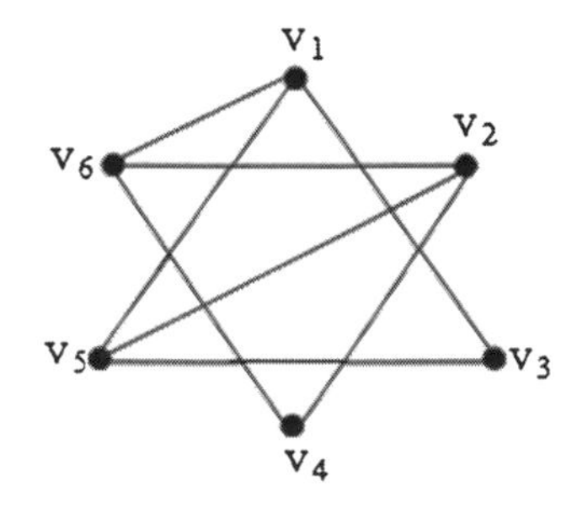

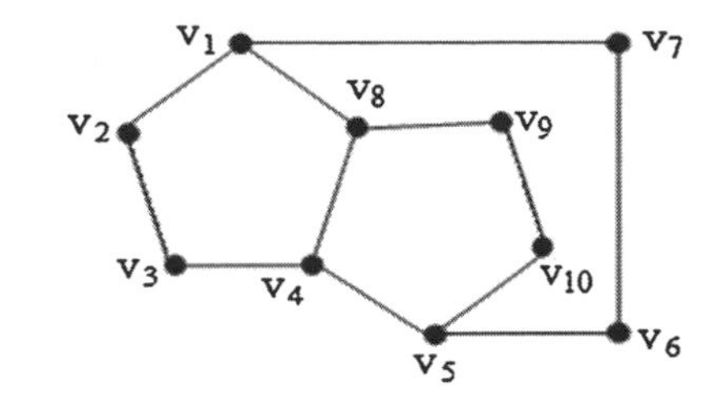

Fig. 2 Non-super strongly perfect graph

Example 2 $\{v_1, v_3, v_6, v_9\}$ is a minimal dominating set which does not meet all maximal cliques of G.

4 Cycle Graph

A closed path is called a cycle. A path is a walk in which all vertices are distinct. A walk on a graph is an alternating series of vertices and edges, beginning and ending with a vertex, in which each edge is incident with the vertex immediately preceding it and the vertex immediately following it. An odd cycle is a cycle with odd length, that is, with an odd number of edges. An even cycle is a cycle with even length, that is, with an even number of edges. The number of vertices in a cycle equals the number of edges.

4.1 Theorem

Let $G = (V, E)$ be a graph with number of vertices n, where $n \geq 5$. Then, G is super strongly perfect if and only if it does not contain an odd cycle as an induced subgraph [6] .

5 Tree

An acyclic graph is a graph which contains no cycle. A tree is a connected acyclic graph. Every tree on n vertices has exactly $n - 1$ edges. Any two vertices of a tree are connected by exactly one path. Figure 3 illustrates the tree structure which is SSP.

Example 3 $\{1\}$ is a minimal dominating set which meets all maximal cliques of G.

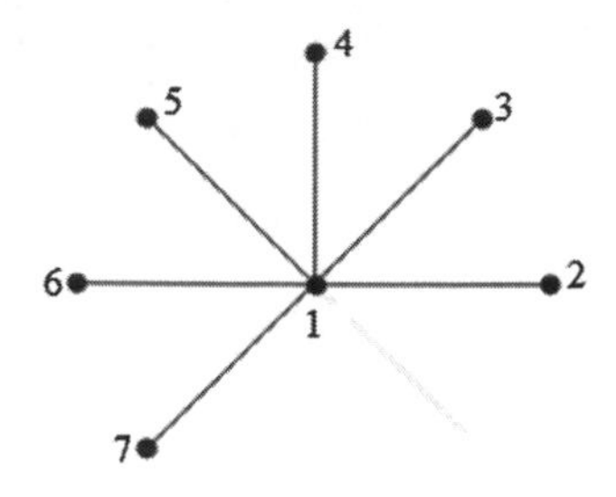

Fig. 3 Tree

5.1 Theorem

Every tree is super strongly perfect.

Proof Let G be a tree.

$\Rightarrow$ G does not contain an odd cycle as an induced subgraph.

Now, by the Theorem 4.1, G is super strongly perfect.

Hence, every tree is super strongly perfect.

5.2 Theorem

Let G be a tree which is super strongly perfect with number of vertices n, where $n \geq 2$, then $\gamma(G) = 1$ if and only if diam $(G) = 2$.

Proof Let G be a tree which is super strongly perfect with $n \geq 2$.

Assume that $\gamma(G) = 1$.

$\Rightarrow$ There exists a vertex $v \in G$ which is adjacent to all the remaining vertices in G.

To prove diam $(G) = 2$.

Suppose diam $(G) > 2$, then there exists at least two vertices a and b with diam $(a, \mathrm{b}) \geq 3$.

$\Rightarrow$ There does not exist a vertex in G which is adjacent to both a and b.

$\Rightarrow \quad \gamma(G) > 1$.

which is a contradiction to the assumption, hence diam $(G) = 2$.

Conversely assume that diam $(G) = 2$.

To prove $\gamma(G) = 1$.

Suppose $\gamma(G) \neq 1$,

$\Rightarrow \quad \gamma(G) \geq 2$, then there does not exist a vertex in G which is adjacent to all the remaining vertices.

$\Rightarrow$ There exists at least two vertices a and b such that diam $(a, b) \geq 3$.

$\Rightarrow$ diam $(G) \geq 3$.

Which is a contradiction to the assumption, hence $\gamma(G) = 1$.

5.3 Theorem

Let G be a tree which is super strongly perfect with number of vertices n, where $n \geq 2$ with $\gamma(G) = 1$, then $\gamma(\overline{G}) = 2$ if and only if diam $(\overline{G})$ is not defined.

Proof Let G be a tree which is super strongly perfect, $n \geq 2$ with $\gamma(G) = 1$.

Assume that $\gamma(\overline{G}) = 2$.

To prove diam $(\overline{G})$ is not defined.

Since $\gamma(G) = 1$, there exists a vertex $v \in G$ which is adjacent to all the vertices in G.

$\Rightarrow$ v is an isolated vertex in $(\overline{G})$.

Since $\gamma(\overline{G}) = 2$ and v is an isolated vertex in $\overline{G}$, hence v must be one of the vertex of the minimum dominating set of $\overline{G}$.

Let u $\in \overline{G}$ such that $\{u, v\}$ be a minimum dominating set of $\overline{G}$.

$\Rightarrow$ diam (u, v) is not defined in $\overline{G}$, since v is an isolated vertex in $\overline{G}$.

$\Rightarrow$ diam $(\overline{G})$ is not defined.

Conversely assume that $\gamma(G) = 1$ and diam $(\overline{G})$ is not defined.

To prove $\gamma(\overline{G}) = 2$.

Since diam $(\overline{G})$ is not defined, there exists a vertex $v \in \overline{G}$ which is an isolated vertex such that diam (u, v) is not defined for some u $\in \overline{G}$.

Since $\gamma(G) = 1$, $\{v\}$ will be the minimum dominating set of G.

Since G is a tree, there exists at least one pendent vertex u^1 incident with a pendent edge, let it be $e = u^1v^1 \in G$.

$\Rightarrow u^1$ is a pendent vertex in G, it is adjacent with all the remaining vertices except v^1 in $\overline{G}$.

$\Rightarrow$ There exists a vertex u^1 which is adjacent with all the remaining vertices in $\overline{G}$ and v is an isolated vertex in $\overline{G}$.

$\Rightarrow \{u^1, v\}$ is a minimum dominating set of $\overline{G}$.

$\Rightarrow \quad \gamma(\overline{G}) = 2$.

Hence proved.

5.4 Theorem

Let G be a tree which is super strongly perfect with number of vertices n, where $n \geq 2$, then $\gamma(G) > 1$ if and only if diam $(G) \geq 3$.

Proof Let G be a tree which is super strongly perfect with $n \geq 2$.

Assume that diam $(G) \geq 3$.

To prove $\gamma(G) > 1$.

By the assumption, diam $(G) \geq 3$.

Then, there exists at least two vertices u and v of G such that diam $(u, v) \geq 3$.

$\Rightarrow$ There does not exist a vertex in G which is adjacent to both u and v.

$\Rightarrow \quad \gamma(G) > 1$.

Conversely assume that $\gamma(G) \geq 1$.
$\Rightarrow \gamma(G) \neq 1$.
By the Theorem 5.2, diam $(G) \neq 2$.
$\Rightarrow$ diam $(G) > 2$.
$\Rightarrow$ diam $(G) \geq 3$.
Hence proved.

5.5 Theorem

Let G be a tree which is super strongly perfect with number of vertices n, where $n \geq 2$ and diam $(G) = 2$ then $\gamma(G) = \delta(G)$.

Proof Let G be a tree which is super strongly perfect, $n \geq 2$ and diam $(G) = 2$.
Since G is a tree, $\delta(G) = 1$.
To prove $\gamma(G) = \delta(G)$.
It is enough to prove $\gamma(G) = 1$.
Suppose $\gamma(G) > 1$
Then, by the Theorem 5.4, diam $(G) \geq 3$.
which is a contradiction to the hypothesis.
Hence, our assumption is wrong.
$\Rightarrow \gamma(G) = 1$.
$\Rightarrow \gamma(G) = \delta(G)$.
Hence proved.

5.6 Theorem

Let G be a tree which is super strongly perfect with number of vertices n, where $n \geq 2$, then diam (G) and diam $(\overline{G})$ cannot be 1.

Proof Let G be a tree which is super strongly perfect with $n \geq 2$.
To prove diam $(G) \neq 1$ and diam $(\overline{G}) \neq 1$.
Suppose diam $(G) = 1$.
$\Rightarrow$ All the vertices are pairwise adjacent in G.
$\Rightarrow$ G is complete, which is a contradiction to the hypothesis.
Hence, our assumption is wrong
$\Rightarrow$ diam $(G) \neq 1$.
Now to prove diam $(\overline{G}) \neq 1$.
Suppose diam $(\overline{G}) = 1$.
$\Rightarrow$ All the vertices are pairwise adjacent in $\overline{G}$.
$\Rightarrow$ All the vertices are isolated in G.
$\Rightarrow$ G is not connected.

Which is a contradiction to the hypothesis.
Hence, our assumption is wrong.
$\Rightarrow$diam $(\overline{G}) \neq 1$.
Hence, diam$(G) \neq 1$ and diam $(\overline{G}) \neq 1$.

5.7 *Proposition*

Let G be a tree which is super strongly perfect with number of vertices n, where $n \geq 2$, then diam $(G) \geq 3$ if and only if $\gamma(\overline{G}) = 2$.

5.8 *Proposition*

Let G be a tree which is super strongly perfect with number of vertices n, where $n \geq 2$, then diam $(\overline{G}) \leq 3$ if and only if $\gamma(\overline{G}) = 2$.

6 Conclusion

We have investigated the characterization of super strongly perfect graphs on trees. Also, we have given the relationship between diameter, domination, and codomination numbers of super strongly perfect graph in trees. This investigation can be applicable for the well known architectures like bipartite graphs, butterfly graphs, Benes butterfly graphs, and chordal graphs.

References

1. Foulds, R.: Graph theory applications. Springer, New York (1994)
2. Cayley, A.: On the theory of analytic forms called trees. Philos. Mag. **13**, 19–30 (1857)
3. Cayley, A.: On the theory of analytic forms called trees. Mathematical Papers (Cambridge) **3**, 242–246 (1891)
4. Bondy, J.A., Murty, U.S.R.: Graph theory with applications. Elsevier Science Publishing Company Inc., New York (1976)
5. Haynes, T.W., Hedetniemi, S.T., Slater, P.J.: Domination in graphs advanced topic. Marcel Dekker, Inc., New York (1998)
6. Amutha, A., Mary Jeya Jothi, R. : Characterization of super strongly perfect graphs. In: Bipartite graphs, proceedings of an international conference on mathematical modelling and scientific computation, 1, 183–185 (2012).

A Novel Approach for Thin Film Flow Problem Via Homotopy Analysis Sumudu Transform Method

Sushila and Y. S. Shishodia

Abstract In this paper, a numerical algorithm based on new homotopy analysis sumudu transform method (HASTM) is proposed to solve a nonlinear boundary value problem arising in the study of thin flow of a third-grade fluid down an inclined plane. The homotopy analysis sumudu transform is a combined form of sumudu transform and homotopy analysis method. The proposed technique finds the solution without any discretization or restrictive assumptions and avoids the round-off errors. The numerical results show that the proposed approach is very efficient and simple and can be applied to other nonlinear problems.

Keywords Thin flow problem · Third-grade fluid · Nonlinear boundary value problem · Homotopy analysis sumudu transform method · Maple code

1 Introduction

Nonlinear phenomena that appear in many areas of scientific fields, such as solid state physics, plasma physics, fluid mechanics, population models, and chemical kinetics, can be modeled by nonlinear differential equations. Most of the science and engineering problems, especially some heat transfer and fluid flow equations, are nonlinear; therefore, some of these problems are solved by computational fluid dynamic (numerical) method, and some are solved by analytical perturbation method.

Sushila (✉)
Department of Physics, Jagan Nath University, Village-Rampura, Tehsil-Chaksu, Jaipur, Rajasthan 303901, India
e-mail: sushila.jag@gmail.com

Y. S. Shishodia
Pro-Vice-Chancellor, Jagan Nath University, Village- Rampura, Tehsil-Chaksu, Jaipur, Rajasthan 303901, India
e-mail: yad.shi@gmail.com

B. V. Babu et al. (eds.), *Proceedings of the Second International Conference on Soft Computing for Problem Solving (SocProS 2012), December 28–30, 2012*, Advances in Intelligent Systems and Computing 236, DOI: 10.1007/978-81-322-1602-5_32,

Recently, for solving the linear and nonlinear boundary value problems, the analytical techniques have become an ever-increasing interest of the scientists and engineers. These techniques have been dominated by the perturbation methods and have found many applications in physical problems. However, perturbation methods, like other analytical techniques, have their own limitations. For example, all perturbation methods require the presence of a small parameter in the equation, and approximate solutions of the equation are expressed in the form of series expansion containing this parameter. Selection of small parameter also requires special skill. Therefore, an analytical method is welcomed which does not require a small parameter in the equation modeling the phenomenon. Unlike perturbation methods, the homotopy analysis method (HAM) [1] does not require the existence of a small parameter in terms of which a perturbation solution is developed and is thus valid for both weakly and strongly nonlinear problems. In recent years, determining approximate analytical solutions using the homotopy analysis method has generated a lot of interest due to its applicability and efficiency, and this technique has been successfully applied to a number of nonlinear problems [2–6]. Very recently, Sushila et al. [7] have introduced a new approximate method, named homotopy analysis sumudu transform method (HASTM) to handle nonlinear partial differential equations. The homotopy analysis sumudu transform method is a combined form of sumudu transform and homotopy analysis method.

In this paper, we apply the homotopy analysis sumudu transform method (HASTM) to obtain the approximate solutions of nonlinear equation governing the thin flow of a third-grade fluid down an inclined plane. The HASTM provides the solution in a rapid convergent series which may lead to the solution in a closed form. The advantage of this method is its capability of combining two powerful methods for obtaining exact and approximate solutions for nonlinear equations.

2 Sumudu Transform

The sumudu transform [8] is defined over the set of functions

$$A = \{f(t) | \exists M, \tau_1, \tau_2 > 0, |f(t)| < M\, e^{|t|/\tau_j}, \text{if } t \in (-1)^j \times [0, \infty)\}$$

by the following formula

$$\bar{f}(u) = \mathrm{S[f(t)]} = \int_0^\infty f(ut)\, e^{-t} \mathrm{d}t, \;\; u \in (-\tau_1, \tau_2). \tag{1}$$

Some of the properties of the sumudu transform were established by Asiru [9]. Further, fundamental properties of this transform were established by Belgacem et al. [10]. The sumudu transform has scale and unit preserving properties, so it can be used to solve problems without resorting to a new frequency domain.

3 Governing Equation

The thin film flow of a third-grade fluid down an inclined plane of inclination $\alpha \neq 0$ is governed by the following nonlinear boundary value problem [11, 12]

$$\frac{d^2U}{dy^2} + \frac{6(\beta_2 + \beta_3)}{\mu}\left(\frac{dU}{dy}\right)^2 \frac{d^2U}{dy^2} + \frac{\rho g \sin\alpha}{\mu} = 0, \tag{2}$$

$$U(0) = 0, \quad \frac{dU}{dy} = 0 \quad at\ y = \delta. \tag{3}$$

Introducing the parameters

$$y = \delta y^*, U = \frac{\delta^2 \rho g \sin\alpha}{\mu} U^*,$$

$$\beta^* = \frac{3\delta^2\rho^2 g^2 \sin^2\alpha}{\mu^3}(\beta_2 + \beta_3) \tag{4}$$

the problem in Eqs. (2) and (3) , after omitting asterisks, takes the following form

$$\frac{d^2U}{dy^2} + 6\beta\left(\frac{dU}{dy}\right)^2 \frac{d^2U}{dy^2} + 1 = 0, \tag{5}$$

$$U(0) = 0, \quad \frac{dU}{dy} = 0 \quad at\ y = 1, \tag{6}$$

where μ is the dynamic viscosity, g is the gravity, ρ is the fluid density, and $\beta > 0$ is the material constant of a third-grade fluid. We note that Eq. (5) is a second-order nonlinear and inhomogeneous differential equation with two boundary conditions; therefore, it is a well-posed problem.

Through integration of Eq. (5), we have

$$\frac{dU}{dy} + 2\beta\left(\frac{dU}{dy}\right)^3 + y = C_1, \tag{7}$$

where C_1 is a constant of integration. Employing the second condition of (6) in Eq. (7), we obtain $C_1 = 1$. Thus, the system (5)–(6) can be written as

$$\frac{dU}{dy} + 2\beta\left(\frac{dU}{dy}\right)^3 + (y - 1) = 0, \tag{8}$$

$$U(0) = 0. \tag{9}$$

It should be noted that for $\beta = 0$, Eq. (5) corresponds to that of Newtonian fluid whose exact solution subjected to the boundary conditions (6) is given by

$$U(y) = -\frac{1}{2}\left[(y-1)^2 - 1\right]. \tag{10}$$

In what follows, we will obtain the approximate analytic solution of the nonlinear system (8)–(9) by using the HASTM.

4 Basic Idea of HASTM

To illustrate the basic idea of this method, we consider an equation $N[U(x,t)] = g(x,t)$, where N represents a general nonlinear ordinary or partial differential operator including both linear and nonlinear terms. The linear terms are decomposed into $L+R$, where L is the highest order linear operator, and R is the remaining of the linear operator. Thus, the equation can be written as

$$LU + RU + NU = g(x,t), \tag{11}$$

where NU indicates the nonlinear terms.

Applying the sumudu transform on both sides of Eq. (11) and using the differentiation property of the sumudu transform, we have

$$S[U] - u^n \sum_{k=0}^{n-1} \frac{U^{(k)}(0)}{u^{(n-k)}} + u^n \left[S[RU] + S[NU] - S[g(x)]\right] = 0. \tag{12}$$

We define the nonlinear operator

$$\begin{aligned} N[\phi(x,t;q)] &= S[\phi(x,t;q)] - u^n \sum_{k=0}^{n-1} \frac{\phi^{(k)}(x,t;q)(0)}{u^{(n-k)}} \\ &\quad + u^n \left[S[R\phi(x,t;q)] + S[N\phi(x,t;q)] - S[g(x)]\right], \end{aligned} \tag{13}$$

where $q \in [0, 1]$ and $\phi(x,t;q)$ is a real function of x, t and q. We construct a homotopy as follows:

$$(1-q)S[\phi(x,t;q) - U_0(x,t)] = \hbar\, q H(x,t)\, N[U(x,t)], \tag{14}$$

where S denotes the sumudu transform, $q \in [0, 1]$ is the embedding parameter, $H(x,t)$ denotes a nonzero auxiliary function, $\hbar \neq 0$ is an auxiliary parameter, $U_0(x,t)$ is an initial guess of $U(x,t)$, and $\phi(x,t;q)$ is a unknown function. Obviously, when the embedding parameter $q = 0$ and $q = 1$, it holds

$$\phi(x,t;0) = U_0(x,t), \qquad \phi(x,t;1) = U(x,t), \tag{15}$$

respectively. Thus, as q increases from 0 to 1, the solution $\phi(x,t;q)$ varies from the initial guess $U_0(x,t)$ to the solution $U(x,t)$. Expanding $\phi(x,t;q)$ in Taylor series with respect to q, we have

$$\phi(x,t;q) = U_0(x,t) + \sum_{m=1}^{\infty} U_m(x,t)q^m, \tag{16}$$

where

$$U_m(x,t) = \frac{1}{m!}\frac{\partial^m \phi(x,t;q)}{\partial q^m}\Big|_{q=0}. \tag{17}$$

If the auxiliary linear operator, the initial guess, the auxiliary parameter $\hbar$, and the auxiliary function are properly chosen, the series (16) converges at $q = 1$, then we have

$$U(x,t) = U_0(x,t) + \sum_{m=1}^{\infty} U_m(x,t), \tag{18}$$

which must be one of the solutions of the original nonlinear equations. According to the definition (18), the governing equation can be deduced from the zero-order deformation (14). Define the vectors

$$\vec{U}_m = \{U_0(x,t), U_1(x,t), \ldots, U_m(x,t)\}. \tag{19}$$

Differentiating the zero-order deformation Eq. (14) m-times with respect to q and then dividing them by $m!$ and finally setting $q = 0$, we get the following mth-order deformation equation:

$$S[U_m(x,t) - \chi_m U_{m-1}(x,t)] = \hbar H(x,t)\Re_m(\vec{U}_{m-1}). \tag{20}$$

Applying the inverse sumudu transform, we have

$$U_m(x,t) = \chi_m U_{m-1}(x,t) + \hbar S^{-1}[H(x,t)\Re_m(\vec{U}_{m-1})], \tag{21}$$

where

$$\Re_m(\vec{U}_{m-1}) = \frac{1}{(m-1)!}\frac{\partial^{m-1} N[\phi(x,t;q)]}{\partial q^{m-1}}\Big|_{q=0}, \tag{22}$$

and

$$\chi_m = \begin{cases} 0, & m \leq 1, \\ 1, & m > 1. \end{cases} \tag{23}$$

5 Solution of the Problem

According to HASTM, we take the initial guess as

$$U_0(y) = y - \frac{y^2}{2}. \tag{24}$$

Applying sumudu transform on both sides of Eq. (8), we have

$$S[U] - u + u^2 + 2\beta u S\left[\left(\frac{\mathrm{d}U}{\mathrm{d}y}\right)^3\right] = 0. \tag{25}$$

We define the nonlinear operator

$$N\left[\phi(y;q)\right] = S\left[\phi(y;q)\right] - u + u^2 + 2\beta u S\left[\left(\frac{\mathrm{d}\phi(y;q)}{\mathrm{d}y}\right)^3\right] \tag{26}$$

and thus

$$\Re_m(\vec{U}_{m-1}) = S\left[\vec{U}_{m-1}\right] - (1-\chi_m)(u-u^2) + 2\beta u S\left[\sum_{i=0}^{m-1}\sum_{l=0}^{i}\frac{\mathrm{d}U_l}{\mathrm{d}y}\frac{\mathrm{d}U_{i-l}}{\mathrm{d}y}\frac{\mathrm{d}U_{m-1-i}}{\mathrm{d}y}\right]. \tag{27}$$

The mth -order deformation equation is given by

$$S\left[U_m(y) - \chi_m U_{m-1}(y)\right] = \hbar\Re_m(\vec{U}_{m-1}). \tag{28}$$

Applying the inverse sumudu transform, we have

$$U_m(y) = \chi_m U_{m-1}(y) + \hbar S^{-1}[\Re_m(\vec{U}_{m-1})]. \tag{29}$$

Solving the above Eq. (29), for $m = 1, 2, 3, \cdots$, we get

$$U_1(y) = -\frac{\beta\hbar}{2}\left[(y-1)^4 - 1\right], \tag{30}$$

$$U_2(y) = -\frac{\beta\hbar}{2}(1+\hbar)\left[(y-1)^4 - 1\right] - 2\beta^2\hbar^2\left[(y-1)^6 - 1\right], \tag{31}$$

$$U_3(y) = -\frac{\beta\hbar}{2}(1+\hbar)^2\left[(y-1)^4-1\right] - 4\beta^2\hbar^2(1+\hbar)\left[(y-1)^6-1\right]$$
$$- 12\beta^3\hbar^3\left[(y-1)^8-1\right], \tag{32}$$
$$\vdots$$

The series solution is given by

$$U(y) = U_0(y) + U_1(y) + U_2(y) + U_3(y) + \cdots. \tag{33}$$

If we put $\hbar = -1$ in Eqs. (30)–(33), we can recover the solutions obtained by using HPM [9], VIM [10], and ADM [10].

In the solution (33), the terms involving the powers of β gives the contribution of the non-Newtonian fluid. It is worth noting that by setting $\beta = 0$ in the above approximations, we recover the exact solution for the case of Newtonian fluid. Thus, the first approximation of the nonlinear system (8)–(9) obtained by HASTM is identical with the exact solutions of the corresponding linear problem. This indicates that the HASTM can be equally applied to linear equations. The effects of the non-Newtonian parameter β on the velocity given in (33) are shown in Fig. 1. It is depicted that as we decrease the non-Newtonian parameter β, the solution converges to the Newtonian case.

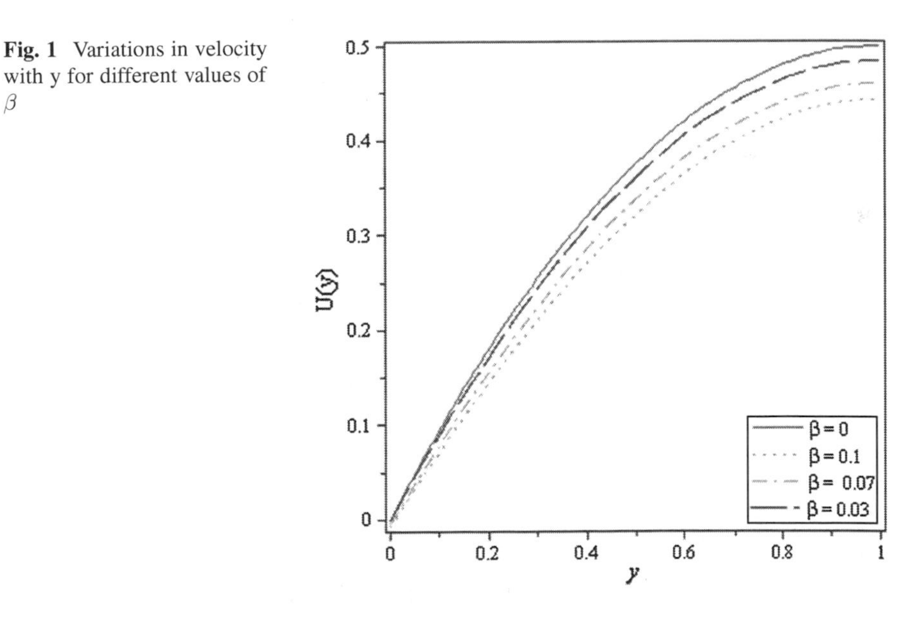

Fig. 1 Variations in velocity with y for different values of β

6 Conclusions

In this paper, the homotopy analysis sumudu transform method (HASTM) is employed for solving the nonlinear equation governing the thin flow of a third-grade fluid down an inclined plane. The numerical results reveal that the HASTM is a very effective method and might find wide applications to solve many types of nonlinear differential equations in science, engineering, and finance. Maple has been used for computation in this paper.

References

1. Liao, S.J.: Beyond Perturbation: Introduction to homotopy analysis method. Chapman and Hall / CRC Press, Boca Raton (2003)
2. Liao, S.J.: On the homotopy analysis method for nonlinear problems. Appl. Math. Comput. **147**, 499–513 (2004)
3. Liao, S.J.: A new branch of solutions of boundary-layer flows over an impermeable stretched plate. Int. J. Heat Mass Trans. **48**, 2529–2539 (2005)
4. Qi, W.: Application of homotopy analysis method to solve relativistic Toda-Lattice system. Commun. Theor. Phys. **53**, 1111–1116 (2010)
5. Shidfar, A., Molabahrami, A.: A weighted algorithm based on the homotopy analysis method: application to inverse heat conduction problems. Commun. Nonlinear Sci. Numer. Simul. **15**, 2908–2915 (2010)
6. Kheiri, H., Alipour, N., Dehgani, R.: Homotopy analysis and Homotopy-Pade methods for the modified Burgers-Korteweg-de-Vries and the Newell Whitehead equation. Math. Sci. **5**(1), 33–50 (2011)
7. Sushila, Kumar, D., Singh, J., Gupta, S.: Homotopy analysis sumudu transform method for nonlinear equations. Int. J. Ind. Math. **4**, 301–314 (2012)
8. Watugala, G.K.: Sumudu transform- a new integral transform to solve differential equations and control engineering problems. Math. Engg. Indust. **6**(4), 319–329 (1998)
9. Asiru, M.A.: Further properties of the Sumudu transform and its applications. Int. J. Math. Educ. Sci. Tech. **33**, 441–449 (2002)
10. Belgacem, F.B.M., Karaballi, A.A., Kalla, S.L.: Analytical investigations of the Sumudu transform and applications to integral production equations. Math. Prob. Engg. **3**, 103–118 (2003)
11. Siddique, A.M., Mahmood, R., Ghori, Q.K.: Homotopy perturbation method for thin film flow of a third grade fluid down an inclined plane. Chaos Solitons Fractals **35**(1), 140–147 (2008)
12. Siddique, A.M., Farooq, A.A., Haroon, T., Rana, M.A., Babcock, B.S.: Application of He's variational iterative method for solving thin flow problem arising in non-Newtonian fluid mechanics. World J. Mech. **2**, 138–142 (2012)

Buckling and Vibration of Non-Homogeneous Orthotropic Rectangular Plates with Variable Thickness Using DQM

Renu Saini and Roshan Lal

Abstract The present work analyzes the buckling and vibration behavior of non-homogeneous orthotropic rectangular plates of variable thickness and subjected to constant in-plane force along two opposite simply supported edges on the basis of classical plate theory. The other two edges may be clamped, simply supported, and free. For non-homogeneity of the plate material, it is assumed that Young's moduli and density vary exponentially along one direction. The governing partial differential equation of motion of such plates has been reduced to an ordinary differential equation using the sine function for mode shapes between the simply supported edges. This has been solved numerically employing DQM. The effect of various parameters has been studied on the natural frequencies for the first three modes of vibration. Critical buckling loads by allowing frequencies to approach zero have been computed. Comparison has been made with the known results.

Keywords Non-homogeneous · Orthotropic · Rectangular · Buckling · DQM

1 Introduction

Plates of various geometries are key components in many structural and machinery applications, particularly in aerospace, civil, mechanical, and automotive industries. In various engineering applications, plates are often subjected to in-plane stresses arising from hydrostatic, centrifugal, and thermal stresses [1–3], which may induce buckling, a phenomenon that is highly undesirable. A lot of work has been carried out to study the vibration of rectangular plates and reported in references [4, 5]. Orthotropic plates are often non-homogeneous either by design or because of the

R. Saini (✉) · R. Lal
Department of Mathematics, Indian Institute of Technology Roorkee,
Roorkee 247667, India
e-mail: renusaini189@gmail.com

B. V. Babu et al. (eds.), *Proceedings of the Second International Conference on Soft Computing for Problem Solving (SocProS 2012), December 28–30, 2012*, Advances in Intelligent Systems and Computing 236, DOI: 10.1007/978-81-322-1602-5_33,

physical composition and imperfection in the underlying materials. Very few models representing the behavior of non-homogeneous materials have been proposed in the literature, and some recent references are [6–8]. The present paper analyzes the behavior of non-homogeneous orthotropic rectangular plates whose two opposite edges are assumed to be simply supported and are subjected to constant in-plane force on the basis of classical plate theory. For non-homogeneity of the plate material, it is assumed that Young's moduli and density vary exponentially along one direction. The governing differential equation for such plates reduces to fourth-order differential equation with variable coefficients whose analytical solution is not feasible. DQM has been employed to obtain the natural frequency for C-C, C-S, and C-F boundary conditions.

2 Formulation

Consider a non-homogeneous orthotropic rectangular plate of dimension $a \times b$ and thickness $h(x, y)$. The x and y axes are taken along the principal directions of orthotropy, and axis of z is perpendicular to the xy plane. The plate that is simply supported at $y = 0$ and b is taken to be under constant in-plane force N_y along these two edges (Fig. 1). The differential equation of motion is given by

$$\begin{aligned} D_x \frac{\partial^4 w}{\partial x^4} &+ D_y \frac{\partial^4 w}{\partial y^4} + 2H \frac{\partial^4 w}{\partial x^2 \partial y^2} + 2\frac{\partial H}{\partial x} \frac{\partial^3 w}{\partial x \, \partial y^2} + 2\frac{\partial H}{\partial y} \frac{\partial^3 w}{\partial y \, \partial x^2} \\ &+ 2\frac{\partial D_x}{\partial x} \frac{\partial^3 w}{\partial x^3} + 2\frac{\partial D_y}{\partial y} \frac{\partial^3 w}{\partial y^3} + \frac{\partial^2 D_x}{\partial x^2} \frac{\partial^2 w}{\partial x^2} + \frac{\partial^2 D_y}{\partial y^2} \frac{\partial^2 w}{\partial y^2} \\ &+ \frac{\partial^2 D_1}{\partial x^2} \frac{\partial^2 w}{\partial y^2} + \frac{\partial^2 D_1}{\partial y^2} \frac{\partial^2 w}{\partial x^2} + 4\frac{\partial^2 D_{xy}}{\partial x \, \partial y} \frac{\partial^2 w}{\partial x \, \partial y} + \rho h \frac{\partial^2 w}{\partial t^2} - N_y \frac{\partial^2 w}{\partial y^2} = 0, \end{aligned} \tag{1}$$

where $D_x = E_x^* h^3/12$, $D_y = E_y^* h^3/12$, $D_{xy} = G_{xy} h^3/12$, $D_1 = E^* h^3/12$,

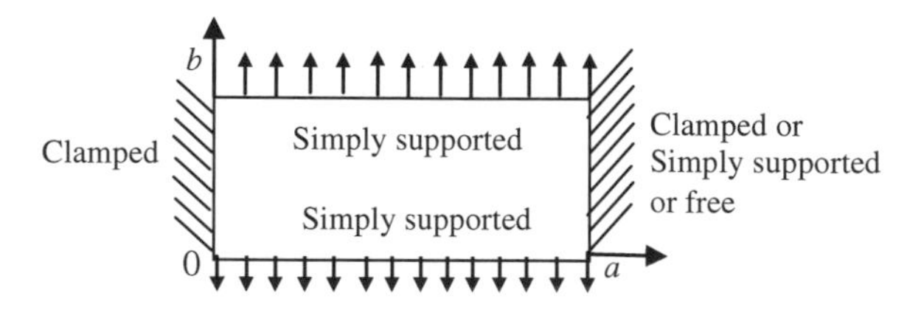

Fig. 1 Geometry of the plate and boundary conditions

$H = D_1 + 2D_{xy}$, $E^* = \nu_y E_x^* = \nu_x E_y^*$, $(E_x^*, E_y^*) = (E_x, E_y)/(1 - \nu_x \nu_y)$, $w(x, y, t)$ is the transverse deflection, t is the time, and $E_x, E_y, \nu_x, \nu_y, G_{xy}$ are material constants.

Let us assume that $h = h(x)$, i.e., independent of y. For harmonic solution, the deflection w is assumed to be

$$w(x, y, z) = \overline{w}(x) \sin(p\pi y/b)e^{i\omega t} \tag{2}$$

where p is a positive number and ω is the frequency in radians. Further, for elastically non-homogeneous material, it is assumed that the E_x, E_y, and ρ are the functions of x only, i.e., $E_x = E_1 e^{\mu X}$, $E_y = E_2 e^{\mu X}$, $\rho = \rho_0 e^{\beta X}$ and the shear modulus is $G_{xy} = \sqrt{E_x E_y}/2(1 + \sqrt{\nu_x \nu_y})$. Using (2) and introducing the non-dimensional variables $X = x/a$, $Y = y/b$, $\overline{h} = h/a$, $W = \overline{w}/a$ where $\overline{h} = h_0 e^{\alpha X}$, Eq. (1) reduces to $A_0 W^{iv} + A_1 W''' + A_2 W'' + A_3 W' + A_4 W = 0$, where $A_0 = 1$

$$A_1 = 2(\mu + 3\alpha),\ A_2 = (\mu + 3\alpha)^2 - 2\sqrt{E_2/E_1}\lambda^2,\ A_3 = -2(\mu + 3\alpha)\sqrt{E_2/E_1}\lambda^2$$
$$A_4 = \lambda^4 E_2/E_1 - \nu_y(\mu + 3\alpha)^2\lambda^2 - \Omega^2 e^{(\beta - \mu - 2\alpha)X} + \lambda^2 \overline{N_y} e^{(-\mu - 3\alpha)X} \tag{3}$$

$\lambda^2 = \pi^2 a^2 p^2/b^2$, $\Omega^2 = 12\,\rho_0(1 - \nu_x \nu_y)a^2\omega^2/E_1 h$, $N_0^* = 12N_y(1 - \nu_x \nu_y)/aE_1 h_0^3$ and prime denotes the differentiation with respect to X. Here, $(h_0, \rho_0) = (h, \rho)_{X=0}$, μ is the non-homogeneity parameter, α is the taper parameter, β is the density parameter, and E_1, E_2 are Young's moduli. Equation (3) is a fourth-order differential equation with variable coefficients whose approximate solution is obtained by DQM.

3 Method of Solution and Boundary Conditions

Let $X_1, X_2 \ldots X_N$ be the N grid points in the applicability range [0, 1] of the plate. According to DQM, the nth-order derivative of $W(X)$ with respect to X can be expressed discretely at the point X_i as

$$\frac{d^n W(X_i)}{dX^n} = \sum_{j=1}^{N} c_{ij}^{(n)} W(X_j),\ i = 1, 2, \ldots N$$

where $c_{ij}^{(n)}$ are the weighting coefficients at X_i and given by

$$c_{ij}^{(n)} = M^{(1)}(X_i)/(X_i - X_j)M^{(1)}(X_j),\quad i = 1, 2 \ldots N (i \neq j),$$
$$M^{(1)}(X_i) = \prod_{j=1, j\neq i}^{N} (X_i - X_j)$$

$$c_{ij}^{(n)} = n\left(c_{ii}^{(n-1)}c_{ij}^{(1)} - \frac{c_{ij}^{(n-1)}}{(X_i - X_j)}\right) \quad i, j = 1, 2 \ldots N, j \neq i \text{ and } n = 2, 3, 4$$

$c_{ii}^{(n)} = -\sum_{j=1, j\neq i}^{N} c_{ij}^{(n)}$, $i = 1, 2, ...N$ and $n = 1, 2, 3, 4$. Discretizing Eq. (3) and substituting for $W(X)$ and its derivative at the ith grid point

$$\sum_{j=1}^{N}\left(A_0c_{ij}^{(4)} + A_1c_{ij}^{(3)} + A_2c_{ij}^{(2)} + A_3c_{ij}^{(1)}\right)W(X_j) + A_{4,i}W(X_i) = 0 \tag{4}$$

For $i = 3, 4 \ldots (N-2)$, ones obtain a set of $(N-4)$ equations in terms of unknowns $W_j (\equiv W(X_j))$, $j = 1, 2, \ldots N$, which can be written in the matrix form as

$$[B][W^*] = [0], \tag{5}$$

where B and W^* are matrices of order $(N-4) \times N$ and $(N \times 1)$, respectively. The $(N-2)$ internal grid points are the zeroes of shifted Chebyshev polynomial in the range [0, 1] given by $X_{k+1} = 1/2[1+\cos((2k-1)\pi/2(N-2))]$, $k = 1, 2, ...N-2$.

The three sets of boundary conditions, namely C-C, C-S, and C-F, have been considered. By satisfying the relations, a set of four homogeneous equations are obtained.

$W = dW/dX = 0$; $W = (d^2W/dX^2) - (E^*/E_x^*)\lambda^2 W = 0$ and
$W = (d^2W/dX^2) - (E^*/E_x^*)\lambda^2 W$
$= (d^3W/dX^3) - \lambda^2(E^* + 4G_{xy})/E_x(dW/dX) = 0.$

This set together with field Eq. (5) gives a complete set of N equations in N unknowns, which is expressed as

$$\begin{bmatrix} A \\ B^{CC} \end{bmatrix}\{B\} = 0 \tag{6}$$

For a non-trivial solution of Eq. (6), the frequency determinant must vanish, and hence,

$$\begin{vmatrix} A \\ B^{CC} \end{vmatrix} = 0, \tag{7}$$

Similarly for C-S and C-F plates

$$\begin{vmatrix} A \\ B^{CS} \end{vmatrix} = 0, \tag{8}$$

$$\begin{vmatrix} A \\ B^{CF} \end{vmatrix} = 0. \tag{9}$$

4 Numerical Results and Discussions

The frequency Eqs. (7–9) provide the values of frequency parameter Ω for various values of $\overline{N}_y = -50\,(20)\,50, 0$, $\beta = -0.5\,(0.2)\,0.5, 0.0, 0.1$, $\alpha = -0.5\,(0.2)\,0.5, 0.0, 0.1$, $\mu = -0.5\,(0.2)\,0.5, 0.0, 0.1$ and $a/b = 0.25\,(0.25)\,2.0$ for $p = 1$. The values of elastic constants for plate material 'ORTHO1' are taken as $(E_1, E_2) = (1 \times 10^{10}, 0.5 \times 10^{10})$ MPa, $\nu_x = 0.2$, $\nu_y = 0.1$. To choose the appropriate number of grid points N, a convergence study has been carried out and graphs are shown in Fig. 2. The value of N has been fixed as 17, for all the three plates.

The results are presented in Figs. (3, 4, 5, 6 and 7) and Table 1. In Fig. 3a, it is observed that Ω increases with the increasing values of $\overline{N}_y$ for all three boundary conditions. The rate of increase in Ω with $\overline{N}_y$ increases in the order of boundary conditions C-C, C-S, and C-F. In Fig. 3b, c, the behavior of Ω with $\overline{N}_y$ is the same except the rate of increase in Ω with $\overline{N}_y$ decreases with the increase in the number of modes.

In Fig. 4a, it is observed that Ω increases with the increasing values of $\mu$for all three plates. The rate of increase in Ω with μ increases in the order of boundary conditions C-F, C-S, and C-C. In Fig. 4b, c, the rate of increase in Ω with increasing values of μincreases with the increase in the number of modes.

From Fig. 5a, it is clear that Ω decreases with increasing value of β. The rate of decrease in Ω with β for a C-S plate is higher than that for a C-F plate but lower for a C-C plate. For II and III modes of vibration, this rate of decrease in Ω further increases in the same order of boundary conditions as shown in Fig. 5b, c.

From Fig. 6a, it is observed that Ω increase with the increasing values of α when $\overline{N}_y < 0$ for all three plates. In Fig. 6b, c, the value of Ω increases with the increasing values of α for all three plates.

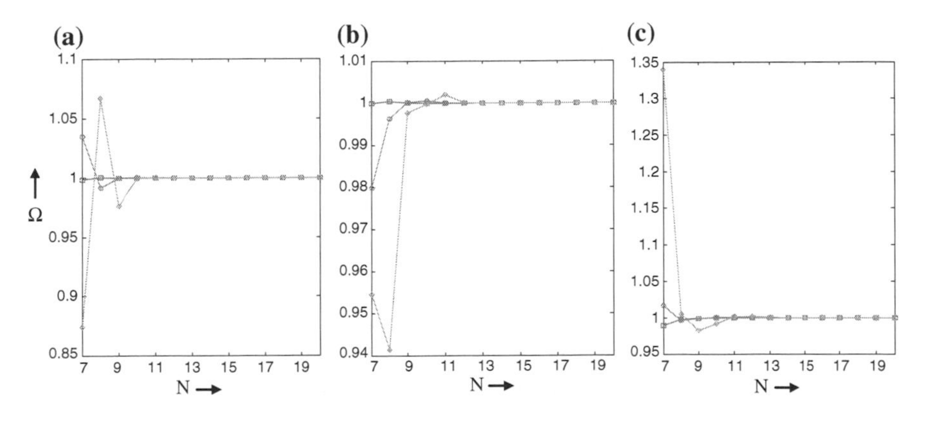

Fig. 2 Normalized frequency parameter Ω/Ω^*: **a** C-C plate, **b** C-S plate, and **c** C-F plate, for $a/b = 1$, $\beta = -0.5$, $\mu = -0.5$, $\overline{N}_y = 30$, *square*, I mode *circle*, II mode *lozenge*, III mode. Ω^*-result using 20 grid points

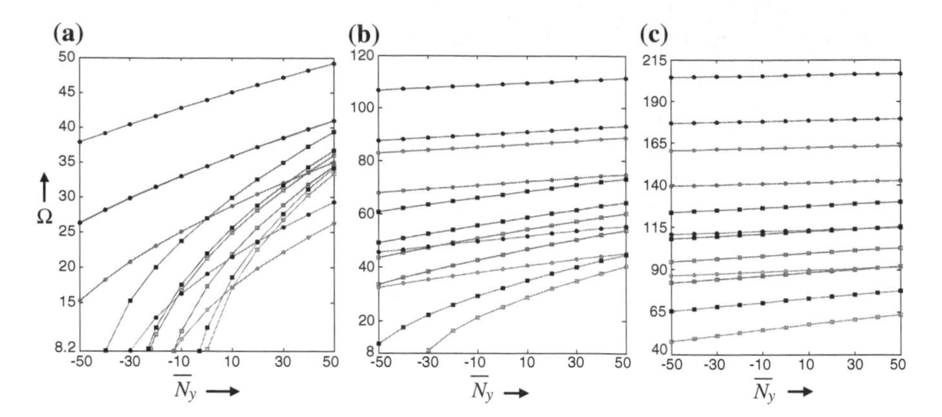

Fig. 3 Frequency parameter: **a** I mode, **b** II mode, and **c** III mode for $\beta = -0.5, a/b = 1$. —C-C; - - -, C-S; -.-., C-F; *square*, $\mu = \alpha = -0.5$; *blacksquare*, $\mu = 0.5, \alpha = -0.5$; *circle*, $\mu = -0.5, \alpha = 0.5$; *blackcircle*, $\mu = \alpha = 0.5$

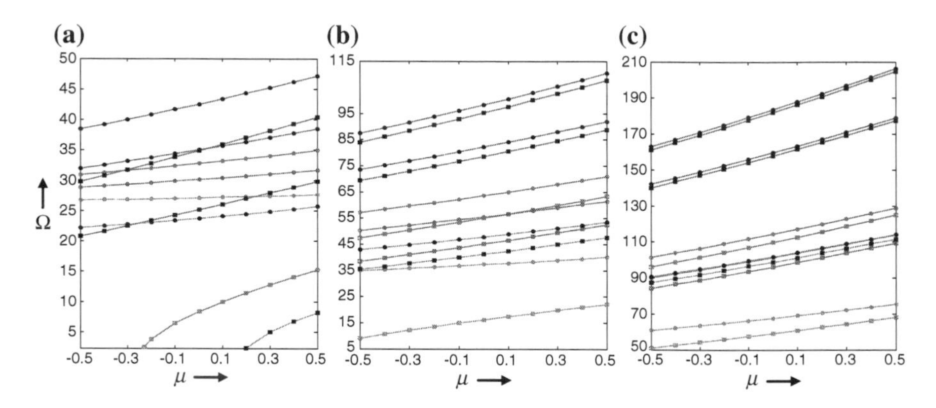

Fig. 4 Frequency parameter: **a** I mode, **b** II mode, and **c** III mode for $\beta = -0.5$,.—C-C; - - -, C-S; -.-. C-F; *square*, $\overline{N}_y = 30, \alpha = -0.5$; *blacksquare*, $\overline{N}_y = 30, \alpha = -0.5$; *circle*, $\overline{N}_y = 30, \alpha = 0.5$; *blackcircle*, $\overline{N}_y = 30, \alpha = 0.5$

In Fig. 7a, it is found that Ω increases with the increasing values of a/b for all three plates. The rate of change of Ω with a/b increases for $\overline{N}_y = -30$ and decreases for $\overline{N}_y = 30$ with the increase in the number of modes as shown in Fig. 7b, c. By allowing the frequency to approach zero, the critical values $\overline{N}_{cr}$ of $\overline{N}_y$ have been computed for all the three plates (Table 1).

A comparison of result by other methods has been given in Tables 2 and 3.

Table 1 Values of critical buckling loads $\overline{N_{cr}}$ for $\beta = -0.5$, $E_2/E_1 = 0.5$, $\nu_y = 0.1$

α	μ/Mode	C-C			C-S			C-F		
		I	II	III	I	II	III	I	II	III
−0.5	−0.5	−26.4422	−163.114	−579.019	−16.6631	−116.046	−445.942	−3.48216	−34.4576	−179.265
	−0.3	−29.3030	−181.174	−643.859	−18.4300	−128.690	−495.404	−3.89020	−38.1689	−198.872
	0	−34.1661	−211.872	−754.077	−21.4265	−150.132	−579.353	−4.59909	−44.4753	−232.124
	0.3	−39.8103	−247.480	−881.885	−24.8952	−174.931	−676.519	−5.44568	−51.7942	−270.589
	0.5	−44.0663	−274.303	−978.119	−27.5056	−193.567	−749.564	−6.10067	−57.3151	−299.498
0	−0.5	−56.7337	−353.962	−1263.60	−35.2560	−248.717	−965.732	−8.13189	−73.7733	−385.074
	−0.3	−62.7361	−391.602	−1398.31	−38.9216	−274.687	−1067.49	−9.13560	−81.5918	−425.390
	0	−72.9123	−455.248	−1625.80	−45.1300	−318.487	−1239.01	−10.8950	−94.8847	−493.419
	0.3	−84.6849	−528.608	−1887.53	−52.3074	−368.817	−1435.90	−13.0172	−110.332	−571.656
	0.5	−93.5380	−583.585	−2083.33	−57.7042	−406.439	−1582.91	−14.6717	−122.002	−630.191
0.5	−0.5	−119.785	−745.634	−2658.80	−73.7176	−516.929	−2013.68	−19.8536	−156.906	−802.402
	−0.3	−132.175	−821.662	−2927.96	−81.2905	−568.588	−2214.57	−22.4353	−173.551	−883.095
	0	−153.122	−949.546	−3379.54	−94.1220	−655.266	−2550.85	−26.9843	−201.947	−1018.77
	0.3	−177.273	−1096.04	−3895.12	−108.971	−754.272	−2933.73	−32.5003	−235.095	−1174.20
	0.5	−195.383	−1205.26	−4278.41	−120.150	−827.922	−3217.70	−36.8162	−260.241	−1290.18

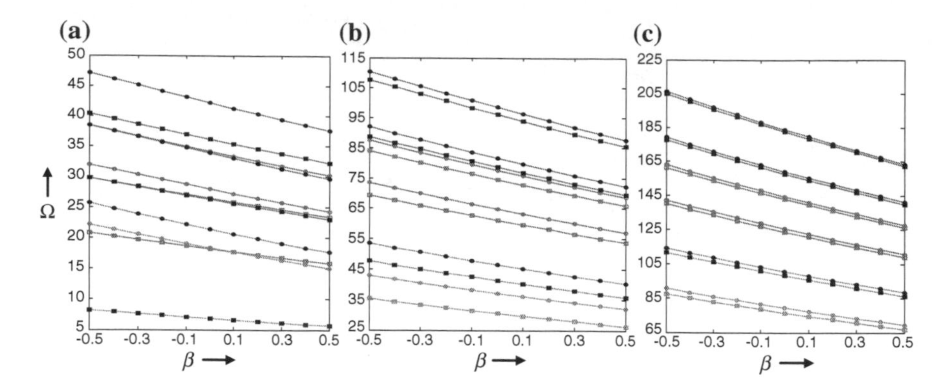

Fig. 5 Frequency parameter: **a** I mode, **b** II mode, and **c** III mode for $a/b = 1$. —C-C; - - -, C-S; -.-.-, C-F; *square*, $\overline{N}_y = -30, \mu = -0.5$; *blacksquare*, $\overline{N}_y = -30, \mu = 0.5$ *circle*, $\overline{N}_y = 30, \mu = -0.5$; *blackcircle*, $\overline{N}_y = 30, \mu = 0.5$

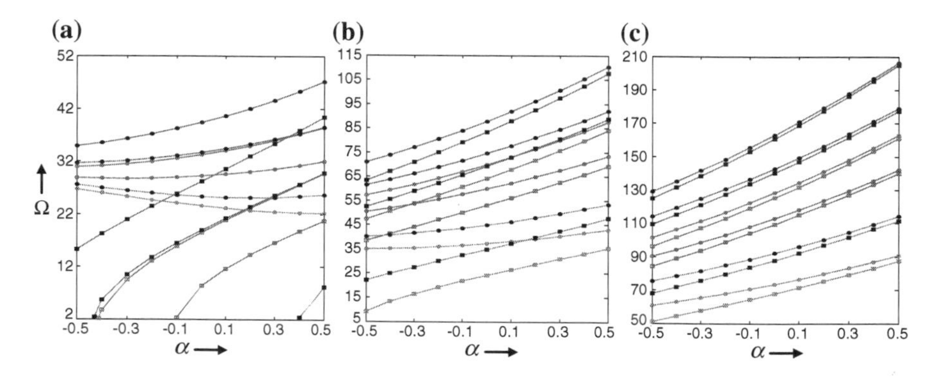

Fig. 6 Frequency parameter: **a** I mode, **b** II mode, and **c** III mode for $\beta = -0.5$.. —C-C; - - -, C-S; -.-. C-F; *square*, $\overline{N}_y = -30, \mu = -0.5$; *blacksquare*, $\overline{N}_y = -30, \mu = 0.5$, *circle*, $\overline{N}_y = 30, \mu = -0.5$; *blackcircle*, $\overline{N}_y = 30, \mu = 0.5$

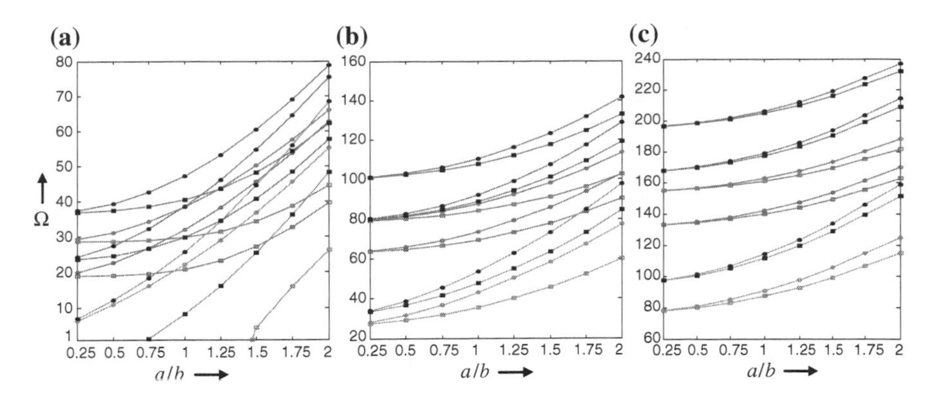

Fig. 7 Frequency parameter: **a** I mode, **b** II mode, and **c** III mode for $\alpha = 0.5$. —C-C; - - -, C-S; -.-. C-F; *square*, $\overline{N}_y = -30, \mu = -0.5$; *blacksquare*, $\overline{N}_y = -30, \mu = 0.5$, *circle*, $\overline{N}_y = 30, \mu = -0.5$; *blackcircle*, $\overline{N}_y = 30, \mu = 0.5$

Table 2 Comparison of frequency parameter Ω for homogeneous ($\mu = 0, \beta = 0$), isotropic ($E_x = E_y = E$), plates for $\nu = 0.3$

	I Mode				II Mode			
a/b	0.5		1		0.5		1	
α	0.0	0.5	0.0	0.5	0.0	0.5	0.0	0.5
C-C	23.8156	30.7594	28.9509	37.2761	63.5345	81.6107	69.3270	88.9922
	23.8204[a]	30.7573[a]	28.9499[a]	37.2675[a]	63.6027[a]	81.6757[a]	69.3796[a]	89.0385[a]
	–	–	28.946[b]	–	–	–	69.320[b]	–
	23.8156[c]	30.7594[c]	28.9508[c]	37.2761[c]	63.6345[c]	81.6101[c]	69.3270[c]	88.9921[c]
C-S	17.3318	21.2621	23.6463	30.1345	52.0979	65.6981	58.6464	74.5075
	17.3350[a]	21.2595[a]	23.6468[a]	30.1282[a]	52.0978[a]	65.7417[a]	58.6880[a]	74.5381[a]
	–	–	23.646[b]	–	–	–	58.641[b]	–
	17.1614[c]	21.2621[c]	23.6463[c]	30.1345[c]	52.0978[c]	65.6999[c]	58.6463[c]	74.5102[c]
C-F	5.70387	6.86986	12.6874	17.2586	24.9438	30.1903	33.0651	42.0484
	5.7031[a]	6.8682[a]	12.6838[a]	17.2545[a]	24.949[a]	30.194[a]	33.064[a]	42.045[a]
	–	–	12.69[b]	–	–	–	33.06[b]	–
	5.7039[c]	6.8677[c]	12.6873[c]	17.2583[c]	24.9438[c]	30.1922[c]	33.0652[c]	42.0566[c]

[a] values from quintic spline [7], [b] exact values [9] and [c] values from Chebyshev collocation technique [10]. A comparison of $\overline{N_{cr}}$ has been presented in Table 3

Table 3 Comparison of critical buckling loads $\overline{N}_{cr}$ for homogeneous ($\mu = 0, \beta = 0$), isotropic ($E_x = E_y = E$), C-C plates for $\nu = 0.3$

Ref. a/b	0.4	0.5	0.6	0.7	0.8	0.9	1
Present	93.247	75.910	69.632	69.095	72.084	77.545	84.922
[7, 11]	93.209	75.887	69.604	69.072	72.067	77.533	75.877
	93.247	75.91	69.632	69.095	72.084	77.545	75.91

5 Conclusions

The DQM has been used to study the effect of in-plane force together with non-homogeneity of the material on the transverse vibrations of orthotropic rectangular plates of exponentially varying thickness on the basis of classical plate theory. The frequency parameter Ω is found to be increased with the increasing values of $\overline{N}_y$, μ, a/b and decreased with the increasing values of β keeping all other parameters fixed. Critical buckling load $\overline{N}_{cr}$ for a C-S plate is higher than that for a C-F plate but less than that for a C-C plate keeping all other plate parameters fixed. An excellent agreement of the results with those obtained by other methods shows the high accuracy and computational efficiency of the present approach.

References

1. Brayan, G.H.: On the stability of a plane plate under thrust in its own plane with application to the buckling of the sides of a ship. Proc. London Math. Soc. **22**, 54–67 (1890)
2. Timoshenko, S., Gere, J.: Theory of Elastic Stability, 2nd ed. McGraw-Hill, New York (1963)
3. Gorman, D.G.: Vibration of thermally stressed polar orthotropic annular plates. Earthquake Eng. Struct. Dynam. **11**, 843–855 (1983)
4. Gorman, D.J.: Free vibration and buckling of in-plane loaded plates with rotational elastic edge support. J. Sound Vib. **229**(4), 755–773 (2000)
5. Dhanpati.: Free transverse vibrations of rectangular and circular orthotropic plates. Ph.D. thesis, Indian Institute of Technology Roorkee, Roorkee, India (2007)
6. Chakraverty, S., Jindal, R., Agarwal, V.K.: Vibration of nonhomogeneous orthotropic elliptic and circular plates with variable thickness. J. Sound Vib. **129**, 256–259 (2007)
7. Lal, R.: Dhanpati.: Quintic splines in the study of buckling and vibration of non-homogeneous orthotropic rectangular plates with variable thickness. Int. J. Appl. Math. Mech. **3**(3), 18–35 (2007)
8. Lal, R., Kumar, Y.: Buckling and vibration of orthotropic nonhomogeneous rectangular plates With bilinear thickness variation. J. Appl. Mech. **78**(06), 1011–1012 (2011)
9. Leissa, A.W.: Vibration of Plates. NASA SP-160 Washington, DC. Govt. Office (1969)
10. Lal, R., Gupta, U.S., Goel, C.: Chebyshev polynomials in the study of transverse vibrations of non-uniform rectangular orthotropic plates. S. V. Digest. **33**(2), 103–112 (2001)
11. Leissa, A.W., Kang, J.H.: Exact solution for vibration and buckling of an SS-C-SS-C rectangular plate loaded by linearly varying in plane stresses. Int. J. Mech. Sci. **44**, 1925–1945 (2002)

A Dual SBM Model with Fuzzy Weights in Fuzzy DEA

Jolly Puri and Shiv Prasad Yadav

Abstract The dual part of a SBM model in data envelopment analysis (DEA) aims to calculate the optimal virtual costs and prices (also known as weights) of inputs and outputs for the concerned decision-making units (DMUs). In conventional dual SBM model, the weights are found as crisp quantities. However, in real-world problems, the weights of inputs and outputs in DEA may have fuzzy essence. In this paper, we propose a dual SBM model with fuzzy weights for input and output data. The proposed model is then reduced to a crisp linear programming problem by using ranking function of a fuzzy number (FN). This model gives the fuzzy efficiencies and the fuzzy weights of inputs and outputs of the concerned DMUs as triangular fuzzy numbers (TFNs). The proposed model is illustrated with a numerical example.

Keywords Fuzzy DEA · Fuzzy SBM model · Fuzzy efficiency · Fuzzy weights

1 Introduction

Data envelopment analysis (DEA) [1] is a nonparametric and linear programming-based technique which evaluates the relative efficiency of homogeneous DMUs on the basis of multiple inputs and multiple outputs. Since the time DEA was proposed, it has got comprehensive attention both in theory and in applications. The beauty of DEA is its ability to measure relative efficiencies of DMUs without assuming prior weights on the inputs and outputs. The first model in DEA is the CCR model [1] which deals with proportional changes in inputs and outputs. The CCR efficiency score reflects the proportional maximum input reduction (or output augmentation)

J. Puri (✉) · S. P. Yadav
Department of Mathematics, I.I.T. Roorkee, Roorkee 247667, India
e-mail: puri.jolly@gmail.com

S. P. Yadav
e-mail: yadavfma@iitr.ernet.in

B. V. Babu et al. (eds.), *Proceedings of the Second International Conference on Soft Computing for Problem Solving (SocProS 2012), December 28–30, 2012*, Advances in Intelligent Systems and Computing 236, DOI: 10.1007/978-81-322-1602-5_34,

rate which is common to all inputs (outputs). But it neglects the slacks corresponding to inputs and outputs. To overcome this shortcoming of CCR model, Tone presented Slack-based Measure (SBM) model [15] in DEA, which puts aside the assumption of proportionate changes in inputs and outputs, and deals with slacks directly. The primal part of the SBM model directly deals with input excesses and output shortfalls of the concerned DMUs. On the other hand, the dual part of the SBM model can be interpreted as profit maximization model and it aims to calculate the optimal virtual costs and prices (also known as weights) of inputs and outputs for the concerned DMUs. Other theoretical extensions of SBM model can be seen in [4, 13].

The conventional DEA models are limited to only crisp input/output data and also their weights take only crisp values. However, in real-world problems, two situations can be possible: (1) Input/output data may have imprecision or fuzziness and (2) the weights of data may have fuzzy essence. To deal with imprecise data, the notion of fuzziness has been introduced in DEA. The DEA is extended to fuzzy DEA (FDEA) in which the imprecision is represented by fuzzy sets or FNs [7, 14]. The SBM efficiency in DEA is extended to fuzzy settings in [6, 11, 12]. Several approaches have been developed to deal with fuzzy data in FDEA. These approaches are as follows: (1) tolerance approach [14], (2) α-cut approach [7], (3) fuzzy ranking approach [5], and (4) possibility approach [8]. However, very less emphasis has been given to FDEA models with fuzzy weights. Mansourirad et al. [10] are the first who introduced fuzzy weights in fuzzy CCR model and proposed a method based on α-cut approach to evaluate weights for outputs in terms of TFNs. In this paper, we propose a dual SBM model with fuzzy weights corresponding to crisp input and output data. We reduce the proposed model into crisp linear programming problem (LPP) by using ranking function of an FN. The proposed model gives the fuzzy efficiencies and the fuzzy weights corresponding to inputs and outputs of the concerned DMUs as TFNs.

The paper is organized as follows: Section 2 presents preliminaries which include basic definitions. Section 3 presents the description of primal and dual parts of the SBM model. Section 4 presents the dual SBM model with fuzzy weights and its reduction to a crisp LPP. Section 5 presents the results and discussion of a numerical example to illustrate the proposed model. The last Section 6 concludes the findings of our study.

2 Preliminaries

The basic definitions in the fuzzy set theory can be seen from [16]. This section includes the definition of TFN and arithmetic operations on TFNs [2]. It also includes ranking function which maps FN to the real line [9].

2.1 Triangular Fuzzy Number

A TFN $\tilde{A}$, denoted by (a_1, a_2, a_3), is defined by the membership function $\mu_{\tilde{A}}$ given by

$$\mu_{\tilde{A}}(x) = \begin{cases} \dfrac{x - a_1}{a_2 - a_1}, & a_1 < x \leq a_2, \\ 1, & x = a_2, \\ \dfrac{x - a_3}{a_2 - a_3}, & a_2 \leq x < a_3, \\ 0, & \text{otherwise}, \end{cases}$$

$\forall x \in R$. In the present study, $\tilde{0} = (0, 0, 0)$, $\tilde{1} = (1, 1, 1)$ and $\tilde{a} = (a, a, a)$ where $a \in R$.

2.2 Arithmetic Operations on TFNs

Let $\tilde{A} = (a_1, a_2, a_3)$ and $\tilde{B} = (b_1, b_2, b_3)$ be two TFNs. Then,
Addition: $\tilde{A} \oplus \tilde{B} = (a_1 + b_1, a_2 + b_2, a_3 + b_3)$.
Subtraction: $\tilde{A} \ominus \tilde{B} = (a_1 - b_3, a_2 - b_2, a_3 - b_1)$.
Scalar multiplication: $k\tilde{A} = \begin{cases} (ka_1, ka_2, ka_3), & k \geq 0, \\ (ka_3, ka_2, ka_1), & k < 0. \end{cases}$
Multiplication: $\tilde{A} \otimes \tilde{B} = (\min(a_1b_1, a_1b_3, a_3b_1, a_3b_3), a_2b_2, \max(a_1b_1, a_1b_3, a_3b_1, a_3b_3))$.

2.3 Ranking Function

Let $F(R)$ be the set of all FNs. A ranking function [9] $\Re$ is a mapping from $F(R)$ to the real line. The FNs can easily be compared by using ranking functions. The rank of TFN $\tilde{A} = (a_1, a_2, a_3)$, represented by $\Re(\tilde{A})$, is defined by $\Re(\tilde{A}) = (a_1 + 2a_2 + a_3)/4$.
Let $\tilde{A} = (a_1, a_2, a_3)$ and $\tilde{B} = (b_1, b_2, b_3)$ be two TFNs in $F(R)$. Then,

1. $\tilde{A}$ is said to be equal to $\tilde{B}$ based on ranking function $\Re$, written as $\tilde{A} \underset{\Re}{=} \tilde{B}$, iff $\Re(\tilde{A}) = \Re(\tilde{B})$.
2. $\tilde{A}$ is said to be less than or equal to $\tilde{B}$ based on ranking function $\Re$, written as $\tilde{A} \underset{\Re}{\leq} \tilde{B}$, iff $\Re(\tilde{A}) \leq \Re(\tilde{B})$.
3. $\tilde{A}$ is said to be greater than or equal to $\tilde{B}$ based on ranking function $\Re$, written as $\tilde{A} \underset{\Re}{\geq} \tilde{B}$, iff $\Re(\tilde{A}) \geq \Re(\tilde{B})$.

4. $\tilde{A}$ is said to be less than or equal to $\tilde{0}$ based on ranking function $\Re$, written as $\tilde{A} \underset{\Re}{\leq} \tilde{0}$, iff $\Re(\tilde{A}) \leq \Re(\tilde{0})$.

Theorem: $\Re(c\tilde{A}+\tilde{B}) = c\Re(\tilde{A})+\Re(\tilde{B})$, c is any constant. (*Linearity property* [11]).

3 Slack-based Measure Model

Assume that the performance of a set of n homogeneous DMUs (DMU$_j$; $j = 1, \ldots, n$) is to be measured. The performance of DMU$_j$ is characterized by a production process of m inputs (x_{ij}; $i = 1, \ldots, m$) to yield s outputs (y_{rj}; $r = 1, \ldots, s$). Let y_{rk} be the amount of the rth output produced by the kth DMU and x_{ik} be the amount of the ith input used by the kth DMU. Assume that input and output data are positive. The primal of SBM model [15] of the kth DMU, represented by SBM-P$_k$, is defined as

$$\text{SBM-P}_k \qquad \rho_k = \min \frac{1-(1/m)\sum_{i=1}^{m} s_{ik}^{-}/x_{ik}}{1+(1/s)\sum_{r=1}^{s} s_{rk}^{+}/y_{rk}}$$

$$\text{subject to } x_{ik} = \textstyle\sum_{j=1}^{n} x_{ij}\lambda_{jk} + s_{ik}^{-}\ \forall i,$$

$$y_{rk} = \textstyle\sum_{j=1}^{n} y_{rj}\lambda_{jk} - s_{rk}^{+}\ \forall r,$$

$$\lambda_{jk} \geq 0\ \forall j,\ s_{ik}^{-} \geq 0\ \forall i,\ s_{rk}^{+} \geq 0\ \forall r,$$

where s_{rk}^{+} is the slack in the rth output of the kth DMU; s_{ik}^{-} is the slack in the ith input of the kth DMU; λ_{jk}'s, i.e., $(\lambda_{j1}, \lambda_{j2}, \ldots, \lambda_{jn})$ are non-negative variables for $j = 1, 2, \ldots, n$. The kth DMU is SBM efficient if $\rho_k = 1$ and all $s_{ik}^{-} = 0$, $s_{rk}^{+} = 0$, i.e., no input excesses and no output shortfalls in any optimal solution.

SBM-P$_k$ can be transformed into LPP using Charnes–Cooper transformation given in [1]. Multiply a scalar $t_k > 0$ to both the denominator and the numerator of SBM-P$_k$. This causes no change in the value of ρ_k. The value of t_k can be adjusted in such a way that the denominator becomes 1. The SBM-P$_k$ model in LPP form becomes

$$\text{LPP-SBM-P}_k \quad \tau_k = \min t_k - \frac{1}{m}\sum_{i=1}^{m} S_{ik}^{-}/x_{ik}$$

$$\text{subject to } 1 = t_k + \frac{1}{s}\sum_{r=1}^{s} S_{rk}^{+}/y_{rk},$$

$$t_k x_{ik} = \sum_{j=1}^{n} x_{ij}\Lambda_{jk} + S_{ik}^{-}\forall i,$$

$$t_k y_{rk} = \sum_{j=1}^{n} y_{rj}\Lambda_{jk} - S_{rk}^{+}\forall r,$$

$$\Lambda_{jk} \geq 0\,\forall j,\, S_{ik}^{-} \geq 0\,\forall i,\, S_{rk}^{+} \geq 0\,\forall r,\, t_k > 0,$$

where $\rho_k = \tau_k, \lambda_{jk} = \Lambda_{jk}/t_k \forall j, s_{ik}^{-} = S_{ik}^{-}/t_k \forall i$ and $s_{rk}^{+} = S_{rk}^{+}/t_k \forall r$.

The dual of LPP-SBM-P_k, represented by SBM-D_k, can be expressed as follows:

$$\begin{aligned}
&\text{SBM-D}_k \qquad E_k = \max \xi_k \\
&\text{subject to } \xi_k + \textstyle\sum_{i=1}^{m} x_{ik} v_{ik} - \sum_{r=1}^{s} y_{rk} u_{rk} = 1, \\
&\textstyle\sum_{r=1}^{s} y_{rj} u_{rk} - \sum_{i=1}^{m} x_{ij} v_{ik} \leq 0\,\forall j, \\
&v_{ik} \geq \tfrac{1}{m x_{ik}}\,\forall i,\, u_{rk} \geq \tfrac{\xi_k}{s y_{rk}}\,\forall r.
\end{aligned}$$

where $\xi_k \in R$, $v_{ik}\,\forall i$, and $u_{rk}\,\forall r$ are the dual variables corresponding to LPP-SBM-P_k. The dual variables v_{ik} and u_{rk} are the weights associated with the ith input and the rth output, respectively. The E_k is the SBM efficiency of the kth DMU.

4 Dual SBM Model with Fuzzy Weights

In conventional SBM-D_k model, the weights of inputs and outputs are found as crisp quantities. However, in real-world problems, the weights may have fuzzy essence. Therefore, in this paper, weights of inputs and outputs are taken as TFNs, and thus, the SBM-D_k model becomes fuzzy SBM-D_k (FSBM-D_k) model given by

$$\begin{aligned}
&\text{FSBM-D}_k \qquad \tilde{E}_k \underset{\Re}{=} \max \tilde{\xi}^k \\
&\text{subject to } \tilde{\xi}^k \oplus \textstyle\sum_{i=1}^{m} x_{ik} \tilde{v}^{ik} \,\Theta \sum_{r=1}^{s} y_{rk} \tilde{u}^{rk} \underset{\Re}{=} \tilde{1}, \\
&\textstyle\sum_{r=1}^{s} y_{rj} \tilde{u}^{rk} \Theta \sum_{i=1}^{m} x_{ij} \tilde{v}^{ik} \underset{\Re}{\leq} \tilde{0}\,\forall j, \\
&\tilde{v}^{ik} \underset{\Re}{\geq} \tfrac{1}{m x_{ik}} \tilde{1}\,\forall i,\, \tilde{u}^{rk} \underset{\Re}{\geq} \tfrac{1}{s y_{rk}} \tilde{\xi}^k\,\forall r, \\
&v_1^{ik} \leq v_2^{ik} \leq v_3^{ik}\,\forall i,\, u_1^{rk} \leq u_2^{rk} \leq u_3^{rk}\,\forall r,\, \xi_1^k \leq \xi_2^k \leq \xi_3^k.
\end{aligned}$$

where $\tilde{v}^{ik}$ and $\tilde{u}^{rk}$ are the triangular fuzzy weights associated with the ith input and the rth output, respectively. The $\tilde{E}_k$ is the fuzzy SBM efficiency of the kth DMU which is also found as a TFN. By using the ranking function of TFN, FSBM-D_k model reduces to Model 1, which is as follows:

$$\begin{aligned}
&\text{Model-1} \quad \Re(\tilde{E}_k) = \max \Re(\tilde{\xi}^k) \\
&\text{subject to } \Re(\tilde{\xi}^k) + \textstyle\sum_{i=1}^{m} x_{ik} \Re(\tilde{v}^{ik}) - \sum_{r=1}^{s} y_{rk} \Re(\tilde{u}^{rk}) = \Re(\tilde{1}), \\
&\textstyle\sum_{r=1}^{s} y_{rj} \Re(\tilde{u}^{rk}) - \sum_{i=1}^{m} x_{ij} \Re(\tilde{v}^{ik}) \leq \Re(\tilde{0})\,\forall j, \\
&\Re(\tilde{v}^{ik}) \geq \tfrac{1}{m x_{ik}} \Re(\tilde{1})\,\forall i,\, \Re(\tilde{u}^{rk}) \geq \tfrac{1}{s y_{rk}} \Re(\tilde{\xi}^k)\,\forall r, \\
&v_1^{ik} \leq v_2^{ik} \leq v_3^{ik}\,\forall i,\, u_1^{rk} \leq u_2^{rk} \leq u_3^{rk}\,\forall r,\, \xi_1^k \leq \xi_2^k \leq \xi_3^k.
\end{aligned}$$

Table 1 Input and output data of six DMUs

Inputs and outputs	A	B	C	D	E	F
I_1	4	14	24	20	48	50
I_2	3	6	3	2	4	7.5
O_1	1	2	3	2	4	5
O_2	2	6	12	6	16	30

Source The input and output data are taken from [3]

Table 2 Fuzzy efficiencies $\tilde{\xi}^k = (\xi_1^k, \xi_2^k, \xi_3^k)$ and $\Re(\tilde{\xi}^k)$

$\tilde{\xi}^k$	A	B	C	D	E	F
ξ_1^k	0.3429	0.2930	0.6720	0.2733	0.2172	0.3247
ξ_2^k	0.8964	0.6113	0.7671	0.6183	0.4434	0.7321
ξ_3^k	1.8644	1.2535	1.7939	1.5758	2.2292	2.2112
$\Re(\tilde{\xi}^k)$	1.0000	0.6923	1.0000	0.7714	0.8333	1.0000

By putting the values of $\Re(\tilde{\xi}^k)$, $\Re(\tilde{v}^{ik})\ \forall i$, and $\Re(\tilde{u}^{rk})\ \forall r$, the Model-1 reduces to Model-2, which is crisp LPP.

$$\begin{aligned}
&\text{Model-2}\quad E_k = \max(\xi_1^k + 2\xi_2^k + \xi_3^k)/4\\
&\text{subject to } (\xi_1^k + 2\xi_2^k + \xi_3^k) + \textstyle\sum_{i=1}^{m} x_{ik}(v_1^{ik} + 2v_2^{ik} + v_3^{ik})\\
&\qquad - \textstyle\sum_{r=1}^{s} y_{rk}(u_1^{rk} + 2u_2^{rk} + u_3^{rk}) = 4,\\
&\textstyle\sum_{r=1}^{s} y_{rj}(u_1^{rk} + 2u_2^{rk} + u_3^{rk}) - \sum_{i=1}^{m} x_{ij}(v_1^{ik} + 2v_2^{ik} + v_3^{ik}) \le 0\ \forall j,\\
&\tfrac{v_1^{ik}+2v_2^{ik}+v_3^{ik}}{4} \ge \tfrac{1}{m x_{ik}}\ \forall i,\ u_1^{rk} + 2u_2^{rk} + u_3^{rk} \ge \tfrac{\xi_1^k+2\xi_2^k+\xi_3^k}{s y_{rk}}\ \forall r,\\
&v_1^{ik} \le v_2^{ik} \le v_3^{ik}\ \forall i,\ u_1^{rk} \le u_2^{rk} \le u_3^{rk}\ \forall r,\ \xi_1^k \le \xi_2^k \le \xi_3^k.
\end{aligned}$$

5 Results and Discussion of a Numerical Example

In this section, we provide a numerical example to illustrate the proposed dual SBM model with fuzzy weights. Table 1 presents the performance evaluation problem of six DMUs with two inputs I_1 and I_2, and two outputs O_1 and O_2.

The fuzzy efficiencies of all DMUs are evaluated from Model-2, which are shown in Table 2. The results reveal that the rank of each fuzzy efficiency score lies between 0 and 1, i.e., $0 < \Re(\tilde{\xi}^k) \le 1$. The fuzzy weights corresponding to inputs and outputs of the concerned DMU are also evaluated by using Model-2, which are shown in Tables 3 and 4, respectively. These fuzzy weights provide additional information to the decision maker, which is not provided by crisp weights in crisp dual SBM model.

Table 3 Fuzzy weights corresponding to inputs

Fuzzy weights		A	B	C	D	E	F
$\tilde{v}^{1k}$	v_1^{1k}	4.4466	0.0220	3.8200	0.0080	0.0034	2.2823
	v_2^{1k}	10.6011	0.0543	8.4805	0.0203	0.0079	6.0048
	v_3^{1k}	35.0495	0.1515	26.6735	0.0514	0.0224	27.9433
$\tilde{v}^{2k}$	v_1^{2k}	4.1217	0.0262	10.2804	16.3350	33.8386	6.0896
	v_2^{2k}	10.0303	0.0645	23.4940	39.3018	78.3524	16.3315
	v_3^{2k}	31.1839	0.1781	61.7977	96.1473	175.6625	54.6475

Table 4 Fuzzy weights corresponding to outputs

Fuzzy weights		A	B	C	D	E	F
$\tilde{u}^{1k}$	u_1^{1k}	28.6998	0.1356	24.3567	16.7974	12.9102	11.6846
	u_2^{1k}	62.6096	0.3327	55.0749	40.0764	32.2595	32.5052
	u_3^{1k}	129.3616	0.8657	115.9649	93.9072	80.8491	89.4276
$\tilde{u}^{2k}$	u_1^{2k}	4.6918	0.0181	5.1196	0.0197	3.5286	3.6698
	u_2^{2k}	11.0246	0.0443	11.1800	0.0490	10.2253	10.0071
	u_3^{2k}	36.0650	0.1241	34.5782	0.1394	28.0860	42.3709

6 Conclusion

In this paper, we proposed a dual SBM model with fuzzy weights (FSBM-D$_k$) for crisp inputs and outputs. The FSBM-D$_k$ model is then reduced to crisp LPP by using ranking function. The proposed model evaluates the components of fuzzy efficiencies and fuzzy weights corresponding to inputs and outputs as TFNs. These fuzzy efficiencies and fuzzy weights provide additional information to the decision maker, which helps to deal with uncertainty in real-life problems.

Acknowledgments The first author is thankful to the University Grants Commission (UGC), Government of India, for financial assistance.

References

1. Charnes, A., Cooper, W.W., Rhodes, E.: Measuring the efficiency of decision making units. Eur. J. Oper. Res. **2**, 429–444 (1978)
2. Chen, S.M.: Fuzzy system reliability analysis using fuzzy number arithmetic operations. Fuzzy Set. Syst. **66**, 31–38 (1994)
3. Cooper, W. W., Seiford, L. M., Zhu, J.: Handbook on Data Envelopment Analysis. 2nd edn. International Series in Operations Research and Management Science, Springer, **164**, p. 200 (2011)
4. Goudarzi, M.R.M.: A slack-based model for estimating returns to scale under weight restrictions. Appl. Math. Sci. **6**(29), 1419–1430 (2012)

5. Hatami-Marbini, A., Saati, S., Makui, A.: An application of fuzzy numbers ranking in performance analysis. J. Appl. Sci. **9**(9), 1770–1775 (2009)
6. Jahanshahloo, G.R., Soleimani-damaneh, M., Nasrabadi, E.: Measure of efficiency in DEA with fuzzy input-output levels: A methodology for assessing, ranking and imposing of weights restrictions. Appl. Math. Comput. **156**, 175–187 (2004)
7. Kao, C., Liu, S.T.: Fuzzy efficiency measures in data envelopment analysis. Fuzzy Sets Syst. **113**, 427–437 (2000)
8. Lertworasirikul, S., Fang, S.C., Joines, J.A., Nuttle, H.L.W.: Fuzzy data envelopment analysis: A possibility approach. Fuzzy Set. Syst. **139**(2), 379–394 (2003)
9. Mahdavi-Amiri, N., Nasseri, S.H.: Duality in fuzzy number linear programming by use of a certain linear ranking function. Appl. Math. Comput. **180**, 206–216 (2006)
10. Mansourirad, E., Rizam, M.R.A.B., Lee, L.S., Jaafar, A.: Fuzzy weights in data envelopment analysis. Int. Math. Forum **5**(38), 1871–1886 (2010)
11. Puri, J., Yadav, S.P.: A concept of fuzzy input mix-efficiency in fuzzy DEA and its application in banking sector. Expert Syst. Appl. **40**, 1437–1450 (2013)
12. Saati, S., Memariani, A.: SBM model with fuzzy input-output levels in DEA. Aust. J. Basic Appl. Sci. **3**(2), 352–357 (2009)
13. Saen, R.F.: Developing a nondiscretionary model of slacks-based measure in data envelopment analysis. Appl. Math. Comput. **169**, 1440–1447 (2005)
14. Sengupta, J.K.: A fuzzy systems approach in data envelopment analysis. Comput. Math. Appl. **24**(8–9), 259–266 (1992)
15. Tone, K.: A slacks-based measure of efficiency in data envelopment analysis. Eur. J. Oper. Res. **130**, 498–509 (2001)
16. Zimmermann, H.J.: Fuzzy Set Theory and its Applications, 3rd edn. Kluwer-Nijhoff Publishing, Boston (1996)

Ball Bearing Fault Diagnosis Using Continuous Wavelet Transforms with Modern Algebraic Function

R. Sharma, A. Kumar and P. K. Kankar

Abstract Ball bearing plays a very crucial part of any rotating machineries, and the fault diagnosis in rotating system can be detected at early states when the fault is still small. In this paper, a ball bearing fault is detected by using continuous wavelet transform (CWT) with modern algebraic function. The reflected vibration signals from ball bearing having single point defect on its inner race, outer race, ball fault, and combination of these faults have been considered for analysis. The features extracted from a non-stationary multi-component ball bearing signal are very difficult. In this paper, a CWT with selected stretching parameters is used to analyze a signal in time–frequency domain and extract the features from non-stationary multi-component signals. The algebraic function norms are calculated from the matrix which can be generated with the help of wavelet transforms. The norms lookup table is used as a reference for fault diagnosis. The experimental results show that this method is simple and robust.

Keywords Bearing faults · Norms · CWT

R. Sharma (✉) · A. Kumar · P. K. Kankar
Department of Electronic and Communication Engineering, PDPM Indian Institute of Information Technology Design and Manufacturing Jabalpur, Jabalpur, Madhya Pradesh 482005, India
e-mail: rahul.ece23@gmail.com

A. Kumar
e-mail: anil.dee@gmail.com

P. K. Kankar
e-mail: pavankankar@gmail.com

B. V. Babu et al. (eds.), *Proceedings of the Second International Conference on Soft Computing for Problem Solving (SocProS 2012), December 28–30, 2012*, Advances in Intelligent Systems and Computing 236, DOI: 10.1007/978-81-322-1602-5_35, © Springer India 2014

1 Introduction

Rolling element bearings are used in wide variety of rotating machineries from small devices to heavy industrial systems. Ball bearing defect may be categorized as point of local defect and distribution defects. These defects are generated at the time of manufacturing due to geometrical imperfection bearing components. Literature review reveals that an extensive work has been done on fault diagnosis in ball bearing system. Chiang et al. [1] have been used Fisher discriminate analysis and support vector machines for fault diagnosis. Lei et al. [2] have proposed an intelligent classification method to mechanical fault diagnosis based on wavelet packet transform (WPT), empirical mode decomposition (EDM), dimensionless parameters, a distance evolution technique, and radial basis function network. It becomes difficult to diagnosis the fault when the amplitude of noise is high and also affects the system performance. Hidden Markov model (HMM), support vector machine (SVM), and artificial neural network (ANN) methods are used for fault classifications [3–5].

Wigner–Ville distribution and wavelet decomposition have been used for their excellent time frequency analysis. Generally, it is difficult to analyze model-based technique of a nonlinear system or non-stationary multi-component signals. Phakde et al. [6] have used a set of coefficients based on the sequence current and voltage phasor components to calculate the apparent impedance. Staszewski [7] have proposed a wavelet-based method for fault detection in mechanical system. The energy-confined DWT is used for fault detection by Prabhakar et al. [8]. Wavelet transform gives the better solution than any other known method; however, the problem is for selecting the parameter for wavelet analysis. The Gaussian correlation of vibration signal and wavelet coefficients for fault diagnosis has been used in [9]. Yuan and Chu [10] have used the particle swarm optimization (PSO) technique for feature selection of wavelet function. Kankar et al. [11] have used the maximum relative wavelet energy criterion and maximum energy to Shannon entropy ratio criterion for fault diagnosis. Adaptive wavelet filter with the selection of wavelet parameter on the basis of amplitude and frequency has been used for fault diagnosis in ball bearing system [12].

In this paper, the algebraic norms are used for fault classification, which can be calculated with the help of continuous wavelet transform. For the selection of the suitable parameter, the sensitive analysis technique is used. The fast Fourier transform (FFT) of healthy as well as non-healthy ball bearing is taken, and only those frequencies will be considered where there is appropriate difference in the magnitude of FFT of the signals and these frequencies are called pseudo-frequencies. These pseudo-frequencies are used to find out the scales of wavelet. With the help of selected scales, the matrix is generated and from that matrix the energy confine norms can be calculated, which are essentially used for fault diagnosis in ball bearing system.

2 Overview of CWT

Wavelet transform is a powerful tool that provides the analysis of signal at transient's state. The continuous wavelet transform (CWT) of $x(t)$ by Ψ (both belong to real domain) is a projection of a function $x(t)$ onto a particular wavelet $\Psi(t)$. The wavelet analogy of the spectrogram is the scalogram since CWT behaves like orthonormal basis decomposition. It is energy preserving transformation [13, 14].

$$(W_{\Psi}^{x})(a,b) = \frac{1}{\sqrt{a}} \int_{-\infty}^{\infty} x(t)\Psi\left(\frac{t-b}{a}\right) dt \tag{1}$$

where $a > 0$ and b are scale and translation parameters, respectively. Ψ is the mother wavelet.

$$(W_{\Psi}^{x})x(a,b) = \int x(l)\Psi_{(a,b)}(l)dl \tag{2}$$

Equation (2) is called wavelet equation or the inner product of $x(t)$ with the scaled and translation versions of the basis function $\psi_{(a,b)}(\mathrm{l})$. The scale 'a' is assumed to be restricted to R^{+}, although tenuously interpreted as a reciprocal of frequency. When 'a' decreases, the oscillation becomes more intense and shows high-frequency behavior. Similarly, when 'a' increases, the oscillations become drawn out and show low-frequency behavior.

$$x(t) = \frac{1}{C_{\Psi}} \int_{0}^{\infty} \int_{-\infty}^{\infty} (W_{\Psi}^{x})(a,b)\Psi(\frac{t-b}{a})\frac{da}{a^2}db \tag{3}$$

Square magnitude of CWT is defined as wavelet spectrogram or scalogram. It is distribution of signal in timescale plane and is expressed in power per frequency unit.

$$\int_{-\infty}^{\infty} |x(t)|^2 dt = \frac{1}{C_{\Psi}} \int_{-\infty}^{\infty} \int_{-\infty}^{\infty} |(W_{\Psi}^{x})(a,b)|^2 \frac{da}{a^2} db \tag{4}$$

C_{Ψ} is the constant that depends on Ψ and $W_{\Psi}^{x}(a, b)$ is the CWT. Equation (4) is called the scalogram of CWT. It is also called an energy preserving transformation as Eq. (4) has only constraint, that is, the mother wavelet ΨC L (R) satisfies:

$$C_{\Psi} = \int_{-\infty}^{\infty} \frac{|\Psi(w)|^2}{w} dw \langle \infty \tag{5}$$

Equation (5) is called admissibility condition, which leads to define the wavelet spectrogram. Sensitivity analysis can be viewed as to remove distracting variance from the dataset. Therefore, in this work, we are analyzing those sensors' data which are most sensitive and discard other.

3 Norms

The norm is used to quantify the size of a matrix or the distance between two matrices. If K denotes the field of real or complex numbers. Let $K^{m \times n}$ denote the vector space containing all matrices with 'm' rows and 'n' columns with entries in K.

Let 'A' be any real matrix then $||A||$ represents the norm of a real matrix 'A' in vector space $K^{m \times n}$. The norm can be defined as follows:

- $\|A\| > 0$ If$A \neq 0$ and $\|A\| = 0$iff$A = 0$, say null matrix.
- $\|\alpha A\| = |\alpha| \, \|A\|$, for all 'α' in $K$and all matrices 'A' in $K^{m \times n}$.
- $\|A + B\| \leq \|A\| + \|B\|$, for all matrices 'A' and'B$'$in$K^{m \times n}$.
  Incase of square matrices ($m = n$), some (but not all) matrix norms satisfy the following condition:
- $\|AB\| \leq \|A\| \, \|B\|$, for all matrices 'A' and 'B' in $K^{m \times n}$ also called a submultiplicative norm.
  Let $\lambda_1, \lambda_{2,.........,} \lambda_n$ be the eigenvalues of 'A', then
- $\frac{1}{\|A^{-1}\|} \leq |\lambda| \leq \|A\|$,

If the vector norms treat a matrix as a vector of size and use one of the familiar vector norms. The norms can be defined as follows:

$$\|A\|_p = \left(\sum_{i=1}^{m} \sum_{j=1}^{m} |a_{ij}|^p \right)^p \tag{6}$$

Equation (6) is called p-norms of a vector. Depending on the value of p, norms can be defined in various way such as 1-norm, 2-norm, and ∞-norm for p=1, 2, and ∞, respectively.

$$\|A\|_p = \sqrt{\sum_{i=1}^{m} \sum_{j=1}^{m} |a_{ij}|^2} = \sqrt{trace(A^* A)} = \sqrt{\sum_{i=1}^{\min\{m,n\}} \sigma_i^2} \tag{7}$$

where $A*$ denoted the conjugate transpose of A and are defined as the singular values of. The 2-norm can also be defined as the square root of confined matrix energy [15, 16].

Table 1 Parameters specification of experimental setup

Parameters	Values
Outer race diameter	28.2 mm
Inner race diameter	18.738 mm
Ball diameter	4.762 mm
Contact angle	0°
Radial clearance	10 μm
Ball number	8

Fig. 1 Experimental setup: *1* Digital encoder; *2* Variable speed control; *3* Motor; *4* Enclosure; *5* Flexible coupling; *6* Accelerometer; *7* Bearing housing; *8* Tested bearing; *9* Rotor; *10* Load disk; *11* Base; *12* Alignment adjustor; *13* Magnetic load system; *14* Gearbox Ref. [11]

4 Experimental Setup

The problem of predicting the degradation of working conditions of bearings before they reach the alarm or failure threshold is extremely important in industries to fully utilize the machine production capacity and to reduce the plant downtime. Figure 1 shows the experimental setup which is used for extracting vibration signals. In the present study, an experimental test rig is used and vibration response for healthy bearing and bearing with faults is obtained. Table 1 shows dimensions of the ball bearings taken for the study. Accelerometers are used for picking up the vibration signals from various stations on the rig. As a first step, the machine was run with healthy bearing to establish the baseline data.

Table 1 shows the test bearing characteristics like diameter of inner race, outer race, ball diameter, number of balls, contact angle, and radial clearance. For the study point of view, collection of vibration data of healthy as well as faulty ball bearing at different loading conditions (no loader, one loader, and two loader) and bearings are simulated on the rig at different rotor speed 1,000, 1,500, and 2,000 rpm Ref. [11].

The following five bearing conditions are considered for the study:

1. Healthy bearings (HB).
2. Bearing with spall on inner race (BSIR).
3. Bearing with spall on outer race (BSOR).
4. Bearing with spall on ball (BSB).
5. Combined bearing component defects (CBD).

5 Methodology

In this methodology, first all reflected signals from ball bearing at different rotating speeds and different loading conditions are collected via the most sensitive sensors, but these signals are contaminated by the noise and other unwanted variance.

The following steps explain the proposed methodology for fault diagnosis in ball bearing system:

Step 1: Collect all the signals which reflect the faults in ball bearing system as an occurrence of fault in a signal introduces distinctive and detectable change in the energy distribution of the sensor data.

Step 2: As the reflected signals are not faulty at all time with respect to non-faulty signal, those part of signal is analyzed, where the probability of fault is more.

Step 3: For applying CWT, the selection of stretching parameters plays a very important roll. In this paper, the sensitivity analysis technique is used for the analysis of noise-contaminated reflected signal features.

Step 4: Fast Fourier transform (FFT) of both faulty as well as non-faulty reflected signals is taken, and only those values of signals will be considered, where there is an appropriate difference in the magnitude of fast Fourier transforms (FFT) of the signals as shown in Fig. 2.

Step 5: Collect all those frequencies at which there are appropriate differences of faulty and non-faulty data and these frequencies are called pseudo-frequencies.

Step 6: Calculate the scales by using Eq. (8).

$$a = \frac{F_c}{F_a} \tag{8}$$

where a scales, F_c center frequency of wavelet, and F_a pseudo-frequency corresponding to scales.

Step 7: Calculate the big matrices of wavelet coefficients corresponding to reference and faulty signals. The first one is called as signature matrix and the other is known as indicator matrix.

Step 8: Estimate the norms of these big matrices and make a lookup table with some tolerance. This is the reference table of norms. The sample version of lookup table is shown in Table 2.

Table 2 Sample version of lookup table

2-Norm	Speed (rpm)	Loading	Fault
0.1219	1,000	No	BFB
0.2622	1,000	One	HEL
0.5211	2,000	One	IRD
0.1912	2,000	Two	HEL
0.2312	1,500	One	MFB
0.2365	1,500	Two	IRD
0.3564	2,000	One	BFB
0.0932	1,500	No	ORD
0.1168	1,000	No	ORD

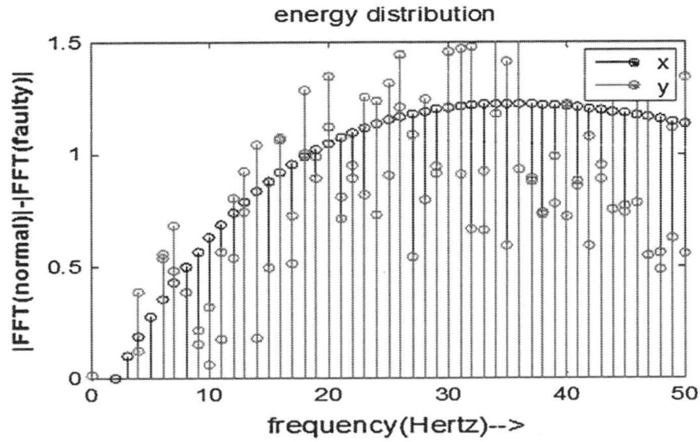

Fig. 2 Variation in amplitude difference between faulty or non-faulty bearing versus frequency

Step 9: Now take a test signal from a faulty test bearing and calculate the norms by using above-mentioned steps and compare that norms with the reference lookup Table 2 to diagnosis the fault that which type of fault occurs in test bearing.

A complete flowchart for the proposed method is depicted in Fig. 3.

6 Results and Discussion

In this paper, the study is carried out on total of 72 instances (36 for horizontal response and 36 for vertical response) at different loading conditions and at different speeds. The square root of energy-confined matrix or second norm can be calculated

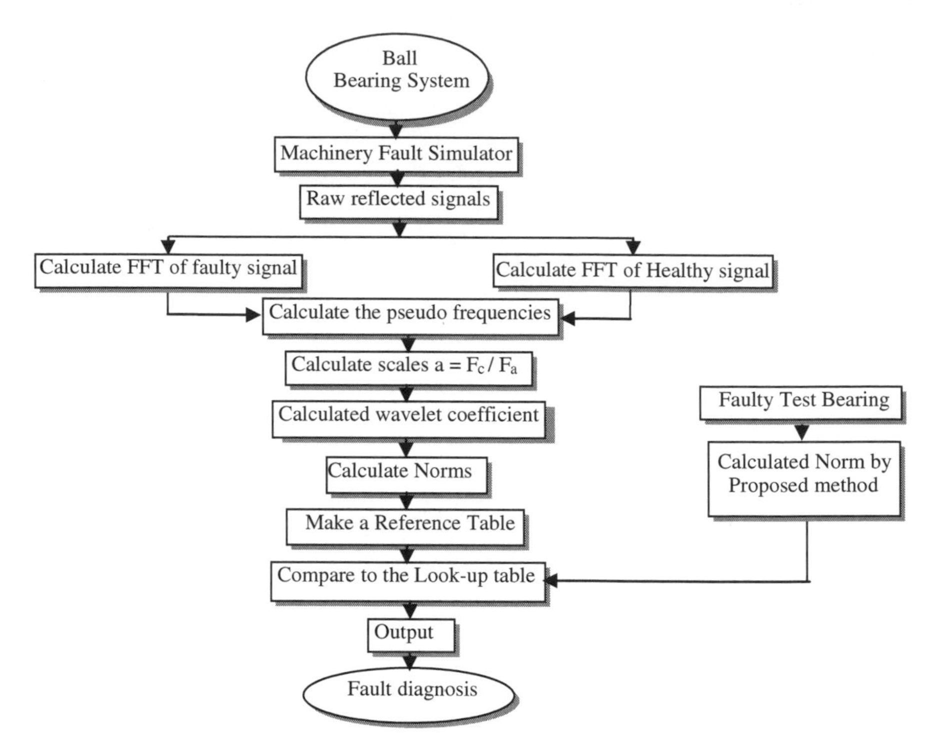

Fig. 3 Flowchart of the proposed method

for horizontal and vertical responses. The sample version of these norms is shown in Table 2. This table is called reference table or lookup table.

Example

1. Take a reflected signal from the faulty test ball bearing (already known that which type of fault occurs in bearing by which one can test the proposed methodology).
2. Calculate FFT of that reflected signal ball bearing.
3. Compare that test signal FFT to the healthy bearing FFT.
4. Calculate those frequencies at which there is an appropriate difference in their FFT magnitude (in this example, 100 frequency points can be taken as the number of points increases, the accuracy will be increased)
5. Calculate the stretching parameters with the help of Eq. (8) (100 stretching parameters can be calculated as the number of stretching parameters is equal to the number of selected frequency points).
6. Apply the wavelet transform on that stretching parameters and calculate the big matrix of wavelet coefficients.
7. Calculate the norms from that coefficient matrix with some ± 0.0010 tolerance.
8. Compare that norm to the reference lookup table norms and predict the type of fault occurring in the test bearing as from the result one can see that the result is 100 % true.

The evaluation of success by using wavelet transform (selected stretching parameters) with modern algebraic function methods for analyzing of non-stationary multi-component faulty reflected signals. With the help of the lookup table, one can figure out the type of fault. The method is very simple and gives out magnificent result.

7 Conclusions

In this paper, the fault diagnosis in ball bearing system is done by using CWT with modern algebraic function. As the complete data are not faulty at all its limit, in this paper only those part of signal has been considered, where the probability of fault is more for which the sensitivity analysis technique is used. The proposed method is useful for extracting features from the original data, and dimension of original data can be reduced by removing irrelevant features. The stretching parameters for the wavelet are calculated with the help of proposed technique. The energy-confined norms are used for the fault classification from the matrix that would be generated by WT at selected stretching parameters. Experimental results included in this paper clearly show the key advantageous features of the proposed methodology. It is evident from the experimental results that the proposed method shows better performance. The example taken in this study shows that this methodology gives the effectively accurate result. This process shows the potential application for developing knowledge-based system. Therefore, the proposed technique can be effectively used for developing online fault diagnosis.

References

1. Chiang, L.H., Kotanchek, M.E., Kordon, A.K.: Fault diagnosis based on Fisher discriminate analysis and support vector machines. **28**, 1389–1401 (2004)
2. Lei, Y., He, Z., Zi, Y.: Application of an intelligent classification method to mechanical fault diagnosis. Expert Syst. Appl. **36**, 9941–9948 (2009)
3. Li, Z., Wu, Z., He, Y., Fulei, C.: Hidden Markov model-based fault diagnostics method in speed-up and speed-down process for rotating machinery. Mech. Syst. Sign. Process. **19**, 329–339 (2005)
4. Widodo, A., Yang, B.S.: Review on a vector support machine in machine condition monitoring and fault diagnosis. Mech. Syst. Sign. Process. **21**, 2560–2574 (2007)
5. Vyas, N.S., Satishkumar, D.: Artificial neural network design for fault identification in a rotor bearing system. Mech. Syst. Sign. Process. **36**, 157–175 (2001)
6. Phadke, A.G., Ibrahim, M., Hlibka, T.: Fundamental basis for distance relaying with symmetrical components. IEEE Trans. Power Apparatus Syst. **96**, 635–646 (1977)
7. Staszewski, W.J.: Structural and mechanical damage detection using wavelets. Shock Vibr. Dig. **30**, 457–472 (1998)
8. Prabhakar, S., Mohanty, A.R., Sekhar, A.S.: Application of discrete wavelet transforms for detection of ball bearing race faults. Tribol. Int. **35**, 793–800 (2002)

9. Galli, A.W., Heyd, G.T.: Comments on power quality assessment using wavelets. The 27 Annual NAPS. Bozeman, MT, U.S.A
10. Yuan, F.S., Chu, F.L.: Fault diagnosis based on particle swarm optimization and support vector machines. Mech. Syst. Sign. Process. **20**, 1787–1798 (2007)
11. Kankar, P.K., Sharma, S.C., Harsha, S.P.: Fault diagnosis of ball bearing using continuous wavelet transforms. Appl. Soft Comput. **11**, 2300–2312 (2011)
12. Lin, J., Zuo, M.J.: Gearbox fault diagnosis using adaptive wavelet filter. Mech. Syst. Sign. Process. **17**, 1259–1269 (2003)
13. https://ewhdbks.mugu.navy.mil/wavelet.htm
14. Daubechies, I.: Ten Lectures on Wavelets. Regional Conference Series in Applied Mathematics. SIAM, Philadelphia (1992)
15. Watrous, J.: Theory of Quantum Information, 2.4 Norms of operators, lecture notes, University of Waterloo (2008)
16. Meyer, C.D.: Matrix Analysis and Applied Linear Algebra. SIAM (2000)

Engineering Optimization Using SOMGA

Kusum Deep and Dipti Singh

Abstract Many real-life problems arising in science, business, engineering, etc. can be modeled as nonlinear constrained optimization problems. To solve these problems, population-based stochastic search methods have been frequently used in literature. In this paper, a population-based constraint-handling technique C-SOMGA is used to solve six engineering optimization problems. To show the efficiency of this algorithm, the results are compared with the previously quoted results.

Keywords SOMGA · C-SOMGA · Optimization

1 Introduction

Constraint handling is considered to be challenging and difficult task in optimization. Many real-life problems in engineering can be modeled as nonlinear constrained optimization problems. In view of their practical utility, there is a need to develop efficient and robust computational algorithms, which can numerically solve problems in different fields irrespective of their size. These days a number of probabilistic techniques are available for obtaining the global optimal solution of nonlinear optimization problems. Though GAs are very efficient at finding the global optimal solution of unconstrained or simply constrained (i.e., box constraints) optimization problems but encounter some difficulties in solving highly constraint nonlinear optimization problems, because the operators used in GAs are not very efficient in dealing with the constraints. Several methodologies have been developed to handle constraints when

K. Deep (✉)
Indian Institute of Technology, Roorkee, India
e-mail: kusumfma@iitr.ernet.in

D. Singh
Gautam Buddha University, Greatar Noida, India
e-mail: diptipma@rediffmail.com

B. V. Babu et al. (eds.), *Proceedings of the Second International Conference on Soft Computing for Problem Solving (SocProS 2012), December 28–30, 2012*, Advances in Intelligent Systems and Computing 236, DOI: 10.1007/978-81-322-1602-5_36, © Springer India 2014

GAs are used to solve constrained optimization problems refer Kim and Myung [12], Michalewicz [13], Myung and Kim [14], Orvosh and Davis [15]. Deep and Dipti [8] proposed a penalty parameter free hybrid approach C-SOMGA for solving the nonlinear constrained optimization problems. It is not only easy to implement but also does not require any parameter to be fine-tuned for constraint handling. It works with a very low population size, hence uses low function evaluations where the term "function evaluations" represents the number of times an objective function is evaluated in the entire run. In this paper, six engineering optimization algorithms has been solved using C-SOMGA. The results obtained are compared with the previously quoted results. On the basis of the results, it is concluded that the C-SOMGA is efficient to solve these problems.

The paper is organized as follows: in Sect. 1, introduction is given; in Sect. 2, methodology of C-SOMGA is presented; in Sect. 3, mathematical models of the problems are given and results obtained using C-SOMGA are discussed and compared with the previously quoted results; and Sect. 4 summarizes the conclusions based on the present study.

2 Methodology of C-SOMGA

The algorithm C-SOMGA is an extension of SOMGA [7] for solving the constraint nonlinear optimization problems in which SOMGA is combined with constraint-handling tournament selection scheme, and as a result of this, C-SOMGA has been proposed. The methodology of C-SOMGA algorithm is as follows:

First, the individuals are generated randomly. These individuals compete with each other through constraint tournament selection method: Create new individuals via single-point crossover and bitwise mutation. Then, the best individual among them is considered as leader and all others are considered as active. For each active individual, a new population of size N is created, where N is the ratio of path length and step size. This population is nothing but the new positions of the active individual proceeds in the direction of the leader in n steps of the defined length. The movement of this individual is given by

$$x_{i,j}^{\text{MLnew}} = x_{i,j,\text{start}}^{\text{ML}} + \left(x_{L,j}^{\text{ML}} - x_{i,j,\text{start}}^{ML}\right) t\text{PRTVector}_j \tag{1}$$

where $t \in < 0, \quad \text{by Step to}, \quad \text{PathLength} >$,

ML	is actual migration loop.
$x_{i,j}^{\text{MLnew}}$	is the new positions of an individual.
$x_{i,j,\text{start}}^{\text{ML}}$	is the positions of active individual.
$x_{L,j}^{\text{ML}}$	is the positions of leader.

PRT vector is created before an individual proceeds toward leader. This parameter has the same effect as mutation in GA. It is defined in the range <0, 1>. Then, sort this population according to the fitness value in decreasing order. Starting from the best one of the new population, evaluate the constraint violation function described by Eq. 2.

$$\psi(x) = \sum_{m=1}^{M} [h_m(x)]^2 + \sum_{k=1}^{K} G_k [g_k(k)]^2 \quad (2)$$

where G_k is the Heaviside operator such that $G_k = 0$ for $g_k(x) \geq 0$ and $G_k = 1$ for $g_k(x) < 0$.

If $\psi(x) = 0$, replace the active individual with the current position and move to the next active individual and if $\psi(x) > 0$, then move to the next best position of the sorted new population. In this way, all the active individuals are replaced by the new updated feasible position. If no feasible solution is available, then active individual remains the same. At last, the best individuals (number equal to population size) from the previous and current generations are selected for the next generation. The computational steps of this approach are given below:

Step 1: Generate the initial population.
Step 2: Evaluate all individuals.
Step 3: Apply tournament selection for constrained optimization on all individuals to select the better individuals for the next generation.
Step 4: Apply crossover operator on all individuals with crossover probability P_c to produce new child individuals.
Step 5: Evaluate the new child individuals.
Step 6: Apply mutation operator on every bit of every individual of the population with mutation probability P_m.
Step 7: Evaluate the mutated individuals.
Step 8: Find leader (best fit individual) of the population and consider all others as active individuals of the population.
Step 9: For each active individual, a new population of size N is created. This population is nothing but the new positions of the active individual toward the leader in n steps of the defined length. The movement of this individual is given in Eq. (1).
Step 10: Sort new population with respect to fitness in decreasing order.
Step 11: For each individual in the sorted population, check feasibility criterion.
Step 12: If feasibility criterion is satisfied, replace the active individual with the new position, else move to next position in sort order, and go to Step 11.
Step 13: Select the best individuals (in fitness) of previous and current generation for the next generation via tournament selection.
Step 14: If termination criterion is satisfied go to 15 else go to Step 3.
Step 15: Report the best chromosome as the final optimal solution.

3 Mathematical Models of Engineering Optimization Problems

In this section, mathematical model of six engineering optimization problems has been given and the results obtained using C-SOMGA are compared with the available results. These models have been taken from the literature to see the performance of the C-SOMGA on constrained optimization problems. Many researchers used these models to demonstrate the performance of their techniques [2, 16–18]. The experimental setup for C-SOMGA is given in Table 1.

3.1 Gas Transmission Compressor Design

This problem is taken from Beightler and Phillips [2]. This is a real-life problem in which the values of design parameters P_1, x_1, x_2, x_3 are to be determined that will deliver 100 million cu. Ft. of gas per day with minimum cost for a gas pipe line transmission system. Here,

P_1 Compressor discharge pressure,
Q Flow rate,
x_1 Length between compressor stations (in miles),
x_2 Compressor ratio $= P_1/P_2$,
x_3 Pipe inside diameter (in inches).

The mathematical model of the problem is

$$\text{Minimize } g_0 = 8.61 \times 10^5 x_1^{1/2} x_2 x_3^{-2/3} x_4^{-1/2} + 3.69 \times 10^4 x_3 + 7.72 \times 10^8 \times 10^8 x_1^{-1} x_2^{0.219} - 765.43 \times 10^6 x_1^{-1}$$

subject to $x_4 x_2^{-2} + x_2^{-2} \leq 1$, where $x_1, x_2, x_3, x_4 > 0$.

Bounds on the variables are as follows:

$$20 \leq x_1 \leq 50, 1 \leq x_2 \leq 10, 20 \leq x_3 \leq 50, 0.1 \leq x_4 \leq 60$$

Table 1 Experimental setup

Population size	20
P_c	0.85
P_m	0.009
Step size	0.31
Path length	3
String length	20

Table 2 Optimal solution to the design of a gas transmission compressor

	Value of objective	Values of variables
Solution obtained by C-SOMGA	296.490×10^4	$x_1 = 49.9996, x_2 = 1.17834,$ $x_3 = 24.5996, x_4 = 0.388482$
Solution given in Pant [16]	296.528×10^4	$x_1 = 50.000, x_2 = 1.183,$ $x_3 = 24.347, x_4 = 0.339$
Solution given in Beightler and Phillips [2]	299×10^4	$x_1 = 28.760, x_2 = 1.109,$ $x_3 = 25.030, x_4 = 0.230$

The problem turns out to be a constrained geometric programming problem. This problem is earlier solved by Beightler and Phillips [2], Verma [18], Thanh [17], and Pant [16]. The results obtained using C-SOMGA and those given in source are shown in Table 2. It is evident with the Table 2 that the cost obtained by C-SOMGA in deliver the gas per day that is 2964900 is lesser than the cost obtained by Pant, i.e., 2965280 and by Beightler and Phillips, i.e., 2990000. In other words, C-SOMGA provides far better results than previously quoted results.

3.2 Optimization of a Riser Design

This problem is taken from Gaindhar et al. [9]. The objective of this problem is to determine the optimal volume of the riser. Any metal will shrink in volume when it is allowed to cool and solidify from a molten state. A riser is a device by which the location of a shrinkage cavity is shifted from within the casting to the riser, which is an extraneous portion cast as an integral but distinct portion of the casting. After the casting is solidified, all extraneous parts are cut off leaving behind the desired casting free of any shrinkage cavity.

The basic requirement for the riser design is that the solidification time of the riser must not be less than the solidification time of the casting. From the practical point of view, it is considered advantageous to have top riser connected to the casting through a neck. The molding sand in the neck region gets up more heated as compared to the rest of the region surrounding the riser. This ensures molten metal in the region of the neck. This also facilitates cutting off of the riser from the casting after the casting has been solidified.

The mathematical modal of the problem, as given in Gaindhar et al. [9] is

$$\text{Minimize } f(x) = (1/4)\,\pi x_1 x_2^2 + (1/12)\,\pi x_4 \left(3 - 3x_4/x_3 + x_4^2/x_3^3\right) x_2$$

$$\text{subject to } 2E\left(5 + (x_4/x_3)\,(2 - x_4/x_3)\left(1 + x_3^2\right)^{1/2}\right) x + 4Ex_2^{-1}$$

$$- (x_4/3)\left(3 - 3x_4/x_3 + x_4^2/x_3^2\right) \le 1$$

$x_1, x_2, x_3, x_4 > 0.$
where

x_1 height of riser
x_2 diameter of the riser
E riser modulus constant ($E = 10/7$)
x_3 $\tan\theta$ and
x_4 height of the neck riser.

The variable bounds are as follows:

$$1 \le x_1 \le 8;\ 1 \le x_2 \le 10;\ 0 \le x_3 \le 1;\ 0 \le x_4 \le 1$$

This problem is earlier solved by Gaindhar et. al [9] and by Pant [16]. The numerical results obtained are compared with the available results and are presented in Table 3. It is evident with the Table 3 that the result obtained by C-SOMGA that is 290.78142 is better than the result obtained by Pant [16] i.e 290.8532 and by Gaindhar et al. [9] i.e. 290.8069.

3.3 *Optimum Design of a Welded Beam*

Optimum design of a welded beam problem is a well-known problem. The formulation of this problem is available in literature with two models. In model (a), the number of constraints is six and in model (b), it is seven. Both the models are described below:

Model (a):
This problem is taken from Beightler and Phillips [2]. In this problem, the assembly of the welded structure as is being considered for mass production. Outside considerations fix the material of the bar A as well as the design parameters F and L. Assuming that the design engineer has fixed the specifications, $F = 6{,}000$ lb, $L = 14$ in and bar $A = 1{,}010$ steel; the objective function is to find a feasible combination of x_1, x_2, x_3 and x_4 such that the total cost assembly construction is minimum.

Table 3 Optimal design of a riser

	Value of objective	Values of variables
Solution obtained by C-SOMGA	290.78142	$x_1 = 4.276, x_2 = 8.7510,$ $x_3 = 1, x_4 = 0.1001$
Solution given in Pant [16].	290.8532	$x_1 = 4.2233, x_2 = 8.6055,$ $x_3 = 1.0000, x_4 = 0.1000$
Solution given in Gaindhar et al. [9]	290.8069	$x_1 = 4.266, x_2 = 8.5710,$ $x_3 = 1.000, x_4 = 0.1000$

The mathematical model of the problem is
Minimize $g_0(X) = 1.1047x_1^2x_2 + 0.6735x_3x_4 + 0.04811x_2x_3x_4$
Subject to

$$g_1(X) = 16.8x_4^{-1}x_3^{-2} \le 1,\ g_2(X) = x_1x_4^{-1} \le 1,\ g_3(X) = 0.125x_1^{-1} \le 1,$$
$$g_4(X) = 9.08x_3^{-3}x_4^{-1} \le 1,\ g_5(X) = 0.09428x_3^{-1}x_4^{-3} + 0.02776x_3 \le 1$$
$$g_6(X) = \left[\left(\begin{array}{c}\frac{F^2}{2}x_1^{-2}x_2^{-2} + \frac{F^2x_2^{-1}(L+x_2/12)}{2\left(\frac{x_2^2}{12}+\frac{(x_3+x_1)^2}{4}\right)^{-1}} \\ +\frac{F^2(L+x_2/2)^2\left(\frac{x_2^2+(x_3+x_1)^2}{4}\right)}{2x_1^2x_2^2\left(\frac{x_2^2}{12}+\frac{(x_3+x_1)^2}{4}\right)^2}\end{array}\right)\right]^{1/2} \le 13,000$$

$(x_1, x_2, x_3, x_4) > 0.$
The variable bounds are as follows:

$$0.1 \le x_1 \le 1;\ 5 \le x_2 \le 7;\ 7 \le x_3 \le 9;\ 0.1 \le x_4 \le 1$$

Model (b):
This model is taken from Xiaohui et al. [19]. The objective is to minimize the cost of a welded beam subject to constraints on shear stress, bending stress in the beam, bucking load on the bar, end deflection of the beam, and side constraints. The problem can be stated as follows:

$$\text{Minimize} \quad f(X) = 1.10471x_1^2x_2 + 0.04811x_3x_4\,(14.0 + x_2)$$

subject to

$$g_1(X) = \tau(X) - \tau_{\max} \le 0$$
$$g_2(X) = \sigma(X) - \sigma_{\max} \le 0$$
$$g_3(X) = x_1 - x_4 \le 0$$
$$g_4(X) = 0.10471x_1^2 + 0.04811x_3x_4\,(14.0 + x_2) - 5.0 \le 0$$
$$g_5(X) = 0.125 - x_1 \le 0,$$
$$g_6(X) = \delta(X) - \delta_{\max} \le 0$$
$$g_7(X) = P - P_c(X) \le 0$$

where

$$\tau(X) = \sqrt{(\tau')^2 + 2\tau'\tau''\frac{x_2}{2R} + (\tau'')^2}$$

$$\tau' = \frac{P}{\sqrt{2}x_1x_2},\quad \tau'' = \frac{MR}{J},\quad M = P\left(L + \frac{x_2}{2}\right),\ R = \sqrt{\frac{x_2^2}{4} + \left(\frac{x_1 + x_3}{2}\right)^2}$$

$$J = 2\left\{\sqrt{2}x_1x_2\left[\frac{x_2^2}{12} + \left(\frac{x_1+x_3}{2}\right)^2\right]\right\}, \sigma(X) = \frac{6PL}{x_4x_3^2}, \quad \delta(X) = \frac{4PL^3}{Ex_3^3x_4}$$

$$P_c(X) = \frac{4.013E\sqrt{x_3^2x_4^6/36}}{L^2}\left(1 - \frac{x_3}{2L}\sqrt{\frac{E}{4G}}\right)$$

$$P = 6{,}000\,lb, \quad L = 14\,in, \quad E = 30 \times 10^6\,psi, \quad G = 12 \times 10^6\,psi,$$
$$\pi_{\max} = 13{,}600\,psi, \quad \sigma_{\max} = 30,000\,psi, \quad \delta_{\max} = 0.25\,in$$

The following ranges of the variables were used:

$$0.1 \le x_1 \le 2, \quad 0.1 \le x_2 \le 10, \quad 0.1 \le x_3 \le 10, \quad 0.1 \le x_4 \le 2$$

Both the models are solved by C-SOMGA. The numerical results obtained and the results given in source are presented in Table 4 for model (a) and Table 5 for model (b).

In Table 4, although the results available in source are lesser than the results obtained by C-SOMGA, but the solutions are not satisfying the feasibility conditions. Hence, these solutions cannot be accepted. The result obtained by C-SOMGA is a feasible solution. Therefore, C-SOMGA is best in this problem.

In Table 5, the result attained by C-SOMGA is superior to Coello [5] and Deb [6] but slightly inferior at fifth place to Xiaohui et al. [19]. It shows that the results are comparable.

3.4 Optimal Capacity of Gas Production Facilities

This problem is taken from Beightler and Phillips [2]. This is the problem of determining the optimum capacity of production facilities that combine to make an oxygen

Table 4 Optimal design of a welded beam based on model (a)

	Value of objective	Value of variables	Feasibility
Solution obtained by C-SOMGA	2.45694	$x_1 = 0.244241, x_2 = 6.4712,$ $x_3 = 8.43726, x_4 = 0.244364$	Satisfied
Solution given in Pant [16]	1.9786	$x_1 = 0.1489, x_2 = 5.000,$ $x_3 = 8.2736, x_4 = 0.2454$	Not Satisfied
Solution given in Beightler and Phillips [2]	2.3860	$x_1 = 0.2455, x_2 = 6.1960,$ $x_3 = 8.2730, x_4 = 0.2455$	Not Satisfied

Table 5 Optimal design of a welded beam based on model (b)

	Value of objective	Value of variables
Solution obtained by C-SOMGA	1.72486	$x_1 = 0.205731, x_2 = 3.47048,$ $x_3 = 9.03669, x_4 = 0.20573$
Solution given in Xiaohui [19]	1.72485084	$x_1 = 0.20573, x_2 = 3.47049,$ $x_3 = 9.03662, x_4 = 0.20573$
Solution given in Coello [5]	1.74830941	$x_1 = 0.2088, x_2 = 3.4205,$ $x_3 = 8.9975, x_4 = .2100$
Solution given in Deb [6]	2.43311600	$x_1 = 0.2489, x_2 = 6.1730,$ $x_3 = 8.1739, x_4 = 0.2533$

Table 6 Optimal capacity of gas production facilities

	Value of objective	Value of variables
Solution obtained by C-SOMGA	169.844	$x_1 = 17.500, x_2 = 600.000,$
Solution given in Pant [16]	169.844	$x_1 = 17.500, x_2 = 600.000,$
Solution given in Beightler and Phillips [2]	173.760	$x_1 = 17.500, x_2 = 465.000,$

producing and storing system. Oxygen for basic oxygen furnace is produced at a steady-state level. The demand for oxygen is cyclic with a period of one hour, which is too short to allow an adjustment of level of production to the demand. Hence, the manager of the plant has two alternatives:

1. He can keep the production at the maximum demand level; excess production is lost in the atmosphere.
2. He can keep the production at lower level; excess production is compressed and stored for use during the high demand period. The mathematical model of the problem is

$$\begin{aligned}\text{Minimize } g_0(\mathrm{X}) = 61.8 + 5.72x_1 + .2623\left[(40 - x_1)\ln\frac{x_2}{200}\right]^{-0.85} \\ + .087\,(40 - x_1)\ln\frac{x_2}{200} + 700.23x_2^{-.75}\end{aligned}$$

subject to $x_1 \geq 17.5, x_2 \geq 200, x_1, x_2 > 0$.

The variable bounds are as follows: $17.5 \leq x_1 \leq 40$; $300 \leq x_2 \leq 600$

The numerical results obtained using C-SOMGA and the numerical results given in source are presented in Table 6. In this problem, C-SOMGA produced better results than Beightler and Philips [2] but similar results as obtained by Pant [16] using GRST.

Table 7 Minimization of the weight of a tension/compression Spring

	Value of objective	Value of variables
Solution obtained by C-SOMGA	.0126656	$x_1 = .0516216, x_2 = 0.355094, x_3 = 11.385$
Solution given in Xiaohui [19]	0.0126661409	$x_1 = 0.05147, x_2 = .35138394, x_3 = 11.60865920$
Solution given in Coello [5]	.0127047834	$x_1 = .051480, x_2 = .351661, x_3 = 11.632201$
Solution given in Arora [1]	.127302737	$x_1 = .053396, x_2 = .399180, x_3 = 9.185400.$

3.5 *Minimization of the Weight of a Tension/Compression Spring*

This problem was described by Arora [1] and Belegundu [3]. The problem consists of minimizing the weight of a tension/compression spring subject to constrains on minimum deflection, shear stress, surge frequency, limits on outside diameter and on design variables. The design variables are the mean coil diameter D, the wire diameter d, and the number of active coils N. The problem can be expressed as follows:

$$Minimize\, f(X) = (N+2)\, Dd^2$$

subject to

$$g_1(X) = 1 - \frac{D^3 N}{71785 d^4} \le 0$$
$$g_2(X) = \frac{4D^2 - dD}{12566\left(Dd^3 - d^4\right)} + \frac{1}{5108 d^2} - 1 \le 0$$
$$g_3(X) = 1 - \frac{140.45 d}{D^2 N} \le 0$$
$$g_4(X) = \frac{D + d}{1.5} - 1 \le 0.$$

The following ranges of the variables were used:

$$0.05 \le x_1 \le 2, \quad 0.25 \le x_2 \le 1.3, \quad 2.0 \le x_3 \le 15.$$

The numerical results of the solution obtained using C-SOMGA and the numerical results given in source are presented in Table 7. The result attained by C-SOMGA is superior to Coello and Mezura [4] and Arora [1] at the fourth place and at the sixth place to Xiaohui et al. [19]. Hence, the results are comparable.

3.6 Himmelblau's Nonlinear Optimization Problem

This problem has been taken from Xiaohui [19]. This problem was proposed by Himmelblau [10], and it has been used before as a benchmark for several evolutionary algorithm-based techniques. In this problem, there are five design variables, six nonlinear inequality constraints, and ten boundary conditions. The problem can be stated as follows:

$$\textit{Minimize}\ f(X) = 5.3578547x_3^2 + 0.8356891x_1x_5 + 37.2932239x_1 - 40792.141$$

subject to

$$0 \le 85.334407 + .0056858x_2x_5 + .00026x_1x_4 - .0022053x_3x_5 \le 92$$
$$90 \le 80.51249 + 0.0071317x_2x_5 + 0.00026x_1x_2 + 0.0021813x_3^2 \le 110$$
$$20 \le 9.300961 + 0.0047026x_3x_5 + 0.0012547x_1x_3 + 0.0019085x_3x_4 \le 25$$
$$78 \le x_1 \le 102,\quad 33 \le x_2 \le 45,\quad 27 \le x_3 \le 45,\quad 27 \le x_4 \le 45,\quad 27 \le x_5 \le 45.$$

The results obtained by C-SOMGA and available from the other source are presented in Table 8. C-SOMGA gives better results than Coello and Mezura [4] and Homaifar et al [11] and results are comparable to Xiaohui et al. [19].

Table 8 Himmelblau's Nonlinear Optimization Problem

	Value of objective	Value of variables
Solution obtained by C-SOMGA	−31025.6	$x_1 = 78$, $x_2 = 33.0001$, $x_3 = 27.071$, $x_4 = 45$, $x_5 = 44.969$
Solution given in Xiaohui [19]	−31025.56142	$x_1 = 78.0$, $x_2 = 33.0$, $x_3 = 27.070997$, $x_4 = 45$, $x_5 = 44.96924255$
Solution given in Coello [5]	−31020.859	$x_1 = 78.0495$, $x_2 = 33.0070$, $x_3 = 27.0810$, $x_4 = 45$, $x_5 = 44.9400$
Solution given in Homaifar et al. [11]	−30665.609	$x_1 = 78.0000$, $x_2 = 33.0000$, $x_3 = 29.9950$, $x_4 = 45$, $x_5 = 36.7760$.

4 Conclusions

In this paper, six real-life constrained optimization problems arising in various fields of engineering have been solved. For solving these constrained optimization problems, a population-based hybridized algorithm C-SOMGA has been used. In four problems, C-SOMGA provides better results than the previously quoted results, and in two problems, results are comparable. The algorithm requires only 20 population size for solving these problems. It is therefore concluded that C-SOMGA is well suited for obtaining the global optimal solution of engineering optimization problems.

References

1. Arora, J.S.: Introduction to Optimum Design, 2nd edn. Academic Press, New Delhi (2006)
2. Beightler, C.S., Phillips, D.T.: Applied Geometric Programming. Wiley, New York (1976)
3. Belegundu, A.D.: A Study of Mathematical Programming for Structural Optimization. University of Iowa, Iowa,Department of Civil and Environmental Engineering(1982)
4. Coello, C.A., Mezura, M.E.: Constraint-handling in genetic algorithms through the use of dominance-based tournament selection. Adv. Eng. Inf. **16**, 193–203 (2002)
5. Coello, C.A.: Use of a self-adaptive penalty approach for engineering optimization problems. Comput. Ind. **41**, 113–127 (2000)
6. Deb, K., Gene AS: A Robust Optimal Design Technique for Mechanical Component Design. In Dasgupta, D., Michalewicz, Z.: (eds). Evolutionary Algorithms in Engineering Applications, pp. 497–514. Springer, Berlin (1997b)
7. Deep, K., and Dipti: A New Hybrid Self Organizing Migrating Genetic Algorithm for Function Optimization. IEEE Congress on Evolutionary Computation, pp. 2796–2803 (2007)
8. Deep, K., and Dipti: A self organizing migrating genetic algorithm for constrained optimization. Appl. Math. Comput. **198**(1), 237–250 (2008)
9. Gaindhar, J.L., Mohan, C., Tyagi, S.: Optimization of riser design in metal casting. J. Eng. Optim. **14**, 1–26 (1988)
10. Himmelblau, D.M.: Applied Nonlinear Programming. McGraw-Hill, New York (1972)
11. Homaifar, A.A., Lai, S.H.Y., Qi, X.: Constrained optimization via genetic algorithms. Simulation **62**, 242–254 (1994)
12. Kim, J.H., Myung, H.: A Two Phase Evolutionary Programming for General Constrained Optimization Problem. Proceedings of the Fifth Annual Conference on Evolutionary Programming, San Diego (1996)
13. Michalewicz, Z.: Genetic Algorithms, Numerical Optimization and Constraints. In: Echelman, L.J. (ed.) Proceedings of the Sixth International Conference on Genetic Algorithms, pp. 151–158 (1995)
14. Myung, H., Kim, J.H.: Hybrid evolutionary programming for heavily constrained problems. Bio-Systems **38**, 29–43 (1996)
15. Orvosh, D., Davis, L.: Using a Genetic Algorithm to Optimize Problems with Feasibility Constraints. In: Echelman, L.J. (ed). Proceedings of the Sixth International Conference on Genetic Algorithms, pp. 548–552 (1995)
16. Pant, M.: Genetic Algorithms for Global Optimization and their Applications. Ph.D. Thesis, Department of Mathematics, IIT Roorkee, Formerly University of Roorkee (2003)
17. Thann, N.H.: Some Global Optimization Techniques and their use in solving Optimization Problems in Crisp and Fuzzy Environments. University of Roorkee, Roorkee, India,Department of Mathematics(1996)

18. Verma, S.K.: Solution of Optimization Problems in Crisp Fuzzy and Stochastic Environments. Ph.D. Thesis, Dept. of Mathematics, University of Roorkee, Roorkee (1997)
19. Xiaohui, H., Eberhart, R.C., Shi, Y.: Engineering Optimization with Particle Swarm. IEEE Swarm Intelligence Symposium, Indianapolis, USA (2003)

Goal Programming Approach to Trans-shipment Problem

Om Prakash Dubey, Kusum Deep and Atulya K. Nagar

Abstract The technocrats put their efforts regularly to minimize the total cost/budget of transportation problem. However, the proper effort has not been put for minimizing the total cost of trans-shipment problem. A goal programming approach has been developed to obtain the minimum budget for trans-shipment problem. The trans-shipment problem is regarded as the extended transportation problem and hence be solved by the transportation techniques. In the present algorithm, trans-shipment problem is transferred to suitable transportation problem and further modified as a proper goal programming problem. The priorities of goal programming explore the wider impact for decision maker. Therefore, the solution obtained is more suitable for decision makers. Hence, it is widely acceptable for any organization. At the end, a numerical example is solved in support of the procedure.

Keywords Trans-shipment problem · Transportation problem · Lexicographic goal programming · Priority level · Decision maker

1 Introduction

A transportation problem (TP) allows only shipments which go directly from a supply/source point, acts only as a shipper of the goods, to a demand point/destination, acts only as receiver of the goods. Transportation models deal with problems

O. P. Dubey (✉)
Department of Mathematics, Bengal College of Engineering and Technology, Durgapur, India
e-mail: omprakashdubeymaths@gmail.com

K. Deep
Department of Mathematics, Indian Institute of Technology, Roorkee, India
e-mail: kusumfma@iitr.ernet.in

A. K. Nagar
Department of Computer Science, Liverpool Hope University, Liverpool L 169 JD, UK
e-mail: nagara@hope.ac.uk

B. V. Babu et al. (eds.), *Proceedings of the Second International Conference on Soft Computing for Problem Solving (SocProS 2012), December 28–30, 2012*, Advances in Intelligent Systems and Computing 236, DOI: 10.1007/978-81-322-1602-5_37, © Springer India 2014

concerned with the effectiveness function when each of a number of origins associates with each of a possibly different number of destinations. In many situations, shipments are allowed between supply points or between demand points. Sometimes there may also be trans-shipment points through which goods can be transshipped on their journey from a supply point to a demand point. Shipping problems with any or all of these characteristics are Trans-shipment Problems. Thus, the optimal solution to a Trans-shipment Problem can be found by solving a Transportation problem [1–3].

Goal programming (GP) is a decision-making technique generally used to solve multiple objective problems, which provides a best compromise solution according to the DM's needs and desires. The concept was originally developed by Charnes and Cooper [4]. In GP, instead of trying to optimize the objective function directly, the deviation between the goals and what can be achieved under a given set of constraints are to be minimized. In the priority-based GP, the priorities/weights are assigned to the goals according to their importance as specified by the DM. Sometimes it happens that the DM is not satisfied with the solution(s) and DM wants some other solutions to suit his desires; then, set of alternate solutions can be provided using interactive GP techniques [5, 6]. The GP formulations ordered the unwanted deviations into a number of priority levels, with the minimization of a deviation in a higher priority level being of infinitely more important than any deviations in lower priority levels, known as Lexicographic or Pre-emptive GP [7]. It should be used when there exists a clear ordering among the decisions.

2 Methodology

The TP assumes that direct routes exist from each source to each destination. However, there are situations in which units may be shipped from one source to another or to other destinations before reaching their final destination, known as Trans-shipment Problem.

In generalized trans-shipment model, items are supplied from different sources to different destination. It is sometimes economical if the shipment passes through some transient nodes in between sources and destinations. Unlike in TP, in trans-shipment problem, the objective is to minimize the total cost of shipments, and thus, the shipment passes through one or more intermediate nodes before it reaches its desired destination.

For the purpose of trans-shipment, the distinction between a source and destination is dropped so that a TP with m sources and n destinations gives rise to a trans-shipment problem with $m + n$ sources and $m + n$ destinations. The basic feasible solution to such a problem will involve $[(m+n)+(m+n)-1]$ or $2m+2n-1$ basic variables, and if we omit the variables appearing in the $(m+n)$ diagonal cells, we have $m+n-1$ basic variables.

Here, as each source or destination is a potential point of supply as well demand, the total supply (say of K units) is added to the actual supply of each source as well as to the actual demand at each destination. Also, the 'demand' at each source and 'supply' at each destination are set equal to K. It may assume the supply and demand of each location to be fictitious one. These quantities (K) may be regarded as buffer stocks, and each of these buffer stocks should at least be equal to the total supply/demand in the given problem.

Therefore, construct a transportation tableau creating a row for each supply point and trans-shipment point, and a column for each demand point and trans-shipment point. Each supply point will have a supply equal to its original supply, and each demand point will have a demand equal to its original demand [3].

The general TP with m production sites and n destinations is given by the cost matrix $[C_{ij}]$, $i = 1, 2, \ldots, m$ and $j = 1, 2, \ldots, n$, together with production capacity a_i and demand b_j. The problem is said to be balanced if total supply = total demand, i.e., $\sum_{i=1}^{m} a_i = \sum_{j=1}^{n} b_j$, otherwise unbalanced. TP can be expressed mathematically as,

$$\text{Minimize } Z = \sum_{i=1}^{m}\sum_{j=1}^{n} C_{ij}x_{ij}$$

$$\text{subject to, } \sum_{j=1}^{n} x_{ij} = a_i \quad i = 1, 2, \ldots, m$$

$$\sum_{i=1}^{m} x_{ij} = b_j \quad j = 1, 2, \ldots, n$$

$$\sum_{i=1}^{m} a_i = \sum_{j=1}^{n} b_j, \quad x_{ij} \geq 0.$$

where x_{ij} is the amount of goods to be transported.

Now, the GP version of the above TP model can be written as,
Minimize $F(d)$
subject to,

$$\text{supply constraints, } \sum_{j=1}^{n} x_{ij} + d_j^- - d_j^+ = a_i, \quad i = 1, 2, \ldots, m; \quad j = 1, 2, \ldots, k.$$

$$\text{demand goals, } \sum_{i=1}^{m} x_{ij} + d_j^- - d_j^+ = b_i, \quad j = 1 + k, 2 + k, \ldots, n + k.$$

$$\text{and, budget goal, } \sum_{i=1}^{m}\sum_{j=1}^{n} C_{ij}x_{ij} + d_{k+n+1}^- - d_{k+n+1}^+ = B; d_j^-.d_j^+ = 0, forallj; x_{ij}, d_j^-, d_j^+ \geq 0$$

where B represents budget aspiration level as fixed by the DM [8].

3 Case Study

To support the algorithm, a sample problem is considered as follows.

A firm has two factories X and Y and three retail stores A, B, and C. The numbers of units of a product available at factories X and Y are 200 and 300, respectively, while demanded at retail stores are 100, 150, and 250, respectively. Rather than shipping directly from sources to destinations, it is decided to investigate the possibility of trans-shipment. Find the optimal shipping schedule. The transportation costs in rupees per unit are given in Table 1.

Table 1 Sample problem

		Factory		Retail store		
		X	Y	A	B	C
Factory	X	0	6	7	8	9
	Y	6	0	5	4	3
Retail store	A	7	2	0	5	1
	B	1	5	1	0	4
	C	8	9	7	6	0

Solution

For this trans-shipment problem, buffer stock = total supply = total demand = 500 units. Adding 500 units to each supply/demand point, we get Table 2.

Table 2 Trans-shipment problem as transportation problem

		Factory		Retail store			Supply
		X	Y	A	B	C	
Factory	X	0 (500)	6	7 (200)	8	9	700
	Y	6	0 (500)	5	4 (50)	3 (250)	800
Retail store	A	7	2	0 (400)	5 (100)	1	500
	B	1	5	1	0 (500)	4	500
	C	8	9	7	6	0 (500)	500
Demand		500	500	600	650	750	

Following is the initial solution obtained by the Vogel's approximation method (Table 3).

Factory X supplies 100 units each to retail stores A and B, whereas factory Y supplies 50 units to retail store B and 250 units to C.

Goal programming formulation of the transportation problem obtained from the given trans-shipment problem is as follows in two models.

Model -1

Minimize $F(d)$

subject to,

Table 3 Solution by Vogel's approximation method

		Factory		Retail store			Supply
		X	Y	A	B	C	
Factory	X	0 (500)	6	7 (100)	8 (100)	9	700
	Y	6	0 (500)	5	4 (50)	3 (250)	800
Retail store	A	7	2	0 (500)	5	1	500
	B	1	5	1	0 (500)	4	500
	C	8	9	7	6	0 (500)	500
Demand		500	500	600	650	750	

$$x_1 + x_6 + x_{11} + x_{16} + x_{21} \geq 500 \tag{1}$$

$$x_2 + x_7 + x_{12} + x_{17} + x_{22} \geq 500 \tag{2}$$

$$x_3 + x_8 + x_{13} + x_{18} + x_{23} \geq 600 \tag{3}$$

$$x_4 + x_9 + x_{14} + x_{19} + x_{24} \geq 650 \tag{4}$$

$$x_5 + x_{10} + x_{15} + x_{20} + x_{25} \geq 750 \tag{5}$$

$$\begin{aligned} 0 \cdot x_1 &+ 6 \cdot x_2 + 7 \cdot x_3 + 8 \cdot x_4 + 9 \cdot x_5 \\ &+ 6 \cdot x_6 + 0 \cdot x_7 + 5 \cdot x_8 + 4 \cdot x_9 + 3 \cdot x_{10} \\ &+ 7 \cdot x_{11} + 2 \cdot x_{12} + 0 \cdot x_{13} + 5 \cdot x_{14} \\ &+ 1 \cdot x_{15} + 1 \cdot x_{16} + 5 \cdot x_{17} + 1 \cdot x_{18} + 0 \cdot x_{19} \\ &+ 4 \cdot x_{20} + 9 \cdot x_{21} + 9 \cdot x_{22} + 7 \cdot x_{23} + 6 \cdot x_{24} + 0 \cdot x_{25} \leq 10,000 \end{aligned} \tag{6}$$

$$x_1 + x_2 + x_3 + x_4 + x_5 \leq 700 \tag{7}$$

$$x_6 + x_7 + x_8 + x_9 + x_{10} \leq 800 \tag{8}$$

$$x_{11} + x_{12} + x_{13} + x_{14} + x_{15} \leq 500 \tag{9}$$

$$x_{16} + x_{17} + x_{18} + x_{19} + x_{20} \leq 500 \tag{10}$$

$$x_{21} + x_{22} + x_{23} + x_{24} + x_{25} \leq 500 \tag{11}$$

$$x_i \geq 0, \; i = 1, 2, 3, \ldots, 25. \tag{12 to 36}$$

where $F(d)$ is a function of deviational variables.

In the present trans-shipment problem, for $F(d)$, the authors consider demand goal as 1st priority level, budget goal as 2nd priority level, supply goal as 3rd priority level, and non-negative constraints as 4th priority level (Table 4).

Here, factory X supplies 100 units to retail store A and B each, whereas factory Y supplies 50 units to retail store B. However, C received 250 units from A in the processing of the solution.

Model - 2

Minimize $F(d)$

subject to,

Table 4 Initial solution obtained by the goal programming technique

		Factory		Retail store			Supply
		X	Y	A	B	C	
Factory	X	0	6	7 (600)	8 (100)	9	700
	Y	6	0 (250)	5	4 (550)	3	800
Retail store	A	7	2 (56.25)	0	5	1 (443.75)	500
	B	1 (500)	5	1	0	4	500
	C	8	9 (193.75)	7	6	0 (306.25)	500
Demand		500	500	600	650	750	

$$x_1 + x_6 + x_{11} + x_{16} + x_{21} \geq 500 \quad (1)$$

$$x_2 + x_7 + x_{12} + x_{17} + x_{22} \geq 500 \quad (2)$$

$$x_3 + x_8 + x_{13} + x_{18} + x_{23} \geq 600 \quad (3)$$

$$x_4 + x_9 + x_{14} + x_{19} + x_{24} \geq 650 \quad (4)$$

$$x_5 + x_{10} + x_{15} + x_{20} + x_{25} \geq 750 \quad (5)$$

$$\begin{aligned} 0 \cdot x_1 + 6 \cdot x_2 + 7 \cdot x_3 + 8 \cdot x_4 + 9 \cdot x_5 + 6 \cdot x_6 + 0 \cdot x_7 \\ + 5 \cdot x_8 + 4 \cdot x_9 + 3 \cdot x_{10} + 7 \cdot x_{11} + 2 \cdot x_{12} + 0 \cdot x_{13} \\ + 5 \cdot x_{14} + 1 \cdot x_{15} + 1 \cdot x_{16} + 5 \cdot x_{17} + 1 \cdot x_{18} + 0 \cdot x_{19} \\ + 4 \cdot x_{20} + 9 \cdot x_{21} + 9 \cdot x_{22} + 7 \cdot x_{23} + 6 \cdot x_{24} + 0 \cdot x_{25} \leq 1000 \end{aligned} \quad (6)$$

$$x_1 + x_2 + x_3 + x_4 + x_5 \leq 700 \quad (7)$$

$$x_6 + x_7 + x_8 + x_9 + x_{10} \leq 800 \quad (8)$$

$$x_{11} + x_{12} + x_{13} + x_{14} + x_{15} \leq 500 \quad (9)$$

$$x_{16} + x_{17} + x_{18} + x_{19} + x_{20} \leq 500 \quad (10)$$

$$x_{21} + x_{22} + x_{23} + x_{24} + x_{25} \leq 500 \quad (11)$$

$$x_i \geq 0, \ i = 1, 2, 3, \ldots, 25. \quad (12 \text{ to } 36)$$

where $F(d)$ is a function of deviational variables.

In the present trans-shipment problem, for $F(d)$, the authors consider demand goal as 1st priority level, budget goal as 2nd priority level, supply goal as 3rd priority level, and non-negative constraints as 4th priority level (Table 5).

Factory X supplies 7.1429 units to retail store A and 92.86 units generated to A in the processing, whereas factory Y supplies 50 units to retail store B and 250 to C. However, B received 100 units from A in the processing of the solution.

These two models show the impact of assigning the budget value. Hence, care should be taken in this situation.

Table 5 Initial solution obtained by the goal programming technique

		Factory		Retail store			Supply
		X	Y	A	B	C	
Factory	X	0 (500)	6	7 (7.1429)	8	9	700
	Y	6	0 (500)	5	4 (50)	3 (250)	800
Retail store	A	7	2	0 (592.86)	5 (100)	1	500
	B	1	5	1	0 (600)	4	500
	C	8	9	7	6	0 (500)	500
Demand		500	500	600	650	750	

4 Conclusion

The beauty of this algorithm is that DM may select priority level as per his/her own choice suitable for the concern organization interchanging demand, budget, supply or different levels of demand, different levels of supply and budget. In the above-mentioned case study, the algorithm minimizes the budget compared to available traditional techniques. Hence, it is better and the obtained results may be more realistic and useful for the organization in view of the DM. Care should be taken in assigning the budget value, because under-budget as well as over-budget explores bad impact on solution. Further, set of solutions may be generated for each model (priority-wise formulation) formulated by the DM.

References

1. Srinivasan, P.: A trans-shipment model for cash management decisions. Manage. Sci. **20**, 1350–1363 (1974)
2. Mohan, C.: Kusum Deep: Optimization techniques. New Age International Publishers, UK (2009)
3. Gupta, P.K., Hira, D.S.: Operations Research. S. Chand and Co. Ltd., India (2007)
4. Charnes, A., Cooper, W.W.: Management models and industrial application of linear programming. Wiley, New York (1961)
5. Dyer, J.S.: Interactive goal programming. Manage. Sci. **19**, 62 (1972)
6. Masud, A.S.M.: Interactive goal Programming- A Survey. Times/ORSA Meeting, Chicago (1983)
7. Ignizio, J.P.: Generalized Goal Programming- An overview. Comput. Oper. Res. **10**, 277–289 (1983)
8. Dubey, O.P., Singh, M.K., Dwivedi, R.K., Singh, S.N.: Interactive Decisions for transport management—Application in the coal transportation sector. IUP J. Oper. Manage. **10**(2), 7–21 (2011)

An Efficient Solution to a Multiple Non-Linear Regression Model with Interaction Effect using TORA and LINDO

Umesh Gupta, Devender Singh Hada and Ankita Mathur

Abstract Goal programming (GP) has been proven a valuable mathematical programming form in a number of venues. GP model serves a valuable purpose of cross-checking answers from other methodologies. Different software packages are used to solve these GP models. Likewise, multiple regression models can also be used to more accurately combine multiple criteria measures that can be used in GP model parameters. Those parameters can include the relative weighting and the goal constraint parameters. A comparative study on the solutions using TORA, LINDO, and least square method has been made in this paper. The objective of this paper is to find out a method that gives most accurate result to a nonlinear multiple regression model.

Keywords Goal programming · Multiple regression · Least square method · TORA · LINDO

1 Introduction

Regression analysis is used to understand the statistical dependence of one variable on other variables. Linear regression is the oldest and most widely used predictive model in decision making in managerial sciences, environmental science, and all the

U. Gupta (✉)
Institute of Engineering and Technology, JK Lakshmipat University, Jaipur, India
e-mail: umeshindian@yahoo.com

D. S. Hada
Kautilya Institute of Technology and Engineering, Jaipur, India
e-mail: dev.singh1978@yahoo.com

A. Mathur
Jaipur Institute of Technology and Group of Institutions, Jaipur, India
e-mail: ankita25jaipur@gmail.com

B. V. Babu et al. (eds.), *Proceedings of the Second International Conference on Soft Computing for Problem Solving (SocProS 2012), December 28–30, 2012*, Advances in Intelligent Systems and Computing 236, DOI: 10.1007/978-81-322-1602-5_38,

areas wherever it is required to describe possible relationships between two or more variables. This technique can show what proportion of variance between variables is due to the dependent variable, and what proportion is due to the independent variables. The earliest form of regression was the method of least squares, which was published by Legendre [1] and by Gauss [2]. The linear regression can be classified into two types, simple linear regression and multiple linear regression (MLR). The simple linear regression describes the relationship between two variables and MLR analysis describes the relationship between several independent variables and a single dependent variable. A number of methods for the estimation of the regression parameters are available in the literature. These include methods of minimizing the sum of absolute residuals, minimizing the maximum of absolute residuals, and minimizing the sum of squares of residuals [3], where the last method of minimizing the sum of squares of residuals popularly known as least square methods is commonly used. Alp et al. [4] explained that linear goal programming (GP) can be proposed as an alternative of the least square method. For this, he took an example of vertical network adjustment. Hassonpour et al. [5] proposed a linear programming model based on GP to calculate regression coefficient.

An interaction occurs when the magnitude of the effect of one independent variable on a dependent variable varies as a function of a second independent variable. This is also known as a moderation effect, although some have more strict criteria for moderation effects than for interactions. Nowadays, interaction effects through regression models are a widely interested area of investigation as there has been a great deal of confusion about the analysis of moderated relationships involving continuous variables. Alken and West [6] have analyzed such interaction effects; further, this method was applied into several models by the researchers, for example, Curran et al. [7] applied into hierarchical linear growth models.

Multiple objective optimization techniques provide more realistic solutions for most of the problems as it deals with multiple objectives, whereas single objective optimization techniques provide solutions to the problems that deals with single objective. GP is a type of multiple objective optimization technique that converts a multi-objective optimization model into a single objective optimization model. GP model has been proven a valuable tool in support of decision making. The first publication using GP as the form of a constrained regression model was used by Charnes et al. [8]. There have been many books devoted to this topic over past years (Ijiri [9]; Lee [10]; Spronk [11]; Ignizio [12]). This tool often represents a substantial improvement in the modeling and analysis of multi-objective problems (Charnes and Cooper [13]; Eiselt et al. [14]; Ignizio [15]). By minimizing deviation, the GP model can generate decision variable values that are the same as the beta values in some types of multiple regression models. Tamiz et al. [16] presents the review of current literature on the branch of multi-criteria decision modeling known as GP. Machiel Kruger [17] proposed a GP approach to efficiently managing a bank's balance sheet while maximizing returns and at the same time taking into account the conflicting goals such as minimizing risk, subject to regulatory and managerial constraints. Gupta et al. [18] solved a multi-objective investment management planning problem using fuzzy min sum weighted fuzzy goal programming technique.

Application of a multi-objective programming model like GP model is an important tool for studying various aspects of management systems (Sen and Nandi [19]). As an extension to the findings of Sharma et al. [20], this paper is focused on comparative study of the results obtained through different software packages like LINDO and TORA.

2 Regression and Goal Programming Formulation

The regression equation used to analyze and interpret a two-way interaction is:

$$y_{ir} = b_0 + b_1 X_i + b_2 Z_i + b_3 X_i^2 + b_4 Z_i^2 + b_5 X_i Z_i + e_i, \quad i = 1, 2, \ldots, m.$$

where b_0, b_1, b_2, b_3, b_4 and b_5 are the parameters to be estimated, and e_i is the error components which are assumed to be normally and independently distributed with zero mean and constant variance. The linear absolute residual method requires us to estimate the values of these unknown parameters so as to minimize $\sum_{i=1}^{m} |y_{io} - y_{ir}|$.

Let y_i be the ith goal, d_i^+ be positive deviation from the ith goal, and d_i^- be the negative deviation from the ith goal. Then, the problem of minimizing $\sum_{i=1}^{m} |y_i - y_{ir}|$ may be reformulated as

Minimize $\sum_{i=1}^{m} \left(d_i^+ + d_i^-\right)$

Subject to:

$$a_0 + a_1 X_{i1} + a_2 X_{i2} + a_3 X_{i3} + a_4 X_{i4} + a_5 X_{i5} + d_i^+ - d_i^- = y_{iG},$$

$$d_i^+ \geq 0$$

$$d_i^- \geq 0$$

and $a_0, a_1, a_2, a_3, a_4, a_5$ are unrestricted.

$$i = 1, 2, \ldots, m.$$

where X_i^2, Z_i^2, and $X_i Z_i$ are taken as X_{i3}, X_{i4}, and X_{i5}, respectively, to formulate the multiple nonlinear regression problem into linear GP model.

3 Mathematical Modeling and Solution

3.1 Mathematical Modeling

Relationship between two methods can be established by taking a simple example. We consider a regression equation of Y on X and Z. The data for illustration are:

y	x	z
7.88	3	2
7.43	2	1
8.38	4	3
7.42	2	1
7.97	3	2
7.49	2	2
8.84	5	3
8.29	4	2

Reformulating the above problem into linear GP model:

Minimize $\sum_{i=1}^{8} (d_i^+ + d_i^-)$

Subject to:

$$a_0 + 3a_1 + 2a_2 + 9a_3 + 4a_4 + 6a_5 + d_1^+ - d_1^- = 7.88$$

$$a_0 + 2a_1 + a_2 + 4a_3 + a_4 + 2a_5 + d_2^+ - d_2^- = 7.43$$

$$a_0 + 4a_1 + 3a_2 + 16a_3 + 9a_4 + 12a_5 + d_3^+ - d_3^- = 8.38$$

$$a_0 + 2a_1 + a_2 + 4a_3 + a_4 + 2a_5 + d_4^+ - d_4^- = 7.42$$

$$a_0 + 3a_1 + 2a_2 + 9a_3 + 4a_4 + 6a_5 + d_5^+ - d_5^- = 7.97$$

$$a_0 + 2a_1 + 2a_2 + 4a_3 + 4a_4 + 4a_5 + d_6^+ - d_6^- = 7.49$$

$$a_0 + 5a_1 + 3a_2 + 25a_3 + 9a_4 + 15a_5 + d_7^+ - d_7^- = 8.84$$

$$a_0 + 4a_1 + 2a_2 + 16a_3 + 4a_4 + 8a_5 + d_8^+ - d_8^- = 8.29$$

$$d_i^+ \geq 0, \quad i = 1, 2, \ldots, 8$$

$$d_i^- \geq 0, \quad i = 1, 2, \ldots, 8$$

a_i are unrestricted, $i = 0, 1, 2, \ldots, 5$.

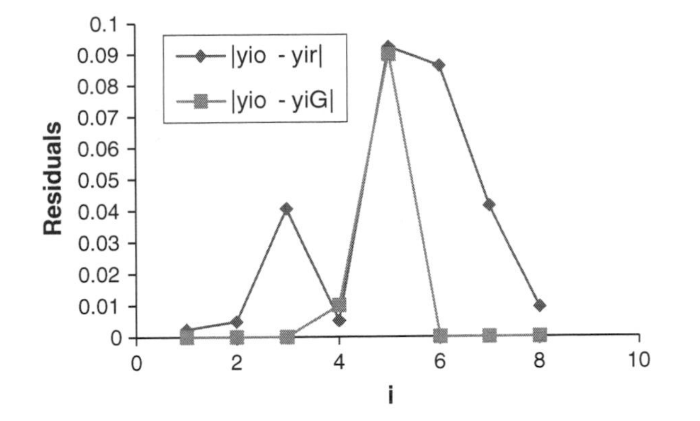

Fig. 1 Comparative results of residuals through different algorithms

3.2 Solution

The values of coefficients in the above problem through different methods are tabulated in Table 1

Final results are tabulated in Table 2:

4 Discussion

It is clear from Table 1 that all software packages give the same results to linear GP formulation with zero difference in the results.

It is observed from the above-tabulated results of Table 2 and Fig. 1 that Minimize $\sum_{i=1}^{m} |y_{io} - y_{iG}| <$ Minimize $\sum_{i=1}^{m} |y_{io} - y_{ir}|$, where y_{iG} be the estimate of the ith response using GP technique, and y_{ir} be the estimate using the least square method. Hence, it is concluded that the GP technique provide better estimate of the multiple nonlinear regression parameters with two-way interaction effect than the least square method.

Table 1 The values of coefficients using different methods

Coefficients	Least square method	TORA	LINDO
a_0	6.9215	6.74	6.74
a_1	0.00001	0.28	0.28
a_2	0.3181	0.045	0.045
a_3	0.0602	0.01	0.01
a_4	−0.0557	−0.015	−0.015
a_5	0.0001	0.03	0.03

Table 2 Values of y for paired values of x and z

y_i	Observed value (y_{io})	Expected values Least square method (y_{ir})	TORA (y_{iG})	LINDO (y_{iG})
y_1	7.88	7.8776	7.88	7.88
y_2	7.43	7.4251	7.43	7.43
y_3	8.38	8.3393	8.38	8.38
y_4	7.42	7.4251	7.43	7.43
y_5	7.97	7.8776	7.88	7.88
y_6	7.49	7.5763	7.49	7.49
y_7	8.84	8.8815	8.84	8.84
y_8	8.29	8.2993	8.29	8.29
	$\sum_{i=1}^{8} \lvert y_i - y_{ir} \rvert$	0.2826	0.1	0.1

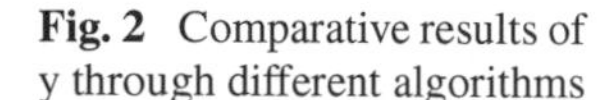

Fig. 2 Comparative results of y through different algorithms

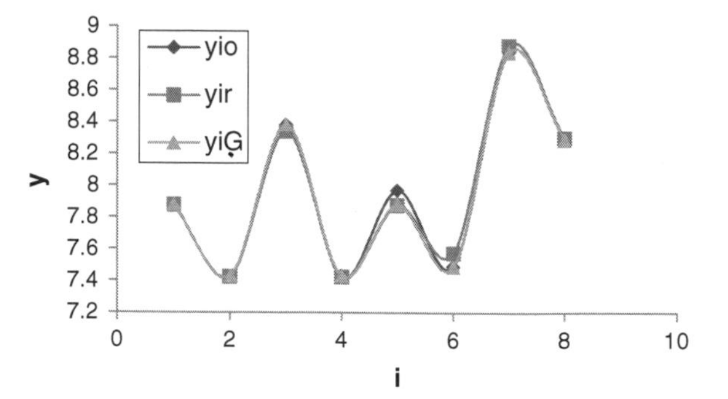

It is clear from Fig. 2 that the data are best fitted into the curve when we get the values of coefficients through solutions of the GP formulation comparative to the solutions using least square method.

5 Conclusion

1. The software packages TORA and LINDO both give similar results to a linear GP problem.
2. GP formulation gives better and best-fitted results than the traditional least square method.
3. The error is minimized when we solve regression model using GP formulation.

References

1. Legendre, A.M.: Nouvelles méthodes pour la détermination des orbites des comètes. “Sur la Méthode des moindres quarrés” appears as an appendix (1805)
2. Gauss, C.F.: Theoria Motus Corporum Coelestium in Sectionibus Conicis Solem Ambientum (1809)
3. Weisberg, s: Applied Linear Regression, 2nd edn. Wiley, Inc, New York (1985)
4. Alp, S., Yavuz, E., Ersoy, N.: Using linear goal programming in surveying engineering for vertical network adjustment. Int. J. Phys. Sci. **6**(8), 1982–1987 (2011)
5. Hassanpour H., Maleki R.H., Yaghoobi A.M.: Fuzzy linear regression model with crisp coefficients: a goal programming approach. Int. J. Fuzzy Syst. **7**(2), 19–39 (2010)
6. Alken, L.S., West, S.G.: Multiple Regression: Testing and Interpreting Interactions. Sage Publications, Thousand Oaks (1991)
7. Curran, P.J., Bauer, D.J., Willoughby, M.T.: Testing main effects and interactions in hierarchical linear growth models. Psychol. Methods **9**(2), 220–237 (2004)
8. Charnes, A., Cooper, W.W., Ferguson, R.: Optimal estimation of executive compensation by linear programming. Manage. Sci. **1**(2), 138–151 (1955)
9. Ijiri, Y.: Management Goals and Accounting for Control. North-Holland Publishing Company, Amsterdam (1965)
10. Lee, S.M.: Goal Programming for Decision Analysis. Auerbach Publishers Inc., Philadelphia (1972)
11. Spronk, J.: Interactive Multiple Goal Programming: Application to Financial Planning. Martinus Nijhoff, Amsterdam (1981)
12. Ignizio, J.P.: Introduction to Linear Goal Programming. Sage Publications, Thousand Oaks, CA (1986)
13. Charnes, A., Cooper, W.W.: Goal programming and multiple objective optimizations. Eur. J. Operat. Res. **1**, 39–54 (1977)
14. Eiselt, H.A., Pederzoli, G., Sandblom, C.L.: Continuous Optimization Models. W De G, New York (1987)
15. Ignizio, J.P.: A review of goal programming—a tool for multiobjective analysis. J. Opl. Res. Soc. **29**(11), 1109–1119 (1978)
16. Tamiz, M., Jones, D., Darzi, E.: A review of goal programming and its applications. Ann. Oper. Res. **58**(1), 39–53 (1995)
17. Machiel, K.: A goal programming approach to strategic bank balance sheet management. SAS Global Forum 2011, Centre for BMI, North-West University, South Africa, Paper 024–2011 (2011)
18. Gupta, M., Bhattacharjee, D.: Min sum weighted fuzzy goal programming model in investment management planning: a case study. Int. Res. J. Fin. Econ. Issue 56 (2010)
19. Sen, N., Nandi, M.: Goal programming, its application in management sectors–special attention into plantation management: a review. Int. J. Sci. Res. Pub. **2**(9) (2012)
20. Sharma, S.C., Hada, D.S., Gupta, U.: A goal programming model for the interaction effects in multiple nonlinear regression. J. Comp. Math. Sci. **1**(4), 477–481 (2010)

On the Fekete–Szegö Problem for Certain Subclass of Analytic Functions

Ritu Agarwal and G. S. Paliwal

Abstract The purpose of the present investigation is to derive several Fekete–Szegö-type coefficient inequalities for certain subclasses of normalized analytic function $f(z)$ defined in the open unit disk. Various applications of our main results involving (for example) the operators defined using generalized fractional differential operator are also considered. Thus, as one of these applications of our result, we obtain the Fekete–Szegö-type inequality for a class of normalized analytic functions, which is defined here by means of the convolution and the fractional differential operators.

Keywords Starlike functions · Fekete–Szegö problem · Fractional derivatives · Generalized Ruscheweyh derivative · Convolution

1 Introduction and Definitions

Let **A** denote the class of functions $f(z)$ of the form

$$f(z) = z + \sum_{n=2}^{\infty} a_n z^n, \tag{1}$$

which are analytic in the open unit disk

R. Agarwal (✉)
Malviya National Institute of Technology, J.L.N. Marg, Jaipur 302017, India
e-mail: ritugoyal.1980@gmail.com

G. S. Paliwal
JECRC UDML College of Engineering, Kukas, Jaipur 302028, India
e-mail: gaurishankarpaliwal@gmail.com

B. V. Babu et al. (eds.), *Proceedings of the Second International Conference on Soft Computing for Problem Solving (SocProS 2012), December 28–30, 2012*, Advances in Intelligent Systems and Computing 236, DOI: 10.1007/978-81-322-1602-5_39,

$$\Delta = \{z : z \in C \quad \text{and} \quad |z| < 1\}. \tag{2}$$

Also, let **S** be the subclass of **A** consisting of all univalent functions in Δ. A function $f(z)$ in **A** is said to be in class $\mathbf{S}^*$ of *starlike functions* of order zero in Δ, if Re $\left\{\frac{zf'(z)}{f(z)}\right\} > 0$ for $z \in \Delta$. Let **K** denote the class of all functions $f \in \mathbf{A}$ that are convex. Further, f is convex if and only if $zf'(z)$ is starlike.

Definition 1.1 Let $(f{*}g)(z)$ denote the *convolution* of two functions f(z) given by (1) and $g(z) = z + \sum_{n=2}^{\infty} b_n z^n$, then

$$(f{*}g)(z) = z + \sum_{n=2}^{\infty} a_n b_n z^n. \tag{3}$$

We shall be requiring the following fractional differential operator in the present investigations:

Definition 1.2 Let f(z) is an analytic function in a simply connected region of the z plane containing the origin, and the multiplicity of $(z-\zeta)^{\lambda}$ is removed by requiring that $\log(z-\zeta)$ to be real when $(z-\zeta) > 0$. Then, the generalized fractional derivative of order λ is defined for a function $f(z)$ by (see, e.g., [16])

$$J_{0,z}^{\lambda,\mu,\nu} f(z) = \begin{cases} \frac{1}{\Gamma(1-\lambda)} \frac{d}{dz} \left\{ \begin{array}{l} z^{\lambda-\mu} \int_0^z (z-\zeta)^{-\lambda} \\ \cdot {}_2F_1\left(\mu-\lambda, -\nu; 1-\lambda; 1-\frac{\zeta}{z}\right) f(\zeta) d\zeta \end{array} \right\}, (0 \le \lambda < 1) \\ \frac{d^n}{dz^n} J_{0,z}^{\lambda-n,\mu,\nu} f(z), (n \le \lambda < n+1, n \in N) \end{cases}$$

and $f(z) = O(|z|^k), (z \to 0, k > \max\{0, \mu-\nu-1\} - 1)$.

It follows at once from the above definition that $J_{0,z}^{\lambda,\lambda,\nu} f(z) := D_z^{\lambda} f(z), (0 \le \lambda < 1)$ which is fractional derivative of f of order λ (see, e.g., [10, 11]). Furthermore, in terms of gamma function, we have:

$$J_{0,z}^{\lambda,\mu,\nu} z^{\rho} := \frac{\Gamma(\rho+1)\Gamma(\rho-\mu+\nu+1)}{\Gamma(\rho-\mu+1)\Gamma(\rho-\lambda+\nu+2)} z^{\rho-\mu},$$
$$(0 \le \lambda; \rho > \max\{0, \mu-\nu-1\} - 1)$$

Definition 1.3 Let f and g analytic in Δ. We say that the function f is subordinate to g if there exists a Schwarz function $w(z)$, analytic in Δ with $w(z) = 0$ and $|w(z)| < 1$, such that $f(z) = g(w(z))$ for $z \in \Delta$. We denote this subordination by $f \prec g$ or $f(z) \prec g(z)$.

Let $\phi(z)$ be an analytic function in Δ with $\phi(0) = 1, \phi'(0) > 0$ and $Re(\phi(z)) > 0, (z \in \Delta)$, which maps the open unit disk Δ onto a region starlike with respect to 1 and is symmetric with respect to the real axis. Motivated by the class $R_{\lambda}^{\tau}(\phi)$ in paper [1], we introduce the following class.

Definition 1.4 Let $0 \leq \alpha < 1, 0 \leq \gamma < 1, 0 \leq \rho < 1, \tau \in C\backslash\{0\}$. A function $f \in A$ is in the class $R_{\alpha,\gamma,\rho}^{\tau}(\phi)$ if

$$1 + \frac{1}{\tau}\left(\rho\left\{(f*S_\alpha)(z)\right\}' + \gamma z\left\{(f*S_\alpha)(z)\right\}'' - \rho\right) \prec \phi(z), \quad (z \in \Delta) \tag{4}$$

where $\phi(z)$ is defined same as above.

The function

$$S_\alpha(z) = \frac{z}{(1-z)^{2(1-\alpha)}} = z + \sum_{n=2}^{\infty} C(\alpha, n) z^n, \tag{5}$$

is the well-known extremal function (see, e.g., [9]). It is observed that $C(\alpha, n)$, $n = 2, 3, \ldots$ is decreasing in α . Also,

$$C(\alpha, n) = \frac{\prod_{k=2}^{n}(k - 2\alpha)}{(n-1)!} \quad \text{and} \quad \lim_{n\to\infty} C(\alpha, n) = \begin{cases} \infty & (\alpha < 1/2) \\ 1 & (\alpha = 1/2) \\ 0 & (\alpha > 1/2) \end{cases} \tag{6}$$

If we set $\phi(z) = \frac{1+Az}{1+Bz}$, $(-1 \leq B < A \leq 1, z \in \Delta)$, in (4), we get

$$R_{\alpha,\gamma,\rho}^{\tau}(A, B) = \left\{ f \in A : \left| \frac{\rho\left\{(f*S_\alpha)(z)\right\}' + \gamma z\left\{(f*S_\alpha)(z)\right\}'' - \rho}{\tau(A-B) - B\left(\rho\left\{(f*S_\alpha)(z)\right\}' + \gamma z\left\{(f*S_\alpha)(z)\right\}'' - \rho\right)} \right| < 1 \right\},$$

which is again a new class. The classes discussed recently by Bansal [1], Swaminathan [17], Ponnusamy and Ronning [12], Ponnusamy [14] and Li [4] follow as special cases of our class.

To prove our main results, we need the following Lemma:

Lemma 1.1 *[5] If $p(z) = 1 + c_1 z + c_2 z^2 + c_3 z^3 + \cdots (z \in \Delta)$ is a function with positive real part, then for any complex number ε, $\left|c_3 - \varepsilon c_2^2\right| \leq 2\max\{1, |2\varepsilon - 1|\}$. The result is sharp for the functions given by $p(z) = \frac{1+z^2}{1-z^2}$ and $p(z) = \frac{1+z}{1-z}$.*

2 Fekete–Szegö Problem

In this section, we shall be finding the Fekete–Szegö coefficient inequalities for the class of functions $R_{\alpha,\gamma,\rho}^{\tau}(\phi)$. Our main result is contained in the following theorem:

Theorem 2.1 *Let* $\phi(z) = 1 + B_1 z + B_2 z^2 + B_3 z^3 + \cdots$, *where* $\phi(0) = 1$ *with* $\phi\prime(0) > 0$. *If* $f(z)$ *given by (1) belongs to* $R^{\tau}_{\alpha,\gamma,\rho}(\phi)(\alpha, \gamma, \rho \in [0, 1], \tau \in C\backslash\{0\}, z \in \Delta)$, *then for any complex number* υ

$$\left|a_3 - \upsilon a_2^2\right| \leq \frac{2B_1 |\tau|}{3(\rho + 2\gamma)(2 - 2\alpha)(3 - 2\alpha)} \max\left\{1, \left|\frac{B_2}{B_1} - \frac{3\upsilon\tau B_1(\rho + 2\gamma)(3 - 2\alpha)}{8(\rho + \gamma)^2(2 - 2\alpha)}\right|\right\} \quad (7)$$

The result is sharp.

Proof If $f(z) \in R^{\tau}_{\alpha,\gamma,\rho}(\phi)$, then there exists a Schwarz function $w(z)$ analytic in Δ with $w(z) = 0$and $|w(z)| < 1$, $(z \in \Delta)$ such that

$$1 + \frac{1}{\tau}\left(\rho\left\{(f{*}S_\alpha)(z)\right\}' + \gamma z\left\{(f{*}S_\alpha)(z)\right\}'' - \rho\right) = \phi(w(z)) \quad (8)$$

Define the function $p_1(z)$ by

$$p_1(z) = \frac{1 + w(z)}{1 - w(z)} = 1 + c_1 z + c_2 z^2 + c_3 z^3 + \cdots \quad (9)$$

Since $w(z)$ is a Schwarz function, we see that $Re(p_1(z)) > 0$ and $p_1(0) = 1$. Define the function $p(z)$ by

$$p(z) = 1 + \frac{1}{\tau}\left(\rho\left\{(f{*}S_\alpha)(z)\right\}' + \gamma z\left\{(f{*}S_\alpha)(z)\right\}'' - \rho\right) \quad (10)$$

In view of (8), (9) and (10),

$$p(z) = \varphi\left(\frac{p_1(z) - 1}{p_1(z) + 1}\right) = \varphi\left(\frac{1}{2}c_1 z + \frac{1}{2}\left(c_2 - \frac{c_1^2}{2}\right)z^2 + \cdots\right) \quad (11)$$

$$= 1 + \frac{B_1 c_1 z}{2} + \frac{B_1}{2}\left(c_2 - \frac{c_1^2}{2}\right)z^2 + \frac{B_2 c_1^2}{4}z^2 + \cdots \quad (12)$$

Using (6) in (5), we get

$$S_\alpha(z) = z + (2 - 2\alpha)z^2 + \frac{(2 - 2\alpha)(3 - 2\alpha)}{2}z^3 + \cdots$$

which on substitution in (10) gives

$$p(z) = 1 + \frac{1}{\tau}\left(2a_2(2 - 2\alpha)(\rho + \gamma)z + 3a_3\frac{(2 - 2\alpha)(3 - 2\alpha)(\rho + 2\gamma)}{2}z^2 + \cdots\right) \quad (13)$$

Comparing (12) and (13)

$$a_2 = \frac{B_1 c_1 \tau}{4(2-2\alpha)(\rho+\gamma)}$$

and

$$a_3 = \frac{\tau}{3(2-2\alpha)(3-2\alpha)(\rho+2\gamma)}\left(B_1\left(c_2 - \frac{c_1^2}{2}\right) + \frac{B_2 c_1^2}{2}\right).$$

Therefore, we have

$$a_3 - \upsilon a_2^2 = \frac{\tau B_1}{3(2-2\alpha)(3-2\alpha)(\rho+2\gamma)}(c_2 - \varepsilon c_1^2)$$

where

$$\varepsilon = \frac{1}{2}\left\{1 - \frac{B_2}{B_1} + \frac{3\upsilon\tau B_1(3-2\alpha)(\rho+2\gamma)}{8(2-2\alpha)(\rho+\gamma)^2}\right\}.$$

Our result is followed by application of Lemma (1.1).

Also, by the application of Lemma (1.1), equality in (7) is obtained when

$$p_1(z) = \frac{1+z^2}{1-z^2} \quad \text{or} \quad p_1(z) = \frac{1+z}{1-z}.$$

For the class $R^{\tau}_{\alpha,\gamma,\rho}(A,B)$, $\phi(z) = \frac{1+Az}{1+Bz} = 1 + (A-B)z - (AB - B^2)z^2 + \cdots$. Thus, putting $B_1 = A\text{-}B$ and $B_2 = -B(A-B)$ in Theorem 2.1, we get the following corollary:

Corollary 2.2 *If $f(z)$ given by (1) belongs to $R^{\tau}_{\alpha,\gamma,\rho}(A,B)$, then*

$$\left|a_3 - \upsilon a_2^2\right| \le \frac{2\,|\tau|\,(A-B)}{3(2-2\alpha)(3-2\alpha)(\rho+2\gamma)} \max\left\{1, \left|B - \frac{3\upsilon\tau(A-B)(3-2\alpha)(\rho+2\gamma)}{8(2-2\alpha)(\rho+\gamma)^2}\right|\right\}.$$

3 Applications to Functions Defined Using Fractional Derivative

For fixed $g \in A$, we define the class $R^{\tau,g}_{\alpha,\gamma,\rho}(\varphi)$ of functions $f \in A$ for which $(f*g) \in R^{\tau}_{\alpha,\gamma,\rho}(\varphi)$. Suppose that $g(z) = z + \sum_{n=2}^{\infty} g_n z^n (g_n > 0)$. Then, $f(z) = z + \sum_{n=2}^{\infty} a_n z^n \in R^{\tau,g}_{\alpha,\gamma,\rho}(\phi)$ if and only if $(f*g)(z) = z + \sum_{n=2}^{\infty} g_n a_n z^n \in R^{\tau}_{\alpha,\gamma,\rho}(\phi)$.

By applying Theorem (2.1) to the convolution $(f*g)(z) = z + g_2 a_2 z^2 + g_3 a_3 z^3 + \cdots$, we get Theorem (3.1) below after an obvious change of the parameter υ.

Theorem 3.1 *Let* $\phi(z) = 1 + B_1 z + B_2 z^2 + B_3 z^3 + \cdots$ *where* $\phi(0) = 1$ *with* $\phi\prime(0) > 0$. *If f(z) given by (1) belongs to* $R^{\tau,g}_{\alpha,\gamma,\rho}(\phi)$ $(\alpha, \gamma, \rho \in [0, 1), \tau \in C\backslash\{0\}, z \in \Delta)$, *then for any complex number* υ

$$\left|a_3 - \upsilon a_2^2\right| \leq \frac{2B_1 |\tau|}{3g_3(\rho + 2\gamma)(2 - 2\alpha)(3 - 2\alpha)} \max\left\{1, \left|\frac{B_2}{B_1} - \frac{g_3}{g_2^2}\frac{3\upsilon\tau B_1(\rho + 2\gamma)(3 - 2\alpha)}{8(\rho + \gamma)^2(2 - 2\alpha)}\right|\right\}$$

The result is sharp.

We, now, discuss some applications of the above theorem to the subclasses defined using fractional derivatives.

1. In terms of generalized Ruscheweyh derivative operator, we now introduce the function class $R^{\tau,\lambda,\mu}_{\alpha,\gamma,\rho}(\phi)$ in the following way:

$$R^{\tau,\lambda,\mu}_{\alpha,\gamma,\rho}(\phi) := \left\{f : f \in A \text{ and } J^{\lambda,\mu} f \in R^{\tau}_{\alpha,\gamma,\rho}(\phi)\right\}, \tag{14}$$

where the generalized Ruscheweyh derivative introduced by Goyal and Goyal [2], Parihar and Agarwal [13] is defined as

$$\begin{aligned} J^{\lambda,\mu} f(z) &= \frac{\Gamma(\mu - \lambda + \nu + 2)}{\Gamma(\mu + 1)\,\Gamma(\nu + 2)} z J^{\lambda,\mu,\nu}_{0,z}\left(z^{\mu-1} f(z)\right) \\ &= z + \sum_{n=2}^{\infty} a_n B^{\lambda,\mu}(n) z^n = (f * g)(z) \end{aligned} \tag{15}$$

where

$$B^{\lambda,\mu}(n) := \frac{\Gamma(n + \mu)\Gamma(\nu + 2 + \mu - \lambda)\Gamma(n + \nu + 1)}{\Gamma(n)\Gamma\,(n + \nu + 1 + \mu - \lambda)\,\Gamma\,(\nu + 2)\,\Gamma\,(1 + \mu)} \tag{16}$$

It is easily seen that the function class $R^{\tau,\lambda,\mu}_{\alpha,\gamma,\rho}(\phi)$ is a special case of the function class $R^{\tau,g}_{\alpha,\gamma,\rho}(\phi)$ when $g(z) = z + \sum_{n=2}^{\infty} B^{\lambda,\mu} z^n = z\cdot{}_2F_1(\mu + 1, \nu + 2; \nu + 2 + \mu - \lambda; z)$.

Thus, we obtain the coefficient estimates for functions in the subclass $R^{\tau,\lambda,\mu}_{\alpha,\gamma,\rho}(\phi)$ from the corresponding estimates for functions in the class $R^{\tau,g}_{\alpha,\gamma,\rho}(\phi)$.

Theorem 3.2 *Let* $\phi(z) = 1 + B_1 z + B_2 z^2 + B_3 z^3 + \cdots$, *where* $\phi(0) = 1$ *with* $\phi'(0) > 0$. *If* $f(z)$ *given by (1) belongs to* $R^{\tau,\lambda,\mu}_{\alpha,\gamma,\rho}(\phi)$ $(\alpha, \gamma, \rho \in [0, 1), \tau \in C\backslash\{0\}, z \in \Delta)$, *then for any complex number* υ

$$\left|a_3 - \upsilon a_2^2\right| \leq \frac{4(\nu+2+\mu-\lambda)(\nu+3+\mu-\lambda)B_1\,|\tau|}{3(\mu+1)(\mu+2)(\nu+2)(\nu+3)(\rho+2\gamma)(2-2\alpha)(3-2\alpha)}$$
$$\max\left\{1, \left|\frac{B_2}{B_1} - \frac{(\mu+2)(\nu+3)(\nu+2+\mu-\lambda)}{2(\mu+1)(\nu+2)(\nu+3+\mu-\lambda)}\frac{3\upsilon\tau B_1(\rho+2\gamma)(3-2\alpha)}{8(\rho+\gamma)^2(2-2\alpha)}\right|\right\}$$

The result is sharp.

2. In terms of generalized Owa–Srivastava operator, we now introduce the function class

$$S_{\alpha,\gamma,\rho}^{\tau,\lambda,\mu}(\phi) := \left\{f : f \in A \quad \text{and} \quad \Omega_\nu^{\lambda,\mu} f \in R_{\alpha,\gamma,\rho}^{\tau}(\phi)\right\} \tag{17}$$

where we define the generalized Owa–Srivastava operator as

$$\Omega_\nu^{\lambda,\mu} f(z) = \frac{\Gamma(2-\mu)\Gamma(3-\lambda+\nu)}{\Gamma(3-\mu+\nu)} z^\mu J_{0,z}^{\lambda,\mu,\nu} f(z)$$
$$= z + \sum_{n=2}^{\infty} \frac{\Gamma(n+1)\Gamma(2-\mu)\Gamma(n-\mu+\nu+2)\Gamma(3-\lambda+\nu)}{\Gamma(n-\mu+1)\Gamma(3-\mu+\nu)\Gamma(n-\lambda+\nu+2)} a_n z^n$$

For $\mu = \lambda$, $\Omega_\nu^{\lambda,\mu}$ reduces to the Owa–Srivastava operators Ω^λ [10].

It is easily seen that the function class $S_{\alpha,\gamma,\rho}^{\tau,\lambda,\mu}(\phi)$ is a special case of the function class $R_{\alpha,\gamma,\rho}^{\tau,g}(\phi)$ when

$$g(z) = z + \sum_{n=2}^{\infty} \frac{\Gamma(n+1)\Gamma(2-\mu)\Gamma(n-\mu+\nu+2)\Gamma(3-\lambda+\nu)}{\Gamma(n-\mu+1)\Gamma(3-\mu+\nu)\Gamma(n-\lambda+\nu+2)} z^n \tag{18}$$

The coefficient estimates for functions in the subclass $S_{\alpha,\gamma,\rho}^{\tau,\lambda,\mu}(\phi)$ are given by

Theorem 3.3 *Let $\phi(z) = 1 + B_1 z + B_2 z^2 + B_3 z^3 + \cdots$, where $\phi(0) = 1$ with $\phi'(0) > 0$. If $f(z)$ given by (1) belongs to $S_{\alpha,\gamma,\rho}^{\tau,\lambda,\mu}(\phi)$ $(\alpha, \gamma, \rho \in [0,1], \tau \in C\backslash\{0\}, z \in \Delta)$, then for any complex number υ*

$$\left|a_3 - \upsilon a_2^2\right| \leq \frac{(3-\mu)(2-\mu)(\nu+3-\lambda)(\nu+4-\lambda)B_1\,|\tau|}{9(\nu+4-\mu)(\nu+3-\mu)(\rho+2\gamma)(2-2\alpha)(3-2\alpha)}$$
$$\max\left\{1, \left|\frac{B_2}{B_1} - \frac{3(\nu+4-\mu)(\nu+3-\lambda)(2-\mu)}{2(3-\mu)(\nu+4-\lambda)(\nu+3-\mu)}\frac{3\upsilon\tau B_1(\rho+2\gamma)(3-2\alpha)}{8(\rho+\gamma)^2(2-2\alpha)}\right|\right\}$$

3. In terms of *Najafzadeh operator*, we now introduce the function class

$$S_{\alpha,\gamma,\rho}^{\tau,\lambda,k}(\phi) := \left\{f : f \in A \quad \text{and} \quad \Omega_\lambda^k f \in R_{\alpha,\gamma,\rho}^{\tau}(\phi)\right\}. \tag{19}$$

For k∈ $N \cup \{0\}$ and $\lambda \geq 0$, the operator $\Omega_\lambda^k f : N \rightarrow N$ is defined by Najafzadeh [8] as

$$\Omega_\lambda^k f(z) = (1-\lambda)S^k f(z) + \lambda R^k f(z), z \in \Delta, \tag{20}$$

where $S^k f$ is the Salagean differential operator [15] and $R^k f$ is the Ruscheweyh differential operator [9].

For f(z) ∈A given by (1), we have respectively

$$S^k f(z) = z + \sum_{n=2}^{\infty} n^k a_n z^n \quad \text{and} \quad R^k f(z) = z + \sum_{n=2}^{\infty} \binom{k+n-1}{k} a_n z^n \tag{21}$$

and hence

$$\Omega_\lambda^k f(z) = \sum_{k=2}^{\infty} \left[(1-\lambda)n^k + \lambda \binom{k+n-1}{k} \right] a_n z^n, z \in \Delta \tag{22}$$

It is easily seen that the function class $S_{\alpha,\gamma,\rho}^{\tau,\lambda,k}(\phi)$ is a special case of the function class $R_{\alpha,\gamma,\rho}^{\tau,g}(\phi)$ when

$$g(z) = z + \sum_{n=2}^{\infty} \left[(1-\lambda)n^k + \lambda \binom{k+n-1}{k} \right] z^n \tag{23}$$

The coefficient estimates for functions in the subclass $S_{\alpha,\gamma,\rho}^{\tau,\lambda,k}(\phi)$ is given by:

Theorem 3.4 *Let $\phi(z) = 1+B_1z+B_2z^2+B_3z^3+\cdots$, where $\phi(0) = 1$ with $\phi\prime(0) > 0$. If f(z) given by (1) belongs to $S_{\alpha,\gamma,\rho}^{\tau,\lambda,k}(\phi)$ $(\alpha, \gamma, \rho \in [0, 1), \tau \in C\backslash\{0\}, z \in \Delta)$, then for any complex number υ*

$$|a_3 - \upsilon a_2^2| \leq \frac{2B_1 |\tau|}{3(3^k(1-\lambda) + \lambda(k+2)(k+1)/2)(\rho+2\gamma)(2-2\alpha)(3-2\alpha)}$$
$$\max\left\{1, \left|\frac{B_2}{B_1} - \frac{(3^k(1-\lambda) + \lambda(k+2)(k+1)/2)}{(2^k(1-\lambda) + \lambda(k+1))^2} \frac{3\upsilon\tau B_1(\rho+2\gamma)(3-2\alpha)}{8(\rho+\gamma)^2(2-2\alpha)}\right|\right\}$$

For $\lambda = 0$ and $\lambda = 1$, Najafzadeh operators $\Omega_\lambda^k f$ reduce to Salagean differentiation of f and Ruscheweyh derivative of f, respectively. Hence, the Fekete–Szegö inequality for these functions follows immediately from the Theorem 3.4.

References

1. Bansal, D.: Fekete -Szegö problem for a new class of analytic function. Int. J. Math. Math. Sci., article ID 143096, 5 pp (2011)
2. Goyal, S.P., Goyal, R.: On a class of multivalent functions defined by generalized Ruscheweyh derivatives involving a general fractional derivative operator. J. Indian Acad. Math. **27**(2), 439–456 (2005)

3. Koepf, W.: On the Fekete-Szegö problem for close-to-convex functions. Archiv derMathematik **49**(5), 420–433 (1987)
4. Li, J.L.: On some classes of analytic functions. Mathematica Japonica **40**(3), 523–529 (1994)
5. Libera, R.J., Złotkiewicz, E.J.: Coefficient bounds for the inverse of a function with derivative in ρ. Proc. Am. Math. Soc. **87**(2), 251–257 (1983)
6. London, R.R.: Fekete-Szegö inequalities for close-to-convex functions. Proc. Am. Math. Soc. **117**(4), 947–950 (1993)
7. Ma, W., Minda, D.: A unified treatment of some special classes of univalent functions. In: Li, Z., Ren, F., Yang, L., Zhang, S. (eds.) Proceedings of the Conference on Complex Analysis, pp. 157–169. Conference Proceedings and Lecture Notes in Analysis, vol. I. International Press, Cambridge, Massachusetts (1994)
8. Najafzadeh, S.: Application of Salagean and Ruscheweyh operators on univalent holomorphic functions with finitely many coefficients. Fractional Calculus Appl. Anal. **13**(5), 517–520 (2010)
9. Owa, S., Uralegaddi, B.A.: A class of functions α- prestarlike of order β. Bull. Korean Math. Soc. **21**(4), 77–85 (1984)
10. Owa, S., Srivastava, H.M.: Univalent and starlike generalized hypergeometric functions. Canad. J. Math. **39**, 1057–1077 (1987)
11. Owa, S.: On the distortion theorems I. Kyungpook Math. J. **18**, 53–58 (1978)
12. Ponnusamy, S., Ronning, F.: Integral transforms of a class of analytic functions. Complex Variables Elliptic Equ. **53**(5), 423–434 (2008)
13. Parihar, H.S., Agarwal, R.: Application of generalized Ruscheweyh derivatives on p-valent functions. J. Math. Appl. **34**, 75–86 (2011)
14. Ponnusamy, S.: Neighborhoods and Carathėodory functions. J. Anal. **4**, 41–51 (1996)
15. Salagean, G.S.: Subclasses of univalent functions. Lect. Notes Math. **1983**, 362–372 (1013)
16. Srivastava, H.M., Saxena, R.K.: Operators of fractional integration and their applications. Appl. Math. Comput. **118**, 1–52 (2001)
17. Swaminathan, A.: Certain sufficiency conditions on Gaussian hypergeometric functions. J. Inequalities Pure Appl. Math. **5**(4), article 83, 6 pp (2004)

Part III
Soft Computing for Operations Management (SCOM)

Bi-Objective Scheduling on Parallel Machines in Fuzzy Environment

Sameer Sharma, Deepak Gupta and Seema Sharma

Abstract The present chapter pertains to a bi-objective scheduling on parallel machines involving total tardiness and number of tardy jobs (NT). The processing time of jobs are uncertain in nature and are represented by triangular fuzzy membership function. The objective of the chapter is to find the optimal sequence of jobs processing on parallel identical machines so as to minimize the secondary criteria of NT with the condition that the primary criteria of total tardiness remains optimized. The bi-objective problem with total tardiness and NT as primary and secondary criteria, respectively, for any number of parallel machines is NP-hard. Following the theoretical treatment, a numerical illustration has also been given to demonstrate the potential efficiency of the proposed algorithm as a valuable analytical tool for the researchers.

Keywords Fuzzy processing time · Average high ranking · Total tardiness · Due date · Tardy job.

1 Introduction

Scheduling is a very common activity in both industry and non-industry settings. Everyday meetings are scheduled, deadlines are set for projects, vacations and work periods are set, maintenance and upgrade operations are planned, operation rooms are

S. Sharma (✉) · S. Sharma
Department of Mathematics, D.A.V. College, Jalandhar, Punjab, India
e-mail: samsharma31@yahoo.com

S. Sharma
e-mail: seemasharma7788@yahoo.com

D. Gupta
Department of Mathematics, M.M. University, Mullana, Ambala, India
e-mail: guptadeepak2003@yahoo.co.in

B. V. Babu et al. (eds.), *Proceedings of the Second International Conference on Soft Computing for Problem Solving (SocProS 2012), December 28–30, 2012*, Advances in Intelligent Systems and Computing 236, DOI: 10.1007/978-81-322-1602-5_40,

booked, and sports games are scheduled and arenas are booked. Proper scheduling allows various activities, jobs or tasks to be executed in an organized manner, while preventing resource conflicts. The parallel machine scheduling problem is a widely studied optimization problem. It is a kind of important multi-machine scheduling in which every machine has same work function and every job can be processed by any available machine. Scheduling problems in real-life applications generally involve optimization of more than one criterion. A large number of deterministic scheduling algorithms have been proposed in last decades to deal with scheduling problems with various objectives and constraints. However, in real-world applications, it is usually difficult to set exact processing times for jobs. More often, the processing time of a job may vary within an interval. Thus, it is natural and realistic to represent this kind of uncertainties by fuzzy numbers.

A survey of the literature has revealed little work reported on the bi-objective scheduling problems on parallel machines. Most of the work done in the bi-objective problems has been on the single machine. Anghinolfi and Paolucci [1] studied total tardiness scheduling problems on parallel machines. Azizoglu et al. [2] discussed bi-criteria scheduling problem involving total tardiness and total earliness penalties. Parkash [9] studied the bi-criteria scheduling problems on parallel machines. Shim and Kim [10] dealt with scheduling on parallel identical machines to minimize the total tardiness. Gupta and Sharma [6] studied the scheduling on parallel machines with bi-objective function NT/$T_{\max}$ in fuzzy environment. Some of the noteworthy approaches are due to Chand and Schneerbrg [4], Moore [8], and Singh and Sunita [11].

The present chapter addresses the bi-objective scheduling problems on identical parallel machines involving total tardiness and number of tardy jobs (NT) with bi-objective function as NT/Total Tardiness. Two approaches can be used to address the bi-objective problems: Both the criteria are optimized simultaneously by using suitable weights for the criteria, and secondly, the criteria are optimized sequentially by first optimizing the primary criterion and then the secondary criterion subject to the value obtained for the primary criterion. In this research paper, we have used the second approach. A practical application of this paper can be taken as to minimize the cost of production or production time given the penalty for delaying the product.

2 Problem Formulation

The following assumptions are made for the problem formulation

1. The jobs are available at time zero.
2. The jobs are independent of each other.
3. No preemption of jobs is allowed.
4. The machines are identical in all respects.
5. No machine can handle more than one job at a time.

The following notations will be used all the way through out the chapter

i: Designate the ith job, i = 1, 2, 3, − − − , n
k: Machine on which ith job is assigned at the jth position
j: Location of ith job on machine k , where j = 1, 2, 3, − − −, n
d_i: Due date of the ith job
c_i: Completion time of ith job
T_i: Tardiness of the ith job= max $(c_i - d_i, 0)$
T: Total tardiness
n: Total number of jobs to be scheduled
NT: Number of tardy jobs.
$X_{\text{ijk}} = 1$; if job i is located at the jth position on kth machine and 0; otherwise.

Chen and Bulfin [5] studied the scheduling on a single machine to minimize the two criteria of maximum tardiness and NT. Akker et al. [3] discussed the minimization of NT. Lawer et al. [7] described the minimization of maximum lateness in a two-machine open shop scheduling. Before formulating the bi-criteria problem, the formulation for the single criterion is represented first. They are as follows:

Criterion: Total Tardiness

Tardiness is given by $\max(0, c_i - d_i)$, where c_i and d_i are the completion and due date of job i. This function is a nonlinear function but can be linearized. The formulation is as follows:

$$\text{Min } Z = \sum_{i=1}^{n} T_i$$

Subject to:

$$\sum_{j=1}^{n}\sum_{k=1}^{n} X_{\text{ijk}} = 1 \quad \forall i \quad \text{(i)}, \qquad \sum_{i=1}^{n} X_{ijk} \leq 1 \quad \forall j, k \quad \text{(ii)},$$

$$X_{\text{ijk}} \text{ is Binary} \quad \forall i, j, k \quad \text{(iii)}, \qquad T_i \geq c_i - d_i \quad \forall i \quad \text{(iv)};$$

along with non-negativity constraint.

Criterion: Number of Tardy Jobs

A job is considered to be late only if its tardiness is strictly positive. Let Y_i be a binary variable that depict whether or not the job i is late. It assumes a value 1 when the job i is late and 0 otherwise. The $\sum_{i=1}^{n} Y_i$ gives the total number of the tardy jobs. The formulation is as follows:

$$\text{Min } Z = \sum_{i=1}^{n} Y_i$$

Subject to: Constraint set (i), (ii), (iii) and (iv)

$$Y_i \geq MT_i \quad \forall i \quad \text{(v)}, \qquad Y_i \text{ binary} \quad \forall i \quad \text{(vi)};$$

where M is a very large number.

The formulation of the bi-criteria problems is similar to that of single criterion problems but with some additional constraints requiring that the optimal value of the primary objective is not violated. The two parts of the bi-criteria problem formulation are as follows:

Primary Objective Function

Subject to: Primary problem constraints

Secondary Objective Function

Subject to:

1. Secondary problem constraint.
2. Primary objective function value constraint.
3. Primary problem constraint.

In the present work, we consider the parallel machines bi-criteria scheduling problem in which the objective is to schedule jobs on parallel identical machines so as to minimize primary and secondary criteria. So here, the problem is distributed in two steps: first, the primary criterion in which total tardiness of jobs is minimized, and in secondary step, the NT is minimized under the objective function value of primary criterion.

3 Algorithm

The following algorithm is proposed to optimize the bi-objective function NT/Total Tardiness by considering total tardiness and number of tardy jobs as primary and secondary criteria.

Step 1: Arrange all the jobs in early due date (EDD) order, and find the tardiness of each job (if any). Let L be the set of late jobs in the current schedule and T be the total tardiness. Initialize a set of jobs, $C = \varphi$. It contains the jobs that cannot be switched.

Step 2: Calculate $T_i, \forall i \in L$. If $T_i < 1 \forall i \in L$, then exit; else go to step 3.

Step 3: Select the first late job $i \in L$ and $i \notin C$. If none exist then exit; else go to step 4.

Step 4: Check if $c_i = d_j$ for some late job $j \in L$. If so, then exchange jobs i and j; Set $L = L - \{j\}$, else set $C = C + \{i\}$. Go to step 3, in any case.

4 Theorems

The following theorems have been developed to optimize the bi-criteria scheduling on parallel machines involving total tardiness and NT.

4.1 Theorem

The proposed algorithm 3 optimizes the bi-objective function NT/Total tardiness.

Proof To prove the optimality of proposed, we need to prove the optimality of steps 2, 3, and 4 of the proposed algorithm. As we know that the EDD rule optimizes total tardiness, further, the completion time of any job i is the location number of that job in the schedule. Hence, the completion time of a job can be determined by its new location after it is switched with another job.

For Optimality of Step 2: By assumption, $T_i < 1 \forall i \in L$. If there exist a better schedule than that of schedule obtained by EDD rule and can be obtained by moving jobs, we shall show that the movement of any job from EDD schedule does not lead to a better schedule.

Case I: If a job i moves to an earlier position from its position in the EDD schedule, then it delays all the intermediate jobs by at least one unit each. Hence, irrespective of whether job i is early or late in the EDD schedule, the value of NT and total tardiness at best remain the same and could possibly deteriorate from their values in the EDD schedule.

Case II: If a job i moves to a late position (delaying) from its position in the EDD schedule, then again two subcases arises:

If job i is late in the EDD schedule, then every unit of time that is delayed, its tardiness increases by a unit, while the tardiness of already late j improves only by the amount $T_j (< 1$, by assumption). Hence, the net effect is an increment in total tardiness.

If job i is not late in the EDD schedule, then two situations may further arise. If job i is delayed to a position where it is still early, then both tardiness and NT remain the same. However, if it is moved to a position where it becomes late, for unit increment in its tardiness, the reduction in tardiness of already late job j is T_j (<1, by assumption). Hence, the net effect is an increment in total tardiness.

Therefore, if $T_i < 1 \forall i \in L$, no improvement in the job scheduling is possible.

For optimality of Steps 3 and 4: Here, we shall show that the conditions mentioned in steps 3 and 4 of the proposed algorithm are the only conditions under which an improved schedule can be obtained.

First, we observe that for these steps $T_i \geq 1$for all $i \in L$with strict inequality holding for at least one i. Further, switching either the early jobs or jobs with the same due date will not result in a better schedule. In fact, it may make the number of late jobs and total tardiness worse. Thus, we consider the switching of a late job with a job that is either late or early job. Pick any two jobs i and j from the EDD schedule. Let job j be the late job. The following cases may arise:

Case I: Job i is late and $d_i < d_j$

In this case, we have either $c_i = c_j$ or $t_i < t_j$.

If $c_i = c_j$, then the switching of these jobs will not improve the solution.

If $c_i < c_j$, then the tardiness T and T' before and after the exchange are

$T_i = \max(0, c_i - d_i) + c_j - d_j, T_i' = \max(0, c_i - d_j) + c_j - d_i$

In case if $c_i > d_j$, then switching job i and job j will worsen the primary criterion.

In case if $c_i < d_j$, then switching job i and job j does not change the total tardiness and NT values. Hence, the only case in which the primary criterion is not violated and NT improves is, if $c_i = d_j$.

Case II: If job *i* is not late and $d_i < d_j$

In this case, the total tardiness before and after switching job i and j is $T = c_j - d_j, T' = c_i - d_j$. Here, we have $T < T'$.

Hence, the primary criterion of total tardiness is violated.

Case III: If job *i* is not late job $d_i > d_j$

In this case, the total tardiness before and after switching job i and job j is $T = c_j - d_j, T' = c_i - d_j$. Here, we have$T < T'$.

Hence, the primary criterion of total tardiness is again violated.

Case IV: If job *i* is late and $d_i > d_j$

In this case, we get the similar result as we get in case I, discussed above.

Hence, we have shown that a switching among any two jobs will worsen the EDD schedule except that made under the exchange condition $c_i = d_j$ as stated in the algorithm. Hence, the proposed algorithm optimizes the bi-objective function NT/Total tardiness.

4.2 Theorem

If the problem of single criterion, total tardiness, is NP-hard, the scheduling problem on parallel machines optimizing the bi-objective function NT/Total Tardiness will also be NP-hard.

Solution: We shall prove the result by the method of contradiction:

Let if possible the bi-objective function NT/Total Tardiness is not NP-hard. Therefore, there must exist a polynomial algorithm which can solve the problem of optimizing the bi-objective function NT/Total Tardiness on parallel processing machines.

This implies that single criterion of total tardiness can be optimized in polynomial time, .i.e., total tardiness is not NP-hard. This is a contradiction as total tardiness is NP-hard.

Hence, the scheduling problem optimizing the bi-objective function NT/Total Tardiness on parallel processing machines is NP-hard.

5 Numerical Illustration

Optimize the NT with condition of total tardiness, whenever the processing times of jobs are in fuzzy environment with due time on parallel machines are as follows (Tables 1, 2):

Solution: The AHR of the processing time by using Yagers [12] formula of the given jobs is as shown in Table 2.

Table 1 Processing time of jobs in fuzzy environment

Jobs (i)	1	2	3	4	5	6
Processing time	(6,7,8)	(5,6,7)	(9,10,11)	(7,8,9)	(5,6,7)	(10,11,12)
Due date (d_i)	20/3	27/3	32/3	26/3	25/3	35/3

Table 2 AHR of processing time of jobs

Jobs (i)	1	2	3	4	5	6
Processing time	29/3	20/3	32/3	26/3	20/3	35/3
Due date (d_i)	20/3	27/3	32/3	26/3	25/3	35/3

On arranging the jobs in EDD order on parallel machines M_1, M_2 and M_3, we have (Table 3).

Therefore, total tardiness = 75/3 units and NT = 4.

Set of late jobs = $L = \{1, 2, 3, 6\}$ and set of jobs that cannot be switched $C = \varphi$.

On considering the 1st late job $i = 1 \in L$ and $1 \notin C$. Here, for the late job $j = 2 \in L$, we have $c_i = d_j$. Therefore, on exchanging jobs $i = 1 \in L$ and $j = 2 \in L$, setting $L = L - \{2\}$, the jobs schedule becomes (Table 4).

Therefore, total tardiness = 75/3 units and NT = 3.

Here, we observe that no further improvement in scheduling of jobs is possible.

Hence, the optimal sequence of jobs processing optimizing the bi-objective function NT/Total Tardiness on parallel machines is 2–5–4–1–3–6 with minimum total tardiness = 75/3 units and minimum NT as 3.

Table 3 Job scheduling with EDD order

Jobs (i)	1	5	4	2	3	6
M_1	0–29/3	–	–	–	–	29/3–64/3
M_2	–	0–20/3	–	20/3–40/3	–	–
M_3	–	–	0–26/3	–	26/3–58/3	–
d_i	20/3	25/3	26/3	29/3	32/3	35/3
T_i	9/3	–	–	11/3	26/3	29/3

Table 4 Reduced job scheduling table

Jobs (i)	2	5	4	1	3	6
M_1	0–20/3	–	–	20/3–49/3	–	–
M_2	–	0–20/3	–	–	20/3–52/3	–
M_3	–	–	0–26/3	–	–	26/3–61/3
d_i	29/3	25/3	26/3	20/3	32/3	35/3
T_i	–	–	–	29/3	20/3	26/3

6 Conclusion

The present chapter is aimed at developing heuristic algorithm to solve the bi-objective problem with total tardiness and NT as primary and secondary criteria more efficiently, in reasonable amount of time and with little conciliation on the optimality of the solution on parallel machines. In past, the processing time for each job was usually assumed to be exactly known. But, in many real-life situations, processing times may vary dynamically due to human factors or operating faults, and hence, the concept of fuzziness in processing time of jobs is introduced. For a given set of jobs initially arranged in EDD order, a late job needs to be considered for being exchanged only with another job or a job having the same due date in order to potentially improve the value of a secondary criteria, given the primary criteria of minimum total tardiness. The study may further be extended by generalizing the number of parallel machines and by introducing trapezoidal fuzzy membership function to represent the fuzziness in processing time.

References

1. Anghinolfi, D., Paolucci, M.: Parallel machine total tardiness scheduling problem. Comput. Oper. Res. **3**, 3471–3490 (2007)
2. Azizoglu, M., Kondacki, S., Omer, K.: Bicriteria scheduling problem involving total tardiness and total earliness penalties. Int. J. Prod. Econ. **23**, 17–24 (1991)
3. Akker Vanden, J.M., Hoogeveen, J.A.: Minimizing the number of tardy jobs. Hand Book of Scheduling. Chap-man and Hall/CRC, Boca Raton, Florida (2004)
4. Chand, S.: Schneerberger: A note on single machine scheduling problem with minimum weighted completion time and maximum allowable tardiness. Naval Res. Logistic **33**, 551–557 (1986)
5. Chen, C.L., Bulfin, R.L.: Scheduling a single machine to minimize two criteria: Maximum tardiness and number of tardy jobs. IIE Trans. **26**, 76–84 (1994)
6. Gupta, D., Sharma, S., Aggarwal, S.: Bi-objective parallel machine scheduling with uncertain processing time. Adv. Appl. Sci. Res. **3**, 1020–1226 (2012)
7. Lawler, E.L., Lenstra, J.K., Rinnooy Kan, A.H.G.: Minimizing maximum lateness in a two machine open shop. Math. Oper. Res. **6**, 153–158 (1981)
8. Moore, J.M.: Sequencing n jobs one machine to minimize the number of tardy jobs. Manage. Sci. **77**, 102–109 (1968)
9. Parkash, D.: Bicriteria scheduling problems on parallel machine. Ph.D. thesis. University of Birekshurg. Virginia (1997)
10. Shin, S.O., Kim, Y.D.: Scheduling on parallel machines to minimize total tardiness. J. Oper. Res. **177**, 629–634 (2007)
11. Singh, T.P.: Bi-objective in fuzzy scheduling on parallel machines. Arya Bhatta J. Math. Inf. **2**, 149–152 (2011)
12. Yager, R.R.: A procedure for ordering fuzzy subsets of unit interval. Inf. Sci. **24**, 143–161 (1981)
13. Zadeh, L.A.: Fuzzy sets. Inf. Control **8**, 78–98 (1965)

Inventory Model for Decaying Item with Continuously Variable Holding Cost and Partial Backlogging

Ankit Prakash Tyagi, Shivraj Singh and Rama Kant pandey

Abstract Holding costs are determined from thc investment in physical stocks and storage facilities for items during a cycle. In most of the research papers, holding cost rate per unit time for perishable inventory is assumed as constant. However, this is not necessarily the case when items in stock are decaying. In this work, paying better attention on the holding cost, we present a deteriorating inventory model in which the unit holding cost is continuously based on the deterioration of the inventory with the time the item is in stock. The deterioration rate is assumed as a Weibull distribution function. Declining market demand is considered in this paper. Shortages are allowed and partial backlogged. The partial backlogging rate is a continuous exponentially decreasing function of waiting time in purchasing the item during stock out period. Conditions for uniquely existence of global minimum value of the average total cost per unit time are carried out. Numerical illustration and sensitivity analysis are presented.

Keywords Inventory · Weibull deterioration · Partial backlogging · Continuously variable holding cost

---

A. P. Tyagi (✉) · R. K. Pandey
Department of Mathematics, D.B.S. (PG) College, Dehradun, Uttarakhand, India
e-mail: ankitprakashtyagi88@gmail.com

R. K. Pandey
e-mail: rkpandey0055@gmail.com

S. Singh
Department of Mathematics, D.N. (PG) College, Meerut, Uttar Pradesh, India
e-mail: shivrajpundir@gmail.com

B. V. Babu et al. (eds.), *Proceedings of the Second International Conference on Soft Computing for Problem Solving (SocProS 2012), December 28–30, 2012*, Advances in Intelligent Systems and Computing 236, DOI: 10.1007/978-81-322-1602-5_41,

1 Introduction

In daily life, deteriorating goods cannot use for long time. Foods, vegetables, fruits and pharmaceuticals are a few examples of such items. So, the loss due to deterioration is needed to give a batter consideration of inventory managers. First, Ghare and Schrader [1] considered continuously deteriorating inventory model with a constant demand. Later, Cover and Philip [2] assumed variable deterioration rate. In this study, they considered two-parameter Weibull distributive deterioration rate. Since then, a great deal of research has focused on variable deterioration rate in models.

The assumption of constant demand rate is not always applicable for decaying inventory. In reality, market demand of an item goes up in the growth phase of its life cycle. On the other hand, on introducing more attractive products consumer's preference may change. This causes demand of some items to decline. The consumer's confidence on quality of such products loses due to the age of the product. Therefore, the age of inventory has a negative pressure on demand. This phenomenon attracted numerous researchers to developed deteriorating models with time varying demand pattern.

Hollier and Mak [3] were first presented the consideration of exponentially decreasing demand. They also developed optimal policies under both conditions where replenishment intervals are constant and variable. In developing such inventory models, Goyal and Giri [4] provided a detail review of deteriorating inventory literatures.

In the mention above, most researchers assumed that shortage are completely backlogged. In practice, during the shortage period a few customers would like to wait for backlogging but the other would not. In daily life, a common observation is that, in the event of shortage, the proportion of consumers who will to purchase the item decreases as the waiting time increases. Therefore, backlogging rate should be variable and dependent on the waiting time for the next replenishment. Chang and Dey [5] investigated an inventory model with shortage. They assumed a variable backlogging rate which depends on the length of waiting for the next replenishment. Recently, Pentico and Drake [6] provide a prominent survey of deterministic models for the EOQ with partial backlogging.

In the consideration above, most researchers assumed that holding cost rate per unit time is constant. However, more sophisticated storage facilities and services may be needed for holding perishable items if they are kept for longer periods of time. So, in holding of decaying items, the assumption of constant holding cost rate is not always suitable. Weiss [7] noted that variable holding cost is appropriate when the value of an item decreases the longer it is in stock. Ferguson et al. [8] indicated that this type of model is suitable for perishable items in which price markdowns and removal of ageing product are necessary. Recently, Mishra and Singh [9] developed the inventory model for deteriorating items with time dependent linear demand and holding cost. To give attention on the concept of variability of the holding cost of decaying item, Tyagi et al. [10] developed an inventory lot-size model for decaying item following the power patterns of demand of item. In that study, shortages were allowed and partial backlogged inversely with the waiting time for the next replenishment.

On the basis of the discussion above, a question crops up that what optimal policy will be adopted by the inventory managers when demand of item follows the exponentially declining path and partial backlogging rate is also a exponentially decreasing function of the waiting time for the next replenishment? To give optimal policy in this situation, in this paper, an Economic Order Quantity (EOQ) inventory model of deteriorating item is considered with declining market demand. To extend such EOQ models, it is assumed continuously variable holding cost rate per unit per unit time based on deterioration. The deterioration rate of item is considered as two-parameter Weibull distribution function. Partial backlogging is allowed. The backlogging rate is an exponentially decreasing function of the waiting time for the next replenishment. In this paper, the primary problem is to minimize the average total cost per unit time by optimizing the shortage point per cycle. We also show that minimized objective function is convex and the obtained solution is uniquely determined. A numerical example is proposed to illustrate the model and the solution procedure. The sensitivity analysis of major parameters is performed in the last, here.

2 Notations and Assumptions

The following notations and assumptions are used throughout the whole paper.

2.1 Notations

$I(t)$ the inventory level at any time t, $t \geq 0$; T constant prescribed scheduling period or cycle length (time units); $I_{\max}$ maximum inventory level at the start of a cycle (units); S maximum amount of demand backlogged per cycle (units); t_1 duration of inventory cycle when there is positive inventory; Q order quantity (units/cycle); c_1 cost of the inventory item ($); c_2 fixed order cost ($/per order); c_3 shortage cost per unit backordered per unit time ($/unit/unit time); c_4 opportunity cost due to lost sales ($/unit).

2.2 Assumptions

In developing the mathematical model of the inventory system, the following assumptions are made: (1) the replenishment rate is infinite; (2) lead time is negligible; (3) the replenishment quantity and cycle length are constant for each cycle; (4) there is no replacement or repair of deteriorated items during a given cycle; (5) the time to deterioration of the item is Weibull distributed. So, the rate of deterioration is

$d(t) = \alpha\beta t^{\beta-1}$, where α and β are shape and scale parameters; (6) the demand rate $R_1(t)$ is known and decreases exponentially as $R_1(t) = De^{-\lambda t}$ for $I(t) > 0$ and $R_1(t) = D$ for $I(t) \leq 0$ where $D(> 0)$ is initial demand and $\lambda(> 0)$ is a constant governing the decreasing rate of the demand; (7) shortage are allowed. Unsatisfied demand is partially backlogged. The backlogging rate $B(t)$ which is a decreasing function of the waiting time t for next replenishment, we here assume that $B(t) = e^{-\delta t}$, where $1 > \delta \geq 0$ is backlogging parameter and t is the waiting time; (8) the holding cost rate per unit time is assumed continuously variable with storage period of the item. The holding cost $h(t)$ consists of fix holding charges and variable handling charges of item due to its deterioration. So, we here assume that $h(t) = R + He^{dt}$ which is increasing function of storage period t of item, where $R(> 0)$ is fix holding charge per unit, $H(> 0)$ is handling charge per unit time and $d(> 0)$ is deterioration rate governing the increasing path of handling charges.

3 Model Formulations

The inventory system goes like this: At $t = 0$, initial replenishment Q units are made, of which S units are delivered towards backorders, leaving a balance of $I_{\max}$ units in the initial inventory. From $t = 0$ and $t = t_1$ time units, the inventory level depletes due to both demand and deterioration. At t_1, the inventory level is zero. During the time $(T - t_1)$ the shortage is partially backlogged at the rate of $B(t)$ after receiving next lot and left part of demand during shortage is lost. That is, only the backlogging items are replaced by the next replenishment.

The inventory function with respect to time can be determined by evaluating the differential equations

$$\frac{dI(t)}{dt} + d(t)I(t) = -R_1(t); \quad 0 \leq t \leq t_1, \tag{1}$$

and

$$\frac{dI(t)}{dt} = -DB(t); \quad t_1 \leq t \leq T. \tag{2}$$

with the boundary conditions $I(0) = I_{\max}$ and $I(t_1) = 0$.

The solutions of (1) and (2) are

$$I(t) = D\left[(t_1 - t) - \lambda\left(\frac{t_1^2}{2} - \frac{t^2}{2}\right) + \frac{\alpha}{\beta+1}\left(t_1^{\beta+1} - t^{\beta+1}\right)\right]\left(1 - \alpha t^{\beta}\right); 0 \leq t \leq t_1, \tag{3}$$

and

$$I(t) = -\frac{D}{\delta}\left[e^{-\delta(T-t)} - e^{-\delta(T-t_1)}\right]; \quad t_1 \leq t \leq T. \tag{4}$$

The maximum inventory level at the starting point of the cycle is

$$I_{\max} = I(0) = I(t) = D\left[t_1 - \frac{\lambda t_1^2}{2} + \frac{\alpha t_1^{\beta+1}}{\beta+1}\right]. \tag{5}$$

The maximum amount of demand backlogged per cycle can be obtained as

$$S = -I(T) = \frac{D}{\delta}\left[1 - e^{-\delta(T-t_1)}\right]. \tag{6}$$

So, from (5) and (6), the order quantity per cycle is

$$Q = I_{\max} + S = D\left(t_1 - \frac{\lambda t_1^2}{2} + \frac{\alpha t_1^{(1+\beta)}}{(1+\beta)}\right) + \frac{D}{\delta}\left[1 - e^{-\delta(T-t_1)}\right]. \tag{7}$$

The average total cost per unit time per cycle consists of ordering cost per cycle, holding cost per cycle, deterioration cost per cycle, shortage cost per cycle and opportunity cost per cycle. Now, the ordering cost (OC) per order is c_2.

The inventory holding cost (HC) per cycle is

$$= \int_0^{t_1} h(t)I(t)dt \tag{8}$$

The deterioration cost (DC) per cycle is

$$= \int_0^{t_1} c_1 d(t)I(t)dt \tag{9}$$

The shortage cost (SC) per cycle is

$$= \int_{t_1}^{T} c_3\{-I(t)\}dt \tag{10}$$

And, the opportunity cost (OPC) due to lost sales per cycle is

$$= \int_{t_1}^{T} [1 - B(t)]Ddt \tag{11}$$

The objective of this model is to determined the optimal value (t_1^*) of t_1 in order to minimize the average total cost per unit time is

$$ATC(t_1) = \frac{[OC + HC + DC + SC + OPC]}{T} \tag{12}$$

The necessary condition for average total cost per unit time $ATC(t_1)$ to be minimized is

$$\frac{dATC(t_1)}{dt_1} = \frac{Df(t_1)}{T} = 0, \tag{13}$$

where

$$\begin{aligned} f(t_1) = &\left[(R+H)\left(t_1 - \lambda t_1^2\right) + \frac{\alpha\beta(R+2H)t_1^{\beta+1}}{(1+\beta)} + \frac{\{R - H(\beta-1)\}\alpha\lambda t_1^{\beta+2}}{(1+\beta)}\right] \\ &+ c_1\alpha\left[t_1^{\beta} - \lambda t_1^{\beta+1}\right] - \frac{\left(c_4\delta - c_3 e^{-\delta(T-t_1)}\right)}{\delta} - (T-t_1)c_3 e^{-\delta(T-t_1)} \\ &- \frac{(c_3 - \delta c_4)\, e^{-\delta(T-t_1)}}{\delta}. \end{aligned}$$

Now, our main concern is to know about the existence of the solution of (13) that will be the inner point in the interval $[0, T]$ at which average total cost per unit time is at minimum value, globally. This is analogous to show that the solution of $f(t_1) = 0$ uniquely exists.

Theorem 1 *If $R > H(\beta - 1)$ and $1 > \lambda T$, then the solution of $f(t_1) = 0$ not only exists but also is uniquely determined as an inner point of $[0, T]$.*

Theorem 2 *If $R + H + \alpha\beta(R + 2H)t_1^{\beta} + \frac{\alpha\lambda R(\beta+2)t_1^{\beta+1}}{(1+\beta)} + (c_3 + \delta c_4)e^{-\delta(T-t_1)}$ $+c_1\alpha[\beta t_1^{\beta-1} - \lambda(1 + \beta)t_1^{\beta}] > \frac{\alpha\lambda H(\beta-1)(\beta+2)t_1^{(\beta+1)}}{(\beta+1)} + 2(R + H)\lambda t_1$ $+\delta(T - t_1)c_3e^{-\delta(T-t_1)}$, the average total cost per unit time $ATC(t_1)$ is convex and reaches its global minimum at point t_1^*.*

Next, by using t_1^*, we can obtain the optimal maximum inventory level, the optimal order quantity and the minimum average total cost per unit time from (5), (7) and (12).

4 Numerical Illustrations

To illustrate the preceding discussion we consider the following example.

Example 1 We consider an inventory system which verifies the assumptions described above. The randomly chosen input data of parameters are $T = 1, \alpha = 0.8$

$$R = 2,\ H = 0.4,\ \delta = 0.1,\ \lambda = 0.4,\ D = 100,\ \beta = 2,\ c_1 = 3,\ c_2 = 10,\ c_3 = 3 \quad \text{and} \quad c_4 = 7.$$

The optimal value $t_1^* = 0.516906$ unit time of t_1 are calculated by MATHEMATICA 8.0 for the proposed model. By using the optimal value t_1^*, the minimum average total cost per unit time $ATC(t_1^*) = \$91.7083$ is obtained and Theorem 2 is satisfied. We can obtain the optimal value $Q^* = 97.1908$ units of ordering quantity Q per cycle.

5 Sensitivity Analysis

Here, taking one parameter at a time and keeping the remaining parameters unchanged, we have studied the effect of changes ($\pm$ 5 % and $\pm$ 10 %) in the values of some parameters α, β, R and H on optimal shortage point t_1^*, optimal order quantity Q^* and

Table 1 Effect of changes in the parameters of the inventory model

Parameters	% Change	% Change in the value of		
		t_1^*	Q^*	$ATC(t_1^*)$
$R = 2$	+10	−2.38	−0.06	+2.67
	+5	−1.20	−0.03	+1.35
	−5	+1.23	+0.03	−1.38
	−10	+2.49	+0.07	−2.80
$H = 0.4$	+10	−0.54	−0.015	+0.57
	+5	−0.27	−0.007	+0.29
	−5	+0.27	+0.007	−0.29
	−10	+0.54	+0.015	−0.58
$\alpha = 0.8$	+10	−1.74	+0.31	+1.29
	+5	−0.88	+0.15	+0.65
	−5	+0.91	−0.16	−0.67
	−10	+1.87	−0.31	−1.36
$\beta = 2$	+10	+2.14	−0.64	−2.33
	+5	+1.09	−0.33	−1.21
	−5	−1.15	+0.35	+1.32
	−10	−2.36	+0.74	+2.75

the minimum average total coast per unit $ATC(t_1^*)$. Example 1 is used and results are shown in Table 1. From Table 1, it is clearly observed that when the values of parameters R and H of holding cost rate increase or decrease, the optimal value of $ATC(t_1^*)$ increase or decrease. But this trend is reversed for the solutions t_1^* and Q^*. The main reason is that when R and H increase, the holding cost will increase. Therefore, inventory managers reduce the order quantity and consequently the shortage point is reduced. Finally, it also increases the average total cost for the inventory system.

6 Conclusion

In this paper we study an inventory model where the inventory level is depleted not only by exponentially decreasing demand but also by Weibull distributive deterioration, in which holding cost per unit time is considered a continuously variable function depends upon item's deterioration nature. Shortages are allowed and partially backlogged. Therefore, the proposed model can be used in inventory controlling of certain perishable items like food items, electronic components and other fashionable products. Moreover, the advantage of the proposed inventory model is that the behavior of the model illustrated by the help of given example is easy to understand by sensitivity analysis due to major parameters. From sensitivity analysis, it is shown that the optimal order quantity is highly sensitive to changes in the value of α and β, on the other hand the optimal value of average total cost per unit time is highly sensitive to changes in the value of R, α and β as well as slightly sensitive to changes in the value of H. According this situation, inventory managers have to take decision to place an order on the basis of decaying nature of goods after setting a justifying level of average total cost that can be accepted by their organization. As far as the future researches biased on this study are concerned, this paper can be extended with stochastic demand and permissible delay in payment.

References

1. Ghare, P.M., Schrader, G.H.: A model for exponentially decaying inventory system. Int. J. Prod. Econ. **21**, 449–460 (1963)
2. Covert, R.B., Philip, G.S.: An EOQ model with weibull distribution deterioration. AIIE Trans. **5**, 323–326 (1973)
3. Hollier, R.H., Mak, K.L.: Inventory replenishment policies for deteriorating items in a declining market. Int. J. Prod. Econ. **21**, 813–826 (1983)
4. Goyal, S.K., Giri, B.C.: Recent trends in modeling of deteriorating inventory. Eur. J. Oper. Res. **134**, 1–16 (2001)
5. Chang, H.J., Dye, C.Y.: An EOQ model for deteriorating items with time varying demand and partial backlogging. J. Oper. Res. Soc. **50**, 1176–1182 (1999)
6. Pentico, D.W., Drake, M.J.: A survey of deterministic models for the EOQ and EPQ with partial backlogging. Eur. J. Oper. Res. **214**(2), 179–198 (2011)

7. Weiss, H.J.: Economic order quantity models with nonlinear holding costs. Eur. J. Oper. Res. **9**(1), 56–60 (1982)
8. Ferguson, M., Hayaraman, V., Souza, G.C.: Note: an application of the EOQ model with nonlinear holding cost to inventory management of perishables. Eur. J. Oper. Res. **180**(1), 485–490 (2007)
9. Mishra, V.K., Singh, L.S.: Deteriorating inventory model for time dependent demand and holding cost with partial backlogging. Int. J. Manage. Sci. Eng. Manage. **6**(4), 267–271 (2011)
10. Tyagi, A.P., Pandey, R.K., Singh, S.R.: Optimization of inventory model for decaying item with variable holding cost and power demand. Proceedings of National Conference on TAME, pp 774–781 (2012). ISBN: 978-93-5087-574-2

The Value of Product Life-Cycle for Deteriorating Items in a Closed Loop Under the Reverse Logistics Operations

S. R. Singh and Neha Saxena

Abstract Owing to its strategic implications, reverse logistics has received much attention in recent years. Growing green concerns and advancement of reverse logistics concepts make it all the more relevant who can be achieved through the End-of-Life (EoL) treatment. In the proposed model, we develop a production inventory model with the reverse flow of the material. Here we determined the value of product life cycle with EoL scenario where the reverse logistics operations deal with the collection, sorting, cleaning, dissembling, and remanufacturing of the buyback products. The purpose of this paper is to develop an effective and efficient management of product remanufacturing. As a result, in this article, we establish a mathematical formulation of the model to determine the optimal payment period and replenishment cycle. Illustrative examples, which explain the application of the theoretical results as well as their numerical verifications, are also given. Finally, the sensitivity analysis is reported.

Keywords Production · Reverse operations · Deterioration · Short life cycle products · Collection investment

1 Introduction

Supply chain management has received remarkable attention both from the business world and from academic researchers. Most of the research concentrates on the forward movement of supply chain and transformation of the materials from the suppliers to the end consumer. However, rapid developments in technology, the

S. R. Singh · N. Saxena (✉)
D. N. College, Meerut, India
e-mail: shivrajpundir@gmail.com

N. Saxena
e-mail: nancineha.saxena@gmail.com

B. V. Babu et al. (eds.), *Proceedings of the Second International Conference on Soft Computing for Problem Solving (SocProS 2012), December 28–30, 2012*, Advances in Intelligent Systems and Computing 236, DOI: 10.1007/978-81-322-1602-5_42,

emergence of new industrial products and shortened product life cycles have resulted in an increasing number of discarded products and caused growing environmental problems in the developed world. Due to the governmental regulations and consumer concerns regarding these environmental issues, an increasing number of companies have focused on reduction efforts in the amount of waste stream, diversion of the discarded products and disposition of the retired products properly. Enforced legislation and customer expectations increasingly force manufacturers to take back their products after use, which can be achieved through the collection investment. The collection investment represents the monetary amount of effort (e.g., promotion, marketing) that the recycled-material supplier applies to the end-user market to create the necessary incentive to receive targeted returns. This subject is related to the concept of reverse logistics. Due to this awareness manufacturers and researchers in many countries have been paid much attention to the reverse flow of products from consumers to upstream businesses interest. The Reverse logistics is the process of retrieving the product from the end consumer for the purposes of or proper disposal. A Reverse Production System includes collection, sorting, and remanufacturing processes for end-of-life products. Reverse distribution can take place through the original forward channel, through a separate reverse channel, or through combinations of the forward and the reverse channel.

The green supply chain, which links the natural environment both with the forward and reverse supply chain, has a growing stream of research and is quickly becoming a well-established field of its own. This development has stimulated a number of companies to explore options for take-back and recovery of their products. There are two types of reverse logistics (RL) classified on the basis of the degree of the openness in its network. One of the two classifications is Open-loop structure and the other one is Closed-loop System. In the open-loop RL system, the products from the end user do not return to the original manufacturers or suppliers. The products are taken away by the third logistics party for the purpose of waste reduction, resale etc. while in case of closed-loop RL system products get returned to the original manufacturers or suppliers, for the purpose of repair, reformation or reuse. In need of repair or renovation it usually points the original source, belonging to the closed-loop structure.

In the past recent years, a growing environmental consciousness enforce the researchers to be more environmental responsible. A lot of work has been done in the field of RL. There are very few models treating forward and reverse distribution simultaneously. Schrady [16] was the first who determined both of the system, reverse flow of material with forward system. He considers the traditional Economic Order Quantity (EOQ) model for repairable items assuming that the manufacturing and recovery (repair) rates are instantaneous. This model was generalized by Nahmias and Rivera [13] for the case of finite repair rate and limited storage in the repair and production shops. Another extension of the model of Schrady [16] was made by Mabini et al. [12]. Ishii et al. [8] developed a model and demonstrate the need of life-cycle design to maximize the life-cycle value of a product at the initial stages of design. Koh et al. [10] generalized the model of Nahmias and Rivera [13] by assuming a limited repair capacity. Dobos and Richter [5] explore a RL inventory system with non instantaneous production and remanufacturing rate. Dobos and Richter [6] generalized their earlier work (2003) to the case of multiple remanufacturing and production cycle. Dobos and Richter [7] extended their previous model and assumed that the quality of collected returned items is not always suitable for further repairing [14]. In a further study a closed-loop supply chain for the returned items is developed by Savaskan et al. [15] assuming that the returned rate depends on the Collection investment. Dekker et al. [4] proposed a quantitative model for closed loop supply chain. He investigated that the amount of returns is highly uncertain and this uncertainty greatly affect the collection and inventory decisions. Bayındır et al. [2] investigated the level of the desired recovery effort with the imperfect recovery process. King et al. [9] defined the term repair as the correction of specified faults in a product, where the quality of repaired products is inferior to those of remanufactured. Srivastava [19] generalized an overview in green supply chain. He showed that RL is a complex process to achieve greater economic benefits. El Saadany and Jaber investigated the model by assuming that the collection rate of returned items is dependent on the purchasing price and the acceptance quality level of these returns. That is, the flow of buyback items increases as the purchasing price increases, and decreases as the corresponding acceptance quality level increases. Konstantaras and Skouri [11] generalized the model by considering a general cycle pattern in which a variable number of reproduction lots of equal size are followed by a variable number of manufacturing lots of equal size. They also have studied the case where shortages are allowed in each manufacturing and reproduction cycle. Alamri [1] proposed a general reverse Logistics inventory model for the optimal returned quantity with deteriorated items. Singh and Saxena [18] developed a RL inventory model for stock out situation. Along the same line Singh et al. [17] developed there model for the flexible manufacturing under the stock out situation. Green supply chain inventory model with short life cycle product is developed by Chung and Wee [3]. This paper differs from the previous research, since in this study a closed loop system in reverse logistics is considered. In this article, we have developed the model for the short life cycle products.

In this paper, a closed loop system for integrated production of new items and remanufacturing of returned items is presented for an infinite planning horizon. The effect of deterioration is taken as under consideration. We developed a model assuming the coordination of joint production and reproduction options by producing new items and reproducing the returned items to quality standards that are "as-good-as" those of new products. In this model, the demand of customer is satisfied by the serviceable stock which is either produced or remanufactured items from the market the buyback products are subjected to the sorting, cleaning, and dissembling from where a constant ratio of the products (repairable stock) that confirms the certain quality standard are collected to be remanufactured and the rest is salvaged. The process is going on. A general framework of such a system is depicted in Fig. 1. The next section is for assumption and notations. Section 3 is for the formulation of the model. In Sect. 4 we have determined the solution procedures for the model. The numerical example to illustrate the model and sensitivity analysis is presented

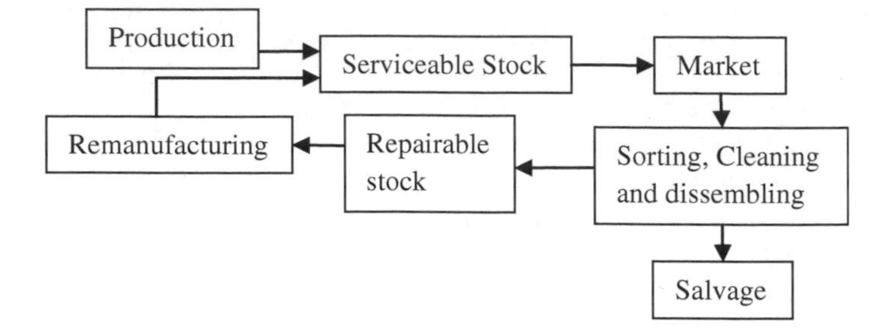

Fig. 1 Material flow in a reverse logistics inventory model

in Sect. 5. Concluding remarks are derived and future research topics are suggested in Sect. 6.

2 Assumption and Notations

Notations for the forward logistics

- $I_m(t)$ = Inventory level at time t in manufacturing stock.
- P_m = The production rate.
- d = The demand rate (satisfied from the newly produced and reproduced items).
- δ_m = Deterioration rate.
- α = Scaling parameter, production formulation.
- Cost parameters for the manufacturing stock are as follows.

  C_m = Unit item cost.
  K_m = Fixed unit production cost.
  S_m = Variable unit production cost.
  H_m = Unit holding cost.

Notations for reverse logistic

- $I_r(t)$ = Inventory level at time t in remanufacturing process.
- $I_m(t)$ = Inventory level at time t in production process.
- P_r = The reproduction rate.
- R = The returned rate.
- β = Scaling parameter, remanufacturing formulation.
- γ = Scaling parameter, collection investment formulation.
- η = Scaling parameter, salvage formulation.
- The cost parameters for the reproduced stock are as follows.

  F_r = Fixed unit reproduction cost.
  S_r = Variable unit reproduction cost.
  H_r = Unit holding cost.

- δ_r = Deterioration cost.
- The cost parameters for the returned stock are as follows.

  C_R = Unit returned item cost.
  H_R = Unit holding cost.

- δ_R = Deterioration cost.
- S_{av} = Salvage.
- M = number of the life cycles before the component is recycled or disposed off.
- CI = Collection investment.
- F_{cl} = fixed cost including cleaning and disassembly cost during the collecting process.
- C_{cl} = variable cost including cleaning and disassembly cost during the collecting process.
- A_{df} = fixed component life-cycle design cost ratio for the green design.
- B_{dv} = variable component life-cycle design cost ratio for the green design.
- C_d = component life-cycle design cost for the green design.
- r_j = reliability of the sub function j.

Assumptions

- Production and remanufacturing rate taken to be demand dependent as

$$P_r = \alpha \mathrm{d}, \quad \alpha > 1$$
$$P_m = \beta \mathrm{d}, \quad \beta > 1$$

- Returned items are collected at a rate R determined by the collection investment CI and demand. The collection investment represents the economical amount of effort (e.g., promotion, marketing) that the supplier applies to the end-user to create the necessary incentive to receive targeted returns. $R = \left(\sqrt{CI/\gamma}\right)\mathrm{d}$ where γ is a scaling parameter and $\sqrt{CI/\gamma} < 1$.

3 Formulation of the General Model

The change in inventory of Production and Remanufacturing house is depicted in Fig. 2. In the remanufacturing house the reproduction starts at time T_0 and the inventory level rises up at a rate $P_r - \mathrm{d} - \delta_r I_r(t)$. At the time T_1 reproduction stops then the stock level wind up at a rate $-\mathrm{d} - \delta_r I_r(t)$ to the time T_2. At the same time due to the production the stock level rises up at a rate $P_m - \mathrm{d} - \delta_m I_m(t)$ in the Production house and at the time T_3 when the production stops the stock level starts to decrease at a rate $-\mathrm{d} - \delta_m I_m(t)$ up to the time T_4. Now the stock level of the returned items assumed to be remanufactured is start to decreasing at a rate $\eta R - P_r - \delta_R I_R(t)$ up

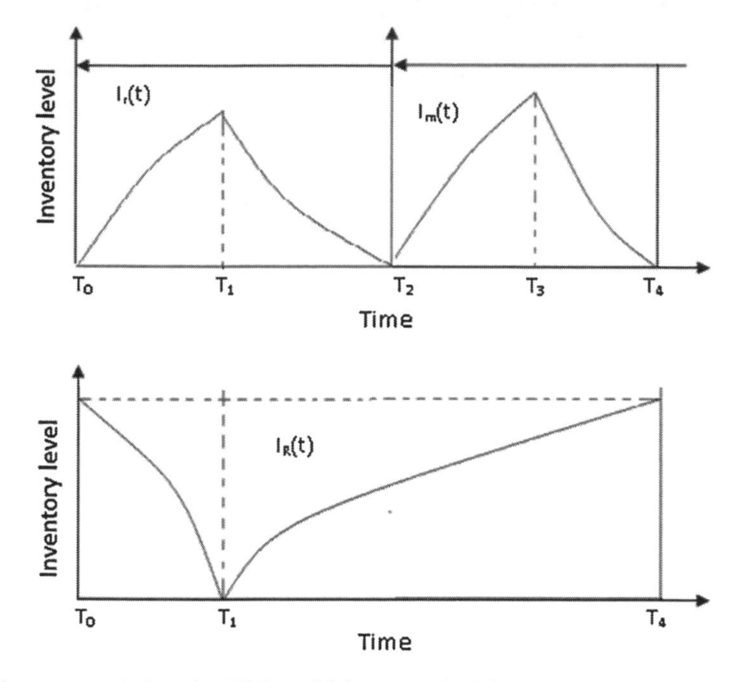

Fig. 2 Inventory variation of an EPQ model for reverse logistics system

to the time T_1 and after the end of the reproduction the stock level is raising at a rate up to the time T_4

The changes in the inventory levels depicted in Fig. 2 are governed by the following differential equations:

$$I_r'(t) + \delta_r I_r(t) = P_r - \mathrm{d}, \quad \text{With the ending condition} \quad I_r(T_0) = 0 \quad 0 \le t \le T_1 \tag{1}$$

$$I_r'(t) + \delta_r I_r(t) = -\mathrm{d}, \quad \text{With the ending condition} \quad I_r(T_2) = 0 \quad T_1 \le t \le T_2 \tag{2}$$

$$I_r'(t) + \delta_m I_m(t) = P_m - \mathrm{d}, \quad \text{With the initial condition} \quad I_m(T_2) = 0 \quad T_2 \le t \le T_3 \tag{3}$$

$$I_m'(t) + \delta_m I_m(t) = -\mathrm{d}, \quad \text{With the ending condition} \quad I_m(T_4) = 0 \quad T_3 \le t \le T_4 \tag{4}$$

$$I_R'(t) + \delta_R I_R(t) = \eta R - P_r, \quad \text{With the ending condition} \quad I_R(T_1) = 0 \quad 0 \le t \le T_1 \tag{5}$$

$$I_R'(t) + \delta_R I_R(t) = \eta R, \quad \text{With the initial condition} \quad I_R(T_1) = 0 \quad T_1 \le t \le T_4 \tag{6}$$

The solution of the above differential equations is as follows:

$$I_r(t) = \left\{ \frac{P_r - \mathrm{d}}{\delta_r} \right\} (1 - e^{-\delta_r t}) \quad 0 \le t \le T_1 \tag{7}$$

$$I_r(t) = \left\{\frac{d}{\delta_r}\right\} (e^{\delta_r(T_2-t)} - 1) \quad T_1 \le t \le T_2 \tag{8}$$

$$I_m(t) = \left\{\frac{P_m - d}{\delta_m}\right\} (1 - e^{\delta_m(T_2-t)}), \quad T_2 \le t \le T_3 \tag{9}$$

$$I_m(t) = \left\{\frac{d}{\delta_m}\right\} (e^{\delta_m(T_4-t)} - 1) \quad T_3 \le t \le T_4 \tag{10}$$

$$I_R(t) = \left\{\frac{P_r - \eta R}{\delta_R}\right\} (e^{\delta_R(T_1-t)} - 1) \quad 0 \le t \le T_1 \tag{11}$$

$$I_R(t) = \left\{\frac{\eta R}{\delta_R}\right\} (1 - e^{\delta_R(T_1-t)}), \quad T_1 \le t \le T_4 \tag{12}$$

Now the per cycle cost components for the given inventory system are as follows.
Item cost: $C_m \int_{T_2}^{T_3} P_m \mathrm{d}u + C_R \int_0^{T_4} R\mathrm{d}u$ this cost includes the deterioration cost

Production cost: $K_m + S_m \int_{T_2}^{T_3} P_m \mathrm{d}u$

Holding cost $= h_r [I_r(0, T_1) + I_r(T_1, T_2)] + h_m [I_m(T_2, T_3) + I_m(T_3, T_4)] + h_R [I_R(0, T_1) + I_R(T_1, T_4)]$

Remanufacturing cost $= K_r + \frac{F_r}{M} + MS_r \int_0^{T_1} P_r \mathrm{d}u + F_{cl} + C_{cl} \int_0^{T_4} R\mathrm{d}u$

Salvage $= \int_0^{T_4} (1 - \eta) R\mathrm{d}u$

Design life cost $= C_D \left\{\frac{A_{DF}}{M} + MB_{DV} \prod_{j=1}^{2} r_j\right\}$
Total cost = cost for the forward supply chain + cost for the reverse supply chain

$$\begin{aligned} Z(T_1, T_2, T_3, T_4, M) = \frac{1}{T_4} \Bigg[& C_m \int_{T_2}^{T_3} P_m \mathrm{d}u + C_R \int_0^{T_4} R\mathrm{d}u + K_m + S_m \int_{T_2}^{T_3} P_m \mathrm{d}u \\ & + K_r + \frac{F_r}{M} + MS_r \int_0^{T_1} P_r \mathrm{d}u + F_{cl} + C_{cl} \int_0^{T_4} R\mathrm{d}u \\ & - \int_0^{T_4} (1 - \eta) R\mathrm{d}u + CI + C_D \left\{\frac{A_{DF}}{M} + MB_{DV} \prod_{j=1}^{2} r_j\right\} \\ & + h_R \left\{ \int_{T_0}^{T_1} \left\{\frac{P_r - \eta R}{\delta_R}\right\} (e^{\delta_R(T_1-t)} - 1)\mathrm{d}t \right. \end{aligned}$$

$$+\int_{T_0}^{T_1}\left\{\frac{\eta R}{\delta_R}\right\}(1-e^{\delta_R(T_1-t)})\mathrm{d}t\Bigg]$$

$$+h_m\Bigg[\int_{T_2}^{T_3}\left\{\frac{P_m-\mathrm{d}}{\delta_m}\right\}(1-e^{\delta_m(T_2-t)})\mathrm{d}t$$

$$+\int_{T_3}^{T_4}\left\{\frac{\mathrm{d}}{\delta_m}\right\}(e^{\delta_m(T_4-t)}-1)\mathrm{d}t\Bigg]$$

$$+h_r\Bigg[\int_{T_0}^{T_1}\left\{\frac{P_r-d}{\delta_r}\right\}(1-e^{-\delta_r t})dt$$

$$+\int_{T_1}^{T_2}\left\{\frac{d}{\delta_r}\right\}(e^{\delta_r(T_2-t)}-1)dt\Bigg]\Bigg]$$

$$\begin{aligned}Z(T_1,T_2,T_3,T_4,M)=\frac{1}{T_4}\Bigg[&\mathrm{d}\left\{\frac{H_r\beta}{\delta_r}-\frac{H_R\beta}{\delta_R}+MS_r\beta\right\}T_1\\&+\mathrm{d}\left\{-C_m\alpha-S_m\alpha-\frac{H_r}{\delta_r}-\frac{H_m(\alpha-1)}{\delta_m}\right\}T_2\\&+\mathrm{d}\left\{C_m\alpha+S_m\alpha+\frac{H_m}{\delta_m}+\frac{H_m(\alpha-1)}{\delta_m}\right\}T_3\\&+d\left\{\left(-1+\eta+C_R+C_{cl}+\frac{H_r}{\delta_r}\right)\sqrt{\frac{CI}{\gamma}}-\frac{H_m}{\delta_m}\right\}T_4\\&+\left\{K_m+K_r+CI+F_{cl}+\frac{F_r}{M}+C_D\left(\frac{A_{DF}}{M}+MB_{DV}r_1r_2\right)\right\}\Bigg]\end{aligned}\tag{13}$$

The total cost can be rewriting the cost function as follows:

$$Z(T_1,T_2,T_3,T_4,M)=\frac{1}{T_4}[AT_1+BT_2+CT_3+DT_4+E]\tag{14}$$

Now we have to find the value of T_1, T_2, T_3 and T_4 that minimize $Z(T_1,T_2,T_3,T_4)$. But there are some relations between the variables as follows:

$$0<T_1<T_2<T_3<T_4\tag{15}$$

$$\left\{\frac{P_r-\mathrm{d}}{\delta_r}\right\}(1-e^{-\delta_r T_1})=\left\{\frac{d}{\delta_r}\right\}(e^{\delta_r(T_2-T_1)}-1)\tag{16}$$

$$\left\{\frac{P_m - \mathrm{d}}{\delta_m}\right\}(1 - e^{\delta_m (T_2 - T_3)}) = \left\{\frac{d}{\delta_m}\right\}(e^{\delta_m (T_4 - T_3)} - 1) \tag{17}$$

$$\left\{\frac{P_r - \eta R}{\delta_R}\right\}(e^{\delta_R (T_1)} - 1) = \left\{\frac{\eta R}{\delta_R}\right\}(1 - e^{\delta_R (T_1 - T_4)}) \tag{18}$$

Thus our purpose is to minimize the total cost $Z\ (T_1, T_2, T_3, T_4, M)$ subject to the Eqs. (15), (16), (17) and (18).

4 Solution Procedure

Let Q be the acceptable returned quantity for used items in the interval $[0, T_4]$ then

$$Q = \int_0^{T_4} R dt \tag{19}$$

From (19), we note that T_4 can be determined as a function of Q, say

$$T_4 = f_4(Q) = \frac{Q}{D\sqrt{\frac{CI}{\gamma}}} \tag{20}$$

From which and (18) we find that T_1 can be determined as a function of T_4, hence of Q, say

$$T_1 = f_1(Q) = \frac{1}{\delta_r}\left[\log\left\{\frac{\beta}{\beta - \eta\sqrt{CI/\gamma}\left(1 - e^{\delta_r T_4}\right)}\right\}\right] \tag{21}$$

From which and (16) , T_2 can be determined as a function of T_1, hence of Q, say

$$T_2 = f_2(Q) = \frac{1}{\delta_r}\left[\log\left\{\beta e^{\delta_r T_1} - \beta + 1\right\}\right] \tag{22}$$

From which, (17), T_3 can be determined as a function of T_2 and T_4 hence of Q, say

$$T_3 = f_3(Q) = \frac{1}{\delta_m}\left[\log\left\{\frac{1}{\alpha}\left(e^{\delta_m T_4} + (\alpha - 1)\, e^{\delta_m T_2}\right)\right\}\right] \tag{23}$$

Thus, if we substitute (20), (21), (22), (23) in (14) then the problem will be converted to the following unconstrained problem with the variable Q.

$$W(Q, M) = \frac{1}{f_4}\left[Af_1 + Bf_2 + Cf_3 + Df_4 + E\right] \tag{24}$$

$$\frac{\partial W(Q,M)}{\partial Q} = \left[A\left(f_4 f_1' - f_1 f_4'\right) + B\left(f_4 f_2' - f_2 f_4'\right) + C\left(f_4 f_3' - f_3 f_4'\right) - E f_4'\right] = 0 \tag{25}$$

By which

$$f_4' = \frac{1}{d}\sqrt{\frac{\gamma}{CI}} \tag{26}$$

$$f_1' = \left\{\frac{\eta\sqrt{CI/\gamma}}{\beta - \eta\sqrt{CI/\gamma}\left(1 - e^{-\delta_R f_4}\right)}\right\} f_4{}' e^{-\delta_R f_4} \tag{27}$$

$$f_2' = f_1'\beta e^{\delta_r (f_1 - f_2)} \tag{28}$$

$$f_3' = \frac{1}{\alpha}\left\{f_4' e^{\delta_m (f_4 - f_3)} + (\alpha - 1) f_2' e^{\delta_m (f_2 - f_3)}\right\} \tag{29}$$

Here f_i' is the differentiation of f_i

Now

$$\frac{\partial W(Q,M)}{\partial M} = \frac{-F_r}{M^2} + S_r d f_1 \beta + C_D\left\{\frac{-A_{DF}}{M^2} + B_{DV}\left(r_1 r_2\right)\right\} = 0 \tag{30}$$

From Eq. (24) we can see that the total cost function is the function of $M$and Q only. While the value of M and Q can be find from the Eqs. (25) and (26). Since M is a discrete variable therefore the optimal value of M is find by satisfying the equation $W(Q^*, M-1) \le W(Q^*, M) \le W(Q^*, M+1)$ where $Q^*$is the optimal value of Q. Therefore, now our aim is to find the optimal values of $M$and Q such that

Minimize$W(Q, M)$

$$\text{Subject to: } W(Q^*, M-1) \le W(Q^*, M) \le W(Q^*, M+1). \tag{31}$$

5 Numerical Example and Sensitivity Analysis

The above theoretical results are illustrated through the numerical verification. The example is based on the following parametric values. We have considered the input parameters in appropriate units:

$d = 2500$, $C_m = 2.5$, $C_R = 1$, $K_m = 800$, $H_m = 1.5$, $H_r = 1$, $H_R = 1$, $S_m = 2$, $\alpha = 1.5$, $\beta = 1.8$, $\gamma = 196$, $\eta = 0.9$, $\delta_m = 0.1$, $\delta_r = 0.1$, $\delta_R = 0.1$, $C_{cl} = 0.25$, $F_{cl} = 500$, $F_r = 100$, $S_r = 1$, $A_{df} = 8$, $B_{dv} = 1$, $C_d = 50$, $r_1 = 0.99$, $r_2 = 0.98$, $S_{av} = 3$, $CI = 100$.

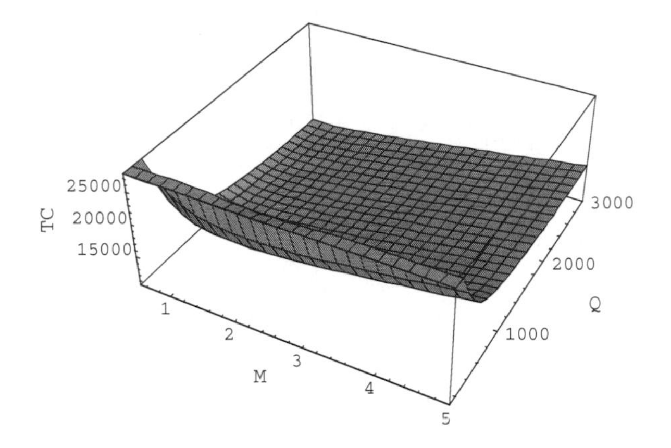

Fig. 3 Convexity of the green supply chain

Table 1 Sensitivity analysis for the scaling parameter (production formulation) α

α	M	T_1	T_2	T_3	T_4	Q	TC
1.3	1	0.569041	1.00205	1.52977	1.6828	3005	10829.3
1.4	1	0.561878	0.989702	1.47332	1.6604	2965	10865.9
1.5	1	0.555961	0.979495	1.42595	1.664192	2932	10897.3
1.6	1	0.550935	0.970821	1.38546	1.62624	2904	10924.5
1.7	1	0.546622	0.963377	1.35048	1.6128	2880	10948.4

Table 2 Sensitivity analysis for the scaling parameter (remanufacturing formulation) β

β	M	T_1	T_2	T_3	T_4	Q	TC
1.6	1	0.65162	1.02314	1.48488	1.708	3050	10747.6
1.7	1	0.599931	0.999445	1.45291	1.67216	2986	10827.8
1.8	1	0.555961	0.979495	1.42595	1.664192	2932	10897.3
1.9	1	0.518284	0.962829	1.40347	1.61672	2887	10958.2
2.0	1	0.485444	0.948409	1.384	1.59488	2848	11012.1

Under the given values of the parameters in above section, the optimal acceptable returned quantity for the used items is $Q* = 2932$ units and the optimal number of life-cycles $M = 1$ which is accumulated by the time $T_1* = 0.555961$, $T_2* = 0.979495$, $T_3* = 1.42595$ and $T_4* = 1.64192$ by which the minimum relevant cost is as $TC* = 10897.3$. The convexity of the problem is shown in Fig. 3.

Sensitivity Analysis

In order to study the effect of various parameters on the optimal policy, a sensitivity analysis is done and results are presented in the following tables.

Observations

When the parameters fluctuate, the following observations have been made during the sensitivity analysis.

Table 3 Sensitivity analysis for the scaling parameter (returned formulation) η

η	M	T_1	T_2	T_3	T_4	Q	TC
0.88	1	0.545717	0.961814	1.42566	1.64976	2946	10896.5
0.89	1	0.552093	0.970678	1.42581	1.64584	2939	10896.9
0.90	1	0.555961	0.979495	1.42595	1.664192	2932	10897.3
0.91	1	0.561045	0.988264	1.42608	1.638	2925	10897.8
0.92	1	0.566286	0.997303	1.42668	1.63464	2919	10898.4

- From Tables 1, 2 and 3 it is observed that as the production, remanufacturing and returned rate increase, total profit increases.
- It is observed from the above tables that as the production, remanufacturing and returned rate increase, the acceptable returned quantity and the cycle length decreases.
- The analysis shows that the deviation in the value of M is negligible according to the deviation of the parameters.
- The cost function is stable for all the parameter but more stable for the returned formulation parameter.

6 Conclusion

This study develops a closed loop supply chain inventory model for the short life cycle products with deterioration. In the development of inventory models, most of the previous researchers have considered that an item can be repaired an indefinite number of times but in practical it is not always possible. We have studied a supply chain system from an End of Life perspective for the decaying items where the whole life-cycle of the product is considering. In this process, returned products are collected, cleaned, dismantling, make some re-use value be re-applications or create "new" products, the "new" products has the performance of original product. This paper develops a model where an item is recovered a finite number of times. Examples, which explain the application of the theoretical results, are given here and these illustrative examples are numerically verified too. Further, the effects of production, remanufacturing and return rates are compared.

References

1. Alamri, A.A.: Theory and methodology on the global optimal solution to a general reverse logistics inventory model for deteriorating items and time-varying rates. Comput. Ind. Eng **60**, 236–247 (2010)

2. Bayındır, Z.P., Dekker, R., Porras, E.: Determination of recovery effort for a probabilistic recovery system under various inventory control policies. Omega **34**, 571–584 (2006)
3. Chung, C.-J., Wee, H.-M.: Short life-cycle deteriorating product remanufacturing in a green supply chain inventory control system. Int. J. Prod. Econ. **129**, 195–203 (2011)
4. Dekker, R., Fleischmann, M., Inderfurth, K.: Reverse logistics: quantitative models for closed-loop supply chains. Springer-Verlag, Heidelberg (2004)
5. Dobos, I., Richter, K.: A production/recycling model with stationary demand and return rates. CEJOR **11**(1), 35–46 (2003)
6. Dobos, I., Richter, K.: An extended production/recycling model with stationary demand and return rates. Int. J. Prod. Econ. **90**(3), 311–323 (2004)
7. Dobos, I., Richter, K.: A production/recycling model with quality considerations. Int. J. Prod. Econ. **104**(2), 571–579 (2006)
8. Ishii, K., Eubanks, C.F., Marco, P.D.: Design for product retirement and material life-cycle. J. Mater. Des. **15**(4), 225–233 (1994)
9. King, A.M., Burgess, S.C., Ijomah, W., McMahon, C.A.: Reducing waste: repair, recondition, remanufacture or recycle? Sustain. Dev. **14**(4), 257–267 (2006)
10. Koh, S.G., Hwang, H., Sohn, K.I., Ko, C.S.: An optimal ordering and recovery policy for reusable items. Comput. Ind. Eng. **43**, 59–73 (2002)
11. Konstantaras, I., Skouri, K.: Lot sizing for a single product recovery system with variable setup numbers. Eur. J. Oper. Res. **203**(2), 326–335 (2010)
12. Mabini, M.C., Pintelon, L.M., Gelders, L.F.: EOQ type formulations for controlling repairable inventories. Int. J. Prod. Econ. **28**(1), 21–33 (1992)
13. Nahmias, N., Rivera, H.: A deterministic model for a repairable item inventory system with a finite repair rate. Int. J. Prod. Res. **17**(3), 215–221 (1979)
14. El Saadany, A.M.A., Jaber, M.Y.: A production/remanufacturing inventory model with price and quality dependant return rate. Comput. Ind. Eng. **58**(3), 352–362 (2010)
15. Savaskan, R.C., Bhattacharya, S., van Wassenhove, L.N.: Closed-loop supply chain models with product remanufacturing. Manage. Sci. **50**(2), 239–252 (2004)
16. Schrady, D.A.: A deterministic inventory model for repairable items. Naval Res Logistics Q. **14**, 391–398 (1967)
17. Singh, S.R., Prasher, L, Saxena, N.: A centralized reverse channel structure with flexible manufacturing under the stock out situation. Int. J. Ind. Eng. Comput.**4**, 559–570 (2013)
18. Singh, S.R,. Saxena N.: An optimal returned policy for a reverse logistics inventory model with backorders. Adv. Decis. Sci. **2012**, 21 (2012). Article ID 386598
19. Srivastava, S.K.: Green supply-chain management: a state-of-the-art literature review. Int. J. Manage. Rev. **9**(1), 53–80 (2007)

A Fuzzified Production Model with Time Varying Demand Under Shortages and Inflation

Shalini Jain and S. R. Singh

Abstract We develop an inventory model with time-dependent demand rate and deterioration, allowing shortages. The production rate is assumed to be finite and proportional to the demand rate. The shortages are partially backlogged with time-dependent rate. Inflation is also taken in this model. Inflation plays a very significant role in inventory policy. We developed the model in both fuzzy and crisp sense. The model is solved logically to obtain the optimal solution of the problem. It is then illustrated with the help of numerical examples. Sensitivity of the optimal solution with respect to changes in the values of the system parameters is also studied.

Keywords Time-dependent demand · Shortages · Deterioration · Fuzzy

1 Introduction

Inventory control involves human capability to deal with uncertainty of future demand of stock items. Hence, the application of fuzzy reasoning models in inventory control systems is quite important as fuzzy inferring procedures are becoming essential in managing uncertainties. In the past, a great deal of research has been done in the areas of inventory control systems. But, only few of the researchers have contributed in the applications of fuzzy logic. [29] first developed a inventory model in fuzzy sense. [24] fuzzified the ordering cost into trapezoidal fuzzy number in the total cost of an inventory without backordering and obtained the fuzzy total cost. Later, they used the centroid method and gained the total cost in the fuzzy sense. [7] fuzzified the

S. Jain (✉)
Centre of Mathematical Sciences, Banasthali University, Banasthali, Rajasthan, India
e-mail: shalini.shalini2706@gmail.com

S. R. Singh
Department of Mathematics, D.N. College, Meerut, India
e-mail: shivrajpundir@gmail.com

B. V. Babu et al. (eds.), *Proceedings of the Second International Conference on Soft Computing for Problem Solving (SocProS 2012), December 28–30, 2012*, Advances in Intelligent Systems and Computing 236, DOI: 10.1007/978-81-322-1602-5_43,

ordering cost, inventory cost, and backordering cost into trapezoidal fuzzy numbers and used the functional principle to obtain the estimate of the total cost in the fuzzy sense. [14] proposed an inventory model without shortages by fuzzifying the order quantity into a triangular fuzzy number. [28] generalized an inventory model without any backlogging for fuzzy order quantity and fuzzy total demand quantity. [5] considered the fuzzy problems for the mixture of backorders and lost sales in inventory model. The total expected annual cost is obtained in the fuzzy sense. [6] considered the mixture inventory model involving a fuzzy variable and obtained the total cost in the fuzzy sense. [22] consider inflation and apply discounted cash flow in a inventory model and formulated the total cost of the system using genetic algorithm. [18] presented solution changed model to a crisp multipurpose problem using defuzzification of fuzzy constraints and fuzzy chance-constrained programming methods.

Production is a process whereby raw material is converted into semifinished products and then converted into finished products. The main purpose of production function is to produce the goods and services demanded by the customers in the most efficient and economical way. So efficient management of the production function is of utmost importance in order to achieve this objective. An optimal production quantity model for a deteriorating item was developed by [15]. [25] proposed economic production quantity (EPQ) deteriorating inventory with partial backordering. [11] introduced an EPQ model with marketing policies and a deteriorating item. [23] developed a model with price and stock-dependent demand considering a production model for deteriorating items. [20] introduced a model that generates an economic run quantity solution and the total production solution simultaneously. Economic run quantity is an extended model of production model. The objective of production planning, and control, like that of all other manufacturing controls, is to contribute to the profits of the industry. As with inventory management and control, this is accomplished by keeping the customers satisfied during the meeting of delivery schedules.

Usually, it is assumed that lifetime of an item is immeasurable when it is in storage. However, there are abundant types of items such as food grains, highly volatile substances, radioactive materials, films, drugs, blood, fashion goods, electronic components, and high-tech products in which there is gradual loss of potential or value with a passage of time. Therefore, the effect of deterioration cannot be ignored in inventory models. [10] were the first to consider deterioration of inventory with constant demand. As time passed, several researchers developed inventory models by assuming either instantaneous or finite production with different assumptions on the patterns of deterioration. In this connection, researchers may refer to work by [4, 8, 9, 13, 15]. Also, some researchers [26, 27] have studied the chances and effect of integration and co-operation between the buyer and the producer of deteriorating items. Interested readers may review the articles by [19] and [12]. Lin and Gong assumed varying production rate and deteriorating inventory.

Inflation plays a very significant role in inventory models. Inflation refers to the movement in the general level of prices. It does not refer to changes in one price relative to other prices. Rather than measure inflation by using the actual rate at which prices are rising, some economists prefer a measure of inflation that reflects primarily only the systematic factors that act to raise prices. [3] developed the first

EOQ model taking inflation into the model. [1] developed the inventory decisions under inflationary condition. [2] proposed an economic order quantity under variable rate of inflation and mark-up prices. [17] criticized a net present value. [16] studied the inflation effects on inventory system. Yang et al. presented a deterministic inventory lot-size models under inflation with shortages and deterioration for fluctuating demand. [21] presented a fuzzy inventory model under inflation.

In the present paper, we assume the production model with time-dependent demand and deterioration. As a result, the finite production rate is also time dependent. Shortages are allowed and are partially backlogged. The model developed in both fuzzy and crisp sense. An analytical solution of the model is discussed and is illustrated with the help of numerical examples. Sensitivity of the optimal solution with respect to changes in different parameter values is also examined.

2 Assumptions and Notations

The following assumptions and notations have been used throughout the paper:

1. The demand rate is deterministic and is a function of time.
2. The rate of production is finite.
3. Production rate depends on the demand rate.
4. Inflation and time value of money are considered.
5. Shortages are allowed.
6. Holding cost is taken to be variable in nature

The following notations are used in our study:

D(t) Demand rate (units/unit time), $D(t) = a+bt$, a and b are positive constants, $a > b$.
P(t) Production rate (units/unit time), k > 1, P(t) = kD(t) for any t.
$\theta(t)$ Rate of deterioration where $\theta(t) = \theta t$, θ is a positive constant.
$I_i(t)$ Inventory level at any time t.
s Per unit selling price of the item.
B Backlogging rate, $B = e^{-\delta t}$, δ is a positive constant.
r Constant representing the difference between the discount rate and inflation rate.
c Production cost per unit item.
c_3 Set up cost per unit item.
c_2 Shortage cost per unit item.
$c_1 + \alpha t$ Inventory holding cost per unit item per unit time, $\alpha > 0$.

3 Model Illustration

The problem has been formulated in two steps. In the first step, we formulate a crisp model and then in the next step, we extend the model into a fuzzy sense. The crisp formulation of the model has been presented here:

$$I_1'(t) + \theta t I_1(t) = k(a + bt) - (a + bt), 0 \le t \le T_1 \tag{1}$$

$$I_2'(t) + \theta t I_2(t) = -(a + bt), T_1 \le t \le T_2 \tag{2}$$

$$I_3'(t) = -e^{-\delta t}(a + bt), T_2 \le t \le T_3 \tag{3}$$

$$I_4'(t) = k(a + bt) - (a + bt), T_3 \le t \le T_4 \tag{4}$$

With the boundary conditions:

$$I_1(0) = 0, I_2(T_2) = 0, I_3(T_2) = 0, I_4(T_4) = 0 \tag{5}$$

Solutions of (1)–(4) are

$$I_1(t) = (k - 1)\left[\left(at + \frac{bt^2}{2} + \frac{a\theta t^3}{6} + \frac{b\theta t^4}{8}\right)\right] e^{-\frac{\theta t^2}{2}} \tag{6}$$

$$I_2(t) = \left[a(T_2 - t) + \frac{b}{2}\left(T_2^2 - t^2\right) + \frac{a\theta}{6}\left(T_2^3 - t^3\right) + \frac{b\theta}{8}\left(T_2^4 - t^4\right)\right] e^{-\frac{\theta t^2}{2}} \tag{7}$$

$$I_3(t) = \left[a(T_2 - t) + \frac{b}{2}\left(T_2^2 - t^2\right)\right] \tag{8}$$

$$I_4(t) = (k - 1)\left[a(t - T_4) + \frac{b}{2}\left(t^2 - T_4^2\right)\right] \tag{9}$$

At $t = T_1$, $I_1(T_1) = I_2(T_1)$. From Eqs. (6) and (7), we get

$$T_1 = f(T_2) \tag{10}$$

At $t = T_3$, $I_3(T_3) = I_4(T_3)$. From Eqs. (8) and (9), we get

$$T_3 = f(T_2, T_4) \tag{11}$$

The present worth of holding cost for the period under consideration

$$HC = \int_0^{T_1} (c_1 + \alpha t) I_1(t) e^{-rt} dt + c_1 \int_{T_1}^{T_2} (c_1 + \alpha t) I_2(t) e^{-rt} dt \tag{12}$$

Production has been taking place in the period $[0, T_1]$ and $[T_3, T_4]$, hence, the present worth of the production cost is:

$$PC = \left(\int_0^{T_1} k(a + bt) e^{-rt} dt + \int_{T_3}^{T_4} k(a + bt) e^{-rt} dt \right) \tag{13}$$

Before the start of a production run, the fixed cost to be borne by the producer is:

$$SPC = c_3 + c_3 e^{-rT_3} \tag{14}$$

Shortages are accumulated in the system during $[T_2, T_4]$. The maximum level of shortages are present at $t = T_4$. The total present worth of shortages during this time is

$$SC = \left(\int_{T_2}^{T_3} -I_3(t) e^{-rt} dt + \int_{T_3}^{T_4} -I_4(t) e^{-rt} dt \right) \tag{15}$$

Inventory is present and sold during $[0, T_4]$. The present worth of the generated revenue is given by

$$SP = s \left(\int_0^{T_1} D(t) e^{-rt} dt + \int_{T_1}^{T_2} D(t) e^{-rt} dt + \int_{T_2}^{T_3} D(t) e^{-\delta t} e^{-rt} dt + \int_{T_3}^{T_4} D(t) e^{-rt} dt \right) \tag{16}$$

So, the net profit of the system is represented by:

$$NP = \frac{s.SP - c_1.HC - c.PC - c_3 - c_2.SC}{T_4} \tag{17}$$

4 Fuzzy Modeling

In order to develop the model in a fuzzy environment, we consider the Production cost per inventory unit per unit time $c^{\vee}$ and Shortage cost per inventory unit per unit time $c_2^{\vee}$ as the triangular fuzzy numbers $\tilde{c} = (c - \Delta_1, c, c + \Delta_2)$ and $\tilde{c}_2 = (c_2 - \Delta_3, c_2, c_2 + \Delta_4)$ such that that $0 < \Delta_1 < c, 0 < \Delta_2, 0 < \Delta_3 < c_2, 0 < \Delta_4$

and where Δ_1, Δ_2, Δ_3, Δ_4 are determined by the decision maker based on the uncertainty of the problem. Thus, the Production cost c and the shortage cost c_2 are considered as the fuzzy numbers $\tilde{c}$ and $\tilde{c}_2$ with membership functions.

$$\mathrm{NP} = \frac{\mathrm{s.SP} - \mathrm{c_1.HC} - \hat{\mathrm{c}}.\mathrm{PC} - \mathrm{c_3.SPC} - \hat{\mathrm{c}}_2.\mathrm{SC}}{T_4} \tag{18}$$

$$\tilde{c} = [c - \Delta_1, c, c + \Delta_2] \tag{19}$$

$$\tilde{c}_2 = [c_2 - \Delta_3, c_2, c_2 + \Delta_4] \tag{20}$$

By Centroid Method, we get

$$\tilde{c} = c + \frac{1}{3}(\Delta_2 - \Delta_1) \tag{21}$$

$$\tilde{c}_2 = c_2 + \frac{1}{3}(\Delta_4 - \Delta_3) \tag{22}$$

$$\mathrm{F_1} = \frac{\mathrm{s.SP} - \mathrm{c_1.HC} - (\mathrm{c} + \Delta_2).\mathrm{PC} - \mathrm{c_3.SPC} - (\mathrm{c_2} + \Delta_4).\mathrm{SC}}{\mathrm{T_4}} \tag{23}$$

$$\mathrm{F_2} = \frac{\mathrm{s.SP} - \mathrm{c_1.HC} - \mathrm{c.PC} - \mathrm{c_3.SPC} - \mathrm{c_2.SC}}{\mathrm{T_4}} \tag{24}$$

$$\mathrm{F_3} = \frac{\mathrm{s.SP} - \mathrm{c_1.HC} - (\mathrm{c} - \Delta_1).\mathrm{PC} - \mathrm{c_3.SPC} - (\mathrm{c_2} - \Delta_3).\mathrm{SC}}{\mathrm{T_4}} \tag{25}$$

By the method of Defuzzification,

$$\mathrm{NP} = \frac{F_1 + 2F_2 + F_3}{4} \tag{26}$$

we find out Net profit in fuzzy sense.

5 Solution Procedure

To maximize total average profit per unit time (NP), the optimal values of t_2 and t_4 can be obtained by solving the following equations simultaneously

$$\frac{\partial NP}{\partial T_2} = 0 \tag{27}$$

and

$$\frac{\partial NP}{\partial T_4} = 0 \tag{28}$$

Provided, they satisfy the following conditions

$$\frac{\partial^2 NP}{\partial T_2^2} < 0, \frac{\partial^2 NP}{\partial T_4^2} < 0 \text{ and} \left(\frac{\partial^2 NP}{\partial T_2^2}\right)\left(\frac{\partial^2 NP}{\partial T_4^2}\right) - \left(\frac{\partial^2 NP}{\partial T_2 \partial T_4}\right)^2 > 0$$

Equation (16) is our objective function which needs to be maximized. For this, we use the classical optimization techniques. The Eqs. (27) and (28) obtained thereafter are highly nonlinear in the continuous variable T_2, T_4 and the discreet variables T_1, T_3. However, if we give particular values to the discreet variable T_1, T_3, our objective function becomes the function of two variables T_2 and T_4. We have used the mathematical software MATHEMATICA 8.0 to arrive at the solution of the system in consideration. We can obtain the optimal values of different values of the time with the help of software. With the use of these optimal values, Eq. (17) provides maximum total average profit per unit time of the system in consideration.

6 Numerical Example

To illustrate the results, let us apply the proposed method to efficiently solve the following numerical example. For convenience, the values of the parameters are selected randomly.

$k = 1.5, a = 80, b = 0.2, \alpha = 0.02, c = 1.6, c_1 = 0.9, c_2 = 10, c_3 = 40, s = 22,$
$r = 0.03, \theta = 0.05, T_1 = 2, T_3 = 26.5472$

Optimal Solution of the Proposed Model

T_2	T_4	Total profit (NP)
4.49286	67.5452	6575.58

The graph shows the variation of the system cost with T_2 and T_4. From the figure, it is very clear that the total profit function is concave with respect to the two variables.

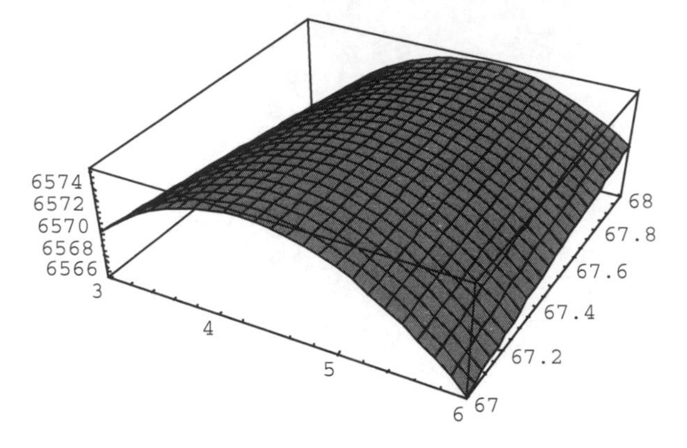

7 Sensitivity Analysis

We now study the effects of changes in the values of the system parameters $a, b, k, \alpha, c, c_1, c_2, c_3, s, r, \theta$ on the total system cost in consideration. The sensitivity analysis is performed by changing each of the parameters by − 10–20 %, taking one parameter at a time and keeping the remaining parameters unchanged. The analysis is based on the familiar results obtained from Example.

Parameter	Change in parameter			
	−20 %	−10 %	10 %	20 %
a	5579.61	6077.65	7073.41	7571.17
b	6255.86	6415.76	6735.3	6894.94
k	2257.36	4105.75	9606.43	13106.3
α	6575.99	6575.99	6575.98	6575.97
c	6583.38	6579.44	6571.67	6571.67
c_1	6578.17	6576.87	6574.28	6584.65
c_2	5409.32	5992.45	7158.71	7741.83
c_3	6575.75	6575.66	6576.49	6575.4
s	6416.06	6495.82	6655.33	6735.09
r	8082.49	7325.72	5831.22	5091.93
θ	6575.37	6575.47	6575.68	6575.78

Graphical representation of the sensitivity results with respect to different system parameters have been plotted in figures shown below.

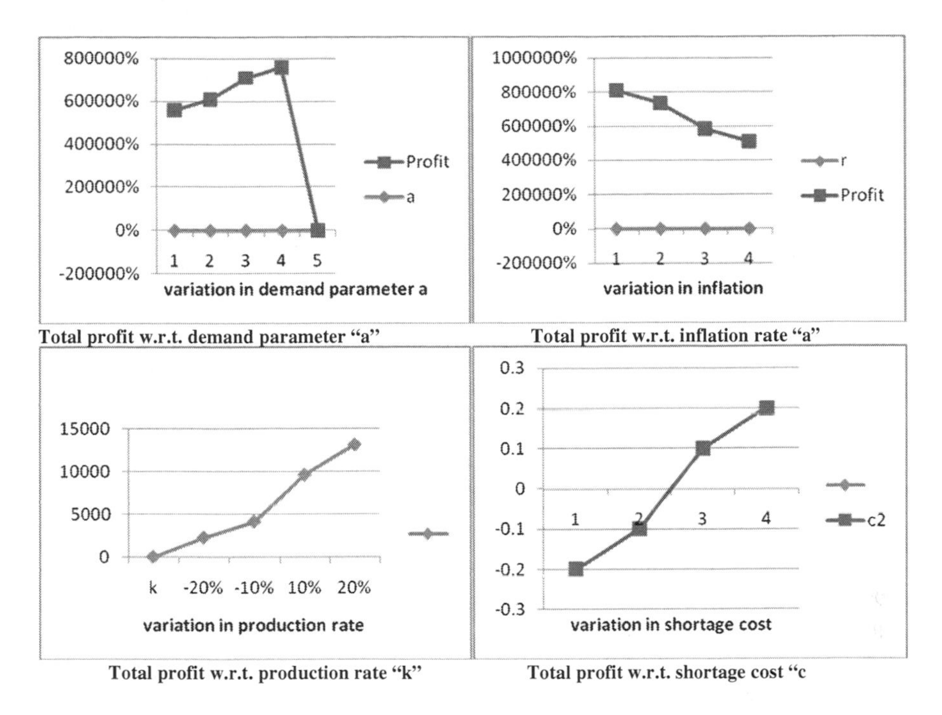

Total profit w.r.t. demand parameter "a"

Total profit w.r.t. inflation rate "a"

Total profit w.r.t. production rate "k"

Total profit w.r.t. shortage cost "c

8 Conclusion

The model has been developed for time-dependent demand with time-dependent deterioration in inventory. In this model, production rate is dependent on the demand rate. The average net profit function is the objective function in the case of crisp model. The same function extends to give the fuzzy model of the situation. This model is later defuzzified to a crisp model. Both the crisp as well as the fuzzy models have been solved numerically. In this study, shortages in inventory are allowed and the backlogging rate is taken as time dependent. The proposed model can be extended in numerous ways. For example, we may extend the inflation-dependent demand to inflation. Also, we could extend the model to incorporate some more features, such as quantity discount, and permissible delay in payment.

References

1. Bierman, H., Thomas, J.: Inventory decisions under inflationary condition. Deci. Sci. **8**(1), 151–155 (1977)
2. Brahmbhatt, A.C.: Economic order quantity under variable rate of inflation and mark-up prices. Productivity **23**, 127–130 (1982)
3. Buzacott, J.A.: Economic order quantities with inflation. Oper. Res. Quart. **26**(3), 553–558 (1975)

4. Chakrabarti, T., Giri, B.C., Chaudhuri, K.S.: An EOQ model for items weibull distribution deterioration shortages and trended demand: an extension of philip's model. Comp. Oper. Res. **25**, 649–657 (1998)
5. Chang, H.C., Yao, J.S., Ouyang, L.Y.: Fuzzy mixture inventory model with variable lead-time based on probabilistic fuzzy set and triangular fuzzy number. Math. Comp. Mode. **39**(2–3), 287–304 (2004)
6. Chang, H.C., Yao, J.S., Ouyang, L.Y.: Fuzzy mixture inventory model involving fuzzy random variable lead time demand and fuzzy total demand. Euro. J. Oper. Res. **169**(1), 65–80 (2006)
7. Chen, S.H., Wang, C.C.: Backorder fuzzy inventory model under functional principle. Inf. Sci. **95**, 71–79 (1996)
8. Covert, R.P., Philip, G.C.: An EOQ model for items with weibull distribution deterioration. AIIE Trans. **5**, 323–326 (1973)
9. Dave, U.: On a discrete-in-time order-level inventory model for deteriorating items. Oper. Res. **30**, 349–354 (1979)
10. Ghare, P.M., Schrader, S.F.: A model for exponentially decaying inventory. J. Ind. Eng. **14**, 238–243 (1963)
11. Goyal, S.K., Gunasekaran, A.: An integrated production-inventory-marketing model for deteriorating items. Comp. Ind. Eng **28**(4), 755–762 (1995)
12. Goyal, S.K., Giri, B.C.: Recent trends in modeling of deteriorating inventory. Euro. J. Oper. Res. **134**, 1–16 (2001)
13. Kang, S., Kim, I.: A study on the price and production level of the deteriorating inventory system. Int. J. Prod. Res. **21**, 899–908 (1983)
14. Lee, H.M., Yao, J.S.: Economic order quantity in fuzzy sense for inventory without backorder model. Fuzzy Sets Syst. **105**, 13–31 (1999)
15. Misra, R.B.: Optimum production lot-size model for a system with deteriorating inventory. Int. J. Prod. Res. **13**(5), 495–505 (1975)
16. Misra, R.B.: A study of inflation effects on inventory system. Logi. Spect. **9**(3), 260–268 (1997)
17. Moon, I., Yun, W.: A note on evaluating inventory systems: a net present value frame work. Eng. Econ. **39**(1), 93–99 (1993)
18. Nayebi, M.A., Sharifi, M., Shahriari, M.R., Zarabadipour, O.: Fuzzy-chance constrained multi-objective programming applications for inventory control model. App. Math. Sci. **6**(5), 209–228 (2012)
19. Raafat, F.: Survey of literature on continuously deteriorating inventory models. J. Oper. Res. Soci. **42**, 27–37 (1991)
20. Simmons, D., Cheng, J.: An alternative approach to computing economic run quantity. Int. J. Prod. Res. **56**(3), 837–847 (2008)
21. Singh, S.R., Bhatia, D.: Fuzzy inventory model for non-instantaneous perishable products under inflation. Int. J. Man. Res. Tech. **4**(2), 261–270 (2010)
22. Singh, S.R., Kumar, T., Gupta, C.B.: Optimal replenishment policy for ameliorating item with shortages under inflation and time value of money using genetic algorithm. Int. J. Comp. App. **27**(1), 5–17 (2011)
23. Teng, J.T., Chang, C.T.: Economic production quantity models for deteriorating itemswith price- and stock-dependent demand. Comp. Oper. Res. **32**(2), 297–308 (2005)
24. Vojosevic, M., Petrovic, D., Petrovic, R.: EOQ formula when inventory cost is fuzzy. Int. J. Prod. Eco. **45**, 499–504 (1996)
25. Wee, H.M.: Economic production lot-size model for deteriorating items with partial backordering. Comp. Ind. Eng. **20**(2), 187–197 (1993)
26. Wee, H.M., Jong, J.F.: An integrated multi-lot-size production inventory modelfor deteriorating items. Manage. Syst. **5**, 97–114 (1998)
27. Yang, P.C., Wee, H.M.: Economic order policy of deteriorated item for vendor and buyer: an integrated approach. Prod. Plan. Cont. **11**, 474–480 (2000)
28. Yao, J.S., Chang, S.C., Su, J.S.: Fuzzy inventory without backorder for fuzzy order quantity and fuzzy total demand quantity. Comp. Oper. Res. **27**, 935–962 (2000)
29. Zadeh, L.: Fuzzy sets. Inf. control **8**, 338–353 (1965)

A Partial Backlogging Inventory Model for Decaying Items: Considering Stock and Price Dependent Consumption Rate in Fuzzy Environment

S. R. Singh and Swati Sharma

Abstract In this article, an inventory model is developed to deal with the impreciseness present in the market demand and the various cost parameters. The presented model is developed in crisp and fuzzy environments. Signed distance method is used for defuzzification. In most of the classical models, constant demand rate is considered. But in practice purchasing deeds of the customers is affected by the selling price and inventory level. In this study, we have considered demand rate as a function of stock-level and selling price. Two parameters Weibull distribution deterioration is considered. It is assumed that shortages are allowed and are partially backordered with the time dependent backlogging rate. A numerical experiment is provided to illustrate the problem. Sensitivity analysis of the optimal solution with respect to the changes in the value of the system parameters is also discussed.

Keywords Inventory model · Triangular fuzzy numbers · Signed distance · Partial backlogging · Stock and price dependent demand rate

1 Introduction

In most of the inventory models, it is assumed that the inventory parameters like demand rate and ordering cost, etc., are precisely known. But in actual living situations, the nature of these parameters is imprecise, so it is important to consider them as fuzzy numbers. The concept of fuzzy set theory first introduced by Zadeh [1], after that Park [2] extended the classical EOQ model by introducing the fuzziness

S. R. Singh · S. Sharma (✉)
Department of Mathematics, D.N. College, Meerut, India
e-mail: shivrajpundir@gmail.com

S. Sharma
e-mail: jmlashi0@gmail.com

B. V. Babu et al. (eds.), *Proceedings of the Second International Conference on Soft Computing for Problem Solving (SocProS 2012), December 28–30, 2012*, Advances in Intelligent Systems and Computing 236, DOI: 10.1007/978-81-322-1602-5_44, © Springer India 2014

of ordering cost and holding cost. Recently, Chang [3] and Wang et al. [4] presented an inventory model under fuzzy demand.

Deterioration is a natural phenomenon in many real situations so it plays an important role in developing an inventory model. Generally, deterioration is defined as damage, spoilage, decay andobsolescence, vaporization, etc., that result in decrease of value of the original one. The first model for decaying items was presented by Ghare and Schrader [5]. It was extended by Covert and Philip [6] considering Weibull distribution deterioration. A complete survey for deteriorating inventory models was presented by Raafat [7]. Some other papers relevant to this topic are Teng et al. [8] and Bakker et al. [9].

Various researchers considered the situation in which shortages are either completely backlogged or completely lost which is not realistic. Many practical experiences disclose that some but not all customers will wait for backlogged items during a shortage period, such as for fashionable supplies and the products with short life cycle. According to such phenomenon, backlogging rate should not be disregarded. Thus, it is necessary to consider the backlogging rate. Researchers, such as Park [10] and Wee [11] developed inventory models with partial backorders. Some recent work in this direction is done by Singh [12] and Hsieh et al. [13].

In many classical research articles, it is assumed that demand rate is constant during the sales period. In real life, the demand may be inspired if there is a large pile of goods displayed on shelf. Gupta and Vrat [14] developed the first models for stock-dependent consumption rate. Mandal and Phaujdar [15] then developed a production inventory model for deteriorating items with uniform rate of production and linearly stock-dependent demand. Other papers related to this research area are Ray et al. [16], Goyal and Giri [17], and Chang et al. [18].

All the above-cited papers reveal that many research articles are developed in which demand is considered as the function of stock level or selling price, shortages are allowed and partially backlogged, but there is no such research paper which is partially backlogged assuming demand rate as the function of selling price and inventory level in fuzzy environment. In lots of business practices, it is observed that several parameters in inventory system are imprecise. Therefore, it is necessary to consider them as fuzzy numbers while developing the inventory model.

In this study, we have developed a partial backlogging inventory model for deteriorating items considering stock and price sensitive demand rate in crisp and fuzzy surroundings. A numerical example to prove that the optimal solution exists and is unique is provided and the sensitivity analysis with respect to system parameters is discussed. The concavity is also shown through the figure (Fig. 1).

2 Assumptions and Notations

The notations and basic assumptions of the model are as follows:

(1) The demand rate is a function of stock and selling price considered as $f(t) = (a + bQ(t) - p)$ where $a > 0, 0 < b < 1, a > b$ and p is selling price.

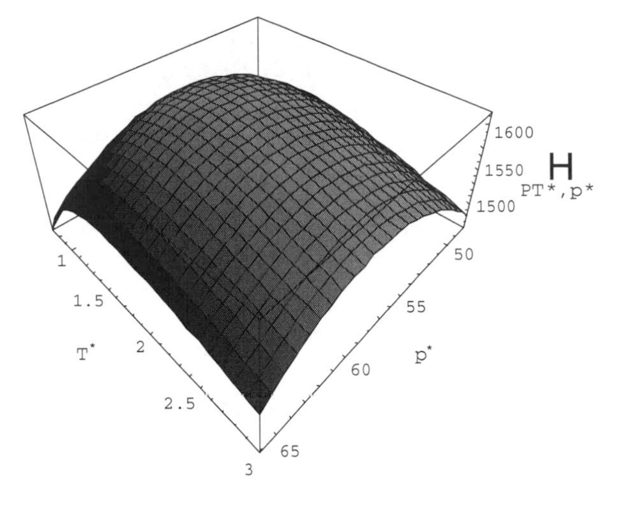

Fig. 1 Concavity of the profit function

(2) Holding cost $h(t)$ per item per time-unit is time dependent and is assumed to be $h(t) = h + \delta t$ where $\delta > 0, h > 0$.
(3) Shortages are allowed and partially backlogged and rate is assumed to be $1/(1 + \eta t)$, which is a decreasing function of time, $\eta \geq 0$.
(4) The deterioration rate is time-dependent.
(5) T is the length of the cycle.
(6) Replenishment is instantaneous and lead time is zero.
(7) The order quantity in one cycle is Q.
(8) The selling price per unit item is p.
(9) A is the ordering cost per order.
(10) c_1 is the purchasing cost per unit per unit time.
(11) c_2 is the backordering cost per unit per unit time.
(12) c_3 is the opportunity cost per unit.
(13) c_4 is the deterioration cost per unit.
(14) $P(T, t_1, p)$ the total profit per unit time.
(15) The deterioration of units follows the two parameter Weibull distribution (say) $\theta(t) = \alpha\beta t^{\beta-1}$ where $0 < \alpha < 1$ is the scale parameter and $\beta > 0$ is the shape parameter.
(16) During time t_1, inventory is depleted due to deterioration and demand of the item. At time t_1 the inventory becomes zero and shortages start going on (Fig. 2).

3 Mathematical Formulation and Solution

Let $Q(t)$ be the inventory level at time $t\,(0 \leq t \leq T)$. During the time interval $[0, t_1]$ inventory level decreases due to the combined effect of demand and deterioration both and at t_1 inventory level depletes up to zero. The differential equation to describe immediate state over $[0, t_1]$ is given by

$$Q'(t) + \alpha\beta t^{\beta-1}Q(t) = -(a + bQ(t) - p) \quad 0 \le t \le t \tag{1}$$

Again, during time interval $[t_1, T]$ shortages stars occurring and at T there are maximum shortages, due to partial backordering some sales are lost. The differential equation to describe instant state over $[t_1, T]$ is given by

$$Q'(t) = -\frac{(a-p)}{[1+\eta(T-t)]} \quad t_1 \le t \le T \quad t_1 \le t \le T \tag{2}$$

With condition $Q(t_1) = 0$. Solving Eqs. (1) and (2) and neglecting higher powers of α, we get

$$Q(t) = (a-p)\left[t_1 - t + \frac{b}{2}(t_1^2 - t^2) + \frac{\alpha}{(\beta+1)}(t_1^{\beta+1} - t^{\beta+1})\right]e^{-bt-\alpha t^{\beta}} 0 \le t \le t_1 \tag{3}$$

$$Q(t) = \frac{(a-p)}{\eta}[\log 1 + \eta(T-t) - \log 1 + \eta(T-t_1)] t_1 \le t \le T \tag{4}$$

At time $t = 0$ inventory level is $Q(0)$ and is given by

$$Q(0) = (a-p)\left(t_1 + \frac{bt_1^2}{2} + \frac{\alpha t_1^{\beta+1}}{\beta+1}\right)$$

At time T maximum shortages (Q_1) occurs and is given by

$$Q_1 = \frac{(a-p)}{\eta}[\log\{1 + \eta(T-t_1)\}]$$

The order quantity is Q and is given by

$$Q = (a-p)\left(t_1 + \frac{bt_1^2}{2} + \frac{\alpha t_1^{\beta+1}}{\beta+1} + \frac{1}{\eta}\log(1 + \eta(T-t_1))\right) \tag{5}$$

The purchasing cost is

$$\text{PC} = c_1(a-p)\left(t_1 + \frac{bt_1^2}{2} + \frac{\alpha t_1^{\beta+1}}{\beta+1} + \frac{1}{\eta}\log(1 + \eta(T-t_1))\right) \tag{6}$$

Ordering cost is given by

$$\text{OC} = A \tag{7}$$

Holding cost during the period $[0, t_1]$ is given by

$$\begin{aligned} \mathrm{IHC} &= \int_0^{t_1} (h + \delta t) Q(t) dt \\ &= \int_0^{t_1} (h + \delta t)(a - p) \left[t_1 - t + \frac{b}{2}(t_1^2 - t^2) + \frac{\alpha}{(\beta + 1)}(t_1^{\beta+1} - t^{\beta+1}) \right] \\ &\quad \times e^{-bt - \alpha t^\beta} dt \end{aligned} \tag{8}$$

Deterioration cost during the period $[0, t_1]$ is given by

$$\mathrm{DC} = c_4 \left\{ Q(0) - \int_0^{t_1} (a + bQ(t) - p) dt \right\} \tag{9}$$

Shortage cost due to backordered is

$$\begin{aligned} \mathrm{BC} &= c_2 \int_{t_1}^{T} [-Q(t)] \, dt \\ &= -c_2 \int_{t_1}^{T} \frac{(a - p)}{\eta} \left[\log\{1 + \eta(T - t)\} - \log\{1 + \eta(T - t_1)\} \right] dt \end{aligned} \tag{10}$$

Lost sales cost due to lost sales is

$$\mathrm{LS} = c_3(a - p) \int_{t_1}^{T} \left[1 - \frac{1}{(1 + \eta(T - t))} \right] dt \tag{11}$$

Sales revenue is given by

$$\mathrm{SR} = p \int_0^{t_1} (a + bQ(t) - p) \, dt + p \int_{t_1}^{T} \frac{(a - p)}{[1 + \eta(T - t)]} dt \tag{12}$$

From Eqs. (6), (7), (8), (9), (10), (11), and (12) total profit per unit time is given by

$$P(T, t_1, p) = \frac{1}{T} \mathrm{SR} - \mathrm{OC} - \mathrm{PC} - \mathrm{IHC} - \mathrm{BC} - \mathrm{LS} - \mathrm{DC} \tag{13}$$

Let $t_1 = \gamma T, 0 < \gamma < 1$. Hence we get the profit function is

$$P(T,p)=\frac{1}{T}\left[p(a-p)\gamma T+b(a-p)pK_1-b(a-p)pK_2+p(a-p)K_3-A\right.$$
$$\left.-c_1(a-p)K_4-\frac{c_2(a-p)K_5}{\eta}-c_3(a-p)K_5-c_4(a-p)K_6-(a-p)\,K_7\right] \tag{14}$$

Where,

$$K_1=\left(\gamma T+\frac{b\gamma^2T^2}{2}+\frac{\alpha\gamma^{\beta+1}T^{\beta+1}}{\beta+1}\right)\left(\gamma T-\frac{b\gamma^2T^2}{2}-\frac{\alpha\gamma^{\beta+1}T^{\beta+1}}{\beta+1}\right)$$

$$K_2=\left[\frac{\gamma^2T^2}{2}+\frac{b\gamma^3T^3}{6}+\frac{\alpha\gamma^{\beta+2}T^{\beta+2}}{(\beta+1)(\beta+2)}-b\left\{\frac{\gamma^3T^3}{3}+\frac{b\gamma^4T^4}{8}\right.\right.$$
$$\left.\left.+\frac{\alpha\gamma^{\beta+3}T^{\beta+3}}{(\beta+1)(\beta+3)}\right\}-\alpha\left\{\frac{\gamma^{\beta+2}T^{\beta+2}}{\beta+2}+\frac{b\gamma^{\beta+3}T^{\beta+3}}{2(\beta+3)}+\frac{\alpha\gamma^{2\beta+2}T^{2\beta+2}}{2(\beta+1)^2}\right\}\right]$$

$$K_3=\frac{\{\log\{1+\eta(T-\gamma T)\}\}}{\eta},\quad K_4=\left(\gamma T+\frac{b\gamma^2T^2}{2}+\frac{\alpha\gamma^{\beta+1}T^{\beta+1}}{\beta+1}\right.$$
$$\left.+\frac{1}{\eta}\log\left(1+\eta(T-\gamma T)\right)\right),\quad K_5=\frac{1}{\eta}\left\{\eta(T-\gamma T)-\log\left(1+\eta(T-\gamma T)\right)\right\},$$

$$K_6=\left(\frac{b\gamma^2T^2}{2}+\frac{\alpha\gamma^{\beta+1}T^{\beta+1}}{\beta+1}-b\left(\frac{\gamma^2T^2}{2}+\frac{b\gamma^3T^3}{6}-\frac{b^2\gamma^4T^4}{8}-\frac{2\alpha\gamma^{\beta+2}T^{\beta+2}}{(\beta+1)(\beta+2)}\right.\right.$$
$$\left.\left.-\frac{b\alpha\gamma^{\beta+3}T^{\beta+3}}{2\,(\beta+1)}-\frac{\alpha^2\gamma^{2\beta+2}T^{2\beta+2}}{2(\beta+1)^2}\right)\right)$$

$$K_7=\left\{\frac{h\gamma^2T^2}{2}+\frac{\delta\gamma^3T^3}{6}+\frac{bh\gamma^3T^3}{6}+\frac{b\delta\gamma^4T^4}{24}-\frac{b^2h\gamma^4T^4}{8}-\frac{b^2\delta\gamma^5T^5}{15}\right.$$
$$+\frac{\alpha h\gamma^{\beta+2}T^{\beta+2}}{\beta+2}-\frac{\alpha h\gamma^{\beta+2}T^{\beta+2}}{(\beta+1)(\beta+2)}+\frac{3\alpha\delta\gamma^{\beta+3}T^{\beta+3}}{2(\beta+3)}-\frac{\alpha bh\gamma^{\beta+3}T^{\beta+3}}{2(\beta+1)}$$
$$-\frac{\alpha\delta\gamma^{\beta+3}T^{\beta+3}}{(\beta+2)}+\frac{\alpha\delta b\gamma^{\beta+4}T^{\beta+4}}{6(\beta+4)}-\frac{\alpha\delta b\gamma^{\beta+4}T^{\beta+4}}{2(\beta+2)}-\frac{\alpha^2h\gamma^{2\beta+2}T^{2\beta+2}}{2(\beta+1)^2}$$
$$\left.-\frac{\alpha^2\delta\gamma^{2\beta+3}T^{2\beta+3}}{(\beta+2)(2\beta+3)}\right\}$$

Our objective is to maximize the profit function $P(T,p)$. The necessary conditions for maximizing the profit are

$$\frac{\partial P(T,p)}{\partial T}=0\quad\text{and}\quad\frac{\partial P(T,p)}{\partial p}=0$$

Using the software Mathematica-8.0, from these two equations, we can determine the optimum values of T^* and p^* simultaneously and the optimal value $P^*(T^*, p^*)$ of the average net profit can be determined by (14) provided they satisfy the sufficiency conditions for maximizing $P^*(T^*, p^*)$ are

$$\frac{\partial^2 P(T,p)}{\partial T^2} < 0, \frac{\partial^2 P(T,p)}{\partial p^2} < 0 \text{ and } \frac{\partial^2 P(T,p)}{\partial T^2}\frac{\partial^2 P(T,p)}{\partial p^2} - \left(\frac{\partial^2 P(T,p)}{\partial T \partial p}\right)^2 > 0$$

4 Numerical Example

To illustrate the theory of the model, we consider the following data on the basis of the previous study.

$A = 200, a = 100, b = 0.02, c_1 = 12, \gamma = 0.6, c_2 = 8, c_3 = 15, c_4 = 0.4, \alpha = 0.1, h = 0.6, \beta - 0.01, \eta = 0.5, \delta = 0.02$.

Based on these input data, the findings are as follows: $p^* = 57.2434, t_1^* = 0.896826, Q^* = 64.8475, T^* = 1.49471$ and $P^*(T^*, p^*) = 1612.65$.

5 Sensitivity Analysis

See Table 1.

6 Fuzzy Mathematical model

In this study, we consider a, A, c_1, c_2, c_3 and c_4 as fuzzy numbers, i.e., $\tilde{a}$, $\tilde{A}$, $\tilde{c}_1$, $\tilde{c}_2$, $\tilde{c}_3$ and $\tilde{c}_4$. Then $\tilde{P}(T, P)$ is regarded as the estimate of total profit per unit time in the

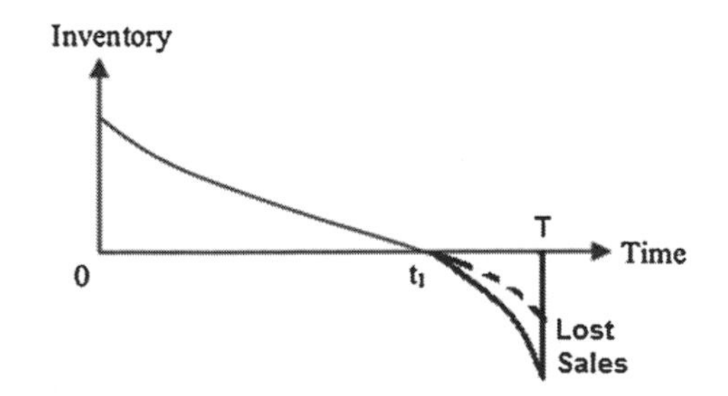

Fig. 2 Graphical representation of the inventory system

Table 1 Sensitivity analysis is performed by changing (increasing or decreasing) the parameters by 10 and 20 % and considering one parameter at a time, keeping the left over parameters at their original values

Changing parameter	% change in system	Change in T^*	Change in p^*	Change in t_1^*	Change in Q^*	Change in $P^*(T^*, p^*)$
A	−20	1.30930	57.1406	0.785580	53.1966	1641.20
	−10	1.40364	57.1932	0.842184	61.1014	1626.45
	+10	1.58302	57.2914	0.949812	68.4592	1599.65
	+20	1.66899	57.3376	1.001390	71.9558	1587.35
a	−20	1.89826	47.4582	1.138960	62.1062	896.308
	−10	1.67136	52.3388	1.002820	63.6073	1229.32
	+10	1.35289	62.1650	0.811734	65.8912	2046.22
	+20	1.23631	67.0994	0.741786	66.7825	2529.98
b	−20	1.48084	57.2373	0.888504	64.2087	1610.57
	−10	1.48771	57.2403	0.892626	64.5252	1611.61
	+10	1.50184	57.2464	0.901104	65.1760	1613.70
	+20	1.50910	57.2496	0.905460	65.5101	1614.76
c_1	−20	1.45761	55.9485	0.874566	65.2108	1718.40
	−10	1.47591	56.5958	0.885546	65.0308	1665.12
	+10	1.51403	57.8912	0.908418	64.6609	1561.00
	+20	1.53389	58.5394	0.920334	64.4704	1510.15
c_2	−20	1.53379	57.1790	0.920274	66.5816	1619.58
	−10	1.51388	57.2115	0.908328	65.6983	1616.09
	+10	1.47623	57.2747	0.885738	64.0271	1609.25
	+20	1.45842	57.3054	0.875052	63.2361	1605.88
c_3	−20	1.53126	57.1831	0.918756	66.4694	1619.14
	−10	1.51266	57.2135	0.907596	65.6442	1615.88
	+10	1.47737	57.2727	0.886422	64.0778	1609.46
	+20	1.46061	57.3016	0.876366	63.3334	1606.30
α	−20	1.49451	57.1739	0.896706	64.1849	1618.36
	−10	1.49461	57.2086	0.896766	64.5166	1615.50
	+10	1.49480	57.2781	0.896880	65.1776	1609.80
	+20	1.49488	57.3128	0.896928	65.5067	1606.95
h	−20	1.50387	57.2314	0.902322	65.2491	1614.03
	−10	1.49926	57.2374	0.899556	65.0318	1613.34
	+10	1.49020	57.2493	0.894120	64.6499	1611.97
	+20	1.48574	57.2552	0.891444	64.4543	1611.28

(continued)

fuzzy sense. In this study, we considered the signed distance method as proposed by Chang [3].

$\tilde{a} = (a - \Delta_1, a, a + \Delta_2)$ where $0 < \Delta_1 < a$ and $\Delta_1\Delta_2 > 0, \tilde{A} = A - \Delta_3, A, A + \Delta_4$ where $0 < \Delta_3 < A$ and $\Delta_3\Delta_4 > 0$.

$\tilde{c}_1 = c_1 - \Delta_5, c_1, c_1 + \Delta_6$ where $0 < \Delta_5 < c_1$ and $\Delta_5\Delta_6 > 0, \tilde{c}_2 = c_2 - \Delta_7, c_2, c_2 + \Delta_8$ where $0 < \Delta_7 < c_2$ and $\Delta_7\Delta_8 > 0$.

Table 1 (continued)

Changing parameter	% change in system	Change in T^*	Change in p^*	Change in t_1^*	Change in Q^*	Change in $P^*(T^*, p^*)$
η	−20	1.59373	57.2295	0.956238	69.6239	1634.32
	−10	1.54094	57.2358	0.924564	67.0794	1623.19
	+10	1.45381	57.2518	0.872286	62.8699	1602.62
	+20	1.41731	57.2609	0.850386	61.1021	1593.04

$\tilde{c}_3 = c_3 - \Delta_9, c_3, c_3 + \Delta_{10}$ where $0 < \Delta_9 < c_3$ and $\Delta_9\Delta_{10} > 0, \tilde{c}_4 = c_4 - \Delta_{11}, c_4, c_4 + \Delta_{12}$ where $0 < \Delta_{11} < c_4$ and $\Delta_{11}\Delta_{12} > 0$.

And the signed distance of $\tilde{a}$ to $\tilde{0}$ is given by the relation $d(\tilde{a}, \tilde{0}) = a + 1/4(\Delta_2 - \Delta_1)$ where $d(\tilde{a}, \tilde{0}) > 0$ and $d(\tilde{a}, \tilde{0}) \in [a - \Delta_1, a + \Delta_2]$. Similarly, for other parameters signed distance can be defined as above.

Now, by the fuzzy triangular rule fuzzy total profit per unit is FP $(\tilde{a}, \tilde{A}, \tilde{c}_1, \tilde{c}_2, \tilde{c}_3, \tilde{c}_4) = (F_1, F_2, F_3)$.

And F_1, F_2, F_3 are obtained as

$$F_1 = \frac{1}{T}[p(a - \Delta_1 - p)\gamma T + b(a - \Delta_1 - p)pK_1 - b(a + \Delta_2 - p)pK_2 + p(a - \Delta_1 - p)K_3 - (A + \Delta_4) - (c_1 + \Delta_6)(a + \Delta_2 - p)K_3 - \frac{(c_2 + \Delta_8)(a + \Delta_2 - p)K_5}{\eta} - (c_3 + \Delta_{10})(a + \Delta_2 - p)K_5 - (c_4 + \Delta_{12})(a + \Delta_2 - p)K_6 - (a + \Delta_2 - p)K_7] \tag{15}$$

$$F_2 = \frac{1}{T}[p(a - p)\gamma T + b(a - p)pK_1 - b(a - p)pK_2 + p(a - p)K_3 - A - c_1(a - p)K_4 - \frac{c_2(a - p)K_5}{\eta} - c_3(a - p)K_5 - c_4(a - p)K_6 - (a - p)K_7] \tag{16}$$

$$F_3 = \frac{1}{T}[p(a + \Delta_2 - p)\gamma T + b(a + \Delta_2 - p)pK_1 - b(a - \Delta_1 - p)pK_2 + p(a + \Delta_2 - p)K_3 - (A - \Delta_3) - (c_1 - \Delta_5)(a - \Delta_1 - p)K_4 - \frac{(c_2 - \Delta_7)(a - \Delta_1 - p)K_5}{\eta} - (c_3 - \Delta_9)(a - \Delta_1 - p)K_5 - (c_4 - \Delta_{11})(a - \Delta_1 - p)K_6 - (a - \Delta_1 - p)K_7] \tag{17}$$

Now, defuzzified average profit using the signed distance method is given by

$$\tilde{P}(T, p) = (F_1 + 2F_2 + F_3)/4 \tag{18}$$

Also, the defuzzified order quantity is Q and is given by

$$Q = \left(a + \frac{(\Delta_2 - \Delta_1)}{4} - p\right)\left(t_1 + \frac{bt_1^2}{2} + \frac{\alpha t_1^{\beta+1}}{\beta+1} + \frac{1}{\eta}\log\left(1 + \eta(T - t_1)\right)\right) \tag{19}$$

The necessary conditions for maximizing the average profit are $\partial\tilde{P}(T, p)/\partial T = 0$ and $\partial\tilde{P}(T, p)/\partial p = 0$.

Using the software Mathematica-8.0, from the above two equations, the optimum values of $\tilde{T}$ and $\tilde{p}$ can be determined simultaneously and the optimal value of the average net profit ($\tilde{P}(T, p)$) can be obtained by (18).

7 Numerical Example

$A = 200, \Delta_3 = 10, \Delta_4 = 20, a = 100, \Delta_1 = 5, \Delta_2 = 10, b = 0.02, c_1 = 12, \Delta_5 = 0.60, \Delta_6 = 1.2, c_2 = 8, \Delta_7 = 0.4, \Delta_8 = 0.8, c_3 = 15, \Delta_9 = 0.75, \Delta_{10} = 1.5, c_4 = 0.4, \Delta_{11} = 0.02, \Delta_{12} = 0.04, \alpha = 0.1, \beta = 0.01, \gamma = 0.6, h = 0.6, \eta = 0.5, \delta = 0.02$.

Based on these input data, the findings are as follows:
$p_f^* = 58.0034, T_f^* = 1.48131, t_{1f}^* = 0.888786, Q_f^* = 65.0234$ and $\tilde{P}(p_f^*, T_f^*) = 1646.53$.

8 Sensitivity Analysis

See Table 2.

8.1 Observations

(1) When $\Delta_1 > \Delta_2$, i.e., demand parameter **a** decreases then the optimal profit $\tilde{P}(p_f^*, T_f^*)$ decreases and when $\Delta_1 < \Delta_2$ i.e., **a** increases then $\tilde{P}(p_f^*, T_f^*)$ increases.

(2) When $\Delta_3 > \Delta_4$, i.e., ordering cost **A** decreases then $\tilde{P}(p_f^*, T_f^*)$ slightly increases and when $\Delta_3 < \Delta_4$ i.e., **A** increases then profit decreases. Similarly, profit increases and decreases as $\Delta_5 > \Delta_6, \Delta_7 > \Delta_8, \Delta_9 > \Delta_{10}$ and $\Delta_5 < \Delta_6, \Delta_7 < \Delta_8, \Delta_9 < \Delta_{10}$ respectively.

(3) There is a small decline in profit $\tilde{P}(p_f^*, T_f^*)$, as $\Delta_{11} < \Delta_{12}$ i.e., deterioration cost c_4 increases and profit increases as $\Delta_{11} > \Delta_{12}$ i.e., c_4 decreases.

Table 2 Sensitivity table with respect to system parameters

Changing parameter	Change in parameter	Change in T_f^*	Change in p_f^*	Change in t_{1f}^*	Change in Q_f^*	Change in $\tilde{P}(p_f^*, T_f^*)$
Δ_1, Δ_2	(40, 20)	1.57895	54.9325	0.947370	64.0666	1387.46
	(10, 5)	1.52093	56.7755	0.912558	64.7378	1545.21
	(5, 10)	1.48131	58.0034	0.888786	65.0234	1646.53
	(20, 40)	1.42152	59.8449	0.852912	65.2461	1792.81
Δ_3, Δ_4	(80, 40)	1.42603	57.9725	0.855618	62.7247	1655.13
	(20, 10)	1.45933	57.9912	0.875598	64.1103	1649.93
	(10, 20)	1.48131	58.0034	0.888786	65.0234	1646.53
	(40, 80)	1.51398	58.0216	0.908388	66.3780	1641.53
Δ_5, Δ_6	(4.8, 2.4)	1.47188	57.5999	0.883128	65.2270	1670.29
	(1.2, 0.6)	1.47660	57.8415	0.885960	65.0666	1660.11
	(0.6, 1.2)	1.48131	58.0034	0.888786	65.0234	1646.53
	(2.4, 4.8)	1.49085	58.2477	0.894510	65.0577	1615.99
Δ_7, Δ_8	(3.3, 1.6)	1.49001	57.9809	0.894006	65.4258	1648.17
	(0.8, 0.4)	1.48605	57.9957	0.891630	65.2357	1647.42
	(0.4, 0.8)	1.48131	58.0034	0.888786	65.0234	1646.53
	(1.6, 3.3)	1.47002	58.0139	0.882012	64.5295	1644.42
Δ_9, Δ_{10}	(6, 3)	1.48900	57.9833	0.893400	65.3794	1647.98
	(1.5, 0.75)	1.48576	57.9961	0.891456	65.2228	1647.36
	(0.75, 1.5)	1.48131	58.0034	0.888786	65.0234	1646.53
	(3, 6)	1.47135	58.0124	0.882810	64.5881	1644.67
Δ_{11}, Δ_{12}	(0.16, 0.08)	1.48134	58.0018	0.888804	65.0271	1646.63
	(0.04, 0.02)	1.48133	58.0028	0.888798	65.0252	1646.59
	(0.02, 0.04)	1.48131	58.0034	0.888786	65.0234	1646.53
	(0.08, 0.16)	1.48128	58.0044	0.888768	65.0206	1646.41

9 Conclusion

In this study, the model is proposed in the following two senses: (1) crisp and (2) fuzzy. In fuzzy situation, demand rate, ordering cost, purchasing cost, deterioration cost, backordering cost, and opportunity cost are considered as triangular fuzzy numbers. It is assumed that the demand rate is a function of price and stock both, shortages are allowed and are partially backlogged. The proposed model is more practical as real life businesses are affected by the stock and price dependent demand. If there is a large pile of stock and a suitable selling price of the products is sustained in a business then it attracts the customers to buy more. In real life situations, all the shortages cannot be fully backlogged as some sales will be lost due to interruption of time, therefore partial backlogging is more realistic. In today's market due to the impreciseness of the inventory costs and demand, it is more useful to consider them as fuzzy numbers since the business strategy will be able to face the upper and down conditions of the market. Moreover, from sensitivity table, it is observed that the model is enough stable toward the changes in the system parameters.

Acknowledgments The second author wish to thank to Council of Scientific and Industrial Research (New Delhi) for providing financial help in the form of JRF vide letter no. 08/017(0017)/2011-EMR-I.

References

1. Zadeh, L.: Fuzzy sets. Inf. Control **8**, 338–353 (1965)
2. Park, K.S.: Fuzzy-set theoretic interpretation of economic order quantity. IEEE Trans. Syst. Man Cybern. SMC-**17**, 1082–1084 (1987)
3. Chang, H.C.: An application of fuzzy sets theory to the EOQ model with imperfect quality items. Comput. Oper. Res. **31**, 2079–2092 (2004)
4. Wang, J., Fu, Q.L., Zeng, Y.R.: Continuous review inventory models with a mixture of backorders and lost sales under fuzzy demand and different decision situations. Expert Syst. Appl. **39**, 4181–4189 (2012)
5. Ghare, P.M., Schrader, G.F.: A model for exponential decaying inventory. J. Ind. Eng. **14**, 238–243 (1963)
6. Covert, R.P., Philip, G.C.: An EOQ model for items with Weibull distribution deteriorating. AIIE Trans. **5**, 323–326 (1973)
7. Raffat, F.: Survey of literature on continuously deteriorating inventory model. Eur. J. Oper. Res. Soc. **42**, 27–37 (1991)
8. Teng, J.T., Chang, H.J., Dye, C.Y., Hung, C.H.: An optimal replenishment policy for deteriorating items with time-varying demand and partial backlogging. Oper. Res. Lett. **30**, 387–393 (2002)
9. Bakker, M., Riezebos, J., Teunter, R.H.: Review of inventory systems with deterioration since 2001. Teunter. **221**, 275–284 (2012)
10. Park, K.S.: Inventory models with partial backorders. Int. J. Syst. Sci. **13**, 1313–1317 (1982)
11. Wee, H.M.: A deterministic lot-size inventory model for deteriorating items with shortages and a declining market. Comput. Oper. Res. **22**, 345–356 (1995)
12. Singh, S.R., Singh, T.J.: Perishable inventory model with quadratic demand, partial backlogging and permissible delay in payments. Int. Rev. Pure Appl. Math. **1**, 53–66 (2008)
13. Hsieh, T.P., Dye, C.Y., Ouyang, L.Y.: Optimal lot size for an item with partial backlogging rate when demand is stimulated by inventory above a certain stock level. Math. Comput. Model. **51**, 13–32 (2010)
14. Gupta, R., Vrat, P.: Inventory model with multi-items under constraint systems for stock dependent consumption rate. Oper. Res. **24**, 41–42 (1986)
15. Mandal, B.N., Phaujdar, S.: An inventory model for deteriorating items and stock-dependent consumption rate. J. Oper. Res. Soc. **40**, 483–488 (1989)
16. Ray, J., Goswami, A., Chaudhuri, K.S.: On an inventory model with two levels of storage and stock-dependent demand rate. Int. J. Syst. Sci. **29**, 249–254 (1998)
17. Goyal, S.K., Giri, B.C.: The production-inventory problem of a product with time varying demand, production and deterioration rates. Eur. J. Oper. Res. **147**, 549–557 (2003)
18. Chang, C.T., Chen, Y.J., Tsai, T.R., Wu, S.J.: Inventory models with stock-and price dependent demand for deteriorating items based on limited shelf space. Yugoslav J. Oper. Res. **20**, 55–69 (2010)

Efficient Protocol Prediction Algorithm for MANET Multimedia Transmission Under JF Periodic Dropping Attack

Avita Katal, Mohammad Wazid and R H Goudar

Abstract Mobile ad hoc network is prone to denial of service attack. Jellyfish is a new denial of service attack and is categorized as JF Reorder Attack, JF Periodic Dropping Attack, JF Delay Variance Attack. In JF Periodic Dropping Attack, intruder node intrudes into forwarding group and starts dropping packets periodically. Due to JF Periodic Dropping attack, the delay in the network increases and throughput decreases. In this paper a comparative performance analysis of three reactive routing protocols i.e. AODV, DSR and TORA used in mobile ad hoc network under JF Periodic Dropping attack is done. This work is specially done for multimedia transmission i.e. video and voice. If we have a mobile ad hoc network in which probability of occurrence of JF Periodic Dropping attack is high and also if it requires time efficient network multimedia service for information exchange then TORA protocol is to be chosen. If it requires high multimedia throughput and consistent service in the network then AODV protocol is recommended. An algorithm has been proposed depending upon the analysis done particularly for multimedia transmission in MANET which will help in choosing the best suited protocol for the required network parameters under JF Periodic Dropping attack.

Keywords JF periodic dropping attack · Multimedia transmission · AODV · DSR · TORA

A. Katal (✉) · M. Wazid · R. H. Goudar
Department of CSE, Graphic Era University, Dehradun, India
e-mail: avita207@gmail.com

M. Wazid
e-mail: wazidkec2005@gmail.com

R. H. Goudar
e-mail: rhgoudar@gmail.com

B. V. Babu et al. (eds.), *Proceedings of the Second International Conference on Soft Computing for Problem Solving (SocProS 2012), December 28–30, 2012*, Advances in Intelligent Systems and Computing 236, DOI: 10.1007/978-81-322-1602-5_45,

1 Introduction

An intruder can easily access mobile ad hoc network because of weak defense mechanism and high mobility of nodes. Multimedia transmission includes streaming of multimedia to an end user by the provider. It mainly includes voice and video transmission. MANET assailable to JF Periodic Dropping attack increases the end to end delay between selected packets in a flow with any lost packets being ignored. This is called Jitter which increases under the above mentioned attack leading to degradation in the quality of the media being transmitted. In this paper the impact of presence of JF Periodic Dropping attack on the performance of network transmitting multimedia is analyzed. Comparative analysis of the three routing protocols i.e. AODV, DSR and TORA used for the transmission of multimedia for various network parameters is done. Section 2 includes the literature review about different kinds of work done by various authors in area related to JellyFish attacks. The novelty of the proposed idea is discussed in Sect. 3. In Sect. 4 a brief introduction to JellyFish attack is given. The methodology and experiment design of this work is discussed in Sect. 5. Section 6 contains performance parameters and results. Section 7 includes the algorithm for protocol prediction designed on the basis of results obtained followed by conclusion, future work and application of the work done in Sect. 8.

2 Literature Review

Paper [1] discusses about techniques for resilience of denial of service attacks on a mobile ad hoc network focusing on JellyFish attacks. The throughput of network under JellyFish attacks introduced here is calculated. Techniques to protect MANET i.e. flow-based route access control (FRAC), Multi-Path Routing Source-Initiated Flow Routing, Sequence Numbers etc are discussed. In [2] authors calculate the performance of MANET under black hole attack using AODV routing protocol with HTTP traffic load. In [4] authors explain various attacks on a mobile ad hoc network corresponding to different MANET layers and they also discuss some available attack detection techniques. They give a brief idea about JellyFish attack. In [5] the performance of different routing protocols for multimedia data transmission over vehicular ad hoc networks is done. The focus was put on the performance evaluation metrics that were used in simulations. Three popular routing protocols were selected for the evaluation: two reactive (AODV, DSR) and one proactive (OLSR). In [6] authors develop an algorithm that detects the Jellyfish attack at a single node and that can be effectively deployed at all other nodes. A novel metric depending on reorder density is proposed and comparison table is given which shows the effectiveness of novel metric which helps protocol designers to develop the counter strategies for JF attack. The main objective of [7] is to analyze and compare the performance of Preemptive DSR and temporarily ordered routing algorithm (TORA). It discusses the effect of variation in number of nodes and average speed on protocol performance.

It concludes that PDSR outperforms TORA in terms of the number of MANET control packets used to maintain or erase routes. TORA is a better choice than PDSR for fast moving highly connected set of nodes. In [8] an attempt has been made to compare the performance of two prominent on demand reactive routing protocols for MANETs: ad hoc on demand distance vector (AODV), dynamic source routing (DSR) protocols. It concludes that if the MANET has to be setup for a small amount of time then AODV should be preferred due to low initial packet loss. If we have to use the MANET for a longer duration then both the protocols can be used, because after sometime both the protocols may have same ratio of packet delivering. But AODV have very good packet receiving ratio in comparison to DSR. In [10] the performance of DSR and TORA routing protocols is calculated using the OPNET simulator. It concludes that delay experienced with mobile nodes employing DSR routing is higher than that of fixed nodes. Delay experienced with fixed nodes employing TORA routing is higher than that of mobile nodes. In [12] authors discuss the most common types of attacks on MANET, namely Rushing attack, Blackhole attack, Neighbor attack and JellyFish attack. They simulate these attacks and calculate parameters such as Average end-to-end delay, Average throughput etc. In Paper [13] JellyFish and Black hole attacks are discussed. Authors calculate the impact of JF on the system performance i.e. Throughput etc. They introduce three factors: mobility, node density and system size and calculate the effect of these factors on fairness to receive packets under the presence of various number of JF attackers. They observe that the effect of mobility is more under the absence of JF attackers and fairness reduces when we increase mobility.

3 Problem Definition and Novelty

Previously many authors have analyzed the performance of various MANET protocols for multimedia transmission. In this paper the performance analysis of the most popular reactive routing protocols in the multimedia transmission under JF Periodic Dropping attack is done followed by an algorithm used for selecting the best suited routing protocol for transmitting multimedia with the desire network parameters.

4 JellyFish Attack

JellyFish attack is related to transport layer of MANET stack. The JF attacker disrupts the TCP connection which is established for communication. JellyFish (JF) attacker needs to intrude into forwarding group and then it delays data packets unnecessarily for some amount of time before forwarding them. Due to JF attack, high end to end delay takes place in the network. So the performance of network (i.e. throughput etc) decreases substantially. JF attacker disrupts the whole functionality of TCP, so performance of real time applications become worse. JF attack is further divided into

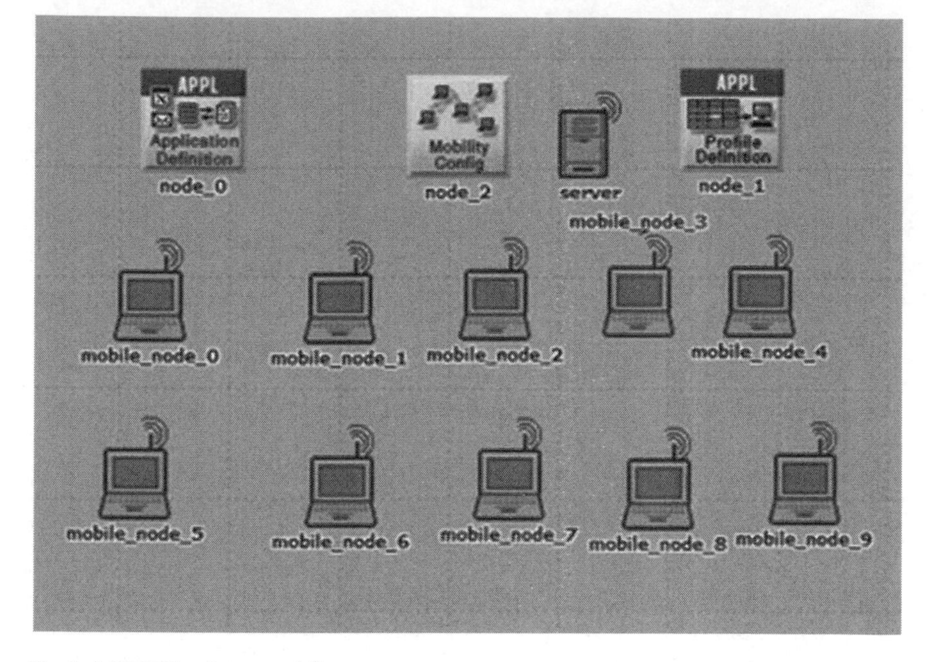

Fig. 1 MANET under normal flow

three categories- JF Reorder Attack, JF Periodic Dropping Attack, JF Delay Variance Attack.

4.1 JF Periodic Dropping Attack

In this attack the JF attackers drop all packets for a short duration of time. Thus JF nodes seem passive in nature and do not generate traffic themselves. JF nodes drop packets for only a small fraction of time due to dropping of packets the performance becomes worse [1].

5 Methodology and Experiment Design

For experimental purpose we simulate a mobile ad hoc network under JF Periodic Dropping attack for three reactive routing protocols i.e. AODV, DSR and TORA using Opnet modeler. We are using the following two simulation scenarios in this paper:

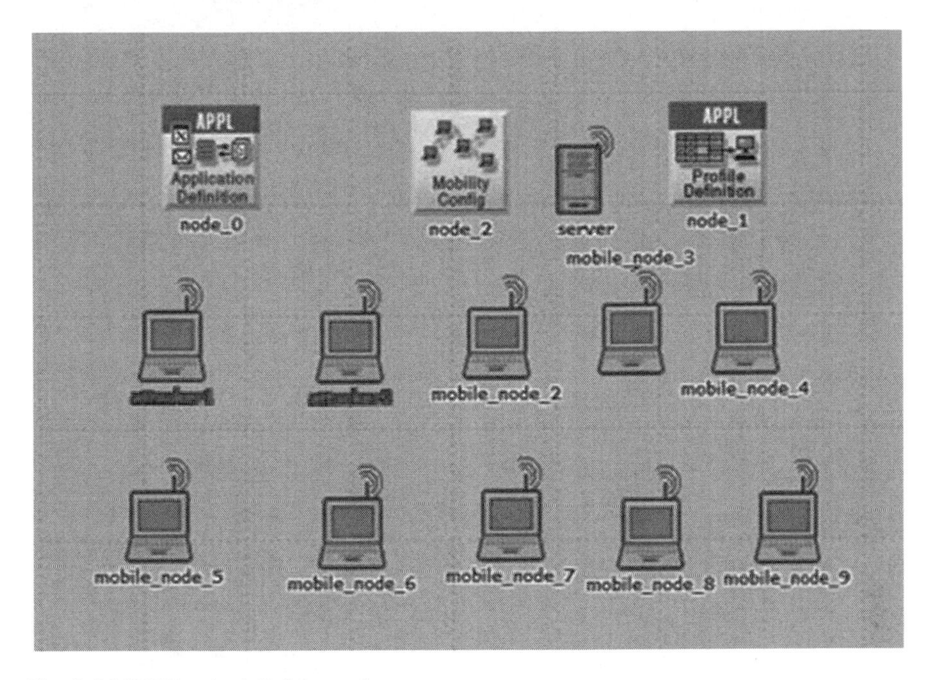

Fig. 2 MANET under jellyfish attack

In Fig. 1 we use 10 mobile nodes and build a scenario without any JF attacker showing a normal flow of traffic. In Fig. 2 we use 10 mobile nodes and build a scenario with two JF Periodic Dropping attackers. JF attackers are shown in red label i.e. attacker 1 and attacker 2. All scenarios are simulated using AODV, DSR and TORA protocols.

The experimental design setup is used to examine the performance of three reactive routing protocols under JF Periodic Dropping attack in MANET transmitting multimedia data.

5.1 Experiment Design Parameters

5.1.1 Common Parameters

Implementations of JF Periodic Dropping Attack

The normal packet forwarding rate is 100,000 packets per second and simulation time is ten minutes. To simulate JF Periodic Dropping attack the time of periodic dropping is taken to five minutes. During other five minutes there is normal flow. Given two scenarios are simulated under routing protocols i.e. AODV, DSR and TORA (Table 1).

Table 1 Common parameters used in simulation

Parameter	Value
Platform	Windows XP SP2
Simulator	Opnet modeler 14.5
Area	500 × 500 m (Fix)
Network size	10 nodes
Mobility model	20 m/s (Fix)
Traffic type	Video and Voice both
Simulation time	10 min
Address mode	Only IPv4
Ad Hoc routing protocol	AODV, DSR, TORA
AODV, DSR, TORA, TCP parameters	Default
JellyFish attackers	Zero attacker for normal flow (Scenario 1)
	Two attackers (Scenario 2)
Attacking scenario	For 5 min, normal flow
	For 5 min, flow under JF packet dropping
Packet size (bits)	Exponential (1024)

Table 2 End-to-end delay and throughput

Parameters	End-to-end delay (s)		Throughput (bps)	
Protocol	Normal flow	Under attack	Normal flow	Under attack
AODV	6.72	8.47	51573.91	41517.39
DSR	7.74	10.32	25832.07	19263.31
TORA	6.87	8.47	21714.92	16903.32

5.2 Results

In simulation we take following statistics of the network: End-to-end Delay (msec), Throughput (bps).

Figure 3 shows end-to-end delay with normal flow (zero attackers) and also in the presence of JF attackers for all given three protocols. Figure 4 shows throughput

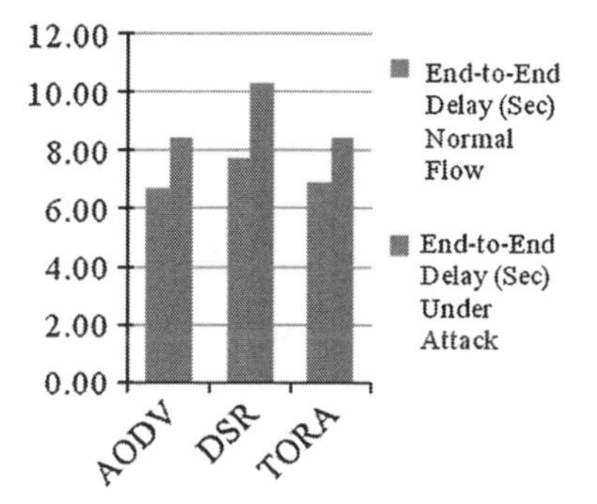

Fig. 3 End-to-end delay

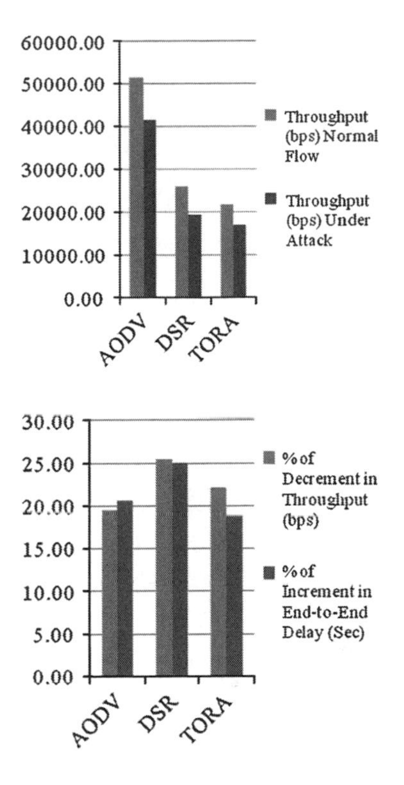

Fig. 4 Throughput (bps)

Fig. 5 Impact of JF periodic dropping attack

Table 3 Impact of JF periodic dropping attack on end-to-end delay and throughput

Protocol	% of Decrement in throughput (bps)	% of Increment in end-to-end delay (Sec)
AODV	19.50	20.66
DSR	25.43	25.00
TORA	22.16	18.89

(bps) with normal flow (zero attackers) and also in the presence of JF attackers for all given three protocols.

Figure 5 shows impact of JF Periodic Dropping attack on end-to-end delay and throughput for all given three protocols.

6 Protocol Prediction Algorithm

enum protocol set => {AODV,TORA,DSR}
enum service set => {Throughput Efficient, In Time}
if service = Throughput Efficient then

select_protocol = AODV
otherwise if service = In Time then
select_protocol = TORA
otherwise
select_protocol = DSR

Complexity Analysis:
The complexity of protocol prediction algorithm comes to be $\ominus$ (1) in combination with the complexity of AODV, TORA or DSR. We can say the complexity of the above algorithm comes in the order of the complexity of the AODV, TORA or DSR.

The above algorithm designed is basically used for choosing the appropriate reactive routing protocol out of the set of three protocols i.e. AODV, DSR and TORA for efficient multimedia transmission depending upon the network parameter requirements that is throughput efficiency and in time delivery.

7 Key Findings

Here, we try to evaluate the performance of three reactive protocols i.e. AODV, DSR and TORA which are implemented in mobile ad hoc network under the presence of JF Periodic Dropping attack for multimedia transmission. Some of the observations are as:

- The performance of DSR is worst for both the network parameters i.e. throughput and delay in multimedia transmission (as shown in Table 2).
- In multimedia transmission a throughput efficient service is provided by AODV protocol as compared to DSR and TORA. The % of decrement in throughput for AODV is 19.50 as compared to DSR and TORA which are having a decrement of 25.43 and 22.16 respectively (as shown in Table 3).
- Multimedia time demanding service must use TORA protocol as TORA proves to be more efficient under JF Periodic Dropping attack with a % increment of end-to-end delay of 18.89 as compared to AODV and DSR which are having 20.66 and 25.00 respectively (as shown in Table 3).

8 Conclusion

If we have a mobile ad hoc network in which probability of occurrence of JF Periodic Dropping attack is high and we want a good time efficient network multimedia service then we have to choose TORA protocol whereas good throughput multimedia service should make use of AODV protocol (as shown in Table 3). Depending upon these results the protocol prediction algorithm is proposed which efficiently chooses the required protocol for multimedia transmission.

Here we take mobility and system size as constant, if we change these two factors then performance may vary. So this work can be further extended to calculate the performance of MANET under varying mobility and system size.

There are certain applications which can bear the time inefficiency but require high throughput service as in case of virtual classrooms where the student can attend lectures even when they are not sitting in the campus class whereas in case of Warfield the time efficiency would be important as compared to throughput. Depending upon these different requirements for different scenarios we recommend AODV for virtual classrooms and TORA for Warfield applications.

References

1. Begum, S. A., Mohan, L., Ranjitha, B.: Techniques for resilience of denial of service attacks in mobile ad hoc networks. In: Proceedings published by International Journal of Electronics Communication and Computer Engineering, Vol. 3(1), NCRTCST, ISSN 2249–071X National Conference on Research Trends in Computer Science and Technology (2012)
2. Barkhodia, E., Singh, P., Walia, G. K.: Performance analysis of AODV using HTTP traffic under Black Hole Attack in MANET. Comput. Sci. Eng. Int. J. (CSEIJ). **2**(3), (2012)
3. Jan, V. M., Welch, I., Seah, W. K. G: Security threats and solutions in MANETs: A case study using AODV and SAODV. Elsevier J. Netw. Comput. Appl. **35**, 1249–1259 (2012)
4. Wazid, M., Singh, R. K., Goudar, R. H.: A survey of attacks happened at different layers of mobile ad-hoc network and some available detection techniques. In: Proceedings published by International Journal of Computer Applications (IJCA) International Conference on Computer Communication and Networks CSI- COMNET, Hawaii, December (2011)
5. Adam, G., Kapoulas, V., Bouras, C., Kioumourtzis, G., Gkamas, A., Tavoularis. N.: Performance evaluation of routing protocols for multimedia transmission over mobile ad hoc networks. In: IEEE International Conference on Wireless and Mobile Networking Conference (WMNC), (2011)
6. Jayasingh, B. B., Swathi, B.: A novel metric for detection of jellyfish reorder attack on ad hoc network. BVICAM'S Int. J. Inf. Technol. (BIJIT) **2**(1), ISSN 0973–5658 (2010)
7. Ramesh, V., Subbaiah, P., Koteswar Rao, N., Subhashini, N., Narayana.D.: Performance comparison and analysis of preemptive-DSR and TORA. Int. J. Ad hoc, Sens. and Ubiquitous Comput. (IJASUC) **1**(4) (2010)
8. Thakare, A. N., Joshi, M. Y.: Performance analysis of AODV & DSR routing protocol in mobile ad hoc networks. IJCA Special Issue on Mobile Ad-hoc Networks MANETs (2010)
9. Cai, J., Yi, P., Tian, Y., Zhou, Y., Liu, N.: The simulation and comparison of routing attacks on DSR protocol. In: IEEE International Conference on Wireless Communications, Networking and Mobile Computing (WiCom), Beijing, September (2009)
10. Amer, S. H., Hamilton, J. A.: DSR and TORA in fixed and mobile wireless networks. In: Proceedings of ACM-SE, ACM, Auburn, March (2008)
11. Alshanyour, A. M., Baroudi, U.: Bypass AODV: improving performance of ad hoc on- demand distance vector (AODV) routing protocol in wireless ad hoc networks. In: Proceedings of ACM Ambi-sys, ACM Digital Library, Quebec city, 11–14 February (2008)

12. Nguyen, H. L., Nguyen, U. T.: A study of different types of attacks on multicast in mobile ad hoc networks. Elsevier J. Ad Hoc Netw. **6**, 32–46 (2008)
13. Aad, I., Hubaux, J.-P., Knightly, E. W.: Denial of service resilience in ad hoc networks. In: Proceedings of ACM MobiCom, Philadelphia, Sept. 26–Oct.1 (2004)

New Placement Strategy for Buffers in Critical Chain

Vibha Saihjpal and S. B. Singh

Abstract With the introduction of Critical chain by Goldratt in 1997, there has been a lot of research in the field of resource constraint project scheduling problems (RCPSP) and Buffer Sizing techniques. This paper suggests a Buffer management technique which aims at reducing the make span time yet maintaining the stability of the project. In the theory of CCPM it is suggested to reduce the duration of all the activities by half to remove the excess safety time in each activity. The trimmed duration is collected and made available at the end of the project in the form of project buffer which could be used if the project gets delayed. Another buffer called the feeding buffer is added whenever a noncritical chain joins a critical chain. This increases the project duration if the slack of the last activity in the feeding chain is smaller than the feeding buffer. In such cases, this paper suggests the division of the project buffer into parts, fitting each part at the junction of critical and noncritical chains so that the delay occurred because of addition of feeding buffer can be utilized and the duration of project buffer is shortened. The use of the proposed technique has reduced the project duration by a significant value.

Keywords Buffer · Feeding buffer · Project buffer · Buffer sizing technique · CCPM

V. Saihjpal (✉)
University College, Chunni Kalan, Punjab, India
e-mail: saihjpal.vibha@gmail.com

S. B. Singh
Department of Mathematics, Punjabi University, Patiala, India
e-mail: sbsingh69@yahoo.com

B. V. Babu et al. (eds.), *Proceedings of the Second International Conference on Soft Computing for Problem Solving (SocProS 2012), December 28–30, 2012*, Advances in Intelligent Systems and Computing 236, DOI: 10.1007/978-81-322-1602-5_46,

1 Introduction

In most simple terms, project scheduling means the planned dates for performing activities and the planned dates for meeting milestones. It is a plan of procedure, usually written for a proposed objective, specially, with reference to the sequence of activities satisfying certain constraints such as the precedence relations and resource constraints and time allotted for each activity or operation necessary to its completion. PERT and CPM have been the most basic tools used for project scheduling for years. Considering the probabilistic time durations, the longest path (called the critical path) in the network of activities is determined which marks the project duration. The critical path however does not recognize the resource constraints. Considering the role of resource constraints in a practical project schedule, Goldratt in 1997 introduced the concept of critical chain project management (CCPM).

1.1 PERT/CPM

PERT was devised in 1958 for the POLARIS missile program by the Program Evaluation Branch of the Special Projects office of the U.S. Navy, helped by the Lockheed Missile Systems division and the Consultant firm of Booz-Allen and Hamilton [1]. One key element to PERT's application is that three estimates are required because of the element of uncertainty and to provide time frames for the PERT network. These three estimates are classed as optimistic, most likely and pessimistic time durations, and are predicted for each activity of the overall project. Generally, the optimistic time estimate is the minimum time the activity would take, considering that all goes right the first time. The reverse is the pessimistic estimate, or maximum time estimate for completing the activity. This estimate takes into account Murphy's Law, i.e., whatever can go wrong will and all possible negative factors are considered when computing this estimate. The third is the most likely estimate, or the normal or realistic time an activity requires. It lies anywhere in the interval (a, b), where a represents the optimistic time and b represents the pessimistic time. Two other elements that comprise the PERT networks are the path, or critical path, and slack time. The critical path is a combination of the critical activities, i.e., the activities which if delayed, would delay the project.

The primary goal of a CPM analysis of a project is the determination of the critical path, which determines the completion time of a project [2, 3]. However, the schedule defined by CPM does not consider the resource constraints, hence giving rise to the development of the theory of critical chain.

1.2 CCPM

In 1997, Dr. Eliyahu Goldratt introduced a new significant approach to project management, in over 30 years, with the publication of his best selling business novel, Critical Chain [4]. Critical chain is the theory of constraints (TOC) philosophy for project management which is considered an innovation that would be useful to organizations [5, 6]. Goldratt's approach introduced a new era in the world of project management by merging, for the first time, both the human side and the algorithmic methodology side of project management in a unified discipline. It considers factors such as

- Parkinson's Law: Work expands to fill the available time.
- Student Syndrome: People start to work in full fledge when deadline is near.
- Bad Multitasking: Bad multitasking can delay start of the successor tasks.

As the activity duration is defined, the project manager tends to keep some safety time in the duration of each activity which increases the project duration unnecessarily and as suggested by Parkinson's Law, each activity consumes all the duration allotted to it. Considering this, it is suggested in CCPM to cut the duration in every activity so as to minimize this wastage. However, buffers are added at specific stages of the project so that any delay in any of the activities does not affect the total project duration. Various methods have been proposed to manage buffer sizing. The C&P, i.e., Cut and Paste Method and RSE, i.e., Root Square Error Method are the most traditional ones. Xie et al. suggested improved root square error (IRSE), a method based on critical chain theory for buffer sizing majorly suitable for software projects and the results claim to have a direct effect in shortening the project duration [7]. Tukel et al. suggested two methods for determining feeding buffers, one incorporates resource tightness while the other uses network complexity resulting in smaller buffer sizes [8]. Fuzzy numbers have also been used in methods to determine the project buffers [9]. It has been observed that through Critical Chain Project Management, projects are completed in significantly shorter time than traditional Critical Path project management techniques. Importantly, Critical Chain Project Management is also simpler to use and requires less work for the project team in both the planning and tracking phases of project [10]. Bevilacqua et al. applied the Theory of Constraints and Risk Assessment to develop a prioritization method for Work Packages, using the critical chain concept and concluded that the proposed method allowed the company to maximize the quality and safety of work and minimized the turnaround time and cost [11].

Another important aspect of CCPM is to start each activity at the latest possible starting time. This helps the project manager gain experience on the project till the time the activity starts and also reduces the WIP time. However it is argued that this makes each activity critical. But the presence of buffers compensates any delays in the activity completion. The following steps are used to modify the critical path into the critical chain [8].

Step 1. Reduce duration of each activity by 50 %.

Step 2. Push all the tasks to as late as possible subject to the precedence relations. (i.e., determine the late finish network).

Step 3. Eliminate the resource constraints by resequencing the tasks. Though Goldratt has not offered any specific procedure to resolve resource constraints this has given rise to a research problem marked as RCPSP (Resource constraint Project Scheduling Problem) and a lot of research has been done on these problems.

Step 4. Identify the critical chain as the longest chain of dependent events for feasible schedule that was identified in step 3.

There could be ties in longest chain, in that case an arbitrary choice can be made between them [12].

Step 5. Add the project buffer to the end of the critical chain.

Step 6. Add the feeding buffers wherever a noncritical chain feeds the critical chain and offset the tasks on the feeding chain by the size of the buffer.

2 Buffers

The reduction of activity durations by 50 % as suggested by Goldratt is compensated by the insertion of buffers at various stages of the project. A buffer is a cushion provided at different stages of a project schedule so as to absorb the delays that occur during the execution of the project up to a maximum possible extent. In the development of the project plan the duration of each activity will be coupled with a lot of security time. Even if a task is ahead of schedule, the security time will not accumulate to the next activity. This problem is tackled by the use of buffers. CCPM suggests that duration of each activity is halved and the duration trimmed from each activity is accumulated and added at the end of the project as project buffer and at various other stages in the form of feeding buffer and resource buffer. A project buffer is inserted to protect the project delivery date. Resource buffers are inserted at every point where work passes from one resource to another on the critical chain. Feeding buffers are inserted to protect the critical chain from delays in the noncritical chain. [13] suggested that the project buffer size is the more appropriate robustness measure regardless of the network complexity.

3 The Placing of Buffers

This paper suggests that whenever the project duration is forced ahead because of the insertion of a feeding buffer (Fig. 1b), there is a void or a gap created in the critical chain. In such a case a part of the project buffer is suggested to be removed from the project buffer and inserted at this point (Fig. 1c), since the activities before this point have got extra time and any delays can be absorbed by this extra time and same is the role of a buffer. This part of the critical chain has got its share of the protection at this point only and hence the project buffer placed in the end of the project need not contribute to its protection. Hence this part of the project buffer is taken out and placed in the gap created in the critical chain. The contribution of project buffer to this part of the chain is given by:

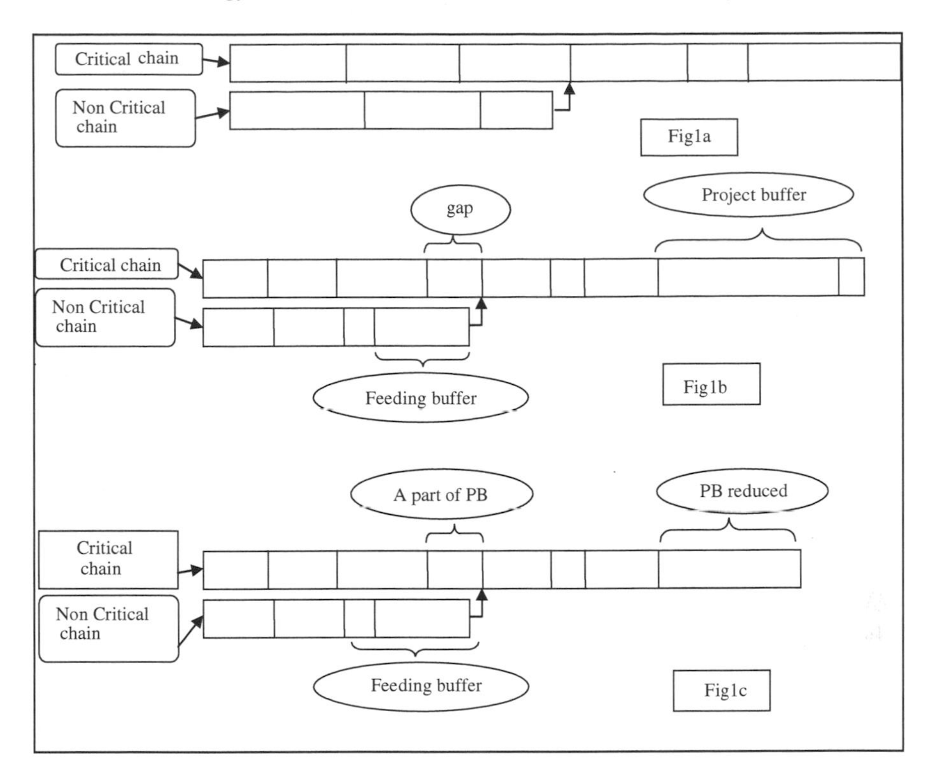

Fig. 1 Use of buffers

$$PB_i = \frac{\sum_{j \in C_i} t_j}{T}(PB) \tag{1}$$

where PB_i is ith part of the project buffer; C_i is the set of activities in PB_i; [t_j is the duration of activities; T is the project duration, i.e., length of the critical chain and PB is the project buffer duration.

If the buffer required by part of the critical chain preceding the gap is less than the gap, the part of project buffer can fit into it. This would decrease the project duration by $\sum_i PB_i$. Else the project would be reduced by difference between the gap and the slack.

4 Numerical Example

The validation of the procedure suggested in Sect. 2 is done here through a numerical example. Consider the project scheduling problem as given in Table 1.

Table 2 shows the results obtained.

Table 1 Numerical example

Activity	Preceded by	Activity durations		
		Optimistic	Most likely	Pessimistic
A		45	51	60
B	A	19	23	30
C	A	27	32	43
D	A	38	42	52
E	A	28	32	39
F	A	23	27	34
G	B	33	40	47
H	B	43	47	54
I	E	31	39	52
J	D, H	24	30	38
K	C, G, F, I	39	43	50
L	C, G, F, I	25	31	42
M	K	16	20	25
N	M, J, L	43	49	60

4.1 Results from the Numerical Example

The critical chain is $A \rightarrow E \rightarrow I \rightarrow K \rightarrow M \rightarrow N$. The feeding chains are $A \rightarrow B \rightarrow G \rightarrow K$, $A \rightarrow C \rightarrow K$, $A \rightarrow F \rightarrow K$, etc. Feeding chains are joining the critical chain at activity K and N. The corresponding feeding buffer sizes are shown in Table 2. Buffer size for the feeding chain $A \rightarrow B \rightarrow G$ is greater than the slack of activity G. Hence the next activity on the critical chain, i.e., activity K is shifted by (34.38749773-8) 27 units (approximately). Hence there exists a gap of 27 units in the critical chain. Hence a part from the project buffer which is the **share of the A → E → I** part of the critical chain given by (1) is added in this gap. If this share is less than the gap, the final project duration is reduced by these many units of time. Similarly share of buffer for the part $K \rightarrow M \rightarrow N$ of the critical chain is added after activity N, i.e., as the project buffer.

Directly adding the project buffer would have increased the project duration by nearly **49.4873** units of time while using the suggested procedure where the project duration is increased by nearly 24 units only.

Also since the buffer consumption is less than 100 % for all the buffers the schedule is stable. The buffer consumption rate is calculated using the formula $(a + b - e)/b$ where a is the most likely time of activity duration, b is the buffer size and e is the estimated duration which is calculated using the Beta distribution.

Table 2 Results of the numerical example

Activity	Expected duration (unit of time)	Slack for activity	Buffer size (units of time)	Buffer utilization	
A	51.5	0			
B	23.5	8			
Buffer for $A \rightarrow C$					
C	33	39	30.104	95.02 %	
D	43	62			
E	32.5	0			
Buffer for $A \rightarrow F$					
F	27.5	44	28.8531	96.53 %	
Buffer for $A \rightarrow B \rightarrow G$					
G	40	8	34.388	97.092 %	
H	47.5	34			
I	39.833	0			
			Buffer for $A \rightarrow B \rightarrow H \rightarrow J$	$A \rightarrow B \rightarrow H \rightarrow J$	$A \rightarrow D \rightarrow J$
J	30.333	34	39.493	44.72 %	95.358 %
K	43.5	0	25.801	92.89 %	
Buffer for $A \rightarrow E \rightarrow I \rightarrow L$					
L	31.833	32	39.074	95.308 %	
M	20.166	0			
N	49.833	0	23.686	93.667 %	
			Project buffer	*Project buffer utilisation*	
			49.4873	93.264 %	

5 Conclusion

The suggested procedure reduces the project duration while maintaining the project stability. The division of the project buffer into parts and each part being used at the gap which otherwise was being wasted has reduced the project duration. Since the buffer is only reshuffled and not reduced, there is no harm to the stability.

References

1. Reference for Business Encyclopedia of Business, 2nd ed. Per-Pro Program Evaluation and Review Technique (PERT) http://www.referenceforbusiness.com/encyclopedia/Per-Pro/Program-Evaluation-and-Review-Technique-PERT.html#ixzz0mVjkM5k2
2. Badiru, A.B.: Activity resource assignments using critical resource diagramming. Proj. Manage. J. **XXIV**, 15–21 (1993)
3. Gemmill, D.D., Tsai, Y.-W.: Using a simulated annealing algorithm to schedule activities of resource constrained projects. Proj. Manage. J., pp. 8–20 (1997)
4. Goldratt, E.: Critical Chain, 1st edn. North River Press, Barrington (1997). ISBN 0-88425-153-6
5. Steyn, H.: An investigation into the fundamentals of critical chain project scheduling. Int. J. Proj. Manage. **19**(6), 363–369 (2001)
6. Steyn, H.: Project management applications of the theory of constraints beyond critical chain scheduling. Int. J. Proj. Manage. **20**(1), 75–80 (2002)
7. Xie, X, Yang, G., Lin, C.: Software development projects IRSE buffer settings and simulation based on critical chain. J .China Univ .Posts Telecommun. **17**(Suppl 1), 100–106 (2010)
8. Tukel, O.I., Rom, W.O., Eksioglu, S.D.: An investigation of buffer sizing techniques in critical chain scheduling. Eur. J. Oper. Res. **172**(2), 401–416 (2006)
9. Long, L.D., Ohsato, A.: Fuzzy critical chain method for project scheduling under resource constraints and uncertainty. Int. J. Proj. Manage. **26**(6), 688–698 (2008)
10. http://www.civiles.org/publi/Gestion/Critical-Chain-Concepts.pdf
11. Bevilacqua, M., Ciarapica, F.E., Giacchetta, G.: critical chain and risk analysis applied to high-risk industry maintenance. A case study. Int. J. Proj. Manage. **27**(4), 419–432 (2009)
12. Herroelen, W., Leus, R.: On the merits and pitfalls of critical chain scheduling Original Research Article. J. Oper. Manage. **19**(5), 559–577 (2001)
13. Hazir, O., Haouari, M., Erel, E.: Robust scheduling and robustness measures for the discrete time/cost trade-off problem. Eur. J. Oper. Res. **207**, 633–643 (2010)

Inducing Fuzzy Association Rules with Multiple Minimum Supports for Time Series Data

Rakesh Rathi, Vinesh Jain and Anshuman Kumar Gautam

Abstract Technological changes have occurred at an exponential rate in recent years leading to the generation of large amount of data in various sectors. Several database and data warehouse is built to store and manage the data. As we know the data which are relevant to us should be extracted from the database for our task. Earlier different mining approaches are proposed in which items are collected at same minimum support value. In this paper we propose a fuzzy data mining algorithm which generates the fuzzy association rules from time series data having different minimum support values. The temperature varying dataset is used to generate fuzzy rules. The proposed algorithm also predicts the variation of temperature. Experiments are also performed to get the desired result.

Keywords Association rule · Data mining · Different minimum support · Fuzzy set

1 Introduction

Data mining plays a vital role in today's application. So, researchers are paying more attention toward the new tricks and techniques which can be evolved in it. It covers a large domain where it is frequently applied such as business, medical, biometrics. Fuzzy concepts have a great impact on data dredging methodology. Various data warehouses are managed to store and use the data efficiently through different domain. Time series data is a collection of data points which has some specific value

R. Rathi (✉) · V. Jain · A. K. Gautam
Government Engineering College Ajmer, Ajmer, Rajasthan, India
e-mail: rakeshrathi4@rediffmail.com

V. Jain
e-mail: vineshjain1280@gmail.com

A. K. Gautam
e-mail: er.anshuman2011@gmail.com

B. V. Babu et al. (eds.), *Proceedings of the Second International Conference on Soft Computing for Problem Solving (SocProS 2012), December 28–30, 2012*, Advances in Intelligent Systems and Computing 236, DOI: 10.1007/978-81-322-1602-5_47,

at that instant of time. It varies with respect to time. This paper proposes an algorithm which induce fuzzy association rule with multiple minimum support value. Earlier many algorithms have been proposed but they follow only single support value condition. Sometimes itemset has different minimum support. To explain this we applied a fuzzy concept on time series data more specifically temperature varying data. As we know that time series data comes under the category of Sequence data which has some trend or pattern in it. So algorithm would predict the near temperature using the trend analysis. The proposed algorithm has two advantages:—First, the result would be easier to understand as we are using fuzzy theory which is quite familiar with natural language. Second, it also helps to determine the sudden change in temperature of a place.

The remaining parts of this paper are assembled as follows: review of fuzzy set theory is given in Sect. 2. The related work of the paper is explained in Sect. 3. The proposed algorithm is explained in given in Sect. 4. Further experimental results are shown in Sect. 5. Finally conclusion and future work is discussed in Sect. 6.

2 Fuzzy Set Theory

Fuzzy set theory was pointed out in 1965 by Zadeh in his seminal paper entitled "Fuzzy sets" which played a vital role in human thinking, focusing in the domains of pattern recognition, communication of information and abstraction. Fuzzy set theory consists of fuzzy membership functions. Fuzzy set expresses the degree to which an element belongs to a set called as characteristic function. For a given crisp set B, the function assigns a value $\mu_B(x)$ to every $x \in X$ such that

$$\mu_B(x) = \{1 \text{ iff } x \in B$$
$$\mu_B(x) = \{0 \text{ iff } x \text{ does not} \in B$$

Assume that x_1 to x_k are the elements in fuzzy set B, and μ_1 to μ_k are respectively their grades of membership function in B. B is usually represented as follows:

$$B = \mu_1/x_1 + \mu_2/x_2 + \cdots + \mu_k/x_k \tag{1}$$

3 Related Work

Data mining is frequently used in inducing association rules from large itemsets. The association rules describes the effects of presence and absence of an item in a transaction with other items in terms of two measures support and confidence.

Hong proposed an algorithm which induces association rules with multiple minimum supports using maximum constraints on general items. Au and Chan proposed a fuzzy dredging approach to find fuzzy rules for time series data. Das proposed a dredging algorithm for time series data prediction. Das used the clustering method to extract basic shapes from time series and applied Apriori method to induce the association rules on it.

4 The Proposed Algorithm with Multiple Minimum Support

Input: A time series TS with n data points, a list of m membership functions for data points, a predefined minimum support threshold for each fuzzy item ms_i, i = 1 to z, a predefined minimum confidence threshold λ, and a sliding window size ws.

Step 1: Convert the time series TS into a list of subsequences W(TS) according to the sliding-window size ws. That is, $W(TS) = \{s_b | s_b = (d_b, d_{b+1}, \ldots, d_{b+ws-1}), b = 1 \text{to } n - ws + 1\}$, where d_b is the value of the b-th data point in TS.

Step 2: Transform the k-th (k = 1 to ws) quantitative value v_{bk} in each subsequence s_b (b = 1 to n-ws + 1) into a fuzzy set f_{bk} represented as $(f_{bk1}/R_{k1} + f_{bk2}/R_{k2} + \ldots + f_{bkn}/R_{kn})$ using the given membership functions, where R_{kl} is the l-th fuzzy region of the k-th data point in each subsequence, m is the number of fuzzy memberships, and f_{bkl} is v_{bk}'s fuzzy membership value in region R_{kl}. Each R_{kl} is called a fuzzy item.

Step 3: Compute the scalar cardinality of each fuzzy item R_{kl} as

$$\text{Count}_{kl} = \sum_{b=1}^{n-ws+1} f_{bkl}$$

Step 4: Check whether the support value (= $\text{count}_{kl}/n - ws + 1$) of each R_{kl} in C_1 is greater than or equal to its predefined minimum support threshold value ms_{Rkl}. If R_{kl} satisfies the above condition, collect it in the set of large 1-itemsets (L_1). That is:

$$L_1 = \{R_{kl} | \text{count}_{kl} \geq ms_{Rkl}, 1 \leq k \leq b + ws - 1 \text{ and } 1 \leq l \leq m\}.$$

Step 5: IF L_1 is not null, then perform the next step; otherwise, terminate the algorithm.

Step 6: Set t = 1, where t is used to represent the number of fuzzy items in the current itemsets to be processed.

Step 7: Join the large t-itemsets L_t to obtain the candidate (t + 1)-itemsets C_{t+1} in the same way as in the Apriori algorithm provided that two items obtained from the same order of data points in subsequences cannot exist in an itemset in C_{t+1} at the same instant provided the minimum support of all the large t-itemsets must be greater than or equal to the maximum of the minimum supports of fuzzy items in theses large t-itemsets.

Step 8: Now, perform the following steps for fuzzy items in C_{t+1}:

(a) Compute the fuzzy value of I in each subsequence s_b as $f_I^{sb} = f_{I1}^{sb} \wedge f_{I2}^{sb} \wedge \ldots \wedge f_{It+1}^{sb}$ where f_{Ik}^{sb} is the membership value of fuzzy item I_k in S_b. If the minimum operator is used for the intersection, then:

$$f_{Ib}^{s} = \mathrm{Min}_{k=1}^{t+1}\, f_{Ip}^{s}.$$

(b) Compute the count of I in all the subsequences as:

$$\mathrm{Count}_I = \sum_{b=1}^{n-ws+1} f_I^{sb}$$

Step 9: If the support ($= \mathrm{count}_I/n-ws+1$) of I is greater than or equal to maximum of the minimum support value, put it in L_{t+1}.

$$L_{t+1} = \{I_k\ |\mathrm{count}_I >= ms_{Ik}, |$$

Step 10: STEP 13: If L_{t+1} is null, then do the next step; otherwise, set $t = t + 1$ and repeat STEPs 6–9.

Step 11: Generate the association rules for each large h-itemset I with items $(I_1, I_2, \ldots, I_h)$, $h \geq 2$, using the following substeps:

(a) Form each possible association rule as follows: $I_1^{\wedge} \ldots {}^{\wedge} I_{n-1}^{\wedge} I_{n+1}^{\wedge} \ldots {}^{\wedge} I_h \rightarrow I_n$, $n = 1$ to h.

(b) Calculate the confidence values of all association rules by the following formula:

$$= \sum_{b=1}^{n-ws+1} f_I^{sb} \backslash \sum_{b=1}^{n-ws+1} \left((f_I^{sb\,\wedge} \ldots {}^{\wedge} f_{IP}^{s} \right)$$

Output: A set of association rules which satisfies the condition of the maximum values of minimum supports.

5 An Example

This section explains the working of the proposed algorithm and generates fuzzy association rule (Table 1).

Assume the membership function used in the example as Fig. 1 (Table 2).

Step 1: The window size is assumed as 5. Using the formula we get $(15-5+1) = 11$ subsequences

Step 2: The data values are then converted into fuzzy item sets using the membership function shown in fig no.

Table 1 Set of data points

Time Series
3, 4, 6, 1, 8, 6, 3, 9, 3, 6, 1, 5, 8, 1, 9

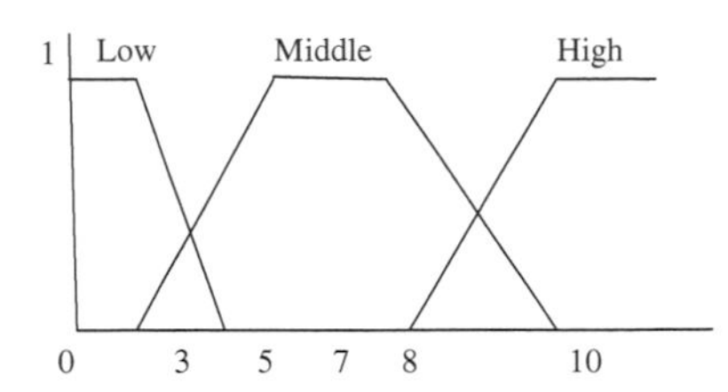

Fig. 1 Membership function used in this example

Table 2 Predefined minimum support value of all fuzzy itemset

Fuzzy item	Minimum support	Fuzzy item	Minimum support
Q1.Low	3.90	Q4.low	4.00
Q1.Middle	3.80	Q4.Middle	3.00
Q1.High	3.50	Q4.High	5.00
Q2.Low	2.00	Q5.Low	4.90
Q2.Middle	2.30	Q5.Middle	3.03
Q2.High	2.00	Q5.High	2.00
Q3.Low	2.10	Q3.High	4.00
Q3.Middle	4.50		

Table 3 Sequence generated, ws = 5

S_b	Subsequence	S_b	Subsequence
S_1	(3, 4, 6, 1, 8)	S_7	(3, 9, 3, 6, 1)
S_2	(4, 6, 1, 8, 6)	S_8	(9, 3, 6, 1, 5)
S_3	(6, 1, 8, 6, 3)	S_9	(3, 6, 1, 5, 8)
S_4	(1, 8, 6, 3, 9)	S_{10}	(6, 1, 5, 8, 1)
S_5	(8, 6, 3, 9, 3)	S_{11}	(1, 5, 8, 1, 9)
S_6	(6, 3, 9, 3, 6)		

Step 3: Add all the value of fuzzy region of the subsequences called as its count. For example-Assume a fuzzy item Q_1.Middle.count is $(0 + 0.33 + 1 + 0 + 0.33 + 1 + 0 + .2 + 0 + 1 + 0) = 3.86$

Step 4: Now,compare the count of all fuzzy item with its individual minimum support count which is predefined. Fuzzy items whose count is greater than minimum support value of itself put the fuzzy item in the table L_1 (Table 3).

Step 5: If L_1 consists of fuzzy item, proceed to step 6, else terminate.

Table 4 Converted fuzzy set

S_t	Q_1			Q_2			Q_3			Q_4			Q_5		
	Low	Mid	High	Low	Mid	High	Low	Mid	High	Low	Mid	High	Low	Mid	High
S_1	0.67	0	0	0.67	0.33	0	0	1	0	1	0	0	0	0.33	0.67
S_2	0.67	0.33	0	0	1	0	1	0	0	0	0.33	0.67	0	1	0
S_3	0	1	0	1	0	0	0	0.33	0.67	0	1	0	0.67	0	0
S_4	1	0	0	0	0.33	0.67	0	1	0	0.67	0	0	0	0.2	1
S_5	0	0.33	0.67	0	1	0	0.67	0	0	0	2	1	0.67	0	0
S_6	0	1	0	0.67	0	0	0	0.2	1	0.67	0	0	0	1	0
S_7	0.67	0	0	0	0.2	1	0.67	0	0	0	1	0	1	0	0
S_8	0	0.2	1	0.67	0	0	0	1	0	1	0	0	0	1	0
S_9	0.67	0	0	0	1	0	1	0	0	0	1	0	0.33	0.33	0.67
S_{10}	0	1	0	1	0	0	0	1	0	0.33	0.33	0.67	1	0	0
S_{11}	1	0	0	0	1	0	0	0.33	0.67	1	0	0	0	0.2	1
Count	4.68	3.86	1.67	4.01	4.86	1.67	3.34	4.68	2.34	4.34	3.68	2.34	2.34	2.70	3.34

Step 6: Candidate set C_{t+1} is generated from L_t. Fuzzy items in $L_1$are (Q1.Low, Q1.Middle, Q2.Low,Q2.Middle, Q3.Low, Q3.Middle, Q4.Low, Q5.Low, Q5.High) .

Step 7: L_1 is joined to generate C_2. The new fuzzy items in C_2 are as follows (Q1.Low,Q2.Mid),(Q1.Low, Q3.Mid), (Q1Low, Q5.High), (Q1.Low, Q2.Mid), (Q1.Low, Q3.Mid),(Q1.Low, Q5.Low), (Q2.Low, Q3.Low), (Q2.Low, Q4.Low), (Q2.low, Q5High), (Q2.low, Q1.Mid), (Q2.Low, Q3.Mid), (Q2.Low, Q5.Low), (Q3.Low, Q4.Low), (Q3.low, Q5.High), (Q3.Low, Q1.Mid), (Q3.low, Q2.Mid), (Q3.low, Q5.low), (Q4.Low, Q1.Mid), (Q4.Low, Q2.Mid), (Q4.low, Q5.High), (Q4.Low, Q5.Low), (Q5.High, Q1.Mid), (Q5.High, Q2.Mid), (Q5.High, Q3.High).

Table 5 Candidate set C_2

Fuzzy itemset	Count	Fuzzy itemset	Count
Q1Low, Q2Mid	4.68	Q2Low, Q5Low	2.34
Q1Low, Q3Mid	4.68	Q3Low, Q5High	3.34
Q1Low, Q5High	3.34	Q3Low, Q1Mid	3.34
Q1Low, Q2Low	4.01	Q3Low, Q2Mid	3.34
Q1Low, Q3Low	3.34	Q3Low, Q5Low	2.34
Q1Low, Q5Low	2.34	Q4Low, Q1Mid	3.86
Q2Low, Q3Low	3.34	Q4Low, Q2Mid	4.34
Q2Low, Q4Low	4.01	Q4Low, Q5High	3.86
Q2Low, Q3Low	3.34	Q4Low, Q2Mid	4.34
Q2Low, Q4Low	4.01	Q4Low, Q5High	3.86
Q2Low, Q5High	3.34	Q4Low, Q5Low	2.34
Q2Low, Q1Mid	3.86	Q5High, Q1Mid	3.34
Q2Low, Q3Mid	4.01	Q5High, Q2Mid	3.34
Q3Low, Q4Low	3.34	Q5High, Q3High	2.34

Table 6 Fuzzy itemset L_2

(Q1Low, Q2Low)(Q1Low, Q2Mid)(Q1Low, Q3Mid)(Q2Low, Q3Low)
(Q2Low, Q4Low)(Q2Low, Q5High)(Q1.Low, Q4.Low)
(Q3.Low, Q2.Mid)(Q4.Low, Q1.Mid)(Q5.High, Q2.Mid)(Q4.Low, Q5.High)

Step 8: Now compute the count of all the fuzzy items of C_2.

Step 9: Compare the C_2 itemset count with minimum support count of Fuzzy itemset. C2 items whose count is greater or equal to minimum support of maximum of the two itemset is stored in L_2.

Step 10: Since L_2 is not null, repeat step no 6–9 until L_t is null (Tables 4, 5, 6).

Step 11: (a) In this example, only (Q3.Low Q2.Mid) exists. It means association rules formed are
If Q3 = Low then Q2 = Mid. If Q2 = Mid then Q3 = Low.
(b) Calculation of confidence of (Q3.Low Q2.Mid) = 3.34\ 3.34 = 1. It means if the value of a data point is mid at time2 then value of a data point is low at time3 with a confidence factor of 1.

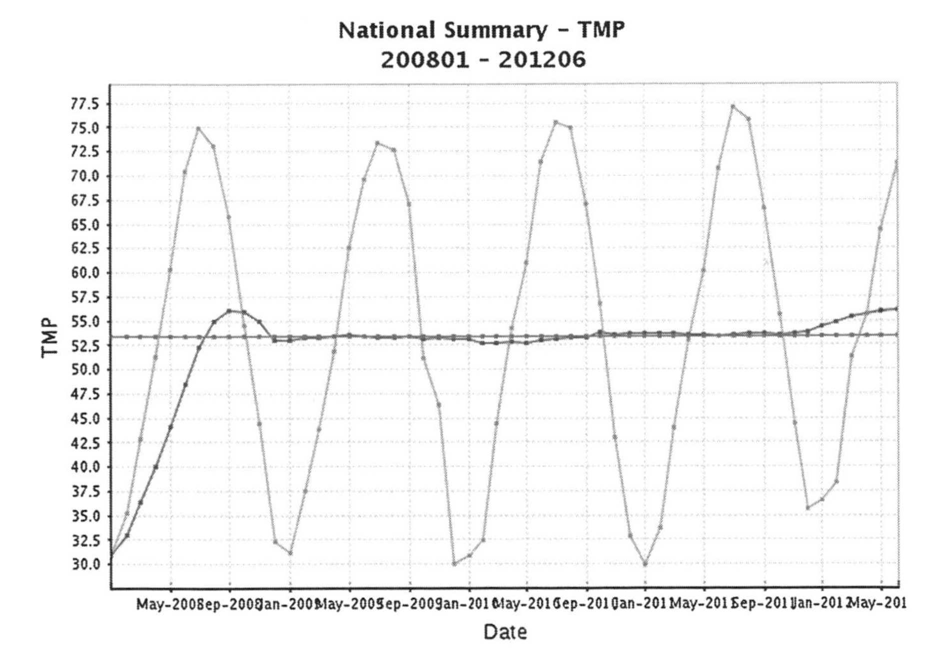

Fig. 2 Temperature varying data

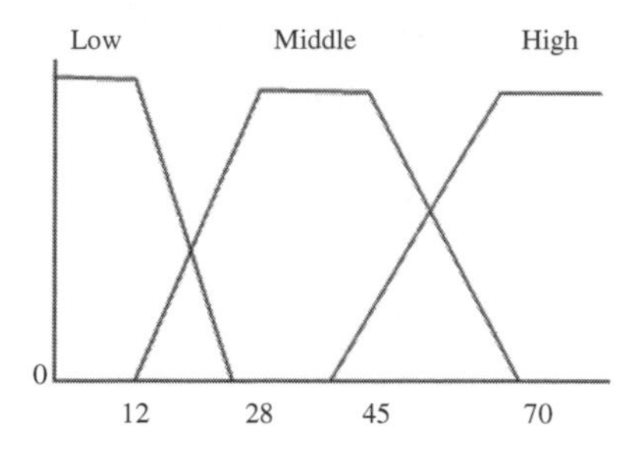

Fig. 3 Membership function used in experiment

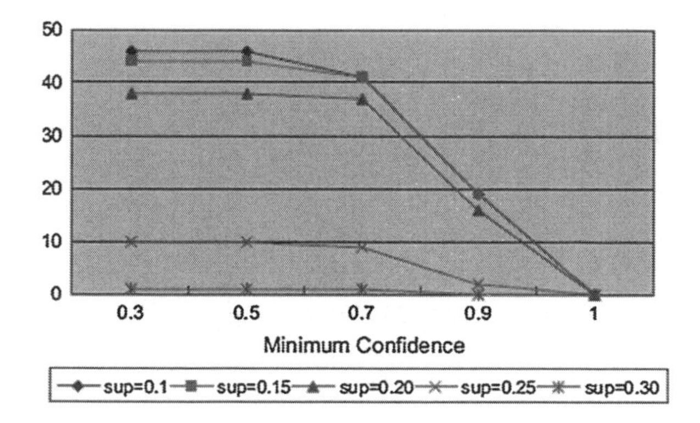

Fig. 4 Relation between support value and confidence

6 Experimental Results

The proposed algorithm is implemented in a programming language C. The dataset points consisted of temperature varying points between year 2008–2012. The dataset is taken from National Data Center (NDC) US (Figs. 2 and 3).

In Fig. 4 as the support value of fuzzy itemset is increased, number of fuzzy association rule decreased. This means change in temperature is effected with change in support value. It means that if the temperature of a day at second day of a month is moderate, then it may be high at third day of the month.

7 Conclusion and Future Work

In this paper, the proposed algorithm provided the best way to induce efficient fuzzy association rule as there is predefined minimum support for all the fuzzy items. The temperature prediction would be more accurate than earlier proposed approaches. Future work suggests that the membership function can be set dynamically. In this paper membership functions are known in advance. More complex operations could be made in near future. It also provided us another view point for defining minimum support of fuzzy items.

References

1. Aggrawal, R.: Mining association rules between sets of items inlarge database ACM
2. Stepnicka, M.: Time series analysis and prediction based on fuzzy rules and the fuzzy transform
3. Dr.Sivatsa S.K.: Inaccuracy minimization by partitioning fuzzy data sets—validation of an analytical methodology(IJCSIS). Int. J. Comput. Sci. Inf. Secur. **8**(1), (2010)
4. Das, G.: Rule discovery from time series. In: Proceedings of the 4 the International Conference
5. Pongracz, R.: Application of fuzzy rule-based modeling technique to regional drought. J. Hydrol. **224**, 100–114 (1999)
6. Mueen, A.A.: Exact primitives for time series data mining. University of California, Riverside (2012)
7. Zhu, Y.: High performance data mining in time series: techniques and case studies. New work University, New York (2004)
8. Herrera, Francisco: Learning the membership function contexts for mining fuzzy association rules by using genetic algorithms. Fuzzy Sets Syst. **160**, 905–921 (2009)
9. Han, J.: Data Mining concepts and Techniques
10. Hong, T.P.: Mining association rules with multiple minimum supports. Int. J. Approximate reasoning. **3**, 38–42 (2005)

RGA Analysis of Dynamic Process Models Under Uncertainty

Amit Jain and B. V. Babu

Abstract The aim of this paper is to gain insights into how process dynamics can affect control configuration decision based on relative gain array (RGA) analysis in the face of model uncertainty. Analytical expressions for worst-case bounds of uncertainty in steady-state and dynamic RGA are derived for two inputs two outputs (TITO) plant models. A simulation example which has been used in several prior studies is considered here to demonstrate the results. The obtained bounds of uncertainty in RGA provide valuable information pertaining to the necessity of robustness and accuracy in the model of decentralized multivariable systems.

Keywords Relative gain array · Parametric uncertainty · Control configuration selection · Worst-case bounds · Multivariable plants

1 Introduction

It has been almost five decades since the introduction of the relative gain array (RGA) in 1966 [1]. It has found to be a promising tool for determination of control loop pairings and in the analysis of the level of the interaction exist in such pairings. Because of its inherent simplicity, the RGA is still popular among industries and academia, particularly after the use of the Niederlinski Index [2] as a complementary tool for analyzing closed-loop stability. The RGA has many useful algebraic properties with

B. V. Babu (✉)
Institute of Engineering and Technology, JK Laxmipat University (JKLU), Jaipur 302026, India
e-mail: director.iet@jklu.edu.in

A. Jain
Chemical Engineering Department, Birla Institute of Technology and Science, Pilani, Rajasthan 333031, India
e-mail: amitjain@bits-pilani.ac.in

B. V. Babu et al. (eds.), *Proceedings of the Second International Conference on Soft Computing for Problem Solving (SocProS 2012), December 28–30, 2012*, Advances in Intelligent Systems and Computing 236, DOI: 10.1007/978-81-322-1602-5_48,

strong control impositions [3]. The applications of RGA are thoroughly discussed in a book written by McAvoy [4] and Shinskey [5]. However, it is clearly indicated [6–13] that the RGA is not just a tool for choosing input/output pairings for decentralized control but also a measure of attainable control quality. It is associated with many elementary properties; robustness, stability, tolerance to the failure of actuators and sensors, decentralized integral controllability (DIC) etc., defined for closed loop systems.

Despite many advantages, the RGA also suffers from certain disadvantages. Since it depends on the steady-state gain information alone, a due consideration has not been given to the process dynamic behavior which leads to incorrect conclusions in certain cases. The dynamic extension of RGA to nonzero frequencies is presented in many ways by different authors [4, 14–19]. The originate work in this field is presented for two inputs two outputs (TITO) plants. Early methods of dynamic RGA employed the transfer function matrix in the definition of RGA as an alternative to the steady-state matrix of gains [14–16]. McAvoy et. al. [18] has offered a novel approach to define a dynamic RGA. It is assumed in this computationally complicated approach that a dynamic model of the process exists and a proportional output optimal controller has been designed using state-space approach. Finally, based on the matrix of consequential controller gains a dynamic RGA is defined. However, these methods [16, 18] have lost the controller independency, which is a valuable property in a pairing criterion. To give a more complete description of control-loop interaction, both the steady-state gain and the process dynamic behavior in terms of bandwidth of the elements of transfer function has been adopted by some authors in a controller independent framework [19, 20].

Although for years, in most of the studies on the analysis of RGA and its properties, the availability of a process model is frequently assumed, still the sensitivity of the RGA analysis to model uncertainty is in nascent stage. However, in practice, the models of real systems always have some uncertainty associated with them. Thus, process models can never be perfect. For plants with uncertain process models, an incorrect pairing decision may result if the RGA analysis carried out based only on a nominal model of the process. The problem further aggravates when a sensitivity analysis of RGA elements to model uncertainty is carried out based on steady-state process model alone.

In this paper, analytical expressions are derived for worst-case bounds of uncertainty in steady-state and dynamic RGA considering TITO plant models. The parametric model uncertainty is considered and is presented here with the aim to identify the possible input–output selection changes resulting from the parameter changes. The transfer function model (2×2) used here to demonstrate the results has also been used by many authors before.

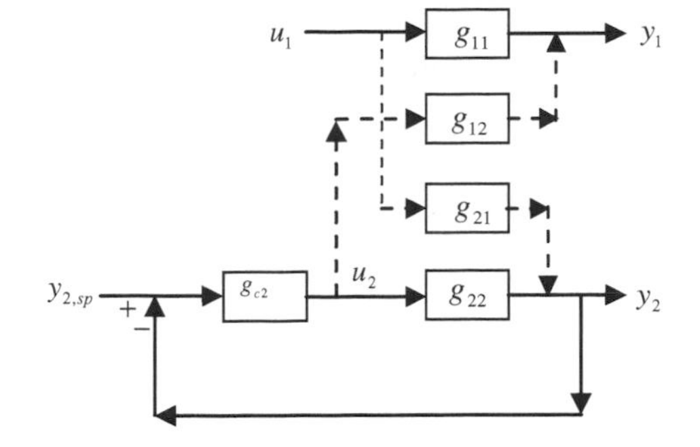

Fig. 1 Block diagram of loop interaction in 2 × 2 plant model

2 Relative Gain Array: Definitions and Properties

The RGA, a measure of interaction for the multivariable control system was formerly proposed by Bristol [1]. Each element in the RGA matrix has been defined as the ratio of gains, i.e., open-loop gain to the gain obtained between the similar variables keeping all other loops under "perfect" control (Fig. 1):

$$\lambda_{ij} = \frac{\left(\partial y_i/\partial u_j\right)_u}{\left(\partial y_i/\partial u_j\right)_y} = \frac{\text{open-loop gain}}{\text{closed-loop gain}} \tag{1}$$

The elements λ_{ij}forms the RGAΛ. Writing definition (1) in the transfer matrix form as,

$$\Lambda\left(G\right) = G\left(s\right) \otimes G^{-1}\left(s\right)' \tag{2}$$

where $\otimes$ denotes element-by-element multiplication or Schur product. The substitution of $(s = j\omega)$ in (2) computes RGA at different frequencies [9]. The steady-state $(s = 0)$ version of the RGA is found by substituting the transfer function matrix $G\left(s\right)$ of (2) with the corresponding matrix of steady-state gains, K. In addition to its numerous algebraic properties, the RGA is also a good gage of sensitivity to uncertainty. The relative gain λ_{ij} provides a direct estimate of the sensitivity of the plant model to independent element-by-element uncertainty. $G\left(j\omega\right)$ turn out to be singular if any element $g_{ij}\left(j\omega\right)$ changes by $-1/\lambda_{ij}\left(j\omega\right)$ [8–10]. Since the steady-state RGA matrix is a function of open loop gains only, thus any uncertainty in open-loop gain leads to the uncertainty in RGA.

3 Analyzing 2 × 2 Plant Model Control Problem

Let a multivariable system $G\left(s\right)$ with two controlled variables and two manipulated variables:

$$G(s) = \begin{bmatrix} g_{11}(s) & g_{12}(s) \\ g_{21}(s) & g_{22}(s) \end{bmatrix} \tag{3}$$

For the given control problem the steady-state gain matrix is,

$$G(0) = K = \begin{bmatrix} g_{11}(0) & g_{12}(0) \\ g_{21}(0) & g_{22}(0) \end{bmatrix} = \begin{bmatrix} K_{11} & K_{12} \\ K_{21} & K_{22} \end{bmatrix} \tag{4}$$

and the subsequent matrix of RGA elements is,

$$\Lambda = \begin{bmatrix} \lambda_{11} & \lambda_{12} \\ \lambda_{21} & \lambda_{22} \end{bmatrix} = \begin{bmatrix} \lambda_{11} & 1-\lambda_{11} \\ 1-\lambda_{11} & \lambda_{11} \end{bmatrix} \tag{5}$$

here, λ_{11} is the relative gain between output-1 and input-1, i.e.,

$$\lambda_{11} = \frac{(\partial y_1/\partial u_1)_{u_2}}{(\partial y_1/\partial u_1)_{y_2}} = \frac{1}{1-\hat{k}}; \tag{6}$$

where,

$$\hat{k} \overset{\Delta}{=} (-1)^n \frac{|K_{12}|\,|K_{21}|}{|K_{11}|\,|K_{22}|} \tag{7}$$

$\hat{k}$, here is referred to as the interaction quotient and n, is the number of negative elements in $G(s)$. Obviously $\hat{k} = 1$ is a singular point for RGA matrix. Note here that for odd number of negative elements (i.e., n = odd number) the RGA matrix can never be singular as $\hat{k} \neq 1$.

3.1 Analyzing Uncertainty Bounds for Steady-State Process Model

Assume lower bounds of uncertainty in all the elements of the steady-state gain matrix K to be symmetrical with that of higher bounds. Thus, the nominal steady-state gain element, K_{ij} varies within $K_{ij} \pm \Delta K_{ij}$. Let $|\Delta K_{ij}| \leq \alpha |K_{ij}|$ represents the uncertainty bound to the every element of K, then,

$$\hat{k}_{uc} = \hat{k}\left(\frac{1 \pm \alpha}{1 \mp \alpha}\right)^2 \tag{8}$$

where $\hat{k}_{uc}$ is the interaction quotient under model uncertainty. The respective limits of the $\hat{k}_{uc}$ are represented as,

$$\hat{k}^l \leq \hat{k}_{uc} \leq \hat{k}^h \tag{9}$$

where,

$$\hat{k}^l\left(=\min\left(\hat{k}_{uc}\right)\right)=\hat{k}\left(\frac{1+\alpha}{1-\alpha}\right)^2 \quad \forall n=\text{odd number} \tag{10}$$

$$=\hat{k}\left(\frac{1-\alpha}{1+\alpha}\right)^2 \quad \forall n=\text{even number} \tag{11}$$

$$\hat{k}^h\left(=\max\left(\hat{k}_{uc}\right)\right)=\hat{k}\left(\frac{1-\alpha}{1+\alpha}\right)^2 \quad \forall n=\text{odd number} \tag{12}$$

$$=\hat{k}\left(\frac{1+\alpha}{1-\alpha}\right)^2 \quad \forall n=\text{even number} \tag{13}$$

For $\hat{k}\neq 1$, the RGA element λ_{11} changes within the uncertainty limits:

$$\frac{1}{1-\hat{k}^l}\leq\lambda_{11}\leq\frac{1}{1-\hat{k}^h} \tag{14}$$

For $\hat{k}=1$ (singular point), the sign of λ_{11}changes. Thus the λ_{11} changes within the uncertainty limits:

$$-\infty\leq\lambda_{11}\leq\frac{1}{1-\hat{k}^h} \tag{15}$$

and

$$\frac{1}{1-\hat{k}^l}\leq\lambda_{11}\leq\infty \tag{16}$$

The application of above methodology will be demonstrated with the help of examples available in open literature.

3.2 Worst-Case Bounds for Dynamic Process Model

In order to assimilate both the static and dynamic behavior of the plant in the interaction measure, we encourage the use of the elective gain, e_{ij} as defined in Xiong et al. [19] for a given element $g_{ij}(\omega)$ of transfer function model at the critical frequencies. Approximating the effective gain integration with a rectangular area, the effective gain e_{ij} can simply be given as,

$$e_{ij}\approx g_{ij}(0)\,\omega_{c.ij} \tag{17}$$

Therefore, the matrix of effective gains is defined as,

$$E = G(0) \otimes \Omega; \tag{18}$$

where,

$$\Omega = \begin{bmatrix} \omega_{c,11} & \omega_{c,12} \\ \omega_{c,21} & \omega_{c,22} \end{bmatrix} \tag{19}$$

Based on (4) and (19), the dynamic interaction quotient can be defined as:

$$\hat{k}_{dy} = (-1)^n \frac{|K_{12}|\left(\omega_{c,12}\right) \times |K_{21}|\left(\omega_{c,21}\right)}{|K_{11}|\left(\omega_{c,11}\right) \times |K_{22}|\left(\omega_{c,22}\right)} = \hat{k}\varpi; \tag{20}$$

where,

$$\varpi = \frac{\left(\omega_{c,12}\right)\left(\omega_{c,21}\right)}{\left(\omega_{c,11}\right)\left(\omega_{c,22}\right)} \tag{21}$$

Equations (9)–(15), defined for steady-state system are all directly applicable to dynamic process model after replacing $\hat{k}$ in (8) by $\hat{k}_{dy}$ of (20).

4 Examples

Example 1 Consider a process model [21]:

$$G(s) = \begin{bmatrix} \frac{5}{4.s+1} & \frac{2.5e^{-5s}}{(2.s+1)(15.s+1)} \\ \frac{-4e^{-6s}}{20.s+1} & \frac{1}{3.s+1} \end{bmatrix} \tag{22}$$

For steady-state system:

The steady-state RGA matrix and interaction quotient $\hat{k}$ for $n = 1$ are:

$$\hat{\Lambda} = \begin{bmatrix} 0.3333 & 0.6667 \\ 0.6667 & 0.3333 \end{bmatrix} \;\&\; \hat{k} = -2 \tag{23}$$

which suggest the 1-2/2-1 (off-diagonal) pairing. Taking into consideration the different values for model uncertainty α. The range of λ_{11} for given α value can be found.

Case 1 $\alpha = 0.01$. According to (10)–(13), the bounds for $\hat{k}$ are $\hat{k}^l = -2.08$ and $\hat{k}^h = -1.92$, the corresponding bounds for λ_{11} are given by (14):

$$0.32 \leq \lambda_{11} \leq 0.34$$

Since, $\lambda_{11} < 0.5$ thus, the suggested control-loop pairing based on the RGA analysis is 1-2/2-1 (off-diagonal).

Case 2 $\alpha = 0.1$. The bounds are $\hat{k}^l = -2.99$ and $\hat{k}^h = -1.34$ and

$$0.25 \leq \lambda_{11} \leq 0.43$$

Thus, suggested pairing is still 1-2/2-1 (off-diagonal).

Case 3 $\alpha = 0.2$. Now $\hat{k}^l = -4.55$ and $\hat{k}^h = -0.88$. The range of λ_{11} can be obtained using (14):

$$0.18 \leq \lambda_{11} \leq 0.53$$

Still in its range, $\lambda_{11} \leq 0.5$. Thus, the suggested control-loop pairing based on the RGA analysis is 1-2/2-1 (off-diagonal).

Case 4 $\alpha = 0.5$. Now $\hat{k}^l = -18.18$ and $\hat{k}^h = -0.22$ and

$$0.05 \leq \lambda_{11} \leq 0.82$$

For the obtained range of λ_{11} the pairing decision is quite ambiguous.

Case 5 $\alpha = 0.9$. Now $\hat{k}^l = -666.7$ and $\hat{k}^h = -0.006$ and

$$0.001 \leq \lambda_{11} \leq 0.99$$

In the obtained range of λ_{11} the pairing decision remained ambiguous. As was discussed, for odd number of negative elements in given matrix of the transfer function the RGA matrix will never be singular as $\hat{k} \neq 1$, within the whole uncertainty range.

For Dynamic Systems:

The critical frequency quotient ϖ and dynamic interaction quotient $\hat{k}_{dy}$ can be obtained using (20) and (21) respectively, as:

$$\varpi = 0.04 \tag{24}$$

and,

$$\hat{k}_{dy} = -0.08 \tag{25}$$

The range of λ_{11} corresponding to the same α values that was used for steady-state analysis could be found.

Case 1 $\alpha = 0.01$. In accordance to (10)–(13), the bounds for $\hat{k}_{dy}$ are $\hat{k}^l_{dy} = -0.083$ and $\hat{k}^h_{dy} = -0.077$, consequently the λ_{11} varies within the bounds as per (14):

$$0.92 \leq \lambda_{11} \leq 0.93$$

Thus, the suggested control-loop pairing based on the RGA analysis is 1-1/2-2, i.e., diagonal.

Case 2 $\alpha = 0.1$ Now $\hat{k}^l_{dy} = -0.12$ and $\hat{k}^h_{dy} = -0.05$ and

$$0.89 \leq \lambda_{11} \leq 0.95$$

Thus, the suggested control-loop pairing based on the RGA analysis is 1-1/2-2, i.e., diagonal.

Case 3 $\alpha = 0.5$; The bounds are given as $\hat{k}^l_{dy} = -0.73$ and $\hat{k}^h_{dy} = -0.01$ and

$$0.58 \leq \lambda_{11} \leq 0.99$$

Thus, the suggested control-loop pairing based on the RGA analysis is 1-1/2-2, i.e., diagonal.

Case 4 $\alpha = 0.9$. Now $\hat{k}^l_{dy} = -28.88$ and $\hat{k}^h_{dy} = -0.0002$ and

$$0.03 \leq \lambda_{11} \leq 0.999$$

For the obtained range of $\lambda_{11}$the pairing decision is quite ambiguous.

Table 1 compares the outcome of uncertainty analysis of steady-state and dynamic RGA for Example 1. As is clearly shown, the uncertainty analysis of steady-state gain matrix gives ambiguous results and no unique control loop-pairing is favored within the whole range of uncertainty whereas dynamic RGA analysis clearly suggest diagonal pairing. The results obtained are in complete agreement with the results of Grosdidier and Morari [21], whose analysis of interaction between the loops was based on frequency response characteristics. The example under considerations illustrates: (i) The uncertainty in process model may have a severe effect on the selection of control configuration if the RGA analysis based on steady-state alone is utilized. (ii) In order to ensure robustness in the system pertaining to decentralized

Table 1 Uncertainty analysis results of Example 1 for different α values

α	Uncertainty range of static RGA Elements	Uncertainty range of dynamic RGA Elements
0.01	$0.32 \leq \lambda_{11} \leq 0.34$	$0.92 \leq \lambda_{11} \leq 0.93$
0.10	$0.25 \leq \lambda_{11} \leq 0.43$	$0.89 \leq \lambda_{11} \leq 0.95$
0.50	$0.05 \leq \lambda_{11} \leq 0.82$	$0.58 \leq \lambda_{11} \leq 0.99$
0.90	$0.001 \leq \lambda_{11} \leq 0.99$	$0.03 \leq \lambda_{11} \leq 0.999$

control, the uncertainty in model parameters should be infused in the RGA analysis under a dynamic framework.

5 Conclusions

In this paper, analytical expressions are derived for bounds of uncertainty (worst-case) in steady-state and dynamic RGA for TITO plant models. On the analysis of considered example it is thus reasoned that the steady-state RGA analysis leads to incorrect conclusions about worst-case bounds and tolerable uncertainty. It is thus recommended to use dynamic RGA uncertainty analysis to approximate the highest degree of uncertainty in the model parameters such that the suggested control-loop pairing will remain unaffected. Such analysis provides an idea about how much change in the operating parameter values are tolerable so as to keep the control-loop pairing decision unchanged.

The method presented is introductory in essence that it is applicable to 2×2 plant models with uncertainty in steady-state gains only.

References

1. Bristol, E.H.: On a new measure of interactions for multivariable process control. IEEE Trans. Autom. Control AC-**11**, 133–134 (1966)
2. Niederlinski, A.: A heuristic approach to the deisgn of linear multivariable interacting control systems. Automatica **7**, 691–701 (1971)
3. Skogestad, S., Postlethwaite, I.: Multivariable feedback control: analysis and design. Wiley, Chichester (2005)
4. McAvoy, T.J.: Interaction Analysis. Instrument Society of America, Research Triangle Park (1983a)
5. Shinskey, F.G.: Distillation Control, 2nd edn. McGraw-Hill, New York (1984)
6. Grosdidier, P., Morari, M., Holt, B.R.: Closed-loop properties from steady-state gain information. Ind. Eng. Chem. Fundam. **24**, 221–235 (1985)
7. Yu, C.C., Luyben, W.L.: Robustness with respect to integral controllability. Ind. Eng. Chem. Res. **26**, 1043–1045 (1987)
8. Skogestad, S., Morari, M.: Implications of large rga elements on control performance. Ind. Eng. Chem. Res. **26**, 2323–2330 (1987)
9. Chiu, M.S., Arkun, Y.: A new result on relative gain array, niederlinski index and decentralized stability condition: 2x2 plant cases. Automatica **27**, 419–421 (1991)
10. Hovd, M., Skogestad, S.: Simple frequency-dependent tools for control system analysis, structure selection and design. Automatica **28**, 989–996 (1992)
11. Chen, J., Freudenberg, J.S., Nett, C.N.: The role of the condition number and the relative gain array in robustness analysis. Automatica **30**, 1029–1035 (1994)
12. Zhu, Z.X., Jutan, A.: Loop decomposition and dynamic interaction analysis of decentralized control systems. Chem. Eng. Sci. **51**, 3325–3335 (1996)
13. Lee, J., Edgar, T.F.: Computational method for decentralized integral controllability of low dimensional processes. Comput. Chem. Eng. **24**, 847–852 (2000)
14. Witcher, M., McAvoy, T.J.: Interacting control systems: steady-state and dynamic measurement of interaction. ISA Trans. **16**, 35–44 (1977)

15. Bristol, E.H.: Recent results on interactions in multivariable process control. Presented at 71st Annual AIChE Meeting, Houston, TX (1979)
16. Tung, L., Edgar, T.: Analysis of control-output interactions in dynamic systems. A.I.Ch.E. J. **27**, 690–693(1981)
17. Gagnepain, J.P., Seborg, D.E.: Analysis of process interactions with application to multiloop control system design. Ind. Eng. Chem. Pro. Des. Dev. **21**, 5–11 (1982)
18. McAvoy, T.J., Arkun, Y., Chen, R., Robinson, D., Schnelle, P.D.: A new approach to defining a dynamic relative gain. Cont. Eng. Pract. **11**, 907–914 (2003)
19. Xiong, Q., Cai, W., He, M.: A practical loop pairing criterion for multivariable processes. J. Process Control **15**, 741–747 (2005)
20. Monshizadeh-Naini, N., Fatehi, A., Khaki-Sedigh, A.: Input-output pairing using effective relative energy array. Ind. Eng. Chem. Res. **48**, 7137–7144 (2009)
21. Grosdidier, P., Morari, M.: Interaction measures for systems under decentralized control. Automatica **22**, 309–319 (1986)

Part IV
Soft Computing in Industrial and Management Applications (SCIMA)

Steps Towards Web Ubiquitous Computing

Manu Ram Pandit, Tushar Bhardwaj and Vikas Khatri

Abstract With evasion of digital convergence [1], computing has by and large pervaded into our environment. WWW has enhanced day-to-day life by utilizing information such as Location awareness, User-context awareness; touch API, mutation observer [2], and many more. The future [3] trends in ubiquitous computing [4] provide a great scope for innovation and value-added services. With approach of "computing being embedded," the future sees its usage more pervasive and appealing. Web is evolving and so are supporting technologies (in terms of hardware technologies). Many real-life examples including augmented-reality, wearable technologies, gesture-based recognition systems, etc., are already in place illustrating its high-end usage. Such diverse future targeting billions of people and devices need streamlined approach. Some steps have already been taken care by World Wide Web consortium (W3C) to provide standards relating to API usage. In this paper, we highlight various aspects of web-ubiquitous computing and how they can be dealt w.r.t to their implementation.

Keywords Web ubiquitous computing · Ubiquitous computing recommendation · WWW pervasive computing · Future trends in web ubiquitous computing

---

M. R. Pandit (✉)
Pursuing M.S. from BITS, Pilani, India
e-mail: manupandit123@gmail.com

T. Bhardwaj
Pursuing M.Tech, Pilani, India
e-mail: tusharbhardwaj19@gmail.com

V. Khatri
M.Tech(K.U.), Pilani, India
e-mail: khatrivikas.mit@gmail.com

B. V. Babu et al. (eds.), *Proceedings of the Second International Conference on Soft Computing for Problem Solving (SocProS 2012), December 28–30, 2012*, Advances in Intelligent Systems and Computing 236, DOI: 10.1007/978-81-322-1602-5_49,

1 Introduction

With evasion of digital convergence [1] computing has by and large pervaded into our environment. WWW has enhanced day-to-day life by utilizing information such as Location awareness, User-context awareness; touch API, mutation observer [2] and many more. The future [3] trends in ubiquitous computing [4] provide a great scope for innovation and value-added services. With approach of "computing being embedded," the future sees its usage more pervasive and appealing.

Today, much of user data can be seen on internet. This provides an opportunity for better services in ubiquitous computing domain. Some strategic approach is already taken in this prospect by World Wide Web consortium (W3c) to make standards for upcoming devices. This paper provides insight of web usage of ubiquitous computing domain. We first describe classifying requirements for enforcement of web ubiquitous domain followed by in-depth strategies to be taken care off during their implementation.

2 Context Scenarios and Recommendations

In lieu of the same, requirements of ubiquitous systems may be categorized into-(Fig. 1)

(a) Need of high-end computing devices
(b) Seamless network integration
(c) Contextual awareness(User, Social, cultural and location-specific)
(d) Security
(e) Policy enforcement

2.1 Need of Computing Devices

Computing devices are important ingredient of this system. Devices may be hand-held devices or might be embedded in others like cloth. Some recommendation for making them suitable w.r.t. web-ubiquitous computing are:

(a) They should conform to Device API specification.
(b) Should expose themselves through <*meta*> [5] tag. This will help in self-discovery of the devices. We recommend every device to have a local information server at its ephemeral port (called as discovery port). Other devices within the device '*X*' periphery would discover with help of : http://www.device-x:portNo/ Similar page would turn up:
(c) Data transfer format should be preferably HTTP.
(d) Conformance to W3.org API specification [6].

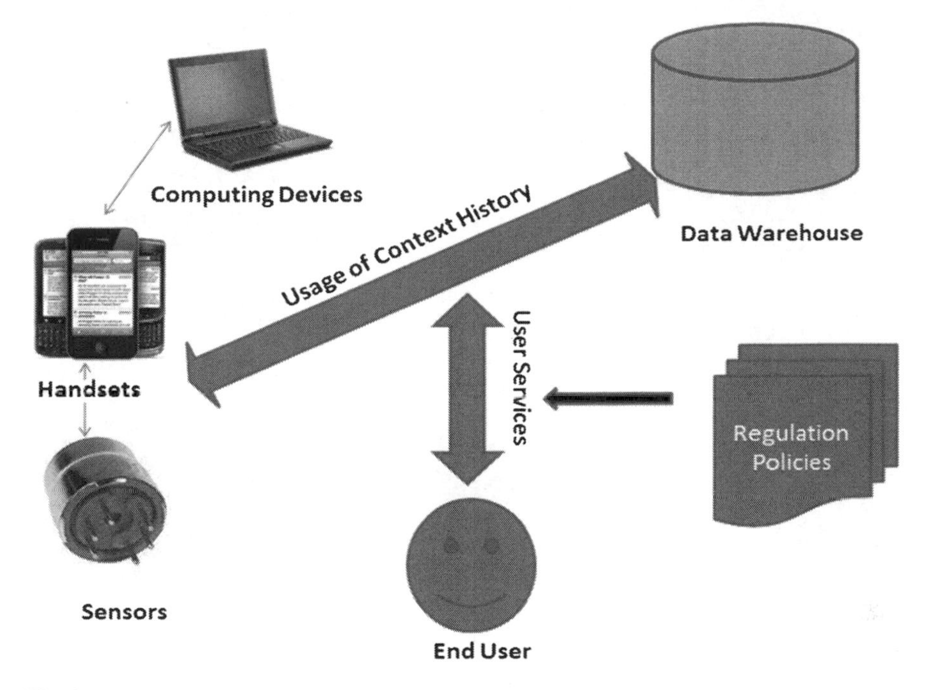

Fig. 1 The complete picture of ubiquitous computing

2.2 Seamless Network Integration

Underlying network technologies should be seamlessly accessible whether it is Bluetooth, wireless, connected or RF-based, etc. Frequency interference should be taken care of which are typically catered by some regulation committee such as FCC (Fig. 2).

2.3 Contextual Awareness

Contextual awareness is one of the most sighted features in ubiquitous computing with the help of which surrounding devices (environment) makes decision thereby adding value services to the subject. This contextual data has to be stored and regularly updated. Since contextual data will be huge, we recommend use of data warehousing technique. To add a further step, we recommend use of *data-warehousing as a service* (**'daas'**) [7] for particular user. This service can be categorized into particular subsections like user travel information, health information, hobbies information, etc. As it involves complex performance requirements, a centralized cloud-based server typically fit into this context (Fig. 3).

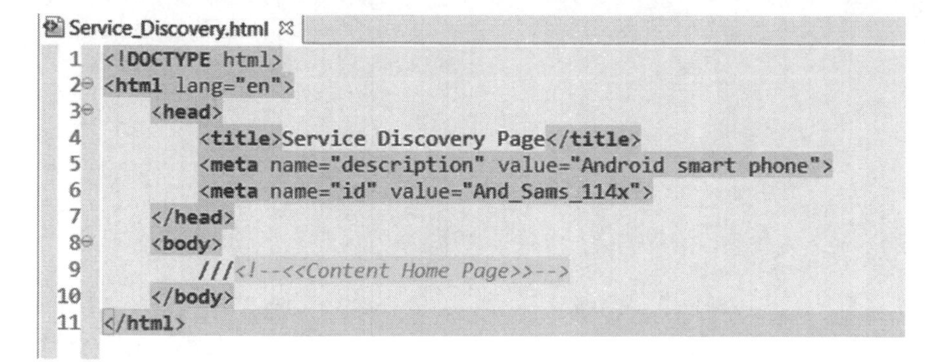

Service_Discovery.html

```
1  <!DOCTYPE html>
2  <html lang="en">
3      <head>
4          <title>Service Discovery Page</title>
5          <meta name="description" value="Android smart phone">
6          <meta name="id" value="And_Sams_114x">
7      </head>
8      <body>
9          ///<!--<<Content Home Page>>-->
10     </body>
11 </html>
```

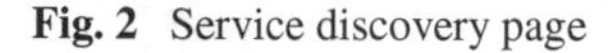

Fig. 2 Service discovery page

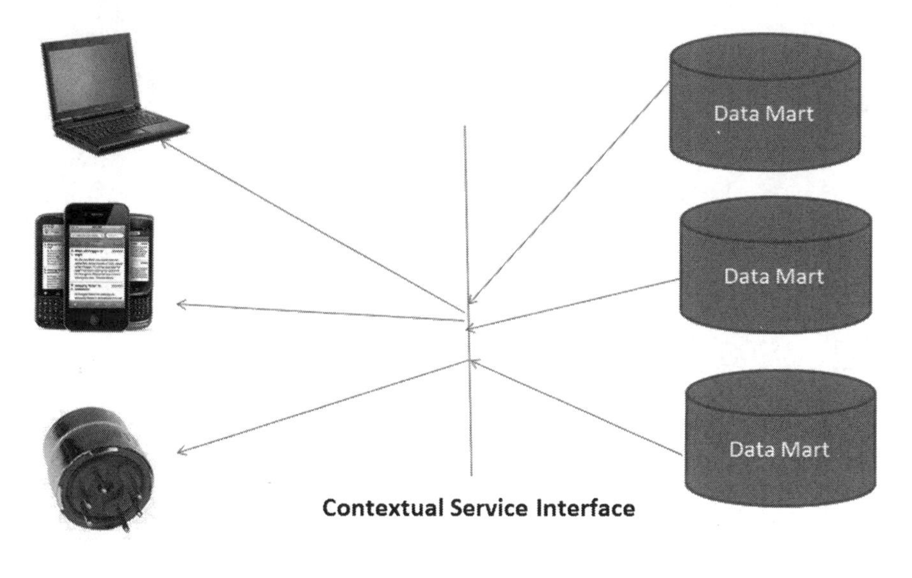

Fig. 3 Contextual service scenario

For fast performance, the data warehouse repository and underlying web-service can be deployed on cloud based server.

2.4 Security

While the above-mentioned services provide lucrative services, it provides vulnerabilities to personal life. User data can be hacked and details may be compromised. Basic cryptographic property viz. confidentiality, integrity, and nonrepudiation needs to be dealt with. One possible recommended strategy to deal with is to allow user data to be shared only with devices/environment which user trust. This may be based on preconfigured security certificates or user passed tokens based on RFID [8] /other master configurable interface (e.g., Open Id-based authentication [9], Facebook/twitter/Google authentication, etc.)

2.5 Regulatory Policies

We highly recommend enforcement of environment policies before the devices provides user some contextual service. For instance, recommending medical aid (drugs that may be banned in some geographies), automatic software downloads (handling IP issues), purchasing some product from customer e-wallet (automatically cater new sales/revenue tax), etc. This can be easily achieved by having some centralized web-service exposed to the devices. Based on the parameters achieved, the devices can make valuable decisions. With the help of a centralized policy server, it will be effective for government/regulatory bodies to make/change decision on-the-fly.

3 Conclusion

With a huge user data on web, web ubiquitous computing provides a great deal of scope for innovation and development. Supporting technologies and standards will provide smoother gateway for its success. We propose some changes in classical model of ubiquitous domain (viz. intelligent devices, network, and contextual information) to suit web followed by two new parameters of security and regulatory policies.

References

1. Infosys SETLabs Briefings. Vol 8, No 5 (2010)
2. Mozilla, Mutation Observers. https://developer.mozilla.org/en-US/docs/DOM/DOM_Mutation_Observers
3. Wireless Future: Ubiquitous Computing. Friedemann Mattern, ETH Zürich
4. Ubiquitous Web Domain group. http://www.w3.org/UbiWeb/
5. World Wide Web Consortium, Meta Tag Description. http://www.w3.org/TR/html401/struct/global.html#h-7.4.4.2
6. Device API Working Group. http://www.w3.org/2009/dap/ (2009)
7. Example of Data-Warehousing as a Service. http://www.kognitio.com/data_warehousing_as_a_service_daas
8. Lee, Y.K., Verbauwhede, I.: Secure and Low-cost RFID Authentication Protocols
9. Open Id Authentication Specification. http://openid.net/specs/openid-authentication-1_1.html

Estimation of Uncertainty Using Entropy on Noise Based Soft Classifiers

Rakesh Dwivedi, Anil Kumar and S. K. Ghosh

Abstract In remote sensing noise is some kind of ambiguous data that occurs due to some inadequacy in the sensing, digitization or data recording process. This paper examines the effect of noise clustering algorithm of image classification. In remotely sensed data the easiest and usual assumption is that each pixel represents a homogeneous area on the ground. However in real world, it is found to be heterogeneous in nature. For this reason, it has been proposed that fuzziness should be accommodated in the classification procedure and preserves the extracted information. Classification of satellite images are complex process and accuracy of the output is dependent on classifier parameters. This paper examines the effect of various parameters like weighted exponent 'm' as well as resolution parameter '∂' for noise clustering (NC) classifier. The prime focus in this work is to select suitable parameters for classification of remotely sensed data which improves the accuracy of classification output to study the behaviour of associated learning parameters for optimization estimation using noise clustering classifier. A concept of "Noise Cluster" is introduced such that noisy data points may be assigned to the noise class. In this research work it has been tried to generate, a fraction outputs of noise clustering based classifier. The remote sensing data used has been from AWiFS, LISS-III and LISS-IV sensors of IRS-P6 satellite. This study proposes the entropy, as a special criterion for visualising and evaluating the uncertainty and it has been used as an absolute uncertainty indicator from output data. From the resultant aspect, while monitoring entropy of fraction images for different values, optimum weighting exponent 'm' and resolution parameter '∂' has been obtained for AWIFS, LIIS-III and LISS-IV images and that is 'm' = 2.9

R. Dwivedi (✉) · S. K. Ghosh
Indian Institute of Technology Roorkee, Roorkee, Uttarakhand, India
e-mail: r_dwivedi2000@yahoo.com

S. K. Ghosh
e-mail: scangfce@iitr.ernet.in

A. Kumar
Indian Institute of Remote Sensing, Dehradun, Uttarakhand, India
e-mail: anil@iirs.gov.in

B. V. Babu et al. (eds.), *Proceedings of the Second International Conference on Soft Computing for Problem Solving (SocProS 2012), December 28–30, 2012*, Advances in Intelligent Systems and Computing 236, DOI: 10.1007/978-81-322-1602-5_50,

and '∂' = 10^6, providing highest degree of membership value with minimum entropy value as shown in Table 1.

Keywords Entropy · Noise clustering (NC) · Fuzzy c-mean (FCM) · Possibilistic c-mean (PCM) · All wide field sensor (AWiFS) · Linear imaging self scanning (LISS).

1 Introduction

A traditional hard classification technique does not take into account that continuous spatial variation in land cover classes. To incorporate the gradual boundary change problem researchers had been proposed the 'soft' classification techniques that decompose the pixel into class proportions. Fuzzy classification is a soft classification technique, which deals with vagueness in class definition. Therefore, it can model the gradual spatial transition between land cover classes.

Fuzzy c-Means (FCM) [4–6] is an unsupervised clustering algorithm which has been widely used to find fuzzy membership grades between zero and one. The aim of FCM is to find cluster centers in the feature space that minimize an objective function. The objective function is associated with the optimization problem, which minimizes within class variation and maximizes variation between two classes. Standard FCM algorithm considers the spectral characteristics.

FCM clustering algorithm has been widely used to classify Satellite images with vague land cover classes. It is a popular fuzzy set theory based soft classifier, which handles the vagueness of a pixel at sub-pixel level. FCM has been successful in assigning the membership (u_{ij}) of a pixel to multiple classes but this assignment is relative to total number of classes defined and not absolute [16, 17]. This is because of the constraint on the membership values as given by Eq. (1)

$$\sum_{i=1}^{C} u_{ij} = 1 \text{ for all } j \tag{1}$$

i.e. the sum of membership values for a pixel in all the classes should be equal to one [5, 17] hence a new variation of FCM , called Possibilistic c-Means(PCM) which relaxes the constraint on membership value in Eq. (1) and gives absolute membership value, as stated by Eq. (2)

$$\max_i \quad u_{ij} > 0 \text{ for all } j \tag{2}$$

In case of PCM, this membership value represents the "degree of belongingness or compatibility or typicality", contrary to that represented by FCM, where it is, "degree of sharing".

The bias due to noise is a classical problem affecting all clustering algorithms. A good solution to this problem does not exist, although the field of clustering has been in existence for decades. An ideal solution would be one where the noise points get automatically identified and removed from the data. The concept of having an approach where one can define one cluster as the noise cluster is also promising, provided there is a way by which all the noise points could be dumped into one single cluster.

A concept of "Noise Cluster" is introduced such that noisy data points may be assigned to the noise class. The approach is developed for objective functional type (K-means or fuzzy K-means) algorithms, and its ability to detect 'good' clusters amongst noisy data is demonstrated. The approach presented is applicable to a variety of fuzzy clustering algorithms as well as regression analysis [9].

It is believed that cluster validity plays a pivotal role in robust clustering because without the concept of validity, it is not possible to separate the good points from the noise points and outliers and verify that our solution is good. The solution to the robust clustering problem requires that a fuzzy subset of the data set is rejected before the parameter estimates are computed. However, it is possible to optimize the objective function very trivially by excluding all points. Therefore, an additional constraint such as, cluster validity to avoid the trivial solution, is required. Hence, the solution to the general clustering problem appears to be inalienable from the notion of validity. Ideally, the objective function should be the same as the cluster validity [10].

The purpose of study of noise clustering without entropy is not only to establish a connection between fuzzy set theory and robust statistics, but also to discuss and compare several popular clustering methods from the point of view of robustness [8].

The aim of this paper is to study the behaviour of associated learning parameters of FCM, PCM and noise clustering without entropy for optimization estimation, with different fuzzy based functions, which is used for classification of multi-spectral remote sensing data in sub-pixel mode. In the next section, the details of parameters considered in FCM, PCM and noise clustering without entropy are provided. After that the remote sensing data used to produce optimization estimation based soft classification. After this experimental setup, results and their analysis are described. In last conclusion of this work is mentioned. This work has been done using in-house developed Sub-Pixel Multi-Spectral Image Classifier (SMIC) package using JAVA programming language [18].

2 Classifiers and Accuracy Assessment Approaches

2.1 Fuzzy C-Means Approach (FCM)

Fuzzy c-means (FCM) was originally introduced by J. C. Bezdek in [7]. In this clustering technique, each data point belongs to a cluster to some degree that is specified by a membership grade, and the sum of the memberships for each pixel must be equal to unity. This can be achieved by minimizing the generalized least-square error objective function,

$$J_m(U, V) = \sum_{i=1}^{N}\sum_{j=1}^{c}(\mu_{ij})^m \left\| X_i - x_j \right\|_A^2 \quad (3)$$

Subject to constraints, that

$$\sum_{j=1}^{c}\mu_{ij} = 1 \text{ for all } i$$

and

$$\sum_{i=1}^{N}\mu_{ij} > 1 \text{ for all } j$$

$$0 \le \mu_{ij} \le 1 \text{ for all } i, j \quad (4)$$

where X_i is the vector denoting spectral response of a pixel i, x is the collection of vector of cluster centers x_j, μ_{ij} are class membership values of a pixel, c and N are number of clusters and pixels respectively, and m is a weighting exponent ($1 < \mathrm{m} < \infty$), which controls the degree of fuzziness, $\left\| X_i - x_j \right\|_A^2$ is the squared distance ($\mathrm{d_{ij}}$) between X_i and x_j, and is given by,

$$d_{ij}^2 = \left\| X_i - x_j \right\|_A^2 = (X_i - x_j)^T A (X_i - x_j) \quad (5)$$

where A is the weight matrix.

Amongst a number of A-norms, three namely Euclidean, Diagonal and Mahalonobis norm, each induced by specific weight matrix, are widely used. The formulations of each norm are given as [7],

$$\begin{aligned} A &= I \quad \text{Euclidean Norm} \\ A &= D_j^{-1} \quad \text{Diagonal Norm} \\ A &= C_j^{-1} \quad \text{Mahalonobis Norm} \end{aligned} \quad (6)$$

where I is the identity matrix, D_j is the diagonal matrix having diagonal elements as the eigen values of the variance covariance matrix, C_j given by,

$$C_j = \sum_{i=1}^{N}(X_i - x_j)(X_i - x_j)^T \quad (7)$$

The class membership matrix μ_{ij} is obtained by:

$$\mu_{ij} = \frac{1}{\sum_{k=1}^{c} \left(\frac{d_{ij}^2}{d_{ik}^2} \right)^{1/(m-1)}} \tag{8}$$

where

$$d_{ik}^2 = \sum_{j=1}^{c} d_{ij}^2; \tag{9}$$

2.2 Possibilistic C-Means Approach (PCM)

The basic change in PCM in comparison to FCM is that one would like the memberships for representative feature points to be as high as possible, while unrepresentative points should have low membership in all clusters [17]. The objective function, which satisfies this requirement, may be formulated as:

$$J_m(U, V) = \sum_{i=1}^{N} \sum_{j=1}^{c} (\mu_{ij})^m \left\| X_i - v_j \right\|_A^2 + \sum_{j=1}^{c} \eta_j \sum_{i=1}^{N} (1 - \mu_{ij})^m \tag{10}$$

Subject to constraints;

$$\max_j \mu_{ij} > 0 \quad \text{for all } i$$

$$\sum_{i=1}^{N} \mu_{ij} > 0 \quad \text{for all } j$$

$$0 \leq \mu_{ij} \leq 1 \quad \text{for all } i, j$$

here μ_{ij} is calculated from Eq. (8).

In Eq. (10) where η_j is the suitable positive number, the first term demands that the distances from the feature vectors to the prototypes be as low as possible, whereas the second term forces the μ_{ij} to be as large as possible, thus avoiding the trivial solution. Generally, η_j depends on the shape and average size of the cluster j and its value may be computed as:

$$\eta_j = K \frac{\sum_{i=1}^{N} \mu_{ij}^m d_{ij}^2}{\sum_{i=1}^{N} \mu_{ij}^m} \tag{11}$$

where K is a constant and is generally kept as one. After this, class memberships, μ_{ij} are obtained as:

$$\mu_{ij} = \frac{1}{1 + \left(\frac{d_{ij}^2}{\eta_j}\right)^{1/(m-1)}} \tag{12}$$

2.3 Noise clustering without Entropy

A concept of "Noise Cluster" is introduced such that all noisy data points may be assigned to the noise class. The approach has been developed for objective functional type (K-means or fuzzy K-means) algorithms, and its ability to detect 'good' clusters amongst noisy data is demonstrated. Clustering methods need to be robust if they are to be useful in practice. Uncertainty is imposed simultaneously with multispectral data acquisition in remote sensing. It grows and propagates in processing, transmitting and classification processes. This uncertainty affects the extracted information quality. Usually, the classification performance is evaluated by criteria such as the accuracy and reliability. These criteria can not show the exact quality and certainty of the classification results. Unlike the correctness, no special criterion has been propounded for evaluation of the certainty and uncertainty of the classification results.

$$\mu_{i,j} = \left[\sum_{k=1}^{c}\left(\frac{d_{ij}^2}{d_{ik}^2}\right)^{\frac{1}{m-1}} + \left(\frac{d_{ij}^2}{\delta}\right)^{\frac{1}{m-1}}\right]^{-1} \tag{13}$$

where $1 \leq k \leq c$
where $1 \leq j \leq c$
and

$$\mu_{i,c+1} = \left[\sum_{j=1}^{c}\left(\frac{\delta}{d_{ik}^2}\right)^{\frac{1}{m-1}} + 1\right]^{-1} \tag{14}$$

Resolution parameter $\delta > 0$, any float value greater than zero.

This study proposes the entropy, as a special criterion for visualizing and evaluating the uncertainty of the results. This paper follows the uncertainty problem in multispectral data classification process. In addition to entropy, several uncertainty criteria are introduced and applied in order to evaluate the classification performance.

The objective function, which satisfies this requirement, may be formulated as:

$$U(u_{ij}|d) = \left[\sum_{i=1}^{N}\sum_{j=1}^{c} u_{ij} d_{ij} + \sum_{i=1}^{N} (u_{k,c+1})^m \partial\right] \tag{15}$$

and

Resolution parameter $\delta > 0$, any float value greater than zero
where $\infty > m > 1$, (any constant float value more than 1)

N = row * column (image size)

i = stands for pixel position at ith location distance between X_i and V_j

$$d_{ij}^2 = \|X_i - v_j\|_A^2 = (X_i - v_j)^T A(X_i - v_j)$$

V_j = Mean Vector for each cluster centers.

2.4 Accuracy Assessment Approach

With the availability of IRS-P6 satellite data, it is possible to acquire spectrally same and spatial different data sets of same area with same acquisition time. Due to the uniqueness of availability of these data sets, soft fraction images generated from coarser resolution data set (e.g. LISS-III, IRS-P6) can be evaluated from fraction images generated from finer resolution data sets (e.g. LISS-IV, IRS-P6) as reference data set is acquired at same time.

In any closed system, the entropy of the system will either remain constant or increase. This is known as the second law of thermodynamics from where the concept of entropy evolves. In information technology, entropy is measure of the uncertainty Dehghan et al. Further, Entropy is also considered to be a measure of disorder, or more precisely unpredictability Shannon.

This study envisages the usage of entropy, as a special criterion for visualizing and evaluating the uncertainty of the classified results of noise clustering classifiers. This criterion is able to purely and completely reflect the uncertainty from the classified image Congalton and Goodchild M. F. For the uncertainty visualization and evaluation of the classification results, the entropy criterion is proposed. This measure expressed by the following equation:

$$Entropy(x) = \sum_{i=1}^{M} -\mu(w_i/x) \log_2(\mu(w_i/x)) \tag{16}$$

where "M" denotes no. of classes and $\left(\mu\left(\frac{w_i}{x}\right)\right)$ is the estimated membership function of class i for pixel x.

For high uncertainty, the calculated value of entropy (Eq. 16) is high and inverse. Therefore, this criterion may be able to visualize the pure uncertainty in the classification results.

3 Study Area And Data Used

The study area for the present research work belongs to Sitarganj Tehsil, Udham Singh Nagar District, Uttarkhand, India. It is located in the southern part of the state. In terms of Geographic lat/long, the area extends from 28°52'29"N to 28°54'20"N and 79°34'25"E to 79°36'34"E. The area consists of agricultural farms with sugarcane and paddy as one of the few major crops with two reservoirs namely, Dhora and Bhagul reservoir.

The images for this research work have been taken from two different sensors namely LISS-III and LISS-IV belonging to satellite IRS-P6. The LISS-III dataset used here for classification and LISS -IV for referencing purposes.

4 Methodology

All three datasets (AWiFS, LISS-III, and LISS-IV) were geometrically corrected with RMSE less than 1/3 of a pixel and resampled using nearest neighbour resample method at 20 m, and 5 m spatial resolution respectively to maintain the correspondence of a LISS-III pixel with specific number of LISS-IV pixels (here four pixels will corresponding to 1 LISS-III pixel) with respect to sampling during accuracy assessment. The flow chart of the methodology adopted is shown in Fig. 1 (Figs. 2 and 3).

The six classes of interest, namely Agriculture land with crop, Sal forest, Eucalyptus plantation, Agriculture dry land without crop, Agriculture moist land without crop, and water body have been used for this study work.

5 Results And Discussions

The ambiguity and uncertainty is one of the major issue in the classification of remote sensing data. The estimation of uncertainity in the classification results is important and is necessary to evaluate the performance of any classifier. This study addresses the evaluation of entropy, based on noise clustering classifier which estimates uncertainty in classification results. In varying spatial resolution of classification and reference soft outputs, entropy gives the true reflectance of uncertainty ratio among various classes. The uncertainty criteria have been estimated from computed entropy based on actual output of classifier.

In this research work it has been tried to generate fraction outputs from NC classifiers. These outputs have been generated from AWIFS, LISS-III and LISS-IV images of IRS-P6 data. Entropy has been used as assessment parameters of accuracy for various land cover classes i. e. water bodies, Sal forest, Eucalyptus plantation,

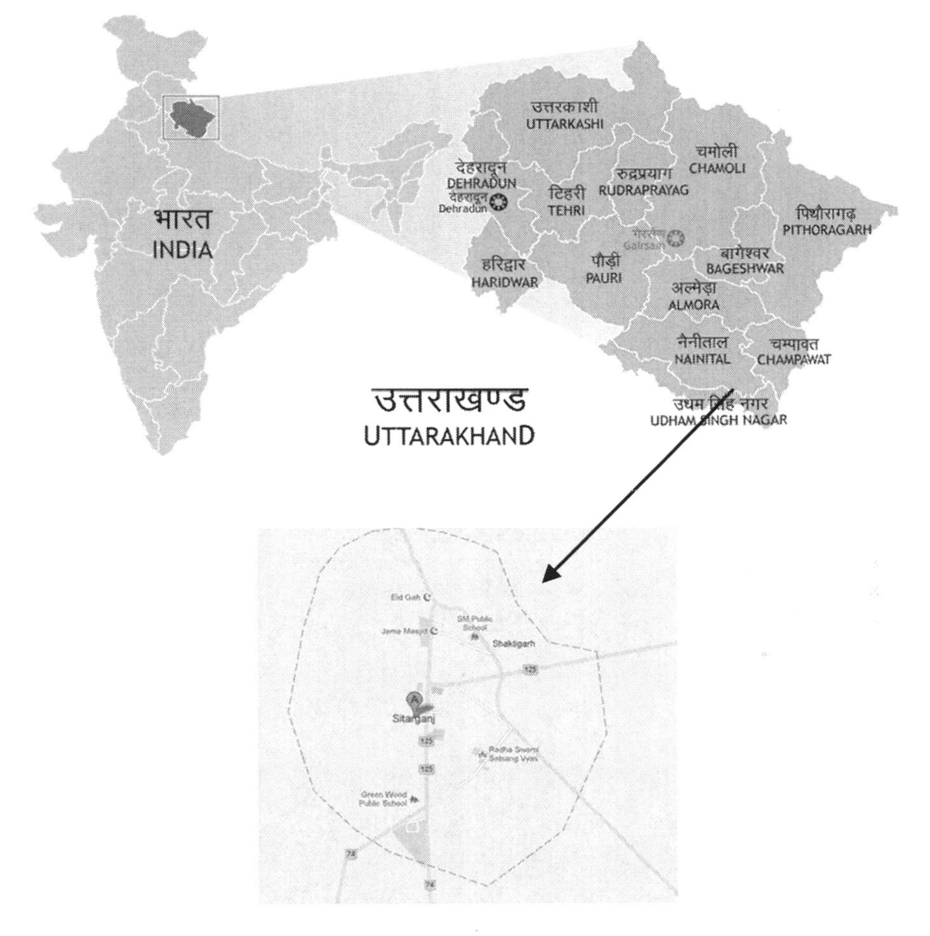

Fig. 1 Location map of study area

agriculture land with crop, agriculture moist land without crop, agriculture dry land without crop.

To investigate the effect of uncertain pixels in noise clustering classifiers , Euclidean norm has been chosen for noise clustering classifiers whereas , for fixed optimized weighting exponent 'm' = 2.9, resolution parameter '∂' varies from 1 to 10^9 for Sal forest, Eucalyptus plantation, water bodies, agriculture land with crop, agriculture moist land without crop, and agriculture dry land without crop. It is observed from the result Tables 1, 2, and 3, that uncertainty ratio is almost less to referential value 2.585, for noise classifiers using Euclidean norm. This reflects that noise based soft classifier is producing higher classification accuracy with minimum level of uncertainty. The computation of entropy is an absolute reflector of an uncertainty and this study identifies that entropy criterion provides stable results for noise, classifier for optimized value of 'm' and '∂'.

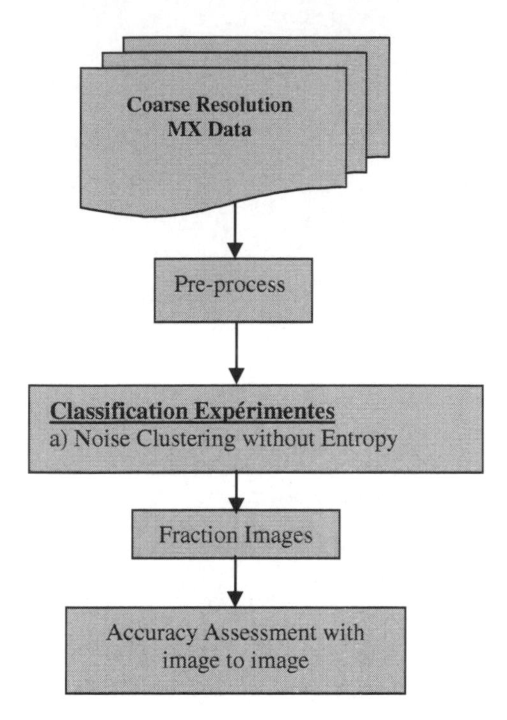

Fig. 2 Methodology adopted

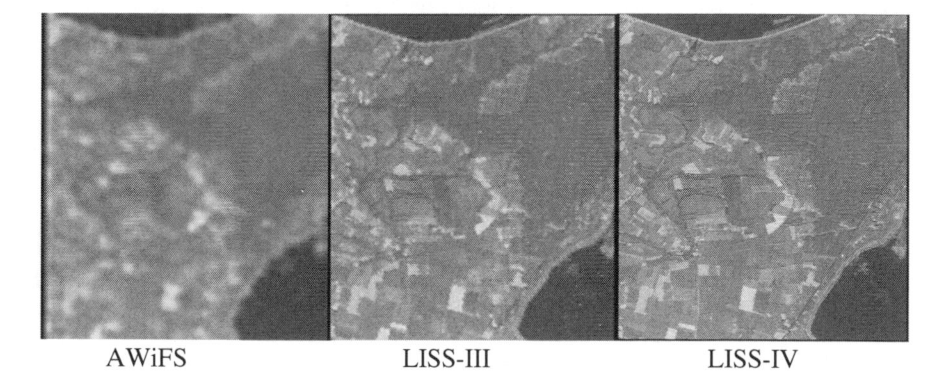

Fig. 3 Location of study area

For setting the optimized value of m and '∂', a number of experiments have been conducted individually for noise classifier for fixed optimized value of 'm' $= 2.9$ and varying '∂' from 1 to 10^9. It has been observed from the resultant Tables 1, 2, and 3 that is for homogenous classes like Agriculture land with crop, Agriculture dry land without crop Agriculture moist land without crop, and Water Body for noise classifiers the optimized value of '∂' is 10^6. Similarly for heterogeneous classes like

Sal forest and Eucalyptus plantation, the optimized value of '∂' for noise classifier is also 10^6. These findings suggest that using these optimized values of weighting exponent 'm' and resolution parameter '∂' for noise classifier on homogenous and heterogeneous land cover classes the range of the computed entropy varies between the range of [0, 2] as shown in resultant Tables 2, 3, and 4. This in turn states that the information uncertainty is not exceeding more than 2 %.

In this research entropy has been used to measure the accuracy in terms of uncertainty without using any kind of ground reference data. This classification accuracy is directly measured by entropy. Measuring the spatial statistics of a satellite image using an entropy, of six land cover classes can be measured using Eq. (16) i. e. $6*(-1/6*\log2 1/6) = 2.585$ [21]. This states that if the computed entropy values of classified images are lying within this range; then indirectly this reflects better classification results. It is shown in Table 2, 3, and 4, where AWIFS, LISS-III and LISS-IV entropy of noise classifiers for six land cover classes have been computed and, found that the entropy values are approximately lying within the specified range wherein the value of weighting exponent 'm' $= 2.9$ and resolution parameter '∂' is varying from 1 to 10^9.

6 Conclusion

In this research work entropy has been used as an assessment parameters of accuracy for various land cover classes i. e. water bodies, Sal forest, Eucalyptus plantation, agriculture land with crop, agriculture moist land without crop, agriculture dry land without crop. Entropy is used as an uniform measure to quantify the total spatial data uncertainty and fuzzy mixture uncertainty. In nutshell this study on spatial variation has identified that total uncertainty which is not exceeding the referential value, 2.585; mentioned in Table 1 for any of the above mentioned six classes of homogenous and heterogeneous categories. This mathematical model of entropy computation is used as an absolute indicator of measuring uncertainty among various land cover classes, without using any ground reference data. Accuracy assessment of a classified image is an integral part of image classification and in this research two things were involved first optimization of weighting exponent 'm' and resolution parameter '∂' and secondly computation of entropy. From the resultant Tables 2, 3, and 4, it shows that the optimum values of 'm' and '∂' for NC classifier on homogenous and heterogeneous classes are 2.9 and 10^6 respectively; wherein the membership values are varying from 0.9 to 1.0 with lesser entropy values, i. e. 0.86. From the Table 1, it can be observed that when the value of weighting exponent 'm' $= 2.9$ and resolution parameter '∂' $= 10^6$, the uncertainty is almost stable for all six land cover classes.

Table 1 Entropy variation for NC classifier

Classifiers used for various land cover classes	NC Classifier					
	AWiFS Entropy		LISS-III Entropy		LISS-IV Entropy	
	Min	Max	Min	Max	Min	Max
Agriculture land with crop	0.69 at m = 2.9 & $\partial = 10^6$	1.21 at m = 2.9 & $\partial = 1$	1.15 at m = 2.9 & $\partial = 10^6$	1.47 at m = 2.9 & $\partial = 1$	1.25 at m = 2.9 & $\partial = 10^6$	1.67 at m = 2.9 & $\partial = 100$
Sal forest	0.16 at m = 2.9 & $\partial = 10^6$	0.87 at m = 2.9 & $\partial = 1$	1.11 at m = 2.9 & $\partial = 10^6$	1.59 at m = 2.9 & $\partial = 10$	0.63 at m = 2.9 & $\partial = 10^6$	1.31 at m = 2.9 & $\partial = 1$
Eucalyptus plantation	0.87 at m = 2.9 & $\partial = 10^6$	1.32 at m = 2.9 & $\partial = 100$	0.55 at m = 2.9 & $\partial = 10^6$	$1.19atm = 2.9\&$ $\partial = 1$	0.75 at m=2.9 & $\partial = 10^6$	1.35 at m = 2.9 & $\partial = 1$
Agriculture dry land without crop	0.84 at m = 2.9 & $\partial = 10^6$	1.33 at m = 2.9 & $\partial = 100$	1.19 at m = 2.9 & $\partial = 10^6$	1.40 at m = 2.9 & $\partial = 100$	1.20 at m = 2.9 & $\partial = 10^6$	1.65 at m = 2.9 & $\partial = 100$
Agriculture moist land without crop	1.13 at m = 2.9 & $\partial = 10^6$	1.52 at m = 2.9 & $\partial = 100$	0.92 at m = 2.9 & $\partial = 10^6$	1.31 at m = 2.9 & $\partial = 10$	1.16 at m = 2.9 & $\partial = 10^6$	1.56 at m = 2.9 & $\partial = 1$
Water Body	0.19 at m = 2.9 & $\partial = 10^6$	1.04 at m = 2.9 & $\partial = 10$	0.67 at m = 2.9 & $\partial = 10^6$	1.22 at m = 2.9 & $\partial = 10$	1.13 at m = 2.9 & $\partial = 10^6$	1.51 at m = 2.9 & $\partial = 1$

Table 2 AWIFS entropy of various land cover classes from NC classification output

Value of 'm' = 2.9 and varying Resolution Parameter '∂'	Agriculture land with crop	Sal forest	Eucalyptus plantation	Agriculture dry land without crop	Agriculture moist land without crop	Water Body
$\partial = 1$	1.21	0.87	0.93	0.51	0.61	0.93
$\partial = 10$	1.25	0.52	1.21	1.06	1.20	1.04
$\partial = 10^2$	0.98	0.30	1.32	1.33	1.52	0.76
$\partial = 10^3$	0.76	0.20	1.08	1.19	1.44	0.45
$\partial = 10^4$	0.70	0.16	0.95	0.99	1.31	0.29
$\partial = 10^5$	0.69	0.16	0.87	0.92	1.22	0.23
$\partial = 10^6$	0.69	0.16	0.87	0.84	1.13	0.19
$\partial = 10^7$	0.69	0.16	0.87	0.84	1.13	0.19
$\partial = 10^8$	0.69	0.16	0.87	0.84	1.13	0.19
$\partial = 10^9$	0.69	0.16	0.87	0.84	1.13	0.19

Table 3 LISS-III entropy of various land cover classes from NC classification output

Value of 'm' = 2.9 and varying Resolution Parameter '∂'	Agriculture land with crop	Sal forest	Eucalyptus plantation	Agriculture dry land without crop	Agriculture moist land without crop	Water Body
$\partial = 1$	1.47	1.45	1.193	0.90	1.109	0.99
$\partial = 10$	1.66	1.59	1.013	1.22	1.31	1.22
$\partial = 10^2$	1.51	1.40	0.80	1.40	1.128	1.09
$\partial = 10^3$	1.33	1.25	0.66	1.37	0.99	0.90
$\partial = 10^4$	1.22	1.174	0.58	1.30	0.91	0.79
$\partial = 10^5$	1.16	1.11	0.55	1.24	0.91	0.68
$\partial = 10^6$	1.15	1.11	0.55	1.19	0.92	0.67
$\partial = 10^7$	1.15	1.11	0.55	1.19	0.92	0.67
$\partial = 10^8$	1.157	1.11	0.55	1.19	0.92	0.67
$\partial = 10^9$	1.157	1.11	0.55	1.19	0.92	0.67

Table 4 LISS-IV entropy of various land cover classes from NC classification output

Value of 'm'=2.9 and varying Resolution Parameter '∂'	Agriculture land with crop	Sal forest	Eucalyptus plantation	Agriculture dry land without crop	Agriculture moist land without crop	Water Body
$\partial = 1$	1.37	1.31	1.35	0.96	1.11	1.03
$\partial = 10$	1.66	1.076	1.19	1.51	1.57	1.49
$\partial = 10^2$	1.67	0.83	1.04	1.65	1.56	1.51
$\partial = 10^3$	1.54	0.74	0.89	1.49	1.35	1.32
$\partial = 10^4$	1.40	0.68	0.81	1.35	1.36	1.19
$\partial = 10^5$	1.30	0.64	0.77	1.29	1.20	1.16
$\partial = 10^6$	1.25	0.63	0.75	1.20	1.16	1.14
$\partial = 10^7$	1.25	0.63	0.75	1.20	1.16	1.13
$\partial = 10^8$	1.25	0.63	0.75	1.20	1.16	1.13
$\partial = 10^9$	1.25	0.63	0.75	1.20	1.16	1.135

References

1. Binaghi, E., Rampini, A.: Fuzzy decision making in the classification of multisource remote sensing data. Opt. Eng. **6**, 1193–1203 (1993)
2. Binaghi, E., Rampini, A., Brivio, P.A., Schowengerdt, R.A. (eds.): Special Issue on Non-conventional Pattern Analysis in Remote Sensing. Pattern Recogn. Lett. **17**(13) (1996)
3. Binaghi, E., Brivio, P.A., Chessi, P., Rampini, A.: A fuzzy set based accuracy assessment of soft classification. Pattern Recogn. Lett. **20**, 935–948 (1999)
4. Bezdek, J.C.: A convergence theorem for the fuzzy ISODATA clustering algorithms. IEEE Trans. on Pattern Anal. Machine Intell. PAMI-**2**(1), 1–8 (1980)
5. Bezdek, J.C., Ehrlich, R., Full, W.: FCM: the fuzzy c-means clustering algorithm. Comput. Geosci. **10**(2–3), 191–203 (1984)
6. Bezdek, J.C., Hathaway, R.J., Sabin, M.J., Tucker, W.T.: Convergence theory for fuzzy c-means: counterexamples and repairs. IEEE Trans. Syst., Man, Cybern. SMC-**17**(5), 873–877 (1987)
7. Bezdek, J.C.: Pattern recognition with fuzzy objective function algorithms. Plenum, New York, USA (1981)
8. Dav'e, R.N.: Fuzzy-shell clustering and applications to circle detection in digital images. Int. J. Gen. Syst. **16**, 343–355 (1990)
9. Dave, R.N.: Characterization and detection of noise in clustering. Pattern Recogn. Lett. **12**, 657–664 (1991)
10. Dav'e, R.N., Krishnapuram, R.: Robust clustering methods: a unified view. IEEE Trans. Fuzzy Syst. **5**(2), 270–293 (1997)
11. Fisher, P.: The pixel: a snare and a delusion. Int. J. Remote Sens. **18**(3), 679–685 (1997)
12. Foody, G.M.: Cross-entropy for the evaluation of the accuracy of a fuzzy land cover classification with fuzzy ground data. ISPRS J. Photogrammetry Remote Sens. **50**, 2–12 (1995)
13. Foody, G.M.: Approaches for the production and evaluation of fuzzy land cover classifications from remotely sensed data. Int. J. Remote Sens. **17**(7), 1317–1340 (1996)
14. Foody, G.M., Arora, M.K.: Incorporating mixed pixels in the training, allocation and testing stages of supervised classification. Pattern Recogn. Lett. **17**, 1389–1398 (1996)
15. Foody, G.M., Lucas, R.M., Curran, P.J., Honzak, M.: Non-linear mixture modelling without end-members using an ANN. Int. J. Remote Sens. **18**(4), 937–953 (1997)
16. Foody, G.M.: Estimation of sub-pixel land cover composition in the presence of untrained classes. Comput. Geosci. **26**, 469–478 (2000)
17. Krishnapuram, R., Keller, J.M.: A possibilistic approach to clustering. IEEE Trans. Fuzzy Syst. **1**(2), 98–110 (1993)
18. Kumar, A., Ghosh, S.K., Dadhwal, V.K.: Study of sub-pixel classification algorithms for high dimensionality data set. In: IEEE International Geoscience and Remote Sensing Symposium and 27th Canadian Symposium on Remote Sensing, Denver, Colorado, USA, 31 July–04 August (accepted), (2006)
19. Okeke, F., Karnieli, A.: Methods for fuzzy classification and accuracy assessment of historical aerial photographs for vegetation change analyses. Part I: algorithm development. Int. J. Remote Sens. **27**(1–2), 153–176 (2006)
20. Pontius, Jr, R.G., Cheuk, M.L.: A generalized cross tabulation marix to compare soft classified maps at multiple resolutions. Int. J. Geogr. Inf. Sci. **20**(1), 1–30 (2006)
21. Verhoeye, J., Robert, D.W.: Sub-pixel mapping of sahelian wetlandsusing multi-temporal SPOT vegetation images. Laboratory of Forest Management and Spatial Information Techniques, Faculty of Agricultural and Applied Biological Sciences, University of Gent, Belgium, (2000)

Location Management in Mobile Computing Using Swarm Intelligence Techniques

Nikhil Goel, J. Senthilnath, S. N. Omkar and V. Mani

Abstract Location management is an important and complex issue in mobile computing. Location management problem can be solved by partitioning the network into location areas such that the total cost, i.e., sum of handoff (update) cost and paging cost is minimum. Finding the optimal number of location areas and the corresponding configuration of the partitioned network is NP-complete problem. In this paper, we present two swarm intelligence algorithms namely genetic algorithm (GA) and artificial bee colony (ABC) to obtain minimum cost in the location management problem. We compare the performance of the swarm intelligence algorithms and the results show that ABC give better optimal solution to locate the optimal solution.

Keywords Location management · Genetic algorithm · Artificial bee colony

1 Introduction

The aim of location management is to track where the subscribers are in order to route incoming calls to appropriate mobile terminals. Location Management strategies can be divided into always-update strategy and never-update strategy. In the

---

N. Goel (✉)
Department of Information Technology, National Institute of Technology Karnataka, Surathkal, India
e-mail: nikhil8877@gmail.com

J. Senthilnath · S. N. Omkar · V. Mani
Department of Aerospace Engineering, Indian Institute of Science, Bangalore, India
e-mail: snrj@aero.iisc.ernet.in

S. N. Omkar
e-mail: omkar@aero.iisc.ernet.in

V. Mani
e-mail: mani@aero.iisc.ernet.in

B. V. Babu et al. (eds.), *Proceedings of the Second International Conference on Soft Computing for Problem Solving (SocProS 2012), December 28–30, 2012*, Advances in Intelligent Systems and Computing 236, DOI: 10.1007/978-81-322-1602-5_51,

always-update strategy, as soon as mobile terminals enter new cells, it performs location update. Using this strategy you can get accurate location information and hence save resources spent on searching (paging). But this frequent location update will require high resource (overhead). In the never-update strategy, location update is never performed. Instead, when an incoming call comes, a search operation is conducted to route the call to appropriate mobile terminals. Here, no resources would be used for location update, overhead for search operation (paging) would be very high. Hence, there is need to have balance between location update and paging so as to minimize the total cost of mobile communication [1–3].

The location area scheme is another location management technique that is commonly used in existing networks [4–6]. In this scheme, the network is partitioned into regions known as location areas (LA), with each region consisting of one or more cells. The update operation is performed only when any user moves from one location area to another. When a call arrives, searching (paging) operation is performed for cells in specific LA. Optimal LA partitioning is NP-complete problem [7].

Genetic algorithm with modified genetic operators was used to solve this optimization problem in [8]. Another swarm intelligence algorithm—ant colony optimization was discussed in [1]. Minimizing the location update subject to a paging bound constraint is studied in [9]. A polynomial time approximate algorithms with the objective of minimizing the sum of handoff traffic cost and paging cost are presented in [10].

Contributions of this paper: In this paper, we consider the location area scheme model and try to minimize the cost per arrival presented in earlier studies [1, 11]. We obtain the optimal cost per arrival using swarm intelligence techniques namely GA and artificial bee colony (ABC). Call per cost arrival is defined as total cost divided by total call arrivals in given network in time T, and obtains optimal number of location areas. Previous study [2] on location management using GA has been carried out using reporting cell model [3]. Here, we present GA for location management problem but using location area scheme [1]. We also apply ABC algorithm to minimize cost per call arrival and compare our results with results of earlier study [1, 2].

2 Problem Formulation

Let us assume a network with n cells. With each cell i we associate call movement weight W_{mi} and call arrival weight W_{ci}. Movement weight represents total number of movement into the cell. Call arrival weight represents total number of call arrivals within a cell. Paging Cost for a cell in Kth location area is

$$a_j = \sum_{j \in k}^{n} W_{cj}, \quad \text{for all } j = 1, 2 \ldots, n \tag{1}$$

Total paging cost for location area k, if there are NLA_k cells in kth location area is

$$\mathrm{PC_k} = \mathrm{NLA}_k{}^{*} a_j \tag{2}$$

Hence total paging cost is:

$$N_P = \sum_{k-1}^{m} \mathrm{PC}_k \tag{3}$$

Whenever there is movement among two adjacent cells in different two location areas, cost is incurred for updating its location. Total handoff cost is:

$$N_{LU} = \sum_{i=1}^{n} \sum_{j=1}^{n} (1 - Y_{ij})^{*} h(i, j) \tag{4}$$

We know from [1] that:

$$\text{Total cost} = N_p + C^{*} N_{LU} \tag{5}$$

Therefore,

$$\text{Total cost} = \sum_{k=1}^{m} \mathrm{NLA_k}^{*} \left\{ \sum_{j \in k}^{n} W_{cj} \right\} + C^{*} \sum_{i=1}^{n} \sum_{j=1}^{n} (1 - Y_{ij})^{*} h(i, j) \tag{6}$$

$Y_{ij} = 1$, if both cell i and cell j are assigned to the same location area;
$Y_{ij} = 0$ otherwise.

where $h(i, j)$ is cost per unit time handoff occur between cell i and cell j and depends upon the movement weight of cell i and cell j. C is a constant representing the cost ratio of location update and paging. It was studied in [12] that cost of location update is much higher than paging cost and it was also concluded in [8] that C value is approximately 10. Cost per call arrival is calculated by dividing total cost by total number of call arrivals. For a given network, total number of call arrivals is calculated using

$$\text{Total number of call arrivals} = \sum_{j \in k}^{n} W_{cj} \tag{7}$$

Hence,

$$\text{Cost per call arrival} = (\text{total Cost}) / \text{total number of call arrivals} \tag{8}$$

3 The Methodology

This section gives brief introduction to swarm intelligence algorithms [13–22] used in our study, the way they are used to solve combinatorial optimization and the pseudocode for the algorithms.

3.1 Genetic Algorithm

Genetic algorithm is a population based evolutionary computation technique [13]. This algorithm takes a predetermined number of random solutions (population) in the search space called chromosomes. At each iteration the chromosomes are made to crossover. At any random point, chromosomes undergo mutation based on mutation rate and the fitness of each chromosome is calculated using Eq. (8). The fitness is calculated and the best solutions carry-on till termination criteria is reached. Thus, optimal distribution of cells among different LA can be determined using this algorithm.

3.2 Artificial Bee Colony

In the ABC algorithm [14], all the population is considered as employed bee and onlooker bee, i.e., they checked for food source in their neighbor region. Solutions consisting of cells belonging to specific LA are changed to get a new solution given by Eq. (9).

$$V[ij] = x[ij] + phi[ij]^{*} (x[kj] - x[ij]) \tag{9}$$

where $x[ij]$ represents LA of jth cell in ith solution and $x[kj]$ represents LA of jth cell in kth solution (k is randomly selected). $Phi[ij]$ ensures that new LA assigned to jth cell in ith solution already exists. New LA value is stored in $V[ij]$.

4 Results and Discussion

In this section, the results and the performance evaluation are shown. A discussion is carried out at the end of this section pertaining to the results obtained. Here, we have considered two networks (i) 6×6 and (ii) 8×8 with the values of movement weight W_{mi} and call arrival weight W_{ci} for all the cells in the networks [2]. For each network we found the optimal number of LA so that cost per call arrivals is minimum using GA and ABC and compare our results with ant colony optimization method given in [1]. We run the code for 10 times for each of the algorithm mentioned above and

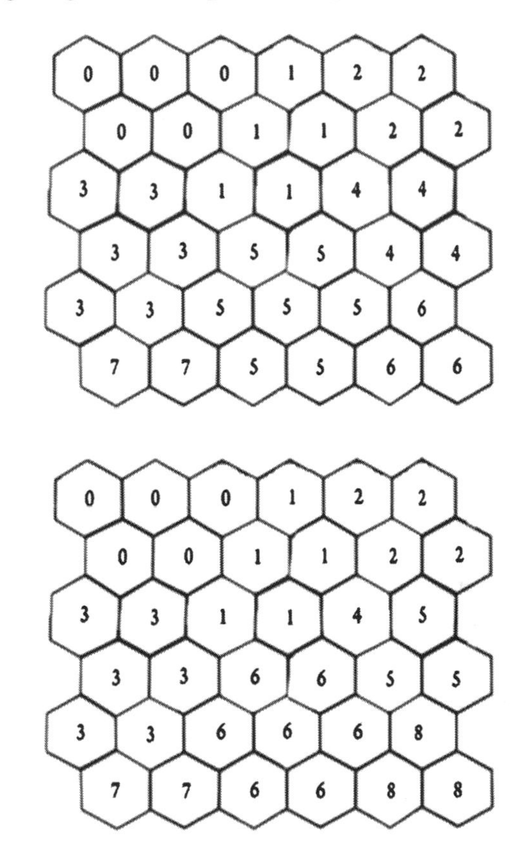

Fig. 1 Configuration with optimal number of LA using GA

Fig. 2 Configuration with optimal number of LA using ABC

Table 1 6×6 network

Method	Minimum fitness	Number of LA	Maximum	Mean	SD
ACO [1]	11.1147	8	–	–	–
GA	11.1147	8	11.586426	11.3098419	0.1554
ABC	11.0943	9	11.545944	11.2829718	0.1393

maximum, minimum, mean, and deviation are calculated. We also made the graph for fitness value versus number of iterations for the run where we get optimal fitness.

In first case, we consider 6×6 (36 cells) network. For 6×6 network we have search space of size $2^{36} = 68{,}719{,}476{,}736$. In all the methods used in this study we use population of 10 and maximum generation of 10,000. In GA, the best 20 % of the parent are reproduced and the remaining 80 % undergoes crossover and mutation for the next generation. Parameter in ABC such as limit value is set to be 50. Best results of GA and ABC have been shown in Figs. 1 and 2, respectively and corresponding regions have been assigned to the same value. From Table 1, we observe after 10 runs for this network, ABC provides better optimal solution compared to ACO and GA.

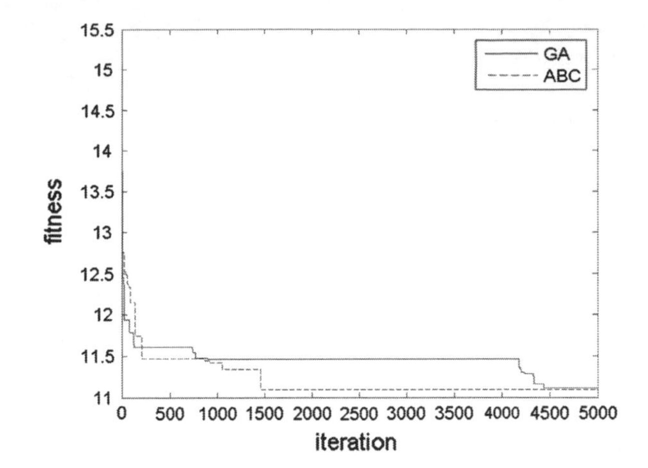

Fig. 3 6 × 6 network

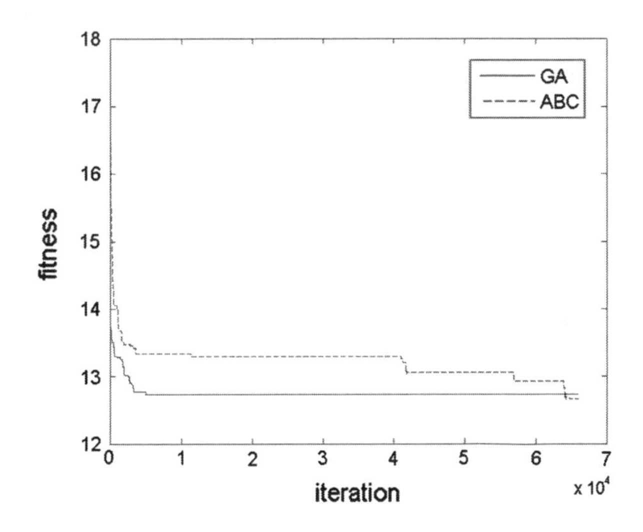

Fig. 4 8 × 8 network

ABC shows minimum standard deviation. From Fig. 3, we can observe GA converges faster in comparison to ABC.

In second case, we consider 8 × 8 (64 cells) network. For 8 × 8 network we have search space of size $2^{64} = 18{,}446{,}744{,}073{,}709{,}551{,}616$. The parameters such as population size and maximum generation are kept constant which is 20 and 1,00,000. Here also in GA, the best 20 % of the parent are reproduced and the remaining 80 % undergoes crossover and mutation for the next generation, whereas in ABC limit value is set to be 1,000. From Fig. 4 we can once again observe GA converges faster in comparison to ABC. Again best result obtained using GA and ABC has been shown in Figs. 5 and 6, respectively and corresponding regions have been assigned

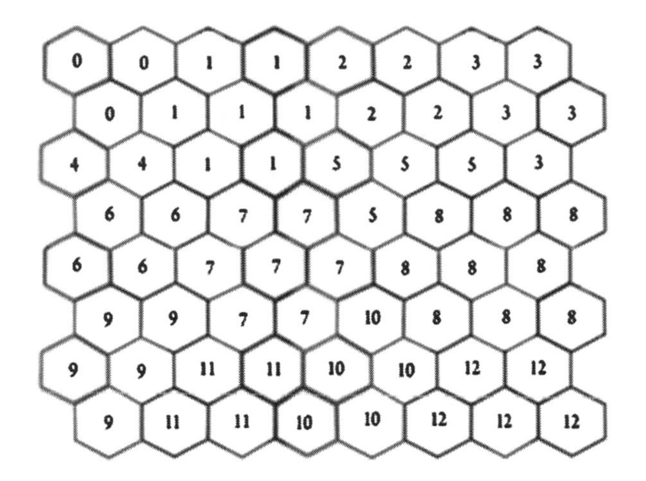

Fig. 5 Configuration with optimal number of LA using GA

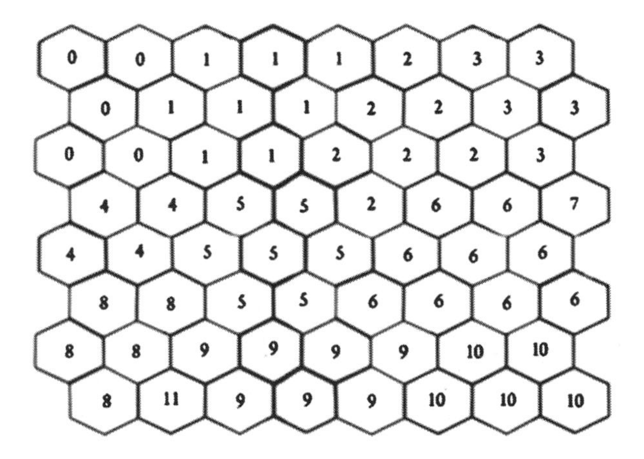

Fig. 6 Configuration with optimal number of LA using ABC

Table 2 8×8 network

Method	Minimum fitness	Number of LA	Maximum	Mean	SD
ACO [1]	12.86895	13	–	–	–
GA	12.731830	13	13.266327	12.926776	0.1596
ABC	12.668662	13	13.158280	12.842936	0.1255

to the same value. From Table 2, we observe that after 10 runs for this network, ABC provides better optimal solution compared to ACO and GA. ABC shows minimum standard deviation.

5 Conclusion

In this paper, we have implemented and analyzed two nature inspired techniques for location management problem and compared it with earlier studied method. The optimal partitions of 6×6 and 8×8 networks are presented. Results of the experiments show that ABC and GA can be effectively implemented for location management scheme. While ABC show best performance in terms of optimal value of cost per call arrival and standard deviation, GA converges faster in both the network.

References

1. Kim, S.S., Kim, I.H., Mani, V., Kim, H.J., Agarwal, D.P.: Partitioning of mobile network into location areas using ant colony optimization. ICIC Express Lett. Part B: Appl. **1**(1), 39–44 (2010)
2. Subrata, R., Zomaya, A.Y.: A comparison of three artificial life techniques for reporting cell planning in mobile computing. IEEE Trans. parallel Distrib. Syst. **14**(2), 142–153 (2003)
3. Bar, N.A., Kessler, I.: Tracking mobile users in wireless communications networks. IEEE Trans. Inf. Theory **39**, 1877–1886 (1993)
4. Okasaka. S., Onoe, S., Yasuda, S., Maebara, A.: A new location updating method for digital cellular systems. In: Proceedings of 41st IEEE Vehicular Technology Conference (1991)
5. Plassmann, D.: Location management strategies for mobile cellular networks of 3rd generation. In: Proceedings of IEEE 44th Vehicular Technology Conference (1994)
6. Yeung, K.L., Yum, T.S.P.: A comparative study on location tracking strategies in cellular mobile radio systems. In: Proceedings of IEEE Global Telecommunication Conference (1995)
7. Gondim, P.R.L.: Genetic algorithms and the location area partitioning problem in cellular networks. In: Proceedings of IEEE 46th Vehicular Technology Conference (1996)
8. Taheri, J., Albert Y.Z.: A genetic algorithm for finding optimal location area configurations for mobility management. IEEE Conference on Local Computer Networks 30th Anniversary (2005)
9. Yannis, M., Magdalene, M., Michael, D., Nikolaos, M., Constantin, Z.: A hybrid stochastic genetic-GRASP algorithm for clustering analysis. Oper. Res. Int. J. **8**, 33–46 (2008). doi: 10.1007/s12351-008-0004-8
10. Bejerano, Y., Smith, M.A., Naor, J.S., Immorlica, N.: Efficient location area planning for personal communication systems. IEEE/ACM Trans. Netw. **14**, 438–450 (2006)
11. Subrata, R., Zomaya, A.Y.: Evolving cellular automata for location management in mobile computing networks. IEEE Trans. Parallel Distrib. Syst. **14**, 13–26 (2003)
12. Imielinski, T., Badrinath, B.R.: Querying locations in wireless environments. In: Proceedings of Wireless Communication and Future Directions (1992)
13. Goldberg, D.E.: Genetic Algorithms in Search Optimization and Machine Learning. Addison-Wesley, Reading (1989)
14. Karaboga, D., Basturk, B.: On the performance of artificial bee colony (ABC) algorithm. Appl. Soft Comput. **8**(1), 687–697 (2008)
15. Senthilnath, J., Omkar, S.N., Mani, V., Tejovanth, N., Diwakar, P.G., Archana Shenoy, B.: Hierarchical clustering algorithm for land cover mapping using satellite images. IEEE J. Sel. Topics Appl. Earth Obs. Remote Sens. **5**(3), 762–768 (2012)
16. Craig, D., Omkar, S.N., Senthilnath, J.: Pickup and delivery problem using metaheuristics. Expert Syst. Appl. **39**(1), 328–334 (2012)
17. Senthilnath, J., Omkar, S.N., Mani, V.: Clustering using firefly algorithm: performance study. Swarm Evol. Comput. **1**(3), 164–171 (2011)

18. Omkar, S.N., Senthilnath, J., Khandelwal, R., Narayana Naik, G., Gopalakrishnan, S.: Artificial Bee Colony (ABC) for multi-objective design optimization of composite structures. Appl. Soft Comput. **11**(1), 489–499 (2011)
19. Omkar, S.N., Senthilnath, J.: Artificial bee colony for classification of acoustic emission signal sources. Int. J. Aerosp. Innov. **1**(3), 129–143 (2009)
20. Omkar, S.N., Senthilnath, J., Suresh, S.: Mathematical model and rule extraction for tool wear monitoring problem using nature inspired techniques. Indian J. Eng. Mater. Sci. **16**, 205–210 (2009)
21. Omkar, S.N., Senthilnath, J.: Mudigere, D., Manoj Kumar, M.: Crop classification using biologically inspired techniques with high resolution satellite image. J. Indian Soc. Remote Sens. **36**(2), 172–182 (2008)
22. Omkar, S.N., Senthilnath J.: In: Dehuri, S., et al. (eds.) Integration of Swarm Intelligence and Artificial Neutral Network, Neural Network and Swarm Intelligence for Data Mining, chap. 2. World Scientific Press, Singapore, pp. 23–65 (2011)

Fingerprint and Minutiae Points Technique

Karun Verma and Ishdeep Singla

Abstract This paper will extensively dictate the whole basic details of fingerprint and its techniques. Moreover one of the most important and widely used techniques that is minutiae point extraction technique is also covered in detail. The minutiae points are extracted with the help of cross-number algorithm. Cross-number algorithm also helps for the rejection of false minutiae point's extraction

Keywords Biometrics · Fingerprint · Minutiae points

1 Introduction

In an increasingly digital world, the control over entry of authorized person has become a vital thing. From the personal computer to National security, there is big use of identity checking that is Authentication. And biometrics provide automated access to the security systems. It's always better to use some automated methods instead of remembering and filling passwords. In biometrics, fingerprint technology is widely used technology, using this technology we need not to carry any identity card. Finger works as identity card, meaning there are no tension of forgetting and losing identity cards [1].

K. Verma · I. Singla (✉)
Computer Science and Engineering Department, Thapar University, Patiala, India
e-mail: ishdeep.singla@gmail.com

I. Singla
e-mail: karun.verma@thapar.edu

B. V. Babu et al. (eds.), *Proceedings of the Second International Conference on Soft Computing for Problem Solving (SocProS 2012), December 28–30, 2012*, Advances in Intelligent Systems and Computing 236, DOI: 10.1007/978-81-322-1602-5_52,

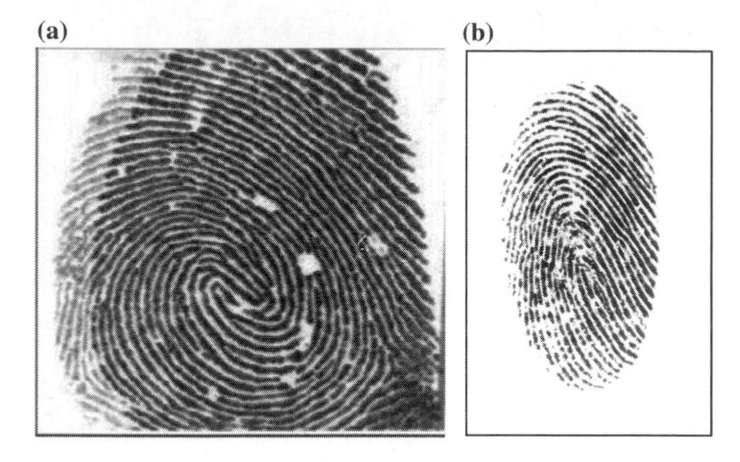

Fig. 1 Fingerprint images: **a** inked fingerprint and **b** live-scan fingerprint [3]

2 What is Fingerprint?

Fingerprint is the graphical flow-like ridges. It is present on each and every finger of every human's fingers as shown in Fig. 1. Ridges are embedded on all fingers from the very first day of our birth and do not change throughout the life. It may only change if a serious accident such as bruises and cuts or surgery on the fingertips occurs. This property makes fingerprints a very attractive biometric identifier and point of research. Basically, there are two resources for getting fingerprint pattern [2]:

(i) Scanning an inked impression of a finger is shown in Fig. 1a.
(ii) Using a live-scan fingerprint scanner shown in Fig. 1b.

In Fig. 1, dark lines are called ridges and the white area that exists between the ridges are called valley or furrow.

3 Fingerprint Features

Fingerprint features are those attributes of a fingerprint that may be useful either to classify or to uniquely identify the fingerprint. There are two main types of features, namely, the local features and the global features. Figure 2a shows the local features and Fig. 2b shows the global features.

3.1 Global Features

The fingerprint global features are identified by means of the local orientation of the fingerprint ridges, that is, the Orientation Field Curves (OFCs). As shown in Fig. 2b the core and the delta are the features which have been located in central position of

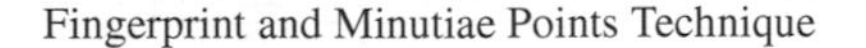

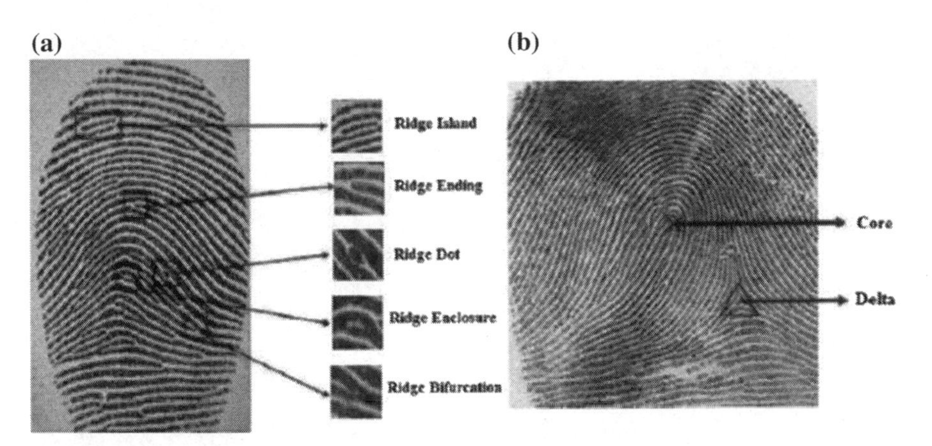

Fig. 2 **a** Local Features: Minutiae; **b** global features: core and delta [3]

fingerprint. A Core is the area around the center of the fingerprint loop and a Delta is the area where the fingerprint ridges tend to triangulate. Due to their unique property, both plays an important role to compare one fingerprint with other fingerprints [4].

3.2 Local Features

The fingerprint local features are those attributes that give the minutiae details about the fingerprint pattern. Minutiae further provide various ways that the ridges can be discontinuous. A ridge can suddenly end (termination), or can divide into two ridges (bifurcation) as shown in Fig. 2a. There are 40–100 minutiae point in a good quality image [8]. And in a fingerprint image of 300 $\times$ 300 pixels the distance between two fingerprints vary between 1-113 pixels. With these features and numerical figures, local features have become more suitable to compare fingerprints [4]. There are many methods like cross number are available to extract the minutiae points.

4 Fingerprint Classification

It is obvious that with the increase database size complexity and automatic comparison time will also increase. So to reduce the search time and computational complexity, there is a need to classify fingerprint in a precise and consistent manner which will help to reduce search time with less number of comparisons. According to Galton–Henry classification (Galton, 1892 and Henry, 1900) classification, we classify fingerprint images into five major classes: plain arch, tented arch, left-loop, right-loop, and whorl (a plain and twin loop, respectively).

Arch: In whole fingerprint arch covers only 5 % of the portion. These consist of ridges that run major in horizontal manner can say from left to right as shown in Fig. 3. There are two types of arches: plain arches and tented arches. Generally, plain arch has no singular points. While tented arch have one core and one delta.

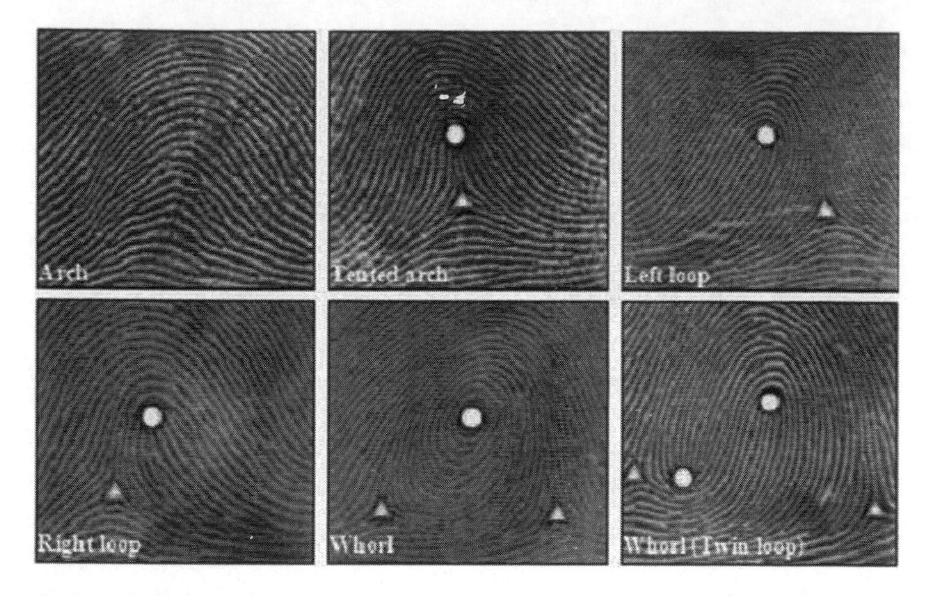

Fig. 3 Fingerprint classes [5]

Loop: Loops cover 60–70 % of whole fingerprint pattern. As the name suggests set of the ridges enters on either side of the fingerprint, bends, touches, or crosses the line running from the delta to the core and run back in same direction of the side where the ridge or ridges entered as shown in Fig. 3. Each loop pattern has one delta and one core. There can be left loop or right loop.

Whorl: 25–35 % of fingerprint pattern is covered by whorl. In a whorl, more than one ridges moves through at least one circuit. A whorl pattern always consists of two or more deltas. There are two types of whorl plain whorl and double whorl. A plain whorl is the pattern which consists of some ridges which make or partially make a complete circuit with two deltas. Double loop whorl consists of two separate and distinct loops.

5 Fingerprint Matching Techniques

There a lot of techniques for matching a fingerprint. There are three most popular methods for matching fingerprints [2] described below.

5.1 Correlation-based Matching

In this method one fingerprint image is superimposed on other. The correlation between corresponding pixels is computed for different alignments and on the basis of these correlations and computations decision is made.

5.2 Pattern-based (or Image-based) Matching

In pattern-based algorithms, the basic fingerprint patterns (arch, whorl, and loop) are used to compare fingerprints, between a previously stored template and a candidate fingerprint. This requires that the images are aligned in the same orientation. To do this, the algorithm finds a central point in the fingerprint image and centers on that. In a pattern-based algorithm, the template contains the type, size, and orientation of patterns within the aligned fingerprint image. The candidate fingerprint image is graphically compared with the template to determine the degree to which they match.

5.3 Minutiae-based Matching

This is the most popular and widely used technique, being the basis of the fingerprint comparison made by fingerprint examiners. Minutiae are extracted from the two fingerprints and stored as sets of points in the two-dimensional plane. Minutiae-based matching essentially consists of finding the alignment between the template and the input minutiae sets that result in the maximum number of minutiae pairings. In this thesis we have implemented a minutiae-based matching technique. This approach has been intensively studied, also is the backbone of the current available fingerprint identification products.

6 Minutiae Extraction

Minutiae points matching is the best approach for the matching of fingerprints. The work of minutiae extraction includes some important steps that are Ridge Thinning, Minutiae Marking, False Minutiae Removal, and Minutiae Representation.

6.1 Ridge Thinning

The main aim of this step is to convert the redundant pixels of ridge into one pixel wide. This will be very helpful in finding minutiae points and to implement minutiae point algorithm. In Matlab there has been one very popular morphological thinning function to perform this task.

bwmorph(binaryImage,'thin',Inf)

The thinned image is then filtered, again using MATLAB's three morphological functions to remove some H breaks, isolated points, and spikes (see Fig. 4).

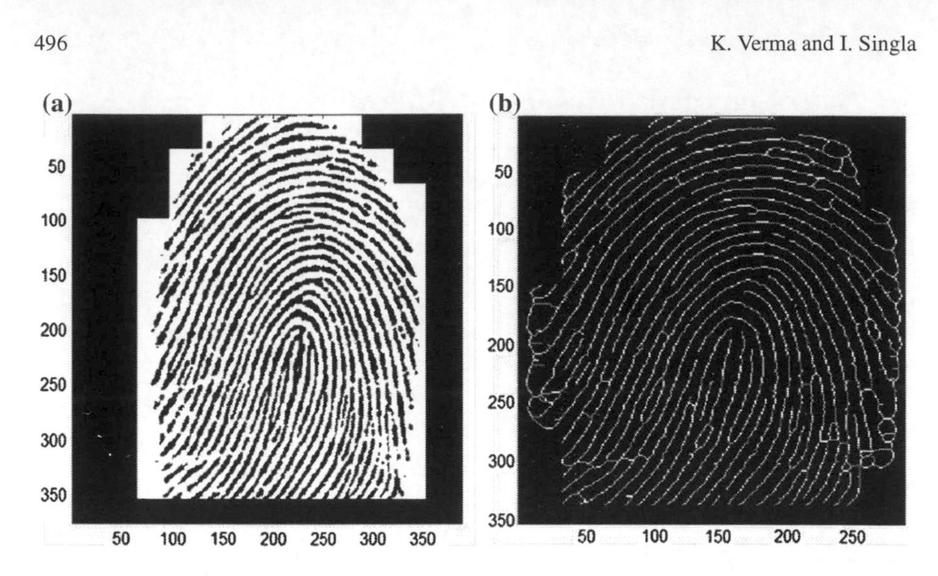

Fig. 4 **a** Binarize image; **b** Thinning image

bwmorph(binaryImage,'hbreak',k) → For H breaks
bwmorph(binaryImage,'clean',k) → For isolated points
bwmorph(binaryImage,'spur',k) → Spikes

6.2 Minutiae Marking

The name of this algorithm is crossing number (CN). It is implemented thinned image. Iteratively a 3×3 pixels wide picture is selected from thinned image then check that if the central pixel is a ridge branch and the central pixel is 1 and has exactly three neighbors of 1's, then its *bifurcation* (see Fig. 5).

If there one central 1 with exactly one 1in its neighbourhood, then its a *ridge ending* (see Fig. 6).

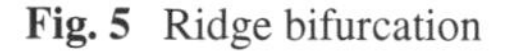

Fig. 5 Ridge bifurcation

0	1	0
0	1	0
1	0	1

Fig. 6 Ridge termination

0	0	0
0	1	0
0	0	1

6.3 False Minutiae Removal

As fingerprint sample maybe taken from ink impression (paper, thing) or it may be taken by using fingerprint scanner. But in case of ink impression the fingerprint quality may suffer. The quality of ink impression fingerprint image may be low which creates false minutiae points. So in most of the cases it need to be eliminated. To remove false minutiae points first calculate the inter ridge distance *D* (say) which is the average distance between two neighboring ridges. For this scan each row calculate the inter ridge distance using the formula [6]:

$$\text{Inter ridge distance} = \frac{\text{sum all the pixels in the row whose value is one}}{\text{row length}}$$

Calculate an average of all the inter ridge distance *D*. A MATLAB morphological operation 'BWLABEL' is helpful in this task. There have been seven different cases studied of false minutiae point's patterns. Follow the steps to remove these seven erogenous patterns (see Fig. 7) [6, 7].

- If d (bifurcation, termination) $< D$ and the two minutia are in the same ridge then remove both of them (case $m1$).
- If d (bifurcation, bifurcation) $< D$ and the two minutia are in the same ridge then remove both of them (cases $m2$, $m3$).

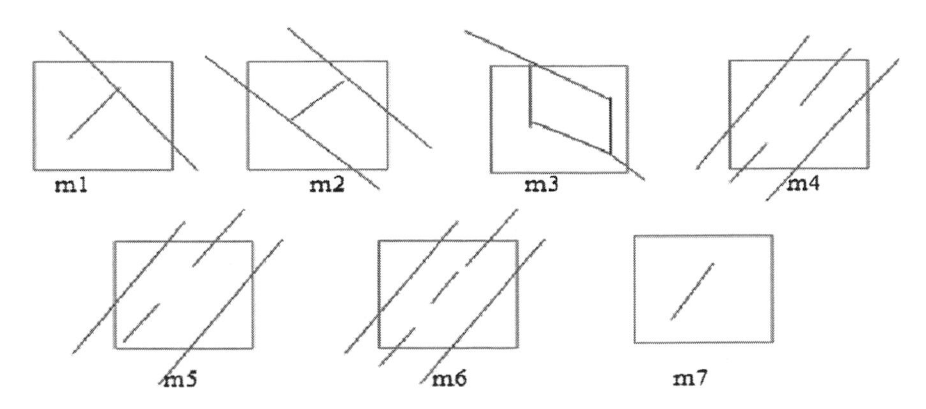

Fig. 7 False minutia structures

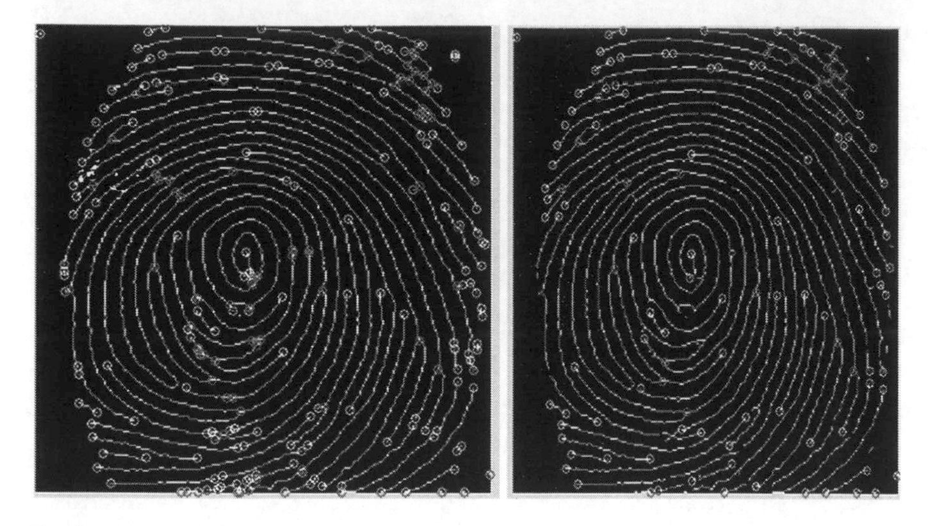

Fig. 8 Remove spurious minutiae

- If d (termination, termination) $\approx D$ and the their directions are coincident with a small angle variation and no any other termination is located between the two terminations then remove both of them (cases $m4$, $m5$, $m6$).
- If d (termination, termination) $<D$ and the two minutia are in the same ridge then remove both of them (case $m7$) (Fig. 8).

7 Conclusion

Fingerprint technique is good biometric technique for identification. Most of the time fingerprint is not of good quality. Due to low quality image false minutiae points get increases. But low quality image needs to preprocessed. Preprocessing to increase contrast, and reduce different types of noises. When some background region is included in the segmented regions of interest, noisy pixels also generate many spurious minutiae because these noisy pixels are also enhanced during preprocessing and enhancement steps. There is a scope of further improvement in terms of efficiency and accuracy, which can be achieved by improving the image enhancement techniques or by improving the hardware to capture the image. So that the input image to the thinning stage could be made, better this could improve the future stages and the outcome.

References

1. Mehtre, B.M., Chatterjee, B.: Segmentation of fingerprint images–a composite method. Pattern Recogn. **22**(4), 381–385 (1989)
2. Maltoni, D., Maio, D., Jain, A.K., Prabhakar, S.: Handbook of Fingerprint Recognition, 2nd edn. Springer, London (2009)
3. Hong, L.: Automatic personal identification using fingerprints. Ph.D. Thesis (1998).
4. Msiza, I.S., Leke-Betechuoh, B., Nelwamondo, F.V., Msimang, N.: A fingerprint pattern classification approach based on the coordinate geometry of singularities. In: Proceedings of the IEEE International Conference on Systems, Man, and Cybernetics, San Antonio, TX, USA, October 2009.
5. Cappelli, R., Lumini, A., Maio, D., Maltoni, D.: Fingerprint classification by directional image partitioning. IEEE Trans. Pattern Anal. Mach. Intell. **21**(5), 402–421 (1999)
6. Ratha, N., Chen, S., Jain, A.K.: Adaptive flow orientation based feature extraction in fingerprint images. Pattern Recogn. **28**, 1657–1672 (1995)
7. Demuth, H., Beale, M.: Neural Network Toolbox for Use with MATLAB. The MathWorks Inc, Natick (1998)

Optimization Problems in Pulp and Paper Industries

Mohar Singh, Ameya Patkar, Ankit Jain and Millie Pant

Abstract Pulp and paper industry plays an important role in Indian as well world economy. These are large scale process industries working round the clock. The focus of the present paper is on optimization problems encountered in pulp and paper industries. Different areas where optimization has been applied are identified and methods available for dealing with such problems are discussed.

Keywords Optimization · Pulp and paper industry · Classical methods · Non-traditional methods

1 Introduction

Pulp and paper industry is one of the most important process industries. It is not just a one industry but is an integration of several processes. Each process is to be carefully dealt for the smooth functioning of the industry. Somefacts about these industries are:

M. Singh (✉) · A. Jain
Department of Paper Technology, Indian Institute of Technology Roorkee, Roorkee, India
e-mail: moharsingh.iitr@gmail.com

A. Jain
e-mail: ankitjain.iitr@gmail.com

A. Patkar
Department of Polymer and Process Engineering, Indian Institute of Technology Roorkee, Roorkee, India
e-mail: ameya1991.iitr@gmail.com

M. Pant
Department of Applied Sciences and Engineering, Indian Institute of Technology Roorkee, Roorkee, India
e-mail: millifpt@iitr.ernet.in

B. V. Babu et al. (eds.), *Proceedings of the Second International Conference on Soft Computing for Problem Solving (SocProS 2012), December 28–30, 2012*, Advances in Intelligent Systems and Computing 236, DOI: 10.1007/978-81-322-1602-5_53,

- Paper industry accounts for nearly 3.5 % of world's industrial production and 2 % of world trade.
- Current annual consumption of paper is of the order of 270 million tones.
- This industry is 10th major section in India.
- Pulp and paper production is a critical part of the global economy with annual revenues of US$ 500 billion from sales of over 300 million tonnes of products [5].
- A modern pulp and paper mill with a production capacity of 300,000 tonnes per year is estimated to cost more than a billion dollars to construct [23].
- In terms of energy use, pulp and paper production accounts for 11 % of the total manufacturing sector, standing in the third place behind the petroleum (24 %) and chemicals (19 %) production industries [6].
- Although the projection of consumption trends towards year 2015 show approximately 2 % annual growth in the demand for pulp and paper products [12].

From the above listed facts it can be easily seen that these are highly capital incentive industries and a proper planning will be beneficial in terms of product/ profit maximization and loss minimization. Mathematically speaking many situations in a pulp and paper industry may be modeled as optimization problems which require suitable methods for their solution. The objective of the present study is to discuss various areas where optimization tools can be applied for improving the various processes of a pulp and paper mill.

Besides introduction, this paper has three more sections. In Sect. 2 we give the general optimization model used in the present study. Section 3, gives a very brief review of optimization problems arising in paper and pulp industries. In Sect. 4, methods available for solving such problems are discussed. The paper concludes with Sect. 4.

2 Definition of an Optimization Problem

Optimization is the art of selecting the best alternative among a given set of options. The process of finding the largest or the smallest possible value, which is a given function can attain in its domain of definitions, is known as optimization. The function to be optimized could be linear, non-linear, fractional or geometric. Sometimes even the explicit mathematical formulation of the function may not be available often the function has to be optimized in a prescribed domain which is specified by a number of constraints in the form of inequalities and equalities.

The general non-linear optimization problem is defined as:

Minimize / Maximize $f(\bar{x})$, where $f : R^n \to R$

Subject to: $x \in S \subset R^n$

where S is defined by:

$$g_j(\bar{x}) \leq 0, j = 1, 2, \ldots, p$$
$$h_k(\bar{x}) = 0, k = 1, 2, \ldots, q$$
$$a_i \leq x_i \leq b_i (i = 1, \ldots\ldots, n).$$

where p and q are the number of inequality and equality constraints respectively, a_i and b_i are lower and upper bounds of the decision variable x_i.

Any vector x satisfying all above constraints is called feasible solution. The best of the feasible solution is called an optimal solution.

- If the objective function and all constraints are linear then the model is called linear programming problem (LPP).
- If the solution has an additional requirement that the decision variables are integers then the model is called integer programming problem (IPP).
- If some variables are integers and other variables are real then the problem is called Mixed integer programming problem (MIPP).
- If the objective function and/or constraints are nonlinear then the problem is called non-Linear programming problem (NLPP).
- If besides nonlinearity, the model also has integer restrictions imposed on it then it called a mixed integer nonlinear programming problem (MINLPP).

3 Common Optimization Problems in Paper and Pulp Industries

In general, industrial processes are defined as large scale, high dimensional, non-linear and highly uncertain. Conventional methods such as 'trial and error' are often used to solve complex optimization problems. These methods often result in sub-optimal solutions due to inherent limitations in representing and exploiting the provided problem information. The exploration of design space is also inhibited. Paper and pulp industry is an integration of several processes that need to be optimized to improve the qualitative and quantitative working of the industry. In pulp and paper industry, optimization is applied during different phases of paper making. There are a number of studies in the literature that address the optimization of the unit operations in pulp and paper mills:
For example:

Some common optimization problems can be identified as:

- Minimizing energy cost and production rate with constrained environment [22].
- Optimizing paper making process [6, 21].
- Controlling the kappa number and pulp yield [16].
- Energy optimization in a pulp and paper mill cogeneration facility [25].
- Energy management technology [17, 19].
- Economic optimization of Kraft pulping process [19].
- Trim loss minimization [4, 10, 11, 14, 19, 26, 27].

- Continuous digester process [13].
- Biological treatment of waste water in pulp and paper industry [15].
- Optimization of a Kraft digester process [2].
- Optimization of refining process for fiber board production [20].
- Optimization of a broke recirculation system [3].

Mathematical model of these problems can be continuous or discrete. It can be as simple as a linear model (optimization of Kraft pulping process) or can be nonlinear, non-convex subject to ordinary or binary variable constraints/restrictions (trim loss optimization). The models can be put under the category of assignment or transshipment. Also, to make the model more realistic, authors have considered multi-objective formulation.

4 Methods Available

The methods available for dealing with optimization problems can be classified as traditional/classical and nontraditional.

4.1 Traditiona/Classical Methods

Traditional methods or classical methods are the ones that rely on the mathematical characteristics of the problem. These methods are often slow and are restricted to a particular class of problems. However these techniques have an advantage of having a strong theoretical proof. Some examples of classical optimization methods include: Simplex Method for linear programming models; Branch and Bound method for integer programming problems etc.

4.2 Nontraditional Methods

Nontraditional methods on the other hand are more user friendly as they do not depend upon the nature of the problem and can thus be applied to a broader range of problems. A major drawback of these algorithms is however that there is no concrete theoretical proof. Some examples are: genetic algorithms (GA) [9], and differential evolution (DE) [24] and algorithms based on systematic and interactive behavior of various species like ant colony optimization (ACO) [7], particle swarm optimization (PSO) [8] and artificial bee colony (ABC) [16] etc. the algorithms mentioned here are also known as nature inspired meta-heuristics as they are based on some natural concept or phenomena.

5 Conclusions

Optimization methods are potential tools that can be applied in almost all the real life scenarios where the situation can be modeled/ formulated in terms of an optimization problem. The objective of the present work is to discuss various areas of a pulp and paper industry where optimization can be applied. This work will help the researchers and industrialists working in this field.

References

1. Borairi, M., Wang, H., Roberts, J.C.: Dynamic modelling of a paper making process based on bilinear system modelling and genetic neural networks. Presented at UKACC international Conference on, control, pp. 1277–1282 (1998)
2. Cristina, H., Aguiar, I.L., Filho, R.M.: Modeling and optimizationof pulp and paper processes using neural networks. Comput. Chem. Eng. 22(Suppl.), S981–S984 (1998)
3. Dabros, M., Perrier, M., Forbes, F., Fairbank, M., Stuart, P.R.: Model based direct search optimization of the broke recirculation system in anewsprint mill. J. Cleaner Prod. **13**, 1416–1423 (2005)
4. Deep, K., Chauhan, P., Bansal, J.C.: Solving nonconvex trim loss problem using an efficient hybrid particle swarm optimization. In: IEEE Xplore Proceedings of the World Congress on Nature and Biologically Inspired Computing Coimbatore, India, pp. 1608–1611, 9–11 Dec 2009
5. DeKing, N. (ed.): Pulp and paper global fact & price book 2003–2004, Boston: Paperloop, Inc. (2004)
6. DoE Annual Energy Review. Report No. DOE/EIA-0384 (2004)
7. Dorigo, M., Maniezzo, V., Colorni, A.: The ant system: optimization by a colony of cooperating agents. IEEE Trans. Syst. Man. Cybern. B. **26**(1), 29–41 (1996)
8. Eberhart, R.C., Kennedy, J.A.: New optimizer using particle swarm theory. In: IEEE Service Center Proceedings of the 6th International Symposium on Micro MMachine and Human Science, pp. 39–43, Piscataway, NJ, Japan, 4–6 Oct 1995
9. Goldberg, D.E.: Genetic Algorithms Search. Optimization and Machine Learning, Addison-Wesley, Reading (1989)
10. Harjunkoski, I., Porn, R., Westerlund, T., Skrifvars, H.: Different transformations for solving noncConvex trim loss problems with MINLP. Eur. J. Oper. Res. **105**, 594–603 (1998)
11. Harjunkosk, i I., Westerlund, T.: Enlarging the trim-loss problem to cover the raw paper mill. Comput. Chem. Eng. 22(Suppl), 1019–1022 (1998)
12. Consulting, Jaakko Poyry Management: World paper markets up to 2015:what drives the global demand? Tappi Solutions **86**(8), 64 (2003)
13. Jin, F.J., Wang, H., Li, P.: Cleaner production for continuous digester processes based on hybrid Pareto genetic algorithm. J. Environ. Sci. **15**(1), 129–135 (2003)
14. Johnson, M.P., Rennick, C., Zak, E.: Skivingaddition to the cuttingstock problem in the paper industry. SIAM Rev. **39**(3), 472–483 (1997)
15. Juuso, E.K.: Hybrid models in dynamic simulation of a biological water treatment process. In: International Conference on Computational Intelligence, Modelling and, Simulation, pp. 30–35 (2009)
16. Karaboga, D.: An idea based on honey bee swarm for numerical optimization.Technical Report. Erciyes University, Engineering Faculty, Kayseri (2005)
17. Kaya, A., Keyes, M.A.: IV. Energy management technology in pulp, paper, and allied industries. Automatica **19**(2), 11–130 (1983)

18. Marshman, D.J., Chmelyk, T., Sidhu, M.S., Gopaluni, R.B., Dumont, G.A.: Energy optimization in a pulp and paper mill cogeneration facility. Appl. Energy **87**, 3514–3525 (2010)
19. Pant, M., Thangaraj, R., Singh, V.P.: The Economic optimization of pulp and paper making processes using computational intelligence. In: Proceedings of AIP Conference, 1146(1), p. 462 Agra July 2009
20. Runklera, T.A., Gerstorfer, E., Schlang, M., Junnemann, E., Hollatz, J.: Modeling and optimisation of a refining process for fibre board production. Control Eng. Pract. **11**, 1229–1241 (2003)
21. Santos, A., Dourado, A.: Global optimization of energy and production in process industries: a genetic algorithm application. Contr. Eng. Pract. **7**, 549–554 (1999)
22. Silva, C.M., Biscaia Jr, E.C.: Multiobjective optimization of a continuous pulp digester. Comput. Aided Chem. Eng. **14**, 1055–1060 (2003)
23. Smook, G.A.: Handbook of pulp and paper Technology, 2nd edn. Angus Wilde Publications, Vancouver (1992)
24. Storn R., Price K: Differential evolution-a simple and efficient adaptive scheme for global optimization over continuous. Techical Report TR-95-012, Spaces, Berkeley 1995
25. Wang, H., Borairi, M., Roberts, J.C., Xiao, H.: Modelling of a paper making process via genetic neural networks and first principle approaches. Presented at the IEEE International Conference on Intelligent Processing Systems, ICIPS, pp. 584–588. Beijing, China (1997)
26. Westerlund, T., Isaksson, J.: Some efficient formulations for the simultaneous solution of trim-loss and scheduling problems in the paper-converting industry. Chem. Eng. Res. Des. **76**, 677–684 (1998)
27. Westerlund, T., Isaksson, J., Harjunkoski, I.: Solving a two-dimensional trim loss problem with MILP. Eur. J. Oper. Res. **104**, 572–581 (1998)

Integrating ERP with E-Commerce: A New Dimension Toward Growth for Micro, Small and Medium-Scale Enterprises in India

Vinamra Nayak and Nitin Jain

Abstract In the developing economy like India Micro, Small and Medium Enterprises (MSMEs) play a very important role as they are the engines of growth in development, upliftment, and transition of economy to the next level. The MSME's in India have played a critical role in generation of employment, providing goods and services at affordable costs by offering innovative solutions in very unstructured and unorganized manner. With the increase in the growing competition and its complexity it is essential for today's business enterprises to work in structured manner by changing its functioning environment dynamically by integrating internal information resourses to the platform of ERP systems which will assist in optimizing potential of utilizing resources efficiently and determining growth. Moreover, with the exponential growth of Internet technology and the emergence of e-business, MSME's can unify the external information resources with internal functional areas within an organization. Use of e-commerce solutions will help enterprises to expand their business through broader product exposure, better customer service, accurate order entry processes and faster product fulfillment. It is expected that in MSME's implementation of an ERP system can facilitate an e-business effort of an organization to optimize its overall functioning. In order to serve as a platform for e-business, it is essential that an ERP system must also be able to be extended to support a range of external constituents for a firm. This paper focuses on that how the challenges faced by MSME's can be overcome by implementing ERP system in an organisation with the efficient integration of e-commerce to optimize its functioning. Further, the paper will highlight the add-on benefits to the companies for using integrated e-commerce platforms with ERP systems. The paper is based on the in-depth study of publically available information collected from the published articles, journals, reports, websites, blogs, and academic literatures in context with the economy of India.

V. Nayak (✉) · N. Jain
Department of MBA, Gyan Ganga Institute of Technology and Sciences, Jabalpur, India
e-mail: vinamra.lucky@gmail.com

N. Jain
e-mail: nitinjain_abc@yahoo.co.in

B. V. Babu et al. (eds.), *Proceedings of the Second International Conference on Soft Computing for Problem Solving (SocProS 2012), December 28–30, 2012*, Advances in Intelligent Systems and Computing 236, DOI: 10.1007/978-81-322-1602-5_54, © Springer India 2014

Keywords ERP · MSME's · E-commerce · Indian economy

1 Introduction

Micro, Small, and Medium-sized Enterprises (MSMEs) are the backbone of all economies and are a key source of economic growth, dynamism, and flexibility in advanced industrialized countries, as well as in emerging and developing economies. Micro, Small, and Medium businesses are particularly very important for bringing innovative products or techniques to the market. For the sustainability of this kind of growth, proper nurturing of MSME sector is imperative. There are around 31 million MSME's units in India, of which 99 % are Micro and small-scale units, contributing nearly 40 % of India domestic production, almost 50 % of total exports, and 45 % of industrial employment.

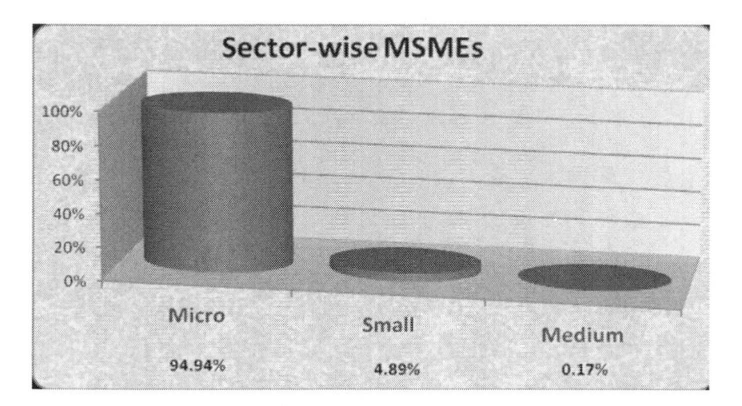

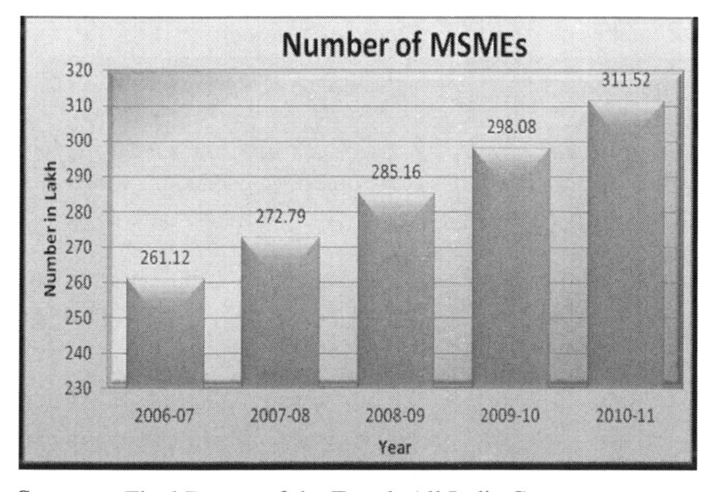

Source: - Final Report of the Fourth All India Census of Micro, Small & Medium Enterprises 2006-07: Registered Sector.

The MSME's in India are mostly in the unorganized sector though have had played a critical role in generation of employment, providing goods and services at affordable costs by offering innovative solutions. The need of the hour is to empower the MSME Sector so that it is able to take its rightful place as the growth engine of the economy.

However, one of the known bottlenecks to the growth of the MSME sector is its lack of adequate access to finance, despite the tremendous efforts laid down by the government and the corporative banks. Focusing on the other challenges will make MSE's more vibrant in managing many new unchartered areas such as infrastructure, nonavailability of suitable technology, ineffective marketing due to limited resources and nonavailability of skilled manpower, and optimum utilization of available resources. These factors are critically affecting the pace at which the micro, small, and medium businesses should prosper and develop.

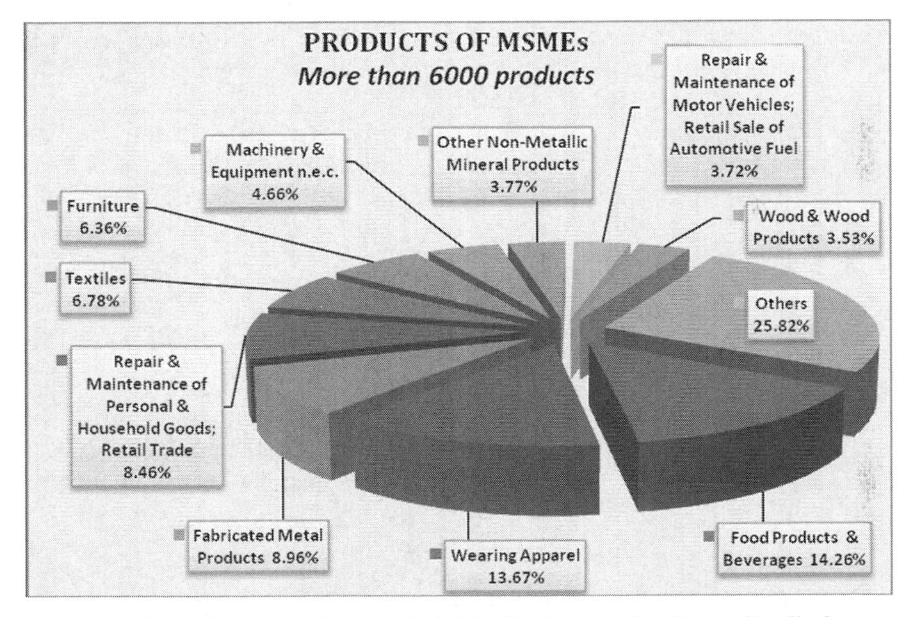

Source: - Final Report of the Fourth All India Census of Micro, Small & Medium Enterprises 2006-07: Registered Sector.

Enterprise Resource Planning (ERP) is an outcome of Information Technology and is a way to integrate the data and processes of an organization into one single system, using subsystems that include hardware, software, and a unified database in order to achieve integration, to store the data for various functions found throughout the organization. Today, ERP is used in almost any type of organization; it does not matter whether it is large, small, or what industry it falls in. The main aim of implementing ERP is to integrate data and processes from all areas of the organization and unify it to provide ease of access and an efficient work flow. ERP systems usually accomplish this through one single database that employs multiple software modules.

According to Karunakaran (2001), Enterprise resource planning or ERP is a method of effective planning of all the resources like money, manpower, materials, and other things that are required to run the enterprise. It comprises management of functional and related areas in SMEs like manufacturing, materials, sales and distribution, human resource, finance, strategy and operations, quality, logistics, and maintenance.

ERP implementation is not just installation and configuration of the software but is the process of defining the way an organization should work; it involves defining the process flows within different departments and across the departments, imposing the checks and controls at different points and implementing the authorization levels in all departments, which requires certain important and stern decisions to be taken for successful execution.

Every business now has the opportunity to take advantage of the Internet's geographical reach and potential for profit. Thus, the Internet has made it possible for MSMEs to compete effectively with the larger organizations. The enabling potential of the Internet guarantees that MSMEs are appropriately represented and active participants in the globalized marketplace.

According to the Internet and Mobile Association of India Report, the total e-commerce market in India was estimated at around Rs. 46,000 crore (US $8.28 billion) at the end of 2011 and is growing at around 40 % year on year.

2 Aims and Objectives

This paper focuses on how the challenges faced by MSME's can be overcomed by implementing the ERP system in an organization with the efficient integration of e-commerce to optimize its functioning. Further, paper will highlight the add-on benefits to the companies for using integrated e-commerce platforms with ERP systems.

3 Rationale

The purpose of the research is to identify the challenges and constraints faced by Indian MSME's, and how by integrating the concept of ERP and its strategies they are overcome efficiently. Moreover, to understand the advantages to MSME's for using e-commerce platform with integrated ERP systems in today's competitive environment.

4 Research Methodology

In order to achieve the research objectives, the blend of deductive and inductive research approach is selected, whereas qualitative research method is utilized. The research is based on only publically available information which has been taken into account. In order to fulfill proposed objectives data on various facts related to the concept of ERP and e-commerce is been presented with the help of the literature review. The data is collected using secondary method to fulfill different issues related to research topic from the published articles, journals, reports, websites, blogs, and academic literatures.

5 Literature Review

The factors like opening up of the economy business era in the postliberalization, ease in international trade barriers, globalization, privatization, disinvestments, and deregulation have thrown several challenges to MSMEs in the fast developing economies like India. Economic downturns, compressed product development cycles, cut-throat domestic and global competition, rapidly changing customer demands, and volatile financial markets have all increased the pressure on MSMEs to come up with effective and competitive capabilities to survive and succeed. ERP is often considered as one of the solutions for their survival [1].

SMEs sector in India had operated under a much-protected economic regime characterized by limited competition and a highly regulated business environment. This business atmosphere had resulted in limited focus on process efficiencies, centralized control structures, highly formalized business settings, and lack of professional business practices [2]. However, following the economic liberalization and opening up of the economy to foreign Multi-National Companies (MNCs), Indian SMEs have been forced to adopt modern business practices and strategies, which in turn can provide SMEs a cutting edge over its competitors.

In recent time, many MSME's implement ERP to increase internal efficiency and external competitiveness. Once ERP is established at all levels of the organization catering all the possible needs necessary to optimize its potentiality to face the challenges of the outside competitive environment ensures desired growth. The enterprise-wide data sharing in SMEs has many benefits like automation of the procedures, availability of high quality information for better decision making, faster response time, and so on Karunakaran (2001).

According to industry reports, the total cost of deploying ERP has ranged between 1 and 2 % of companies' gross sales. Lower cost solutions are available for comparatively smaller-sized companies. Though the market seems to be very encouraging for ERP implementation, the time frame for deployment may be an issue.

While many new MSMEs start each year, nearly 50 % cease to exist in the first 3 years of business itself. Though it is assumed that all MSMEs desire growth, only 40 % survive beyond 10 years. Majority of the firms do not think of long-term business strategy but focus only on survival. They think of change only when the business begins to fail as a result of not keeping track of the changing market scenario. The firms who survive and grow are the ones who have the ability to take risks and respond to the changing circumstances [3].

An ERP system would allow MSMEs to integrate their business functions. MSMEs would be able to increase their efficiency and productivity by implementing a suitable ERP system. Over the next 5 years, the ERP market in India is expected to reach Rs. 1,550 crore ($341 million), according to International Data Corporation (IDC), a market research and analysis firm. Of this, the SME potential in India for the enterprise class is projected to be Rs. 728 crore ($160 million) 47 % of the overall market [4] (Table 1).

According to Gattiker and Goodhue (2005), ERP vendors have developed and customized the ERP software for the use of all types of industries. This has created a great demand on the use of ERP among business entities to integrate and maximize their resources. The growing demand for ERP applications among business firms has several reasons:

- Competitive pressures to become a low cost producer.
- To increase the revenue growth.
- Ability to compete globally.
- Maximizing the resources and the desire to re-engineer the business to respond to market challenges.

According to Gattiker and Goodhue (2005), ERP implementation failure rate is from 40 %, yet companies try to implement these systems because they are absolutely essential to responsive planning and communication. The competitive pressure unleashed by the process of globalization is driving implementation of ERP projects in increasingly large numbers, so a methodological framework for dealing with complex problem of evaluating ERP projects is required. It has been found that, unique risks in ERP implementation arises due to tightly linked interdependencies of

Table 1 Projected growth rate overall industrial sector

Year	Growth rates of SSI sector (2001–2002) base IIP (%)	Over all industrial sector growth rate (%)[a]
2004–2005	10.88	8.40
2005–2006	12.32	8.00
2006–2007	12.60	11.90
2007–2008	13.00[b]	8.70
2008–2009	[c]	3.20
2009–2010	[c]	10.50
2010–2011	[c]	7.80

[a] *Source* M/o Statistics and PI website www.mospi.nic.in
[b] Projected, IIP Index of Industrial Production
[c] Due to revised definition of MSMEs Sector, figures not available

business processes, relational databases, and process reengineering [5]. According to Gordon [6], three main factors that can be held responsible for failure of ERP system: (a) Poor planning or poor management, (b) Change in business goals during project, and (c) Lack of business management support.

According to Gordon [6], a careful use of communication and change management procedures is required to handle the often business process re-engineering impact of ERP systems which can alleviate some of the problems, but a more fundamental issue of concern is the cost feasibility of system integration, training and user licenses, system utilization, etc. need to be checked. A design interface with a process plan is an essential part of the system integration process in ERP.

The pressure of staying ahead in a competitive environment by reduction of costs and product development time is compelling SMEs to leverage technology and enterprise applications like ERP application, as adapted by their large counterparts [7].

Sharma and Bhagat [8] in their study of information system (IS)-related practices in 210 MSMEs of western part of India concluded that though MSMEs understand and acknowledge the importance of the IS in day-to-day operations management in the present dynamic and heterogeneous business environment, but these are yet to implement, operate, and exploit it fully in a formal and professional manner so as to enable them to derive maximum business gains out of it.

According to Mishra et al. [9], in an increasingly competitive and globalized world, MSMEs need to compete more effectively to boost domestic economic activities and contribute toward increasing export earnings. E-commerce is emerging as a new way of helping business enterprises to compete in the marketplace and thereby contributing to their economic success. MSMEs also continue to play an important role in increasing employment and thus contributing to poverty reduction on a sustainable basis. With spread of technology and infrastructure, rural businesses will be the biggest beneficiaries of e-commerce.

6 Research Analysis

The boom in the ERP business segment is accompanied by a lot of challenges. One of the key challenges faced by solution providers is that the MSME sector is poor paymasters. Considering the concept is just a decade old in India, MSMEs are reluctant to invest in an ERP replacement as they lack domain-specific knowledge, technical know-how, and monetary resources to implement solutions.

Moreover, the MSME's are fighting with the problems and challenges faced to survive in this competitive environment but for sure adopting the strategies to optimize its overall business through ERP which will be the key to success. On the basis of the literature review, critical success factors are highlighted which assist in identifying and stating the key elements required for the success of a business operation. Furthermore, these critical success factors and key variables can be described in more details as a small number of easily identifiable operational goals shaped by the industry, the firm, the manager, and the environment that assures the success of an organization. In this research paper, some key success factors and variables are identified which will be helpful to MSME's after integrating ERP systems with e-commerce (Table 2):

Looking into the growing need of business enterprises to optimize its resources and meeting their upcoming problems and challenges for the better functioning in the present competitive environment, ERP tool is one of the most important and complete business software for any organization today, but if you do business through e-commerce and your ERP is not integrated with your e-commerce website then it would not consider that you are using a complete package.

Here are the most important benefits of integrated ERP software with your e-commerce website which will assist MSMEs to overcome the business constraints in present scenario:

1. **Improves Inventory Management**: ERP e-commerce integration allows you to check your inventory in real time, Products uploaded in the ERP system can be automatically published on your e-commerce website which helps enterprises to maintain level of inventory.
2. **Increases Self-Service Functionality**: The availability of real-time data from the ERP system on to the store front, allows customers to view available inventory, latest order status, and track shipments with tracking numbers. This helps in reducing the cost of operations and improves customer experience with store front of the enterprises.
3. **Quick Updates**: On purchases or sales of inventory in business ERP system as well as on online stores, which further helps in making quick decisions and necessary corrections.

Table 2 Success criteria and key variables

ERP	E-commerce
Business process re-engineering	Customer satisfaction
Reduced inventory level	Online product cataloge
Reduced logistic cost	Tight integration with ERP system
Reduced procurement cost	Secure electronic payment
Order fullfillment performance	Reduced cost
Increased productivity and flexibility	Online customer service
Standardization of computing platforms	
Global sharing of information	
Improved responsiveness to the customer	
Key variables	
Top management commitment	
Consultant skills	
Schedule reliability	
Budget reliability	
Implementation team skills	

4. **Less Manual Process**: ERP and e-commerce integration allows businesses to have less manual intervention as every part or entry is taken care with the help of system integration. Automatically uploading of all the entries in the ERP system or vice versa help to reduce re-entry.
5. **Increased Internal Productivity**: ERP integrated system streamlines multiple business processes, helps in reducing human resource involvement in these processes. Web sales orders will be integrated to the ERP system in real time, back office ERP user can instantly track the order, and start the further processing. Thus, the order fulfillment cycle is reduced through this integration.
6. **Reduced Human Involvement, Data Redundancy and Error**: By integration, web customer details, web orders, payment and shipping information to ERP system, similarly Item and Inventory details can be uploaded from ERP to e-commerce portal, eliminating the need of re-entering the data. Thus, the integration solution will reduce human involvement, data redundancy, and error over two platforms.
7. **No Duplication of Entries**: ERP integration with e-commerce website eliminates the re-entering or rekeying process which further allows to have more accuracies in accounting, data processing, billing, and other modules of business enterprises.
8. **Generating Financial Reports in ERP, based on Web Transactions**: E-commerce applications are able to generate financial reports on sales. But integration with ERP provides the merchant the ability to produce Balance Sheet, P/L Statement, Trial Balance, Cash Flow, etc. which gives the transparency in financial information across the organization.
9. **Increase Customer Satisfaction**: Easy availability of up to date product information, inventory availability details, order tracking detail, etc in the web from ERP system; customer satisfaction level raises a lot and it reduces operational hassle for the business.
10. **Reduced Overall Cost**: Adopting the platform of ERP system with e-commerce helps business to cut down its logistics and procurement cost whereas also eliminates the possibility of underutilization of resources by efficiently streamlining the distribution process.
11. **Reduces Process Time**: E-commerce platform gives an opportunity to the business to serve multiple customers by streamlining the efficient distribution process through integrated ERP implementation.
12. **Better Control of Business Enterprise**: Integration of e-commerce and ERP business processes provides the business owners with a better control of their business and there by getting competitive advantage.

7 Conclusion

E-commerce solutions are a big support for business today. The demands and the expectations of the customers are high and so is the level of competition in the industry. E-commerce solution provides assistance to streamline business processes efficiently and covers all the gaps that might have gone unseen. ERP Investments has been top IT spending priorities for MSME's off late. Most business organizations which are MSME in nature have complex ERP environments consisting of customized packages from multiple vendors, as well as an array of internally developed software that must integrate with the packages.

Any information system implementation is a complex task and hence complex ERP environments present several solutions to overcome challenges which are met by the MSMEs. The findings hold significance for any organization in the micro, small, and medium-scale sector which wishes to leverage the benefits of integration of business processes by implementing an ERP system in their organization.

The data analysis presented in this paper clearly specifies the way in which the constraints and challenges commonly encountered by MSMEs in managing the business house can be effectively dealt with the efficient use of integrated ERP system platform with e-commerce to gain competitive position over the conditions in which they are operating.

References

1. Rao, S.S.: Enterprise resource planning: Business needs and technologies. Ind. Manage. Data. Syst. **100**(2), 81–86 (2000)
2. Ranganathan, C., Kannabiran, J.: Effective management of information systems function: An exploratory study of Indian organizations. Int. J. Inf. Manage. **25**, 247–266 (2004)
3. Levy, M., Powell, P.: Strategies for growth in SMEs: The role of information and information systems. Inf. Proc. Manage. Int. J. 42 (2006).
4. Munjal, S.: Small is beautiful: ERP for SMEs. http://www.domainb.com/infotech/itfeature/20060601_beautiful.htm (2006)
5. Wright, S., Wright, A.M.: Information system assurance for enterprise resource planning systems : Unique risk considerations. J. Inf. Sci. **16**, 99–113 (2002)
6. Gordon, A.: ERP applications: Myth and misconceptions, ezinearticles. www.ezinearticles.com (2006)
7. Goswami, S., Sarangdevot S.S.: Study of critical success factors for enterprise system in Indian SMEs. Refeered Q. J., **4**(1) (2010).
8. Sharma, K.M., Bhagat, R.: Practice of information systems evidence from select Indian SMEs. J. Manufact. Technol. Manage. **17**(2), 199–223 (2006)
9. Mishra, B.B., Mishra, U.S., Mishra, P.K., et al.: Perception and adoption of e-commerce in Indian SMEs : A study in the state of Orissa. Int. J. Adv. Comput. Math. Sci. **3**(2), 227–236 (2012) (ISSN 2230–9624)
10. Anonymous., Benefits of having an integrated ERP with an e-commerce store front. Online http://www.sboeconnect.com/blog/e-commerce/benefits-of-having-an-integrated-erp-with-an-e-commerce-store-front.html
11. Anonymous., Companies are benefiting from integrating ERP with e-commerce. Online http://www.erpsoftwareblog.com/2012/03/companies-are-benefiting-from-integrating-erp-with-e-commerce/
12. Anonymous., ERP and e-commerce. Online http://www.managementstudyguide.com/erp-and-ecommerce.htm
13. Anonymous., What is the difference between e-commerce and e-business? Online http://www.eresourceerp.com/What-is-the-difference-between-E-commerce-and-E-Business.html

14. Bajaj, K.K., Nag, D.: E-Commerece-the cutting edge of business, 2nd edn. Tata McGraw Hill Publishing Company Ltd., New Delhi (2009)
15. Dixit, A., Prakash, O.: A study of issues affecting ERP implementation in SME's. J. Arts Sci. Commer., **2**(2), 77–85 (2011) (E-ISSN 2229–4686, ISSN 2231–4172) Avaliable at www.researchersworlld.com
16. Final report of the fourth all India census of micro, small and medium enterprises 2006–07: Registered sector
17. Kale, P.T., Banwait, S. S., Laroiya S. C.: Enterprise resource planning implementation in Indian SMEs: issues and challenges
18. Kale, P.T., Banwait, S.S.: Evaluation of ERP implementation in an Indian company: a case department of mechanical engineering, national institute of technical teachers, training and research. Available Online
19. Leon, A.: Enterprise resource planning, 2nd edn. Tata McGraw Hill Publishing Company Ltd., New Delhi (2008)
20. Lewis, A.: E-commerce application in India : an emperical study. Welingkar Institute of Management Development and Research, Mumbai (2008)
21. Rayport, J.F., Jaworski, B.J.: Introduction to e-commerce, 2nd edn. Tata McGraw Hill Publishing Company Ltd., New Delhi (2007)
22. Thapliyal, M.P., Vashishta. P.: ERP software implementation in Indian small and medium enterprises. Int. J. Emerg. Trends Technol. Comput. Sci. **1**(2), 107 (2012) www.ijettcs.org Accessed 15 Oct 2012
23. The integration strategy of E-commerce platform and ERP based on cooperative application. http://ieeexplore.ieee.org/xpl/articleDetails.jsp?reload=true&arnumber=5593163&contentType=Conference+Publications
24. Upadhyay, P.: ERP in Indian SME's: a post implementation study of the underlying critical success factors. Int. J. Manage. Innovation Syst. **1**(2), E1 (2009). ISSN 1943–1384
25. Whiteley, D.: E-commerce-strategy, technologies and applications, 1st edn. Tata McGraw Hill Publishing Company Ltd., New Delhi (2009)

A New Heuristic for Disassembly Line Balancing Problems with AND/OR Precedence Relations

Shwetank Avikal, Rajeev Jain, Harish Yadav and P. K. Mishra

Abstract Disassembly operations are inevitable elements of product recovery with the disassembly line as the best choice to carry out the same. The product recovery operations are fully based on disassembly line balancing and because of this, disassembly lines have become the chaise of automated disassembly of returned product. It is difficult to find the optimal balance of a disassembly line because of its N-P hard nature. In this paper a new heuristic is proposed to assign the parts to the disassembly workstations under AND/OR precedence constraints. The heuristic solutions are known as intrinsically optimal/suboptimal solutions of the N-P hard problems. The solution obtained by proposed heuristic has been compared with other heuristic solutions. The heuristic tries to minimize the number of workstations and the cycle time of the line while addressing the different criteria. The methodology of the proposed heuristics has been illustrated with the help of examples and it has been observed that the heuristic generates significantly better result.

Keywords Line balancing · Disassembly · Product recovery · Heuristic

S. Avikal (✉) · R. Jain · H. Yadav · P. K. Mishra
Department of Mechanical Engineering, Motilal Nehru National Institute of Technology, Allahabad, India
e-mail: avi.shwetank@gmail.com

R. Jain
e-mail: jainrajeev@rediffmail.com

H. Yadav
e-mail: harishivri@gmail.com

P. K. Mishra
e-mail: pkm@mnnit.ac.in

B. V. Babu et al. (eds.), *Proceedings of the Second International Conference on Soft Computing for Problem Solving (SocProS 2012), December 28–30, 2012*, Advances in Intelligent Systems and Computing 236, DOI: 10.1007/978-81-322-1602-5_55,

1 Introductions

The impact of industrial and domestic waste on the environment has been overwhelming. Extensive diffusion of consumer goods and reduction in product lifecycles have led to a large number of used products being discarded due to chaotic transition and drastic changes in demand for a product during the past decade, which is due to globalization [1].

Disassembly has proven its role in material and product recovery by allowing selective separation of desired parts and materials [2]. Disassembly is defined as the methodical extraction of valuable parts/subassemblies and materials from discarded products through a series of operations. After disassembly, reusable parts/subassemblies are cleaned, refurbished, tested, and directed to inventory for remanufacturing operations. The recyclable materials can be sold to raw-material suppliers, while the residuals are sent to landfills [3].

End-of-life processing of complex products such as electronic products is becoming increasingly important, because they contain a large variety of hazardous, useful, as well as valuable components and materials. Disassembly is often used to separate such components and materials. A disassembly precedence graph (DPG) is frequently used to describe a disassembly process. The box in this graph refers to operations, typically the detachments of components. Arcs represent the precedence relationships. Both yield and costs are associated with every operation [4].

In order to minimize the amount of waste sent to landfills, product recovery seeks to obtain materials and components from old or outdated products through recycling and remanufacturing—this includes the reuse of components and products. There are many attributes of a product that enhance product recovery; examples include ease of disassembly, modularity, type and compatibility of materials used, material identification markings, and efficient cross-industrial reuse of common parts/materials. The first crucial step of product recovery is disassembly [3].

In this paper an analysis of U-shaped disassembly line has been carried out using proposed heuristic. In Sect. 2, the relevant literature has reviewed. The proposed heuristic has been described in Sect. 3, a practical example and computational results have been shown in Sect. 4. while the conclusions are drawn in Sect. 5.

2 Literature Review

The problem of disassembly line balancing (DLBP) can be defined as the assignment of disassembly tasks to workstations such that all the disassembly precedence relations are satisfied and some measure of effectiveness is optimized. Gungor and Gupta [5–7] presented the first introduction to the disassembly line-balancing problem and developed an algorithm for solving the DLBP in the presence of failures with the goal of assigning tasks to workstations in a way that probabilistically minimizes the cost of defective parts. Tiwari et al. [8] presents a Petri Net-based approach to determine the disassembly strategy of a product; it was a cost-based heuristic analysis for the circuit board. McGovern and Gupta [3] presented a disassembly solution which was found first by greedy modal, then improving the solution with the help of 2-opt heuristic. Later, various combinatorial optimization techniques to solve DLBP were compared by McGovern and Gupta [9]. Recently, the fact that even a simple disassembly line balancing problem is NP-complete that has been proven in the literature and a genetic algorithm was presented by McGovern and Gupta [10] for obtaining optimal solutions for DLBP. In order to solve the profit-oriented DLBP, Altekin et al. [11] developed the mixed integer programming algorithm for the DLBP. Koc et al. [12] proposed two exact formulations for disassembly line balancing problems with task precedence diagram construction using an AND/OR graph Ding et al. [13] proposed a multiobjective disassembly line balancing problem and then solved this by an ant colony algorithm.

3 Proposed Heuristic

In this paper a heuristic has been proposed for the assignment of tasks to the workstations under consideration of some criteria such as task time, part demand, part hazardousness, and part removal cost. The proposed heuristic has been developed to achieve some objectives such as minimize the total minimum number of disassembly workstations to decrease total idle time and balance the disassembly line. Remove hazardous parts/components early in the disassembly sequence, remove high demand components before low demand components, remove low disassembly cost components before high disassembly cost components, and remove the parts which have large part removal time before the parts which have small part removal time. In this proposed heuristic a rank has been provided to all the tasks according to the described objectives which has been done by taking the sum score of their criteria. The ranking of all tasks has been done by normalizing the data for each criteria and adding them together. Now assign the rank to all the tasks according to summation of their criteria, and then assign the task to the workstations on the bases of their rank and precedence relations and cycle time constraints.

4 Computational Example

The developed algorithm has been investigated on a variety of test cases to confirm its performance and to optimize parameters. The proposed heuristic has been used to provide a solution to the disassembly line balancing problem based on the modified disassembly sequencing problem presented by McGovern and Gupta [3], where the objective is to completely disassemble a given product (see Fig. 1) consisting of $n = 10$ components and several precedence relationships. The problem and its data were modified with a disassembly line operating at a speed which allows $CT = 40$ s

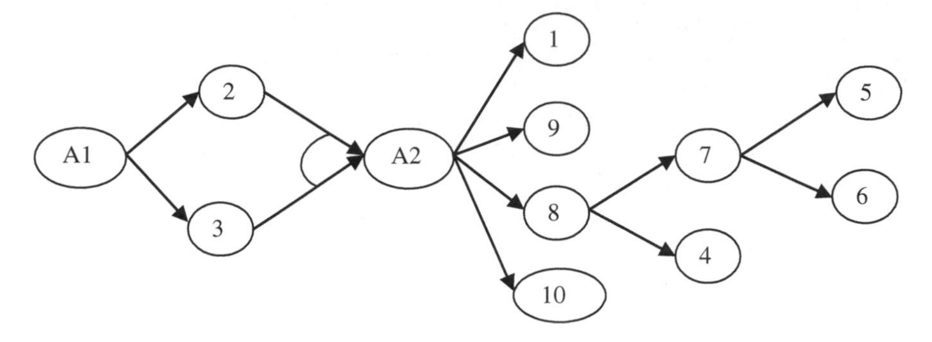

Fig. 1 Modified precedence diagram of McGovern and Gupta [3] example

Table 1 Tasks data set of McGovern and Gupta [3]modified example

Part	Time	Hazardous	Demand	Cost
1	14	0	0	27
2	10	0	500	63
3	12	0	0	48
4	18	0	0	62
5	23	0	0	24
6	16	0	485	18
7	20	1	295	83
8	36	0	0	77
9	14	0	360	93
10	10	0	0	10

for each workstation to perform its required disassembly tasks with AND/OR precedence relations. This provided an application to a previously explored disassembly problem. This practical and relevant example consists of the data for the disassembly of a product as shown in Table 1. It consists of ten subassemblies with part removal times of Tk = {14, 10, 12, 18, 23, 16, 20, 36, 14, 10}, hazardousness as hk = {0, 0, 0, 0, 0, 0, 1, 0, 0, 0}, part demand dk = {0, 500, 0, 0, 0, 750, 295, 0, 360, 0}, and part removal cost as Ck = {27, 63, 48, 62, 24, 18, 83, 77, 93, 10}, The disassembly line is operated at a speed that allows 40 s for each workstation.

4.1 Determination the Rank of Tasks

After applying proposed heuristic, the normalization of data (see Table 2) has been done and, by adding the normalized value of each criterion, the final ranking of tasks has been done. The final ranking of tasks is given in Table 2.

Table 2 Ranking of tasks

Part	Time	Hazardous	Demand	Cost	Total	Rank
1	0.1111	0	0	0.7097	0.8208	8
2	0.0000	0	1	0.3226	1.3226	3
3	0.0556	0	0	0.4839	0.5394	10
4	0.2222	0	0	0.3333	0.5556	9
5	0.3611	0	0	0.7419	1.1030	4
6	0.1667	0	0.97	0.8065	1.9431	2
7	0.2778	1	0.59	0.1075	1.9753	1
8	0.7222	0	0	0.1720	0.8943	5
9	0.1111	0	0.72	0.0000	0.8311	7
10	0.0000	0	0	0.8925	0.8925	6

In this list part number 7 is ranked first because it has hazardous property and also, it has some demand; after this part number 6 is assigned at rank 2 because it has no hazardous property but has maximum demand over all the parts. Then part number 2 is assigned at rank 3 and part number 9 at rank 7, because part number 2 has much more demand and requires a minimum of 8 removed parts to remove it; part 9 does not require any removed part to remove it. Parts 10 and 5 are also assigned at ranks 6 and 4, as these tasks also do not have hazardous property and demand, so here, their part removal cost is considered as decision variable and according to their part removal cost they are assigned the ranking. This process is also repeated for part numbers 1, 2, 3, 4, and 8 and they are assigned their respective ranks.

4.2 Assignment of Tasks to the Workstations

The assignment of tasks to the workstation is performed with the help of ranks assigned to the tasks according to the proposed heuristic. In the assignment procedure the cycle time of the workstation is taken as 40 s and only four workstations are required for the complete disassembly of the product. The tasks assigned to the workstation are as follows: $W1 = \{2, 10, 9\}$; $W2 = \{8\}$; $W3 = \{7, 6\}$; $W4 = \{5, 1\}$; and $W5 = \{4, 3\}$. The idle times of the workstations are: $I_1 = 6$ s; $I_2 = 4$ s; $I_3 = 4$ s; $I_4 = 3$ s; $I_5 = 10$ s. The overall idle time of the disassembly line $I = 27$ s. The performance of the heuristic in terms of minimization of number of workstation required and idle time per workstation is examined because the heuristic is designed for the minimization of the number of workstation and cycle time of the line for complete disassembly. The proposed heuristic balances the line for the given problem and reduces the cycle time up to 3 s and minimizes it up to 37 s, while the compared heuristic can reduce the cycle time by only one unit and minimize up to only 39 s. Now it can be said that the proposed heuristic performs better than the compared heuristic. The results of tasks assignment are given in Table 3.

Table 3 Solution by proposed heuristic

W/S No.	Part assigned	Part removal time	Idle time
1	2	10	30
	10	10	20
	9	14	6
2	8	36	4
3	7	20	20
	6	16	4
4	5	23	17
	1	14	3
5	4	18	22
	3	12	10

5 Conclusion

However, the use of disassembly lines generates the demand to balance the disassembly line [11]. As environmental regulations come into lime light and producers are gratified to collect their end-of-life products and recover the parts and materials, the problem of disassembling them in large volumes arises. In the presence of costs of performing disassembly associated with disposal, and the presence of hazardous material, the discarded products should be disassembled safely, efficiently, environment friendly, and cost-effectively [7]. An efficient, near optimal, multiobjective heuristic is presented for deterministic disassembly line balancing.

References

1. Zhang, H.C., Kuo, T.C., Lu, H., Huang, S.H.: Environmentally conscious Design and Manufacturing: a state-of-the-art survey. J. Manuf. Syst. **16**(5), 352–371 (1997)
2. Gupta, S.M., Taleb, K.N.: Scheduling disassembly. Int. J. Prod. Res. **32**, 1857–1866 (1994)
3. McGovern, S.M., Gupta, S.M.: 2-opt heuristic for the disassembly line balancing problem. Iris Northeastern University, Boston (2003)
4. Gupta, S.M., Lambert, A.J.D.: A heuristic solution for disassembly line balancing problem incorporating sequence dependent costs. Iris Northeastern University, Boston (2005)
5. Gungor, A., Gupta, S.M.: Issues in environmentally conscious manufacturing and product recovery: a survey. Comput. Ind. Eng. **36**(4), 811–853 (1999a)
6. Gungor, A., Gupta, S. M.: Disassembly line balancing. In: Proceedings of the 1999 Annual Meeting of the Northeast Decision Sciences Institute, Newport, Rhode Island, 193–195 March 1999b
7. Gungor, A., Gupta, S.M.: Disassembly line in product recovery. Int. J. Prod. Res. **40**(11), 2569–2589 (2002)
8. Tiwari, M.K., Sinha, N., Kumar, S., Rai, R., Mukhopadhyay, S.K.: A petri net based approach to determine strategy of a product. Int. J. Prod. Res. **40**(5), 1113–1129 (2001)
9. McGovern, S.M., Gupta, S.M.: Combinatorial optimization methods for disassembly line balancing. In: Proceedings of the 2004 SPIE International Conference on Environmentally Conscious Manufacturing, Philadelphia, Pennsylvania, 53–66 (2004)

10. McGovern, S.M., Gupta, S.M.: A balancing method and genetic algorithm for disassembly line balancing. Eur. J. Oper. Res. **179**, 692–708 (2007)
11. Altekin, F.T., et al.: Profit-oriented disassembly-line balancing. Int. J. Prod. Res. **46**, 2675–2693 (2008)
12. Koc, A., Sabuncuoglu, I., Erel, E.: Two exact formulations for disassembly line balancing problems with task precedence diagram construction using an AND/OR graph. IIE Trans. Oper. Eng. **41**, 866–881 (2009)
13. Ding, L.-P., et al.: A new multi-objective ant colony algorithm for solving the disassembly line balancing problems. Int. J. Adv. Manuf. Technol. **48**, 761–771 (2010)

A New Approach to Rework in Merge Production Systems

S. Kannan and Sadia Samar Ali

Abstract This paper introduces and incorporates the concept of rework in modelling of a merge production system under random conditions. Merge and split production stages are common in assembly lines. Merging of components can be performed correctly or otherwise. If not done properly, a merging operation can be redone, i.e. the merging operation can be reworked. This paper explains the modelling of a two stage merge production system subject to rework using semi-regenerative stochastic processes. The modelling has been done to obtain various busy period durations over finite time duration for transient state analysis. Also, the modelling and analysis has been carried out without any particular assumption on the distributions of processing times. All the processing times involved have been assumed to be arbitrarily distributed.

Keywords System performance · Expected duration · Design stage of transfer-line production systems · System over a finite horizon of time

1 Introduction

Production system modelling and analysis is a field as old as industrial engineering itself. Research is being carried out for a long time, over the years and decades, under varied assumptions [10]. In the initial stages, research has been carried out under deterministic assumptions. Later, randomness has been incorporated but confined to

S. Kannan (✉)
Management Department, BITS, Pilani 333031, India
e-mail: rnskannan@gmail.com

S. S. Ali
Operations Management, Fortune Institute of International Business, Vasant Vihar, New Delhi 110057, India
e-mail: sadiasamarali@gmail.com

B. V. Babu et al. (eds.), *Proceedings of the Second International Conference on Soft Computing for Problem Solving (SocProS 2012), December 28–30, 2012*, Advances in Intelligent Systems and Computing 236, DOI: 10.1007/978-81-322-1602-5_56,

exponential distribution only. There is quite an amount of literature on modelling and analysis of production systems under Markovian assumption, that is, the processing times involved in the analysis are assumed to follow an exponential distribution. The present model makes no assumption about the distributions of any processing times involved and provides transient state solution and analysis. A general framework is developed and for particular cases one can substitute desired distributions for various processing times involved. A system can be defined as a collection of interacting elements or processes that operate in a coordinated and combined fashion to achieve a predefined common goal. A production system can be defined as the means by which resource inputs are transformed to create goods and services. In manufacturing industries, the inputs are various raw materials such as energy, labour, machine, etc. In service industries, the inputs are likely to be dominated by labour [2]. The focus of analysis of this paper is discrete part manufacturing systems, where each item processed is distinct and the processing times are non-Markovian. Such systems can be found in mechanical, electrical and electronic industries making components possibly for *cars, refrigerator, computers*, etc. While the 'production line' plays an important role in a manufacturing industry, very little research on the interactions between the stages in a line is reported [3, 11]. The analysis of productions systems, though not given importance to the extent it deserves, is one of the oldest problems in industrial engineering [1, 6]. Analysis of two-stage systems provides useful hints to describe generalised n-stage systems and such an analysis is getting a great deal of attention over the last few years. This is because any multi *"n-stage"* system can be analysed by formulating the system as a series of two-stage systems [4].

A variety of applications particularly those arising out of computers and communication systems lead to the consideration of an input queue of unlimited capacity. This concept is applied in many a serial production systems [5]. Though the assumption of initial buffer of infinite capacity sounds artificial in production lines, it only implies that an input is always available, meaning that the initial stage is never starved [7]. Recently, [4] have analysed a production system taking into account the concept of rework but the model is of deterministic type. Merge production systems have been analysed without taking into account the concept of rework in [8] and [9].

Consider the following quality control problem in a Merge production system. In the first stage, components are made/ processed. In the second stage, the components are assembled. Assembled products coming out of stage II are inspected at an inspection point and the good ones are transferred out of the system while the bad ones are further classified as products that can be reworked (say minor defects) and otherwise (i.e. scraps). This is because not all the rejects can be reworked (major defects). A diagram of this system is given in Fig. 1.

The production system under study is modelled using a semi-regenerative process. This modelling requires knowledge over the concept of renewal processes. These equations, which are of convolution type, have been solved using the numerical method suggested by Jones [6]. Usage of this numerical method provides means to use distributions like Erlang (two-stage), which can be used in practical situations, unlike the traditional Laplace Transform technique, which can only be used when the processing times follow an exponential distribution.

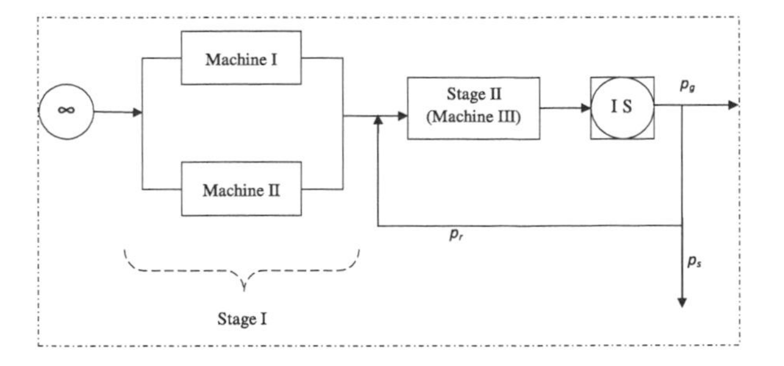

Fig. 1 Schematic diagram of the production system I.S.: inspection station

2 Assumptions

Following is a list of assumptions made to model the system under consideration.

1. Transfer of units from the initial buffer to Machines in stage I to Stage II is of instantaneous type.
2. Inspection is instantaneous.
3. Whenever a merge operation is to be reworked, then the Stage II machine will immediately start reworking the merge operation.
4. Processing times at both the stages are independent, random and arbitrarily distributed.
5. Products from Stage I will be inspected at the inspection station only when Stage II is free.
6. Stage II (i.e. machine 3) is never blocked.
7. Reworked jobs are always perfect.
8. Machine 1/2/3 is perfect (i.e. reliable)
9. Setup time is of instantaneous type.

3 Notations

pdf : Probability density function
cdf : Cumulative distribution function
sf : Survivor function
$f(\cdot)/g(\cdot)$: pdf of processing time of machine I/machine II in Stage I
$F(\cdot)/G(\cdot)$: cdf of processing time of machine I/machine II in Stage I
$\bar{F}(\cdot)/\bar{G}(\cdot)$: sf of processing time of machine I/machine II in Stage I
$h(\cdot)/r(\cdot)$: pdf of merge processing time and that of rework time in Stage II
$H(\cdot)/R(\cdot)$: cdf of merge processing time and that of rework time in Stage II
$\bar{H}(\cdot)/\bar{R}(\cdot)$: sf of merge processing time and that of rework time in Stage II
p_g : Probability of a merge operation completed by Stage II is good

p_r : Probability of a merge operation completed by Stage II is not good but can be reworked

p_s : Probability of a merge operation completed by Stage II is neither good nor can be reworked. Clearly, $p_{g,} + p_r + p_s = 1$

: Convolution : $f(t) * g(t) = \mathbf{R}_0^t f(u)g(t-u)du = g(t) * f(t)$

4 System Modelling

The system under consideration is modelled by identifying the state of the system at any instant t. The exhaustive list of possible states of the system is given in Table 1.

5 Evolution of System Characteristics

In this Section, analytical expressions for various measures of system performance such as busy period durations have been obtained. See Fig. 1.

The following system characteristics have been obtained under the assumption that the distributions of all the random variables involved in the analysis (including the processing times of rework) are arbitrary.

1. Expected duration Machine 1 in Stage I is busy in $[0, t]$.
2. Expected duration Machine 2 in Stage I is busy in $[0, t]$.
3. Expected duration both Machines in Stage I is busy in $[0, t]$.
4. Expected duration Machine 1 in Stage I is blocked in $[0, t]$.
5. Expected duration Machine 2 in Stage I is blocked in $[0, t]$.
6. Expected duration Stage II (i.e. Machine 3) is busy in $[0, t]$.

Table 1 State space

State	Machine 1 in Stage I	Machine 2 in Stage I	Machine 3 in Stage II
1	Busy	Busy	Free
2	Blocked	Busy	Free
3	Busy	Blocked	Free
4	Busy	Busy	Busy
5	Blocked	Busy	Busy
6	Busy	Blocked	Busy
7	Blocked	Blocked	Busy
8	Busy	Busy	Busy with rework
9	Blocked	Busy	Busy with rework
10	Busy	Blocked	Busy with rework
11	Blocked	Blocked	Busy with rework

7. Expected duration Stage II is idle in $[0, t]$.
8. Expected duration Stage II is busy with rework in $[0, t]$.
9. Expected duration Stage II is busy with rework of type "i" in $[0, t]$.

In this paper, the modelling of a two stage merge production system subject to rework using semi-regenerative stochastic processes is explained where all the processing times involved have been assumed to be arbitrarily distributed. The finished components from machines I and II are transferred to machine III for merging). The expected duration for various machines in both stages during $[0, t]$, are found following the expected duration of Stage II is busy with rework and idle, in $[0, t]$.

In case of Expected duration of Machine I in Stage I is busy in $[0, t]$ when there are two types of rework, we shall extend the number of types of defects from one to two. Later, we shall generalise the number of defects to "M" ($M = 2$). Assume that a merging operation can result in two types of defects. The assembled product has a defect of type 1. The expected duration for type 1 and type 2 in Stage II, is busy with rework is then found during $[0,t\ ]$. We shall extend and generalise the number of types of defects from two to "M" ($M = 2$). The completed merging operation is of type 1 defect with probability p_{r1} and is of type 2 defect with probability p_{r2} and so on. Clearly, $p_g + p_s + p_{r1} + p_{r2} + \cdots + p_{rM} = 1$, i.e. after merging in Stage II, products are inspected at the inspection station and the merging operation clears inspection with probability p_g, could be defective of type 1 with probability p_{r1} or could be defective of type 2 with probability $p_r 2, \ldots$, could be defective of type M with probability p_{rM} and the operation is neither good nor can be reworked (i.e. a scrap) with probability p_s. Finally, expected duration of Stage II is busy with rework of type "i" in $[0, t]$ is expressed with:

$$Av_1^{Ri}(t) = [f(t)G(t) + g(t)F(t)] * Av_4^{Ri}(t) \tag{1}$$

$$\begin{aligned} Av_4^{Ri}(t) = \{&f(t)[h(t)G(t) + g(t)H(t)] + g(t)[f(t)H(t) \\ &+ h(t)F(t)] + h(t)[f(t)G(t) + g(t)F(t)]\} \\ &\left(p_g Av_4^{Ri}(t) + \sum\nolimits_{i=1}^{M} p_{r1} Av_8^{Ri}(t) + p_s Av_4^{Ri}(t)\right) \end{aligned} \tag{2}$$

$$\begin{aligned} Av_{8(i)}^{Ri}(t) = \{&f(t)[r_i(t)G(t) + g(t)R_i(t)] + g(t)[f(t)R_i(t) + r_i(t)F(t)] \\ &+ r_i(t)[f(t)G(t) + g(t)F(t)]\} \\ &Av_4^{Ri}(t) + \bar{F}(t);\ 1 \le i \le M; \end{aligned} \tag{3}$$

The expected duration Stage II is busy with rework of type 1, is then given by, in $[0, t]$,

$$\mu^{Ri}(t) = \int_0^t Av_1^{Ri}(u)du \tag{4}$$

6 Conclusion

When the number of types of defects is taken to be 2, i.e. merging operation, when not correct, can result in an item having one of two types of defect. As the time progresses the proportion of blocked duration also increases. If these proportions are measured in terms of % value of the time, the blocked durations are actually decreasing. Similarly it happens for busy, blocked/idle durations of Machines 2 and 3. One can test the sensitivity of various durations with respect to changes in processing rates of Machines 1, 2 and 3. Such a sensitivity analysis can be useful in two ways. First is to fix the desired busy/blocked/idle durations (including that of rework) and experiment on processing rates of various machines until the desired durations are obtained. Another way is to fix the processing times of machines and experiment on the sensitivity of changes in busy/blocked/idle durations. Such a transient state analysis will be highly useful at the design stage of transfer-line production systems under random conditions and also when it is decided to monitor the system over a finite time horizon.

References

1. Birolini, A.: On the use of stochastic processes in modeling reliability problems. Springer-Verlag, Lecture Notes in Economics and Mathematical Systems (1985)
2. Buffa, E.S.: Modern production/operations management, 7th edn. Wiley Eastern Ltd., New Delhi (1983)
3. Buzacott, J.A., Shanthikumar, J.G.: Stochastic models of manufacturing sys-tems. Prentice Hall Inc., New Jersey (1993)
4. De Koster, M.B.M., Wigngaard, J.: On the equivalence of multi-stage production lines and two-stage lines. I.I.E Trans. 19, 351–354 (1987).
5. Yang, L., Jingshan, L.: Performance analysis of split and merge production systems. In: Proceedings of 2009 American Control Conference, pp. 2190–2195 (2009).
6. Jones, J.G.: On the numerical solution of convolution integral equations and systems of such equations. Math. Comput. **15**, 131–142 (1961)
7. Neuts, M.F.: Matrix-geometric solutions in stochastic models: an algorithmic approach. The John Hopkins University Press, Baltimore and London (1981)
8. Gopalan, M.N.: Dinesh Kumar, U.: On the production rate of a merge production system. Int. J. Qual. Reliab. Manage. **11**(1994), 66–72 (1994)
9. Koneigsberg, E.: Production lines and internal storages : a review. Manage. Sci. **5**, 410–433 (1959)
10. Shanthikumar, J.G., Tien, C.C.: An algorithmic solution to two-stage transfer lines with possible scrapping of units. Manage. Sci. **29**, 1069–1086 (1993)
11. Buscher, U., Lindner, G.: Optimizing a production system with rework and equal sized batch shipments. Comput. Oper. Res. **34**, 515–535 (2007)

Part V
Applications of Soft Computing in Image Analysis and Pattern Recognition (SCIAPR)

An Iterative Marching with Correctness Criterion Algorithm for Shape from Shading Under Oblique Light Source

Gaurav Gupta and Manoj Kumar

Abstract In this paper, a fast and robust Shape from Shading (SfS) algorithm by iterative marching with corrections criterion under oblique light source is presented. Usually, SfS algorithms are based on the assumption that image radiance is a function of normal surface alone. SfS algorithms solve first-order nonlinear Hamilton Jacobi equation called image irradiance equation. Both Fast Marching Method (FMM) and Marching with Correctness Criterion (MCC) basically work for the frontal light illumination direction, in which the image irradiance equation is an Eikonal equation. The problem task is to recover the surface from the image—which amounts to finding a solution to the Eikonal equation. FMM copes better the image irradiance iteratively under oblique light sources with the cost of computational complexity $O(NlogN)$. One prominent solution is the Marching with MCC of Mauch which solves the Eikonal equation with computational complexity $O(N)$. Here, we present a new iterative variant of the MCC which copes better with images taken under oblique light sources. The proposed approach is evaluated on two synthetic real images and compared with the iterative variant of FMM. The experimental results show that the proposed approach, iterative variant of MCC is more efficient than the iterative variant of FMM.

Keywords SfS · FMM · MCC · Complexity

G. Gupta (✉)
ITM University, Gurgaon, India
e-mail: guptagaurav.19821@gmail.com

M. Kumar
Babasaheb Bhimrao Ambedkar University, Lucknow, India
e-mail: mkjnuiitr@gmail.com

B. V. Babu et al. (eds.), *Proceedings of the Second International Conference on Soft Computing for Problem Solving (SocProS 2012), December 28–30, 2012*, Advances in Intelligent Systems and Computing 236, DOI: 10.1007/978-81-322-1602-5_57, © Springer India 2014

1 Introduction

Shape from Shading (SfS) is a well-known problem of Computer Vision. SfS was first introduced by Horn [11] in 1970s, where the 3D shape of the object is reconstructed by exploiting the shading information contained in its single shaded image. After the introduction of this field, various techniques have been reported to solve the SfS problems. Zhang et al. [16] categorized the reported techniques on SfS into four categories, viz. *local*, *linear*, *minimization*, and *propagation* approach.

Rouy et al. [10] introduced a viscosity solution and used dynamic programming to solve the SfS problem. Dupis et al. [14] formulated SfS as an optimal control problem and solved it by numerical methods. Bichel et al. [12] improved Dupis and Oliensis approach and proposed a fast converging, minimum downhill approach for SfS. All these propagation methods are guaranteed to converge; however, they are iterative and the convergence speed is uncertain. To deal with this problem, Kimmel and Sethian [15] reported a deterministic, non-iterative single pass algorithm known as Fast Marching Method of complexity $O(NlogN)$ for unique surface reconstruction, where N is the number of pixels in the image. Various algorithms [1–4, 9, 8, 17, 18] have been derived from Fast marching method to solve the SfS problem. Tankus [3, 4] and Yuen et al. [18] have reported the FMM-based methods for perspective SfS. However, Sean Mauch [19] observed that more than one point can be selected as the best solution at each iteration and hence reported a greedier algorithm namely Marching with Correctness Criterion (MCC) to solve Hamilton Jacobi equation. The MCC is faster and more efficient algorithm than FMM algorithm as its computational complexity is $O(N)$. FMM uses binary heap to store and search the minimum data points but MCC uses the array or list, so the space complexity of MCC is also less than FMM. The MCC algorithm produces almost same result as FMM but with less space and time complexity. Both FMM and MCC basically work for the frontal light illumination direction, in which the image irradiance equation is an Eikonal equation.

In order to solve the SfS problem using FMM approach for oblique light source direction, Sethian [15] suggested to rotate the image coordinate system to the light source coordinate system. The rotation of image coordinate system yields the image irradiance equation as an almost Eikonal equation and hence the problem is solved similar to the case of vertical light source direction. However, the rotation of image coordinate system requires the depth values of the surface which are not known in priori and basically the goal of the algorithm is to obtain them. Hence to accomplish this, the way is to use the initial approximation of the depth values, which leads toward the erroneous shape reconstruction of the surface. To deal with this problem, Tankus et al. [4] proposed an iterative scheme, in which the rotation of co-ordinate system is not used. In this scheme, the image irradiance equation is approximated into the series of Eikonal equations and each equation approximation refines the approximation of previous equation. Numerical tests performed in [6, 7, 13] clearly show that the PSFS algorithm is not able to catch discontinuities of the surface. In fact, it tries to reconstruct a continuous surface with the same brightness function as the original one.

In this paper, an iterative Marching with Correctness Criterion algorithm for SfS problem under oblique light source direction has been reported. The Lambertian reflectance map is used to model the SfS problem as it is simple and almost fair approximation of most of the real world objects. Rest of the paper is organized as follows. Section 2.1 describes the mathematical formulation of SfS problem and its solution using FMM for oblique light source direction. The solution of SfS problem of oblique light source directions by MCC is described in Sect. 3. Experimental results, error analysis, and time complexity are discussed in Sect. 4. The proposed work is concluded in Sect. 5.

2 Shape from Shading

Let $z(x, y)$ denotes the depth function in a real world Cartesian coordinate system whose origin is at camera plane. The real world coordinate system $(x, y, z(x, y))$ of a surface is projected orthographically onto image point (x, y). The surface normal and the image intensity at the point (x, y) are denoted by $\overrightarrow{N}(x, y)$ and $I(x, y)$ respectively. If the object is assumed to be Lambertian and illuminated by a point light source at infinity with the direction $\overrightarrow{L}(p_s, q_s, -1)$, then the orientation of the object surface and image intensity are related by the image irradiance equation which is given as follows:

$$I(x, y) = \overrightarrow{L} \cdot \overrightarrow{N}(x, y) = \frac{p_s p + q_s q + 1}{\sqrt{(p_s^2 + q_s^2 + 1)(p^2 + q^2 + 1)}} \tag{1}$$

where $p = \frac{\partial z}{\partial x}$ and $q = \frac{\partial z}{\partial y}$.

If the light source direction is vertical, i.e., $\overrightarrow{L} = (0, 0, -1)$, then the above equation becomes an Eikonal equation which can be written as:

$$p^2 + q^2 = f^2 \text{ or } |\nabla z(x, y)| = f \tag{2}$$

where $f = \sqrt{\frac{1}{I^2(x,y)} - 1}$.

2.1 *Fast Marching Method*

The previously existing methods to solve the Eikonal equation are boundary value methods and are iterative. The Marching methods allow to solve the boundary value problem in a single pass, i.e., without iterations. Therefore these methods are based on an optimal ordering of the grid points in the solution domain. The FMM produces a solution which satisfies the discrete version of the Eikonal equation. The first-order approximations of $\frac{\partial z}{\partial x}$ and $\frac{\partial z}{\partial y}$ in forward and backward directions can be written as:

$$D_{i,j}^{+x} z = \frac{z_{i+1,j} - z_{i,j}}{\Delta x}, \quad D_{i,j}^{-x} z = \frac{z_{i,j} - z_{i-1,j}}{\Delta x}$$
$$D_{i,j}^{+y} z = \frac{z_{i,j+1} - z_{i,j}}{\Delta y}, \quad D_{i,j}^{-y} z = \frac{z_{i,j} - z_{i,j-1}}{\Delta y}$$

The Fast Marching Method is a upwind direction scheme and information always flows from small values to large values of the solution z. This information is encoded into the first-order, adjacent difference scheme:

$$\left(\max(D_{i,j}^{-x} z, -D_{i,j}^{+x} z, 0)\right)^2 + \left(\max(D_{i,j}^{-y} z, -D_{i,j}^{+y} z, 0)\right)^2 = f_{i,j}^2 \quad (3)$$

In this scheme, the z values at all grid points are set to ∞ and the correct z values at the local minimum points are assigned. The solution propagates from these points to all other points. Assuming without loss of generality $\Delta x = \Delta y = 1$, the solution of the above Equation at point (i, j) is given as:

$$z_{i,j} = \begin{cases} \frac{z_1+z_2+\sqrt{2f_{i,j}^2-(z_1-z_2)^2}}{2}, & \text{if} |z_1 - z_2| < f_{i,j} \\ \min\{z_1, z_2\} + f_{i,j}, & \text{otherwise} \end{cases} \quad (4)$$

Here two input arrays namely solution array and status array of size N (size of the image) are used. Initially, in status array, the minimum singular points are set to *Known* and at the corresponding points in solution array, the true depth values are assigned. The rest points in status array are set to be *Unknown* and their values are set as ∞ in solution array. The trial (neighbors of *Known*) points are pushed in a minimum binary heap. The minimum z value vertex with its indices is removed from heap and the corresponding status in status array is set to be *Known*. While updating the value at *Trial* points, value of a point can be decreased which increases its priority in the heap. On the basis of priority, the position of this point is adjusted in the heap.

For oblique light source (non-vertical), Eq. (2) takes the following form

$$p^2 + q^2 = f^2(p, q) \quad (5)$$

where, $f(p, q) = \sqrt{1 - (\frac{p_s p + q_s q + 1}{(p_s^2 + q_s^2 + 1) I(x, y)})}$.

Thus the difference between vertical and oblique light source is the dependency of f on p and q in the right-hand sides of Eqs. (2) and (5). Sethian's Fast Marching Method [15] solves these Equations in "upwind" fashion in a single pass. Tankus et al. [4] analyzed that due to the dependence of f on p and q in case of oblique light source direction, the Sethian's single pass FMM method may not solve this case efficiently and proposed an iterative use of Fast Marching Method. The Eq. 5

is written in iterative fashion as follows:

$$p_{n+1}^2 + q_{n+1}^2 = f^2(p_n, q_n) \tag{6}$$

where p_n and q_n are the values of p and q at the nth iteration. The algorithm is as follows:

1. Initialize (p_0, q_0) on the basis of Sethian's FMM.
2. Calculate the value of $f^2(p_n, q_n)$ using the approximated (p_n, q_n).
3. As the $f^2(p_n, q_n)$ is obtained, Eq. 6 becomes an Eikonal equation. Use Eq. (4) to calculate the z values and hence the values of p_{n+1}, q_{n+1}.
4. $n = n + 1$, go to the step 2 and repeat the above process for n steps.

FMM uses a finite difference scheme to label the adjacent neighbors of the *Known* grid point. If there are N grid points, the computational complexity of labeling operation of these points is $O(N)$. Since there may be atmost N *Trial* grid points, the maintenance of the binary heap and choosing the minimum labeled grid points add a cost of $O(NlogN)$. Hence the total complexity of the FMM is $O(NlogN)$. Sethian proved that the solution by FMM satisfies the discrete version of Eikonal. In this iterative scheme, series of Eikonal equations are used. Therefore the iterative version of FMM is also convergent.

3 Marching with Correctness Criterion (MCC)

Sean Mauch [19] improved the Fast Marching Method and proposed a faster method to solve the Hamilton Jacobi equation named as Marching with Correctness Criterion (MCC). The MCC is similar to FMM as both are based on Dijkstra's single source shortest path algorithm but dissimilar to FMM in the way of storing and freezing the vertices. In MCC, the labeled adjacent neighbors are not stored in the heap and instead of selecting a single vertex from the heap to become *Known* at each iteration, it tries to make the multiple vertices to known at each iteration. To examine whether a *Trial* point could be converted to a *Known* point, the future labeling operations are observed. If the future labeling does not decrease the value of a *Trial* point then the point can be converted to the known point at the same label. To accomplish this task, a lower bound is defined in which the future labeling operations are incorporated. If the value of a *Trial* vertex is less than or equal to the lower bound then the Trial vertex is moved from *Trial* set to the *Known* set at the same step. Let $z_{i,j}$ be the value of a trial vertex then the lower bound is defined as follows: lowerbound $=$ $\min(z_{i,j}, \min(z_{k,l} + f_{k,l}))$ where (k, l) are the unknown neighbors of (i, j). MCC is also used for ordering an upwind finite difference scheme to solve the static Hamilton Jacobi equation. In this method, difference scheme is used in coordination as well as in diagonal directions. Hence a *Trial* point can be possible in three directions of a *Known* point viz. adjacent, diagonal, and adjacent-diagonal directions. The first-order approximation of $\frac{\partial z}{\partial x}$ and $\frac{\partial z}{\partial y}$ is obtained also in diagonal direction and is written as follows:

$$\frac{z_{i,j} - z_{i-1,j-1}}{\Delta x} \approx \frac{\partial z}{\partial x} + \frac{\partial z}{\partial y} \tag{7}$$

If $\frac{\partial z}{\partial x} + \frac{\partial z}{\partial y}$ is obtained from diagonal directions and $\frac{\partial z}{\partial x}$ from differencing in horizontal direction, then $\frac{\partial z}{\partial y}$ can be obtained from these two. Thus $\frac{\partial z}{\partial x}$ and $\frac{\partial z}{\partial y}$ are calculated by differencing in horizontal and diagonal direction. Without loss of generality, we can assume that $\Delta x = \Delta y = 1$ throughout the grid. If $z_{i,j}$ is a trial point then the value at this point is updated as follows:

1. If the points at adjacent position to $z_{i,j}$ are *Known* then $z_{i,j} = z_a + f_{i,j}$, where z_a is the minimum of all *Known* points adjacent to $z_{i,j}$.
2. If the points at diagonal position to $z_{i,j}$ are *Known* then $z_{i,j} = z_d + f_{i,j}$, where z_d is the minimum of all *Known* points diagonal to $Z_{i,j}$.
3. If the points at both adjacent and diagonal positions to $z_{i,j}$ are *Known* then

$$z_{i,j} = \begin{cases} z_{i,j} = Z_d + \sqrt{(2)} f_{i,j}, & \text{if } 0 \leq D \leq \sqrt{2} f_{i,j} \\ \infty \text{ (large value)}, & \text{otherwise;} \end{cases}$$

where, $D = z_a - z_d$.

The corresponding algorithms for vertical and oblique light sources are written as follows:

For Vertical Light Source

1. Label the minimum singular point as *Known*.
2. Label all other grid points as *Far*.
3. Label all the neighbors of *Known* points as *Trial*, compute the z value on these points and add them in the Trial set.

Marching Forward:

1. If a point is in *Trial* set, examine it for its possible conversion into the *Known* point on the basis of lower bound.
2. If the *Trial* point converts into the *Known* point, label the neighbors of this *Known* point as *Trial*, compute the z value, and add them in the New Trial set.
3. Remove the converted *Known* points from the Trial set.
4. Add the points of New *Trial* set in the *Trial* set.
5. Make the New *Trial* set empty.
6. Return to step 1.

The marching forward steps are repeated until all the *Trial* points are converted to the *Known* points.

For Oblique Light Source

In case of oblique light source direction, image irradiance equation takes the form of Eq. (5), in which the right-hand side is depending on p and q. We have used the same iterative fashion of Eq. (5) as by Tankus et al. [4] and which is shown in Eq. (6).

1. After one iteration of the algorithm of MCC for Vertical Light Source, obtain the values of p and q, and initialize them as p_0 and q_0 for this algorithm.
2. Calculate the value of $f^2(p_n, q_n)$ as in Eq. (5), using the above values of p and q.
3. As the $f^2(p_n, q_n)$ is obtained, Eq. 6 becomes an Eikonal Equation. Use the algorithm of MCC for Vertical Light Source to calculate the z values and hence the values of p_{n+1}, q_{n+1}.
4. increase n by one unit and go to the step 2. Repeat the above process for n steps.

The MCC algorithm is convergent because at each iteration at least one vertex (minimum value vertex) converts to a *Known* point. In SfS formulation, the MCC algorithm is basically used to solve the Eikonal equation $| \bigtriangledown z(x, y)| = f$. Let $f_{\min} \leq f \leq f_{\max}$ and $K = f_{\max}/f_{\min}$. Let z_m be the minimum solution of the *Trial* points and the solution at all *Trial* points lie in the range $[z_m, z_m + \frac{\sqrt{2}\Delta x}{f_{\min}}]$. On applying the correctness criterion, vertices with z value less than $z_m + \frac{\Delta x}{\sqrt{2} f_{\max}}$ become *Known*. At each step, the minimum solution at *Trial* points increase by at least $\frac{\Delta x}{\sqrt{2} f_{\max}}$. This implies that a point can stay in the *Trial* set by at most $\frac{\sqrt{2}\Delta x}{f_{\min}} / \frac{\Delta x}{\sqrt{2} f_{\max}} = \frac{2 f_{\max}}{f_{\min}} = 2K$ steps. Hence, if there are N grid points then the computational complexity of correctness criterion is $O(KN)$. The cost of labeling is $O(N)$, the cost of adding and removing *Trial* points is $O(N)$. Hence the total computational complexity of MCC algorithm is $O(KN)$.

4 Results and Discussions

4.1 Synthetic Images

The proposed algorithm is tested on two synthetic images: sphere and vase are used to test the proposed scheme. The first test image is sphere, as the sphere has only one minimum depth point. The sphere is generated by using the formula.

$$z(x, y) = \sqrt{r^2 - (x - 64)^2 - (y - 64)^2} \tag{8}$$

where r is the radius and $0 \leq x, y \leq 127$. The surface of the sphere is shown in Fig. 1a. The second test image is the synthetic vase. The vase is generated using the formula provided by Ascher et al. [5] as follows:

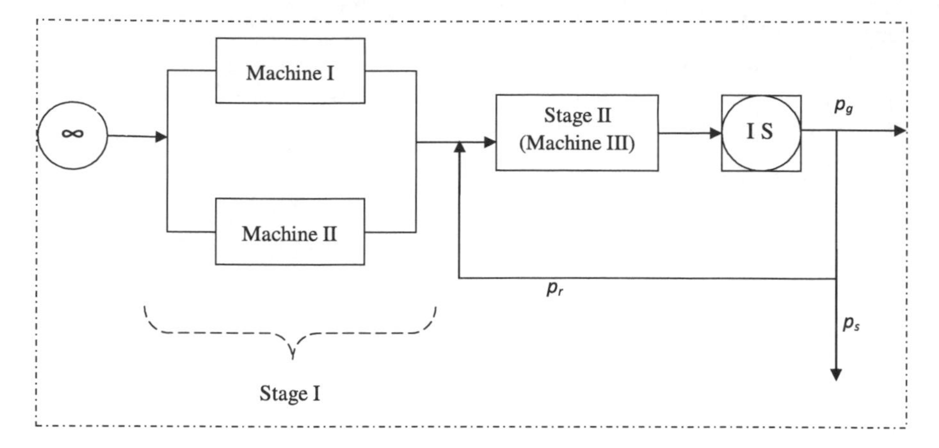

Fig. 1 **a** The original depth map **b** Shaded image generated from true depth map in Oblique light source direction **c** Reconstructed depth map **d** Error depth map for oblique light source

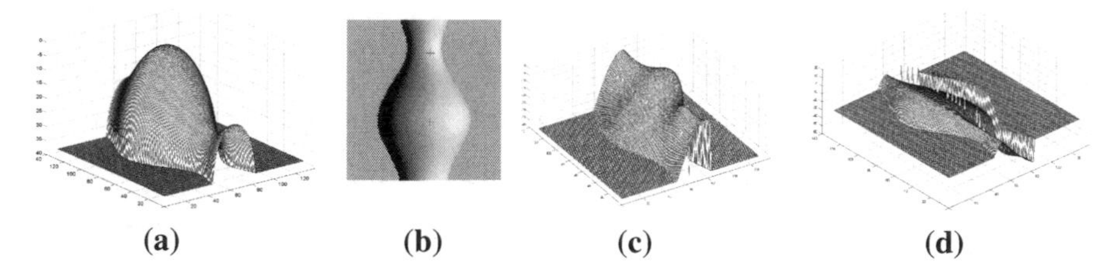

Fig. 2 **a** The original depth map **b** Shaded image generated from true depth map in Oblique light source direction **c** Reconstructed depth map **d** Error depth map for oblique light source

$$z(x, y) = \sqrt{(f(y))^2 - x^2}, \tag{9}$$

where $f(y) = 0.15 - 0.1y(6y + 1)^2(y - 1)^2(3y - 2)$, $-0.5 \leq x \leq 0.5$ and $0 \leq y \leq 1.0$. For the proper size and scale of the depth map, x and y are mapped in the range [0, 127] and z is scaled by a factor of 128. The synthetic surface of vase is shown in Fig. 2a.

In order to obtain the shaded images (of these surfaces), the surfaces are assumed to be Lambertian, and illuminated by a point light source in the direction (0, 0, 1) (vertical) and (0, −1, 1) (oblique). The surfaces gradients p and q are obtained by the discrete approximation of depth values in x and y directions. The surfaces are projected orthographically on $z = 0$ plane and the intensities at each pixel in the image are calculated by using Eq. (1). The obtained shaded images of sphere and vase under oblique light source direction are shown in Figs. 1b and 2b. The obtained shaded images under oblique light source direction are used as the input in MCC algorithm for the possible 3D shape reconstruction. However, the images are synthetic and the known true depth values are assigned on the minimum singular points. In sphere image, only one minimum singular point (center point) and vase image for two minimum singular points are chosen respectively. The minimum singular points in the shaded images are shown by + marks. These minimum singular points are taken as the initial points and the z values at all the grid points are calculated.The minimum singular points in the image are shown by + marks as shown in Fig. 1b and 2b.

The iterative MCC algorithm is used under oblique light source direction and the z values are computed using three iterations for each image. The reconstructed surfaces of sphere and vase under oblique light source direction are shown in Figs. 1c and 2c. Reconstructed depth values are normalized in the range of true data and the error maps are calculated which is the difference of true z values and reconstructed z values. The error maps of synthetic images are shown in Figs. 1d and 2d.

4.2 Real Images

We have tested the proposed scheme on two real images: pepper and real vase. The size of both the images is 256×256. However, the light source directions for real images are not known, by perception the light source directions for both the images are assumed as vertical. We have taken two minimum singular points for pepper image and one minimum singular point for vase image. The minimum singular points are taken on the basis of human perception. The minimum singular points in the image are shown by + marks as shown in Figs. 3a and 4a. On these points, the constant values are assigned according to the perception of relative heights. Algorithm starts from these points and calculates the z value at each point of the image. The reconstructed depth maps of pepper and vase are shown in the Figs. 3b and 4b. As the true data were not available for real images, is not available hence to show the efficiency of the proposed scheme, the reconstructed depth maps are projected orthographically on $z = 0$ plane, surface gradients p and q are calculated from the obtained z values, and intensity at each point is calculated using Eq. (1). Thus the shaded images from

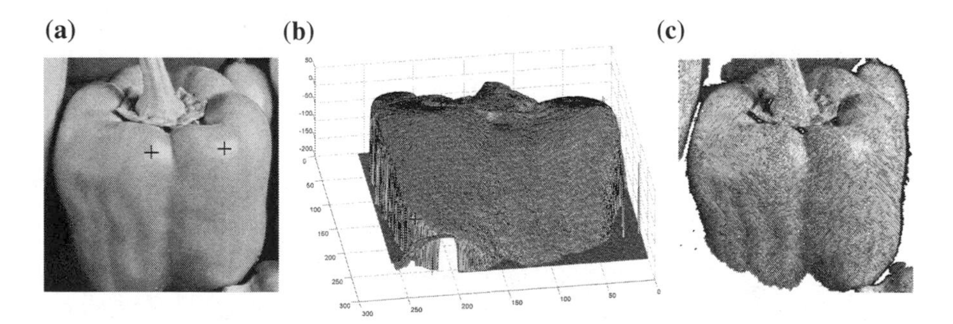

Fig. 3 **a** Original shaded image with minimum singular points. **b** Reconstructed depth map. **c** Shaded image generated from reconstructed depth map

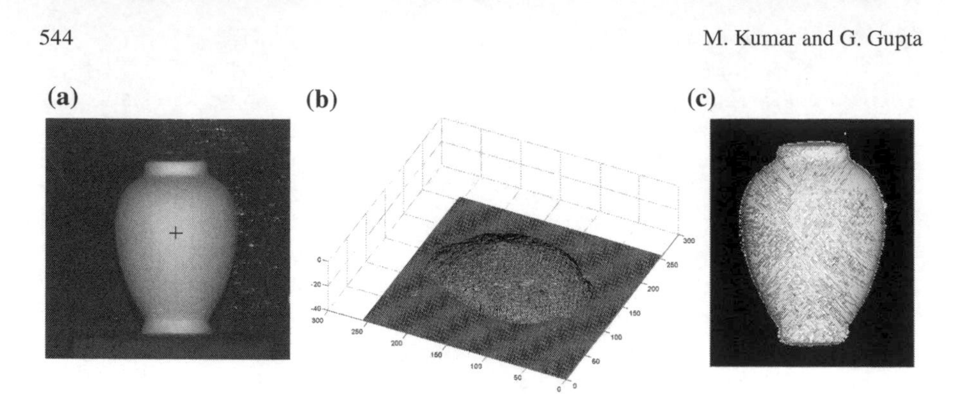

Fig. 4 **a** Original shaded image with minimum singular points. **b** Reconstructed depth map. **c** Shaded image generated from reconstructed depth map

the reconstructed depth maps are rendered as shown in Figs. 3c and 4c. The obtained shaded images are quite similar to the original images.

4.3 Error Analysis and Efficiency

In case of synthetic images, the error map is calculated by the absolute difference of true depth map and reconstructed depth map. The mean error and standard deviation error are calculated as shown in Table 1. All reconstruction processes are performed on a PC with Pentium Core 2 Duo 1.86 GHz and 1.00 GB memory. The algorithm is coded in C++ and MATLAB is used as a tool. The computation time is recorded for the sphere image of size 8×8, 16×16, 32×32, 64×64, 128×128, 256×256, and 320×320 in milliseconds as shown in Table 2.

Table 1 Computed errors

Surfaces	Oblique	
	mean error	Std. error
Sphere	8.63	6.37
Vase	18.04	9.14

Table 2 Computational time (in milliseconds) for sphere image

Size	FMM	MCC
8×8	0	2
16×16	15	31
32×32	31	49
64×64	94	212
128×128	344	710
256×256	1281	2809
320×320	2046	4392

5 Conclusions

This paper has presented a fast and robust algorithm for SfS under oblique light source. The MCC method is used to solve the Hamilton Jaccobi equation and the iterative variant of MCC is applied to deal with the images taken under the oblique light source direction. The orthographic projection and the most widely used Lambertian reflectance map have been used in SfS problem formulations. It can be observed that the present algorithm reconstructs the shapes in less computation time than the method proposed by Tankus et al. [4] which reconstructs the shape of the objects using Fast Marching Method. The computational time table indicates that as the size of image increases, the computing time in MCC is less as compared to FMM. The reported algorithm solves the problem with the time complexity of $O(KN)$, where, K is the ratio of maximum to minimum z values and N is the total number of pixels in the image. The method is guaranteed to obtain the solution provided the depth at local minimum singular points are given as it always freezes at least one grid point to the correct solution at each step. The effectiveness of the proposed algorithm has been demonstrated by experimental results. Experiments are conducted on various synthetic surfaces computed by formulae. The results suggested that the proposed algorithm successfully reduced the error and the reconstructed shapes are closely matched with the true data. However, the proposed scheme does not deal with the perspective projection and discontinuities due to occlusion in the surfaces. Furthermore, these problems may be discussed with the development of theory proposed in this paper and will be further refined in the future.

References

1. Tankus, A., Sochen, N., Yeshurun, Y.: A new prespective on shape from shading. In: Proceeding of 9th ICCV, **2**, 862–869 (2003)
2. Tankus, A., Sochen, N., Yeshurun, Y.: Perspective shape from shading by fast marching. In: Proceedings of IEE Computer Society Conference on Computer Vision and, Pattern Recognition, (2004)
3. Tankus, A., Sochen, N., Yeshurun, Y.: Shape from shading under perspective projection. Int. J. Comp. Vision **63**(1), 21–43 (2005)
4. Tankus, A., Sochen, N., Yeshurun, Y.: Shape from shading by iterative fast marching for vertical and oblique light source, geometric properties for incomplete data, Springer, Netherlands **1**, 237–258 (2006)
5. Ascher, U.M., Carter, P.M.: A multigrid method for shape from shading. SIAM J. Numer. Anal. **30**(1), 102–115 (1993)
6. Cristiani, E., Fast marching and semi-lagrangian methods for hamilton-jacobi equations with applications, Ph.D. thesis, Dipartimento di Metodi e Modelli Matematici per le Scienze Applicate, SAPIENZA Universit'a di Roma, Rome, Italy (2007) www.iac.rm.cnr.it/_cristiani/attach/ECristianiPhDthesis.pdf
7. Cristiani, E., Falcone, M., Seghini, A.: Some remarks on perspective shape-from-shading models. In: Proceedings 1st International conference Scale Space and Variational Methods in Computer Vision, Ischia, Italy, 2007, In: Sgallari, F., Murli, A., Paragios, N. (eds.) vol. 4485 of LNCS, 276–287

8. Prados, E., Faugeras, O.: Perspective shape from shading and viscosity solution. In: Proceedings of 9th ICCV, **2**, 826–831 (2003)
9. Prados, E., Soatto, S.: Fast marching method for generic shape from shading. In: Proceedings of VLSM'05. Third International Workshop on Variational Geometry and Level Set Method in Computer Vision, LNCS **3752**, 320–331 (2005)
10. Rouy, E., Tourin, A.: A viscosity solution approach to shape from shading. SIAM J. Numer. Anal. **29**(3), 867–884 (1992)
11. Horn, B.K.P., Brooks, M.J.: Shape from shading. The MIT Press, Cambridge (1989)
12. Bichsel, M., Pentland, A.P.: A simple algorithm for shape from shading. In: Proceedings of CVPR, 459–465 (1992)
13. Breu, M., Cristiani, E., Durou, J.-D., Falcone, M., Vogel, O.: Numerical algorithms for Perspective Shape from Shading, Kybernetika, **46**, 207–225 (2010). http://www.kybernetika.cz/content/2010/2/207
14. Dupis, P., Oliensis, J., Direct method for reconstructing shape from shading. In: Proceedings of CVPR, 453–458 (1992)
15. Kimmel, R., Sethian, J.A.: Optimal algorithm for shape from shading and path planning. J. Math. Imaging Vision **14**(3), 237–244 (2001)
16. Zhang, R., Tsai, P.S., Cryer, j.E., Shah, M.: Shape from shading: a survey. IEEE Trans. Pattern Anal. Mach. Intell. **21**(8), 690–706 (1999)
17. Yuen, S.Y., Tsui, Y.Y., Leung, Y.W., Chen, R.M.M.: Fast marching method for shape from shading under perspective projection. In: 2nd international Conference Visualization, Imaging and Image Processing, 584–589 (2002)
18. Yuen, S.Y., Tsui, Y.Y., Chow, C.K.: A fast marching formulation of perspective shape from shading under frontal illumination. Pattern Recogn. Lett. **28**, 806–824 (2007)
19. Sean, M.: Efficient algorithms for solving static Hamilton-Jacobi equations, Ph.D. thesis, California Institute of Technology Pasadena, California (2003)

Enhancement of Mean Shift Tracking Through Joint Histogram of Color and Color Coherence Vector

M. H. Sidram and N. U. Bhajantri

Abstract Tracking of an object in a scene, especially through visual appearance is weighing much relevance in the context of recent research trend. In this work, we are extending the one of the approaches through which visual features are erected to reveal the motion of the object in a captured video. One such strategy is a mean shift due to its unfussiness and sturdiness with respect to tracking functionality. Here we made an attempt to judiciously exploit the tracking potentiality of mean shift to provide elite solution for various applications such as object tracking. Subsequently, in view of proposing more robust strategy with large pixel grouping is possible through mean shift. The mean shift approach has utilized the neighborhood minima of a similarity measure through bhattacharyya coefficient (BC) between the kernel density estimate of the target model and candidate. However, similar capability is quite possible through color coherence vectors (CCV). The CCV are derived in addition to color histogram of target model and target candidate. Further, joint histogram of color model and CCV is added. Thus, the resultant histograms are empirically less sensitive to variance of background which is not ensured through traditional mean shift alone. Experimental results proved to be better and seen changes in tracking especially in similar color background. This work explores the contribution and paves the way for different applications to track object in varied dataset.

Keywords Object tracking · CCV · Mean shift · Kernel · BC

M. H. Sidram (✉)
JSS Research Foundation, JSS Technical Institutions Campus,
Mysore, Karnataka, India
e-mail: mhsidram@gmail.com

N. U. Bhajantri
Department of CS and E, Government Engineering College Chamarajanagar,
Chamarajanagar, Karnataka, India
e-mail: bhajan3nu@gmail.com

B. V. Babu et al. (eds.), *Proceedings of the Second International Conference on Soft Computing for Problem Solving (SocProS 2012), December 28–30, 2012*, Advances in Intelligent Systems and Computing 236, DOI: 10.1007/978-81-322-1602-5_58, © Springer India 2014

1 Introductiion

In an itinerary of tracking of the moving object in a capture sequence of real time is a blistering theme of research in the broad present scenario. Most of tracking algorithms assume that the object motion is smooth with no abrupt changes. Researchers can further mention the constraints of the object motion to be of invariable velocity or steady acceleration based on a priori information. In other words prior knowledge about the quantity and the size of objects, sometimes object appearance and shape, can also be used to simplify the predicament.

The plentiful approaches for object tracking have been proposed [1, 12, 15]. These are principally contradictory from each other based on the means they loom. Sometime tracking is accurate due to suitability of object representation or for a moment modeled motion, appearance, and shape of the object is appropriate. However, object tracking is posed with many problems like change in illumination, scale of object, occlusion of object, change in dynamism in the background, sudden movement of the objects, and cost of computation [15]. A large number of tracking methods have been proposed which attempt to counter these questions for a diversity of scenarios [2, 9, 10, 14]. The goal of this research work is to tracking the object based on appearance based features.

The suitability of apt strategy to track the object in surveillance is quite encouraging. In this work, we have attempted to do so through Mean shift tracking. It has an adequate amount of capability through its appearance-based features using probability density function (PDF) with nonparametric approach.

This approach proceeding with postulation such as the appearance and position of the object will not change drastically. The subsequent elite part of approach carried with brute force advance where previous spatial information available for estimating the objet location in the next frame.

A suitable kernel is used to establish the relation between the color PDF and spatial information. In this path, joint histogram of color and CCV of the target model and target candidate are perhaps essential to supplement the precise results, which is shown with experimental evidence in Sect. 4. The following milestones are used to assist the flow of the research article. Section 2 presents comprehension of contemporary effort and in Sect. 3 substantiates the proposed criteria and also enhanced version of algorithm is illustrated. Experimental results are elaborated in Sect. 4. The conclusion is given in Sect. 5.

2 Comprehension of Related Effort

In the greater part of circumferences, selecting the appropriate features plays a decisive role in tracking. Thus, the most desirable property of a visual feature is its exceptionality so that the objects can be straightforwardly eminent in the feature space. On other hand feature assortment is closely related to the object depiction such as color is used as a feature for histogram-based appearance representations. In this view, the following are the literature elaborated briefly.

Comaniciu et al. demonstrated the establishment of target model and candidate model to track the object [6, 7]. Ning et al. Experimented using color and texture jointly for target model [5, 10]. In this case texture features are exploited in addition to color histogram. The object is represented by texture using local binary pattern (LBP) [4, 11, 12]. Wang Yagi et al. worked with color and shape features jointly to create the target model and using mean shift object is tracked [2, 13, 14].

We propose more robust criteria to track the object especially in homogeneous background. However, we are motivated by the observation of above literature to erect the suitable model which is capable to produce accurate results. Thus, we have taken assistance of statistical attribute such as CCV. On the other hand, traditional mean shift strategy outcome also encouraged and along with the joint histogram of color and CCV supplemented our motivation.

3 Proposed Methodology

3.1 Basic Mean Shift Tracking Method

Object tracking using mean shift is a nonparametric method and robust to the change in appearance of object. Object under consideration is cropped as a rectangle. In other words it is Template of object or Target model. Its weighted RGB color histogram using suitable number of bins is determined using Eq. 2, q_u as notation where $u = 1 \ldots m$ and $\sum_{u=1}^{m} q_u = 1$. A similar histogram of Candidate of object or Candidate model around the spatial information of previous frame is created using Eq. 3, $p_u(y)$ as notation and $\sum_{u=1}^{m} p_u(y) = 1$. Weights are extracted using Eq. 7. The shift in the mean is calculated by using Eq. 8 and the spatial information y_1 will be used as y_0 for computing the next frame mean shift.

3.2 RGB Color Histogram

To facilitate smooth weights which calculated through the similarity between the target PDF and candidate PDF, a kernel is used. The Epanecknikov kernel is one such approach. Due to its ellipsoidal region it performs better and gives more weights to the pixels nearer to the center of the kernel expressed in Eq. 1.

$$k(x) = \begin{Bmatrix} 1 - r & r \leq 1 \\ 0 & \text{otherwise} \end{Bmatrix} \tag{1}$$

q_u is a PDF of the target model and $p(y)$ is a PDF of candidate model at y location and same concept glimpses through Eq. 2 and Eq. 3.

$$q_u = C_1 \sum_{i=1}^{n} k(\|x_i^*\|)^2 \, \delta\left[c\left(x_i^*\right) - u\right] \tag{2}$$

$$p_u(y) = C_2 \sum_{i=1}^{n} k\left(\left\|\frac{(y - x_i)}{h}\right\|\right)^2 \delta\left[c\left(x_i^*\right) - u\right] \tag{3}$$

3.3 Color Coherence Vectors Histogram

The Color histogram is less capable to discrimi nate the objects clearly because dissimilar images can have similar histograms. Even though its computation is linear, determination of CCV resembles color histogram [1, 3, 4, 8, 13]. The color space of image is converted in to the number of bins using connected component analysis. The pixels of particular bin are classified based on threshold τ Coherent and Incoherent pixels. The larger pixels group of a bin is called Coherent pixels and incoherence noticed by smaller group.

Let there be n number of bins. Therefore, number of Coherent pixels α_i and Incoherent pixels β_i of ith bin. Hence CCV pair is given by (α_i, β_i) and also Color Coherence vector is denoted by

$$(\alpha_1, \beta_1)\,(\alpha_2, \beta_2)\ldots(\alpha_n, \beta_n) \tag{4}$$

Weighted histograms of CCV are c & d calculated using the kernel.

3.4 Joint Histogram

A joint histogram of color and CCV for target model and Target candidate calculated [14]. Joint histograms of target model q_u and target candidate p_u are as follows.

$$q_u = q_u .* \mathrm{c} \tag{5a}$$

$$p_u = p_u .* \mathrm{c} \tag{5b}$$

Bhattacharyya coefficient is extended to measure similarity as shown through Eqs. 6, 7.

$$\rho = \sum_{i=1}^{m} \sqrt{p_u(y), q_u} \tag{6}$$

$$w_i = \sum_{u=1}^{m} \sqrt{\frac{q_u}{p_u(y)}} \delta\left[b\left(x_i^*\right) - u\right] \tag{7}$$

$$Y_1 \frac{\sum_{i=1}^{\text{nk}} x_i w_i g\left(\frac{\|y_0 - x_0\|}{h}^2\right)}{\sum_{i=1}^{\text{nk}} \left(\frac{\|y_0 - x_0\|}{h}^2\right)} \tag{8}$$

3.5 Proposed Algorithm

1. Convert the given avi video into number of frames. Declare variables and Kernel (Epanecknikov).
2. Read the first frame, select the object (Target Model or Template) to be tracked and store its spatial information such as X_0, Y_0, W and H.
3. Create the weighted histogram of color q_u and Compute the weighted histo grams of CCV *(c & d).*
4. Determine the Joint histogram of color and CCV ($q_u = q_u . * c$).
5. Read the next frame and crop the object (Target Candidate) as per the spatial information from previous step (X_0, Y_0, *W and H*).
6. Create the weighted histogram of color p_u and Compute the weighted histo grams of CCV *(c & d).*
7. Determine the Joint histogram of color and CCV ($p_u = p_u . * c$) and w_i using Eq. 7.
8. Calculate the Target's new location Y_1 using Eq. 8. Repeated five times for convergence by substituting $Y_0 = Y_1$.
9. Put the bounding box over the object using the spatial information from previous step (X_0, Y_0, *W and H*).
10. If number of frames not equal to zero-proceed with step 5.

3.6 Computation of Weighted Histogram of CCV

1. Read the object (Target Model or Target Candidate or Template).
2. Color space of object is converted into number of bins.
3. Label the image using connected components theory.
4. Determine the centroids of the labeled regions and estimate the number of pixels in each labeled region.
5. Separate coherent pixels and incoherent pixels (α and β) for each bins using threshold.
6. Calculate weighted histograms (c & d) using the above said Kernel. c is weighted histogram of coherent pixels and d is Weighted histogram of Incoherent pixels.
7. Normalize the histogram (c & d).

Table 1 Color histogram

SN	Dataset	TP*	FP**	FN***	DR+	FAR^^	Remarks
1	PETS2001	15	33	2	88	69	50 frames
2	dtneuWinter	21	16	3	87	43	40 frames

* TP = True Positive, **FP = False Positive, ***FN = False Negative, + DR = Detection Rate in %, ^^FAR = False Alarm Rate in %. Here, we attempted to narrate typical computation for the above said terminology. DR = (TP/(TP + FN))* 100, FAR = (FP/(TP + FP))* 100, DR = (15/(15 + 2))* 100 = 88.23 % ≈ 88 %, FAR = (33/(15 + 33))* 100 = 68.75 % ≈ 69

4 Experimental Results

Here, we address the difficulties by presenting a tracking algorithm that uses a simple symmetric similarity function between kernel density estimates of the template and target distributions in a joint spatial-feature space. Given sufficient samples, the kernel density estimation works well with both in low and high dimensions. The similarity measure is symmetric and the expectation of the density estimates centered on the model image over the target image. Using this similarity measure, we can derive a mean shift tracking algorithm with the histogram-based similarity measures. The proposed approach employed the information-theoretic similarity measure is more robust and more discriminative. An experiment on video sequence is conducted first with color histogram, second with CCV, and finally with joint histogram of color and CCV. We have conducted an experiment with different dataset to confirm the performance efficacy of the proposed mean shift approach. The computational aspects of the method occurred as the quadratic. The same is tested over the available machine Dual-core CPU of 2 GHz with MATLAB periphery.

In the first instance, we have conducted an experiment on PETS2001 (3) video sequence. Here the vehicle such as black car object is attempted to track as it has the dark background. It is observed that with color as feature the tracking fails and shown in Fig. 1a. With CCV histogram the result is better but with joint histogram of color and CCV outcome is best and robust which is portrayed through Fig. 1(b–d). Second an experiment leading via dtneuWinter (Traffic) multiple object video has been accomplished. Due to space limitation we are unable to include. Normally, single component such as color histogram is unable to discriminate the object of its same background color. As it can be observed in Fig. 1b, the bounding box is located other than the ground truth of the object. In other words, it may lead to mistrack. With CCV histogram the result (Fig. 1c) is better further with joint histogram the tracking is efficient. The bounding box is located over the object itself and can be seen in Fig. 1d.

Hence, it is quite possible through enhancement of traditional mean shift with the effort of statistical characteristics as exhibited by CCV which has extensively remarked through the experimental outcomes. In order to reveal the tracking performance, the following empirical metrics are considered such as True Positives and False Positives as shown in Tables 1, 2 and 3.

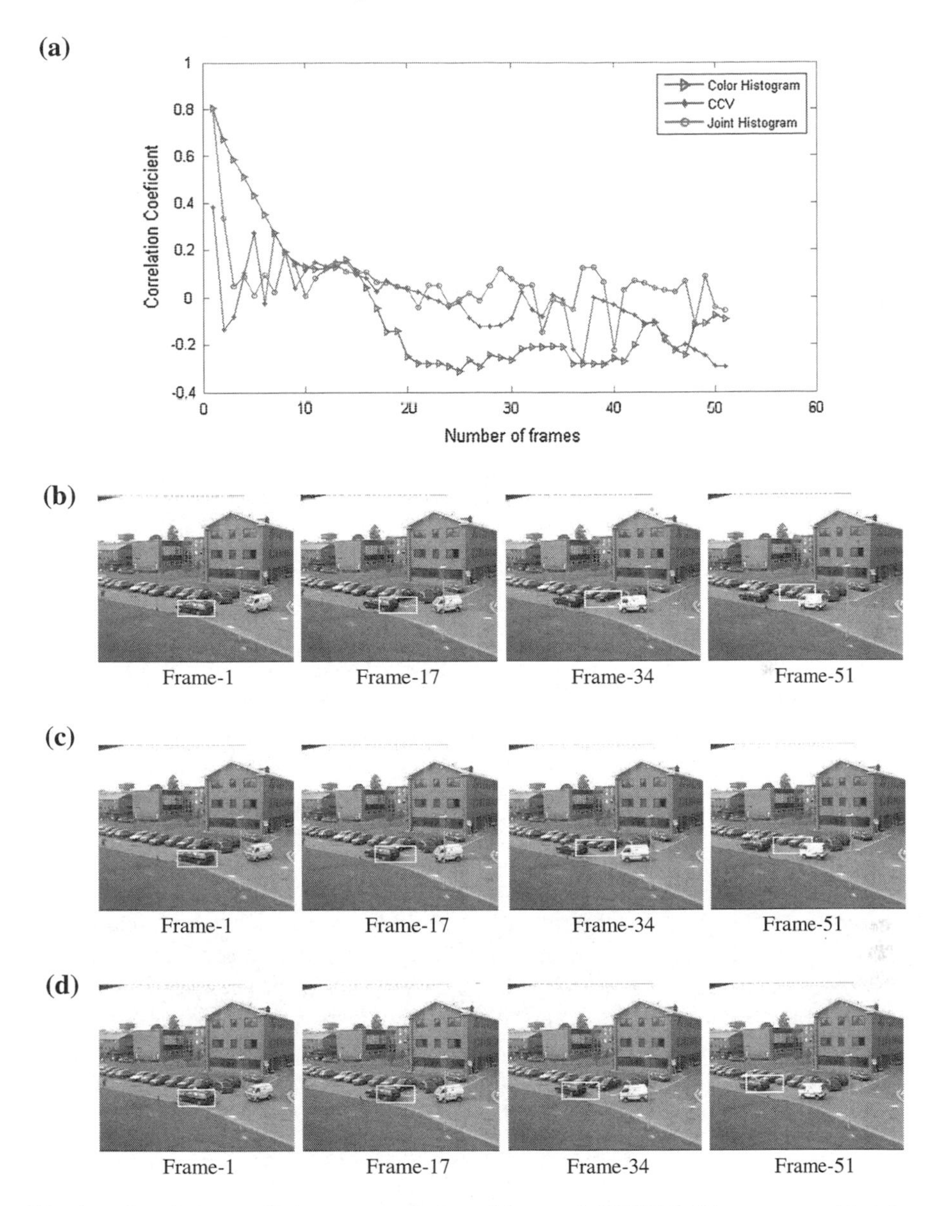

Fig. 1 **a** Correlation coefficient versus number of frames. **b** PETS2001(3) sequence with color histogram. **c** PETS2001(3) sequence with CCV histogram. **d** PETS2001(3) sequence with joint histogram of color and CCV

Table 2 CCV histogram

SN	Dataset	TP	FP	FN	DR	FAR	Remarks
1	PETS2001	46	1	3	94	2	50 frames
2	dtneuWinter	34	2	4	89	6	40 frames

Table 3 Joint histogram

SN	Dataset	TP	FP	FN	DR	FAR	Remarks
1	PETS2001	48	1	1	98	2	50 frames
2	dtneuWinter	36	2	2	95	5	40 frames

Table 4 Detection rate (DR)

SN	Dataset	DR for color hist	DR for CCV hist	DR for joint hist
1	PETS2001	88	94	98
2	dtneuWinter	87	89	95

From the Table 4 results such as DR helping to remark and increased monotonically through Color, CCV, and Joint histograms as an experimental evidences.The plot for CCV versus number of frames for PETS2001(3) dataset is shown in Fig. 1a. It witnessed the determination of correlation coefficient of template and candidate. On the other hand template and candidate extracted from first and subsequent frames, respectively. Further, color histogram, CCV histogram, and joint histogram of color and CCV are considered as the remarkable parameters to unearth the tracking outcome. In order to discriminate the correlation coefficient of three strategies through the appropriate distinct legends are quite essential. Hence, the experimental results of correlation coefficient represented through legends such as arrow, star, and circle are the correspondingly denoted. The three criteria said in the previous Para.

The performance of tracking algorithm through correlation coefficient for joint histogram of color and CCV has shown little supremacy over the color histogram and CCV histogram, which is exhibiting the accuracy of object tracking outcome. Furthermore, the same is demonstrated through the Fig. 1(b–d).

5 Conclusion

The proposed approach is simple symmetric similarity function between spatially smoothed kernel density estimates of the model and target distributions for object tracking. The resemblance measure is based on the expectation of the density estimates over the model or target image. Experimental results prove that they are precise, and reliable as compared to traditional methods.

To track the objects, the similarity function is maximized using the mean shift algorithm to iteratively find the local mode of the function. Since the similarity measure is an expectation taken over all pairs of the pixel between two distributions, the computational complexity is quadratic. The proposed method leads to a very efficient and robust nonparametric tracking algorithm has achieved better performance especially through the correlation coefficient of Target model and candidate.

References

1. Ajay, M., Sanjeev, S., Venkatesh, M.: A novel color coherence vector based obstacle detection algorithm for textured environments. The 3rd international conference on machine vision (2010)
2. Birchfield, S.: Elliptical head tracking using intensity gradients and color histograms. In: Proceedings of the IEEE conference on computer vision and pattern recognition, 232–237(1998)
3. Brian, V.: Funt, Graham, D., Finlayson.: Color constant color indexing. IEEE Trans. Pattern Anal. Mach. Intell. **17**(5), 522–529 (1995)
4. Chan Nguyen, V.: An efficient obstacle detection algorithm using color and texture. World Acad. Sci. Technol. **36**, 132–137 (2009)
5. Cheng, Y.: Mean shift, mode seeking, and clustering. IEEE Trans. Pattern Anal. Ma chine Intell. **17**, 790–799 (1995)
6. Comaniciu, D., Ramesh, V., Meer, P.: Real-time tracking of non-rigid objects using mean shift. In: Proceedings IEEE conference computer vision and pattern recognition **II**s 142–149 (2000)
7. Comaniciu, D., Ramesh, V., Meer, P.: PAMI. Kernel based object tracking **25**(5), 564–575 (2003)
8. Greg, P., Ramin, Z., Justin, M.: Comparing color images using color cohe rence vector. In: Proceedings 4th ACM international conference on multimedia (1997)
9. Han, B., Davis, L.: Object tracking by adaptive feature extraction. In: Proceedings of the IEEE conference on image processing, **3**, 1501–1504 (2004)
10. Ning, J., Zhang, L., Chengke W.: Robust object tracking using joint Co- lor_Texture histogram. Int. J. Pattern Recogn. Artif. Intell. **23**(7), 1245–1263 (2009)
11. Li, X., Wu, F.: Convergence of a mean shift Algorithm. J. Softw. **16**(3), 365–374 (2005)
12. Ning, J., Zhang, L., Zhang, D.: Robust mean shift tracking with corrected background- weighted histogram. IET-CV, (2010)
13. Swain, M., Ballard, D.: Color indexing. Int. J. Comput. Vision **7**(1), 11–32 (1991)
14. Wang, J., Yagi, Y.: Integrating shape and color features for adaptive real-time object tracking. In: Proceedings international conference on robotics and biomimetics, Kunming. 1–6 (2006)
15. Yilmaz, A., Javed, O., Shah, M.: Object tracking. A survey. ACM Comput. Surv. **38**(4), 13, (2006). doi:10.1145/1177352.1177355

A Rubric Based Assessment of Student Performance Using Fuzzy Logic

Meenakshi G. and Manisharma V.

Abstract Assessment of student performance is one of the important tasks in the teaching and learning process. It has great impact on the student approach to learning and their outcomes. Evaluation of student learning in different activities is the process of determining the level of performance of students in relation to individual activities. A rubric is a systematic scoring guideline to evaluate student performance (assignment, quiz, paper presentation, open and closed book test etc.) through the application of detailed description of performance standards. After giving the specific task the student should be explained about the criteria and evaluation points. This allows the students to be aware of the performance and they will try to improve the performance. In this paper the main focus is on student centered learning activities which are mainly based on multiple intelligences. This paper presents an integrated fuzzy set approach Lotfi et al. [2] to assess the outcomes of student-centered learning. It uses fuzzy set principles to represent the imprecise concepts for subjective judgment and applies a fuzzy set method to determine the assessment criteria and their corresponding weights.

Keywords Student centered learning · Fuzzy logic · Active learning · Passive learning · Multiple intelligences

G. Meenakshi (✉) · V. Manisharma
Department of Computer Science, Nalla Malla Reddy Engineering College, Hyderabad, Andhra Pradesh, India
e-mail: meena_ganti@yahoo.com

V. Manisharma
e-mail: manisharmavittapu@gmail.com

B. V. Babu et al. (eds.), *Proceedings of the Second International Conference on Soft Computing for Problem Solving (SocProS 2012), December 28–30, 2012*, Advances in Intelligent Systems and Computing 236, DOI: 10.1007/978-81-322-1602-5_59, © Springer India 2014

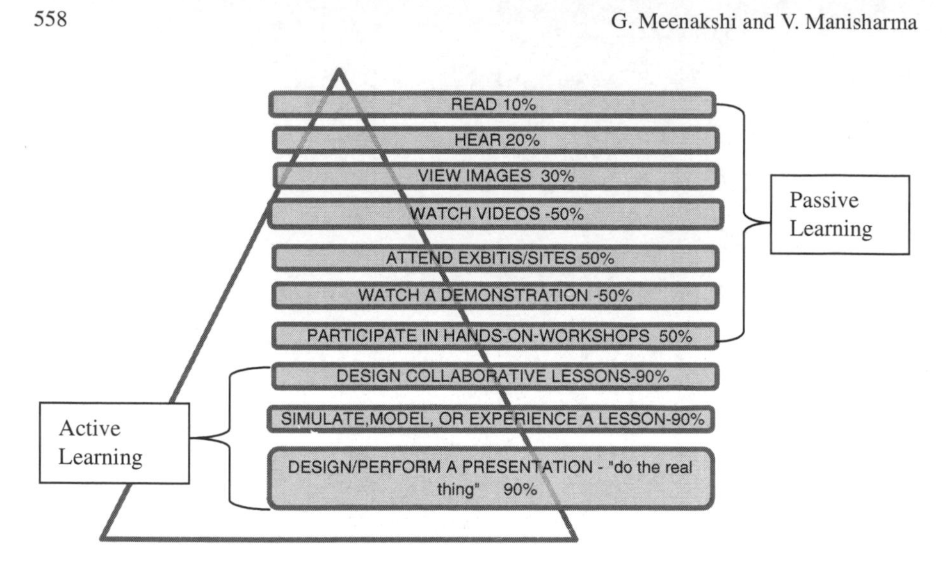

Fig. 1 The Cone of active and passive learning

1 Introduction

Currently the shortage of skilled people at work place is one of the major problems faced by the Indian economy. It was found that there was a large gap between the current education standards and employability skills. Several studies have shown that students prefer strategies promoting active learning to traditional lectures [5].

This is a passive method of teaching which results in less retention of knowledge by students as indicated in the cone of learning shown in the Fig. 1.

2 Proposed Methodology

Howard Gardner's Multiple Intelligences (MI) theory has sparked a revolution in the educational field. According to him each person is unique and a blend of intelligences namely Linguistic, Logical -Mathematical, Spatial, Musical, Interpersonal, Intrapersonal, Naturalistic. People differ in the intelligences. According to multiple intelligence the course contents must be designed to meet the activity based activities so that the students can actively participate in the classroom [4] (Fig. 2).

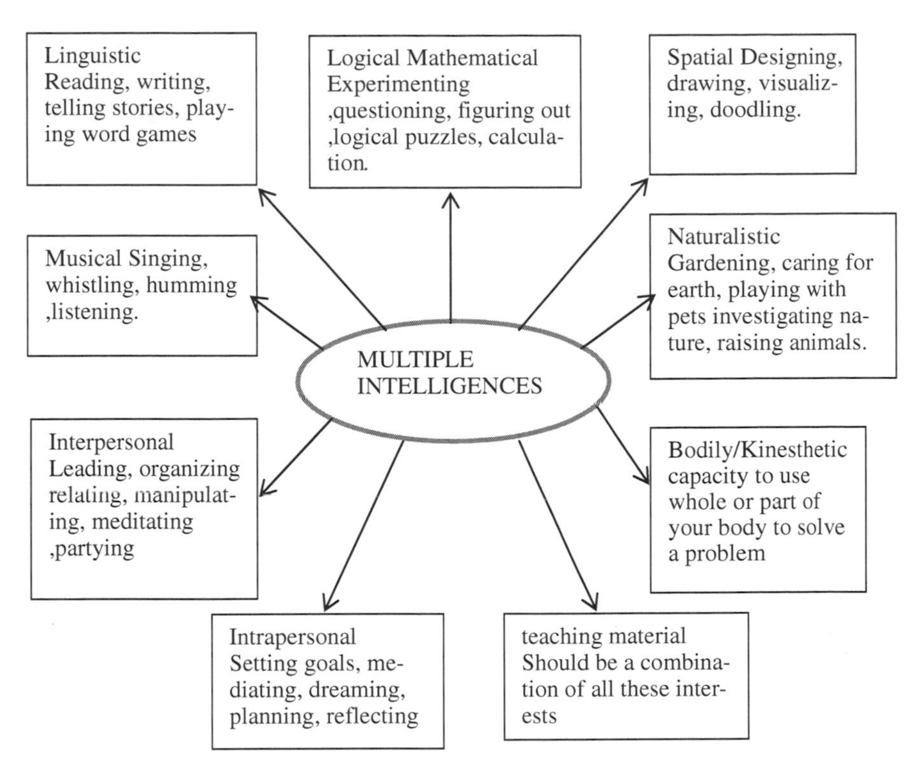

Fig. 2 Multiple intelligences

2.1 Student Assessment Model

To improve and develop the student learning [3], staff are required to prepare a list of tasks to be implemented are prepared, different tasks are assigned to faculty members at the commencement of the year/semester the evaluation of the students in that particular period is evaluated [1].

2.2 Evaluating Steps: The Following Steps are Followed for Evaluating the Student Performance

Student assessment model can be implemented in academics also as shown in [6] Fig. 3.

Step 1: Normalize the marks

The marks obtained by each of the student have to be converted to the normalized values. Normalized value is referred to a value in a range of [0, 1]. It can be obtained

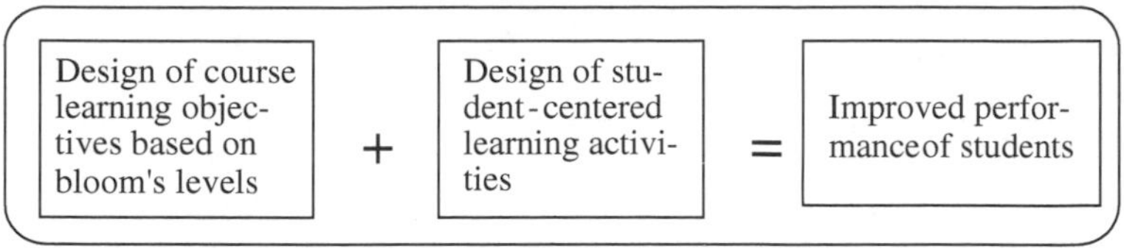

Design of course learning objectives based on bloom's levels + Design of student-centered learning activities = Improved performanceof students

Fig. 3 Student evaluation method

Table 1 Normalized value

Criteria	Total marks	Mark obtained	Normalized value
Poster/chart presentation ($C1$)	100	70	0.70
Subject based quiz ($C2$)	100	80	0.80
Open book/closed book test ($C3$)	100	60	0.60
Power point presentation/seminars ($C4$)	100	80	0.80
Assignments ($C5$)	100	90	0.90

by dividing the mark for each criterion with the total mark. This will be the input value of this evaluation. Table 1 tabulates the example marks and the normalized values obtained by a student for all the criteria.

Step 2: Developed the graph of the Fuzzy Membership Function [2]

The graph of membership function is developed in order to execute the fuzzification process. In this process, the input value is mapped into the graph of membership function to obtain the fuzzy membership value of that particular input value. Each membership value will represent the level of satisfaction.

Table 2 shows five satisfaction levels that have proposed in this study. It is based on the linguistic term. The degree of satisfaction shows the range of marks for each satisfaction level which are based on some grading system. The maximum degree of satisfaction denoted by T(xi) describes the mapping function for corresponding satisfaction level where T(xi) ->[0,1].

Step 3: calculate the degree of satisfaction

The degree of satisfaction of jth criteria which denoted by $D(C_j)$ is evaluated by x

$$\frac{D(C_j) = y_1 * T(x_1) + y_2 * T\ (x_2) + \cdots + y_5 * T(X_4)}{y_1 + y_2 + \cdots y_5} \tag{1}$$

where y_i = degree of membership value for each satisfaction level, y_1

Step 4: Compute the final rank

The final mark for kth student denoted by $F(\mathrm{Sk})$ is calculated using the formula below:

$$\frac{F(\mathrm{Sk}) = w_1 * D\ (C1) + w_2 * D\ (C2) + \cdots w_5 * D\ (C5)}{w_1 + w_2 + \cdots w_5} \tag{2}$$

Table 2 Satisfaction levels

Satisfaction levels (Xi)	Degrees of satisfaction	Maximum degrees of satisfaction T(xj)
Exemplary	80–100 % (0.8–1.0)	1.0
Accomplished	70–79 % (0.70–0.79)	0.79
Delveloping	60–69 % (0.6–0.69)	0.69
Beginning	40–59 % (0.4–0.59)	0.59
Fail	0–39 % (0–0.39)	0.39

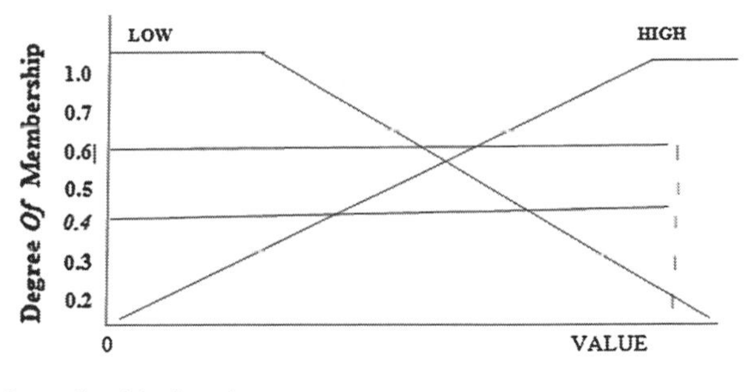

Fig. 4 Graph of membership function

where w_i = the total marks of ith criteria for $i = 1, 2, \ldots 5$.

The result obtained is put into the fuzzy grade sheet (Table 3) in the appropriate columns.

NUMERICAL EXAMPLE: As an illustration, the example mark of a student is taken (as in Table 1). The student is evaluated based on procedure mentioned earlier. The graph of membership function is generated to execute the fuzzification process in step 2 is shown in Fig. 4.

Based on the Fig. 1 we can see the satisfaction level of ACCOMPLISHED and DEVELOPING that represent the degree of membership 0.4 and 0.6 respectively. The degree of satisfaction regarding criterion 1 is calculated as follows:

$$D\,(Ci) = (0.4 * 1.0 + 0.6 * 0.79)\,/\,(0.4 + 0.6) = 0.874$$

The same procedure is applied for calculating the $D(C1)$, $D(C2)$, $\ldots D(C5)$
Finally, the final mark earned by the student for all criteria is computed using (2).

$$\begin{aligned} F\,(S1) &= 100 * 0.874 + 100 * 0.752 + 100 * 0.960 + 100 * 0.704 \\ &\quad + 100 * 0.672/500 \\ &= 0.7924 = 0.79 \end{aligned}$$

Based on the final mark obtained, the student is awarded by the fuzzy linguistic terms of Accomplished at 0.19 (UAc) Exemplary at 0.83 (Uex = 0.81). These values are obtained from the graph of the membership function (as in Fig. 1). Besides that the

Table 3 Final marks of a student

Student	Criteria	Fuzzy membership value					Degree of satisfaction	Final mark
1		E	A	D	B	F		0.801
	$C1$	0.4	0.6	0	0	0	0.874	
	$C2$	0	0.62	0.38	0	0	0.752	
	$C3$	0.81	0.19	0	0	0	0.960	
	$C4$	0	0.57	0.43	0	0	0.747	
	$C5$	0.2	0	0	0.8	0	0.672	

Table 4 Comparison of fuzzy and non fuzzy

	Non-fuzzy method		Fuzzy-evaluation method	
	Final mark	Linguistic term	Final mark	Linguistic term
1	70	Accomplished	0.79	Exemplary at 0.4 Accomplished at 0.6
2	80	Exemplary	0.90	Accomplished at 0.62 Developing at 0.38
3	70	Accomplished	0.73	Exemplary at 0.17 and Accomplished at 0.83
4	40	Beginner	0.59	Developing at 0.43 and Accomplished at 0.57
5	60	Developing	0.71	Exemplary at 0.2 and Beginner at 0.8

final mark also can be valued as 78.69 (by multiplying with 100 %) which represent the linguistic term of Exemplary. The details of the fuzzy marks obtained from this evaluation procedure as shown in Table 3.

Since most of the computation used in fuzzy evaluation method is made as fuzzy sets with the range of [0, 1], therefore the results obtained from this method as shown in the Table 3 are in the range of [0, 1] only. The final mark earned by student 1 is 0.79, but after converted in terms of percentage, it will be 79 % Table 4.

3 Conclusion and Future Direction

Overdependence on lecturing and other instructor-centered learning strategies often fosters a passive learning attitude and mental disengagement during class. Effective instructors regularly utilize more student-centered strategies that are benefitial to students.

Acknowledgments I am heartily thankful to Dr. Mrs. Divya Nalla , Principal, Nalla Mallareddy Engineering College Whose encouragement and support resulted in the preparation of paper.I am thankful to our research Director Prof. Vara Prasasd, Dept HOD and ASSOC. Prof. P.V.S Siva Prasad and Mr. Mani Sarma Assoc. Prof for their encouragement guidance and support from the initial to the final level.

References

1. Jian, M., Duanning, Z.: Fuzzy set approach to the assessment of student-centered learning. IEEE Trans. Educ. **43**(2), 237 (2000)
2. Lotfi A. Zadeh, L.A.: Fuzzy logic. Berkeley
3. Chickering, A.W., Gamson, Z.F.: Seven principles of good practice. AAHE Bull. **39**, 3–7 (1987). (ED 282 491. 6 pp. MF-01; PC-01)
4. Lowman, J.: Mastering the techniques of teaching. Jossey-Bass, San Francisco (1984)
5. McKeachie, W.J., Pintrich P.R., Lin Y.-G., Smith, D.A.F.: Teaching and learning in the college classroom: a review of the research literature. Regents of The University of Michigan, Ann Arbor (1986). (ED 314 999. 124 pp. MF-01; PC-05)
6. Penner, J.G.: Why many college teachers cannot lecture. Charles C. Thomas, Springfield (1984)

Representation and Classification of Medicinal Plants: A Symbolic Approach Based on Fuzzy Inference Technique

H. S. Nagendraswamy and Y. G. Naresh

Abstract In this paper, a method of representing shape of medicinal plant leaves in terms of interval-valued type symbolic features is proposed. Axis of least inertia of a shape and the fuzzy equilateral triangle membership function is exploited to extract features for shape representation. Multiple class representatives are used to handle intra class variations in each species and the concept of clustering is used to choose multiple class representatives. A simple nearest neighbor classifier is used to perform the task of classification. Experiments are conducted on the standard flavia leaf dataset to demonstrate the efficacy of the proposed representation scheme in classifying medicinal plant leaves. Results of the experiments have shown that the method is effective and has achieved significant improvement in classification accuracy when compared to the contemporary work related to leaf classification.

Keywords Axis of least inertia · Fuzzy equilateral triangle · Symbolic data · Shape classification · Shape representation

1 Introduction

Medicinal plants play an important role in Ayurvedic system of medicine. This system emphasizes not only on curing but also on the prevention of the recurrence of diseases. Certain diseases can be treated effectively using Ayurvedic medicine and it has been practiced since ancient years. Though several medicinal plants are available in our surrounding environment, we are not able identify and make use of them for preparation of simple house-hold medicines for common diseases in day-to-day life.

H. S. Nagendraswamy · Y. G. Naresh (✉)
DoS in Computer Science, University of Mysore, Mysore, India
e-mail: swamy_hsn@yahoo.com

Y. G. Naresh
e-mail: naresh.yg@gmail.com

B. V. Babu et al. (eds.), *Proceedings of the Second International Conference on Soft Computing for Problem Solving (SocProS 2012), December 28–30, 2012*, Advances in Intelligent Systems and Computing 236, DOI: 10.1007/978-81-322-1602-5_60, © Springer India 2014

Since the taxonomy of medicinal plants is very vast, even the experts may find it difficult to identify and classify them effectively. Even though there are many biological methods to identify the plant species, it is worth exploring image processing and pattern recognition techniques to create a knowledge base of medicinal plants and to provide an automated support for identification and classification of medicinal plants by analyzing the visual properties such as shape, texture, internal vein patterns, and color of the leaves.

2 Related Work

Shape of a plant leaf plays a major role in identifying and classifying the plant species. In the literature, both contour-based and region-based methods have been proposed to characterize the shape and experimented on leaf datasets.

In [1], the region-based morphological features of a leaf viz., aspect ratio, rectangularity, area ratio of convex hull, perimeter ratio of convex hull, sphericity, circularity, eccentricity, form factor, and invariant moments are used for its description. The concept of physiological width and physiological length of a plant leaf along with other geometric and region-based morphological features have been proposed in [2] for characterizing shape of a plant leaf. In [3], landmark points are considered for polygon approximation. Given any two points in the set of landmark points, Inner-distance between those two points as a replacement to Euclidean distance is computed to build shape descriptors, for classification. In [4], a method of characterizing a shape by modeling shape contour as a complex network through multiscale fractal dimension is proposed. In [5], Fuzzy equilateral triangle membership values have been proposed to describe a shape in terms of multiinterval-valued type features. In [6], Contour-based symbolic representation method has been proposed to describe the shape curve using string of symbols. In [7], symbolic representation in terms of multiinterval-valued type features has been proposed to characterize two dimensional shapes. In [9], Histogram of gradients is used as shape descriptors for characterizing leaf shape.

From the literature survey, it has been observed that several methods have been proposed to effectively represent shape of a plant leaf and to improve the accuracy of classification results. Despite several methods available for the task of classification of leaves, still there is a scope for exploring new methods to improve the accuracy of classification results through realistic and effective representation techniques. It has also been observed from the survey that the representation techniques based on the concept of fuzzy symbolic data have shown good performance when compared to the representation techniques based on conventional data. This is due to the fact that the symbolic data are more unified in terms of relationship and can capture the variations more effectively which is very much essential for most of the real-life objects and the concept of fuzzy theory is a natural choice for handling vagueness and uncertainty in representation. However, only few attempts [5, 7] have been made in this direction which motivated us to think of exploring the unconventional representation technique for representing shape of medicinal plant leaves for their classification. Thus, in this work, we made an attempt to propose a representation technique based on the Fuzzy Inference Technique and the concept of symbolic data. The experimental study has revealed that the proposed representation technique has shown significant improvements in the accuracy of classification and is comparable with the contemporary work in this direction.

3 Proposed Method

Extracting relevant and discriminative features for characterizing a leaf shape is an important step in any recognition or classification task. In this section, a method of extracting interval-valued features adopting axis of least inertia of a shape and the concept of fuzzy equilateral triangle is explained.

3.1 Feature Extraction

The color images of plant leaves are first converted into binary images. Then the contour of binary images is obtained by using a suitable contour extraction algorithm. The extracted closed contour serves as a shape curve of a plant leaf image.

The proposed feature extraction technique adopts the axis of least inertia of a shape to preserve the orientation of a shape curve which is very important to extract features invariant to geometric transformations (Rotation, translation, and scaling). The details regarding the computation of the axis of least inertia of a shape curve can be found in [7].

Once the axis of least inertia of a shape is computed, all the points of shape contour are projected onto the axis and the two farthest points are obtained. Figure 1 shows a shape with axis of least inertia and two extreme points E_1 and E_2. Let PE_1 and PE_2 be the two contour points on the shape curve obtained by projecting the two extreme points E_1 and E_2 of the axis of least inertia as shown in the Fig. 1. The shape contour is traversed in clockwise direction keeping either PE_1 or PE_2 as the starting point. In order to identify this starting point, the Euclidean distance between the PE_1 and the shape centroid and the distance between PE_2 and the shape centroid are computed. The shortest distance among the two is considered as a starting point. In some cases, there is a possibility that these two distances may be same and leads to ambiguity in selecting the starting point. In such case, we resolve the conflict by considering the horizontal width of the shape at subsequent points on the axis starting from the two extreme points.

Keeping either PE_1 or PE_2 as a starting point and traversing in clockwise direction, the contour is split into 'k' equal number of segments. The centroid of each curve segment is obtained by taking the average of all the pixels of the curve segment. The centroids so obtained serve as feature points on the shape curve.

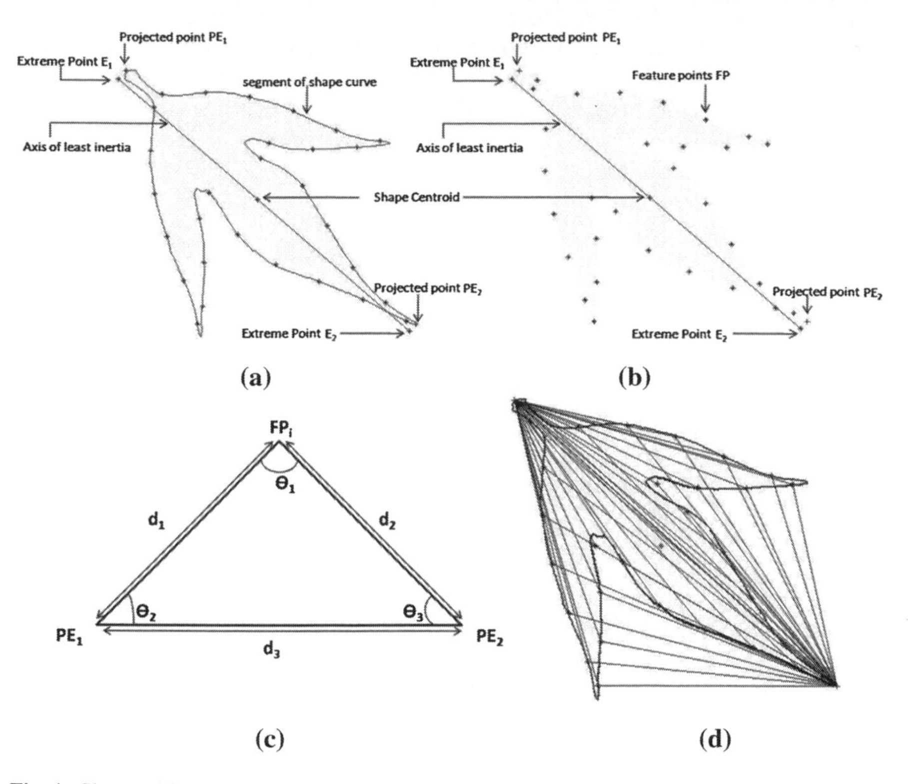

Fig. 1 Shape with axis of least inertia and two extreme points

The idea behind obtaining the local centroids of a shape curve is to take into consideration all the pixels information to define a feature point rather than considering a corner points or dominant points on the shape curve, which are not consistent and robust due to noise and shape transformation. Figure 1b shows the feature points obtained for a shape curve shown in Fig. 1a.

Once the feature points on the shape curve are obtained, we traverse the shape curve in clockwise direction and at every feature point FP_i for $i = \{1, 2, 3, \ldots, k\}$, we form a triangle considering the two extreme points PE_1 or PE_2. We use the fuzzy inference technique [10] to approximate the triangle as equilateral and the corresponding membership value is computed as follows.

Let θ_1, θ_2 and θ_3 be the inner angles of the triangle in the order $\theta_1 > \theta_2 > \theta_3$. Let U be the universe of the triangle.

$$\mathrm{U} = \left\{(\theta_1, \theta_2, \theta_3) / \theta_1 \geq \theta_2 \geq \theta_3 \geq 0; \theta_1 + \theta_2 + \theta_3 = 180^\circ\right\} \tag{1}$$

Let d_1, d_2, d_3 be the Euclidean distances between PE_1 and FP_i, PE_2 and FP_i and PE_1 and PE_2, respectively as shown in the Fig. 1c. The inner angles of a triangle are computed as follows:

$$\theta_1 = \cos^{-1}\left(\frac{d_2^2 + d_3^2 - d_1^2}{2 * d_2 * d_3}\right) \tag{2}$$

$$\theta_2 = \cos^{-1}\left(\frac{d_1^2 + d_3^2 - d_2^2}{2 * d_1 * d_3}\right) \tag{3}$$

$$\theta_3 = \cos^{-1}\left(\frac{d_1^2 + d_2^2 - d_3^2}{2 * d_1 * d_2}\right) \tag{4}$$

The membership value of the triangle which approximates equilateral triangle is computed as

$$\mu\left(\theta_1, \theta_2, \theta_3\right) = 1 - \frac{1}{180}\left(\theta_1 - \theta_3\right) \tag{5}$$

The membership values μ_i (for $i = 1$ to k) are computed and considered as feature values to describe a shape.

3.2 Feature Representation

Since the shape of leaves in a class may vary due to size, maturity, and other biological facts, there will be significant intraclass variations among the shapes. To handle this case, we propose to have multiple representatives for a class by grouping similar shapes into one group and choose a representative for that group within the class. We have used hierarchical clustering technique to group similar shapes in a class by utilizing inbuilt Matlab function. The natural groups in a class are identified using the inconsistency coefficient for each link of the hierarchical cluster tree yielded by the respective linkage. Once the natural groups are obtained for each plant species, we consider 60 % of the samples from each group to form a class representative to be stored in the knowledge base during training and remaining 40 % of the samples for testing. A representative feature vector for a particular group within the class is obtained by aggregating the corresponding features of the shapes in the group to form an interval-valued type feature. Thus, the feature vector representing a group within a class is of interval-valued type rather than crisp. Thus in the proposed approach, the shape of plant leaves in the knowledge base are represented in the form k-dimensional Interval-valued type feature vector.

Lower the intraclass variations in the class C_i, fewer the number of groups obtained and higher the intraclass variations, more the number of groups. Hence the number of groups in a class C_i is directly proportional to its intraclass variations.

Let $[G_1, G_2, G_3 \ldots, G_m]$ be the 'm' number of group formed in a class. Let $[S_1, S_2, S_3 \ldots, S_P]$ be the leaf shapes in the group. Each shape in S_i for $i = \{1, 2, 3, ..., P\}$ in a group is represented by feature vector of dimension k i.e., $S_i = \{F_{i1}, F_{i2}, F_{i3} \ldots, F_{ik}\}$. To form a group representative vector, we aggregate the

Table 1 An example of k-dimensional Interval-valued feature vector representing a group in the knowledge base.

Shapes in a group	1	2	…	…	k
S_1	0.2204	0.1793	…	…	0.1542
S_2	0.1216	0.1503	…	…	0.1536
..	…	…	…	…	…
S_p	0.1246	0.1263	…	…	0.0899
Interval type feature value	[0.1216, 0.2204]	[0.1263, 0.1793]	…	…	[0.0899, 0.1542]

corresponding features of the shape S_i to form an interval by choosing the minimum and maximum of corresponding values feature values. Table 1 shows an example of k-dimensional interval-valued feature vector representing a group in the knowledge base as explained earlier.

3.3 Matching and Classification

Let TS $= \{F_1, F_2, F_3, \ldots, F_k\}$ be the k-dimensional feature vector representing the shape of a plant species to be classified. Let MS $= \{[F_{si}^-, F_{si}^+]\}$ for $s = \{1, 2, 3, \ldots, N\}$ where N is the number of model shapes and $i = \{1, 2, 3, \ldots, k\}$ be the k-dimensional interval-valued feature vector representing jth model shape in the knowledge base pertaining to a particular class of a plant species.

We use the similarity measure proposed in [7] to find the similarity score between the test shape and the model shape as

$$\text{Sim}\,(\text{TS},\ \text{MS}) = \frac{1}{K}\sum_{i=1}^{K} \text{SV}_i \tag{6}$$

where

$$\text{SV}_i = \left\{ \begin{array}{l} 1 \quad if(F_{Si}^- \le F_i \le F_{Si}^+) \\ else \\ \frac{1}{2}\left[\frac{1}{1+abs(F_{Si}^- - F_i)} + \frac{1}{1+abs(F_{Si}^+ - F_i)}\right] \end{array} \right\}$$

The test leaf shape is compared with all the model (reference) shape of the plant species in the knowledge base. The label of model (reference) shape which possesses the highest similarity value is assigned to the test leaf shape.

4 Experimental Results

We have implemented our method using Matlab R2011b. We have conducted experiments on the dataset provided by [2]. All the plant species in the dataset have medicinal values. Thus we have called them as medicinal plant species. We considered 30 plant species for our experiments from the dataset [2]. In our experiments, the 'complete' linkage is chosen empirically for finding the compact groups in every class of leaf shape. The training and testing set of samples are considered as explained in the Sect. 3.2. The performance of the proposed method is evaluated in terms of accuracy, precision, recall, and F-measure for varying values of k and the values are tabulated in Table 2. From Table 2, it is observed that the proposed method has shown highest average Accuracy, Precision, Recall, and F-measure respectively for $k = 40$. The following Table 2 gives Average Accuracy, Precision, Recall, and F-measure, respectively, for various values of k.

Since we have used the same dataset used by [2], we are comparing the performance of the proposed methodology with that of [2]. The authors in [2] have obtained the overall classification accuracy of 90.13 % by observing the number of leaves correctly classified over the total number of leaves considered for experiment. Accordingly, the proposed method has shown the overall classification accuracy of 92.29 %. This reveals that the proposed methodology shows better performance compared to [2] Table 3.

Figure 2 shows example leaves belonging to different species but possess similar shape structures. Thus it is very difficult to classify such leaves just by considering only shape information.

Table 2 Performance measures of the proposed method

No. of features	Average accuracy	Average precision	Average recall	Average F-measure
30	0.9919	0.9062	0.8792	0.8853
40	**0.9949**	**0.9315**	**0.9229**	**0.9246**
50	**0.9943**	**0.9242**	**0.9146**	**0.9157**
60	**0.9943**	**0.9275**	**0.9146**	**0.9172**
70	0.9928	0.9413	0.8917	0.9047
80	0.9931	0.9386	0.8958	0.9082
90	0.9924	0.9307	0.8854	0.8985

Table 3 Comparison for the method proposed in [2]

Scheme	Classification accuracy %
Proposed scheme	92.29
Stephen et al. [2]	90.13

Fig. 2 Leaves belonging to different species possessing similar shape structure

5 Conclusion

In this work, we proposed a novel method of representing medicinal plant leaves for classification. The method exploits the concept of fuzzy inference technique and symbolic data analysis for effective representation. Experiments are conducted on the standard database of considerably large size and the performance of the proposed method is evaluated in terms of standard performance evaluation measures. The results are more encouraging and comparable with that of state of the art work. However, the approach may fail to classify accurately the plant leaves belonging to different species possessing similar shape structure as shown in Fig. 2. So, we shall explore the texture-based representation techniques in such cases. Also, the multistage classifier techniques incorporating shape and texture features will be explored to further improve the classification accuracy.

References

1. Du, Xiao-Feng Wang,Ji-Xiang, : Zhang, Guo-Jun: Leaf shape based plant species recognition. Appl. Math. Comput. **185**, 883–893 (2007)
2. Wu, S.G., Bao, F.S., Xu, E.Y., Wang, Y.-X., Chang, Y.-F., Shiang, C.-L.: A leaf recognition algorithm for plant classification using probabilistic neural network. In: IEEE 7th International Symposium on Signal Processing and Information Technology, Cario, Egypt (2007).
3. Ling, H., Jacobs, D.W.: Shape classification using the inner distance. IEEE Trans. Pattern Anal. Mach. Intell. **29**(2), 286–299 (2007)

4. Backes, A.R., Bruno, O.M.: Shape classification using complex network and multi-scale fractal dimension. Pattern Recogn. Lett. **31**, 44–51 (2010)
5. Nagendraswamy, H.S., Guru, D.S.: A new method of representing and matching two dimensional shapes. Int. J. Image Graphics **7**(2), 337–405 (2007)
6. Daliri, M.R., Torre, V.: Robust symbolic representation for shape recognition and retrieval. Pattern Recogn. **41**, 1782–1798 (2008)
7. Guru, D.S., Nagendraswamy, H.S.: Symbolic representation of two dimensional shapes. Pattern Recogn. Lett. **28**, 144–155 (2007)
8. Tsai, D.M., Chen, M.F.: Object recognition by linear weight classifier. Pattern Recogn. Lett. **16**, 591–600 (1995)
9. Xiao, X., Hu, R., Zhang, S., Wang, X.: HOG-based approach for leaf classification. In: ICIC, vol. 2, pp. 149–155 (2010)
10. Ross, T.J.: Fuzzy Logic with Engineering Applications, 2nd edn. Wiley, Hoboken (2004)

Fractal Image Compression Using Dynamically Pipelined GPU Clusters

Munesh Singh Chauhan, Ashish Negi and Prashant Singh Rana

Abstract The main advantage of image compression is the rapid transmission of data. The conventional compression techniques exploit redundancy in images that can be encoded. The main idea is to remove redundancies when the image is to be stored and replace it back when the image is reconstructed. But the compression ratio of this technique is quite insignificant, and hence is not a suitable candidate for an efficient encoding technique. Other methods involve removing high frequency Fourier coefficients and retaining low frequency ones. This method uses discrete cosine transforms(DCT) and is used extensively in different flavors pertaining to the JPEG standards. Fractal compression provides resolution-independent encoding based on the contractive function concept. This concept is implemented using attractors (seed) that are encoded/copied using affine transformations of the plane. This transformation allows operations such as, skew, rotate, scale, and translate an input image which is in turn is extremely difficult or impossible to perform in JPEG images without having the problem of pixelization. Further, while decoding the fractal image, there exist no natural size, and thus the decoded image can be scaled to any output size without losing on the detail. A few years back fractal image was a purely a mathematical concept but with availability of cheap computing power like graphical processor units (GPUs) from Nvidia Corporation its realization is now possible graphically. The fractal compression is implemented using MatLab programming interface that runs on GPU clusters. The GPUs consist of many cores that together give a very high computing speed of over 24 GFLOPS.The advantage of fractal compression

M. S. Chauhan
Research Scholar, Pacific University, Udaipur, Rajasthan, India
e-mail: muneshchauhan@gmail.com

A. Negi
Department of CSE, G.B. Pant Engineering College, Uttarakhand, India
e-mail: ashish.ne@gmail.com

P. S. Rana (✉)
Research Scholar, Department of ICT, IIITM, Gwalior, MP, India
e-mail: psrana@gmail.com

B. V. Babu et al. (eds.), *Proceedings of the Second International Conference on Soft Computing for Problem Solving (SocProS 2012), December 28–30, 2012*, Advances in Intelligent Systems and Computing 236, DOI: 10.1007/978-81-322-1602-5_61,

can have varied usage in satellite surveillance and reconnaissance, medical imaging, meteorology, oceanography, flight simulators, extra-terrestrial planets terrain mapping, aircraft body frame design and testing, film, gaming and animation media, and besides many other allied areas.

Keywords GPU · Fractal · Attractor · Affine transformation · PIFS

1 Introduction

Image/video data constitute a major chunk that is transmitted over networks. Since image and video data consume more memory as compared to text data, they are generally transmitted in compressed form. The main aim for an efficient multimedia data transfer is to retrieve almost the same quality of the data at the receiver-end that was originally transmitted. Considering the availability of the present algorithms for multimedia data compression, it still remains a challenge. Some of the standards that are in vogue are JPEG [14] (still images), MPEG [20] (motion video images), H.261 [15] (Video telephony on ISDN lines), and H.263 [12, 18] (Video telephony on PSTN lines). All these formats compress data signals using Discrete Cosine Transform (DCT) [7]. Since all of these compressions standards are lossy, the image and video data do not retain the original quality. This is acceptable to an extent as human eye is not able to discern the pixels loss if it is limited to a certain threshold. The major issue arises, if the compressed image is scaled up. This leads to pixelization. Fractal images till recently have been researched in the domain of mathematics. Fractal images are derived exploiting a common property of a real image, i.e., self-similarity [8, 11]. Most partitions of a real image show extensive self-similarity. As a result, the original image can be represented as a finite set of contractive affine transformations using partitioned iterated function systems (PIFS).

2 Nature of Fractals

Fractal images are found in nature in great abundance [3]. Undulated coastlines, mountains chains, ferns, and also galaxies too represent a striking amount of self-similarity, which can be usually represented by fractals. Fractals have applications in gaming, animation, and science-fiction films where the terrains/landscapes are artificially created [16].

The major disadvantage of a fractal image is its encoding time. The time complexity of encoding a fractal image is $O(n^4)$ [14, 19]. The decoding part is very efficient and almost instantaneous [17]. Moreover, the fractal images are resolution-independent and they do not show pixelization unlike JPEG images. The scaled up image contains the same level of details as in the original image.

3 Encoding Images Using Partitioned Iterated Function System (PIFS)

The mathematical equivalent of partitioned copying machine is termed as PIFS. The transformations in PIFS are affine transformations. These affine transformations consist of two spatial dimensions and a third dimension that is the grey level. In sum, the transformations w_i may be defined as in Eq. 1 where s_i denotes contrast and o_i the brightness.

$$w_i \begin{bmatrix} x \\ y \\ z \end{bmatrix} = \begin{bmatrix} a_i & b_i & 0 \\ c_i & d_i & 0 \\ 0 & 0 & s_i \end{bmatrix} \begin{bmatrix} x \\ y \\ z \end{bmatrix} + \begin{bmatrix} e_i \\ f_i \\ o_i \end{bmatrix} \tag{1}$$

Thus the spatial part v_i of the image may be defined as in Eq. 2.

$$v_i(x, y) = \begin{bmatrix} a_i & b_i \\ c_i & d_i \end{bmatrix} \begin{bmatrix} x \\ y \end{bmatrix} + \begin{bmatrix} e_i \\ f_i \end{bmatrix} \tag{2}$$

Different partitions of the image as shown in Fig. 1 are considered as D_i and R_i lying in the plane above the image. Each w_i is restricted to $D_i \times I$ (the vertical space above D_i).

Hence $vi(Di = Ri)$. Thus, we need $W(f)$ as the image with the conditions as given in Eq. 3.

$$\begin{aligned} &\bigcup R_i = I^2 \\ &R_i \bigcap R_j = \emptyset \text{ when } i \neq j \end{aligned} \tag{3}$$

As shown in Eq. 3, we obtain single-valued function above each point of the square I^2. In addition to it each copy is nonoverlapping in terms of the copying machine metaphor.

According to Contractive Mapping Theorem [5], a fixed point (attractor) of a given image f is described as $W(f) = f$. We must guarantee that W has a unique fixed point in order for an image to be encoded. It has been found that for $s_i < 1.2$ (where s_i is defined as the contrast factor) can be considered as a safe limit for an

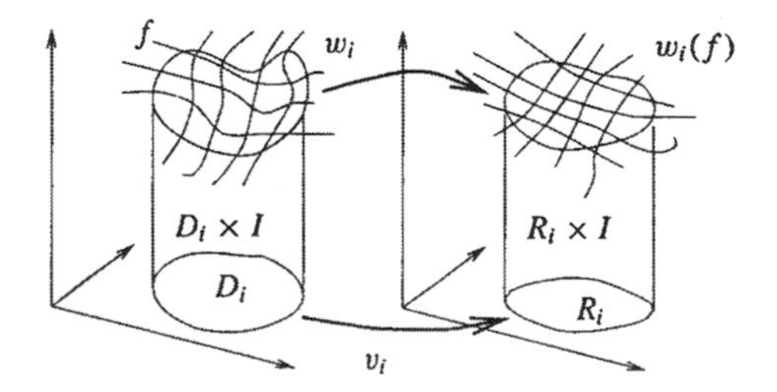

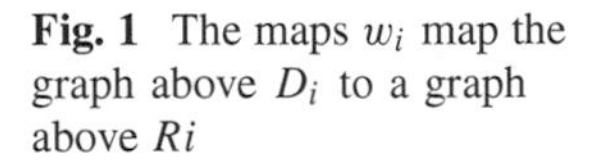
Fig. 1 The maps w_i map the graph above D_i to a graph above Ri

image to be contractive. Though there exists no surety of the contractiveness of an image for any given values of s_i.

Further, the copying machine metaphor, as defined for fractal image compression reduces contrast (s_i) of a given image during each iteration of the feedback loop. This does not mean that the final image through successive iterations will be a homogenous grey. On the flip side, this produces increased levels of details in each of the successive iterations as found in fractals.

Though there are many views of fractal image compression as outlined below but we shall delve on a special case of IFS, i.e., PIFS.

1. Partitioned iterated function system (PIFS)
2. Self-vector quantization: It is based on VQ codebook approach.
3. Self-quantized wavelet subtrees: It uses a type of wavelet transform coding (Haar wavelet coefficient).
4. Convolution transform coding: Uses convolution operation while searching a matching image.

4 Applying PIFS on a Sample Image (Lenna Image)

The Lenna image consists of large chunks of homogeneous regions [4]. As a result, appropriate collage may be derived to encode large blocks. For regions of high contrast smaller blocks are selected for high quality. We apply *variable quadtree partition* approach to encode *Lenna* image. Variable partitioning provide users an option to specify image size and quality parameters according to their needs. The variable blocks in our experiment shall be of 4, 8, 16, and 32 pixels wide.

The image encoding is simulated on a GPU cluster simulated using 24 GPU machines with the host machine having ATI Radeon graphics hardware [6, 10]. The topology followed is ***Dynamic Allocation with Circulating Pipeline Processing*** as outlined in [1, 9].

As depicted in Fig. 2 each range block is circulated through a pipeline that traverses through nodes comprising of PEs (Processing Elements- GPUs [13]). Each PE is represented by a domain and thus each range is matched with it. Once a match is found the range exits the pipeline. If no match is obtained, the host PE_0 subdivides the range into four sub-blocks and the process of matching continues through the pipeline till an appropriate match is found. The advantage of the pipeline approach is that the work is spread evenly among slave processors. In addition to this, less memory is used as each slave processor is required to save only part of the image (domain) and not the entire image. The pipeline topology scale evenly (linearly) as number of slave processors is increased. It is noted that the processor utilization efficiency is 95 % for the sample *Lenna* image.

The experimental results for encoding the *Lenna* image over the pipelined topology are outlined in Table 1.

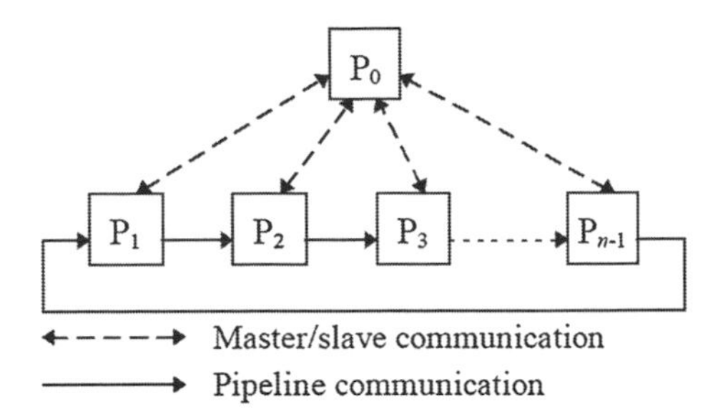

Fig. 2 Circulating pipeline configuration

Table 1 Encoding and compression time chart

Range	PSNR (dB)		Encoding time (s)	Compression ratio
	Collage	Attractor		
4 × 4	34.78	34.56	56.07	6.6
8 × 8	27.95	27.34	34.29	19.67
16 × 16	23.03	23.78	28.55	89.34
32 × 32	19.66	19.54	25.02	341.89

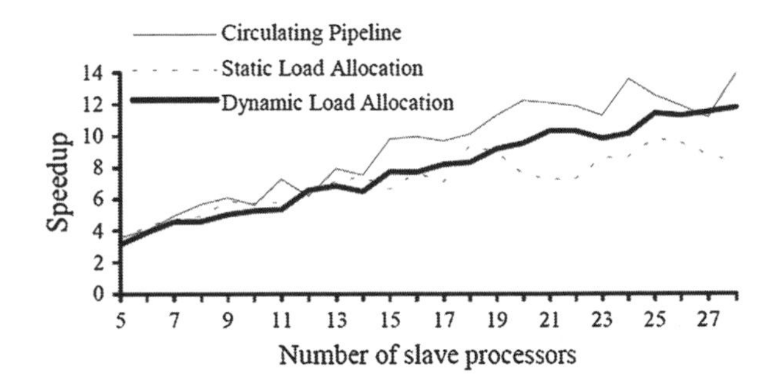

Fig. 3 Performance simulation for a GPU cluster

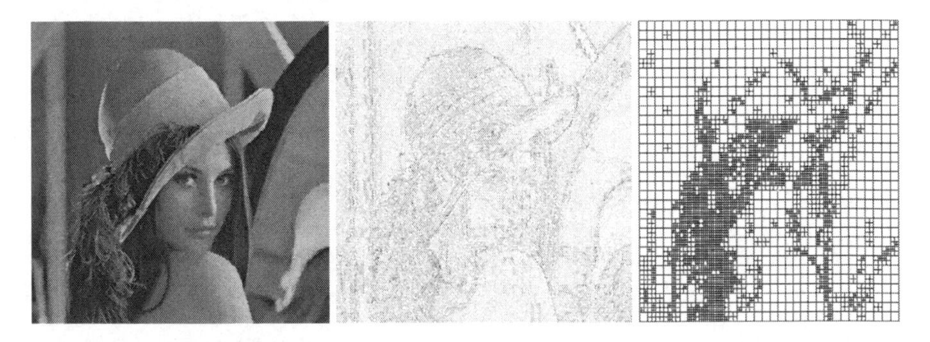

Fig. 4 Encoding results using PIFS partitioning scheme on GPU cluster (PSNR: 19.54, Compression ratio: 341.89 %)

The performance simulation graph [2] shown in Fig. 3 clearly shows performance gain in terms of speedup factor against the number of GPU PEs. The circulating pipeline approach is compared with the other conventional schemes like ***static load allocation and dynamic load allocation***.

The results of the encoding using the PIFS approach with *PSNR* value 19.54 shows an impressive compression ratio of 341.89 % (refer Fig. 4).

5 Conclusion

The results obtained using PIFS scheme on a set of GPU cluster are promising. Though the computing time for this scheme still remains $O(n^4)$ but the pipelining of slave GPU processing cores open new avenues of multithreading (inherent inside GPU cores) the image compression algorithms which till now have been implemented in uniprocessor architectures or selective simulations on SIMD machines. The pipelining approach is scalable and considering the GPU availability in the market on commodity rates it can be implemented using cheap processing power for computer intensive fractal image processing logic. The quality of the image (fidelity) still remains an open question as it has been observed that the PIFS approach creates degradation in the image fidelity in case the image is heterogenous and thus the range blocks need to be partitioned more frequently in each of the pipeline stage.

References

1. Ahmed, N., Natarajan, T.: Discrete cosine transform. IEEE Trans. Comput. **23**, 90–93 (1974)
2. An Introduction to Fractal Image Compression. Literature Number: BPRA065, Texas Instruments Europe, October 1997.
3. Barnsley, M.F., Sloan, A.: Chaotic compression. Comput. Graphics. World **5**, 107–108 (1987)
4. Chady M.: Application of the Bulk Synchronous Parallel Model in Fractal Image Compression. School of Computer, Science (2004).
5. Erra, U.: Toward real time fractal image compression using graphics hardware. In: Proceedings of Advances in Visual Computing, LNCS, vol. 3804, Springer, pp. 723–728. 2005, doi:10.1007/11595755_92.
6. Galabov, M.: Fractal image compression. In: International Conference on Computer System and Technologies - CompSysTech'2003 (1990).
7. Hockney, R.W.: Performance parameters and benchmarking of supercomputers. Parallel Comput. **17**, 1111–1130 (1991)
8. Hurtgen, B., Mols, P., Simon, S.F.: Fractal transform coding of color images. SPIE Conference on Visual Communications and Image Processing, Chicago, In (1994)
9. Jackson, D.J., Blom, T.: Fractal image compression using a circulating pipeline computation model. Int. J. Comput. Appl. (in review).
10. Jacquin, A.: Image coding based on a fractal theory of iterated contractive image transforms. In: SPIE, Visual Communications and Image Processing, vol. 1360 (1990).
11. Kominek, J.: Advances in fractal compression for multimedia applications. Multimedia Systems. Springer, New York (1997).

12. Line Transmission of Non-Telephone Signals. Video CODEC for AudioVisual Services, ITU-T Recommendation (H.261).
13. Mandelbrot, B.B.: The Fractal Geometry of Nature. W.H. Freeman and Company, San Francisco (1982). ISBN 0-7167-1186-9
14. Monro, D.M., Wooley, S.J.: Fractal image compression without searching. In: Proceedings of ICASSP, vol. 5, pp. 557–560 (1994).
15. MPEG-7 Whitepaper. Sonera MediLab, 13 October 2003.
16. Owens, J.D. et al.: A survey of general-purpose computation on graphics hardware. In: Eurographics 2005, State of the Art Reports, pp. 21–51, August 2005.
17. Ruhl, M., Hartenstein, H.: Optimal fractal coding is NP-hard. Proceeding of IEEE Computer Society Washington, DC, USA, In (1997)
18. Series H: Audiovisual and multimedia systems, infrastructure of audiovisual services - Coding of moving video. Video coding for low bit rate communication ITU-T, ITU-T Recommendation (H.263).
19. Sodora, A.: Fractal Image Compression on the Graphics Card. Johns Hopkins University, Baltimore (2010)
20. Wallace, G.K.: The JPEG still picture compression standard. IEEE Trans. Consum. Electron. **38**(1), 18–34 (1992)

Person Identification Using Components of Average Silhouette Image

Rohit Katiyar, K. V. Arya and Vinay Kumar Pathak

Abstract Gait biometrics is one of the non-cooperative biometrics traits particularly in the situation of video surveillance. In the proposed method human knowledge is combined with gait information to get the better recognition performance. Here, individual contributions of different human components, namely head, arm, trunk, thigh, front-leg, back-leg and feet are numerically analyzed. The performance of the proposed method is evaluated by experimentally with CASIA dataset B and C. The effectiveness and impact of seven human gait components is analyzed by using Average Silhouette Image (ASI) under wide range of circumstances.

Keywords Biometrics · Human gait recognition · Average silhouette image · Human gait modeling · Visual surveillance.

1 Introduction

In visual surveillance some kind of biometric information is required to be extracted for human identification and verification. Majority of researchers focused on biometrics traits such as face, fingerprints, iris, handwriting etc but some of the most prominent are discussed below.

R. Katiyar (✉)
CSE Department, H.B.T.I, Kanpur, India
e-mail: rohit.katiyar@rediffmail.com

K. V. Arya
AVB-Indian Institute of Information Technology and Management, Gwalior, India
e-mail: kvarya@iiitm.ac.in

V. K. Pathak
Uttarakhand Open University, Haldwani, India
e-mail: vinaypathak.hbti@gmail.com

B. V. Babu et al. (eds.), *Proceedings of the Second International Conference on Soft Computing for Problem Solving (SocProS 2012), December 28–30, 2012*, Advances in Intelligent Systems and Computing 236, DOI: 10.1007/978-81-322-1602-5_62,

(i) In visual surveillance, the distance between the cameras and the people under surveillance are often large and hence, it is very difficult (almost impossible) to get the detailed and accurate biometric information in above mention conventional biometric systems.
(ii) The subject's cooperation often required to capture conventional biometric information. Therefore, the quality of input image is highly dependent on cooperation.
(iii) People's attention in authentication and authorization: Human gait provides an interesting alternative for visual surveillance applications. A gait describes by the walking pattern of a person. It can be acquired at a distance and even without the cooperation or knowledge of the person.

Human gait is affected by the person's physical characteristics and some factors related to ambient. Tao et al. [1] have presented that carrying status of walking person and its clothes, shoes etc. affect the performance of human gait. In [2] Phillips et al. demonstrated that the effect of camera view point and elapsed time on the gait measurement and demonstrated that for visual surveillance the elapsed time should be as small as possible. It has been shown by Tanawongsuwan and Bobick [4] that walking speed, rhythm and surface bounciness are the inherent characteristics associated with gait and do affect the performance of gait recognition system. As identified by Liu et al. [5], the efforts are still required to resolve the issues like image quality, lighting condition, subject's familiarity of walking surface.

In this work, an effort is made to resolve some of the above mentioned issues. Here, human gait is first decomposed into seven components, and then, the person gait identification is carried out using average silhouette image of gait for each component. During experiments we have observed that the component based gait recognition system is capable of recognizing the people efficiently. Human gait shows the distinctive moving silhouette of a human body which also indicates the physical situation as well as psychological state of the walking person. Hence, the study of various components of the human silhouette along with their effect on human gait recognition process is studied in this work.

The rest of the paper is organized as follows. In Sect. 2 gait modeling scheme is presented along with ways of human silhouette decomposition into seven components also the process of Average Silhouette Image (ASI) generation is discussed. The details of gait dataset used for the experiments are given in Sect. 3. The proposed human gait recognition system is presented in Sects. 4 and 5 experimental results are described. Finally the paper is concluded in Sect. 6.

2 Scheme to Model Human Gait

In the module the human gait the averaged silhouette image partitioned into the seven components as described below using the partition shown in Fig. 1. These components are: (i) head (ii) arm includes shoulder, (iii) the trunk without chest, (iv) thigh including hip, (v) front leg (vi) back leg and (vii) the feet. The front-leg

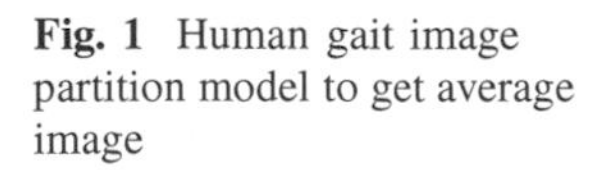

Fig. 1 Human gait image partition model to get average image

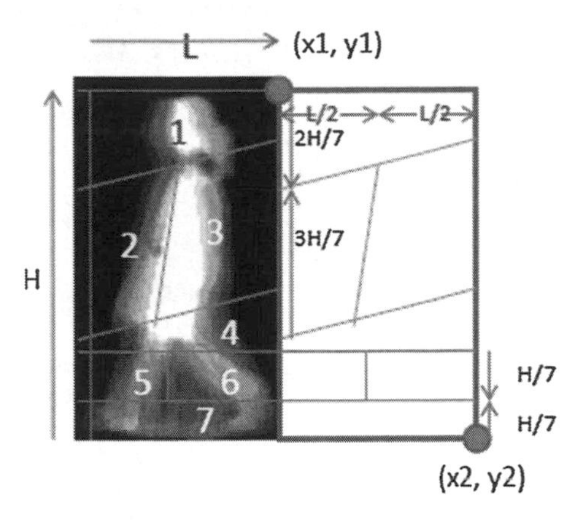

and back-leg are included as separate components because of the bipedal walking style. During walking, the left-leg and the right-leg come to back and front one by one.

The three steps for preparing the averaged silhouette image described in [6] and reproduced below for sake of completing the study. A sample set of images for average gait is shown in Fig. 2.

Step 1: For all the averaged silhouette images in a gallery set compute the average image (as shown in Fig. 1).

Step 2: Mark all the six control points to locate head (2 points), the arm include shoulder (1 points), trunk (1 point) and one point is used to find thigh and feet location respectively reach."

Step 3: Use lines to connect the relevant pairs of points to partition the mean image into seven parts as shown in Fig. 1.

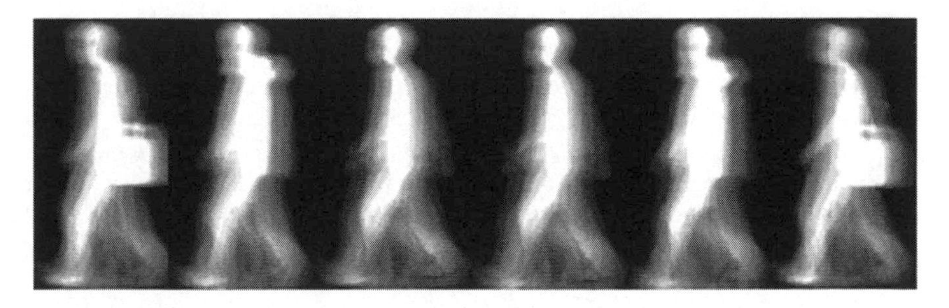

Fig. 2 Average Silhouette Images (ASIs) in the gallery set.

3 The Proposed Human Gait Recognition Method

It is observed that different persons have different average silhouette images (ASIs), hence they can be used for gait recognition. Some examples for a gallery set of ASIs are shown in Fig. 2. The ASIs obtained by taking the average of silhouette images over a gait cycle within a series of images Liu and Sarkar [7] demonstrated that based on the gate period of length N_{Gait} sequence of images is partitioned into a series of cycles (subsequence). The ASIs are achieved by averaging the silhouette (binary images) over one subsequence and represented as follows [7].

$$\text{ASI}_i|_{i=1}^{\lfloor \text{T}/N_{\text{Gait}} \rfloor} = \left(\sum_{k=i N_{\text{Gait}}}^{k=(i+1)N_{\text{Gait}}-1} S(k)/N_{\text{Gait}} \right) \tag{1}$$

where $S(k)$ represents the k silhouette, as stated earlier, a binary image. ASI is very robust against errors in individual frames; the ASI is used to represent a gait cycle. Many ASIs are resulted from one sequence and their number depends the number of cycles contained in a sequence. In the proposed work the ASI act as the input data for gait recognition system. We have used the same definition as given in [7] for computer the distance between training sequence and test sequence and represented by (2).

$$\text{Dist}\left(\text{ASI}_P^{\text{Method}}, \text{ASI}_G^{\text{Method}}\right) = \text{Median}_{i=1}^{N_P}\left(\min_{j=1}^{N_c}\left\|\text{ASI}_P^{\text{Method}}(i) - \text{ASI}_G^{\text{Method}}(j)\right\|\right) \tag{2}$$

where $\text{ASI}_P^{\text{Method}}(i)\Big|_{i=1}^{N_P}$ is the ith projected ASI in the probe data and $\text{ASI}_G^{\text{Method}}(j)\Big|_{j=1}^{N_G}$ is the jth projected ASI in the gallery. Equation (2) uses the median of the Euclidean distances between the averaged silhouettes from the probe and the gallery sequences. The difference between (2) and the gait recognition measure developed in [7] is that here we select a template for recognition which is most closely related. The algorithm is described below and demonstration with respect to dataset images is shown in Fig. 3. First each sequence in the training set is segmented into a few subsequences consisting of a complete gait cycle. Then the averaged silhouette image is calculated for each subsequence and a component or its negative (or complement) is selected from each averaged silhouette image through a predefined template and the similar procedure is adopted for test image dataset. Finally, the acceptance or rejection decision is taken based on the similarity measure between stored training information from gallery and test sample of average silhouette image (ASI) from the probe set. We simply find the Euclidean distance between the test data and training data set and arrange the distances in descending order to find out the matching person.

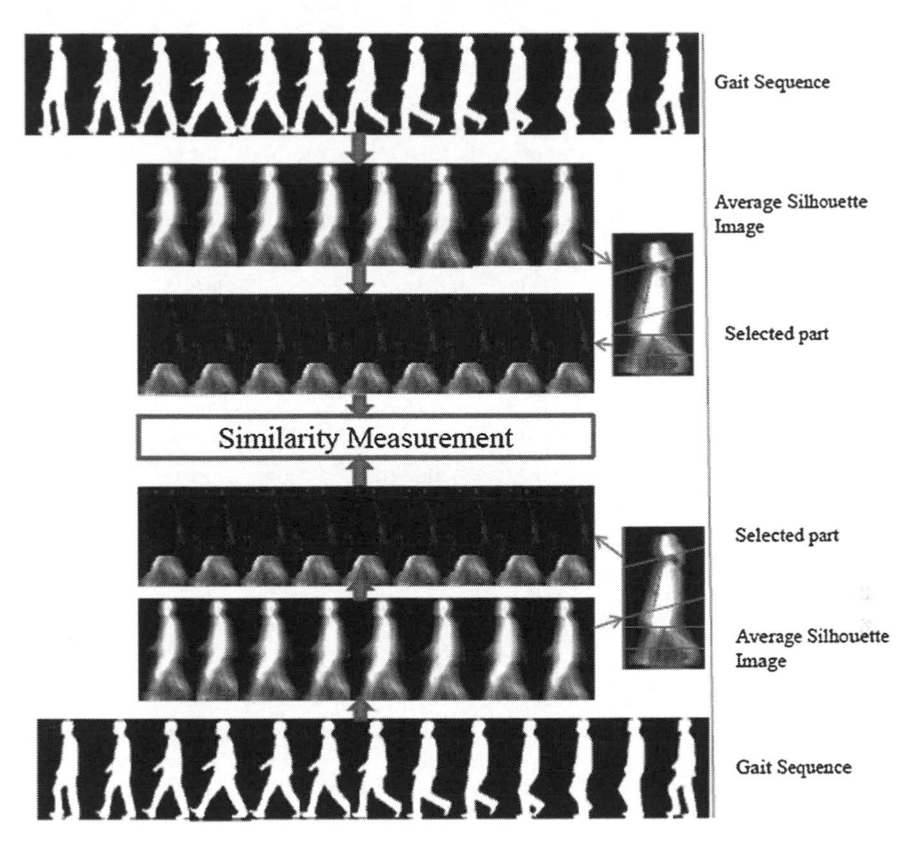

Fig. 3 Construction of the average silhouette image along with the similarity measure *vis-a-vis* the steps of the proposed method

4 Brief Review of CASIA Dataset

The performance of the proposed method is evaluated using the CASIA dataset B [8]. It is a large gait dataset that contains 124 subjects where 6 video sequences are recorded for each subject. Thus, a total of 744 (124 × 6) gait sequences are used in this experiment.

The dataset is partitioned into 2 groups. The first group contains 24 subjects and is used to construct gallery set for training the system are remaining 100 subjects behave as the probe set and used as training images in performance evaluation.

5 Experimental Results

The effectiveness of the averaged silhouette image as well as the components of the averaged silhouette image for gait recognition has been studied through experimental results. Here the contribution of different component of human Average Silhouette Image ASI in human gait recognition system is observed. The effect of individual component on gait recognition is evaluated and results of the experiments are shown in Table 1. The first column contains the identification of 12 subjects from training dataset and the remaining columns contain the recognition rates for using the different parts of the human body for gait recognition where M presents the averaged/mean silhouette images and h, a, t, th, fl, bl, and f respectively indicates the recognition of gait using head, arm, trunk, thigh, front-leg, back-leg and feet. Similarly the human gait recognition employing the averaged silhouette image without head, arm, trunk, thigh, front-leg, back-leg and feet is indicated by −h, −a, −t, −th, −fl, −bl, and −f respectively. In this analysis effect of the individual component is considered as positive if there is a significant reduction in the recognition rate after removing that component from the averaged silhouette image and similarly the effect is considered negative if the recognition rate increases after the component removal.

We presume that the component has little effect on the gait recognition in other case. As an example we can observe from Table 1 that the head has a positive effect for the training subject 1, as the recognition rate is reduced to 79 from 85 by removing the head from the averaged silhouette and the front-leg has a negative effect for the training subject 3, because the recognition rate is increasing to 70 from 67 without considering the trunk in averaged silhouette image. However, the trunk has little effect on the recognition rate for training subject 4 as the recognition rate is unchanged in case of with or without trunk.

Table 1 Rank 1 experimental result for human gait recognition

IDs	M	h	−h	a	−a	t	−t	th	−th	fl	−fl	bl	−bl	f	−f
1	85	33	79	29	80	45	82	33	79	33	84	35	78	27	82
2	86	55	84	53	84	73	84	73	81	48	86	37	85	36	86
3	67	25	66	16	65	28	70	29	62	12	68	21	68	20	68
4	23	7	22	8	23	11	23	5	33	10	23	5	22	7	13
5	22	4	19	5	24	15	20	6	28	6	29	9	26	4	19
6	10	5	11	4	9	5	13	6	10	6	13	8	10	3	13
7	15	7	18	6	11	11	17	4	17	8	17	4	13	6	11
8	53	54	47	37	44	62	42	3	80	32	49	14	51	44	38
9	49	44	43	23	48	58	40	4	67	16	51	11	45	32	39
10	41	30	34	10	37	36	38	4	48	12	46	14	38	13	31
11	2	6	3	9	3	16	3	11	2	2	5	3	5	3	13
12	7	3	4	11	5	8	3	4	7	2	7	3	7	4	14

6 Conclusion

This paper presents a gait recognition method which uses the concept of average silhouette image for efficient gait recognition. The method is developed keeping in mind that the various parts of the human body play a significant role in the human gait recognition. After analyzing the experimental results it can be concluded that in human gait recognition system is not only affected by legs but also head, arm, trunk and hands plays the vital role for person recognition and consequently may contribute in improving the recognition accuracy. We find out the contribution of each body part in recognition rate and observed that dynamic areas affect more than the static parts of the human body. If we go through the Average Silhouette Image (Figs. 2 and 3) then it is almost clear that body part containing gray region which represents the dynamic information have the positive effect on recognition rate.

Our future work will focus on to derive such a image which only contains all the dynamic areas of whole body in a complete gait cycle, so that space required need to store it in gallery will be reduced and will find easy to extract the limited features from it.

References

1. Tao, D., Li, X., Wu, X., Maybank, S.: Human carrying status in visual surveillance. IEEE Comput. Vis. Pattern Recog. **2**, 1670–1677 (2006)
2. Phillips, P., Sarkar, S., Robledo, I., Grother, P., Bowyer, K.: The gait identification challenge problem: datasets and baseline algorithm. IEEE Int. Conf. Pattern Recog. **1**, 385–388 (2002)
3. Tao, D., Li, X., Wu, X., Maybank, S.: Elapsed time in human gait recognition: A new approach. In: Proceedings of IEEE ICASSP, pp. 177–180 (2006)
4. Tanawongsuwan, R., Bobick, A.: Modelling the effects of walking speed on appearance-based gait recognition. IEEE Comput. Vis. Pattern Recog. **2**, 783–790 (2004)
5. Liu, Z., Malave, L., Sarkar, S.: Studies on silhouette quality and gait recognition. IEEE Comput. Vis. Pattern Recog. **2**, 704–711 (2004)
6. Sarkar, S., Phillips, P., Liu, Z., Vega, I., Grother, P., Bowyer, K.: The humanid gait challenge problem: data sets, performance, and analysis. IEEE Trans. Pattern Anal. Mach. Intell. **27**(2), 162–177 (2005)
7. Liu, Z., Sarkar, S.: Simplest representation yet for gait recognition: averaged silhouette. In: Proceedings of International Conference on Pattern Recognition, Cambridge, U.K, vol. 4, pp. 211–214 (2004)
8. CASIA Gait Database available at http://www.cbsr.ia.ac.cn/english/Databases.asp

Modified LSB Method Using New Cryptographic Algorithm for Steganography

R. Boopathy, M. Ramakrishnan and S. P. Victor

Abstract Steganography is different from Cryptography, Steganography is the process of hiding the information so that no one will try to decrypt the information, where as in Cryptography it is obvious that the message is encrypted, so that any one will try decrypting the message. In this paper, we are suggesting new methods to improve the security in data hiding, perhaps by combining steganography and cryptography. In this work, we propose a new encryption method that provides the cipher text as the same size of the plain text. We also presented an extensive classification of various steganographic methods that have used in the field of Data Security. We analyze both security and performance aspects of the proposed methods by PSNR values and proved that in the cryptographic point of view. The proposed method is feasible in such a way that it makes to intricate the steganalyst to retrieve the original information from the Stego-image even if he detect the presence of digital steganography. An embedded message in this method is perceptually indiscernible under normal observation and thus our proposed method achieves the imperceptibility. The volume of data or message to be embedded in this method is comparatively large and proved in Experimental Results hence the high capacity is also achieved.

Keywords Encryption · Steganography · Data hiding · Cryptography

R. Boopathy (✉)
Research Scholar, Manonmaniam Sundaranar University, Tirunelveli, India
e-mail: boopathyr123@gmail.com

M. Ramakrishnan
Department of IT, Velammal Engineering College, Chennai, India
e-mail: ramkrishod@gmail.com

S. P. Victor
Research Director, St. Xavier College, Palyamkottai, India
e-mail: victorsp@rediffmail.com

B. V. Babu et al. (eds.), *Proceedings of the Second International Conference on Soft Computing for Problem Solving (SocProS 2012), December 28–30, 2012*, Advances in Intelligent Systems and Computing 236, DOI: 10.1007/978-81-322-1602-5_63,

1 Introduction

Steganography and watermarking [6] are the cardinal components of the fast escalating area of information security. Steganography and watermarking bring a wide range of very important techniques as to how one can hide important information in an imperceptible and/or irremovable way in an image and video data. Both Steganography and watermarking work toward secret communication. In the other hand, Cryptography hides the contents of the message from an attacker, but not the existence of the message. Steganography/watermarking even hide from the view of the message in communicating data. Cryptography is not concerned with hiding the existence of a message, but rather its meaning by a process called encryption. Generally, AES algorithms are not suitable for many digital video applications [18, 20]. In order to avoid many paper have been proposed [19].

Recently, a research paper was published [1] on how to improve existing methods of hiding secret messages, by combining Steganography and Cryptography in such a way to make it harder for the Steganalyst [7] to retrieve the plain text of a secret message from a Stego-object if cryptanalysis were not used. The prime intention of this paper is the combined approach of Cryptography and Steganography. Very often, a message is encrypted before being hidden in an image in order to achieve a better level of secrecy (which provides a basic model on how to combine Cryptography and Steganography). Steganography embeds the secret message in a harmless looking cover, such as a digital image file [5].

In [21] Esra Satir and Hakan Isik proposed a compression-based text Steganography method to improve the hiding capacity.

2 Related Works

Extensive Research has been carried out on Steganography and Steganalysis [8, 11]. Analysis of Least Significant Bit (LSB) and DCT methods were already proposed [2]. Jessica Fridrichetal [9] has discussed a reliable and accurate method for detecting LSB non-sequential embedding in digital images. The images can be hidden in DCT domain [10] also. The length of the messages can be acquired by inspecting the lossless capacity in the LSB and shifted LSB plane. The most accepted method for Steganography is the LSB encoding [4]. Using any digital image, LSB replaces the least significant bits of each byte by the hidden message bits. The message may also be dispersed at random throughout the image. There are numerous means of hiding information in digital media. Most common approaches are

- Least significant bit insertion
- Masking and filtering
- Redundant Pattern Encoding
- Encrypt and Scatter
- Algorithms and transformations

Every one of these techniques can be applied by varying the degrees of success.

Least significant bit insertion

In this paper, we have used LSB insertion method with little modification. It is a common and simple approach to embed information in an image file. In LSB insertion, the LSB of a byte is altered with an M's bit. This method is commonly used for any image, audio as well as video Steganography. When viewed by an HVS the resulting image resembles the cover object [16].

For example, in [4] an image Steganography, the letter A can be concealed in three pixels, assuming no compression. The raster data for 3 pixels, i.e., 9 bytes may be drawn as

00100111 11101001 11001000
00100111 11001000 11101001
11001000 00100111 11101001

The binary value for B is 01000011. Inserting the binary value for B in the three pixels would result in

(0010011$\underline{0}$ 11101001 11001000)
(0010011$\underline{0}$ 11001000 1110100$\underline{0}$)
(1100100$\underline{1}$ 00100111 11101001)

The underlined bits are the only three actually changed data of the 8 bytes used. On an average, the LSB requires only half the bits in an image to be changed. We can hide data in the least and second least significant bits and still the human eye would not be able to discern it [15]. In [14] they have proposed method to verify whether the secret message was deleted or changed by hackers.

Our first implication in this paper is to get better Steganographic techniques by combining them to Cryptographic ones in a new way that is, as far as we know, not available in the literature. Indeed, most of the techniques that combine Cryptography and Steganography aid in encrypting the secret message before hiding its existence in a cover object. Usually applying an entrenched, common-purpose symmetric-key encryption algorithm to guarantee the privacy during video/image transmission [3] is a high-quality idea from a security point of view. Although numerous classifications of image encryption algorithms [17] have been formerly presented, we provide an extended and more comprehensive such classification. Finally, we show that our method allows for a new type of digital image Steganography where a given message is camouflaged with jpeg/bitmap cover image. We evaluated in both security and performance aspects of the proposed method and found that the method is efficient and adequate from a Cryptographic point of view. Our proposed method satisfies the following requirements.

Imperceptibility—The invisibility of a Steganographic algorithm is the most important and basic need, since the strength of Steganography lies in its ability to be unnoticed by the human eye. The moment one can see that an image has been altered, the algorithm is compromised.

Payload capacity—Steganography requires enormous embedding capacity, unlike watermarking which needs only a small capacity of copyright information.

Robustness—Our method is Robust against statistical attacks—Statistical Steganalysis is the practice of detecting hidden information through applying statistical tests on image data. Many Steganographic algorithms leave an impression when embedding information that can be easily detected through statistical analysis.

3 Proposed Model

High-quality encryption is a process that produces randomized information by the way compression efficiency [12] and is directly reliant on the presence of source data redundancy. The more the data is correlated, better the compression and vice versa [13]. This paper introduces a method of encrypting the text and image files in an image in order to test the accuracy and efficiency of encryption. This process enhances the transfer of information to the intended receiver without any potential risk. In this paper, the proposed method will help to secure the text content within the image and encryption of image file within the image will help to make the document much securer because, though the unauthorized person succeeds in hacking the image, the person will not able to read the message as well as acquire the information in the image file. The proposed approach find the suitable algorithm for embedding the data in an image using Steganography which provides a better security pattern for sending messages through a network. This paper deals with digital images acting as a cover medium to store secret data in such a manner that it becomes invisible. The Java software is used to extensively analyze the functions of the LSB algorithm in Steganography. Mat lab software is used to evaluate the PSNR values and to evaluate the performance of the proposed method.

3.1 Overall View

The data hiding patterns using the Steganography technique in this paper can be explained using this simple block diagram. The block diagram for Steganography technique is as follows (Fig. 1).

3.2 Implementation

Implementation section discusses about the different modules of this work and shows the methods of implementation. This work is implemented by using the following step by step procedure.

The Steganography process is implemented in different stages like:

- Encryption section
- Data transmission section
- Decryption section

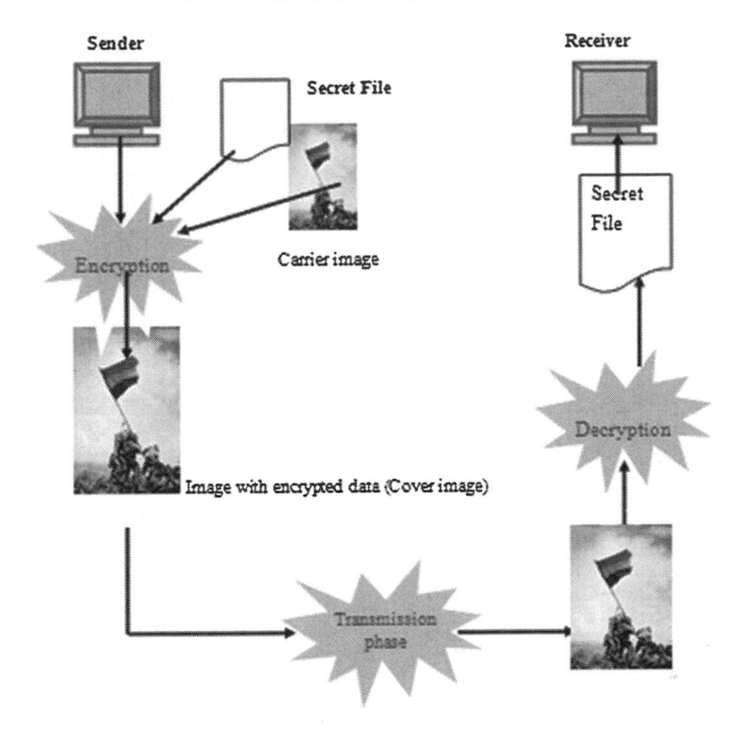

Fig. 1 Overall architecture diagram for proposed model

3.3 The Encryption Section

The encryption section of Steganography is the primary stage. In this stage, the sender sends the data as well as the image file which acts as a carrier image to transfer the data to the destination. In this paper, different images are used as carriers because all images are highly resistant for Steganalysis. In the encryption section, the text message (secret message) is encrypted by using a proposed new algorithm and then it will be embedded into the image file. The data is embedded into the image and the encryption is implemented using Java.

Proposed New Encryption Algorithm

This algorithm was written and coded by us specifically for this paper. The main advantage of this algorithm is that it provides the encrypted text, the same size as the original text.

Encryption Algorithm

Step 1: Generate the ASCII value for every letter.
Step 2: Generate the equivalent binary value of it.

[Binary value should be 8 digits, e.g., for decimal 14 binary number should be 1110]

Step 3: The resultant binary number should be reversed.
Step 4: Take a divisor having 4 digits length (>= 1,000) as the Key.
Step 5: Divide the resultant number already reversed with the divisor.
Step 6: Store the remainder in first 3 digits and quotient in next 5 digits (remainder and quotient wouldn't be more than 3 digits and 5 digits long respectively. If we find any of these digits are less than 3 and 5 respectively then we need to add required number of 0s (zeros) in the left-hand side. So, this would be the preprocessed text to be embedded. Now store the remnants in first 3 digits and quotient in next 5 digits.

Steps to embed a message into a Master file:

- Convert a Secret message into a character array.
- Create a byte array. The length of the binary array is equal to size of input file.
- Open input file and read all bytes into byte Array.
- Skip past OFFSET_JPG bytes.
- Skip past OFFSET_PNG bytes.
- Skip past OFFSET_GIF_BMP_TIF bytes.
- The 32 bit input file size should be converted into byte array.
- 4 byte input File size array should be embedded into the master file.
- Write the remaining bytes.
- 3 byte version information should be embedded into the file.
- Write 1 byte for features.
- Compress the message from level 0 to 9.
- Get the result of compressed message byte array.
- Embed the 1 byte compression ratio into the output file.
- Encrypt the message based on proposed new Encryption algorithm.
- Convert the 32 bit message size into byte array.
- Embed 4 byte message size array into the master file.
- Embed the message.

3.4 Data Embedding Section

The embedding is done based on the well-known Least Significant Bit (LSB) algorithm with little modification. We used the last two significant bits of each pixel and replaced with the significant bits of the text document. The encrypted data is embedded into the cover image. Now the compression is done. The level of compression can be from 0 to 9. Using the modified approach the message bits are embedded properly in the place of Least Significant Bits of cover image, such that the image doesn't lose its resolution and hence the security will be robust. The encrypted image is protected with a secured symmetric key such that is used to avoid the damages caused due to hackers or unauthorized persons.

3.5 Decryption Section

In the decryption phase, the intended receiver receives the carrier image from sender through the transference medium. The intended receiver then sends the carrier image to the decryption phase. In the decryption phase, the same 'Least Significant Algorithm' is implemented for decrypting the LSB from the image and merge in an order to frame the original message bits. After successful completion of the process, the file is decrypted from the carrier file and accessed as an original text document.

3.6 Proposed Decryption Algorithm

Step 1: Multiply last 5 digits of the encrypted text by Key value.
Step 2: Add first 3 digits of the encrypted text with the result produced in the previous step.
Step 3: If the result produced in the step 2 is not an 8-bit number then we need to make it as 8-bit number
Step 4: Reverse the number to get the encrypted text.

4 Results and Discussion

We have chosen image Steganography because it is a simple and user friendly application. Though there are various applications for image hiding but the proposed approach is created using Java which is efficient for coding and the performance is better compared to other languages.

Comparing with paper [1] we have used two-bit stego instead of conventional one-bit stego, and even though the various encryption algorithms have been proposed already we have used our own encryption algorithm as proposed in Sect. 3. The advantage of algorithm is it will produce the cipher text with the same size of that plain text. The algorithm used in this paper falls under the classification of spatial domain.

This paper gave us good experience in dealing with the data security issues in theoretical as well as in technical domain and in Java programming. I performed the paper in a satisfactory level with the help and good guidance from my supervisor (Table 1).

PSNR results
PSNR—phrase peak signal-to-noise ratio, is defined as the ratio between the maximum possible power of a signal and the power of corrupting noise that affects the fidelity of its representation. As most signals have a wide and dynamic range, PSNR is usually denoted in terms of the logarithmic decibel scale. Imperceptibility has been calculated using PSNR between Original cover image (I) and Stego-Image (Js).

Table 1 Comparison of different Steganography algorithms

Steganography algorithm	Speed	Quality of hiding	Security
LSB	High	Good	Strong
F5	Low	High up to 13.4 %	Strong
JSteg	Moderate	Embedding capacity up to 12 %	Less

Table 2 Comparison of different sizes in jpeg images

Cover image name	Cover image size (KB)	Text file (KB)	Compressed file(KB)	Stego-image size(KB)
Cameraman.jpg	7.17	10.4	7.96	7.97
Cutepuppy.jpg	28.8	65.6	56.6	52.1
Flag.jpg	26.5	60.6	49.2	47.1
Windows.jpg	51.3	120	101.2	99.4
Barsilona.jpg	63.6	57.9	55.9	111
Cell.jpg	6.13	104	96.1	46.5

$$\text{PSNR} = 10 * \log(255 * 255/\text{MSE}) / \log(10);$$

In the equation, MSE is the mean square error between the original and the denoised image. The higher the PSNR in the restored image, the better is its quality. In testing, few images were experimented. The quantitative results have been given in the table for the six images (Table 2).

5 Conclusion

In this world, data transfers using Internet is rapidly growing because it is so easier as well as faster to transfer the data to the destination. Security is an important issue and transferring the data using Internet because any unauthorized individual can hack the data and make it useless or obtain information unintended to him.

The proposed approach in this paper uses a new combined approach of Encryption and Steganography. This creates a Stego-image in which the personal data is embedded and is secured with a symmetric key which is highly protective. The main contribution in this paper is the introduction of new encryption algorithm/approach for Steganography application that provides commendable security. The main advantage of this algorithm is that it provides the encrypted text in the same size as the plain text.

The proposed approach provides higher PSNR Value of 41db as an average. It has three levels of security. In the first level, the text file is compressed and zipped. In the second level, it is encrypted using proposed algorithm. In the final level a secret key is used to protect the message from Stego-attacks. The change in image resolution is negligible when we embed the message into the image and the image is concealed or secured with the personal password. And hence the data is protected from damage by any unintended user.

As future work, we plan to use the same concept in Frequency domain to increase the hiding capacity. We are aiming to use the different combination of frequency domain.

SAMPLE SCREEN SHOTS

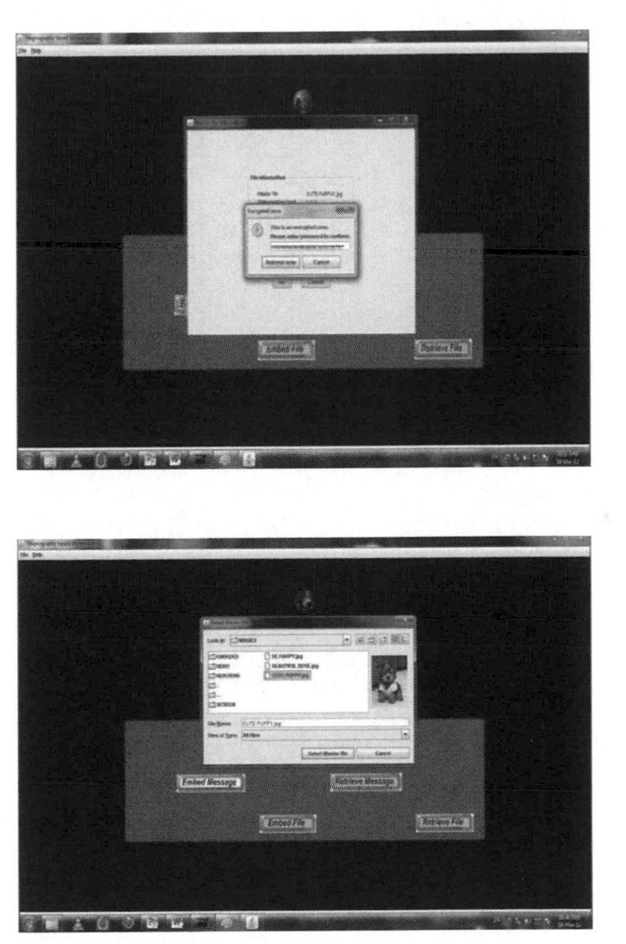

Acknowledgments We would like to express our gratitude to Velammal Engineering College in which where I am working, for providing to access all the resources . I would like to acknowledge Ms. Mary Elizabeth for her great support.

References

1. Challita, K., Farhat, H.: Combining steganography and cryptography: new directions. In: IJCNAA **1**(1), 199–208 (2012). The Society of Digital Information and Wireless Communication
2. Walia, E., Jain, P., Navdeep.: An analysis of LSB & DCT based steganography. Glob. J. Comput. Sci. Technol. **10**(1), 4–8 (2010)
3. Socek, D., Kalva, H., Magliveras, S.S, Marques, O., Culibrk, D., Furht, B.: New approaches to encryption and steganography for digital videos. Multimedia Syst. **13**(3), 191–204 (2007). (Springer)
4. Bandyopadhyay, S.K., Bhattacharyya, D., Ganguly, D., Mukherjee, S., Das: A tutorial review of steganography. In: IC3, pp. 105–114 (2008)
5. Dunbar B.: A detailed look at Steganographic techniques and their use in an open systems environment. Sans Institute InfoSec Reading Room (2002)
6. Katzenbeisser, S., Peticolas, F.A.P.: Information hiding techniques for steganography and watermarking. Artech House, Boston (2000)
7. Upham, D.: Steganographic algorithm JSteg. http://zooid.org/paul/crpto/jsteg
8. Zahedi Kermani, M.J.Z.: A roust steganography algorithm based on texture similarity using gabor filter. In: IEEE 5th International Symposium on Signal Processing and Information Technology (2005)
9. Fridrich, J., Goljan, M.: Detecting LSB steganography in color and grey scale images. IEEE Multimedia Secur **8**, 22–28 (2001)
10. Ashourian, M., Jain, R.C., Ho, Y.-S.:Dithered quantization for image data hiding in the DCT domain. In: Proceeding of IST, pp. 171–175 (2003)
11. Krenn, J.R.: Steganography and Steganalysis, January 2004
12. Hwang, R.-J., Shih, T.K., Kao, C.-H.: A lossy compression tolerant data hiding method based on JPEG and VQ. J. Internet Technol. **5**, 171–178 (2004)
13. Tseng, H.-W., Chang, C.-C.: High capacity data hiding in JPEG compressed images. Informatica, **15**, 127–142 (2004)
14. Park, Y., Kang, H., Yamaguchi, K., Kobayashi, K.: Integrity verification of secret information in image steganography. In: Symposium on Information Theory and its Applications. Hakodate, Hokkaido, Japan (2006)
15. Neeta, D., Snehal, K., Jacobs, D.: Implementation of LSB steganography and its evaluation for various bits. In: 2006 1st International Conference on Digital Information Management on 6 Jan 2007. doi:10.11.09/ICDIM.2007.369349 (2007)
16. Jain, A., Gupta, I.S.: A JPEG compression resistant steganography scheme for raster graphics images. In: IEEE Region 10 Conference TENCON 2007, vol. 2 (2007)
17. Furht, B., Muharemagic, E.A., Socek, D.: Multimedia security: encryption and watermarking. Multimedia Systems and Applications, vol. 28. Springer, Berlin (2005)
18. Li, S., Chen, G., Zheng, X.: Chaos-based encryption for digital images and videos. In: Furht, B., Kirovski, D (eds.) Multimedia Security Handbook. Internet and Communications Series, vol. 4, pp. 133–167. CRC Press, West Palm Beach (2004)
19. Li, S., Zheng, X., Mou, X., Cai, Y.: Chaotic encryption scheme for real-time digital video. In: Real Time Imaging VI, Proceedings of SPIE, vol. 4666, pp. 149–160. SPIE Publishers (2002)
20. Uhl, A., Pommer, A.: Image and video encryption: from digital rights management to secured personal communication. In: Advances in Information Security, vol. 15. Springer, New York (2005)
21. Esra S., Hakan I.: A compression-based text steganography method (2012). doi:10.1016/j.jss.2012.05.027

TDAC: Co-Expressed Gene Pattern Finding Using Attribute Clustering

Tahleen A Rahman and Dhruba K Bhattacharyya

Abstract An effective unsupervised method (TDAC) is proposed for identification of biologically relevant co-expressed patterns. Effectiveness of TDAC is established in comparison to its other competing algorithms over four publicly available benchmark gene expression datasets in terms of both internal and external validity measures.

Keywords Cluster · Outlier · Core · Neighbour · Co-expressed

1 Introduction

Many clustering algorithms have been evolved and applied on gene expression data. The existing approaches for gene data clustering are categorised into three types: (a) Gene-based, where genes are treated as the objects, while the samples are as features; (b) Sample-based, where samples are generally related to various diseases or drug effects within a gene expression matrix; and, finally (c) Subspace clustering, which attempts to find subset of objects such that the objects emerge as a cluster in a subspace created by a subset of the features. Subspace clustering techniques are further classified into two subcategories, i.e. biclustering and triclustering. In biclustering, it attempts to cluster the gene expression data both row-wise as well as column-wise simultaneously [1]. Whereas, triclustering [2] aims to mine biologically relevant coherent clusters over a gene sample time (GST) domain for any gene expression datasets. In this paper, we propose a cost-effective attribute clustering method for finding co-expressed gene patterns that does not require discretisation. To avoid the restrictions caused due to the use of any proximity measure while expanding the cluster, it exploits the regulation information computed over the expression values.

T. A. Rahman (✉) · D. K. Bhattacharyya
Tezpur University, Napaam, Assam 784028, India
e-mail: dkb@tezu.ernet.in

B. V. Babu et al. (eds.), *Proceedings of the Second International Conference on Soft Computing for Problem Solving (SocProS 2012), December 28–30, 2012*, Advances in Intelligent Systems and Computing 236, DOI: 10.1007/978-81-322-1602-5_64,

2 TDAC: The Proposed Attribute Clustering Method

TDAC is basically a three step method. In *step* 1, the gene expression data matrix, i.e. $G_{m\times n}$ of order $m \times n$, is normalised to have a mean 0, and standard deviation 1. In *step* 2, we find condition-wise neighbourhood for each expression value based on regulation information and proximity measure with the neighbouring expression values. To find neighbourhood based on expression value proximity, we use a linear density-based clustering that works based on L_1 norm with reference to β, a user defined threshold. Similarly, to find similarity between a pair of genes based on regulation information, we use the angular deviation (i.e +ve, −ve or *neutral*) computed based on the *arccos* formula given in [3]. It identifies the core gene groups for each condition based on the regulation information and proximity-based neighbourhood information with reference to β. *Step* 3 performs two major tasks, i.e. identification of (i) outlier genes and (ii) co-expressed gene groups. The co-expressed gene group is a subset of genes having common neighbours ≥ 2, over at least k conditions. Here, we assume that to form a co-expressed gene group or cluster, there must be at least 'two' neighbour genes over at least k conditions. An outlier gene is defined as a gene having neighbourhood <2 over $\geq k$ conditions.

3 Algorithm

TDAC operates on a preprocessed gene expression dataset for simultaneous identification of both outlier as well as co-expressed genes based on regulation information and attribute/condition level proximity. The proposed TDAC is free from the restrictions of using (i) discretisation and (ii) specific proximity measure. Based on the regulation information and the expression-level proximity for each condition computed over the pre-processed gene matrix, a faster attribute clustering technique, i.e. *attrib_clus* identifies the core genes. Here, *attrib_clus* finds core genes based on the regulation information using the arccos expression given in [3] and finds expression-level dissimilarity using L_1 norm. However, it is free from the restriction of using any proximity measure. Based on regulation, core genes and their connectivity information and by using the concepts given in the Definitions k-neighbour and Cluster, TDAC can identify the co-expressed gene groups as well as the outlier genes, with reference to a given user defined threshold. The basic steps of TDAC for finding co-expressed gene groups are stated next.

1. Preprocess $G_{m\times n}$ with z-score normalization to obtain $G'_{m\times n}$.
2. Apply attrib_clus() on $G'_{m\times n}$ to obtain core_gene groups for each condition;
   a. Find neighbour gene (s) for a given gene gi at condition, say Ca based on regulation information (with reference to its previous condition, i.e. Ca-1) and L1 proximity with reference to β.
   b. Identify gi as core at Ca if satisfies the core gene conditions.

3. Identify the co-expressed gene groups across n conditions;
   a. Find genes which are core over atleast k conditions.
   b. From the subset of neighbour genes obtained, find genes which have the same nearest neighbours across atleast k conditions.
   c. Find the set of k—neighbors for each gene based on the subsets obtained above.
   d. Each gene in the list of genes obtained along with its respective nearest k—neighbours is assigned cluster ids and form clusters.
   e. If there are common nearest neighbours between two such genes, i.e. some gene is assigned to more than one cluster, then the respective clusters are merged.

4 Performance Evaluation

Our technique was implemented in MATLAB running on a 1.73 GHz(2CPUs) Intel Pentium processor with 16 GB RAM. We have compared the results of our work in terms of several internal and external validity measures with several other competing techniques Table 1.

Homogeneity and Separation We have used CVAP [4] to test the performance of TDAC in terms of homogeneity and separation measures. The clusters detected by our technique in Datasets 1, 2 and 3 are shown in Figs. 1 and 2 respectively. Average homogeneity [5] reflects the compactness of a cluster given by a clustering algorithm. It can be seen from the Table 2 that the proposed TDAC shows a better homogeneity than its other competing algorithms. Average separation [5] reflects the overall distance among the clusters given by a clustering algorithm. As it can be seen from the Table 2 that the proposed TDAC shows superior performance in terms of separation than its other competing algorithms. Silhoutte Index [6] reflects the compactness and separation of clusters. As can be seen from the Table 3 that our work shows significant improvement in performance in comparison to its other competing algorithms [7] like MOGA-SVM, MOGA (without SVM), FCM, SOM, Average-linkage, k-means and DGC for most of the datasets.

Table 1 List of datasets used and their sources

Dataset	Genes	Samples	Source
1. Yeast sporulation [9]	474	7	http://cmgm.stanford.edu/pbrown/sporulation/index.html
2. Human fibroblasts [10]	517	19	http://genomewww.stanford.edu/serum/data/
3. Arabidopsis thaliana	138	8	http://homes.esat.kuleuven.be/thijs/Work/Clustering.html
4. Subset of yeast cell cycle [11]	384	17	http://yscdp.stanford.edu/yeast-cellcycle/fulldata.html

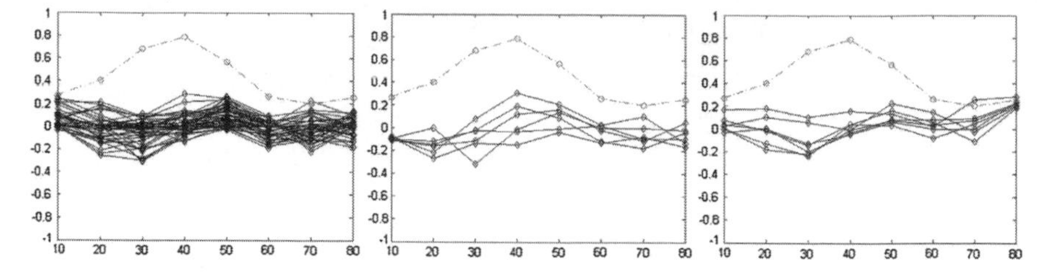

Fig. 1 An outlier gene with a cluster obtained from Dataset 3

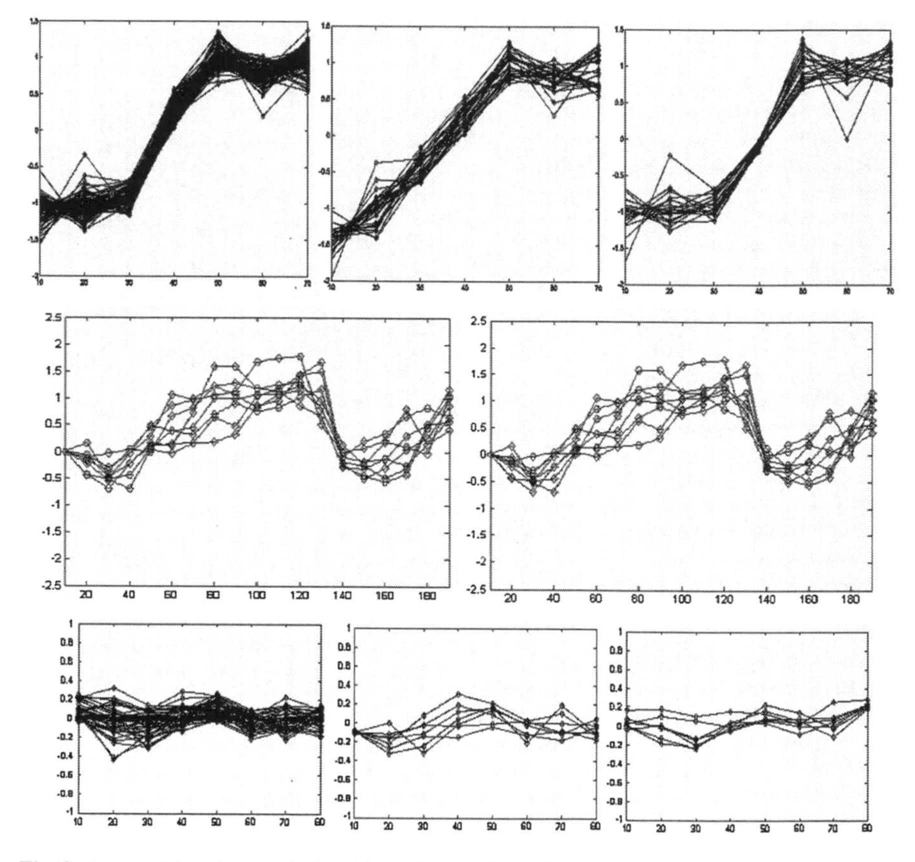

Fig. 2 Some of the clusters obtained from Dataset 1 and 2

External Quality p-value: We have obtained p-value [8] using the software FuncAssociate [8], which is a web-based tool that accepts as input a list of genes, and returns a list of GO attributes that are over-(or under-) represented among the genes in the input list. A low p-value indicates that the genes belonging to the enriched

Table 2 Performance comparison in terms of Homogeneity and Separation

Dataset used	Method applied	Number of clusters	Homogeneity	Separation
Dataset 1	*k*-means	9	0.11047	0.40366
	TDAC	9	0.96915	0.12266
	SOM	9	0.11804	0.40898
Dataset 2	*k*-means	3	0.29568	0.46473
	TDAC	3	0.77574	0.18869
	SOM	3	0.29973	0.47346
	CLICK	3	0.750	0.107
	CAST	3	0.831	0.0166
Dataset 3	*k*-means	5	0.24113	0.39542
	TDAC	5	0.2243	0.20996
	SOM	5	0.24425	0.38723
Dataset 4	*k*-means	3	0.302	0.490
	TDAC	3	0.482	0.171
	SOM	3	0.300	0.512
	CLICK	3	0.349	0.212
	CAST	3	0.360	0.164

Table 3 Performance comparison in terms of Silhoutte Index

Dataset used	Method applied	Number of clusters	Silhoutte index
Dataset 1	*k*-means	9	0.24752
	SOM	9	0.27703
	TDAC	4	0.29877
Dataset 2	SOM	6	0.3235
	MOGA-SVM (RBF)	6	0.4154
	MOGA (without SVM)	6	0.3947
	FCM	8	0.2995
	Average Linkage	4	0.3562
	DGC	16	0.6880
	TDAC	6	0.6909
Dataset 4	DGC	17	0.7307
	SOM	6	0.3682
	MOGA-SVM (RBF)	5	0.4426
	MOGA (without SVM)	5	0.4392
	FCM	6	0.3872
	Average Linkage	4	0.4388
	TDAC	6	0.7248

functional categories are biologically significant in the corresponding clusters. The enriched functional categories for Dataset 1 and Dataset 3 are listed in Tables 4 and 5 respectively. The cluster C1 in Dataset 1 contains several enriched GO categories. Two examples of highly enriched categories in C1 are 'anatomical structure

Table 4 *p*-values of dataset 1

Cluster	*p* value	GO number	GO category
C1	1.187e-22	GO:0048646	Anatomical structure formation involved in morphogenesis
	1.395e-21	GO:0010927	Cellular component assembly involved in mor phogenesis
	6.871e-19	GO:0030476	Developmental process
	2.295e-18	GO:0032502	Ascospore wall assembly
	2.295e-18	GO:0042244	Spore wall assembly
	2.295e-18	GO:0071940	Fungal-type cell
	3.373e-18	GO:0070726	Cell wall assembly
	2.039e-17	GO:0048869	Cellular developmental process
	2.464e-16	GO:0030435	Sporulation resulting in formation of a cellular spore
	2.464e-16	GO:0030154	Cell differentiation
	2.464e-16	GO:0043934	Sporulation
	1.338e-15	GO:0003006	Developmental process involved in reproduction
	1.534e-14	GO:0048610	Cellular process involved in reproduction
	6.201e-13	GO:0022414	Reproductive process
C2	1.298e-21	GO:0002181	Cytoplasmic translation
	9.073e-20	GO:0003735	Structural constituent of ribosome
	5.303e-17	GO:0022625	Cytosolic large ribosomal subunit
	7.408e-19	GO:0005840	Ribosome
	1.641e-16	GO:0005198	Structural molecule activity
	1.856e-16	GO:0006412	Translation
	5.072e-13	GO:0030529	Ribonucleoprotein complex
	6.726e-11	GO:0044267	Cellular protein metabolic process
	1.321e-10	GO:0043228	Nonmembrane-bounded organelle
	1.321e-10	GO:0043232	Intracellular nonmembrane-bounded organelle
	7.389e-10	GO:0005622	Intracellular
C3	4.230e-06	GO:0031145	Anaphase-promoting complex-dependent proteasomal ubiquitin-dependent protein catabolic process

formation involved in morphogenesis' and 'cellular component assembly involved in morphogenesis' with *p*-values 1.187e-22 and 1.395e-21, respectively. Similarly, for Dataset 3, TDAC identifies several clusters with highly enriched GO categories of very low *p*-values. Cluster C1 in this dataset includes GO categories like 'response to stimulus' and 'response to chemical stimulus' with *p*-values 6.12e-08 and 1.19e-07, respectively.

Table 5 *p*-values of dataset 3

Cluster	*p* value	GO number	GO category
C1	6.12e-08	GO:0050896	Response to stimulus
	1.19e-07	GO:0042221	Response to chemical stimulus
	2.04e-04	GO:0070887	Cellular response to chemical stimulus
	2.37e-04	GO:0010033	Response to organic substance
C2	1.68e-05	GO:0003857	05 3-hydroxyacyl-CoA dehydrogenase activity
	3.35e-05	GO:0004300	Enoyl-CoA hydratase activity
	2.09e-04	GO:0006629	Lipid metabolic process
	6.50e-04	GO:0006631	Fatty acid metabolic process

5 Conclusion and Future Work

A method for identification of biologically relevant co-expressed patterns based on regulation information and attribute clustering is reported. The proposed TDAC is established over four publicly available gene datasets in terms of both internal and external validity measures. Work is going on to extend the present TDAC toward handling of large number of gene expression datasets +ve, −ve and mixed-correlated gene patterns.

References

1. Jiang, D., Tang, C., Zhang, A.: Cluster analysis for gene expression data: a survey. IEEE Trans. Knowl. Data Eng. **16**(11), 1370–1386 (2004)
2. Mahanta, P., Ahmed, H.A., Bhattacharyya, D.K., Kalita, J.K.: Triclustering in gene expression data analysis: A selected survey. In: Proceedings of IEEE NCETACS, pp 1–6 (2011)
3. Sarma, S., Sarma, R., Bhattacharyya, D.K.: An effective density based hierarchical clustering technique to identify coherent patterns from gene expression data. In: Proceedings of PAKDD 2011, vol. 6634, pp. 225–236 (2011)
4. Kaijun, Wang, Wang, Baijie, Peng, Liuqing: CVAP: validation for cluster analyses. Data Sci. J. **8**, 88–93 (2009)
5. Sharan, R., Shamir, R.: CLICK: A clustering algorithm with applications to gene expression analysis. In: Proceedings of the International Conference on Intelligent Systems for Molecular Biology, pp. 307–316. AAAI Press, Menlo Park (2000)
6. Rousseeuw, P.: Silhouettes: a graphical aid to the interpretation and validation of cluster analysis. J. Comput. Appl. Math. **20**, 153–165 (1987)
7. Das, R., Bhattacharyya, D.K., Kalita, J.K.: Clustering gene expression data using an effective dissimilarity measure. Int. J. Comput. Bio Sci. (Special Issue) **1**(1), 55–68 (2011)
8. Berriz, F.G., et al.: Characterizing gene sets with Funcassociate. Bioinformatics **19**, 2502–2504 (2003)
9. Chu, S., DeRisi, J., Eisen, M., et. al.: The transcriptional program of sporulation in budding yeast. Science **282**, 699–705 (1998)
10. Iyer, V.R., Eisen, M.B., Ross, D.T., et. al.: The transcriptional program in the response of the human fibroblasts to serum. Science **283**, 83–87 (1999)
11. Cho, R.J., Campbell, M., Winzeler, E., Steinmetz, L., et al.: A genome-wide transcriptional analysis of the mitotic cell cycle. Mol. Cell **2**(1), 65–73 (1998)

An Introduction to Back Propagation Learning and its Application in Classification of Genome Data Sequence

Medha J. Patel, Devarshi Mehta, Patrick Paterson and Rakesh Rawal

Abstract The gene classification problem is still active area of research because of the attributes of the genome data, high dimensionality and small sample size. Furthermore, the underlying data distribution is also unknown, so nonparametric methods must be used to solve such problems. Learning techniques are efficient in solving complex biological problems due to characteristics such as robustness, fault tolerances, adaptive learning and massively parallel analysis capabilities, and for a biological system it may be employed as tool for data-driven discovery. In this paper, some concepts related to cognition by examples are discussed. A classification technique is proposed in which DNA sequence is analyzed on the basis of sequence characteristics near breakpoint that occur in leukemia. The training dataset is built for supervised classifier and on the basis of that back propagation learning classifier is employed on hypothetical data. Our intension is to employ such techniques for further analysis and research in this domain. The future scope and investigation is also suggested.

Keywords Supervised classifier · Artificial neural network · Cancer classification

M. J. Patel (✉)
Gujarat Technological University, Ahmadabad, India
e-mail: medhapate0@gmail.com

D. Mehta
GTU, Gujarat, India

P. Paterson
Industrial Engineering Department, Texas Tech University, Lubbock, Texas, USA

R. Rawal
Gujarat Cancer Research Institute, Gujarat, India

B. V. Babu et al. (eds.), *Proceedings of the Second International Conference on Soft Computing for Problem Solving (SocProS 2012), December 28–30, 2012*, Advances in Intelligent Systems and Computing 236, DOI: 10.1007/978-81-322-1602-5_65,

1 Introduction

To solve a problem on a computer, we need algorithm, but for some applications, we do not have algorithms. We know the inputs, sometimes we know what the output should be, but do not know how to transform input to the output. Human brain is adaptable to get some insight, i.e., cognition by examples for this kind of applications. But in modern scenario, the pressing need of rapid transformation from data to information and from information to knowledge and above all repeating this task a large number of times are ideally suited for machine—not humans. And so we want some combination of machine and learning. Learning is somewhat very subjective—can we make our machine to learn? The easier approach is to mechanize the process of learning. We can develop some learning algorithm which can make computers (machine) to extract automatically the hypothesis/rules from examples, than it is called machine learning [1]. Machine learning is programming computers to optimize performance criterion using example data or past experience. Machine learning uses the theory of statistics in building mathematical models, because the core task is making inference from samples. Artificial neural network (ANN) mimics the learning or adaptability of the biological system and thus provides a kind of machinery of learning.

2 Artificial Neural Network

ANN is an information processing system that has been developed as a generalization of the mathematical model of human cognition (ability to know). It consists of simple computational units called neurons that are highly interconnected and each connection has a strength that is expressed by a positive or negative number called weight. The connection of neurons are normally arranged in layers and executed in parallel. The connections are categorized as network topology. The size of the weight controls the influence that one neuron has on other, with a positive weight excite an element and negative weight inhibit. Overall the activation of an element is determined by a combination (summation) of excitatory and inhibitory influence it receives from its neighbors. The weights of the net are adaptable which store the experimental knowledge from task example through a process of learning. The information is stored in the connections and distributed throughout, so the network can function as a memory of brain. The memory is content addressable, in the sense that the information may be recalled by providing partial or even erroneous input pattern. The information is stored in association of other stored data, hence it is adaptable.

The network architecture determines how and which type of neurons can be connected and in which topology. The way nodes are connected determines how computations proceed. On the basis of the connection patterns (architecture) the ANNs can be grouped as (i) *feed forward*, in which there are no loops. Examples single layer perceptron, multi-layer perceptron (MLP), radial basis function (RBF) (ii) *recurrent* (feedback) in which loop occurs because of feedback connections. Examples are Hopfield network and adaptive resonance theory (ART) models [1].

Learning Process The issue of learning is central to the study and design of artificial neural network. Learning encodes pattern information into interneuronal connection strengths, i.e., free parameters or weights of network. The algorithms are developed according to well-defined learning rules which simulate the learning methodology of brain's mathematical models. The basic learning rules can be broadly categorized as error correction learning, memory based learning, Hebb postulate learning, competitive learning, and Boltzmann machine learning [2]. The learning paradigm can be supervised, unsupervised, and reinforcement according to the way by which the network is trained. If desired output is already available with sample data than the main work is to find the patterns in that data.

Error correcting rules The error correcting rules are fundamental building blocks of supervised learning. The general philosophy underlying most supervised learning is based on ***principle of minimal disturbance.*** 'Adapt to reduce the output error for the current training pattern, with minimum disturbances to the patterns already learnt.' The weights of the network can be altered by either presenting the linear error to reduce the error or gradient information to reduce the mean square error (MSE), usually averaged over all training patterns. Let us consider a single neuron (k) of output layer which produce the output y_k for a particular sample n (or discrete time step for real time system). Now this output is compared with desired output d_k and consequently an error signal e_k is produced. Thus $e_k = d_k - y_k$ error actuate a control mechanism, the purpose of which is to apply a sequence of corrective adjustment to the synaptic weights of neuron k. The corrective adjustments are designed to make the output signal y_k come closer to the desired response in a step by step manner. The objective here is to minimize the cost function defined in terms of error signal. This kind of learning process is used in delta rule, generalized delta rule (α-LMS or widrow-hoff rule). According to delta rule the adjustment Δw_{kj} applied to weight w_{kj} is defined as

$$\Delta w_{kj}(n) = \eta \cdot e_k(n)\, xj(n) \tag{1}$$

Here η is a positive constant that determines rate of learning. The rule can be stated as:

The adjustment made to a synaptic weight of a neuron is proportional to the product of the error signal and the input signal [2].

Back propagation is similar to LMS algorithm and is based on gradient descent: weights are modified in the direction that corresponds to the negative gradient of error measure. For successful application of this method differentiable node activation function is required. The major advance of back propagation over LMS and perceptron learning is in expressing how an error at a higher (outer) layer of a multilayer network can be propagated backwards to node at lower (inner) layers of the network. Back propagation learning has emerged as standard algorithm for the training of multi layer perceptron against which other networks are benchmarked. This algorithm has had a major impact on field of neural network and has been applied to a large number of problems in many disciplines. These ANNs have been applied to virtually all pattern recognition problems and are typically the first network tried on a new problem. The reason for this is the simplicity of the algorithm, and vast body of research that has studied these networks.

2.1 *Why Machine Learning for Genome Data*

Biological data is high dimensional, complex, not fully annotated, noisy, and voluminous. There are number of reasons why machine learning approaches are widely used in practice, especially in bioinformatics.

1. Systems often produce results different from desired ones, due to unknown properties of the inputs during designing the computational system. However, with the capacity to improve/learn dynamically, the machine learning can cope with such types of problems, often occurs in biological world.
2. Missing and noisy data is one of the characteristics of biological data. Though conventional techniques is unable to handle this, machine learning can do well.
3. In molecular biology research, new data and concepts are generated very often and replaces old one. Machine learning can easily adapt this change.
4. For biological data it is possible that some hidden relationship or correlation exist in the data. Machine learning techniques are able to extract such relationship voluminous data and supports for data driven knowledge discovery.
5. There are some biological problems in which experts can specify input/output pairs, but not the relationships between inputs and outputs. This can be addressed by machine learning to predict outputs for new inputs introduced to program, by generalization capabilities.

There are many practical issues to deal with when performing machine learning. This is especially true when it is applied to computational biology because the data sets are complex, high dimensional and not annotated, so inferring meaningful results require high accuracy and deep insight. Some points are narrated here. (1) Using more parameters may lead to more over fitting. For biological purpose we have high dimensional input data with comparatively less samples. So it is very difficult to learn all the parameters by given training sets. If features are more in data than it may be more likely that the classifier finds something that separates data just by chance. It also make difficult to draw conclusions from the parameters of a learned model. (2) If we have not enough input samples we cannot separate training sets and test sets, so cross validation technique is used.

3 A Classification Problem

Philadelphia chromosome or Philadelphia **translocation** is a specific chromosomal abnormality that is associated with chronic myelogenous leukemia (CML). It is the result of a reciprocal translocation between chromosome 9 and 22, and is specifically designated *t* (9; 22) (q34; q11) [3]. Chromosome translocations are very important in the initiation and/or progression of cancer; and consequently high numbers of translocation events have been reported in human genome. If high throughput data available than analyzing sequence features in the vicinity of translocation breakpoints, may have major clinical role as: (1) Diagnostic markers (2) Response markers (3) Therapeutic markers.

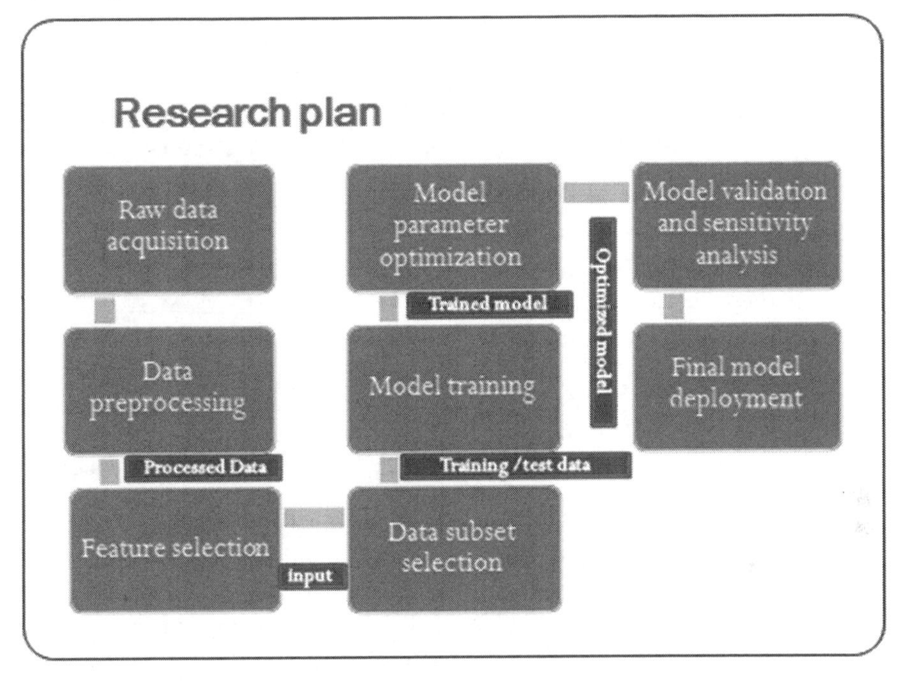

Reasons for selecting this problem are (i) In 2007, cancers caused about 13 % of all human death worldwide (7.9 million) rates are rising as more people live to an old age. And (ii) In 2000 approximately 256,000 children and adults around world developed some form of leukemia and 209,000 died from it. About 90 % of all leukemia is diagnosed in adults [3].

3.1 Materials and Methods

To address this issue the DNA sequence of BCR and ABL chromosomes with precise location of the break point which is involved in reciprocal translocation of leukemia were taken from various databases and further analysis were done. Data retrieval is done from publicly available database. Following databases:

1. Mitelman Database of chromosome aberration and gene fusion in cancer- (http://cgap.nci.nih.gov/chromosomes/Mitelman)
2. NCBI

The bioinformatics analyses are done by following method.

1. The 50 nucleotide base pair upstream and 50 downstream from breakpoint location of BCR and ABL is taken.

2. The occurrence of base pair in a sliding window of window size 3 is calculated and the factor of occurrence to window size is taken as frequency of the position. The graph of it with respect to position is as shown in figure.
3. The mean of frequency is taken as characteristics data for further analysis [4]. The above exercise is done for four nucleotide base pairs “A”, “C”, “G”, and “T” for 10 positive data samples and 5 negative data sample of housekeeping genes. The 15×4 data matrix with 15 sample and four attributes are prepared.
4. For positive data the desired output is ranked as 0.95 instead of 1 and for negative data 0.05 instead of 0, because the activation functions is sigmoid. For sigmoid node function the output value will be 0 only if net input is $-\infty$ and 1 only if net input is $+\infty$ [5]. Since input signal is finite, we need weights of infinitely large magnitude. To avoid this desired out [2] put is taken as 0.95 and 0.05.
5. Data is introduced to the network for several epochs until the error is within desired tolerance.
6. The back propagation algorithm is implemented in MATLAB. A 4–3–1 network with hidden layer of three nodes is designed. Only three layers with one hidden layer are considered as per universal approximation theorem. Input layer neurons are linear while hidden and output layer neuron is sigmoid [6]. The weights are initialized randomly and values between -1 to 1. The frequency of the weight update is either “per pattern”. The value of “***learning rate*** η” is primarily taken as 1. The performance index used is MSE and the goal is to minimize it. The fast convergence of BP algorithm can be obtained by introducing momentum.
7. For testing purpose the weight matrix generated by training program for input to hidden nodes and hidden to output nodes are used to check the validity of the program. A new input vector is introduced in the output is checked.
8. The outputs of the data taken from training set give the correct answer. The output of unseen data not given in training set gives 95 % correct results.

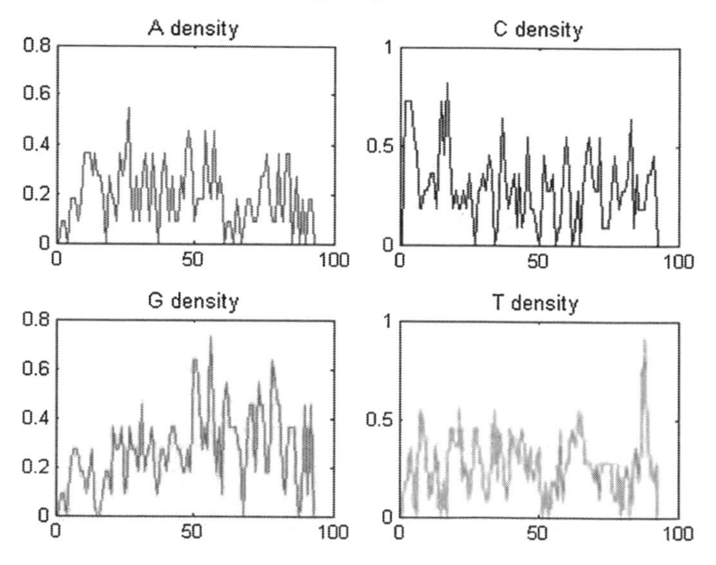

3.2 *Future Scope and Conclusion*

With more sample data available, the result efficiency can be improved. Training model with cross validation technique for both positive and negative examples can be used for better efficiency. The design of architecture is optimized by network pruning. Optimizing the number of hidden nodes is still under process. The genome sequence can be characterized in two ways, mathematically and with biological perception. The sequence can be rich in content of its nucleotides A, C, T or G, it may have more GC content, fractal dimension of ACTG can be calculated. All this characteristics if related with biological characteristics like DNA bend ability, torsion of sequence and other parameters which can cause fragility is analyzed than gene annotation and classification or prognosis phase diagnosis of cancer can also be obtained. The learning rate can also be made adaptive by comparing the errors in two consecutive presentations of input samples. The motivation behind this work is to develop some supervised learners–artificial neural network, support vector machine, and thereby build an ensemble classifier. The gene annotation and diagnosis of cancer at prognosis phase is really a helpful outcome for medical field.

References

1. Sushmita Mitra, E.: Introduction to Machine Learning and Bioinformatics. CRC Press, Boca Raton (2008)
2. Haykins, S.: Neural Networks a Comprehensive Foundation. Pearson Prentice Hall, Hamilton (2008)
3. Wikipedia, (n.d.)
4. Santosh, R., Uma, M.: Back propagation neural network method for predecting lac gene structures. Int. J. Biotechnol. Mol. Biol. Res. **2**(4), 61–72 (2011)
5. Mehrotra, K., Mohan, C.K., Ranka, S.: Elements of Artificial Neural Network. MIT Press, Cambridge (1997)
6. Kumar, S.: Neural Network a Classroom Approach. Tata McGraw Hill Education Private Ltd., Agra (2012)
7. Thompson, J.M.T.: An introduction to the mechanics of DNA. The Roy. Soc. vol. 362, (2004)

Sobel-Fuzzy Technique to Enhance the Detection of Edges in Grayscale Images Using Auto-Thresholding

Jesal Vasavada and Shamik Tiwari

Abstract Images have always been very important in human life because humans are very much adapted in understanding images. Feature points or pixels play very important role in image analysis. These feature points include edge pixels. Edges on the image are strong intensity variations which show the difference between an object and the background. Edge detection is one of the most important operations in image analysis as it helps to reduce the amount of data by filtering out the less relevant information and if edge can be identified, basic properties of object such as area, perimeter, shape, etc can be measured. In this paper, a Sobel-Fuzzy technique using auto-thresholding is proposed by fuzzifying the results of first derivatives of Sobel in *x, y* and *xy* directions. The technique automatically finds the six threshold values using local thresholding. Comparative study has been done on the basis of visual perception and edgel counts. The experimental results show the proposed Sobel-Fuzzy approach is more efficient in comparison to Roberts, Prewitt, Sobel, and LoG and produces better results.

Keywords Edge detection · Sobel edge detection · Image processing · Fuzzy logic

1 Introduction

Edge is defined by discontinuity in the gray levels of pixels. Edge detection is one of the most frequently used techniques in digital image processing [1]. Edge detection is

J. Vasavada (✉) · S. Tiwari
Department of Computer Science and Engineering, Faculty of Engineering and Technology, Mody Institute of Technology and Science (Deemed University), Lakshmangarh, Sikar, Rajasthan 332311, India
e-mail: jesal.vasavada@gmail.com

S. Tiwari
e-mail: tiwari@rediffmail.com

B. V. Babu et al. (eds.), *Proceedings of the Second International Conference on Soft Computing for Problem Solving (SocProS 2012), December 28–30, 2012*, Advances in Intelligent Systems and Computing 236, DOI: 10.1007/978-81-322-1602-5_66,

a preprocessing step to extract some low-level boundary features of an image, which are then used for higher level processing such as object finding and recognition. The Edge detection contains three steps namely Filtering, Enhancement, and Detection. The objective of Filtering is to remove noise from an image so that noise free image is obtained. Edge detection becomes difficult task in noisy images because both edges and noise contain high frequency content. Edge detection in noisy images sometimes give rise to the problems like missing true edges, false edge detection, false edge localization, etc. To filter the noise from an image, the nature and type of noise must be known in prior that helps to choose the correct filtering method [2]. Quality of digital image can be improved by Enhancement techniques. Edge detection identifies the edges by using thresholding. Thresholding can be categorized into global thresholding and local thresholding. Global thresholding is more appropriate in images with uniform contrast distribution of background and foreground like document images. Local thresholding is used when the background illumination is highly nonuniform. The main idea behind edge detection in an image is to find the places where the intensity of pixel changes rapidly, using one of the two general criterias. One of which is finding the places where the first derivative of the intensity is greater in magnitude than a specified threshold. Second is finding the places where the second derivative of the intensity has a zero crossing [3]. Edge detection using fuzzy logic has the advantage that fuzzy set theory and Fuzzy logic offer powerful tools to represent and process human knowledge in the form of fuzzy-if-then rules

Section 2 explains in brief fuzzy image processing. Section 3 explains fuzzy logic-based proposed method. Section 4 gives comparison between Roberts, Prewitt, Sobel, LoG, and proposed method. Finally the Sect. 5 is concluding the paper.

2 Fuzzy Image Processing

Fuzzy logic, one of the decision-making techniques of AI has many application areas. Fuzzy means "unclear". Image data contains a lot of vagueness. Fuzzy logic helps to represent the uncertainties that exist in the image data. Fuzzy image processing is a collection of all approaches that understand, represent, and process images, their segments and features as fuzzy sets. Fuzzy image processing has three main steps:

- Image fuzzification
- Modification of membership values
- Image defuzzification as shown in Fig. 1

Fuzzy image processing is important to represent uncertainty in data. Fuzzy set theory and Fuzzy logic offer powerful tools to represent and process human knowledge in the form of fuzzy-if-then rules. Fuzzy logic is tolerant of imprecise data and can deal with uncertain data which helps to create a model for edge detection in image as presented in [4]. Several approaches on fuzzy logic-based edge detection have been proposed on fuzzy-if-then rules [5–8].

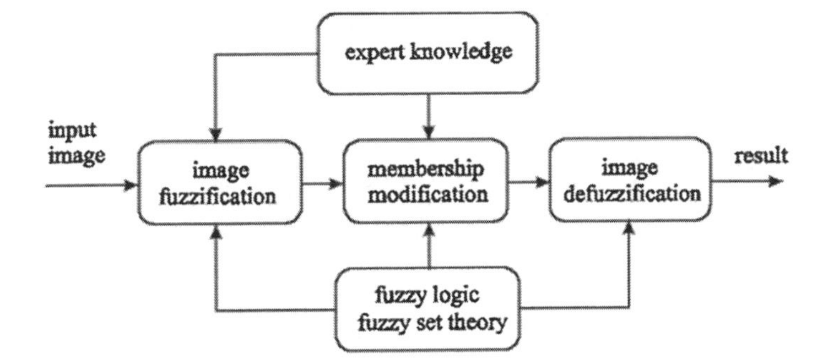

Fig. 1 General structure of fuzzy image processing

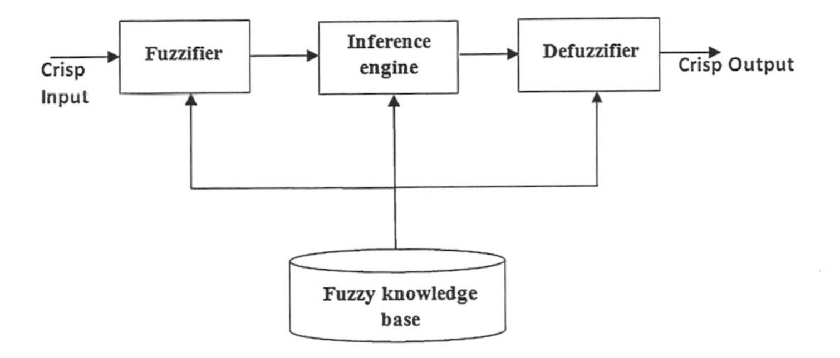

Fig. 2 Block diagram of fuzzy inference system

A fuzzy inference system (FIS) is a system that uses fuzzy set theory to map inputs (*features* in the case of fuzzy classification) to outputs (*classes* in the case of fuzzy classification). The block diagram is shown in Fig. 2.

The function of Fuzzifier is to convert the crisp input to a linguistic variable using the membership functions stored in the fuzzy knowledge base. The function of Inference Engine using If-Then type fuzzy rules is to convert the fuzzy input to the fuzzy output. Functioning of Inference Engine is shown in Fig. 3.

Defuzzifier converts the fuzzy output of the inference engine to crisp using membership functions analogous to the ones used by the fuzzifier. In an FIS, defuzzification is applied after aggregation. Five commonly used defuzzifying methods are Centroid of Area (COA), Bisector of Area (BOA), Mean of Maxima (MOM), Smallest of Maximum (SOM), and Largest of Maximum (LOM). Popular defuzzification methods include the Centroid of Area defuzzifier [9], and the Mean-of-Maxima defuzzifier [9]. The Centroid of Area defuzzifier is the best-known method, which is used to find the centroid of the area surrounded by the MF and the horizontal axis. There are three Fuzzy Models namely Mamdani Fuzzy models, Sugeno Fuzzy Models, and Tsukamoto Fuzzy models. All the models have same style for antecedents but different style for consequence. In the paper, Mamdani Fuzzy Model and COA defuzzifying method are used.

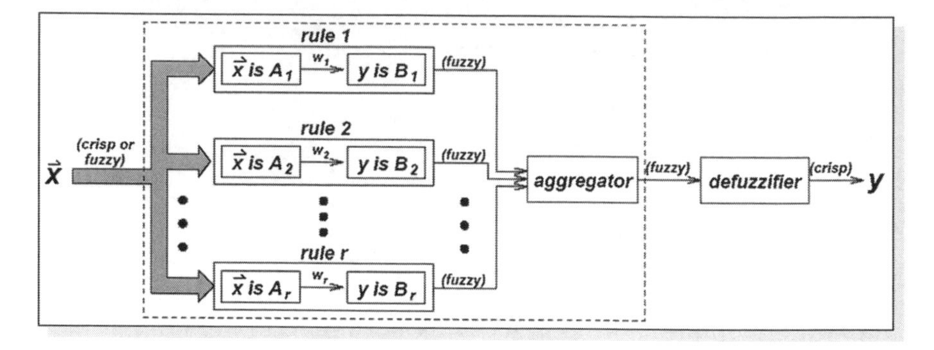

Fig. 3 Functioning of fuzzy inference engine

3 The Proposed Methodology

The block diagram of the algorithm to identify edges in image is given in Fig. 13. In the existing world, almost all the researchers use combined magnitude of Sobel gradients in x and y directions as one input along with one or more inputs like standard deviation, high pass filters, etc. These inputs are then fuzzified. In the paper, first derivatives of Sobel i.e., Gx, Gy, and Gxy are calculated in three different directions i.e., in x, y and xy directions. The three inputs are then fuzzified. The proposed method contains four steps as follows:

3.1 Noise Identification and Filtering

The paper performs noise identification in the image under process and after knowing what type of noise is present, filters the noise and then detects edges using proposed method. At first the image to be processed goes through noise identification as given in [2]. If the image contains no noise then the proposed fuzzy method for edge detection is directly applied to image and if image contains it is filtered after knowing what type of noise it is [2].

3.2 Fuzzification

Derivatives of Sobel Gx, Gy, and Gxy are calculated in three different directions i.e., in x, y and xy directions respectively. The first and the second inputs are calculated using the Sobel masks given in Fig. 4. The third input that is Gxy in 45 degree direction is calculated by the following equation given below where Y and

-1	0	+1
-2	0	+2
-1	0	+1

Gx

+1	+2	+1
0	0	0
-1	-2	-1

Gy

Fig. 4 Sobel convolution masks

X are derivatives in the y and x direction. The equation allows the calculation of the derivative in any direction given by alpha.

$$\text{Gxy} = \sqrt{(\text{Y}\sin(\text{alpha}))^2 + (\text{X}\cos(\text{alpha}))^2}$$

Three computed values are used as fuzzy system inputs. Appropriate membership functions are defined for fuzzy system inputs. To apply these functions, first of all the three inputs are mapped to the range of [0–100]. The classification of mapped values is done into one of the following classes that is low, medium, or high. The Gx classes are shown by GX_L, GX_M, and GX_H symbols, Gy classes are shown by GY_L, GY_M and GY_H and The Gxy classes are shown by GXY_L, GXY_M, and GXY_H symbols. To separate different Gx, Gy and Gxy classes four different thresholds a_1, c_1, a_2, and c_2 are used are used. Such that if values of GX, GY, or GXY lies in the range of $[0– c_1]$, then, the corresponding pixel is classified to GX_L or GY_L or GXY_L for the range of $[a_1–c_2]$, the pixel is classified to GX_M or GY_M or GXY_M, and finally for the range of $[a_2–100]$, pixel is classified to GX_H or GY_H or GXY_H. The defined classes and membership functions are shown in Fig. 5.

If GX value of a pixel is equal to P, GY is equal to Q and GXY value of the pixel is R the fuzzy rules are defined as given in Table 1.

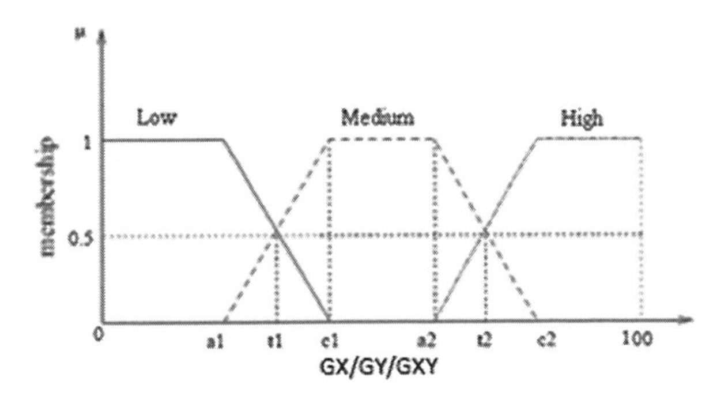

Fig. 5 The defined classes and membership functions

Table 1 Defined Fuzzy Rules

P	Q	R	P_{edge}
GX_L	GY_L	GXY_L	E_L
GX_L	GY_L	GXY_M	E_L
GX_L	GY_L	GXY_H	E_L
GX_L	GY_L	GXY_L	E_L
GX_L	GY_L	GXY_M	E_M
GX_L	GY_M	GXY_H	E_H
GX_L	GD_H	GXY_L	E_L
GX_L	GY_H	GXY_M	E_H
GX_L	GY_H	GXY_H	E_H
SX_L	GY_L	GXY_L	E_L
SX_L	GY_L	GXY_M	E_M
SX_L	GY_L	GXY_H	E_H
SX_M	GY_M	GXY_L	E_M
SX_M	GY_M	GXY_M	E_H
SX_M	GY_M	GXY_H	E_H
SX_M	GD_H	GXY_L	E_H
SX_M	GY_H	GXY_M	E_H
SX_M	GY_H	GXY_H	E_H
GX_H	GY_L	GXY_L	E_L
GX_H	GY_L	GXY_M	E_H
GX_H	GY_L	GXY_H	E_H
GX_H	GY_M	GXY_L	E_H
GX_H	GY_M	GXY_M	E_H
GX_H	GY_M	GXY_H	E_H
GX_H	GD_H	GXY_L	E_H
GX_H	GY_H	GXY_M	E_H
GX_H	GY_H	GXY_H	E_H

On the basis of fuzzy rules described in Table 1, the output of this fuzzy system is classified to one of the three classes E_L, E_M and E_H. The classes E_L, E_M, and E_H corresponds to pixels with low, medium, and high probability value, respectively, to belong to edge pixels set. Output membership functions are shown in Fig.6.

3.3 Defuzzification

Defuzzification is done by Centroid of Area method. This method was developed by Sugeno in 1985. This is the most common and accurate technique. This technique can be expressed as:

$$x^* = \frac{\int \mu_i(x)\ x\ dx}{\int \mu_i(x)\, dx}$$

where x^* is defuzzified output, $\mu_i(x)$ is aggregated membership function and x is the output variable.

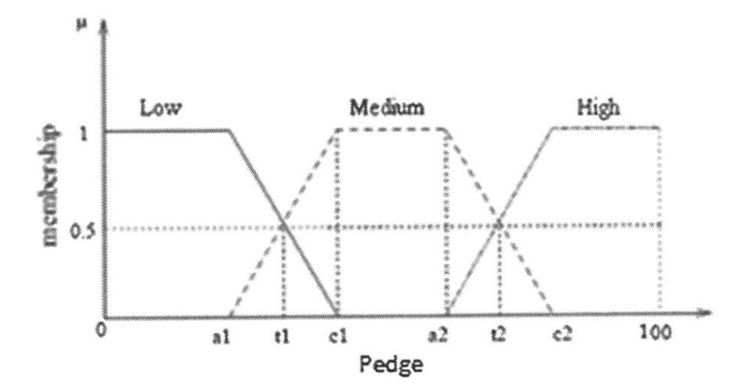

Fig. 6 Output membership functions

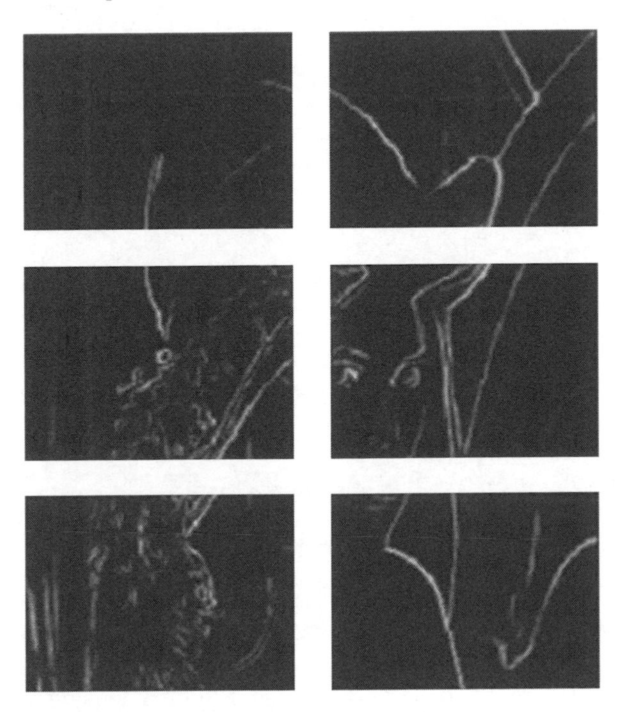

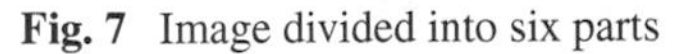

Fig. 7 Image divided into six parts

3.4 Auto-Thresholding

The concept of local thresholding is applied in the paper. The problem with global thresholding is that changes in illumination across the scene may cause some parts to be brighter (in the light) and some parts darker (in shadow) in ways that have nothing to do with the objects in the image. We can deal, at least in part, with such uneven illumination by determining thresholds locally. That is, instead of having a single

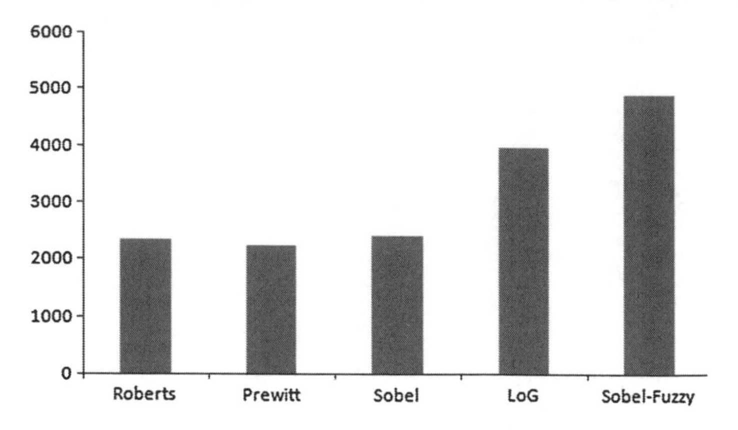

Fig. 8 Edgel counts of the image without noise using different operators

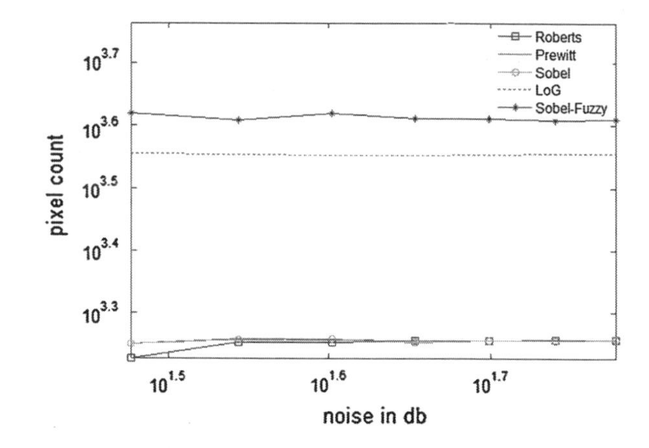

Fig. 9 Edgel counts after filtering speckle noise for different operators

global threshold, we allow the threshold itself to smoothly vary across the image. In the paper, image is divided into six parts as shown in Fig. 7 and then thresholding is applied to each part by taking mean of each part. In the first part if the pixel value is greater than T1 (threshold of 1st part) then it is considered as edge pixel else not. Similarly for the other parts T2, T3, T4, T5, T6 are calculated and edge pixels are found for each part, in the end all the parts are merged to form one complete image.

4 Performance Evaluation

This section compares the proposed method discussed above with the Roberts, Prewitt, Sobel, and LoG on the basis of their edge pixel count. The Fig. 8 compares the edgel count of Roberts, Prewitt, Sobel and LoG with proposed Sobel-fuzzy

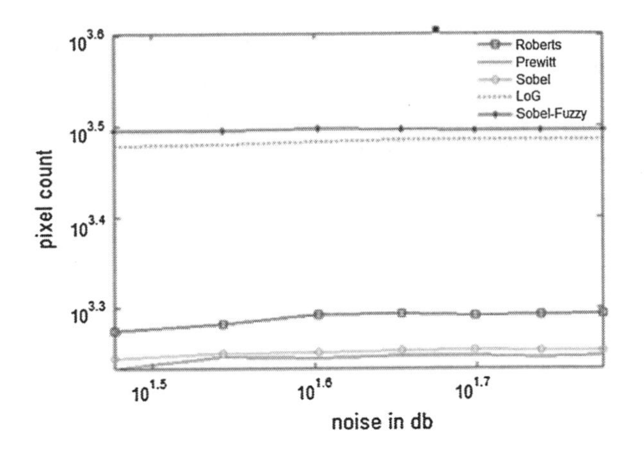

Fig. 10 Edgel counts after filtering Gaussian noise for different operators

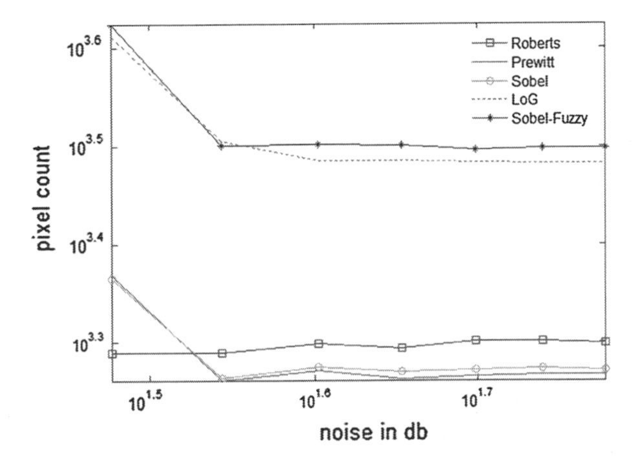

Fig. 11 Edgel counts after filtering salt and pepper for different operators

method for the image without noise. To know how the proposed method performs at different levels of noise, experiments are done with three types of noise, Gaussian noise, Speckle noise, and Salt and Pepper noise at noise levels from 30 to 60 db. Noise identification and filtering are done using the method given in [2]. For Salt and Pepper noise median filter is used, for Gaussian noise weiner filter is used, and for Speckle noise lee filter is used. Then all the techniques considered for comparison are applied to filtered image and edge pixels are counted. The results are shown in Figs. 9, 10 and 11. The evidence for the best detector type is judged by studying the edge maps relative to each other through statistical evaluation. For each edge map the number of edge pixels is count and compared. The proposed Sobel-fuzzy based method reports the higher detected edge pixels as shown in Figs. 8, 9, 10, and 11.

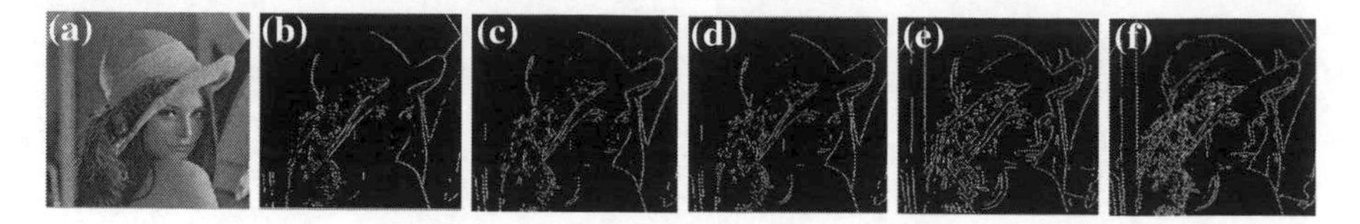

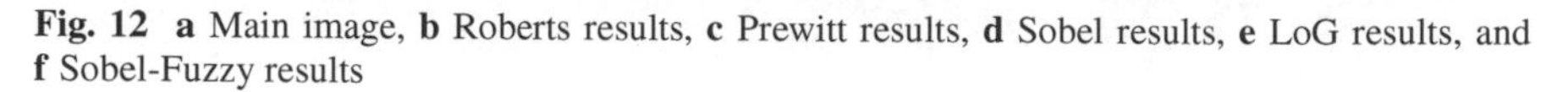

Fig. 12 **a** Main image, **b** Roberts results, **c** Prewitt results, **d** Sobel results, **e** LoG results, and **f** Sobel-Fuzzy results

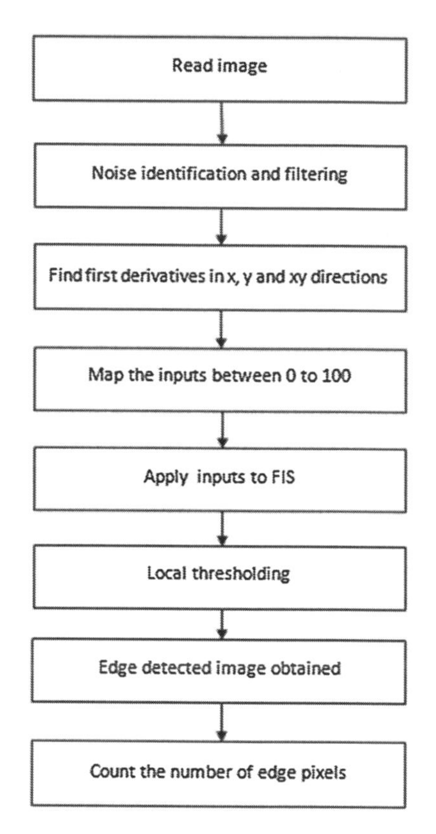

Fig. 13 Block diagram of proposed method

Figure 12 shows the visual comparison of edge detection techniques. Although Sobel provides both differencing and smoothing but it detects only some part of edges. The Roberts operator fails to detect thin/fine edges. LoG is better than gradient-based operators because smoothing is performed before the application of the Laplacian in order to remove sensitivity to noise but still it is sensitive to noise and sometimes produces double edges. We find that Sobel and Prewitt give almost the same results. These operators are highly sensitive to noise. Sobel-Fuzzy based method produces best edge map which is proved in pixel count analysis of all operators with each other in Figs. 8, 9, 10, and 11.

5 Conclusion

In this paper, first different edge detection operators to the Lena image with different levels of noise and without noise are applied, afterward results are compared quantitatively and qualitatively. Performance is measured in presence of three types of noise, Salt and Pepper noise, Gaussian noise, and Speckle noise from 30 to 60 database are introduced and then image is filtered using appropriate filters namely median, weiner, and lee filters after noise classification. The classical edge detectors chosen for the comparison are Roberts, Sobel, and Prewitt and LoG. Analysis is done by comparing the edgel count of all the edge detectors with the proposed Sobel-Fuzzy based method. So the paper concludes that given Sobel-Fuzzy based method performs well in detecting the edges, when compared to other edge detectors.

References

1. Senthilkumaran, N., Rajesh, R.: Edge detection techniques for image segmentation- a survey of soft computing approaches. Int. J. Recent. Trends. Eng. **1**(2), 250–254 (2009)
2. Tiwari, S., Kumar Singh, A., Shukla, V.P.: Statistical moments based noise classification using feedforward back propogation neural network. Int. J. Comput. Appl. **18**(2), 0975–8887 (2011)
3. Maini, R., Aggarwal, H.: Study and comparison of various image edge detection techniques. Int. J. Image Process. **3**(1), 1–12 (2010)
4. Stefno, B.D., Fuk's, H., Lawniczak A.T.: Application of fuzzy logic in CA/LGCA models as a way of dealing with imprecise and vague data. Can. Conf. Electr. Comput. Eng. **1**, 212–217 (2000)
5. Mendoza, O., Melin, P., Licea, G.: A new method for Edge Detection in Image Processing Using Interval Type-2 Fuzzy Logic. In: IEEE International Conference Granular Computing, 151–156 (2007)
6. Ur Rahman khan, A., Thakur, K.: An efficient fuzzy logic based edge detection algorithm for gray scale image. Int.J. Emerg, Technol. Adv. Eng. **2**(8), 2250–2459. http://www.ijetae.com (2012)
7. Ching-Yu, T., Wang, P.P.: Image processing-enhancement, filtering and edge detection using the fuzzy logic approach. In: Second IEEE Conference Fuzzy Systems **1**, 600–605 (1993)
8. Aborisade, D.O.: Novel fuzzy logic based edge detection technique. Int. J. Adv. Sci. Technol. **29**, 75–82 (2011)
9. Mamdani, E.H.: Application of fuzzy algorithms for control of a simple dynamic plant. Proc. Inst. Electr. Eng. **121**, 1585–1588 (1974)

Part VI
Soft Computing for Classification (SCC)

Hesitant k-Nearest Neighbor (HK-nn) Classifier for Document Classification and Numerical Result Analysis

Neeraj Sahu, R. S. Thakur and G. S. Thakur

Abstract This paper presents new approach Hesitant k-nearest neighbor (HK-nn)-based document classification and numerical results analysis. The proposed classification HK-nn approach is based on hesitant distance. In this paper, we have used hesitant distance calculations for document classification results. The following steps are used for classification: data collection, data pre-processing, data selection, presentation, analysis, classification process and results. The experimental results are evaluated using MATLAB 7.14. The Experimental results show proposed approach that is efficient and accurate compared to other classification approach.

Keywords Hesitant k-nearest neighbor · Hesitant distance · Classification · Data mining

1 Introduction

Document classification is the recent issue in text mining. Document classification areas are science, technology, social science, biology, economics, medicine, and stock market, etc. In past recent years, lot of research work has been done decodes some best contributions on Document classification are as follows: An Algorithm for a Selective Nearest Neighbor Decision Rule [1], Gradient-Based Learning Applied to Document Recognition [2], Condensed Nearest Neighbor Rule [3], Fast Nearest-Neighbor

N. Sahu (✉)
Singhania University Rajasthan, Jhunjhunu, India
e-mail: neerajsahu79@gmail.com

R. S. Thakur · G. S. Thakur
MANIT, Bhopal, India
e-mail: ramthakur2000@yahoo.com

G. S. Thakur
e-mail: ghanshyamthakur@gmail.com

B. V. Babu et al. (eds.), *Proceedings of the Second International Conference on Soft Computing for Problem Solving (SocProS 2012), December 28–30, 2012*, Advances in Intelligent Systems and Computing 236, DOI: 10.1007/978-81-322-1602-5_67, © Springer India 2014

Search in Dissimilarity Spaces [4], Branch and Bound Algorithm for Computing k-Nearest Neighbors [5], Finding Prototypes for Nearest Neighbor Decision Rule [6], An Algorithm for Finding Nearest Neighbors in Constant Average Time [7], Strategies for Efficient Incremental Nearest Neighbor Search [8], Accelerated Template Matching Using Template Trees Grown by Condensation [9], An Algorithm for Finding Nearest Neighbors [10], A Simple Algorithm for Nearest-Neighbor Search in High Dimension [11], A Fast k Nearest Neighbor Finding Algorithm Based on the Ordered Partition [12], Multidimensional Binary Search Trees Used for Associative Searching [13], Discriminant Adaptive Nearest-Neighbor Classification [14], Comparing Images Using Hausdorff Distance [15], Empirical Evaluation of Dissimilarity Measures for Color and Textures [16], A Multiple Feature/Resolution Approach to Hand printed Character/Digit Recognition [17], Representation and Reconstruction of Handwritten Digits Using Deformable Templates [18], Sparse Representations for Image Decompositions with Occlusions [19], A Note on Binary Template Matching [20], Classification with Non-metric Distances: Image Retrieval and Class Representation [21], Properties of Binary Vector Dissimilarity Measures [22], Nearest Neighbor Pattern Classification [23], Hesitant Distance Similarity Measures for Document Clustering [24], Classification of Document Clustering Approaches [25], Architecture-Based Users and Administrator Login Data Processing [26]. The above mentioned work suffers from lack of efficiency and accuracy. The low accuracy is still issue and challenge in the Classification. This motivates us to construct the new method for Classification. New Document Classification method we called Hesitant k-nearest neighbor (HK-nn). Hence we proposed new document classification approach HK-nn. The remaining paper is organized as follows: Sect. 1 describes introduction and review of literatures. Section 2 describes HK-nn and k-nn. In Sect. 3, methodology of document classification steps are described. In Sect. 4, experimental results are described. In Sect. 5, results Evaluation and measurement are described. Finally, we concluded and proposed some future directions in Conclusion Section (Fig. 1).

2 Calculations for Hesitant k-nn and General k-nn Classifier

In this calculation, we find k-nearest neighbor based on Hesitant distance (Hd) and General distance (Gd). Hesitant distance and General distance of each P_i to P_j (Tables 1, 2). Represent all distance calculated by Hd, Gd. Hesitant distance and General distance Cluster Point shown in Tables 3 and 4 with ascending order. This calculation shows hesitant distance-based accuracy percentages and General distance-based accuracy percentages Cluster Point show in Tables 5 and 6 (Fig. 2).

For computational model, we give tabulation form from Tables 1, 2, 3, 4, 5, and 6.

3 Methodology

In the classification of different document the steps are used. The steps are as follows:

(A) **Data collection.** In this phase, collect relevant documents like e-mail, news, web pages, etc., from various heterogeneous sources. These text documents are stored in a variety of formats depending on the nature of the data. The datasets are downloaded from UCI KDD Archive. This is an online repository of large datasets and has wide variety of data types.

(B) **Classification method.** Initial step is to complete review of literature in the field of data mining. Next step is a detailed survey of data mining and existing Algorithms for Classification. In this area some works are done by various researchers. After studying their work, it would be attempted to find the Classification algorithm.

(C) **Classification process.** Algorithms develop for Classification Process. Classification Process means transform documents into a suitable determined in classes for the Classification task. In Classification Process, we performed Different tasks. Optimized classification will also be studied. The real data may be great source for the Classification.

(D) **Classification results.** In this Experiment we calculate k-nearest neighbor based on Hesitant distance and General distance. Hesitant distance and General distance from Cluster Points P_i to P_j calculated and gives ascending order of the hesitant distance and General distance for tabulation. Hesitant Distance

Table 1 Hesitant distance from cluster point

Clusters points	Hesitant distance from cluster point $P_1(7, 4)$
$P_1(7, 4)$	0.00
$P_2(9, 6)$	0.16
$P_3(11, 4)$	0.31
$P_4(2, 3)$	0.21
$P_5(4, 5)$	0.14
$P_6(5, 6)$	0.07
$P_7(7, 9)$	0.11
$P_8(9, 8)$	0.05

Table 2 General distance from cluster point

Clusters points	General distance from cluster point $P_1(7, 4)$
$P_1(7, 4)$	0.00
$P_2(9, 6)$	2.82
$P_3(11, 4)$	4.00
$P_4(2, 3)$	5.09
$P_5(4, 5)$	3.16
$P_6(5, 6)$	2.82
$P_7(7, 9)$	5.00
$P_8(9, 8)$	4.47

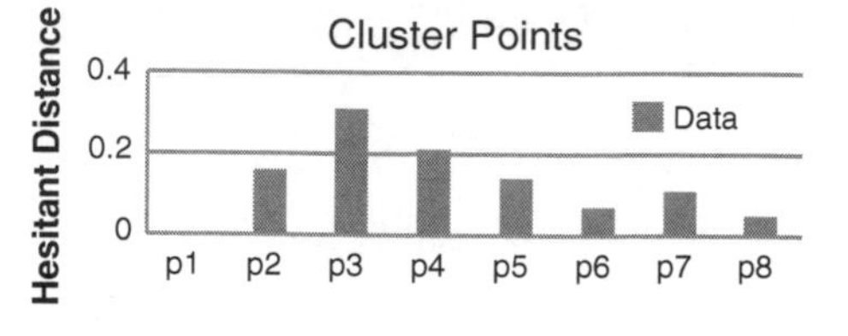

Fig. 1 Hesitant distance from cluster point

accuracy percentages and General distance accuracy percentages from Cluster Point show in tabulation. This Experiment show hesitant distance-based accuracy percentages is efficient and accurate compared with General distance-based accuracy percentages (Fig. 3).

Algorithm 1: This Algorithm obtains Hesitant distance of a cluster from each cluster.

Step 1: Input eight clusters points.
Step 2: Initialize x_1, y_1 for cluster point and x_2, y_2 for each clusters points.
Step 3: Produce and compare hesitant distance one by one.
Step 4: Find minimum Hesitant distance Hd from clusters points say first.
Step 5: Arrange all hesitant distance in ascending order.

Algorithm 2: This Algorithm obtains General distance of a cluster from each cluster.

Step 1: Input eight clusters points.
Step 2: Initialize x_1, y_1 for cluster point and x_2, y_2 for each clusters points.
Step 3: Produce and compare General distance one by one.
Step 4: Find minimum General distance Gd from clusters points say first.
Step 5: Arrange all General distance in ascending order.

4 Experimental Results

In this Experiment we calculate k-nearest neighbor based on Hesitant distance and General distance. Hesitant distance and General distance from Cluster Points P_1 to P_8 calculated and gives ascending order of the hesitant distance and General distance for tabulation describe in Tables 3 and 4. Hesitant Distance accuracy percentages and General distance accuracy percentages from Cluster Point are shown in Tables 5 and 6. This Experiment which shows hesitant distance-based accuracy percentages is efficient and accurate compared with General distance-based accuracy percentages (Fig. 4).

Figures 5 and 6 describe document classification results and accuracy % of classification process.

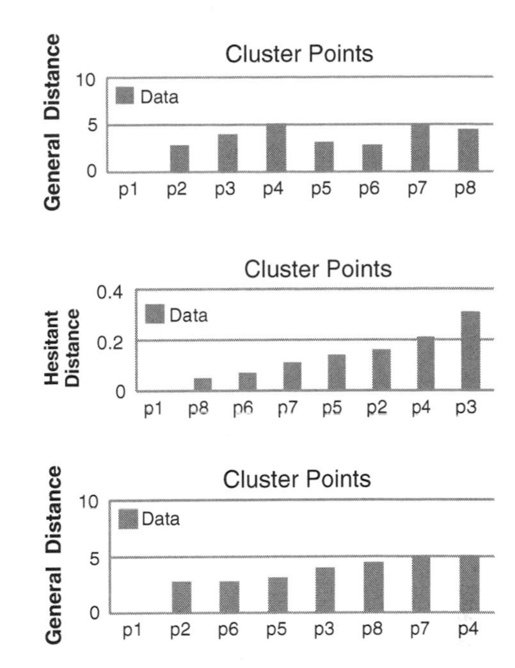

Fig. 2 General distance from cluster point

Fig. 3 Hesitant distance from cluster point in ascending order

Fig. 4 General distance from cluster point in ascending order

Table 3 Hesitant distance from Cluster Point in ascending order

Clusters points	Hesitant distance from cluster point $P_1(7, 4)$
$P_1(7, 4)$	0.00
$P_8(9, 8)$	0.05
$P_6(5, 6)$	0.07
$P_7(7, 9)$	0.11
$P_5(4, 5)$	0.14
$P_2(9, 6)$	0.16
$P_4(2, 3)$	0.21
$P_3(11, 4)$	0.31

Table 4 General distance from cluster point in ascending order

Clusters points	General distance from cluster point $P_1(7, 4)$
$P_1(7, 4)$	0.00
$P_2(9, 6)$	2.82
$P_6(5, 6)$	2.82
$P_5(4, 5)$	3.16
$P_3(11, 4)$	4.00
$P_8(9, 8)$	4.47
$P_7(7, 9)$	5.00
$P_4(2, 3)$	5.09

Table 5 Hesitant distance accuracy percentages from cluster point

Clusters points	Accuracy percentages from cluster point %
$P_1(7,4)$	12.48
$P_2(9,6)$	25.63
$P_3(11,4)$	37.59
$P_4(2,3)$	50.37
$P_5(4,5)$	62.22
$P_6(5,6)$	75.29
$P_7(7,9)$	87.67
$P_8(9,8)$	97.99

Table 6 General distance accuracy percentages from cluster point

Clusters points	Accuracy percentages from cluster point %
$P_1(7,4)$	10.34
$P_2(9,6)$	22.34
$P_3(11,4)$	32.45
$P_4(2,3)$	47.45
$P_5(4,5)$	55.45
$P_6(5,6)$	64.34
$P_7(7,9)$	79.45
$P_8(9,8)$	86.45

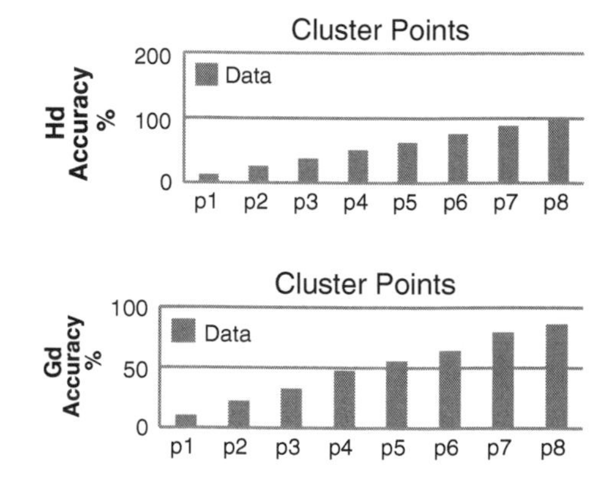

Fig. 5 Accuracy % from cluster point for Hesitant distance

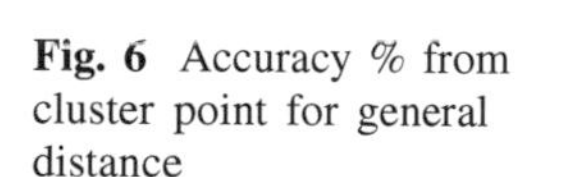

Fig. 6 Accuracy % from cluster point for general distance

Acknowledgments This work is supported by research grant from MPCST, Bhopal M.P., India, Endt.No. 2427/CST/R&D/2011 dated 22/09/2011.

References

1. Ritter, G.L., Woodruff, H.B., Lowry, S.R., Isenhour, T.L.: An algorithm for a selective nearest neighbor decision rule. IEEE Trans. Inf. Theory **21**, 665–669 (1975)
2. LeCun, Y., Bottou, L., Bengio, Y., Haffner, P.: Gradient-based learning applied to document recognition. Proc. IEEE **81**(11), 2278–2324 (1998)
3. Hart, P.E.: Condensed nearest neighbor rule. IEEE Trans. Inf. Theory **14**, 515–516 (1968)
4. Farago, A., Linder, T., Lugosi, G.: Fast nearest-neighbor search in dissimilarity spaces. IEEE Trans. Pattern Anal. Mach. Intell. **15**(9), 957–962 (1993)
5. Fukunaga, K., Narendra, P.M.: A branch and bound algorithm for computing k-nearest neighbors. IEEE Trans. Comput. **24**(7), 750–753 (1975)
6. Chang, C.L.: Finding prototypes for nearest neighbor decision rule. IEEE Trans. Comput. **23**(11), 1179–1184 (1974)
7. Vidal, E.: An algorithm for finding nearest neighbors in (approximately) constant average time. Pattern Recogn. Lett. **4**(3), 145–157 (1986)
8. Broder, A.J.: Strategies for efficient incremental nearest neighbor search. Pattern Recogn. **23**(1/2), 171–178 (1986)
9. Brown, R.L.: Accelerated template matching using template trees grown by condensation. IEEE Trans. Syst Man Cybern. **25**(3), 523–528 (Mar. 1995)
10. Friedman, J.H., Baskett, F., Shustek, L.J.: An algorithm for finding nearest neighbors. IEEE Trans. Comput. **24**(10), 1000–1006 (1975)
11. Nene, S.A., Nayar, S.K.: A simple algorithm for nearest-neighbor search in high dimension. IEEE Trans. Pattern Anal. Mach. Intell. **19**(9), 989–1003 (1997)
12. Kim, B.S., Park, S.B.: A fast k nearest neighbor finding algorithm based on the ordered partition. IEEE Trans. Pattern Anal. Mach. Intell. **8**(6), 761–766 (1986)
13. Bentley, J.L.: Multidimensional binary search trees used for associative searching. Commun. ACM **18**(9), 509–517 (1975)
14. Hastie, T., Tibshirani, R.: Discriminant adaptive nearest-neighbor classification. IEEE Trans. Pattern Anal. Mach. Intell. **18**(6), 607–615 (1996)
15. Hunttenlocher, D., Klanderman, G., Rucklidge, W.: Comparing images using Hausdorff distance. IEEE Trans. Pattern Anal. Mach. Intell. **19**(1), 1–14 (1997)
16. Puzicha, J., Buhmann, J., Rubner, Y., Tomasi, C.: Empirical evaluation of dissimilarity measures for color and textures. In: Proceedings of International Conference on Computer Vision, pp. 1165–1172 (1999)
17. Favata, J.T., Srikantan, G.: A multiple feature/resolution approach to hand printed character/digit recognition. Proc. Intl J. Imaging Syst. Technol. **7**, 304–311 (1996)
18. Jain, A.K., Zongker, D.: Representation and reconstruction of handwritten digits using deformable templates. IEEE Trans. Pattern Anal. Mach. Intell. **19**(12), 1386–1391 (1997)
19. Donahue, M., Geiger, D., Hummel, R., Liu, T.: Sparse representations for image decompositions with occlusions. In: Proceedings of the IEEE Conference on Computer Vision and Pattern Recognition, pp. 7–12 (1996)
20. Tubbs, J.D.: A note on binary template matching. Pattern Recogn. **22**(4), 359–365 (1989)
21. Jacobs, D.W., Weinshall, D.: Classification with non-metric distances: image retrieval and class representation. IEEE Trans. Pattern Anal. Mach. Intell. **22**(6), 583–600 (June 2000)
22. Zhang, B., Srihari, S.N.: Properties of binary vector dissimilarity measures. In: Proceedings of JCIS International Conference on Computer Vision, Pattern Recognition, and Image Processing (2003)
23. Cover, T.M., Hart, P.E.: Nearest neighbor pattern classification. IEEE Trans. Inf. Theory **13**, 21–27 (1968)
24. Neeraj, S., Thakur, G.S.: Hesitant distance similarity measures for document clustering. In: IEEE Conference World Congress on Information and Communication Technologies Mumbai, India. ISBN: 978-1-4673-0125-1, 11–14 December 2011

25. Sahu, S.K., Sahu, N, Thakur, G.S.: Classification of document clustering approaches. Int. J. Adv. Res. Comput. Sci. Softw. Eng. (IJARCSSE) ISSN (ONLINE): 2277 128X, **2**(5), 509–513 (2012)
26. Sahu, B., Sahu, N., Thakur , G.S.: Architecture based users and administrator login data processing In: International Conference on Intelligent Computing and Information System (ICICIS-2012), Pachmarhi, Piparia (MP). ISSN (ONLINE): 2249–071X, Oct. 27–28 2012

Lower Bound on Naïve Bayes Classifier Accuracy in Case of Noisy Data

Karan Rawat, Abhishek Kumar and Anshuman Kumar Gautam

Abstract Classification is usually the final and one of the most important steps in most of the tasks involving machine learning, computer vision, etc., for e.g., face detection, optical character recognition, etc. This paper gives a novel technique for estimating the performance of Naïve Bayes Classifier in noisy data. It also talks about removing those attributes that cause the classifier to be biased toward a particular class.

Keywords Optimal statistical classification · Confusion matrix · ROC · Discrminant functions · Bayes theorem

1 Introduction

Naïve Bayes [1] classification is a generative machine learning algorithm that exploits the possible independence between attributes of data points. It is a special case of more general "**Optimal Statistical Classifier**" [2]. Although other techniques such as **Logistic Regression**, **Neural Networks**, **Gaussian Discriminant Analysis**, its special case, **Linear Discrminant Analysis** [3] are available, but the simplicity in design and implementation makes **NB** classification a popular choice for classification, provided its underlying assumption holds true to some extent. It has proved to be useful in large number of classification problems involving multivariate feature vectors.

K. Rawat (✉)
IIT Allahabad, Allahabad, India
e-mail: rawatkaran4@hotmail.com

A. Kumar · A. K. Gautam
Government Engineering College, Ajmer, Rajasthan, India

B. V. Babu et al. (eds.), *Proceedings of the Second International Conference on Soft Computing for Problem Solving (SocProS 2012), December 28–30, 2012*, Advances in Intelligent Systems and Computing 236, DOI: 10.1007/978-81-322-1602-5_68, © Springer India 2014

2 Related Work

Accuracy of statistical classifiers can be measured using **Confusion Matrix** [4], **ROC** [5] curve, etc. However, these measures (a) estimate how well the assumed model fits the underlying data and (b) this too requires testing the classifier on another data set called "Test Data". But, this paper does something different. It assumes that the data model chosen is correct. It checks the accuracy of the classifier against noisy inputs, i.e., to what extent noise in the input data vector is tolerable. The above-mentioned measures, however, test the accuracy of the data against a given test data and does not give any information about the noise handling capacity of the classifier.

3 Bayes Theorem

The concept of **Inverse probability** was first given by Thomas Bayes but published by Laplace who reached the same result independently and published it in a paper titled "**Théorie analytique des probabilities**" in 1812. It calculates the probability of the event that occurred on performing an experiment given the result (outcome) of the experiment. This is mathematically given as:

$$P(A|B) = P(B|A)P(A)/P(B).$$

In case the experiment is composed of n events, $P(B)$ is given as:

$$P(B) = \sum_i (P(B|A_i)P(A_i)).$$

4 Discriminant Functions

These are the functions used that that determine the class of a data point using one discriminant function [4] corresponding to each class.

The predicted class of a data vector $\mathbf{x}$ is i if

$$d_i(\mathbf{x}) > d_j(\mathbf{x}) \quad \text{for all } j \neq i.$$

Decision Boundary between classes i and j is given by

$$d_{ij}(\mathbf{x}) = d_i(\mathbf{x}) - d_j(\mathbf{x}) = 0.$$

5 Optimal Statistical Classification

If a data vector **x** is predicted to belong to class j when it actually belongs to class I, the loss incurred is, say, L_{ij}. Now, i could have been any possible class, therefore, the prediction that **x** belongs to j incurs an expected loss:

$$r_j(\mathbf{x}) = \sum_i P(i|\mathbf{x})L_{ij}.$$

This is known as **Conditional Average Risk**.

Using Bayes Theorem, we have

$$r_j(\mathbf{x}) = (1/P(\mathbf{x}))\sum_i L_{ij}P(\mathbf{x}|i)P(i).$$

Since, $P(\mathbf{x})$ is same for all i, we can define a new conditional average risk variable,

$$s_i(\mathbf{x}) = \sum_i P(\mathbf{x}|i)L_{ij}P(i).$$

Optimal Statistical Classification calculates $s_1(\mathbf{x}), s_2(\mathbf{x}), \ldots, s_w(\mathbf{x})$ and predicts the class with minimum CAR. Bayes Classifier minimizes the total CAR over all predictions by assigning a data vector **x** to class i if

$$\begin{aligned} &s_i(\mathbf{x}) < s_j(\mathbf{x}) \\ &\sum_p P(\mathbf{x}|p)L_{pi}P(q) < \sum_q P(\mathbf{x}|q)L_{qj}P(q) \quad \text{for all } j \neq i. \end{aligned}$$

Now, for a correct prediction, loss incurred should be 0 and a loss of unity is incurred in case of incorrect prediction.

$$L_{ij} = 1 - \delta(i, j)$$

where δ is the Kronecker delta function.

$$\begin{aligned} &\sum_p P(\mathbf{x}|p)P(q)(1-\delta(p,i)) < \sum_q P(\mathbf{x}|q)P(q)(1-\delta(q,j)). \\ &P(\mathbf{x}) - P(\mathbf{x}|i)P(i) < P(\mathbf{x}) - P(\mathbf{x}|j)P(\mathbf{x}) \\ &P(\mathbf{x}|j)P(j) < P(\mathbf{x}|i)P(i) \quad \text{for all } j \neq i. \end{aligned}$$

Hence, for Bayesian classification,

$$d_i(\mathbf{x}) = P(\mathbf{x}|i)P(i)$$

6 Naïve Bayes Assumption

The classifier assumes that the attributes of data are independent of each other, and hence, contribution of an attribute to the probability that a given data point belongs to a given class does not depend on the values assumed by other attributes. Formulating this mathematically:

$$P(x_1, x_2 \ldots, x_n|C) = P(x_1|C)P(x_2|C)\ldots P(x_n|C)$$

where C is a class label and x_i is the random variable corresponding to ith attribute.

7 Algorithm

1. After the calculation of means and variances for all the classes and for each attribute (to be used in NB classification), for each pair of classes, i, j and for each attribute, k, $x^*_{ij,k}$ is calculated using the following equation :

   $$(\varphi_i)^{1/n} f_{i,k}(x^*_{ij,k}) = (\varphi_j)^{1/n} f_{j,k}(x^*_{ij,k})$$

   where φ_p is the prior probability of occurrence of class p and $f_{p,q}(z)$ is the probability density function for class p and attribute q with random variable Z. Often, when f is not known, Z is assumed to follow Gaussian distribution and hence,

   $$f_{p,q}(z) = (1/\sigma_{p,q}\sqrt{2\pi})\exp(-(z-\mu_{p,q})^2/2\sigma^2_{p,q})$$

   where $\mu_{p,q}$ and $\sigma_{p,q}$ are mean and variance for class p and attribute q, respectively.
2. Using $x^*_{ij,k}$, find the intervals for which $f_{i,k}(z) > f_{j,k}(z)$. Find the c.d.f of $f_{i,k}(z)$ over these intervals and obtain a value $D_{i,k}$.
3. Calculate $D_{i,k}$ for all k.
4. The probability that a data vector belonging to class i comes from the intervals determined in step 2 is

   $$\prod_k D_{i,k}.$$

8 Working

By determining the value of t for classifier boundary between each pair of classes, we can calculate the confusion matrix and hence, find the accuracy. The principle behind the correctness of this method is that for any two classes, i, j a data vector is

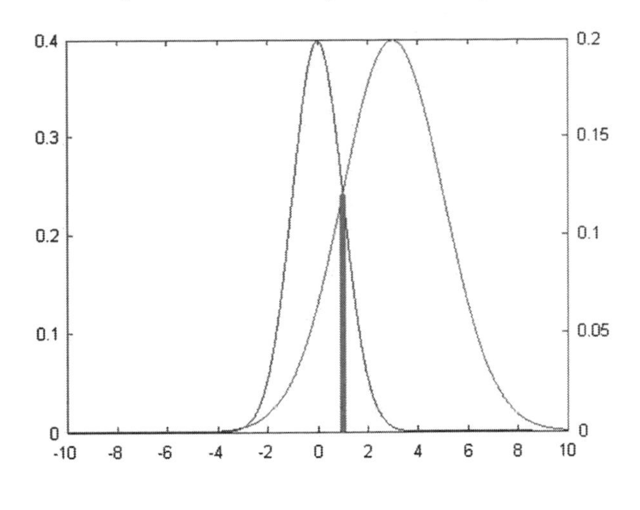

Fig. 1

assigned to class i if

$$d_i(\mathbf{x}) > d_j(\mathbf{x}).$$

And, this will happen for sure if

$$(\varphi_i)^{1/n} f_{i,k}(x_{ij,k}) > (\varphi_j)^{1/n} f_{j,k}(x_{ij,k}), \quad \text{for all } k.$$

By determining the set of values for which these will hold and estimate of the relative frequency of data vectors that will fall within this set using c.d.f. of standard normal distribution.

9 Graphical Approach

Taking a two-class problem in which data vectors are two- dimensional, two graphs are shown corresponding to each attribute.

Prior probability of each class is taken to be 0.5. So, in Fig. 1 the red line shown is the dividing line $x = x^*$. There exists one more point because of the quadratic nature of Gaussian decision boundaries, however, Gaussian value at that point is quite low, therefore, it is not of concern.

Figure 2 tells us that Gaussian for class 1 has higher value than that for class 2 and that too also includes large amount of neighborhood of mean of class 2. This tells that using second attribute for classification will not be a good idea as classifier would be biased toward class 1. Hence, we should reduce the dimensionality to 1 taking only first attribute.

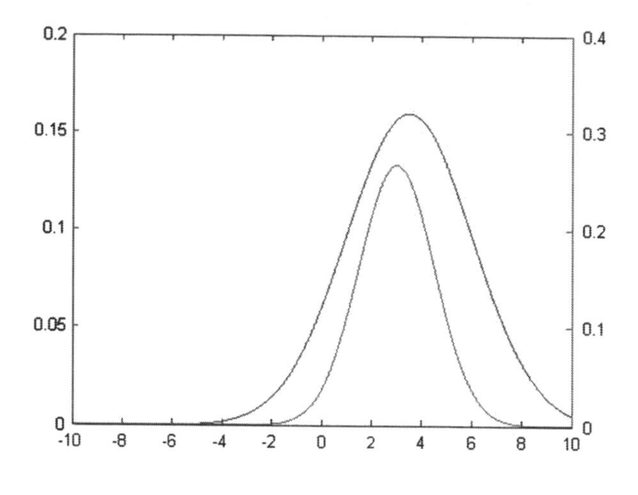

Fig. 2

10 Conclusion

The accuracy of **NB** classifier can be estimated only from the parameters of discriminant functions, determined using training data. Hence, no further test data is required and thus, prior to training, data set does not need to be cross-validated and a larger data set can be used, resulting in higher accuracy. This method, however, assumes that NB assumption holds true and determines how the accuracy of prediction is affected when data points that are less likely to belong to a given class actually belong to that class.

References

1. John, G.H., Langley, P.: Estimating continuous distributions in Bayesian classifiers. In: Proceedings of the Eleventh Conference on Uncertainty in Artificial Intelligence. Morgan Kaufmann Publishers, San Mateo (1995)
2. Gonzalez, R.C., Woods, R.E.: Digital Image Processing. Prentice Hall, New Jersery. ISBN-10:013168728X (2007)
3. Fisher, R.A.: The use of multiple measurements in taxonomic problems. Ann. Eugenics **7**, 179–188 (1936)
4. Ting, K.M.: Confusion Matrix. Encyclopedia of Machine Learning, 1st edn. Springer, Berlin, p. 209. ISBN-10: 0387307680 (2011)
5. Fawcett, T.: An introduction to ROC analysis. Pattern Recogn. Lett. **27**(2006), 861–874 (2006)

A Neuro-Fuzzy Approach to Diagnose and Classify Learning Disability

Kavita Jain, Pooja Manghirmalani Mishra and Sushil Kulkarni

Abstract The aim of this study is to compare two supervised artificial neural network models for diagnosing a child with learning disability. Once diagnosed, then a fuzzy expert system is applied to correctly classify the type of learning disability in a child. The endeavor is to support the special education community in their quest to be with the mainstream. The initial part of the paper gives a comprehensive study of the different mechanisms of diagnosing learning disability. Models are designed by implementing two soft computing techniques called Single-Layer Perceptron and Learning Vector Quantization. These models classify a child as learning disabled or nonlearning disabled. Once diagnosed with learning disability, fuzzy-based approach is used further to classify them into types of learning disability that is Dyslexia, Dysgraphia, and Dyscalculia. The models are trained using the parameters of curriculum-based test. The paper proposes a methodology of not only detecting learning disability but also the type of learning disability.

Keywords Learning disability · Single-layer perceptron · Learning vector quantization · Fuzzy expert system

K. Jain (✉) · P. M. Mishra
Department of Computer Science, University of Mumbai, Mumbai-98, India
e-mail: cavita_jain@yahoo.com

P. M. Mishra
e-mail: pmanghirmalani@gmail.com

S. Kulkarni
Department of Mathematics, Jai Hind College, Mumbai-20, India
e-mail: sushiltry@gmail.com

B. V. Babu et al. (eds.), *Proceedings of the Second International Conference on Soft Computing for Problem Solving (SocProS 2012), December 28–30, 2012*, Advances in Intelligent Systems and Computing 236, DOI: 10.1007/978-81-322-1602-5_69,

1 Introduction

Learning disability [LD] denotes to a neurological condition which disturbs an individual's ability to think and remember. It is established in disorders of listening, thinking, reading, writing, spelling, or arithmetic [1]. These individuals are not attributed to medical, emotional, or environmental causes despite having normal intellectual abilities [2].

LD can be broadly classified into three types. They are difficulties in learning with respect to read (Dyslexia), to write (Dysgraphia) or to do simple mathematical calculations (Dyscalculia) [4] which are often termed as special learning disabilities. Kirk [3] stated that, children with special learning disabilities exhibit a disorder in one or more of the basic psychological processes involved in understanding or in using spoken or written language. These may be manifested in disorders of listening, thinking, talking, reading, writing, spelling, or arithmetic. They include conditions which have been referred to as perceptual handicaps, brain injury, minimal brain dysfunction, dyslexia, developmental aphasia, etc., and they do not include learning problems which are due primarily to visual, hearing, or motor handicaps, to mental retardation, emotional disturbance or to environmental deprivation.

For diagnosing LD, there does not exist a global method. Mostly detection is done using Wechsler Intelligence Scale for Children (WISC) test [5], conducted in the supervision of special educators and with the observation of parent and teachers. In this context, computational approach to detect LD is quite significant. Once LD is successfully diagnosed, there is no substantial work done to classify LD into its three types.

This paper proposes a model for diagnosis and classification of LD. Section 2 of this paper explores in detail different computational methods and models applied in or diagnosing LD. Having elaborately explored different approaches, we have found that there are still possible ways to improvise the entire diagnosing process. Section 3 describes the system parameters and Sect. 4 introduces the proposed models and compares their diagnosis results. Section 5 gives the system results based on accuracy, sensitivity, specificity for the diagnosing phase. Section 6 discusses the classification of LD which is done by applying a fuzzy-based approach and Sect. 7 gives the result for classification phase of the experiment. Finally Sect. 8 discusses the conclusion and future works by the authors.

2 Taxonomy of Computational Diagnosis of Learning Disability

The different computational methods and models used in detecting learning disability can be classified into four groups. The grouping is done based on the broader theoretical foundation and computational characteristics of the models and methods applied. The following subsections deal with such models and methods.

2.1 Digital Signal Processing Methodologies

Reitano [6] analyses and compares spoken words with prerecorded and properly pronounced phonemes. The mispronounced phonemes are indicated which led to the diagnosis of LD. Fonseca et al. [7] conducted electroencephalograms (EEG) to detect abnormalities related to electrical activity of the brain by studying different brainwaves. He concluded that there is a significant difference in brainwaves of normal and learning disabled children. Assecondi et al. [8] carried out Electroencephalograms while awake and resting and the values for absolute and relative powers were calculated. Children with severe reading/ writing disabilities had more delta activity in frontal-temporal regions and those with less intense disabilities had more theta activity and less relative alpha activity. A study of the relationship between quantitative EEG variables and IQ provided greater knowledge about the biological aspects related to LDs. Bonte et al. [9] compared the values of amplitude and latency of some peaks in Reading Related Potential (RRP) of the dyslexia children with the reference template of normal children. The amplitude and latencies in RRPs are based on *Dynamic Time Warping Technique* and this technique is used in the speech processing to match 1-Dimentional signals [10]. This comparison indicated a valuable tool in understanding of dyslexia [11].

2.2 Digital Image Processing (DIP) Methodologies

Mico-Tormos et al. [12] inferred that eye movements of even an infant could indicate LD by analyzing the responses of the movement of eye through oculographic signals. From computer analysis of records, high correlation between the neural activity and eye response, as well as linear dependency of eye movement on stimulus velocity has been documented by numerous studies [13, 14]. Pavlidis [15] observed that erratic and strikingly large number of regressive eye movements pointed to dyslexia.

2.3 Soft Computing Methodologies

The various components of soft computing used to diagnose LD can be viewed as follows:

2.3.1 Artificial Neural Network (ANN)

Jain et al. [16] proposed a simple perceptron based ANN model and it comprised of a single input layer with eleven units which correspond to different sections of a curriculum-based test and one output unit for diagnosing LD. Bullinaria [17] on the other hand applied a multilayer feed forward perceptron to diagnose dyslexia where letter strings were mapped to phoneme strings in multisyllabic words. Wu et al. [18] proved that a multilayer perceptron with back propagation gave better results in diagnosing LD. He later attempted to diagnose LD using support vector machines [19]. Manghirmalani et al. [20] proposed a soft computing technique called Learning Vector Quantization. The model classifies a child as learning disabled or nonlearning disabled. Once diagnosed with learning disability, rule-based approach is used further to classify them into types of learning disability that is dyslexia, dysgraphia, and dyscalculia.

2.3.2 Fuzzy Systems (FS)

Sanchez et al. [21] proposed a technique to deal with the vagueness of the data by analyzing the set of different granularities based on more flexible representation of polygons. Manghirmalani et al. [39] used fuzzy expert systems to classify the type of LD out of the correctly identified LD cases. Fuzzy expert system uses rules which are the most fundamental part for classification purpose and thus successfully classifying the LD cases into one of the seven types of LD.

2.3.3 Genetic Algorithms with Fuzzy Systems

Georgopoulos et al. [22] put forth a strong mechanism to deal with the inputs whose information is not adequate by combining Fuzzy Cognitive Maps (FCM) and genetic algorithm (GA). This combination of FCM and GA led to better accuracy of diagnosing LD.

2.4 Hybridized Computational Techniques

Hybridized approaches in LD include attempts to apply video and signal processing techniques along with soft computing techniques. Salhi et al. [23] used both wavelet transforms and neural networks to diagnose LD from Pathological Voices. The results using the multilayer neural network (MNN) classifier gives the best correct classification. A feature vector based on wavelet coefficients was used for classification of normal and pathological speech data. The MNN with BP used as a classifier has been proved to be more efficient and more precise than the time-frequency analysis method. The MNN classifier represents a low cost, accurate, and automatic tool for pathological voice classification using wavelet coefficients. Using the MNN with BP the system gave the classification rate between 70–100 %. learner is as close as possible to learner is as close as possible toNovak et al. [24] have calculated a set of features from signals of horizontal and vertical eye movement using self-organizing map and genetic algorithm (GA). They concluded that the reading speed increased with the probability of the patient being healthy. Wu et al. [25] combined different feature selection algorithms like brute-force, greedy, and GA along with ANN to improve the identification rate of LD. Macaš et al. [26] developed a system for extracting the features of eye movements from time and frequency domain. They concluded that back propagation-based classification gave better results than that offered by Bayes' and Kohonnen network.

3 Collection of Parameters

A curriculum-based test was designed with respect to the syllabus of primary-level school going children. This test was conducted in schools for collecting non-LD and testing datasets. Historic data for LD cases were collected from LD Clinics of Government hospitals where the tests were conducted in real-time medical environments. The system was fed with 11 input units which correspond to 11 different sections of the curriculum-based test [16]. Dataset consists of 240 cases, out of which 160 are of normal children and 80 of LD children. The system was trained using 120 data items and the remaining was used to test the network.

4 Diagnosis Model

Single layer Perceptron and LVQ are supervised neural network algorithms. Both the soft computing models have been tested and trained using the same dataset and hence their comparison is of relevance with respect to the problem of diagnosing LD using artificial neural networks.

4.1 Single Layer Perceptron

The proposed system is called Perceptron based Learning Disability Detector (PLEDDOR). PLEDDOR consists of one input layer and one output layer. The goal of the Perceptron algorithm is to find a combination of expert predictions such that the performance of the learner is as close as possible to the best combination of experts. The Perceptron maps an input vector $x = [x_1, x_2, x_3, \ldots, x_n]^T$ to a binary output y. The value of x and y is taken to be 0 and 1. Thus, the model could be looked as a simple two-class classifier [27]. In this manuscript, it is classifying the data set into 'Normal' and 'Learning Disabled,' that is, for taking the value 1 for Normal children and 0 for Learning Disabled children. A weight vector $w = [w_1, w_2, w_3, \ldots, w_n]^T$, is used to train the network [28]. The model consists of one input layer and one output layer. In the input layer, the proposed model is fed in with 11 inputs corresponding

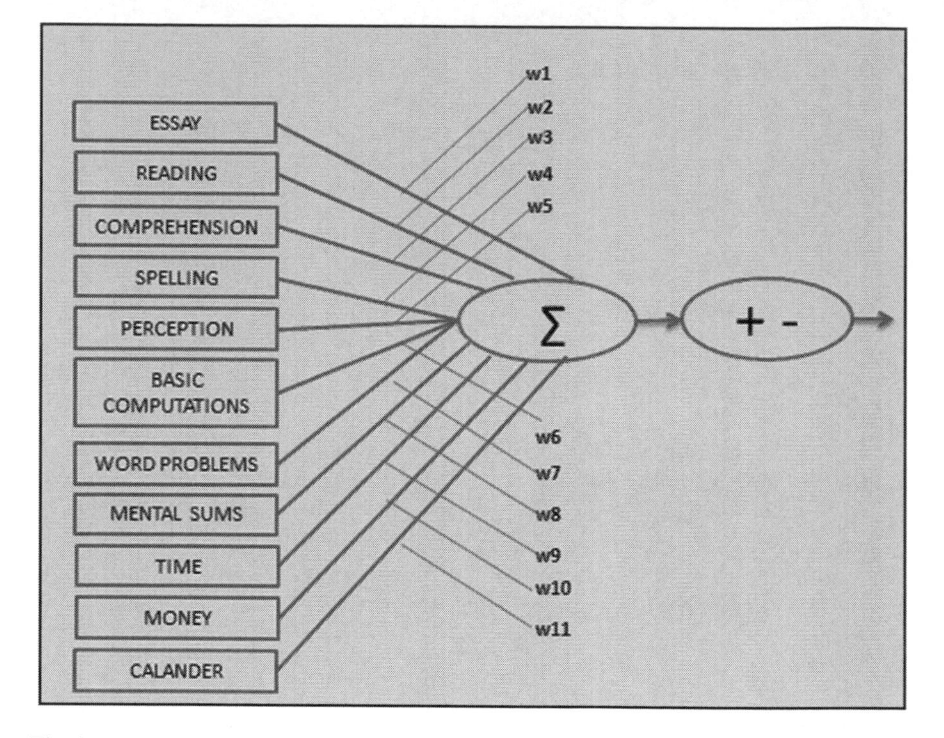

Fig. 1 Perceptron model with system parameters

to the 11 sections of the curriculum-based test as shown in Fig. 1. On these 11 inputs, 11 weight vectors ($w_1 - w_{11}$) are applied.

Algorithm

1. Create a Perceptron with n inputs and n weights.
2. Initialize the weights to random real values.
3. Iterate through the training set, collecting all data misclassified by the current set of weights.
4. If all the data is classified correctly, output the weights and quit.
5. If output is low, increment weights for all inputs which are 1.
6. If output is high, decrement weights for all the inputs which are 1.

---

1. Applying Activation function

The activation function used in the system is given by: $f(z) = 1$ if $z - \theta \geq 0$; $f(z) = 0$ if $z - \theta < 0$ where $z = w^T x$ *and* θ is a threshold, which is called as 'LD threshold'. After experimental results, the value of LD threshold is fixed as 30.

2. Training of the network

The PLEDDOR is trained using the training pair $\{x^1, d^1\}$, $\{x^2, d^2\}, \ldots$ $\{x_m, d_m\}$; where x^k is the kth input vector and $d^k = \{1, 0\}$ where $1 \rightarrow$ Normal; $0 \rightarrow$ Learning Disabled The system is trained by adjusting the weight vector to make the output y^k corresponding to the input x^k match with the desired output d^k.

4.2 Learning Vector Quantization (LVQ)

The second proposed system is called Diagnosis & Classification using Learning Vector Quantization for Learning Disability (DICQOLD) applying the scheme proposed by Kohonen [29]. LVQ is a special case of artificial neural network (ANN) which applies a winner-take all Hebbian Learning-based approach [30]. Here one regulates the pattern for each data which is closest to the input vector agreeing to a Eucledian distance. The location of the winner prototype is thus computed.

This learning technique uses the class information to reposition the weight vectors (w_j) slightly, so as to increase the quality of the classifier. The basic LVQ approach is quite intuitive. This allows us to use the known classification labels of the inputs to find the best classification label for each (w_j). The objective of LVQ is the representation of a set of feature vectors $x \in \chi \subset IR^n$ by a set of prototypes $\nu = \{v_1, v_2 \ldots v_c\} \subset IR^n$. Thus, vector quantization can also be seen as a mapping from a dimensional Euclidean space into the finite set $\nu \in IR^n$, also referred to as the codebook [31].The network consists of an input layer and an output layer. Each node in the input layer is connected directly to the cells, or nodes, in the output layer. A prototype vector is associated with each cell in the output layer [32].

Algorithm:

1. Initialize weights vectors (for the system it is $w_1 - w_{11}$).
2. Initialize learning rate (α), (for the system it is 0.8).
3. For each training input vector x (for the system it is x_1 to x_{11}), do steps 4 to 5.
4. Compute J using squared Euclidean distance: $D(j) = \sum \left(w_{ij} - x_j\right) * D(j) = \sum \left(w_{ij} - x_j\right)$ Find j when $D(j)$ is minimum. This leads to the classification of the entire dataset into two classes, one pertaining to LD and the other pertaining to the normal class. Cj indicates the class of LD; $Cj = 0$ indicates that the child is suffering from LD and $C_j = 1$ indicates the child is normal.
5. Update W_j as follows: If $t = C_j$, then W_j (new) $= W_j$ (old) $+ \alpha \left[x - W_j \text{ (old) }\right]$ If $t \neq C_j$, then W_j (new) $= W_j$ (old) $- \alpha \left[x - W_j \text{ (old) }\right]$
6. Reduce α by a small amount 0.01.
7. System is tested for every new value of α till the result shows redundancy. This is the stopping condition.

5 Results and Discussions–Diagnosing Phase

5.1 Detection Measures

The paper uses the following conventional detection measures as benchmarks to compare the performance of the proposed model with that of similar models.

- *Accuracy*: ratio of the number of correct classification of input-output data to the total number of training data.
- *Sensitivity*: ratio of the correctly classified LD cases to the total number of LD cases.
- *Specificity*: ratio of the correctly classified non-LD cases to the total number of non-LD cases.

a. System performance measures for PLEDDOR: Accuracy = 84 %, Sensitivity = 80 %, Specificity = 80 %.
b. System performance measures for DICQOLD: Accuracy = 91.8 %, Sensitivity = 84.5 %, Specificity = 84.5 %.

5.2 Comparison

It is observed that as the number of samples increases, the accuracy of the system also increases and then it tends to stabilize from the training dataset size 90 onwards. Testing is done using 100 samples. For PLEDDOR the accuracy of 84 % is achieved whereas for DICQOLD an accuracy of almost 92 % is achieved.

6 Classification Model

Once LD has been detected, the system further classifies the case into the type/s of LD. Classification is done using the fuzzy logic [33]. It consists of crisp input stage, fuzzy inference system, and a crisp output stage. The input stage maps inputs to apt membership functions and truth values. The processing stage invokes each suitable rule and generates a result for each, then combines the results of the rules. Finally, the output stage converts the combined result back into a specific control output value [34, 35]. This is done using 'IF variable IS property THEN action' rule [36].

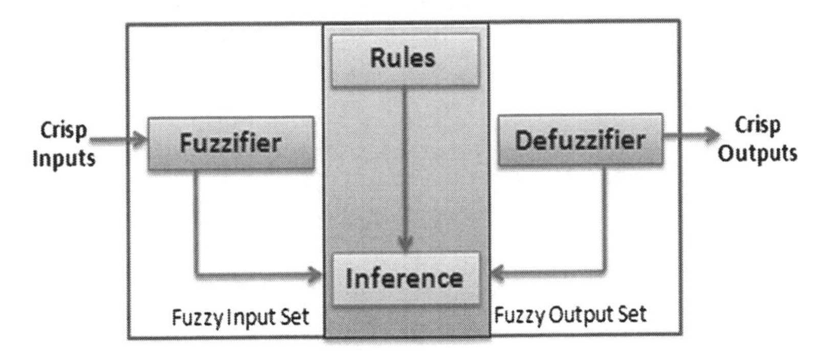

Fig. 2 Fuzzy inference system

6.1 Fuzzy Logic System

A fuzzy logic system (FLS) can be defined as the nonlinear mapping of an input data set to a scalar output data. A FLS consists of four main parts: fuzzifier, rules, inference engine, and defuzzifier. FLS is one of the most famous applications of fuzzy logic and fuzzy sets theory. They can be helpful to achieve classification tasks, offline process simulation, and diagnosis, online decision support tools and process control [37]. As shown in Fig. 2, a crisp set of input data are gathered and converted to a fuzzy set using fuzzy linguistic variables, fuzzy linguistic terms, and membership functions. This step is known as fuzzification. Afterward, an inference is made based on a set of rules. And lastly, the resulting fuzzy output is mapped to a crisp output using the membership functions, in the defuzzification step.

6.2 Fuzzy Modeling

As show in the Fig. 3, given below are the seven steps of fuzzy model applied to the data [38].

1. Define the linguistic variables and terms.
2. Construct the membership functions.
3. Construct the rule base.
4. Convert crisp input data to fuzzy values using membership functions.
5. Applying implication method.
6. Aggregating all output.
7. Convert the output data to nonfuzzy values.

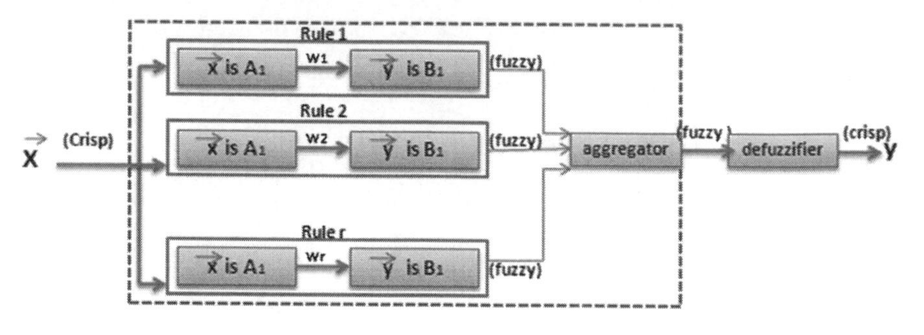

Fig. 3 Fuzzy expert system steps

6.3 Rules for Diagnosis

Case-1: IF score low in essay, reading, comprehension, spelling, perception, solve, word problems, mental sums, time, money; THEN Dyslexia, Dysgraphia and Dyscalculia

Case-2: IF low in of reading, comprehension, perception, word problem; THEN Dyslexia

Case-3: IF low in of spelling, comprehension, essay; THEN Dysgraphia

Case-4: IF low in of solve, mental sums, word problems, sums related to time, calendar, money; THEN Dyscalculia

Case-5: IF low in of reading, comprehension, perception, word problem, spelling, essay; THEN Dyslexia and Dysgraphia

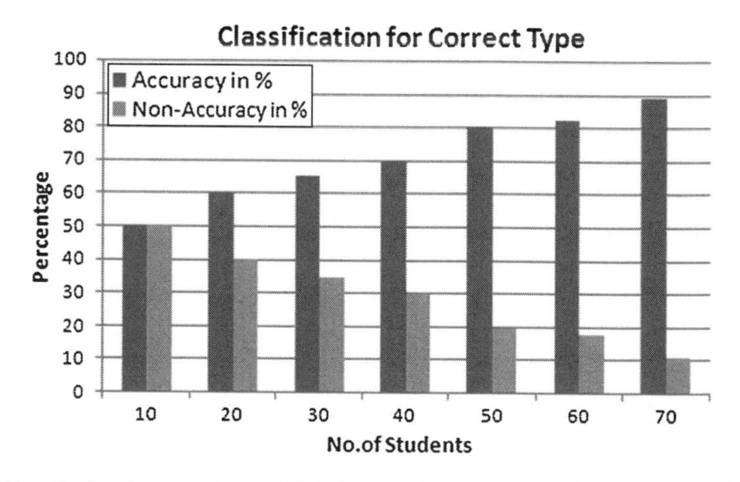

Fig. 4 Classification for correct type of LD showing the accuracy and the nonaccuracy during the training phase. accuracy and the nonaccuracy during the training phase

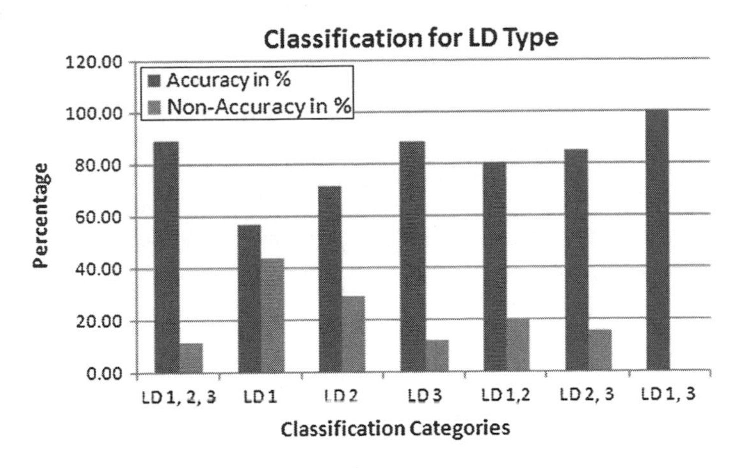

Fig. 5 Classification for specific LD showing the accuracy and the nonaccuracy during the training phase. accuracy and the nonaccuracy during the training phase

Case-6: IF low in of reading, comprehension, perception, solve, mental sums, word problems, sums related to time, calendar and money; THEN Dyslexia and Dyscalculia

Case-7: IF low in of spelling, comprehension essay, solve, mental sums, word problems, sums related to time, calendar, money; THEN Dysgraphia and Dyscalculia.

7 Results and Discussions—Classification Phase

The classification system gives an accuracy of 90 % Figs. 4 and 5.

8 Conclusion

We have presented a comparison between two supervised artificial neural network algorithms names Single-Layer Perceptron and Learning Vector Quantization. We have observe, when trained and tested with the same dataset, the accuracy measure between both these approaches is competitive. However, the model of LVQ has proved to be slightly more accurate with respect to the Single-Layer Perceptron. However, there is scope for further enhancement of system by finding a combination of algorithms so as to build up a model that is satisfactorily accurate. It is also seen that on increasing the number of data in the training set of the system, the overall accuracy in both the models shows a promising growth.

In future we intend to explore the possibility of parameter classification in order to distinguish irrelevant and superfluous variables which might lead to decrease in diagnosis and classification process time and increase in accuracy. This can be beneficial for the special educators, doctors, and teachers by providing suggestions that lead to the exclusion of redundant tests and saving of time needed for diagnosing LD. On the whole, the focus of our research is to identify early diagnosis of LD and to support special education community in their quest to be with mainstream.

References

1. Kirk, S.A.: Educating exceptional children book. Wadsworth Publishing, ISBN: 0547124139
2. Weyandt, L.L.: The physiological bases of cognitive and behavioral disorders. Blausen Medical Communications, United States
3. Lerner, J.W.: Learning disabilities: theories, diagnosis, and teaching strategies. Houghton Mifflin, Boston ISBN 0395961149
4. Fletcher, J.M., Francis, D.J., Rourke, B.P., Shaywitz, B.A., Shaywitz, S.E.: Classification of learning disabilities: an evidencebased evaluation. (1993)
5. Kaplan, R.M., Saccuzzo, D.P.: Psychological testing: principles, applications, and issues, 7th edn. p. 262. Wadsworth, Belmont ISBN 978-0- 495-09555-2 (citing Wechsler (1958) The Measurement and Appraisal of Adult Intelligence) (2009)
6. Carmen T.R.: System and method for dyslexia detection by analyzing spoken and written words. US Patent 6,535,853 B1, 18 Mar 2003
7. Fonseca, L.C., Tedrus, G.M.A.S., Chiodi, M.G., Cerqueira, J.N. Tonelotto J.M.F.: Quantitative EEG in children with learning disabilities. Anal.Band power. **64**(2-B):376–381 (2006)
8. Assecondi, S., Casarotto, S., Bianchi, A.M., Chiarenza, G.A., D'Asseler, Y., Lemahieu, I.: Automatic measurement of reading related potentials in dyslexia. In: IEEE Benelux EMBS Symposium, Belgian Day on Biomedical Engineering, 7–8 Dec 2006
9. Bonte, M.L.: Developmental dyslexia: ERP correlates of anomalous phonological processing during spoken word recognition. Cognitive Brain Res. **21**, 360–376 (2004)
10. Rath, T.M., Manmatha, R.: Word image matching using dynamic time warping. Natl. Sci. Found.
11. Chiarenza, G.A.: Motor-perceptual functions in children with developmental reading disorders: neuropsychophysiological analysis. J. Learn. Disabil. **23**(6), 375–385 (1990)
12. Pau, M-T., David, C-F., Daniel, N.: Early Dyslexia Detection Techniques by means of Oculographic Signals. In: 2nd European Medical and Biological Engineering Conference, Austria, Vienna (2002)
13. Merrill, C.E., Jordan, D.R.: Dyslexia in the classroom. pp.1–4. publishing Company, New York (1977)
14. Sheppard, R.: Why kids can't read. Maclean's. **111**(36), 40–48 (1998)
15. George T.P.: Eye movement in dyslexia: their diagnostic significance. J. Learn. Disabil. (2001)
16. Jain, K., Manghirmalani, P., Dongardive, J., Abraham, S.: Computational diagnosis of learning Disability. Int. J. Recent Trends Eng. **2**(3), (2009)
17. Bullinaria, J.A.: Neural network models of reading multi-syllabic words. In: 1993 International Joint Conference on Neural Networks
18. Wu, T-K., Meng, Y-R., Huang, S-C.: Application of ANN to the identification of students with LD. IC-AI: 162–168 (2006)
19. Wu, T-K., Meng, Y-R., Huang, S-C.: Identifying and diagnosing students with LD using ANN and SVM. In: IEEE International Joint Conference on Neural Networks, Vancouver (2006)
20. Manghirmalani, P., Panthaky, Z., Jain, K.: Learning disability diagnosis and classiication-a soft computing approach. In: IEEE World Congress on Information and Communication Technologies (WICT), 2011 (doi:10.1109/WICT.2011.6141292)

21. Sanchez, L., Palacios, A., Suarez, M.R., Couso, I.: Graphical exploratory analysis of vague data in the early diagnosis of dyslexia.
22. Georgopoulos, V., Stylios, C.: Genetic algorithm enhanced fuzzy cognitive maps for medical diagnosis. In: IEEE 2008
23. Salhi, L., Talbi, M., Cherif, A.: Voice disorders identification using hybrid approach: wavelet analysis and multilayer neural networks.Proceedings of World Academy of Science, Engineering and Technology, **35**, ISSN 2070–374 (2008)
24. Novak, D., Kordk, P., Macas, M., Vyhnalek, M., Brzezny, R., Lhotska, L.: School children dyslexia analysis using self organizing maps. In: IEEE 26th Annual International Conference, San Francisco 2004
25. Wu, T-K., Meng, Y-R., Huang, S-C.: Effects of feature selection on the identification of students with LD Using ANN.Springer-Verlag, Heidelberg (2006)
26. Macaš, M., Lhotská, L., Novák, D.: Bioinspired methods for analysis and classification of reading eye movements of dyslexic children. In: Czech Republic NiSls Symposium, Czech Technical University, PragueDepartment of Cybernetics(2005)
27. Chiarenza,G.A.: Motor-perceptual functions in children with developmental reading disorders: neuro psycho physiological analysis. J. Learn. Disabil. **23**(6), 375–385 (1990)
28. Bonte, M.L.: Developmental dyslexia: ERP correlates of anomalous phonological processing during spoken word recognition. Cognitive Brain Research **21**, 360–376 (2004)
29. Kohonen, T.: Self-Organization and Associative Memory, 3rd edn. Springer-Verlag, Berlin (1989)
30. Lerner, J.W.: Learning disabilities: theories, diagnosis, and teaching strategies. Houghton Mifflin, Boston (2000) ISBN 0395961149
31. Sivamanbam, S.N., Sumathi, S., Deepa, S.N.: Introduction to neural networks using MATLAB 6.0. ISBN 9780070591127
32. Karayiannis, N.B.: A Methodology for Constructing fuzzy algorithms for learning vector quantization. IEEE Trans. Neural Networks **8**(3), (1997). doi:1045-9227(97)00466-9
33. Yegnanarayana, B.: Artificial Neural Networks. Prentice Hall India, Publication (2006)
34. Zadeh, L.A.: Fuzzy sets and systems. In: Fox, J. (ed.) System Theory, pp. 29–39. Polytechnic Press, Brooklyn, NY (1965)
35. Zadeh, L.A.: Fuzzy sets. Information and Control **8**(3), 338–353 (1965)
36. Gerla, G.: Fuzzy Logic Programming and fuzzy control. Studia Logica **79**, 231–254 (2005)
37. Arjona, M.A.L., Escarela-Perez, R., Melgoza- 'Vázquezc, E., Hernández, C. F.: Convergence improvement in two-dimensional finite element nonlinear magnetic problems–a fuzzy logic approach. Finite Elements in Analysis and Design, Science, Direct (2004) doi:10.1016/S0168-874X(02)00165-8.
38. Yuan, G.J.K.: Fuzzy sets and fuzzy logic: theory and applications
39. Manghirmalani, P., More, D., Jain, K.: A fuzzy approach to classify learning disability. Int. J. Adv. Res. Artif. Intell. **1**(2), U.S ISSN:2165–4069(Online), U.S ISSN : 2165–4050(Print) (2012)

A Review of Hybrid Machine Learning Approaches in Cognitive Classification

Shantipriya Parida and Satchidananda Dehuri

Abstract The classification of functional magnetic imaging resonance (fMRI) data involves many challenges due to the problem of high dimensionality, noise, and limited training samples. In particular, mental states classification, decoding brain activation, and finding the variable of interest by using fMRI data was one of the focused research topics among machine learning researchers during past few decades. In the context of classification, algorithms have biases, i.e., an algorithm perform better in one dataset may become worse in other dataset. To overcome the limitations of individual techniques, hybridization or fusion of these machine learning techniques proposed in recent years which have shown promising result and open up new direction of research. This paper reviews the hybrid machine learning techniques used in cognitive classification by giving proper attention to their performance and limitations. As a result, various domain specific techniques are identified in addition to many open research challenges.

Keywords Functional magnetic resonance imaging · Machine learning · General linear model · Genetic algorithm · Particle swarm optimization

1 Introduction

Since decade after decade, the complexity of brain has been the primary research of many studies and experiment. Although the advancement of technologies improve our understanding about brain, still far from being completely understood.

S. Parida (✉)
Carrier Software and Core Network, Huawei Technologies India Pvt Ltd, The Leela Palace, 23 Airport Road, Bangalore, Karnataka 560008, India
e-mail: shantipriya.parida@gmail.com

S. Dehuri
Department of Systems Engineering, Ajou University, San 5, Woncheon-dong, Yeongtong-gu suwon 443-749, South Korea
e-mail: satchi@ajou.ac.kr

B. V. Babu et al. (eds.), *Proceedings of the Second International Conference on Soft Computing for Problem Solving (SocProS 2012), December 28–30, 2012*, Advances in Intelligent Systems and Computing 236, DOI: 10.1007/978-81-322-1602-5_70, © Springer India 2014

The cognitive neuroscience is evolving with the latest neuroimaging techniques combined with experimental methodologies which provide us images of the structure or function of the brain. The magnetic resonance imaging (MRI) uses a powerful magnetic field and radio waves to produce highly transparent images of the human body which shows injury, diseases process, or abnormal condition [1]. The fMRI is a technology to detect the localized changes in blood flow and blood oxygenation which occur in the brain in response to neural activity [2]. The classification techniques can identify many types of activation patterns within or shared across subjects [3].

As a part of this review, this study focus on the challenges involved in the classification task and different machine learning approaches performing the classification activities and the requirement of hybrid classification. The various classification approaches for cognitive states discrimination developed under the umbrella of hybrid machine learning techniques have been reviewed.

The rest of the paper is set out as follows. In Sect. 2, the challenges in cognitive classification have been studied. Some of the popular machine learning approaches used for cognitive classification is discussed in Sect. 3. Hybrid techniques and their applications are reviewed in Sect. 4. Future perspectives and conclusions are derived in Sects. 5 and 6 respectively.

2 Challenges in Cognitive Classification

The fMRI involves many challenges starting from image acquisition to final data analysis, which the neuroscientists trying to make best use of the technique [4]. In cognitive neuroscience the key challenge is to identify the mapping between neural activity and mental representations [5]. The challenge in fMRI data analysis is due to the noisy and weakly activated voxel in the acquired images and may therefore go undetected [6]. The data obtained from fMRI neuroimaging are especially rich and complex [7] and fMRI generates vast amount of data which are handled using computer-based methods [8]. The main factors cause the fMRI data analysis so complex is as follows [9]: (i) extremely large feature to instance ratio; (ii) spatial relationship between features; (iii) low contrast to noise ratio; and (iv) redundancy in the feature set.

One of the major challenges is using multivariate classification techniques to decode cognitive states from fMRI images. The problem involved in this scenario is high intersubject variability in the spatial locations and functional activation degrees [10]. The multivariate pattern recognition (MPR) methods are becoming popular rapidly for fMRI data analysis as these methods detect activity patterns in brain regions which can collectively discriminate one cognitive or participant group from another. The performance of these methods often limited due to the challenge that the number of regions considered in the fMRI data ("features") analysis is larger compared to the number of observations ("trials or participants") [11]. The evolved of real-time fMRI classification opens new direction for interactive self-regulation, as in this technique, the brain response can be modeled better. The major challenges

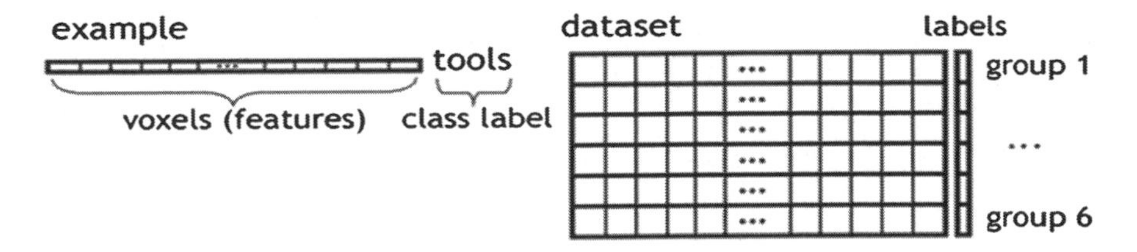

Fig. 1 An example where voxels are features as a row vector (*left*) and a dataset as matrix of such vectors (*right*) (extracted from [14])

in case of real-time fMRI is that the number of training instances comparative less and the classifier must be capable of perform within the time constraint [12].

3 Machine Learning Approaches in Cognitive Classification

One of the popular techniques used in recent development is machine learning classifiers for classifying cognitive states by analyzing the fMRI data. The trained classifier represents a function of the form [13]:

$$\Phi : fMRI\,(t, t+n) \rightarrow Y, \tag{1}$$

where $fMRI\,(t, t+n)$ is the observed fMRI data during the interval from time t to $t+n$, and Y is a finite set of cognitive states to be discriminated.

$\Phi\,(fMRI\,(t, t+n))$ the classifier prediction regarding which cognitive state gave rise to the observed fMRI data $fMRI\,(t, t+n)$.

In case of neuroimaging [14], the voxel represent features and activation of subject denoted as class is shown in Fig. 1.

4 Hybrid Techniques and Application

The hybrid classifier tries to use the desirable properties of each individual classifier and emphasizes to improve the overall accuracy in the combined approach.

Anderson et al. [15], have proposed a hybrid SVM-Markov model to predict accurately the real-time cognitive states such as the subject is viewing a video while either resting or nicotine craving. The markov transition matrix contributed in improving the classification accuracy by effectively removing unlikely state transition and the transition matrix acts as high pass filter for SVM probability prediction.

Wang [16] has combined the advantages of both general linear model (GLM) and SVM to develop a hybrid technique for fMRI data analysis. The proposed technique used the power of SVM for data derive reference function and GLM for statistical inference. His experimental study confirms that the combined approach shown better sensitive than the regular GLM for detecting the sensorimotor task.

Yang et al. [17], have proposed hybrid machine learning approach for classifying the schizophrenia and healthy control subjects. Their experimental result shows that the hybrid approach classification accuracy is promising and able to extract the discriminating information to classify schizophrenia efficiently.

Kharrat et al. [18] have proposed a hybrid machine learning approach based on genetic algorithm (GA) and SVM for fMRI brain tissues classification. The feature extraction is performed using spatial gray-level dependence method (SGLDM) and feature selection by GA-based global search method. The selected feature subset is given as input to the SVM for classification. The proposed method has shown very promising result for classifying the healthy and pathological brain.

Dahshan et al. [19], have proposed a hybrid method for classifying subjects as normal or abnormal using MRI images. The proposed technique consists of discrete wavelet transforms (DWT), the principal component analysis (PCA), feed forward artificial neural network (FP-ANN), and k-nearest neighbor (k-NN). The experimental result confirms that it is efficient in classifying the normal and abnormal brain.

Similarly, Zhang et al. [20], have developed a hybrid classifier to distinguish between the normal and abnormal of the brain using the MR images. In this approach the feature extraction performed using the wavelet transform and subsequently it applied the PCA to reduce the dimensions of the features. The reduced features input to the neural network (NN) and the parameters optimized by the chaotic particle swarm optimization (ACPSO). The k-fold cross validation has been used for enhancing the generalization.

Roussos et al. [21], have extended the ICA and propose a hybrid wavelet-ICA (W-ICA) model for transforming the signals into domain and apply the Bayesian inference. The directed acyclic graph (DAG) which is also known as Bayesian network structure represents the probabilistic dependence relationships. They applied this WICA model to fMRI data. This new approach showing great interest but further analysis required to find the comparative study with other hybrid techniques and its usefulness applying on neuroscientific data.

Kim et al. [22], have proposed a hybrid ICA-Bayesian approach for identifying the differences between the brain regions of normal and abnormal subject. The correlation between brain regions in one group versus another measured by the approximate conditional score (ACL) and its observed that ICA-filtered data significantly improved the magnitude of the ACL score. The ICA for filtering noise components from fMRI datasets is recommended as it plays more significant in determining higher level correlations which are more sensitive to noisy datasets.

Li et al. [23] have presented a hybrid approach for brain state classification based on both statistical data analysis and graphical information modeling (model-based). The proposed framework consists of three components. The first component includes image enhancement, event prediction and detection where the event prediction is performed using the temporal cluster analysis (TCA). The second component includes feature extractions and modeling where the linear predictive coding (LPC) used for feature extraction and variational bayesian gaussian mixture model (VBGMM) used for event classification. The third component includes graphical model based inference using the probabilistic graphical model. The experimental result has shown that the neural activities can be detected and classified using this hybrid approach. As part of further study, other complex neural activities need to test with this hybrid approach.

Tohka et al. [24], have proposed a hybrid-GA algorithm for solving the challenging finite mixture model (FMM) optimization which arises in the neuroimaging classification. The hybrid algorithm tested with T1-weighted MR brain scans of healthy and mental disorder (Alzheimer's diseases (AD), schizophrenia) subject. The hybrid-GA algorithm compared with the other FMM optimization algorithm and found most consistent and accurate although in some case other algorithms recommended for computational reasons. It is recommended to use this hybrid algorithm in many FMM optimization task and automatic image analysis procedure for brain imaging.

Abraham et al. [25], have proposed hybrid schemes which involves particle swarm optimization (PSO) algorithm-based rough set reduction for SVM applied to cognitive state classification. The PSO algorithm can find optimal regions of complex search spaces through interaction of particle in a population of studies and it is converging fast. Based on fMRI experiment they shown that the proposed scheme is feasible for classification for single or multiple subjects.

Brodersen et al. [26] have proposed a hybrid generative-embedding approach to address the two major challenges with fMRI classification, i.e., first challenge due to high dimensionality and low sample size and second challenge is that popular discriminative method like SVM rarely affords mechanistic interpretability. The proposed approach combined the dynamic causal models (DCMs) and SVMs. It shows that the generative-embedding approach achieves 98 % accurate classification and performs better compared with conventional activation-based and correlation-based methods.

Cheng et al. [27], have proposed a hybrid generative/discriminative framework for brain region classification. The hybrid framework consists of two major parts such as "generative part" and "discriminative part". The idea of generative part is to choose a generative model capable of considering all ROI at the same time along with the relations between them. The discriminative part contains the "fisher kernels" which allows an effective general way of mixing generative and discriminative models for classification. In this approach, the data is represented using Fisher Score Space and classification using SVM. The experimental result shows that the hybrid approaches successfully discriminating between the healthy and schizophrenic patients based on analysis of brain MRI scan.

Zhang et al. [28] (after inspired from their earlier model [20], have proposed a hybrid model based on forward neural network (FNN) for classifying the normal and abnormal states based on MR brain image. The proposed approach consists of five stages: (1) in the first stage, it use DWT to extract feature; (2) then in the second stage, use PCA to reduce feature size; (3) third stage contain k-fold stratified cross validation to prevent over-fitting; (4) in the fourth stage, it contain the FNN to construct the classifier; (5) in the final stage, it uses scaled chaotic artificial bee colony

(SCABC) to train the FNN. The SCABC-FNN classifier obtains 100 % classification accuracy to distinguish between normal and abnormal MRIs of the brain.

Jafarpour et al. [29], have proposed a robust classification technique for classifying normal and abnormal brain MRI. The proposed approach consists of the following techniques: gray level co-occurrence matrix (GLCM), PCA, linear discriminant analysis (LDA), ANN, and K-NN. The three stages of their work includes: feature extraction, feature reduction, and classification. The methodology effectively used the image features and employed hybrid feature reduction technique to distinguish the normal, multiple sclerosis (MS), and tumorial brain MRIs. The experimental study shows 100 % accuracy for normal, 100 % accuracy for tumoral, and 92.86 % for MS images with both ANN and KNN classifiers.

5 Future Perspectives

The hybrid feature reduction [29] applied on T2-weighted images and future research can extend to other type of MRI and more kind of brain disease. The SCABC-FNN classifier [28] which successfully distinguishes normal and abnormal MRIs of the brain can further explore for multi class and specific MRIs disorder and advanced wavelet transforms can be applied for improved features. One of the future directions for hybrid generative/discriminative method [27] is to explore more complex probabilistic model and introducing clinical data with more variability. The proposed hybrid GA algorithm [24] for solving FMM optimization problem arise in neuroimaging can be further use to study how to fully automate many FMM optimization task to develop automatic brain image analysis procedure. The hybrid framework which is based on statistical data analysis and graphical information modeling [23] can be further extending to dealing with complex neural activities. The hybrid wavelet-ICA [21] model can be further extended to consider the residual dependencies in the wavelet coefficient to enhance separation and to better match the structure of signals it can explore to adapt the dictionary itself. The hybrid classification method based on ACPSO [20] can be further extended to apply on other type of MR images, reducing computation time, focusing specific brain MRI disorder. The hybrid technique classifying the normal and abnormal classes [19] can be further extended to processing pathological brain tissues (e.g., lesion, tumors).

6 Conclusion

In this paper, we have walked through the recent developments of the hybrid techniques for brain state classification and discussed their efficiency and limitations. It has been observed that the hybrid techniques outperform the non-hybrid one. Hence, this paper prescribes to use hybrid machine learning techniques for fMRI classification. Alongside it recommends toward fine tuning of parameters for enhancing user understandability, and interestingness.

References

1. McGowan, J.C.: Basic principles of magnetic resonance imaging. Neuroimaging Clin. N. Am. **18**(4), 623–636 (2008)
2. Savoy, R.L.: Functional magnetic resonance imaging (fMRI), Encyclopedia of Neuroscience, 2nd edn. Birkhauser, Boston, MA (1996)
3. Etzel, J.A., Gazzola, V., Keysers, C.: An introduction to anatomical ROI-based fMRI classification analysis. Brain Res. **1282**, 114–125 (2009)
4. Amaro Jr, E., Barker, G.J.: Study design in fMRI: basic principles. Brain Cogn. **60**(3), 220–232 (2006)
5. Norman, K.A., Polyn, S.M., Detre, G.J., Haxby, J.V.: Beyond mind-reading: multi-voxel pattern analysis of fMRI data. Trends Cogn. Sci. **10**(9), 424–430 (2006)
6. Friman, O., Borga, M., Lundberg, P., Knutsson, H.: Detection and detrending in fMRI data analysis. NeuroImage **22**(2), 645–655 (2004)
7. Horwitz, B., Tagamets, M., McIntosh, A.R.: Neural modeling, functional brain imaging, and cognition. Trends Cogn. Sci. **3**(3), 91–98 (1999)
8. Nielsen, F.A., Christensen, M.S., Madsen, K.H., Lund, T.E., Hansen, L.K.: fMRI Neuroinformatics. IEEE Eng. Med. Biol. Mag. **25**(2), 112–119 (2006)
9. Kuncheva, L.I., Plumpton, C.O.: Choosing parameters for random subspace ensembles for fMRI classification. Lect. Notes Comput. Sci. **2010**(5997), 54–63 (2010)
10. Fan, Y., Shen, D., Davatzikos, C.: Detecting cognitive states from fMRI Images by machine learning and multivariate classification. In: Proceeings of Conference on Computer Vision and Pattern Recognition Workshop, IEEE Computer Society, Washington, DC, USA (2006)
11. Ryali, S., Supekar, K., Abrams, D.A., Menon, V.: Sparse logistic regression for whole-brain classification of fMRI data. NeuroImage **51**(2), 752–764 (2010)
12. Plumpton, C.O., Kuncheva, L.I.: On-line fMRI data classification using linear and ensemble classifier". In: 20th International Conference Pattern Recognition, pp. 4312–4315 (2010)
13. Mitchell, T.M., Hutchinson, R., Just, M.A., Niculescu, R.S., Wang, X.: Classifying instantaneous cognitive states from fMRI data. In: American Medical Informatics Association, Symposium, pp. 465–469 (2003)
14. Pereira, F., Mitchell, T., Botvinick, M.: Machine learning classifiers and fMRI: a tutorial overview. NeuroImage **45**(1), S199–S209 (2009)
15. Anderson, A., Han, D., Douglas, P.K., Bramen, J., Cohen, M.S.: Real-time functional MRI classification of brain states using Markov-SVM hybrid models: Peering inside the rt-fMRI black box. In: Neural Information Processing Systems (2011)
16. Wang, Z.: A hybrid SVM-GLM approach for fMRI data analysis. NeuroImage **46**(3), 608–615 (2009)
17. Yang, H., Liu, J., Sui, J., Pearlson, G., Calhoun, V.D.: A hybrid machine learning method for fusing fMRI and genetic data: combining both improves classification for schizophrenia. Frontiers Hum. Neurosci. **4**, (2010)
18. Kharrat, A., Gasmi, K., Messaoud, M.B., Benamrane, N., Abid, M.: A hybrid approach for automatic classification of brain MRI using genetic algorithm and support vector machine. Leonardo J. Sci. **9**(17), 71–82 (2010)
19. Sayed, EL., Dahshan, EL., Salem, Abdul- Badeeh. M., Yousin, T.H.: A hybrid technique for automatic MRI brain images classification, vol. LIV. Studia Univ. Babes. Bolyai, Romania (2009)
20. Zhang, Y., Wang, S., Wu, L.: A novel method for magnetic resonance brain image classification based on adaptive chaotic PSO. Prog. Electromagnet. Res. **109**, 325–343 (2010)

21. Roussos, E., Roberts, S., Daubechies, I.: Variational Bayesian learning for wavelet independent component analysis. In: 25th Int. Workshop on Bayesian Inference and Maximum Entropy Methods in Science and Engineering, vol. 803, 274–281 (2005)
22. Kim, D., Burge, J., Lane, T., Pearlson, G.D., Kiehl, K.A., Calhoun, V.D.: Hybrid ICA-Bayesian network approach reveals distinct effective connectivity differences in schizophrenia. NeoroImage **42**(4), 1560–1568 (2008)
23. Li, C., Hao, Q., Guo, W., Hu, F.: A hybrid approach for compressive neural activity detection with functional MR images. In: Proceedings of 31th IEEE Conference on Engineering in Medicine and Biology Society, pp. 4787–4790 (2009)
24. Tohka, J., Krestyannikov, E., Dinov, I.D., Graham, A.M., Shattuck, D.W., Ruotsalainen, U., Toga, A.W.: Genetic algorithms for finite mixture model based voxel classification in neuroimaging. IEEE Trans. Med. Imaging **26**(5), 696–711 (2007)
25. Abraham, A., Liu, H.: Swarm intelligence based rough set reduction scheme for support vector machines, In: IEEE Internationl Conference on Intelligence and Security Informatics, pp. 200–202 (2008)
26. Brodersen, K.H., Schofield, T.M., Leff, A.P., Ong, C.S., Lomakina, E.I., Buhmann, J.M., Stephan, K.E.: Generative embedding for model-based classification of fMRI data". PLoS Comput. Biol. **7**(6), e1002079 (2011)
27. Cheng, D. S., Bicego, M., Castellani, U., Cristani, M., Cerruti, S., Bellani, M., Rambaldelli, G., Aztori, M., Brambilla, P., Murino, V.: A hybrid generative/discriminative method for classification of regions of interest in schizophrenia brain MRI, In: Proceedings of MICCAI09 workshop on Probabilistic Models for, Medical Image Analysis (2009)
28. Zhang, Y., Wu, L., Wang, S.: Magnetic resonance brain image classification by an improved artificial bee colony algorithm. Prog. Electromagnet. Res. **116**, 65–79 (2011)
29. Jafarpour, S., Sedghi, Z., Amirani, M.C.: A robust brain MRI classification with GLCM features. Int. J. Comput. Appl. **37**(12), 1–5 (2012)

"A Safer Cloud", Data Isolation and Security by Tus-Man Protocol

Tushar Bhardwaj, Manu Ram Pandit and Tarun Kumar Sharma

Abstract Today cloud computing is well-known for touching all periphery of technology with its on-demand and elastic capability. Ever since it has come into picture, security has remained a major concern. VM model is already known to be vulnerable to various issues. We introduce Tus-Man protocol which will act in addition to existing system to make computing secure enough for both service provider as well as service consumers. In this protocol, we suggest a tunnel-based protocol to make data transfer not only secure but also safe enough against any malicious attack.

Keywords Cloud computing · Security in cloud computing · Tus-Man algorithm (Cloud computing security protocol)

1 Introduction

Today cloud computing is well-known for touching all periphery of technology with its on-demand and elastic capability. Ever since it has come into picture, security has remained a major concern. VM model is already known to be vulnerable to various issues. We introduce Tus-Man protocol which will act in addition to existing system to make computing secure enough for both service provider as well as service consumers. In this protocol, we suggest a tunnel-based protocol to make data transfer

T. Bhardwaj (✉)
M.T.U Noida, Noida, India
e-mail: tusharbhardwaj19@gmail.com

M. R. Pandit
BITS Pilani, Pilani, India
e-mail: manupandit123@gmail.com

T. K. Sharma
IIT Roorkee, Roorkee, India
e-mail: taruniitr1@gmail.com

B. V. Babu et al. (eds.), *Proceedings of the Second International Conference on Soft Computing for Problem Solving (SocProS 2012), December 28–30, 2012*, Advances in Intelligent Systems and Computing 236, DOI: 10.1007/978-81-322-1602-5_71,

not only secure but also safe enough against any malicious attack. This paper is divided into two main sections viz. Top threats to cloud computing followed by description of Tus-Man protocol.

2 Top Threats to Cloud Computing

Before getting into the business to secure the public cloud let us look into the security threats to the world of cloud computing. We have enlisted the threats which we have been undertaken before writing this paper [1].

1. *Insecure interfaces and APIs*: The interfaces and the different APIs are open to the users for the communication among the different services. The security and the availability of the services highly depend on these interfaces and APIs. There are many activities like encryption, authentication, and access control mechanism that are monitored by these APIs.
2. *Malicious Insiders*: This is a very common attack for any organization. The main cause is the lack of transparency among the customers and the system. The introduction of IT and the management tool under the same roof leads to this problem.
3. *Shared Technology Issues*: Several vendors deliver their services in a highly scalable way by sharing the infrastructure. It is quite obvious that the elements that incorporate this so-called infrastructure (e.g., CPU and GPUs) are not kept isolated to match with the multitenancy architecture.
4. *Data Loss or Leakage*: The data is open for all in the cloud. As this is the main advantage of cloud computing, it leads to the data compromise. It can be done due to several reasons like lack of authentication, authorization, and audit (AAA) control. It also amplifies due to the increase in the number of users day-to-day.
5. *Account or Server Hijacking*: There are many methods like phishing, fraud, and exploitation that still prevail to hijack any server and hence its services. Passwords and credentials are mostly repeated which supports this kind of attack.

3 Tus-Man Protocol

Earlier we have studied that are many threats to the world of cloud computing like malicious insiders, data loss or leakage, account or server hijacking, and insecure interfaces and APIs. All the above- mentioned loopholes of the existing system lead us to focus on this highly increasing demanding field of security. So, we have proposed a protocol for the same. This protocol is assumed to be in injected to the transport layer of the cloud provider communication system. Our aim is to secure the access mechanism and authentication system.

So, this protocol has two distinct but important features:

1. A secure access mechanism.
2. Authentication system (Fig. 1).

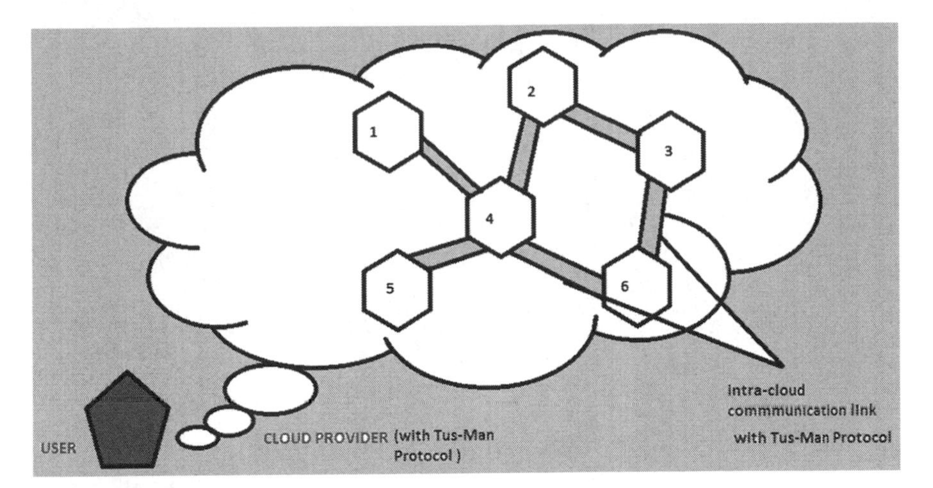

Fig. 1 Cloud communication channel injected with Tus-Man protocol

3.1 Introduction

When we talk about the access mechanism it is the user interaction with the system. It tells that how the user is able to access the data from the cloud. If the access mechanism is weak then any unauthorized user may enter the system and compromise the data. On the other authentication system means that only the authorized user should be allowed to enter the cloud and have access to it.

3.2 Basics

Before entering into the detail of the protocol let us brush up the basics of cloud computing. Whenever a user logs on with his/her user name and password, then he/she is able to access both the public and private data of his own cloud. In addition to that he/she can visualize the shared data (read only mode) of other users in that particular cloud. Consider the case of an organization where many employees are working and having their own private clouds. So, the main aim is to secure the access mechanism and authentication system as to prevent the malicious insiders and data loss attack of cloud computing [2, 3].

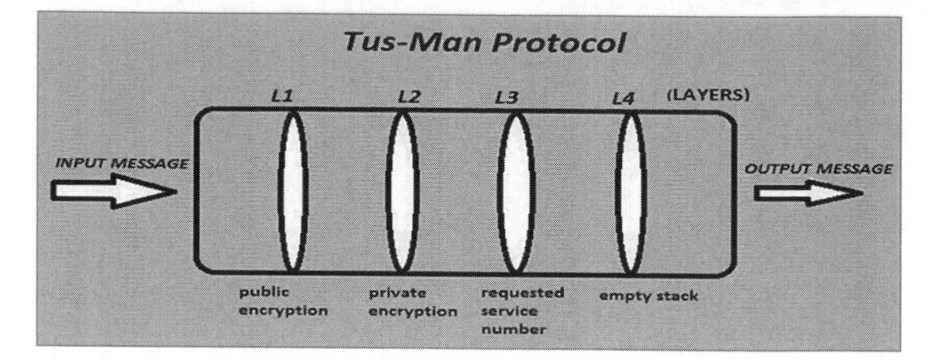

Fig. 2 Tus-Man protocol design

3.3 Design

Consider a tunnel with several layers. Each layer has a specific purpose attached with it. The tunnel has two ends one for input and other for output. There is a flow of data that have to pass from each subsequent layer to the destination, i.e., cloud. The system design is such that if at any layer the data halts or failed to fulfill the given criteria then the data flow will not be done. In other words the message that is traversing in the tunnel has to pass and qualify the necessary condition of each layer.

3.4 Working

In this section we will look into the structure of the protocol in detail (Fig. 2):

Layer 1 (*public key Encryption*): This is the first layer which contains the public information of the user. This layer authenticates the public credentials of the user and opens the gateway to the public data.

Layer 2 (*private key Encryption*): This is the second layer that deals with the private data. When the message gets traversed and passes the Layer 1, it means that an authentic user has entered the cloud. Next is the authentication for the private data. For this layer the password is required. Layer 2 has the collection of all the passwords for matching with the incoming request (message + credentials). This authentication is being done with a highly secure passion. This method of authentication will be discussed later (Fig. 3) [4].

Layer 3 (*Stack: requested service numbers*): All the services present in a particular cloud are given a random number. When a user passes the two layer of authentication, then he/she have to demand for that particular service. The user is unaware (as in most of cases today) of this random number methodology; he/she have performed the required operations. The main function of this layer is that whenever a user requests a service, the corresponding service number is passed to the layer 4 (Fig. 4).

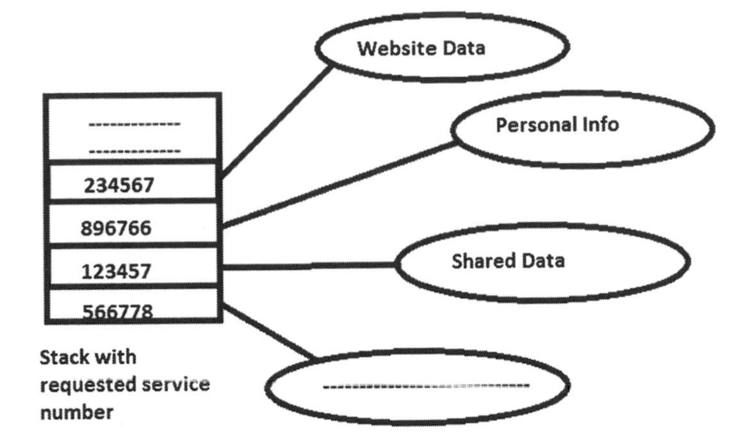

Fig. 3 Requested service mapping with random numbers

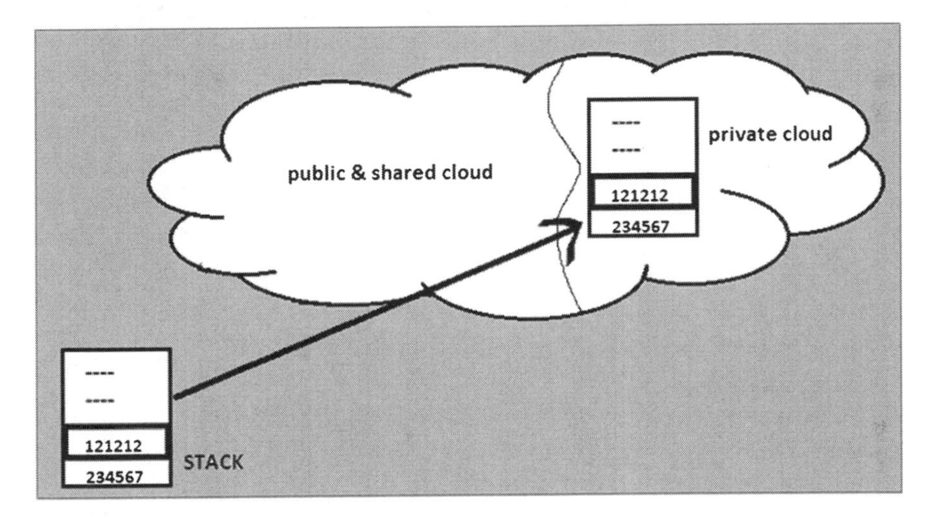

Fig. 4 Operation of layer 4

Layer 4 (*empty stack*): Next the message having the service number is passed to the layer 4. Here, we are having an empty stack on which the random number is updated. These stacks have all the services that have been requested by the user with a time stamp. This information is updated to the private area of that particular cloud. This feature helps for monitoring the activities done by the user.

Monitoring: The basic benefit of layer 3 and 4 is that the user can keep an eye on the service being used and accessed. In the private area of user's cloud the stack is shown in the form of a database where instead of random numbers the corresponding service name is given [5].

3.5 Authentication and Key Exchange

The basic aim of Tus-Man protocol is to provide:

1. A secure access mechanism.
2. Authentication system.

In this section, we deal with the authentication and the key exchange mechanism used in the Tus-Man protocol.

3.5.1 Security Model

The main aspects we have to guarantee are: (**Authentication and confidentiality**). Here we have **third party token provider (TPTP)**. It generates a TOKEN on demand of the client. The **TOKEN** is useful in making the transaction more secure and reliable between the client and the service provider.

1. Theme

This algorithm has been designed on the basis of locking system used in the luxury cars now-a-days. In which there is a microchip built in the car's lock and the key used to unlock that key. Whenever the key is being put into the lock the data in the two chips are matched and car will be opened. On the other hand, when the car is being turned off, the data will be changed and stored both in the car's and key's microchip.

2. Salient features

- The value of the token will be changed every time the user requests for the token from the service provider.
- It provides an extraordinary level of security to the data stored on the cloud.
- The regular change of the TOKEN value makes all attacks on the data just next to impossible.
- Encryption and decryption depends on the key's present values, which makes it very secure.
- The current value of TOKEN is not known to the user himself.

3. Working

Step 1: The User's ID [ID_{USER}] and TUS-MAN CODE is send to the third party token provider.
Step 2: The third party token provider generates a Token [T] and delivers it to both the client and the service provider.
Step 3: Client generates a three way hand shake protocol with the service provider to check that the data is being available and accessible to communicate.
Step 4: After confirmation, user sends the TOKEN and its TUS-MAN CODE to the service provider.

Step 5: The service provider than matches the two TOKEN values and maps it with the particular TUS-MAN CODE.
Step 6: When the two values of TOKEN are matched, the service provider fetches the data from the particular cloud and returns to the client.
Step 7: When the data is being fetched than the user synchronizes with the third party token provider to inform that he/she has to sign out.
Step 8: Than the values of the token will be changed and kept secret with the service provider.
Step 9: Next time the same user logins with the same credentials the stored token value is now active and the same procedure is repeated.

4 Conclusion

Tus-Man protocol will provide capability to service providers and consumers to act seamlessly keeping security context in mind. This will act as a protocol which is to be used for providing secure services (viz. confidentiality, Access control) to both the service provider and service consumer. Salient feature that can be summarized are:

1. Tunnel-based security to hide network traffic.
2. Data Isolation from intercloud and intracloud communication.
3. A highly sensitive monitoring system for keeping track of requested services.

All the above-mentioned features help the secure transmission of data among the clouds as well as cloud-consumer network. The Tus-Man algorithm provides certain protocol that isolates and secure the personal data in the clouds. We are looking forward in making the cloud computing a better place to bank upon.

References

1. Cloud computing security threats [http://www.cloudsecurityalliance.org/topthreats/csathreats.v1.0.pdf]
2. Basics of public key encryption [http://en.wikipedia.org/wiki/Public-key_cryptography]
3. Top security thread to cloud [http://h30458.www3.hp.com/ww/en/ent/954867.html]
4. Basics of private key encryption [http://searchsecurity.techtarget.com/definition/private-key]
5. Working of kerberos [http://searchsecurity.techtarget.com/definition/Kerberos]

Poem Classification Using Machine Learning Approach

Vipin Kumar and Sonajharia Minz

Abstract The collection of poems is ever increasing on the Internet. Therefore, classification of poems is an important task along with their labels. The work in this paper is aimed to find the best classification algorithms among the K-nearest neighbor (KNN), Naïve Bayesian (NB) and Support Vector Machine (SVM) with reduced features. Information Gain Ratio is used for feature selection. The results show that SVM has maximum accuracy (93.25 %) using 20 % top ranked features.

Keywords Poem · Classification · Ranked feature

1 Introduction

A poem is a piece of writing in which the expression of feelings and ideas is given intensity by particular attention to diction (sometimes involving rhyme), rhythm, and imagery [5]. It is generally meant to deliver expressions such as love, happiness, success, etc. Thus, poems of many category are available. However, an effort in automatic poem classification is rare. Usually, poems are as short textual paragraphs with little discriminative value of word features for automatic classification. Therefore, poem classification is a challenging task. Some text classification algorithms have been developed to categorize news [6], patent [7], etc. Many machine learning algorithms have been attempted for automatic text classification.

A poet may use any word for the poem. In fact, poems are often structured differently from normal text documents. Therefore, there is a necessity to identify an effective machine learning algorithm for poem classification. Three machine learning

V. Kumar (✉) · S. Minz
JNU, New Delhi, India
e-mail: rt.vipink@gmail.com

S. Minz
e-mail: sona.minz@gmail.com

B. V. Babu et al. (eds.), *Proceedings of the Second International Conference on Soft Computing for Problem Solving (SocProS 2012), December 28–30, 2012*, Advances in Intelligent Systems and Computing 236, DOI: 10.1007/978-81-322-1602-5_72,

algorithms Naïve Bayesian (NB), k-Nearest Neighbor (KNN), and Support Vector Machine (SVM) are considered. For feature selection, Gain Ratio is used. The comparison of three machine learning algorithms with respect to poem classification is considered to identify the best suited one. The paper is structured as follows: Sect. 2 is on related work. Section 3 introduces poem data set. Implementation is described in Sect. 4. Section 5 contains experiment results and analysis. Finally, we present the conclusions and future work.

2 Related Work

Many types of statistical and machine learning algorithms are available to classify text documents such as k-Nearest Neighbor, Naïve Bayesian [8], and Support Vector Machine [9]. The effort in automatic classification of literary texts such as poetry is less. Malay poetry is classified using support vector machine. In this, Radial Basic Function (RBF) and linear kernel function are implemented to classify pantun by theme, as well as poetry and none poetry [10]. Logan and Kositsky [11] have done a comparison with the acoustic similarity technique and semantic text analysis technique for collecting lyrics from the Web to analyze artistic similarity.

3 Data Set

A collection of 400 text documents from popular sites such as *http://www.poetseer.org*, *http://www.poetry.org*, and *http://www.poemhunter.com* has been considered for experiment. More than 225 numbers of labels of poem are available on the Internet. It is difficult to consider all labels for this research. Therefore, the data set includes only 8 labels as alone, childhood, god, hope, love, success, valentine, and world. Each label has 50 text documents and each text document has nearly the same length.

4 Implementation

Poem classification framework is presented in Fig. 1. RapidMiner5 [2], an open source data mining package has been used to model the NB, k-KNN, and SVM [4]-based classifiers. The preprocessing steps include two main components such as text preprocessing and feature selection. In the text preprocessing task, all poems are input and processed by transformation of upper case to lower case, tokenization, stop word removal, and stemming using WordNet [1]. The text preprocessing task transforms poems into term-document vector, where the vector represents the weight of terms in the documents. Therefore, in feature selection the weight of each term is based on the *tf-idf* weighting scheme as shown below:

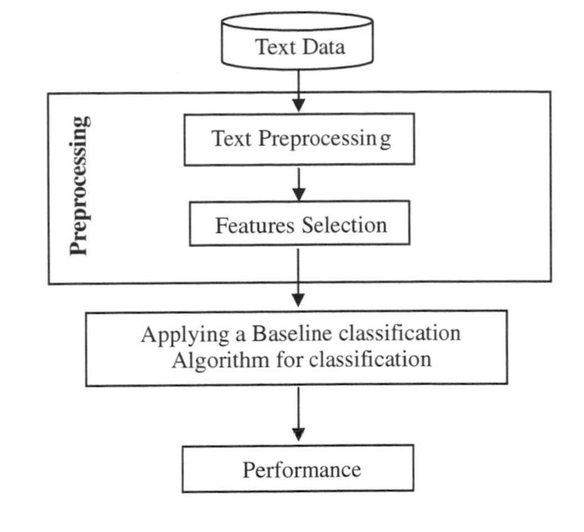

Fig. 1 Poem classification frameworks

$$w_{ij} = tf_{ij} \times \log \frac{N}{n_j}$$

where w_{ij} is the weight of term j, N is the total number of poems, and n_j is the number of poems.

The feature selection scheme Gain Ratio (GR) is applied to the feature vector of the document to rank the features for feature reduction. It applies a kind of normalization to information gain using splitting information value defined analogously with Info(D) as

$$SplitInfo_A(D) = -\sum_{j=i}^{v} \frac{|D_j|}{|D|} \times \log_2\left(\frac{|D_j|}{|D|}\right)$$

where D is data set into v partitions on attribute A. The Gain ratio is defined as

$$\text{Gain Ratio}(A) = \frac{\text{Gain}(A)}{\text{SplitInfo}(A)}$$

The attribute with the higher gain ratio is considered to have more relevance for classification.

After the preprocessing task, baseline classifiers KNN, NB, and SVM are applied for respective percentage of ranked features, where accuracy is a key measure to evaluate the performance of the classifiers. It can be calculated from Table 1 using the formula:

$$\text{Accuracy} = \frac{(TP + TN)}{(TO + TN + FP + FN)}$$

The true positive, true negative, false positive, and false negative are also useful in assessing the costs benefits (or risk ad gain) associated with the classification model.

Table 1 Confusion matrix for 2-class classification

Actual class	Predicted class	
	Positive	Negative
Positive	True positive (TP)	False positive (FP)
Negative	False negative (FN)	True negative (TN)

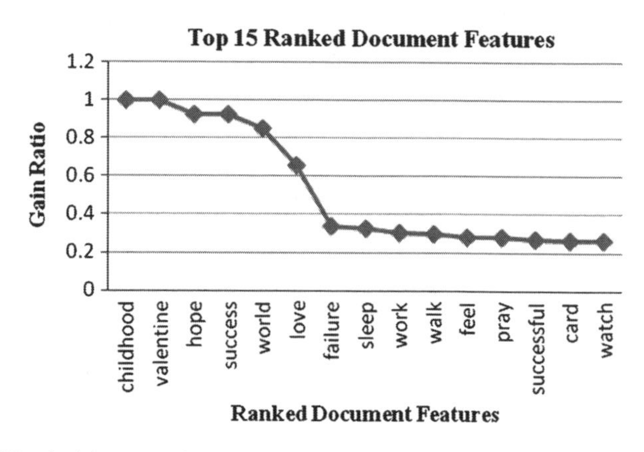

Fig. 2 Top 15 Ranked features of documents

5 Results and Analysis

The objective of the research is to find out suitable machine learning classification algorithms. NB, KNN, and SVM, three classifiers are chosen for the desired objective. Evaluation of classifiers is based on the classification accuracy. The poem data set was partitioned into training and test data. Gain Ratio was used to rank the features of the documents. Classifiers were developed using the 10, 20, 30...90 % and all features using the NB, KNN, and SVM. Ten-fold cross validation method was used to estimate the performances of all the classifiers.

5.1 Results

Gain Ratio was used to rank the document features in the experiment. To form the 3707 features, 15 top ranked features are shown in Fig. 2. It shows that childhood and Valentine features have 1 gain ratio. Hope, Success, World, Love, and Failure features are between 1.0 to 0.25 gain ratio.

Table 2 shows average accuracy of the classifiers using 10-fold cross validation with top ranked features. NB classier has maximum accuracy using 10 % top ranked features. Using 40 % top ranked features, KNN yields maximum accuracy of 87.50 %. SVM -based classifier has maximum 93.50 % accuracy by using 20 % top ranked features.

Table 2 Accuracy of NB, KNN, and SVM classifiers

% of top ranked attribute	Classification accuracy (%)		
	NB	KNN	SVM
10	**80.00**	85.50	93.00
20	79.75	86.00	**93.25**
30	76.50	84.00	93.25
40	71.25	**87.50**	92.75
50	71.25	87.00	93.25
60	68.50	83.25	92.75
70	66.25	80.00	93.00
80	66.25	76.25	92.75
90	65.75	75.50	93.00
100	79.50	85.50	92.25

Table 3 Accuracy of NB, KNN, and SVM classifiers using less than 10 % ranked features

% of top ranked attribute	Classifiers accuracy (%)		
	NB	KNN	SVM
1	**82.00**	86.00	90.00
2	81.00	85.75	90.00
3	80.00	84.00	89.75
4	80.50	86.00	91.25
5	79.50	86.26	91.50
6	79.50	85.25	92.00
7	78.50	**86.50**	91.00
8	78.00	86.00	92.50
9	78.75	85.25	92.75
10	80.00	85.50	**93.00**

From Table 2, it can be observed that SVM does not have significant difference in accuracy using 10 and 20 % top ranked features. The accuracy of KNN classifier does not have significant difference, with using 10 and 40 % top ranked features. Therefore, 10 % top ranked features are considered for the further reduced set of features.

Table 3 shows average accuracy of classifiers using 1, 2, 3... 10 % top ranked selected features. NB, KNN, and SVM have maximum accuracy of 82.00, 86.50, and 93.00 % using 1, 7 and 10 % selected features respectively.

5.2 Analysis

The objective of this research is to identify an efficient machine learning algorithm (NB, KNN, and SVM) for poem classification with reduced features. From Fig. 2, it is observed that only 6 features have Gain Ratio more than 0.6 and others have

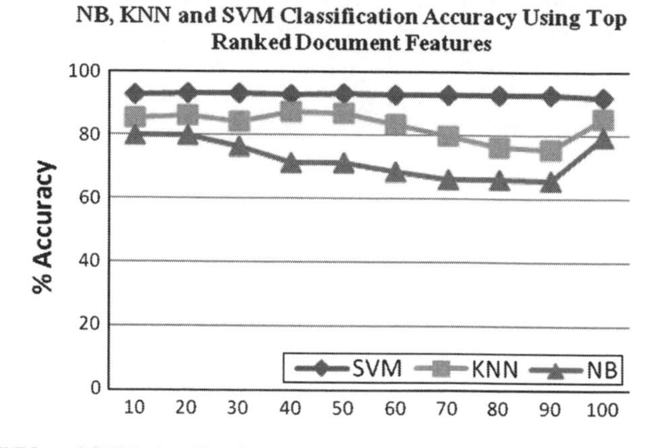

Fig. 3 NB, KNN, and SVM classification accuracy using top ranked document features

Gain Ratio less than 0.34. The top 6 features (childhood, valentine, hope, success, world, and love) are the same as labels mentioned in the data set. It means that these features (words) directly play a role in the poem classification task.

From Fig. 3 it can be easily analyzed that SVM classifier has the best performance for all percentage of ranked features and KNN has the second best performance. The performance of the classifiers displays decrease in accuracy as size of feature set increases except when 100 % features are used to model the classifiers. NB, KNN, and SVM achieved best accuracies of 80.00, 87.50, and 93.25 %, using 10, 40, and 20 % of selected features. The best performances of KNN and SVM classifiers do not indicate a significant difference using 10 % ranked features. Therefore, performance of the classifiers using 10 % ranked features has been considered for comparison of the three machine learning algorithms.

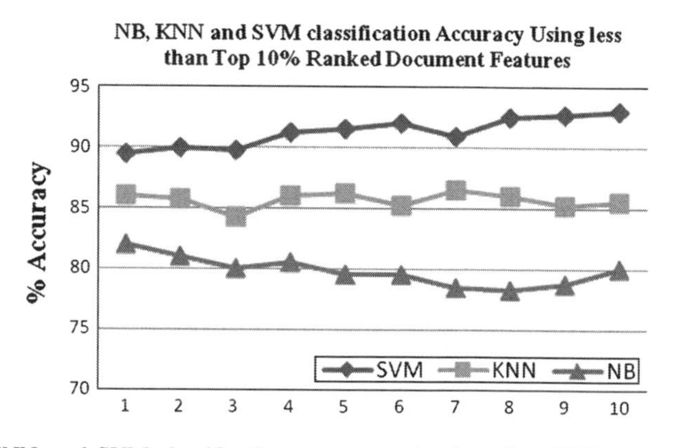

Fig. 4 NB, KNN, and SVM classification accuracy using less than 10 % top ranked document features

Figure 4 exhibits that the performance of NB-based classifiers decreases whereas the size of feature set increases. It achieves best performance 82.00 %, using 1 % of top ranked features. SVM performance is decreasing with 10 to 1 % of top ranked; therefore SVM has best performance using 10 % top ranked features. KNN has best performance 86.50 %, using 7 % of ranked features.

6 Conclusion and Future Work

A large number of poem data sets are available on Internet. Therefore, labeling poems is an important task. This study has attempted to identify an effective machine learning algorithms (NB, KNN, and SVM). The experiment results show that SVM have best performance compared to the NB and KNN. It has 93.5 % using 20 % ranked features. The top 6 features (childhood, valentine, hope, success, world, and love) have played an important role in classifying the respective poem labels.

Gain Ratio has been used for features selection of the documents. Therefore, other feature selection techniques (Information gain, Gini Index…etc) can be used in the future for features ranking. SentiWordNet [3] is a lexical resource, which gives three sentiment scores of positivity, negativity, and objectivity at each synset. Extracted document features, using SentiWordNet can play the important role for poem classification task.

References

1. Andrea, E., Sebastiani F.: SentiwordNet: a publicly available lexical resource for opinion mining. In: Proceedings of the 5th Conference on Language Resources and Evaluation (LREC 2006), Genova, IT, 2006, pp. 417–422 (2006).
2. http://www.rapidminer.com
3. Esuli., Sebastiani, F.: Sentiwordnet: a publicly available lexical resource for opinion mining. In: Proceedings from International Conference on Language Resources and Evaluation (LREC), Genoa (2006).
4. Chih-Chung, C., Chih-Jen, L.: LIBSVM: a library for support vector machines. In: ACM Transactions on Intelligent Systems and Technology, 2:27:127:27, (2011).
5. http://oxforddictionaries.com/definition/english/poem
6. Shih, L.K., Karger, D.R.: Learning classifiers: using URLs and table layout for web classification tasks. In: Proceedings of the 13th International Conference on World Wide Web, New York, NY, pp. 193–202 (2004).
7. Richter, G., MacFarlane, A.: The impact of metadata on the accuracy of automated patent classification. World Patent Inf. **37**(3), 13–26 (March 2005)
8. Wang, B., Zhou, S., Hu, Y.: Naive bayes-based garual Chinese documents categorization. In: Proceedings of World Multi conference on Systems, Cybernetics and Informatics, 2, July, Orlando, pp. 516–521 (2001).

9. Noraini, J., Masnizah, M.: Shahrul Azman, N.: Poetry classification using support vector machines. J. Comput. sci. **8**(9), 1441–1446 (2012)
10. Logan, B., Kositsky, A., Moreno, P.: Semantic analysis of song lyrics. In: the Proceeding of IEEE Int. Conf. on Multimedia and Expo, 2, pp. 827–830 (2004).
11. Tizhoosh, H.R., Dara, R.: On poem recognition. Pattern Anal. Appl. **9**(4), 325–338 (2006)

Novel Class Detection in Data Streams

Vahida Attar and Gargi Pingale

Abstract Data stream classification is challenging process as it involves consideration of many practical aspects associated with efficient processing and temporal behavior of the stream. Two such aspects which are well studied and addressed by many present data stream classification techniques are infinite length and concept drift. Another very important characteristic of data streams, namely, concept-evolution is rarely being addressed in literature. Concept-evolution occurs as a result of new classes evolving in the stream. Handling concept evolution involves detecting novel classes and training the model with the same. It is a significant technique to mine the data where an important class is under-represented in the training set. This paper is an attempt to study and discuss the technique to handle this issue. We implement one of such state-of-art techniques and also modify for better performance.

Keywords Novel class detection · Data stream classification · Concept evolution

1 Introduction

Advances in electronics and hardware technology have highly increased the capability to generate as well as to store enormous data in various forms. Mobile phones, laptops, easily available camera, the Internet are few among the various devices which generate huge amount of digital data. Though data mining provides with many good applications to mine this huge data, data stream classification is a major challenge for data mining community. With increasing volume of data it is no longer possible to process the data in multiple passes. Also the data stream may evolve over the time

V. Attar (✉) · G. Pingale
College of Engineering Pune Shivaji Nagar, Pune, India
e-mail: vahida.comp@coep.ac.in

G.Pingale
e-mail: gargipingale04@gmail.com

B. V. Babu et al. (eds.), *Proceedings of the Second International Conference on Soft Computing for Problem Solving (SocProS 2012), December 28–30, 2012*, Advances in Intelligent Systems and Computing 236, DOI: 10.1007/978-81-322-1602-5_73,

period. Concept-evolution occurs when new classes evolve in the stream. Handling concept evolution involves detecting novel classes and training model with the same. When we are using continuous data stream for classification it is not possible to predict the class or type of data in advance that system is going to encounter. System should be able to detect an emergence of new class of data in stream.

Novel class detection is concerned with recognizing inputs that differ in some way from those that are usually seen or which already exist in the trained model. It is a significant technique to mine the data where an important class is underrepresented in the training set. Traditional classifiers cannot detect presence of novel class. All the instances which belong to such class are misclassified by the algorithm unless there is some kind of manual intervention and model is trained with such a novel class. Novel class detection technique is immensely useful in medical sciences, intrusion detection, fraud detection, signal processing, and image analysis. Novel class detection refers to learning algorithms being able to identify and learn new concepts. In this case, the concepts to be detected and those to be learned correspond to an emerging pattern that is different from noise, or drift in previously known concepts. In simple words, systems detecting novel class must be able to identify new concept emerging in the stream and train the existing models with it so that in future instances belonging to that novel class can be correctly classified. Novelty class detection is now being under constant attention from researchers and academicians. Various approaches to detect the new concept as well as to learn the new concept have been devised.

In rest of the paper Sect. 2 describes about Novel class Detection. In Sect. 3, we explore different technique to handle this issue, Sect. 4 is about the implemented techniques and improvements over the same. Results of both the techniques are presented and compared in Sect. 5 and conclusion in Sect. 6.

2 Overview of Concept Evolution and Novel Class Detection

There are many algorithms in literature that address novelty detection. These algorithms can be divided into one-class approach and multi-class approach. In one class method [6, 7, 9], one class is able to detect a single class of data instances that is different from classes with which system is trained, however, multi-class method can detect more than one new class in training data set. These algorithms show different results for different data sets because their efficiency and accuracy depends on underlying method used and properties of data taken into consideration. Major challenge is to detect novel classes in presence of concept drift in data stream. There are broadly two approaches for novelty detection in data stream, Statistical and Neural Network approach.

Neural networks [6] are immensely used in Novelty Detection. In comparison to statistical methods some critical issues in neural networks are generalization of computational expense during training and expense involved in retraining. Some of these methods are multi-layer perceptions, support vector machines, radial basis function networks, self-organizing maps, oscillatory networks, etc.

Statistical Approach uses the statistical properties of data for creating models. It further uses these models to estimate whether the given instance belongs to the existing class or not. There exist various techniques for novelty detection using this approach. To mention a few, building a density function for data of known class and then calculate the probability of the coming test instance belonging to existing class. Another technique can be finding mean distance of test instance under consideration from the center of nearest cluster of existing class first to detect it as outlier and if there are several such outliers close to each other considering some threshold value for closeness as a novel class. The distance measure can be a Euclidian distance or some other probabilistic distance. Further down there are two types of statistical approach, Parametric and Non-parametric approaches.

In **parametric approach** data distributions are assumed to be known and then the parameters such as mean, variance of model are estimated, the test data falling outside the estimated parameters of distribution are declared as novel. But the parametric approach does not have much practical implications as the distribution of real data is already not known. Some of the parametric methods for novelty detection are Hypothesis testing, Probabilistic/Gaussian mixture modeling.

Non-parametric approach involves estimation of density of data for training for example, Parzen window and K-Nearest Neighbor. The instances which fall out of certain threshold density are regarded as novel. These methods do not make any assumption regarding data distribution functions. Thus in a way they are more flexible. But they do have shortcomings in handling of noise in data. In KNN method, the normal data distribution is defined by a few numbers of spherical clusters formed by k-nearest neighbor technique. Novel class is identified by calculating distance of data point from the center of the clusters which fall beyond its radius. Parzen windows method is used for estimation of data density. It uses Gaussian function. In this method a threshold for detecting novelty is set which is being applied on the probability of test pattern.

3 Techniques for Novel Class Detection

Various papers to deal with the novel class detection are noted in the literature. Smola [5], proposes approach to this problem by trying to estimate a function f which is positive on S and negative on the complement. The functional form of f is given by a kernel expansion in terms of a potentially small subset of the training data; it is regularized by controlling the length of the weight vector in an associated feature space. It provides a theoretical analysis of the statistical performance of algorithm. The algorithm is a natural extension of the support vector algorithm to the case of unlabeled data. Given a small class of sets, the simplest estimator accomplishing this task is the empirical measure, which simply looks at how many training points fall into the region of interest. This algorithm does the opposite. It proposes an algorithm which computes a binary function which is supposed to capture regions in input space where the probability density lives (its support), i.e., a function such that most of the data will live in the region where the function is nonzero. In doing so, it is in line with Vapnik's principle never to solve a problem which is more general than the one we actually need to solve. Moreover, it is applicable also in cases where the density of the data's distribution is not even well-defined, e.g., if there are singular components. It starts with the number of training points that are supposed to fall into the region, and then estimates a region with the desired property.

In [6], a new single class classification technique has been proposed that can detect novelty and handle concept drift. The proposed method uses clustering algorithm to produce the normal model. It relies on Discrete Cosine Transform (DCT) to build compact and effective generative models such that the closest model to a new instance will be an approximation of its K nearest neighbors. Also using the DCT coefficients it presents an effective method for discriminating normal concepts as well as detecting novelty and concept drift. The proposed method referred to as Discrete Cosine Transform Based Novelty and Drift Detection (DETECTNOD), consists of two phases. In the first phase, based on the normal data, it tries to generate an initial model with an effective and compact knowledge about the clusters. At the second phase, the testing data is divided into equal sized blocks whose size is limited only by the storage space. In this phase, using the previously obtained generative models, normal data is discriminated from novel classes and concept drift.

D. Martinez [8] introduced a neural competitive learning tree as a computationally attractive scheme for adaptive density estimation and novelty detection. This approach combines the unsupervised learning property of competitive neural networks with a binary tree-type structure. The initialization process can be performed with input data sampled either randomly from the training set (random initialization) or sequentially as data become available (sequential initialization) in case of an Independently Identically Distributed (IID) sequence and then nodes are splitting each at one time. To avoid this dependency of initialization process on particular data tree is built by taking into account entire dataset and splitting all nodes once. Thus, the learning rule provides an adaptive focusing mechanism capable of tracking time-varying distributions. The constructed tree from the training data serves as a reference tree. Another tree is built for the testing data and a novelty is detected when it differs too much from the reference tree.

Markos Markou [9] proposes a new model of "novelty detection" for image sequence analysis using neural networks. This model uses the concept of artificially generated negative data to form closed decision boundaries using a multilayer perceptron. It uses novelty filter to classify data as known and unknown. One neural network is trained per class where samples are labeled as belonging (positive) or not belonging (negative) to class. Negative data is used for novelty detection. Neural Network with Random Rejects (NN-RR) novelty filter works on thresholding the neural network output activation in response to an input test pattern. Rejected samples are then collected in data storage called bin. Clustering is done using k-means clustering method. This helps in deciding which clusters should be used for retraining.

4 Implemented Technique and Proposed Up-Gradation

In real streaming environment where new classes evolve it is not the case that total number of classes is fixed for classification purpose. Masud et al. [4] propose an algorithm to detect emergence of novel class in the presence of concept drift by quantifying cohesion among unlabeled instances and separation of them from training instances. Traditional novelty detection schemes assume or build a model of normal data and identify outliers that deviate from normal points. This scheme not only detects single point which deviates from normal data but also to find if there is any strong bond among the points. This technique uses ensemble approach to handle concept drift. Data stream is divided into equal sized chunks. Each chunk is accommodated in memory and processed online and classification model is trained from each chunk. Newly trained model replaces original model. And the ensemble evolves representing most up-to-date concepts in the stream. This paper forms the platform of our work. We implement this technique and also propose some improvements in the algorithm.

Input data stream is divided into equal sized chunks. Each unlabeled chunk is given as input to the algorithm. It first detects presence of novel class. Instances belonging to the novel class are separated from the chunk and the remaining instances are classified normally. A new model is trained using the instances of the latest chunk. Finally the ensemble is updated by choosing the best M classifiers from M + 1 classifiers. The base learners used are K Nearest Neighbors and Decision tree. Clusters of the training instances are built and hence store only cluster summaries. These clusters are called as Pseudo points. Any strong cohesion among the instances falling in the unused space indicates presence of a novel class. Novel class detection is a two step process. Initially the training data is clustered and stored as cluster summaries called Pseudopoints which are used to keep the track of used spaces. Later these Pseudopoints are used to detect outliers and if a strong cohesion exists among the outliers a novel class is declared. Every time the data chunk is clustered and the cluster centroid and relevant information is stored as Pseudopoints. Clustering is specific to each base learner. In case of decision tree at its each leaf node where as for KNN classifier the already existing Pseudopoints are used. K Means clustering approach is adapted for the same. The desired value of K parameter in K Means algorithm should be determined experimentally. We build $K_i = (t_i/S)*K$ clusters in l_i, where t_i denoted number of training instances in the leaf node l_i, S is the chunk size.

Cluster summary is stored in Pseudopoints which consist of Weight (W)—total number of instances in the cluster. Centroid (C). Radius(R)—maximum distance between the centroid and the data instances belonging to the cluster. Mean distance—it is the mean distance from each data instance to the cluster centroid. Once the Pseudopoints are formed the raw data is discarded. The union of regions covered by all Pseudopoints represents union of all the used spaces which forms a decision boundary.

R-Outlier is a data instance such that the distance between the centroid of the nearest Pseudopoint and the instance is greater than the radius of this Pseudopoint.

For KNN approach R-Outliers are determined by testing each data instance against all the Pseudopoints. For decision tree each data point is tested against only the Pseudopoints stored at the leaf node where the instance belongs. So any data instance outside the decision boundary is an R-Outlier for that classifier.

Up-Gradations: Early Novel Class Detection (ENCD)

Sometimes test data instance may be considered as an R-outlier because of one or more reasons like: The test instance belongs to an existing class but it is a noise, shift in the decision boundary, Insufficient training data. To avoid an ordinary instance being declared as a novel class instance filtering is done. If a test instance is an R-Outlier to all the classifiers in the ensemble only then it is considered as filtered outlier that is F-Outlier. Rest all are filtered out. Hence being an F-Outlier is a necessary condition for being in a new class. Detection of novel class basically means to verify whether F-Outliers satisfy the two properties of a novel class that is separation and cohesion. λ_C neighborhood is the set of η nearest neighbours of x belonging to class C where η is user defined parameter. In the existing algorithm a new classifier is built only when the test chunk is completed. As soon as a new class is found a new classifier must be built, instead of waiting for the test chunk to finish, and then to train a classifier. Now we don't need to wait for the test chunk to complete, instead we train the new classifier with whatever part of test chunk at hand and this puts us in a position to detect and identify the forthcoming novel instances in that present chunk itself. To achieve this, we put an additional condition for building a classifier. During testing, whenever a new class is detected, a flag is set. Building or training of new classifier is done when this flag is set or when the test chunk is completed. We call this as Early Novel Class Detection (ENCD).

5 Experiments and Results

5.1 Datasets

10 % of KDDCup 99 network intrusion detection [5] contains around 4,90,000 instances. Here different classes appear and disappear frequently, making the new class detection very challenging. There are 22 types of attacks, each record consists of 42 attributes. We have also used synthetically generated dataset from [4] which simulates both concept-drift and novel-class. The data size varies from 100 to 1000 k instances, class label varies from 5 to 40 and data attributes from 20 to 80 (Fig. 1).

5.2 Implementation Environment

The proposed algorithm is implemented in Java programming language on Linux platform. We have used MOA-Massive Online Analysis tool for all the experimen-

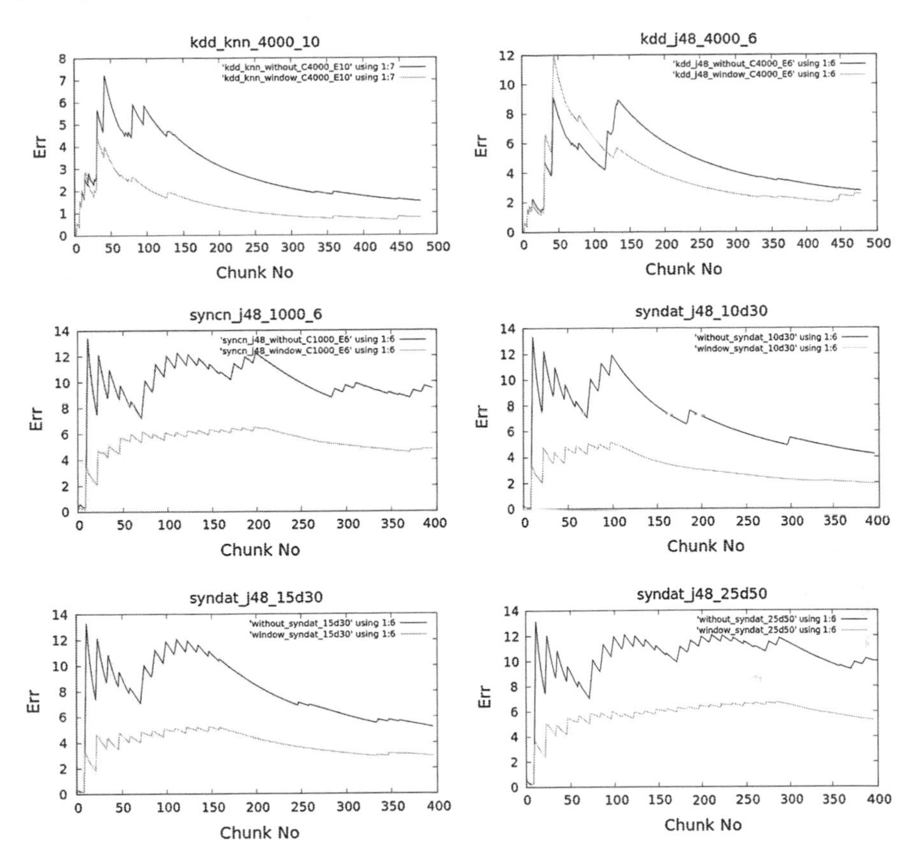

Fig. 1 Error rate versus Chunk No. for Datasets: KDD (KNN, DT), SyncCN ()

tation of the proposed approach. Performance existing algorithm is compared with ETNCD proposed by us. For plotting the graphs of chunk no. and error rate of the two algorithms, we used GNU plot version 4.2.

Parameters-Error: Whenever predicted class value of the instance under consideration is different from its actual class value error is noted. If an existing class instance is misclassified as a novel class instance error is incremented.

$$\text{Global Error} = (100 * \text{Err}) / \text{total no. of instances in dataset.}$$

Chunk Size: We have experimented with chunk size from 500 to 5,000 and selected 4,000 for real and 1,000 for synthetic dataset.

Ensemble Size: 10 for KNN and 6 for Decision tree classifier.

Minimum Number of Points required to declare a novel class: 50.

5.3 Results Based on Up-Gradations

Initially models are built with the first N chunks. From the N + 1 chunk performance of each method is evaluated for that chunk and then the same chunk is used to update the existing models. We compared the proposed approach that is ENCD and the existing algorithm in [4]. We perform comparisons for all the mentioned datasets.

6 Conclusion and Future Work

Various techniques in novel class detection have been studied and analyzed. We propose early detection and identification of every new class. This improvement gives us an edge to classify the novel instances correctly. Experimental results show that the proposed up gradations improve the existing algorithm. The current technique does not take into consideration multiple-label instances. So we would also like to apply our technique to multiple-label instances.

References

1. Mohammad, M.M., Jing, G., Latifur, K., Jiawei, H., Bhavani, M.T.: Classification and novel class detection in concept drifting data streams under time constraints.In: Preprints, IEEE Transactions on knowledge and Data Engineering (TKDE), 2010.
2. Mohammad, M., Masud, J.G., Latifur, K., Jiawei, H.: Bhavani. Classification and novel class detection in data streams in a dynamic feature space, M.T. (2010)
3. Mohammad, M.M., Jing, G., Latifur, K., Jiawei, H.: Bhavani. Classification and novel class detection in data streams with active mining, M.T. (2010)
4. Mohammad, M.M., Jing, G., Latifur, K., Jiawei, H.: Bhavani. M.T , Integrating novel class detection with classification for concept drifting data streams (2009)
5. Smola, A.J., Shawe,-T., Scholkopf, B.P., Williamson R.C.. Advances in neural information processing systems (1999).
6. Zi, M.H., Hashemi M.R.: A DCT based approach for detecting novelty and concept drift in data streams (2010).
7. Martinez, D.: Neural tree density estimation for novelty detection (1998).
8. Markos, M.: Sameer. A neural network-based novelty detector for image sequence, Analysis, S. (2006)
9. Stephen, D.B.: The UCI KDD archive. http://kdd.ics.edu 1999
10. Markou, M., Singh S.: Novelty detection: a review-part 1: statistical approaches part 2: neural network based approaches (2003).

Analyzing Random Forest Classifier with Different Split Measures

Vrushali Y. Kulkarni, Manisha Petare and P. K. Sinha

Abstract Random forest is an ensemble supervised machine learning technique. The principle of ensemble suggests that to yield better accuracy, the base classifiers in the ensemble should be diverse and accurate. Random forest uses decision tree as base classifier. In this paper, we have done theoretical and empirical comparison of different split measures for induction of decision tree in Random forest and tested if there is any effect on the accuracy of Random forest.

Keywords Classification · Split measures · Random forest · Decision tree

1 Introduction

Random forest is an ensemble supervised machine learning technique. Ensembles use multiple classifiers and are always more accurate than individual classifiers. Random forest uses decision tree as base classifier. It is based on principle of bagging [1] and uses randomization in two ways: one in the selection of training data samples to train each base tree and another in the selection of attributes for induction of tree. The principle of ensemble suggests that to yield better accuracy, the base classifiers in the ensemble should be diverse and accurate [2].

V. Y. Kulkarni (✉)
COEP, Pune, India
e-mail: kulkarnivy@rediffmail.com

V. Y. Kulkarni
MIT, Pune, India

M. Petare
MIT, Pune, India

P. K. Sinha
HPC and R&D, CDAC, Pune, India

B. V. Babu et al. (eds.), *Proceedings of the Second International Conference on Soft Computing for Problem Solving (SocProS 2012), December 28–30, 2012*, Advances in Intelligent Systems and Computing 236, DOI: 10.1007/978-81-322-1602-5_74,

In the original Random forest algorithm, Brieman has used Gini index as split measure to induce decision trees. Gini index particularly is not able to detect strong conditional dependencies among attributes [4]. There are various top-down decision tree inducers like ID3, C4.5 which work with split measures other than Gini index [5]. Robnik and Sikonja experimented with Random forest using five different attribute measures; each fifth of the trees in the forest is generated using different split measure (Gini index, Gain ratio, MDL, Myopic ReliefF, or ReliefF). This helped in decreasing correlation between the trees while retaining their strengths. In the literature survey related to Random forest [12], we found that there is very less work done related to attribute split measures in Random forest.

There are 29 different attribute split measures [6], we have selected the commonly used split measures for our work. In this paper we tested use of different split measures (Information gain, Gain ratio, Gini index, Chi square, and ReliefF) to induce decision trees for Random forest, how it is affecting the accuracy, is the strength of individual tree is increased by a specific split measure. For this, we have first done theoretical study of different measures and generated a comparison matrix. In [13] the authors have mentioned that for research related to decision tree induction, empirical comparisons are to be preferred. Hence these results are verified by empirical study and conclusions are derived.

This paper is organized in following way: Sect. 2 introduces in brief the Theoretical Foundations for this research work. They include Random forest and split measures used for experimentation. Section 3 explains methods and results. Section 4 gives concluding remarks.

2 Theoretical Foundations

2.1 Random Forest

Definition: A Random forest is a classifier consisting of a collection of tree-structured classifiers $\{h(x, \theta_k)\, k = 1,\ 2, \ldots.\}$ where the $\{\theta_k\}$ are independent identically distributed random vectors and each tree casts a unit vote for the most popular class at input x [3]. Random forest generates an ensemble of decision trees. To classify a new object from an input vector, the input vector is run down each of the trees in the forest. Each tree gives a classification and each tree votes for the class. The forest chooses the classification having maximum votes. Each tree is grown in the following way: If the number of records in the training set is N, then N records are sampled at random but with replacement, from the original data, this is bootstrap sample. This sample will be the training set for growing the tree. If there are M input variables, a number $m << M$ is selected such that at each node, m variables are selected at random out of M and the best split on these m is used to split the node. The value of m is held constant during forest growing. Each tree is grown to the largest extent possible. There is no pruning. The generalization error of Random forest is given as,

$$PE^* = P_{x,y}(mg(X, Y)) < 0$$

The margin function is given as,

$$mg(X, Y) = av_k I(h_k(X) = Y) - \max_{j \neq Y} av_k I(h_k(X) = j)$$

The margin function measures the extent to which the average number of votes at (X, Y) for the right class exceeds the average vote for any other class. Strength of Random forest is given in terms of the expected value of margin function as,

$$S = E_{X,Y}(mg(X, Y))$$

If ρ is mean value of correlation between base trees, an upper bound for generalization error is given by,

$$PE^* \leq \rho\left(1 - s^2\right) s^2$$

Hence, to yield better accuracy in Random forest, the base decision trees are to be diverse and accurate.

2.2 *Split Measures for Decision Tree*

In decision tree induction, split measures are used to generate splitting rules. A splitting rule is a one-ply look ahead heuristic used to guess the best test to make at the current node in the tree [11].

2.2.1 Information Gain

This attribute selection measure is based on Shannon's information theory, which finds the value or information content of message. Hence for each splitting attribute, Information gain is calculated and the attribute with highest gain is chosen as splitting attribute. This attribute is such that it creates minimum impurity or randomness in the generated splits and hence it minimizes the information needed to classify the tuples. If D is the entire dataset, then information content of D or entropy of D is given as

$$\text{Info}(D) = -\sum_{i=1}^{m} (p_i \log p_i)$$

Here $i = 1, 2, \ldots, m$ are classes in dataset D. P_i is probability that an arbitrary tuple in D belongs to class C_i. Let A be an attribute in D and $\{a_1, a_2, \ldots, a_v\}$ are different values of attribute A in D such that $\{D_1, D_2, \ldots, D_v\}$ are partitions generated based on these values. These partitions are likely to be impure. How much more information is still needed to arrive at an exact classification or pure partition is given as [10]:

$$\text{Info}_A(D)=\sum_{j=1}^{v}(|D_j|/|D|) * \text{Info}(D_j)$$

The smaller this additional information, the greater the purity of the partition.

$$\text{Gain }(A) = \text{Info }(D) - \text{Info}_A\ (D)$$

The attribute with highest Information gain is selected for splitting. Information gain is biased toward choosing attributes with a large number of values.

2.2.2 Gain Ratio

The Gain ratio attempts to overcome the bias toward multi-valued attributes. The less evenly spread its value, the less information in the attribute [7].

$$\text{SplitInfo}_A(D) = -\sum_{j=1}^{v}(|D_j|\ /\ |D|)\ *\ \log\frac{|D_j|}{|D|}$$

$$\text{Gain Ratio}(A) = \text{Gain }(A)\ /\text{SplitInfo}_A\ (A)$$

The attribute with maximum Gain ratio is selected for splitting.

2.2.3 Gini Index

It is known as generalized inequality index. If a dataset D contains examples from n classes, Gini Index, Gini(D) is defined as [10]:

$$\text{Gini}(D) = 1 - \sum_{j=1}^{n} p_j^2$$

p_j is the relative frequency of class j in D. Gini index considers binary split for each attribute. For attributes having more than two distinct values, subsets of attributes are considered. For binary split on partitions D into D_1 and D_2, the Gini index of G is;

$$\text{Gini}_A(D) = \frac{|D_1|}{|D|}\text{Gini }(D_1) + \frac{|D_2|}{|D|}\text{Gini }(D_2)$$

The attribute with minimum Gini index is selected for splitting. The best split is the one with the largest decrease in diversity. Gini prefers split that put the largest class into one pure node, and all remaining classes into the other [9].

Table 1 Comparison of different split measures

Measures issues	Information gain	Information gain ratio	Gini index	Chi square	Relief
Split type	Multiway	Multiway	Binary	Multiway	Binary
Splitting function	Entropy	Entropy	Gini coefficient	X^2 statistics	Diff$(A, I1, I2)$
Target	Continuous, discrete	Continuous, discrete	Continuous, discrete	Discrete	Continuous, discrete
Predictors	Continuous, discrete	Continuous, discrete	Continuous, discrete	Discrete	Continuous, discrete
Algorithm	ID3	C4.5	CART	CHAID	RELIEFF family
Biasing	Biased toward multi-valued attributes	Helps to reduce biasing	Biased toward multi-valued attributes	Adjustment is possible to reduce biasing	Relief exhibits an implicit normalization effect to reduce biasing

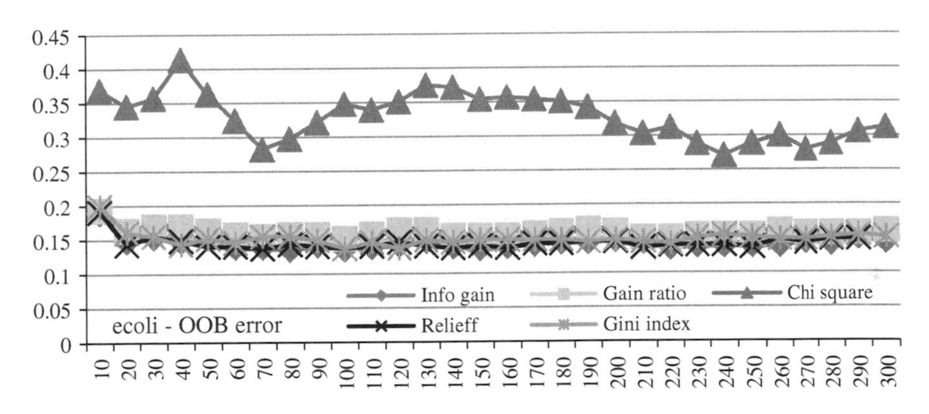

Fig. 1 Graph showing comparative OOB error for ecoli dataset

2.2.4 Chi Square

Chi square test uses contingency matrix to test distribution of available data tuples into different classes. If $(A_{i,}\ B_i)$ denote the events that attribute A takes on value a_i and attribute B takes on value b_j then Chi square value is computed as [10]:

Table 2 Datasets used for analysis

Dataset	Instances	Attributes	Classes	Imbalanced ?	Attribute type
Hypothyroid	3772	30	4	yes	Numeric/nominal
Ionosphere	351	35	24	no	Numeric
kr-vs-kp	3196	37	2	no	Nomial
Sick	3772	37	2	yes	Numeric/nominal
Sonar	208	61	2	no	Numeric
Soybean	683	36	19	no	Nomial
Vehicle	846	19	4	no	Numeric
Anneal	898	39	5	yes	Numeric/nominal
Vote	435	17	2	no	Nomial
Audiology	226	70	24	no	Nomial
Vowel	990	14	11	no	Numeric/nominal
Waveform	5000	41	3	no	Numeric
Diebetes	768	9	2	no	Numeric
Breast cancer	286	10	2	no	Nomial
credit-g	1000	21	2	no	Numeric/nominal
segment	2310	20	7	no	Numeric/nominal
Splice	3190	62	3	no	Nomial
Car	1728	7	4	no	Nomial
Ecoli	336	8	8	no	Numeric
Glass	214	10	7	no	Numeric

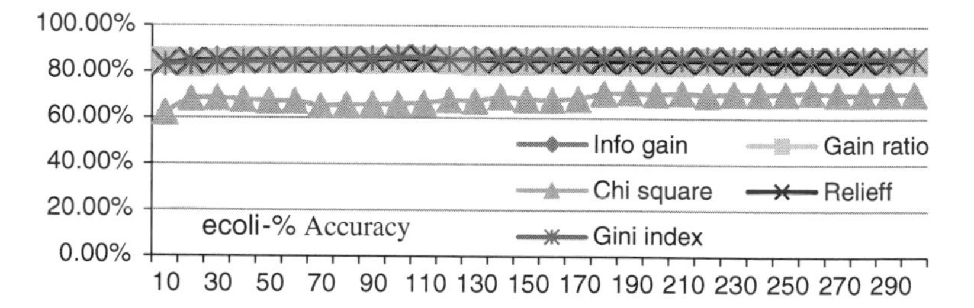

Fig. 2 Graph showing comparative % accuracy for ecoli dataset

$$X^2 = \sum_{i=1}^{c}\sum_{j=1}^{r}(o_{ij} - e_{ij})^2/e_{ij}$$

where o_{ij} is observed frequency and e_{ij} is expected frequency, N is number of data tuples. The attribute that gives highest Chi square value is used as the best split.

Table 3 Percentage accuracy values for different split measures

Trees	Measure	Sonar	Soybean	Vote	Vowel	BC	Diabetis	Ecoli
50	Info gain	87.50	94.00	96.09	98.18	69.23	75.26	84.82
	Gain ratio	85.10	93.70	96.09	97.98	70.63	75.00	84.82
	Chi square	85.10	91.51	96.32	97.78	67.83	74.61	67.26
	ReliefF	86.06	93.70	96.32	98.38	68.53	74.48	84.52
	Gini index	84.62	92.39	96.32	98.59	65.38	75.78	84.52
100	Info gain	85.58	93.41	96.32	98.48	69.58	75.91	85.12
	Gain ratio	86.54	93.85	96.32	98.38	70.98	75.65	85.12
	Chi square	86.54	91.65	96.78	98.59	66.78	74.87	66.07
	ReliefF	83.65	94.00	96.09	98.38	67.48	75.26	85.12
	Gini index	84.62	92.83	96.32	98.69	66.43	75.52	86.31
150	Info gain	85.58	93.41	96.55	98.48	70.63	75.91	85.12
	Gain ratio	87.02	93.27	96.55	98.48	70.28	75.78	84.52
	Chi square	86.06	91.36	96.32	98.59	66.70	74.87	68.15
	ReliefF	86.54	94.00	96.32	98.48	67.48	74.87	84.82
	Gini index	83.17	92.53	96.32	98.89	66.43	76.04	85.71
200	Info gain	87.98	93.56	96.55	98.48	70.98	75.26	85.12
	Gain ratio	87.02	93.12	96.55	98.59	70.63	75.26	84.23
	Chi square	87.02	91.51	96.55	98.89	66.43	75.13	70.54
	ReliefF	87.02	93.56	96.55	98.38	67.13	75.00	84.82
	Gini index	84.62	92.53	96.32	98.79	65.04	75.39	86.01
250	Info gain	86.06	93.70	96.55	98.59	69.23	75.00	85.12
	Gain ratio	87.98	93.12	96.55	98.59	70.63	75.13	84.23
	Chi square	88.46	91.22	96.55	98.69	65.38	75.13	70.83
	ReliefF	87.02	94.00	96.55	98.48	67.13	74.61	84.82
	Gini index	86.54	92.53	96.32	98.69	65.38	74.87	86.31
300	Info gain	84.62	93.70	96.55	98.59	69.23	75.30	85.42
	Gain ratio	87.02	93.12	96.55	98.59	70.98	75.00	84.82
	Chi square	88.46	91.36	96.55	98.38	65.38	75.13	70.83
	ReliefF	86.54	94.00	96.55	98.48	67.83	74.74	85.42
	Gini index	86.54	92.68	96.32	98.69	65.38	75.13	86.01

2.2.5 ReliefF

ReliefF is an attribute selection method that is based on attribute estimation. ReliefF is useful when there are strong inter dependencies among the attributes [8]. It assigns a relevance weight to each attribute. The attributes are estimated based on how well their values distinguish between the instances that are near to each other.

Given a randomly selected instance *R*, ReliefF searches for its two nearest neighbors: One from the same class called nearest hit H, and the other from a different class called nearest miss M. Then a quality estimation $W[A]$ is updated for all attributes depending on their values for R, M, and H. If the values of attribute A at instances R and H are different then attribute A separates two instances of the same class which is not desirable, and hence negative updation is added to quality estimation

$W[A]$. If the values of attribute A at instances R and M are different then attribute A separates two instances with different class values which is desirable, and hence a positive update is done to quality estimate $W[A]$.

The process is repeated for m times where m is user-defined parameter.

Diff (A, $I1$, $I2$) is a function which calculates difference between the values of attribute A for instances $I1$ and $I2$.

For discrete attribute:

If Diff (A, $I1$, $I2$) $= 0$: value(A, $I1$) $=$ value(A, $I2$), 1 : Otherwise

For continuous attribute:

Diff (A, $I1$, $I2$) $=$ [value(A, $I1$) $-$ value(A, $I2$)] / (max(A) $-$ min(A)).

3 Methods and Results

The aim behind performing this experiment was to observe the effect of variation in attribute split measure on the accuracy of Random forest classifier. For this, we have generated five different Random forest classifiers each using different split measures (Info gain, Gain ratio, Chi square, Gini index, and ReliefF). For each of this measure, Random forest is generated with varying size, i.e., from 10 to 300 trees with a step of 10. The datasets used are all selected from UCI machine learning repository, and the selection is such that they are already used in different experiments related to Random forest classifier. The details of datasets used for analysis are given in Table 2. The accuracy values and OOB error [3] values are noted down. The accuracy reflects overall performance of Random forest classifier and the OOB error estimation is a measure of strength of individual tree in the forest. The OOB readings recorded here are average of OOB error values over all trees in the forest. The experimentation is done using 10-fold cross validation. The experiments were conducted on 20 datasets with variation in their size, but due to space limit we are presenting here the readings for 7 datasets (given in Table 3) and graphs for one of them. For all the graphs, values on X-axis show number of trees and values on Y axis show either OOB error or percentage accuracy.

4 Conclusion

The empirical results show that there is not much / significant variation in accuracy obtained except Chi square. Information gain and Gain ratio give comparable results for almost all datasets. Chi square is not suitable for Random forest classifier. ReliefF gives slightly better results as compared to Information gain or Gain ratio, but the time taken by Random forest with ReliefF is more. We are at present working on the time aspect of all the split measures where we will exclude Chi square from the experimentation. With most of the datasets, Gini index slightly lags in the results. Taking into consideration both theoretical and empirical comparisons, we conclude that Gain ratio and ReliefF are better options, and considering the time aspect, Gain ratio is the best option.

As a future scope, we are working on generating Hybrid Decision Tree where individual decision tree of Random forest will be generated using different split measures and the effect on accuracy of Random forest will be observed.

References

1. Breiman, L.: Bagging predictors, Technical report No 421, September (1994).
2. Opitz, David: Maclin, richard: popular ensemble methods: an empirical study. J. Arti. Intel. **11**, 169–198 (1999)
3. Brieman, Leo: Random forests. Machine Learning. **45**, 5–32 (2001)
4. Sikonja, M.R.: Improving random forests. In: Boulicaut, J.F., et al. (eds): Machine Learning, ECML 2004 Proceedings, LNCS, vol. 3201, PP. 359–370, Springer, Berlin (2004).
5. Rokach, Lior: Maimon, oded: top-down induction of decision trees classifiers-a survey. IEEE trans. syst. man. cyber. part c: appli. rev. 35(4), 476–487 (2005).
6. Badulescu, L.A.: The choice of the best attribute selection measure in DecisionTree induction, Annals of University of Craiova, Math. Comp. Sci. Ser. Vol. 34 (1) (2007).
7. Mingers, J.: An empirical comparison of selection measures for decision tree induction. Mach. Learn. **3**, 319–342 (1989)
8. Robnik-Sikonja, M., Kononenko, I.: Attribute dependencies, understandability, and split selection in tree based models, Machine Learning: Proceedings of the 6th International Conference (ICML), 344–353 (1999).
9. Brieman, Leo: Technical note-some properties of splitting criteria. Mach. Learn. **24**, 41–47 (1996)
10. Han, J., Kamber, M.: Data mining: concepts and techniques, 2nd edn. Morgan Kaufmann Publisher, San Francisco (2006)
11. Buntine, Wray: Niblet, tim: a further comparison of splitting rules for decision tree induction. Mach. learn. **8**, 75–85 (1992)
12. Kulkarni, V.Y., pradeep, K.S.: Random forest classifiers: a survey and future research directions. Int. J. Adv. Comput. ISSN 2051–0845. 36(1), 1144–1153 (2013).
13. Liu, W.Z., White, A.P.: The importance of attribute selection measures in decision tree induction. Mach. Learn. **15**, 25–41 (1994)

Text Classification Using Machine Learning Methods-A Survey

Basant Agarwal and Namita Mittal

Abstract Text classification is used to organize documents in a predefined set of classes. It is very useful in Web content management, search engines; email filtering, etc. Text classification is a difficult task due to high- dimensional feature vector comprising noisy and irrelevant features. Various feature reduction methods have been proposed for eliminating irrelevant features as well as for reducing the dimension of feature vector. Relevant and reduced feature vector is used by machine learning model for better classification results. This paper presents various text classification approaches using machine learning techniques, and feature selection techniques for reducing the high-dimensional feature vector.

Keywords Text classification · Feature selection · Machine learning Algorithms

1 Introduction

Text mining means to extract relevant information from text and to search for interesting relationships between extracted entities. Text classification is one of the basic and important tasks of text mining. Text classification means automatically assign a document in some predefined categories of documents based on their contents. Text classification is a supervised learning model that can classify text documents according to their predefined categories. Web content for a search engine can be organized properly using text classification for efficient retrieval of Web documents. Text classification techniques are be used for automatically email filtering, medical diagnosis,

B. Agarwal (✉) · N. Mittal
Malaviya National Institute of Technology, Jaipur, India
e-mail: thebasant@gmail.com

N. Mittal
e-mail: mittalnamita@gmail.com

B. V. Babu et al. (eds.), *Proceedings of the Second International Conference on Soft Computing for Problem Solving (SocProS 2012), December 28–30, 2012*, Advances in Intelligent Systems and Computing 236, DOI: 10.1007/978-81-322-1602-5_75,

news group filtering, documents organization, indexing for document retrieval, word sense disambiguation by detecting the topics a document covers.

Main challenges for text classification are following:

1. High dimensionality, due to which it is difficult to create a classifier model because performance of the classifier degrades as feature vector increases for a classifier [1].
2. Not all features are important for classification, some features may be redundant or irrelevant and some may even misguide the classification result [1].
3. To remove redundancy and noisy features from the data.

In text classification, feature vector generally consist of thousands of attributes/features, that is why feature reduction methods has to be used for removing irrelevant features, in such a way that classifier accuracy does not affected. Efficiency and success of any machine learning algorithm depends on the quality of data. Automatic feature reduction methods are used for reducing the size of feature vector and removing irrelevant features. There are two methods for this purpose, (i) feature selection and (ii) feature transformation. In feature selection important features are identified and used for classification. In feature transformation feature vector is transformed into a new feature vector with selected lower dimensions.

The objective of this paper is to discuss, (i) filter-based feature selection methods, (ii) feature transformation techniques, and (iii) machine learning techniques used for text classification.

The remainder of the paper is organized as follows. Section 2 describes the text classification process, Sect. 2.4 discusses the evaluation methods used for text classification, and Sect. 3 concludes the paper.

2 Text Classification Process

In text classification process, initially documents are read from the collection, then preprocessing like stemming, removal of stop words takes place. After that, important features are selected from the feature vector. Lower dimensional feature vector is fed to the classifier. Common text classification methods include both supervised and unsupervised machine learning methods like Support Vector Machine (SVM) [9], K-Nearest Neighbour (KNN), Neural Network (NN), Naive Bayes [19] etc.

2.1 *Preprocessing*

The most common preprocessing task for text classification is that of stop-word removal and stemming. In stop-word removal, the common words in the documents which are not discriminatory to the different classes are removed from feature vector.

For example “a”, “the”, “that”, etc., are frequent words that do not help in classification, which occurs almost equally in all the documents.

In stemming, different forms of the same word are converted into a single word. For example, singular, plural, and different tenses are converted into a single word. Port stemmer algorithm is well-known algorithm for stemming [7].

2.2 Text Representation

For text classification using machine learning methods, each document should be represented in the form so that learning algorithm can be applied. So each document is represented as a vector of words/terms/feature. The values in the feature vectors are weighted to reflect the frequency of words in the documents and the distribution of words across the collection. The more times a word/term occurs in a document, the more it is relevant to the document. The more times the word occurs throughout all documents in the collection, the more poorly it discriminates between documents [15]. A popular weighting scheme is Term Frequency–Inverse Document Frequency (TF-IDF): $wij = tf(ij)*idf(i)$, where $tf(ij)$ is the frequency of term i in document j, and $idfi$ is the inverse document frequency, it measures if a term is common or rare across documents. IDF can be calculated by $\log(N/F)$, where N is total number of documents in the corpus, F is number of documents where term I appears.

The $tf*idf$ weighting scheme does not consider the length of document, tfc weighting is similar to $tf*idf$ weighting except, length normalization is used in tfc weighting. In addition, a logarithm-based weighting scheme is log-weighted term frequency that uses a logarithm of word frequency, reducing the effect of large number of term frequency in a document with big document length [19]. One method is word frequency weighting, i.e., to use the frequency of the term in the document [19]. Another method for text representation is to simply calculate binary feature values, i.e., a term either present or not in the document [14].

2.3 Feature Reduction

Feature reductions methods are used to remove the irrelevant features and reduce the dimensionality of feature space. Basically, there are two methods for feature reductions (1) Feature selection, and (2) Feature extraction/Feature transformation. Feature extraction means reduce the dimensionality by transformation/projection of all the features in subset features. It maps the high-dimensional data on lower dimensional space. New attributes are obtained by the combination of all the features, for e.g., Principal Component Analysis (PCA) [22], Singular Value Decomposition (SVD) [12]. Feature selection technique selects the important features/attributes from the high-dimensional feature vector using certain criteria for e.g. Information Gain (IG). Its main purpose is to reduce the dimensionality of the feature space, remove the irrelevant features so that performance and accuracy of the machine leaning algorithm can be improved and also algorithm can run faster.

Feature selection methods are basically of three types depending on how they selects feature from the feature vector, i.e., filter approach, wrapper approach, and embedded approach [14, 22].

In filter approach [14, 22], all the features are treated independent to each other. Features are ranked according to their importance score of each feature, which is calculated by using some function. Filter approach-based methods does not depend on the classifier. Advantages of this approach are that they are computationally simple, fast, and independent to the classifier. Feature selection step is performed once and then reduced feature vector is used with any classifier can be used. Disadvantage of this approach is that it does not interact with the classifier. It assumes features are independent; it is possible that a feature performs well but performs worse with the combination of other feature, and similarly a lower scoring attribute can show good performance with the combination to other features [22]. However filter approach with some modification, included features dependency in multivariate filtering approach.

In wrapper approach [14, 22], a search procedure is defined to search the feature subset, and various subsets of features are generated and evaluated for a specific classifier. In wrapper approach features are treated dependant to each other, and model interacts with the classifier. As the number of features subset grows exponentially with increase in the number of features, hence heuristic search methods can be used for selecting feature subsets. Advantages of this approach are that it interacts with the classifier, and features dependencies are considered. Disadvantages are that there is a risk of over fitting, slow and classifier-dependant.

Filter approach is very fast compared to wrapper approach, wrapper approach is very efficient but specific for a classifier algorithm. It is time consuming. If size of dataset is high than it is very difficult to create wrapper.

2.3.1 Filter-based Feature Selection Methods

Document Frequency

Document Frequency (DF) is the number of documents in which a term appears. In document frequency thresholding, those terms are removed whose document frequency is less than a predefined value. This is an unsupervised feature selection method; it can be computed without class labels. Assumption is that rare terms are less informative for learning algorithms [3, 4], and frequent words have more chances that they will be present in future test cases.

Information Gain

Information gain measures decrease in entropy when the feature value is given, means number of bits of information obtained due to knowing the presence or absence of a term for prediction [4].

First, Information gain for each term is computed. Further, terms are removed from the feature vector whose value is below predefined threshold value [4].

Mutual Information
Mutual information of a term and class attribute is used for feature selection methods. Mutual information is used to quantitatively analyze the relationship between any two features or between a feature and a class variable. Mutual information compares the probability of occurring term t and class c together and probability of term and class individually [6, 22]. The mutual information of between term t and class c is defined as

$$I(t,c) = \log \frac{P(t,c)}{P(t)^{*}\mathrm{P}(c)} = \log \frac{P(t \wedge c)}{P(t)^{*}P(c)} \quad (1)$$

If there is a relationship between term and class then joint probability $P(t,c)$ will be greater than the $P(t)^{*}\mathrm{P}(c)$, and $I(t,c) >> 0$. High value of mutual information of a feature with the class indicates higher importance of feature for classification. Threshold value can be set for selecting the features.

Chi Square
The chi squared measures the lack of independence between term *t* and class c. it can be used for testing independence or association between two variables. Chi squared statistic test tries to identify the best terms for the class c and are the ones which are distributed most differently in the sets of positive and negative examples of class c [1, 2].

Odds Ratio
Odds Ratio is a fraction of the word occurring in the positive class normalized by that of the negative class. It has been used for relevance ranking in information retrieval. It is based on the assumption that the distribution of features on the relevant documents is different from the distribution of features on the nonrelevant documents [17].

2.3.2 Feature Transformation

Feature transformation techniques are used to reduce the feature vector size, it does not rank the features according to their importance but it transforms higher dimensional feature space on the lower dimensional feature space.

Singular value decomposition can be used for feature reduction for text classification. Latent Semantic Analysis (LSA) uses singular value decomposition method for mapping high-dimensional features to lower dimensional space that is latent semantic space [12].

Principal Component Analysis (PCA) is a common method for feature transformation. PCA seeks a linear projection of high-dimensional data into lower dimensional space in such way that maximum variance is extracted from the variables. These extracted variables are called principal components those are orthogonal to each other and uncorrelated. Principal Component Analysis rejects data with small variance [11].

Linear Discriminant Analysis (LDA) is one of the popular dimension reduction methods. It finds out the feature that has high-class discriminant capability. Discriminant features are identified by maximizing the ratio of the between-class to the within-class variance of a given data set. So a feature scattered more among different classes and less scattered within class is important for the classification. A novel text classification method is proposed which is based on LDA and SVM. High-dimensional feature vector is transformed into lower dimensional feature vector by LDA feature reduction technique. Then SVM classifier is used for text classification [15].

Independent Component Analysis (ICA) [16], on the other hand, is to identify independent components. ICA transforms the original high dimensional data into lower dimensional components that are maximally independent from each other. These independent components are not necessarily orthogonal to each other like PCA. For dimension reduction ICA finds k components that effectively contain maximum variability of the original data.

2.4 Classifier Models

There has been active research in text classification over the past few years. Most of the research work in text classification has focused on applying machine-learning methods to classify text based on words from a training set [1, 18, 19]. These approaches include Naïve Bayes (NB) classifiers, SVM, K-Nearest Neighbor (KNN), Decision Tree, Rocchio algorithm, etc., and also by combining approaches.

Naïve Bayes classifier assumes independence among attributes. NB approach's implementation is simple and learning time is less, however, its performance is not good for categories defined with very few features [21, 25]. It gives a good classification result of a text document provided there are a sufficient number of training instances of each category. Gini index-based weighted features is combined with NB classifier, this approach improved the performance for text classification [10]. Bayesian classifier is modified to handle one hundred thousand of variables. Experiment result shows that modified tree-like Bayesian classifier works with sufficient speed and accuracy [2]. Maximum entropy is used for a new text classifier proposed in [8], resulting in better performance in contrast to bayes classifier.

SVM produces good results for two class classification problems like text document belongs to a particular category or not, but it is difficult to extend to multiclass classification. To solve multiclass problems of SVM more efficiently, class incremental approach is proposed in [23]. SVM outperforms with KNN and naïve Bayesian classifier for text classification as proposed in [28]. Naïve Bayesian method was used as a preprocessor for dimensionality reduction followed by the SVM method for text classification [5].

A modified k-NN-based text classification is proposed, in which variants of the k-NN method with different decision functions, k values, and feature sets were evaluated to increase the performance of the algorithm [9]. An improved k-NN algorithm is proposed in which unimportant documents are not considered, to increase the performance of the classification [13].

Decision Tree-based text classification does not assume independence among its features as in Naïve Baysian. Decision tree performs well as a text classifier when there are very less number of features; however it becomes difficult to create a classifier for large number of feature [19].

In Rocchio Algorithm, text is indicated as an N-dimensional vector. N is the total number of features, and each feature item is weighted by TF-IDF algorithm. Training text dataset is expressed as a feature vector, and then generated the prototype vector for each class. At the time of classification, similarity between different class features vectors and feature vector of unknown text document is calculated, and the text is assigned to the class which has highest similarity [19].

Boosting and Bagging are two voting-based classifiers. In voting classifier, training samples are taken randomly from the collection multiple times, and different classifiers are learned. To classify a new sample, each classifier gives a different class label; the result of voting classifier is decided by the maximum votes earned for a particular class [29]. Main difference between bagging and boosting is in the way, they take the samples for training a classifier. In bagging, training samples are taken with equal weights randomly, and in boosting, more weightage are given to those samples which have been misclassified by previous classifiers. AdaBoost which is a boosting classifier outperforms rocchio when the training dataset contains a very large number of relevant documents [20].

Feature vector is fed to the inputs of the neural network and classification results come from the output of the network. Problem with the neural network is its slow learning. Performance of neural network-based text classification was improved by assigning the probabilities derived from Naïve Bayesian method as initial weights [24]. In [27], three neural networks, i.e. (i) the Competitive, (ii) the Back Propagation (BP), and (iii) the Radial Basis Function (RBF), in text classification are compared. The competitive network is an unsupervised and BP and RBF are supervised methods for learning. Experimental results show that BP works effectively for text classification, RBF network learns faster compare to others. BP and RBF perform better than competitive network. A modified back propagation neural network is proposed to improve the performance of traditional algorithm. SVD technique is used for reducing the dimension of the feature vector. Experimental results show that the modified neural network outperforms traditional back propagation NN [26].

There is a need to experiment with more such hybrid techniques in order to derive the maximum benefits from machine learning algorithms and to achieve better classification results. Different feature selection and reduction techniques are used in combination with different machine learning algorithm to increase the performance and accuracy of the classifier.

3 Conclusion

The commercial importance of automatic text classification applications has increased due to the number of blogs, Web contents, growth rate of Internet access. Therefore, much research is currently focused in this area. Performance of text classification can be increased using machine learning techniques. However preprocessing plays important role due to high- dimensional data, and feature selection and reduction techniques enhances the quality of training data for the classifier, resulting into improved classifier accuracy.

Text classification for regional language documents can be useful for several governmental and commercial projects. Multitopic text classification, identify contextual use of terms on blogs and use of semantics for better classifiers are some of the areas, where future research can be done.

References

1. Sebastiani, F.: Machine learning in automated text categorization. ACM Comput. Surv. **34**(1), 1–47 (2002)
2. Al-Harbi, S., Almuhareb, A., Al-Thubaity, A., Khorsheed, M., Al-Rajeh, A.: Automatic Arabic text classification. In: JADT'08, France, pp. 77–83 (2008)
3. Forman, George: An extensive empirical study of feature selection metrics for text classification. J. Mach. Learn. Res. **3**, 1289–1305 (2003)
4. Yang, Y., Pedersen, J.O.: A Comparative study on feature selection in text categorization. In: Proceedings of the Fourteenth International Conference on Machine Learning, pp. 412–420, 08–12 July 1997
5. Isa, D., Lee, L.H., Kallimani, V.P., RajKumar, R.: Text document pre-processing with the Bayes formula for classification using the support vector machine. IEEE Trans. Knowl. Data Eng. **20**(9), 1264–1272 (2008)
6. Yan, X., Gareth J., Li J.T., Wang, B., Sun, C.M.: A study on mutual information-based feature selection for text categorization'. J. Comput. Inf. Syst. **3**(3), 1007–1012 (2007)
7. Porter, M.F.: An algorithm for suffix stripping. Program **14**(3). 130–137 (1980)
8. Nigam, K., Mccallum, A.K., Thrun, S., Mitchell, T.: Text classification from labeled and unlabeled documents using EM. Mach. Learn. **39**, 103–134 (2000)
9. Joachims, T.: A statistical learning model for text classification for support vector machines. In: 24th ACM International Conference on Research and Development in Information Retrieval (SIGIR) (2001)
10. Dong, Tao, Shang, Wenqian, Zhu, Haibin: An improved algorithm of Bayesian text categorization. J. Softw. **6**(9), 1837–1843 (September 2011)
11. Kumar, C.A.: Analysis of unsupervised dimensionality reduction techniques. Comput. Sci. Inf. Syst. **6**(2), 217–227 (Dec. 2009)
12. Soon, C.P.: Neural network for text classification based on singular value decomposition. In: 7th International conference on Computer and Information Technology, pp. 47–52 (2007)
13. Muhammed, M.: Improved k-NN algorithm for text classification. Department of Computer Science and Engineering University of Texas at Arlington, TX, USA
14. Ikonomakis, M., Kotsiantis, S., Tampakas, V.: Text classification using machine learning techniques. IEEE Trans. Comput. **4**(8) 966–974 (2005)
15. Wang, Z, Qian, X.: Text categorization based on LDA and SVM. In: Computer Science and Software Engineering, 2008 International Conference, vol. 1, pp. 674–677 (2008)

16. Kolenda, T., Hansen, L.K., Sigurdsson, S.: Independent components in text. In: Girolami, M. (ed.) Advances in Independent Component Analysis, Springer-Verlag, New York (2000)
17. Jia-ni, H.U., Wei-Ran, X.U. Jun, G., Wei-Hong, D.: Study on feature methods in chinese text categorization. Study Opt. Commun. **3**, 44–46 (2005)
18. Aggarwal, C.C., Zhai, C-X.: A survey of text classification algorithms. Mining Text Data. pp. 163–222, Springer (2012)
19. Aas, K., Eikvil, L.: Text categorisation: A survey"m Tech. rep. 941. Norwegian Computing Center, Oslo, Norway (1999)
20. Schapire, R.E., Singer, Y., Singhal, A.: Boosting and Rocchio applied to text filtering. In: Proceedings of SIGIR-98 21st ACM International Conference on Research and Development in Information Retrieval, pp. 215–223, ACM Press, New York US (1998)
21. Kim, S.B., Rim, H.C., Yook, D.S., Lim, H.S.: Effective Methods for Improving Naive Bayes Text Classifiers. LNAI **2417**, 414–423 (2002)
22. Saeys, Y., Inza, I., Larranaga, P.: A review of feature selection techniques in bioinformatics. Bioinformatics **23**(19), 2507–2517 (2007)
23. Zhang, B., Su, J., Xu, X.: A class-incremental learning method for multi-class support vector machines in text classification. In: Proceedings of the 5th IEEE international conference on Machine Learning and, Cybernetics, pp. 2581–2585 (2006)
24. Goyal, R.D.: Knowledge based neural network for text classification. In: Proceedings of the IEEE international conference on Granular, Computing, pp. 542–547 (2007)
25. Meena, M.J., Chandran, K.R.: Naïve bayes text classification with positive features selected by statistical method. In: Proceedings of the IEEE international conference on Advanced, Computing, pp. 28–33 (2009)
26. Li, C.H, Park, S.C.: An efficient document classification model using an improved back propagation neural network and singular value decomposition. J. Expert Syst. Appl. **36**(2), pp. 3208–3215 (2009)
27. Wang, Z., He, Y., Jiang, M.: A comparison among three neural networks for text classification. In: 8th IEEE International Conference on, Signal Processing (2006)
28. Zhijie, L., Lv, X., Liu, K., Shi, S.: Study on SVM compared with other text classification methods. In: 2nd International workshop on education technology and computer, science (2010)
29. Freund, Y., Shapire, R.R.: Experiments with a new boosting algorithm. In: Proceedings of 13th International Conference on, Machine learning, pp. 148–156 (1996)

Weka-Based Classification Techniques for Offline Handwritten Gurmukhi Character Recognition

Munish Kumar, M. K. Jindal and R. K. Sharma

Abstract In this paper, we deal with weka-based classification methods for offline handwritten Gurmukhi character recognition. This paper presents an experimental assessment of the effectiveness of various weka-based classifiers. Here, we have used two efficient feature extraction techniques, namely, parabola curve fitting based features, and power curve fitting based features. For recognition, we have used 18 different classifiers for our experiment. In this work, we have collected 3,500 samples of isolated offline handwritten Gurmukhi characters from 100 different writers. We have taken 60 % data as training data and 40 % data as testing data. This paper presents a novel framework for offline handwritten Gurmukhi character recognition using weka classification methods and provides innovative benchmark for future research. We have achieved a maximum recognition accuracy of about 82.92 % with parabola curve fitting based features and the multilayer perceptron model classifier. In this work, we have used C programming language and weka classification software tool. At this point, we have also reported comparative study weka classification methods for offline handwritten Gurmukhi character recognition.

Keywords Handwritten character recognition · Feature extraction · Classification · Weka · Tool

M. Kumar (✉)
Department of Computer Science, Panjab University Rural Centre,
Kauni, Muktsar, Punjab, India
e-mail: munishcse@gmail.com

M. K. Jindal
Department of Computer Science and Applications, Panjab University Regional Centre,
Muktsar, Punjab, India
e-mail: manishphd@rediffmail.com

R. K. Sharma
School of Mathematics and Computer Applications, Thapar University,
Patiala, Punjab, India
e-mail: rksharma@thapar.edu

B. V. Babu et al. (eds.), *Proceedings of the Second International Conference on Soft Computing for Problem Solving (SocProS 2012), December 28–30, 2012*, Advances in Intelligent Systems and Computing 236, DOI: 10.1007/978-81-322-1602-5_76,

1 Introduction

Offline Handwritten Character Recognition usually abbreviated as Offline HCR. Offline HCR is one of the oldest ideas in the history of pattern recognition by using the computer. In character recognition, the process commences by reading of a scanned image of character, determining its meaning, and finally, translates the image into a computer written text document. In recent times, Gurmukhi character recognition has become one of the fields of practical usage. OHCR involves activities like digitization, preprocessing, feature extraction, classification, and recognition. Recognition rate depends on the quality of features extracted from characters and effectiveness of the classifiers. For the past several years, many academic laboratories and companies are occupied with research on handwriting recognition. In the character recognition system, we need three things, i.e., preprocessing on digitized data, feature extraction, and decision-making algorithms. Preprocessing is the initial stage of character recognition. In this phase, the character image is normalized into a window of size 100×100. After normalization, we produce bitmap image of the normalized image. Afterwards, the bitmap image is transformed into a skeletonized image. In this work, we have used two efficient feature extraction techniques, namely, parabola curve fitting based features and power curve fitting based features for character recognition. Aradhya et al. [1] have presented a multilingual OCR system for South Indian scripts based on PCA. Bansal and Sinha [2, 3] have presented a technique for complete Devanagari script recognition. In this technique, they have recognized the character in two steps. In first step, that recognize the unknown characters and in the second step they recognize the character based on the strokes. Chaudhary et al. [4] have represented a technique for recognition of connected handwritten numerals. Gader et al. [6] have presented a handwritten word recognition system using neural network. Hanmandlu et al. [8] have presented a handwritten Hindi numeral recognition system using Fuzzy logic. Kumar [11] has proposed a AI based approach for handwritten Devanagari script recognition. Kumar et al. [12] have presented a review on OCR for handwritten Indian scripts. They have also proposed two efficient feature extraction techniques for offline handwritten Gurmukhi character recognition [13]. Lehal and Singh [14] have presented a printed Gurmukhi script recognition system, where connected components are initially segmented using thinning based approach. Pal et al. [18] have assimilated a comparative study of handwritten Devanagari character recognition using twelve different classifiers and four sets of features. Rajashekaradhya and Ranjan [20] have proposed zoning based feature extraction technique for Kannada script recognition. Roy et al. [21] have presented a script identification system for Persion and Roman script. Sharma et al. [22] have proposed a offline handwritten character recognition system using quadratic classifier. Pal et al. [16, 17] have come up with a technique for offline Bangla handwritten compound characters recognition. They have used modified quadratic discriminant function for feature extraction. They have also presented a technique for feature computation of numeral images. Classification is the most significant activity for character recognition. In the classification process, we required decision making algorithms. There

The Consonants

ਸ ਹ ਕ ਖ ਗ ਘ ਙ ਚ ਛ ਜ ਝ ਞ ਟ ਠ ਡ ਢ ਣ ਤ ਥ ਦ ਧ ਨ ਪ ਫ

ਬ ਭ ਮ ਯ ਰ ਲ ਵ ੜ

The Vowel Bearers

ੳ ਅ ੲ

The Additional Consonants (Multi Component Characters)

ਸ਼ ਜ਼ ਖ਼ ਫ਼ ਗ਼ ਲ਼

The Vowel Modifiers

◌ੋ ◌ੌ ◌ੇ ◌ੈ ਿ◌ ◌ੀ ◌ਾ ◌ੁ ◌ੂ

Auxiliary Signs

◌ੱ ◌ੰ ◌ਂ

The Half Characters

◌੍ ਵ੍

Fig. 1 *Gurmukhi* script character set

have presented various kinds of decision making algorithms as: Baye's Net, DMNB Text, Naïve Baye's, multilayer perceptron model, etc [5, 7, 9, 10, 15, 19]. We have applied 18 different weka classification methods for offline handwritten Gurmukhi character recognition.

2 Gurmukhi Script and Data Collection

Gurmukhi script is the script used for writing in the Punjabi language and is derived from the old *Punjabi* term *Guramukhi*, which means "from the mouth of the Guru". *Gurmukhi* script has three vowel bearers, thirty two consonants, six additional consonants, nine vowel modifiers, three auxiliary signs, and three half characters. The *Gurmukhi* script is the 12th most widely used script in the world. Writing style of *Gurmukhi* script is from top to bottom and left to right. In the *Gurmukhi* script, there is no case sensitivity. The character set of *Gurmukhi* script is given in Fig. 1. In the *Gurmukhi* script, most of the characters have a horizontal line at the upper part called, headline and the characters are connected with one another through this line.

For the present work, we have collected the data from 100 different writers. These writers were requested to writer each Gurmukhi character. A sample of handwritten Gurmukhi characters by five different writers ($W1$, $W2$,…,$W5$) is given in Fig. 2.

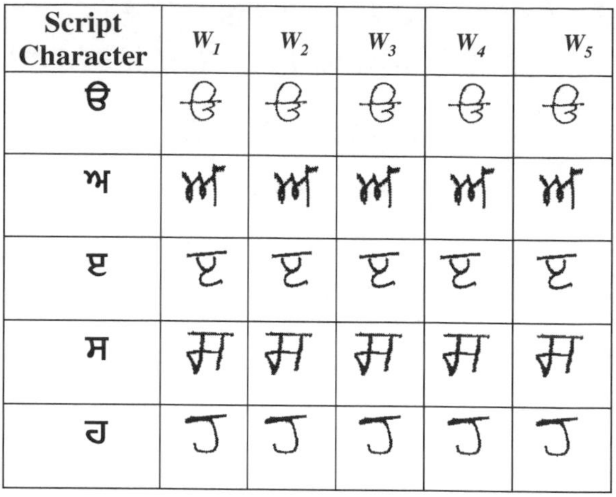

Script Character	W_1	W_2	W_3	W_4	W_5
ੳ	ੳ	ੳ	ੳ	ੳ	ੳ
ਅ	ਅ	ਅ	ਅ	ਅ	ਅ
ੲ	ੲ	ੲ	ੲ	ੲ	ੲ
ਸ	ਸ	ਸ	ਸ	ਸ	ਸ
ਰ	ਰ	ਰ	ਰ	ਰ	ਰ

Fig. 2 Samples of handwritten *Gurmukhi* characters

3 Feature Extraction

In this phase, the features of input character are extracted. The performance of Offline HCR system depends on features, which are being extracted. The extracted features should be able to uniquely classify a character. In this work, we have used two efficient feature extraction techniques, namely, parabola curve fitting based features and power curve fitting based features.

3.1 Parabola Curve Fitting Based Features

In this technique, initially, we have divided the thinned image of a character into n (=100) zones. A parabola is fitted to the series of *ON* pixels in every zone by using the least square method. A parabola $y = a + bx + c$ is uniquely defined by three parameters: a, b, and c. This will give $3n$ features for a given bitmap.

The steps that have been used to extract these features are given below.

Step I: Divide the thinned image into n (=100) number of equal sized zones.

Step II: For each zone, fit a parabola using the least square method and calculate the values of a, b and c.

Step III: Corresponding to the zones that do not have a foreground pixel, the values of a, b, and c are taken as zero.

3.2 Power Curve Fitting Based Features

In this technique also, we have divided the thinned image of a character into n (=100) zones. A power curve is fitted to the series of *ON* pixels in every zone, using the least square method. A power curve of the form $y = a$ is uniquely defined by two parameters: a and b. This will give $2n$ features for a given bitmap.

The steps that have been used to extract these features are given below.

Step I: Divide the thinned image into n (=100) number of equal sized zones.
Step II: In each zone, fit a power curve using least square method and calculate the values of a and b.
Step III: Corresponding to the zones that do not have a foreground pixel, the value of a and b is taken as zero.

Table 1 Experimental results of parabola curve fitting based features

Classifier	Accuracy (%)	Root mean squared error (%)	Weighted average precision (%)	False rate (%)	Rejection rate (%)	Weighted F-measure average (%)
Baye's Net	72.78	11.18	74	0.80	26.20	73
Complement naïve bays	57.35	15.61	65.80	1.20	41.40	55.50
DMNB text	72.86	11.58	73.80	0.80	26.25	73
Naïve bays	71.29	12.41	74.30	0.80	27.90	71.70
Multi-layer perceptronl	82.92	8.88	83.40	0.50	16.60	82.90
Multi class classifier	60.86	16.64	65.70	1.10	38	61.80
Classification Via regression	65.57	12.03	66.70	1	33.40	65.60
J48 (decision tree)	55.64	15.11	56.7	1.30	43.10	55.60
IBI	77.85	11.25	79	0.60	20.40	78
IBK	77.86	11.16	80.00	0.60	21.40	78
K-Star	72.07	12.32	74.70	0.80	27.10	72.40
Attribute selected	57.07	14.81	58.80	1.20	41.70	57.30
Ensemble selection	58.07	13.03	59.40	1.20	40.70	57.60
LWL	66.07	12.53	68.70	1	32.90	66.60
PART	53.36	15.66	54.60	1.40	45.20	53.40
Bagging	63.42	12.50	64.60	1.10	35.50	63.30
LogitBoost	62.14	12.11	63.30	1.10	36.80	62.30
Voting feature interval	67.85	12.03	68.10	0.40	31.70	67

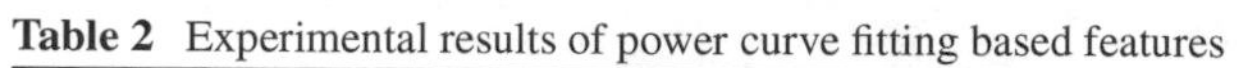

Table 2 Experimental results of power curve fitting based features

Classifier	Accuracy (%)	Root mean squared error (%)	Weighted average precision (%)	False rate (%)	Rejection rate (%)	Weighted F-measure average (%)
Baye's Net	72.07	11.50	72.90	0.80	27.10	72.20
Complement naïve bays	61.28	14.87	66.80	1.10	37.60	60.40
DMNB text	72.86	11.58	73.80	0.80	26.30	72.80
Naïve bays	72.86	12.07	75.30	0.80	26.30	73.30
Multi-layer perceptron	82.86	8.80	83.50	0.50	16.60	82.60
Multi class classifier	65.78	16.64	69.30	1.00	33.20	66.60
Classification Via regression	67.85	11.68	68.10	1.00	31.10	67.70
J48 (decision tree)	55.50	15.04	57.10	1.30	43.20	55.60
IBI	77.57	11.32	79.40	0.60	21.80	77.60
IBK	77.57	11.23	79.40	0.60	21.85	77.60
K-Star	69.42	12.74	72.20	0.90	29.70	69.60
Attribute selected	53.14	15.32	54.60	1.40	45.50	53
Ensemble selection	59.42	13.04	60.60	1.20	39.40	59
LWL	64.50	13.05	66.90	1.00	34.50	64.60
PART	55.64	15.18	57.30	1.30	43.10	55.70
Bagging	64.07	12.47	65.10	1.00	34.90	63.80
LogitBoost	64.50	13.05	66.90	1.00	34.50	64.60
Voting feature interval	60.78	16.71	62.60	1.20	38.00	60.00

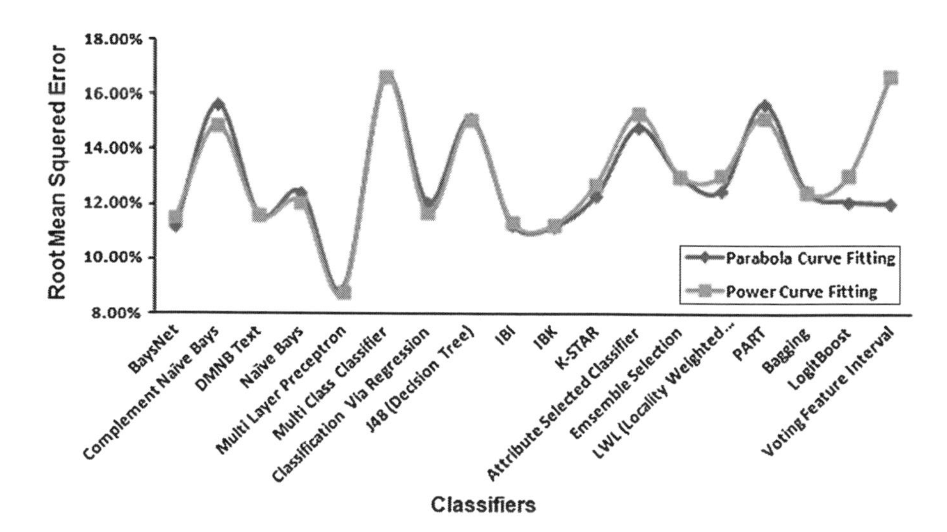

Fig. 3 Recognition accuracy of diverse classification techniques

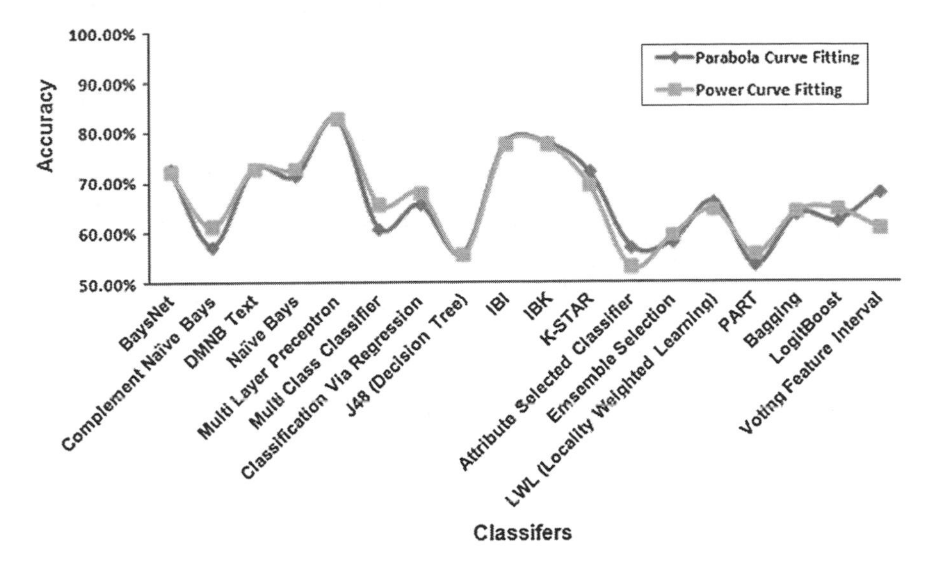

Fig. 4 Comparison of RMSE in various classification techniques

4 Classification

Classification phase is the decision-making phase of an OHCR engine. This phase uses the features extracted in the previous stage for deciding class membership. In this work, we have used 18 different classifiers based on *weka* namely, Baye's Net, Complement Naïve Baye's, Discriminative Multinominal Naïve Baye's Text (DMNB

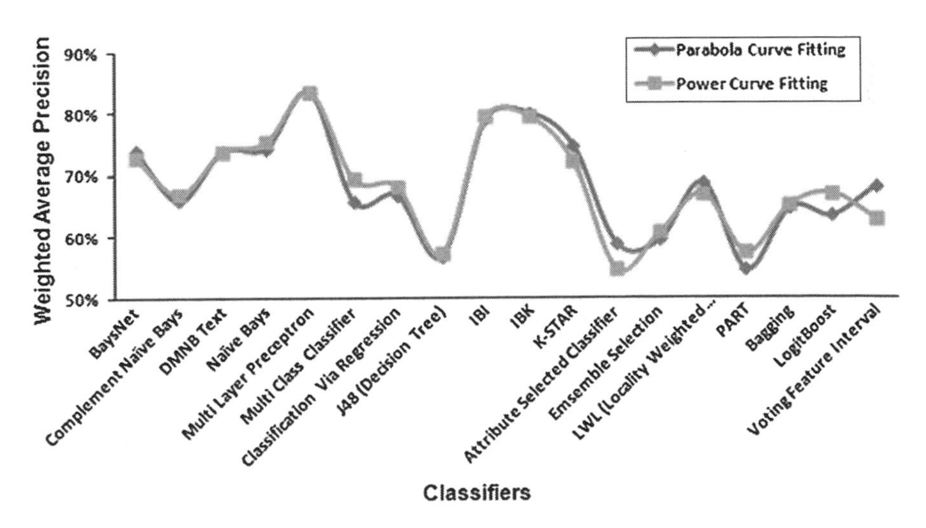

Fig. 5 Weighted average precision of divergent classification techniques

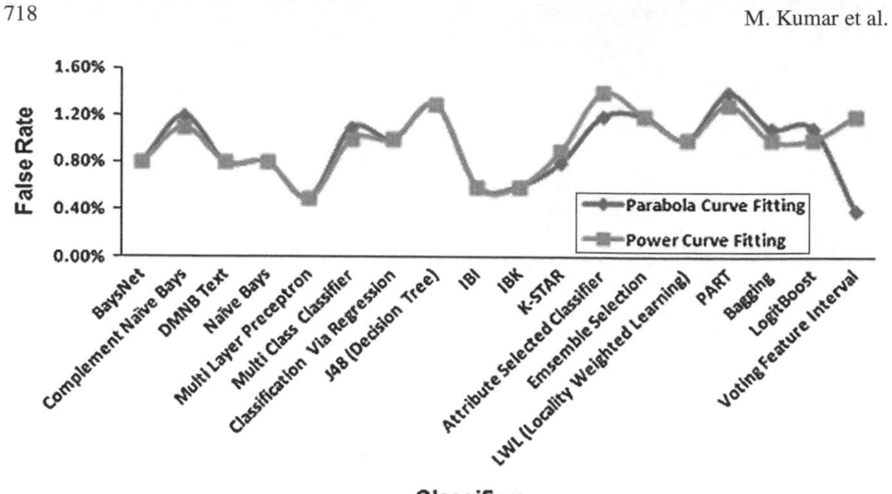

Fig. 6 Comparison of FPRate of different classifiers

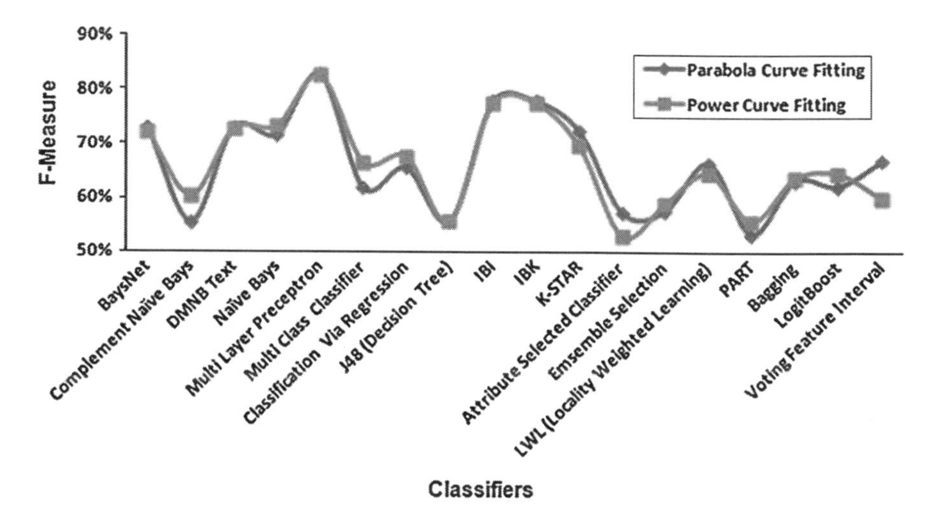

Fig. 7 F-Measure of diverse classification methods

Text), Naïve Baye's, Multi-Layer Perceptron, Multi Class Classifier, Classification via Regression, Decision Tree, IBI, IBK, K-Star, Attribute Selected Classifier, Ensemble Selection, Locality Weighted Learning (LWL), PART, Bagging, LogitBoost, and Voting Feature Interval classifier for offline handwritten *Gurmukhi* character recognition. Experimental results of these classifiers are presented in the next section.

5 Experimental Results

In this section, we have presented results of various *weka*-based classification methods. Table 1, depicts the results based on parabola curve fitting based feature extraction technique and Table 2 shows the power curve fitting based feature extraction results. In Fig. 3, we have presented recognition accuracy of various classification techniques for offline handwritten *Gurmukhi* character recognition. As such, we have seen that the multilayer perceptron model is the preeminent classifier for offline handwritten character recognition. We have achieved a maximum accuracy of about 82.92 % with parabola curve fitting based features and the multilayer perceptron model classifier.

In Fig. 4, we have presented the root mean squared error (RMSE) of various classification techniques, based on parabola curve fitting features and power curve fitting features. Figure 5, signifies the precision of different classification techniques. In Fig. 6, false rate (FP) of different classification techniques is depicted, graphically. Figure 7, describes the F-measure of various classifiers, considered in this work.

6 Conclusion

In present work, we have illustrated the effectiveness of various *weka*-based classification techniques for offline handwritten *Gurmukhi* character recognition. We have experimented on our own data set. We have collected 3,500 samples of isolated handwritten Gurmukhi characters from 100 different writers. After making a comparison of various classifiers for the recognition of character, we drew conclusion that the most appropriate classification technique is multilayer perceptron model. We obtained a maximum accuracy of about 82.92 % with parabola curve fitting based features and multilayer perceptron model classifier. The results of power curve fitting are also promising. A further study of the benefits of this technique can also be extended to offline handwritten character recognition of other Indian scripts.

References

1. Aradhya, V.N.M., Kumar, G.H., Noushath, S.: Multilingual OCR system for south Indian scripts and English documents: an approach based on Fourier transform and principal component analysis. Eng. Appl. Artif. Intell. **21**, 658–668 (2008)
2. Bansal, V., Sinha, R.M.K.: Integrating knowledge sources in devanagari text recognition. IEEE Trans. Syst. Man Cybern. **30**(4), 500–505 (2000)
3. Bansal, V., Sinha, R.M.K.: Segmentation of touching and fused Devanagri characters. Pattern Recogn. **35**(4), 875–893 (2002)
4. Chaudhuri, B.B., Majumder, D.D., Parui, S.K.: A procedure for recognition of connected hand written numerals. Int. J. Syst. Sci. **13**, 1019–1029 (1982)
5. Friedman, N., Geiger, D., Goldszmidt, M.: Bayesian network classifiers. Mach. Learn. **29**, 131–163 (1997)

6. Gader, P.D., Mohamed, M., Chaing, J.H.: Handwritten word recognition with character and inter-character neural networks. IEEE Trans. Syst. Man Cybern. Part B Cybern. **27**(1), 158–164 (1997)
7. Genkin, A., Lewis, D., Madigan, D.: Large-scale Bayesian logistic regression for text categorization. TECHNOME TRICS. **49**(3), 291–304 (2004)
8. Hanmandlu, M., Grover, J., Madasu, V. K. and Vasikarla, S.: Input fuzzy for the recognition of handwritten Hindi numeral. In: Proceedings of ITNG, pp. 208–213 (2007)
9. Jomy, J., Parmod, K.V., Kannan, B.: Handwritten character recognition of south Indian scripts: a review. In: Proceedings of Indian Language Computing, pp. 1–6 (2011)
10. Kala, R., Vazirani, H., Shukla, A. and Tiwari, R.: Offline handwriting recognition using genetic algorithm. Int. J. Comput. Sci. Issues **7**(2,1), 16–25 (2010)
11. Kumar, D.: AI approach to hand written Devanagri script recognition. In: Proceedings of 10th International conference on EC3-Energy, Computer, Communication and Control Systems, **2**, 229–237 (2008)
12. Kumar, M., Jindal, M. K. and Sharma, R. K.: Review on OCR for handwritten Indian scripts character recognition. In: Proceedings of DPPR, pp. 268–276 (2011)
13. Kumar, M., Jindal, M.K., Sharma, R.K.: Efficient feature extraction technique for offline handwritten Gurmukhi character recognition. Chaing Mai J. Sci. (2012)
14. Lehal, G.S., Singh, C.: A Gurmukhi script recognition system. In: Proceedings of 15th ICPR, pp. 557–560 (2000)
15. McCallum, A., Nigam, K.: A comparison of event models for Naive Baye's text classification. Paper presented at the workshop on learning for text categorization (1998)
16. Pal, U., Wakabayashi, T., Kimura, F.: Handwritten numeral recognition of six popular scripts. In: Proceedings of ICDAR, vol. **2**, pp. 749–753 (2007a)
17. Pal, U., Wakabayashi, T., Kimura, F.: A system for off-line Oriya handwritten character recognition using curvature feature. In: Proceedings of 10th ICIT, pp. 227–229 (2007b)
18. Pal, U., Wakabayashi, T., Kimura, F.: Comparative study of devanagari handwritten character recognition using different feature and classifiers. In: Proceedings of 10th ICDAR, pp. 1111–1115 (2009)
19. Patel, C.I., Patel, R., Patel, P.: Handwritten character recognition using neural network. Int. J. Sci. Eng. Res. **2**(4), 1–5 (2011)
20. Rajashekararadhya, S.V., Ranjan, S. V.: Zone based feature extraction algorithm for handwritten numeral recognition of Kannada script. In: Proceedings of IACC, pp. 525–528 (2009)
21. Roy, K., Alaei, A., Pal, U.: Word-wise handwritten Persian and Roman script identification. In: Proceedings of ICFHR, pp. 628–633 (2010)
22. Sharma, N., Pal, U., Kimura, F., Pal, S.: Recognition of off-line handwritten devanagari characters using quadratic classifier. In: Proceedings of ICVGIP, pp. 805–816 (2006)

Part VII
Soft Computing for Security (SCS)

An Efficient Fingerprint Indexing Scheme

Arjun Reddy, Umarani Jayaraman, Vandana Dixit Kaushik and P. Gupta

Abstract This paper proposes an efficient geometric-based indexing scheme for fingerprints. Unlike other geometric-based indexing schemes, the proposed indexing scheme reduces both memory and computational costs. It has been tested on IITK database containing 2,120 fingerprints of 530 subjects. Correct Recognition Rate is found to be 86.79 % at top 10 best matches. Experiments prove its superiority against well-known geometric-based indexing schemes.

Keywords Biometrics · Fingerprint indexing · Core point · Minutiae

1 Introduction

Fingerprint recognition system is used to recognize the identity of a subject. Identification can be done by searching all images in the database (henceforth termed as *models*) against a image (henceforth termed as *query*). To make an efficient process, there is a need of an efficient indexing technique. A fingerprint has the following characteristics:

A. Reddy (✉) · U. Jayaraman · P. Gupta
Indian Institute of Technology, Kanpur, India
e-mail: areddy@cse.iitk.ac.in

U. Jayaraman
e-mail: umarani@cse.iitk.ac.in

P. Gupta
e-mail: pg@cse.iitk.ac.in

V. D. Kaushik
Harcourt Butler Technological Institute, Kanpur, India
e-mail: vandanadixitk@yahoo.com

B. V. Babu et al. (eds.), *Proceedings of the Second International Conference on Soft Computing for Problem Solving (SocProS 2012), December 28–30, 2012*, Advances in Intelligent Systems and Computing 236, DOI: 10.1007/978-81-322-1602-5_77,

- Number of minutiae extracted from a fingerprint of a subject at any two time instants may not be same.
- There are too many minutiae in a fingerprint; some may be false.
- Number of minutiae of any two fingerprints may not be same.
- There may be partial occlusion in a fingerprint of a subject and it may overlap with some other subjects that are not present in the database.
- A query fingerprint may be rotated and translated with respect to the corresponding model fingerprints in the database.

Most of the available fingerprint indexing schemes can be classified on the basis of the features such as singular points [1], directional field [2], local ridge-line orientations [3], orientation image [4], minutiae [5], minutiae descriptor, multiple features, and SIFT features. Since most matching algorithms use minutiae, the use of minutiae is especially beneficial. These schemes derive robust geometric features from triplets of minutiae and use hashing techniques to perform the search.

A prominent geometric-based indexing technique for fingerprints is proposed by Germain et al. [6] in which geometric features from triplets are used with the help of Fast Look up Algorithm for String Homology (FLASH) hashing technique. It does clustering using transformation parameter where all the fingerprints in that bin represent a hypothesis match between the three points in the query fingerprint and those in the fingerprints of database. The best coordinate transformation that matches query triplet and model triplet is calculated with the information. The transformation should be such that squared distance between the points of query triplet and model triplet is minimum. This transformation parameter reduces false matches.

The scheme in [7] uses geometric features from minutiae triplets where the triplet features are maximum length of three sides, median and minimum angles, triangle handedness, type, direction and ridge count minutiae density. Since triangles are formed using all possible minutiae, this increases both memory and computational cost. A fast and robust projective matching for fingerprints has been proposed in [8]. It performs a fast match using a Geometric Hashing [9] which needs large computational time and memory.

Bebis et al. [10] have used the geometric features from Delaunay triangles formed on the minutiae for indexing the fingerprints, instead of all possible combination of triangles. It can be shown that for a given set of minutiae, the Delaunay triangulation produces linear number of triangles. This compares favorably to the number of all possible combinations of triangles/ bases pairs considered in approaches [6–8]. However, the major issue with Delaunay triangulation is that it is more sensitive to noise and distortion. For example, if some minutiae are missed or added (spurious minutiae), the structure of Delaunay triangulation gets affected.

This paper presents an efficient indexing scheme which uses geometric information from triangles formed on minutiae to index model fingerprints. It assumes that the uncertainty of feature locations associated with minutiae feature and shear does not affect the angles of a triangle arbitrarily. Triangles are invariant to translation and rotation. It reduces the number of possible triangles by taking minutiae to form triangles within the specified region R from its core point C and is inserted exactly once into a hash table. So it effectively removes the use of all triangles/bases pairs used in [6–8] reducing memory and computational complexity.

The paper is organized as follows. Next section discusses feature extraction technique from a fingerprint image. Section 3 discusses the proposed indexing scheme. Experimental results are analyzed in the next section. Conclusions are given in the last section.

2 Feature Extraction

Feature extraction is a series of steps such as minutiae detection and core point detection [11]. Let $M = \{m_1, m_2, \ldots, m_o\}$ be the detected minutiae from each model fingerprint image. Each minutia m_i is a 4-tuple $(x_{mi}, y_{mi}, \alpha_{mi}, T_{mi})$ which denotes their coordinates, direction, and type. Let $C = (x_c, y_c, \alpha_c)$ be the core point, detected through [11] where (x_c, y_c) denote the coordinates and α_c is the direction of the core point. The proposed indexing scheme overcomes some of the issues and constraints in [6, 7] by considering small number of possible triangles instead of all possible triangles. In a model fingerprint, it considers the minutiae to form triangles within the specific range *R* from its core point *C* reducing computational and memory cost. Further, it introduces some additional features to reduce false correspondences. The triplet features used in the proposed indexing scheme are:

1. Sides of the triangles s_1, s_2, s_3: Sides of the triangle is considered in certain order as the longest side, the medium side and the smaller side. If any two or three sides are similar, this system does not consider them because this type of triplets is negligible in number and their exclusion does not affect the results.
2. Regions of vertices r_1, r_2, r_3: Considering core point *C* as the center, concentric circles are drawn with radius $r, 2r, 3r, 4r$... till the circle's outer line is completely outside fingerprint boundary. The optimal value of r is found out with the help of experiments. Let p_1, p_2, p_3 be three vertices of triplet and r_1, r_2, r_3 be their respective regions of vertices and d_1, d_2, d_3 be their respective distances from core point C. Then the value of r_i is given by $r_i = \frac{di}{r} + 1, i = 1, 2, 3$.
3. Triangle type λ: Each minutia is either termination or bifurcation. If $\lambda_1, \lambda_2, \lambda_3$ are three vertices to indicate whether they are termination or bifurcation point, then λ for the triplet can be used as one more attribute for indexing component. Since λ_1, λ_2 and λ_3 can have values either 0 or 1, based on termination or bifurcation, λ can be any value between 0 and 7 and is calculated by $\lambda = 4\lambda_1 + 2\lambda_2 + \lambda_3$.
4. Orientation φ : Sometime all the above features may be same for triplets of two triangles. They can be differentiated by their orientation which can be calculated using cross-product between the longest side and the medium side. Orientation has two values, $+1$ or -1.

3 Proposed Indexing Scheme

The proposed indexing technique consists of two stages known as indexing and searching. During indexing, the model fingerprints in the database are indexed into a hash table. For any new fingerprint image, it can be added into a hash table without affecting the performance of the searching algorithm and without modifying the existing hash table. During searching, it recognizes the query fingerprint by searching only indexed hash bins.

3.1 Indexing

For each model fingerprint consisting of core point C and minutiae, concentric circles are drawn with radius r, $2r$, $3r$, $4r$... till the circle's outer line is completely outside fingerprint boundary. The value of r is found experimentally. All possible triangles within various circles are determined. The index $I = (s_1, s_2, s_3, r_1, r_2, r_3\lambda, \varphi)$ is generated from each triangle to select an appropriate bin in hash table. At this bin, model fingerprint ID is added. At the time of indexing, it keeps track of number of triangles generated from each model fingerprint. Let T_i be the total number of triangles generated from a model M_i used for indexing. Steps for indexing are given in Algorithm 1.

Algorithm 1: Indexing

For each model fingerprint M_i do the following
1. Generate all possible triangles for minutiae of fingerprint.
2. Consider only those triangles whose all sides are in the specifies range R.
3. Store numbers of triangles considered in indexing and store it in T_i where T_i indicates total number of triangles of fingerprint M_i considered for indexing.
4. For each triplet do the following
   a) Generate index $I = (s_1, s_2, s_3, r_1, r_2, r_3, \lambda, \varphi)$.
   b) Access appropriate bin of the hash table.
   c) Add model fingerprint ID of the image to this bin.

Algorithm 2: Searching

For the given query Q do the following
1. Generate all possible triangles from the minutiae of the query image.
2. Consider only those triangles whose all sides are in the specified range R. Let there be T such triangles.
3. For each triangle, do the following
   a) Generate index $I = (s_1, s_2, s_3, r_1, r_2, r_3, \lambda, \varphi)$.
   b) Access appropriate bin of the hash table.
   c) Increment count C_i for all model fingerprints whose ID is in that bin.
   d) Calculate S_i, score of image M_i, by formula $S_i = \frac{Ci}{Ti}$.
   e) Obtain largest t S_i's that are to be considered for critical search.

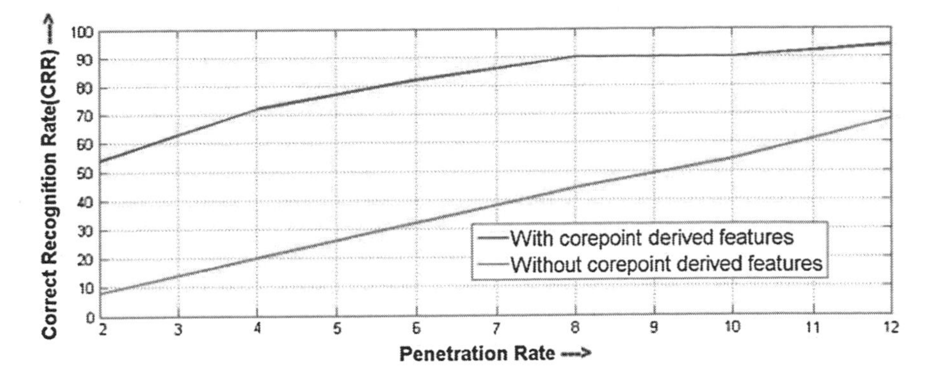

Fig. 1 Performance graph: effect of core point generated features

3.2 Searching

For a given query fingerprint, all possible triangles are generated using minutiae similar to indexing stage. Then only those triangles are considered whose all sides of triangle fall within the range R. Let there be T such triangles. From these triangles, an index $I = (s_1, s_2, s_3, r_1, r_2, r_3, \lambda, \varphi)$ is generated to find an appropriate bin of hash table and to count each model fingerprint ID in that bin C_i. Same process is repeated for remaining query's triangles. Finally, score S_i is calculated by $S_i = \frac{C_i}{T_i}, i = 1, 2, ..., N$where N denotes the number of fingerprints in the database. Largest t S_i's which are considered for top t best matches are used for critical search to find out the exact match. Steps for searching are given in Algorithm 2.

4 Experimental Results

The proposed indexing scheme has been tested on IITK fingerprint database of 2,120 images acquired from 530 subjects. Every person has given four fingerprints of the same finger, at different instant of times. Four datasets have been created to carry out various experiments. DB1 contains 2120 fingerprints, DB2 has 1,336 fingerprints while DB3 and DB4 contains 668 fingerprints and 200 fingerprints respectively. In all these datasets, three impressions of each finger are used for indexing and remaining one impression is used for searching.

Figure 1 shows the performance of the proposed indexing scheme with respect to core point and database sizes. DB4 dataset is used for performing this experiment. The *CRR* is very high when core point is considered to obtain additional features from triangles. This *CRR*, when core point is used, is close to the *CRR* of the existing schemes like [6] and [7] . But the proposed scheme has achieved this performance without using transformation parameter cluster and imposing any geometrical constraints. Experiments have also been conducted on all datasets DB1, DB2, DB3, DB4. It is found that there is no drastic difference in *CRR* for different datasets and as database size is increasing, *CRR* is coming down for a given *Penetration Rate*, slightly.

5 Conclusion

This paper has proposed a fingerprint-based indexing scheme which uses core point. This has helped to overcome various issues and constraints of well-known schemes. It has been tested on IITK database containing 2,120 fingerprints of 530 subjects. Accuracy is found to be 86.79 % at top 10 best matches. This accuracy has been achieved without using clustering based on transformation parameter and without imposing geometrical constraints as in existing schemes. However, it may not perform well when a fingerprint does not contain any core point.

References

1. Liu, T., Zhu, G., Zhang, C., Hao, P.: Fingerprint indexing based on singular points. In: Proceedings of International Conference on Image Processing, pp. 293–296 (2005)
2. Cappelli, R., Maio, D., Maltoni, D.: Indexing fingerprint databases for efficient 1:N matching. In: Proceedings of International Conference on Control, Automation, Robotics and Vision (2000)
3. Alessandra, L., Dario, M., Davide, M.: Continuous versus exclusive classification for fingerprint retrieval. Pattern Recogn. Lett. **18**(10), 1027–1034 (1997)
4. Li, J., Yau, W.Y., Wang, H.: Fingerprint indexing based on symmetrical measurement In: Proceedings of International Conference on Pattern Recognition, pp. 1038–1041 (2006)
5. Liang, X., Asano, T., Bishnu, A.: Distorted fingerprint indexing using minutiae detail and delaunay triangle. In: Proceedings of the 3rd International Symposium on Voronoi Diagrams in Science and, Engineering, pp. 217–223 (2006)
6. Germain, R.S., Califano, A., Colville, S.: Fingerprint matching using transformation parameter clustering. IEEE Comput. Sci. Eng. **4**(4), 42–49 (1997)
7. Bhanu, Bir, Tan, Xuejun: Fingerprint indexing based on novel features of minutiae triplets. IEEE Trans. Pattern Anal. Mach. Intell. **25**(5), 616–622 (2003)
8. Boro, R., Roy, S.D.: Fast and robust projective matching for fingerprints using geometric hashing. In: Proceedings of the 4th Indian Conference on Computer Vision, Graphics and Image Processing, pp. 681–686 (2004)
9. Haim, J.: Wolfson and Isidore Rigoutsos. Geometric Hashing: An overview, IEEE Computational Science and Engineering **4**(4), 10–21 (1997)
10. Bebis, G., Deaconu, T., Georgiopoulos, M.: Fingerprint identification using delaunay triangulation. In: Proceeding of IEEE International Conference on Intelligence, Information, and Systems, pp. 452–459 (1999)
11. Jain, A., Prabhakar, S., Hong, L.: A multichannel approach to fingerprint classification. IEEE Trans. Pattern Anal. Mach. Intell. **21**(4), 348–359 (1999)

Gait Biometrics: An Approach to Speed Invariant Human Gait Analysis for Person Identification

Anup Nandy, Soumabha Bhowmick, Pavan Chakraborty and G. C. Nandi

Abstract A simple and a common human gait can provide an interesting behavioral biometric feature for robust human identification. The human gait data can be obtained without the subject's knowledge through remote video imaging of people walking. In this paper we apply a computer vision-based technique to identify a person at various walking speeds, varying from 2 km/hr to 10 km/hr. We attempt to construct a speed invariance human gait classifier. Gait signatures are derived from the sequence of silhouette frames at different gait speeds. The OU-ISIR Treadmill Gait Databases has been used. We apply a dynamic edge orientation histogram on silhouette images at different speeds, as feature vector for classification. This orientation histogram offers the advantage of accumulating translation and orientation invariant gait signatures. This leads to a choice of the best features for gait classification. A statistical technique based on Naïve Bayesian approach has been applied to classify the same person at different gait speeds. The classifier performance has been evaluated by estimating the maximum likelihood of occurrences of the subject.

Keywords Human gait · Orientation histogram · Naïve Bayesian · Speed invariance gait · OU-ISIR gait database

A. Nandy (✉) · S. Bhowmick · P. Chakraborty · G. C. Nandi
Robotics and AI Lab,Indian Institute of Information Technology Allahabad, Allahabad, India
e-mail: nandy.anup@gmail.com

S. Bhomick
e-mail: iro2011005@iiita.ac.in

P. Chakraborty
e-mail: pavan@iiita.ac.in

G. C. Nandi
e-mail: gcnandi@iiita.ac.in

B. V. Babu et al. (eds.), *Proceedings of the Second International Conference on Soft Computing for Problem Solving (SocProS 2012), December 28–30, 2012*, Advances in Intelligent Systems and Computing 236, DOI: 10.1007/978-81-322-1602-5_78,

1 Introduction

Bi-pedal locomotion is a complex task for a human. It requires a strong coordination of different joints of the human body which generates the rhythmic motion. A normal walking is involved with balancing ability and proper stability through the synchronous oscillations of different body joints of a person. The rhythmic motion [1] is called as gait which holds biometric signatures of the human behavioral walking pattern. Gait biometric has brought an enormous attention in security-related issues for detecting threats in controlled environments like Airports, Banks, Big Malls, and military installations. Johansson [2] has extracted the biological pattern of human gait by mounting Moving Lights Display (MLD) markers onto different major body parts of human subjects. Gait recognition from a video sequence signifies the gait as a potential biometric. It extends its advantages over the other biometrics traits like face, iris-scans, fingerprints, and hand scans for its unobtrusiveness properties, distance identification, and dealing with low resolution videos. Several gait review articles [3, 4] offer a general outline of the gait identification process. Morris [5] first started the wearable sensor-based gait recognition technique where motion recording sensors [6, 7] were attached on different locations of human body to record the acceleration of the gait which is utilized for identification purposes. The disadvantage of this system is that the full cooperation of the person is extremely required. Orr and Abowd [8] have shown the floor sensor-based gait recognition approach. It deals with force plates and an array of sensors deployed on the floor [8, 9] which enable to extract gait features from a person's walking on the floor. The Motion Vision-based gait recognition applies image and video processing techniques to extract gait features for identification from the video camera at a far distance. BenAbdelkader et al. [10] investigates individual identification and verification by calculating stride and cadence. Johnson and Bobick [11] applied a technique to measure static body parameters pertaining to distances among head and pelvis, height of the subject, maximum distance obtained between pelvis and feet, and distance among both the feet. These parameters are used for person recognition. It has been observed that maximum MV-based gait recognition works are based on the human silhouette [12, 13]. Liu and Sarkar [12] applied a technique to estimate the average silhouettes of a gait cycle and adopted a classification method called Euclidean distance to compare two averaged silhouettes for similarity measurements. A model-based approach is applied to determine gait features by evaluating the static and dynamic body parameters [14–16]. It is generally view and scale invariant which provides the advantage over the holistic approaches. The main disadvantage of this approach is low robustness and needs very high quality gait images and difficult capturing which requires proper camera calibration. Holistic approaches directly concentrate on gait sequences rather than any specific model for human body or parts of the human body. They are robust to the quality of human silhouettes and have low computational complexity [17, 18] which offers great advantages comparing to model-based approaches and are easy to implement with acceptable recognition rate. The major disadvantage of this model free approach is changing the shape of the silhouette with the effect of occlusions, wearing different clothes and carrying of objects. Sarkar et al. [19] proposed the baseline algorithm which is directly used on gait silhouette images to extract features which are scaled and aligned before applying on classification techniques. A Radon Transform was applied by Boulgouris and Chi [20] on binary silhouette gait images to construct a template from binary gait sequences. The feature vector was constituted by Radon coefficients derived by applying subspace projections and linear discriminate analysis (LDA). Dimensionality reduction method was used in the context of gait analysis to capture most discriminative features.

The advantage of vision-based technique exposes that no person's physical cooperation is required. It also attracts the person's identification from a large distance with low resolution image where other biometrics modality perhaps fails to obtain good classification result. The overview of this paper is stated as follows:

In Sect. 2, the process of deriving gait signatures as a feature has been explained. Section 3 investigates the classification technique using Naïve Bayesian rule for speed invariant human identification. In Sect. 4, gait identification results together with the distribution of gait features using histogram bins have been addressed accordingly. The conclusion and the future work for enhancement of this research work have been added at the end of this paper.

2 Method for Feature Extraction

As the human gait is purely a nonlinear time varying signal, hence selecting the best feature for speed invariant person identification is indeed a challenging job. The dynamic edge orientation histogram is chosen as a feature vector which appeals its robustness in orientation and translation invariant gait speeds. The orientation histogram computes the gradient by applying three tap derivative filter and subsequently produces the histogram in the desired orientations [21]. It has been investigated that the same person with different gait speeds will generate the similar feature vector. The algorithm for finding out the dynamic edge orientation histogram has been described in [22]. We have used OU-ISIR Gait Databases [23] for analyzing the human gait of 10 people at different gait speeds. The process of obtaining gait signatures at 10 km/hr gait speed has been depicted in Fig. 1.

3 Method for Gait Classification

In gait classification process, a Naïve Bayesian-based technique has been applied to identify a person walking at different speeds. Since the gait signal carries nonlinear characteristics, uncertainties are involved in classifying persons at various gait speeds. The statistical-based approach will allow us to differentiate different gaits at various speeds in order to achieve more accurate results for person identification. The Naïve Bayesian technique has been explained in the following section.

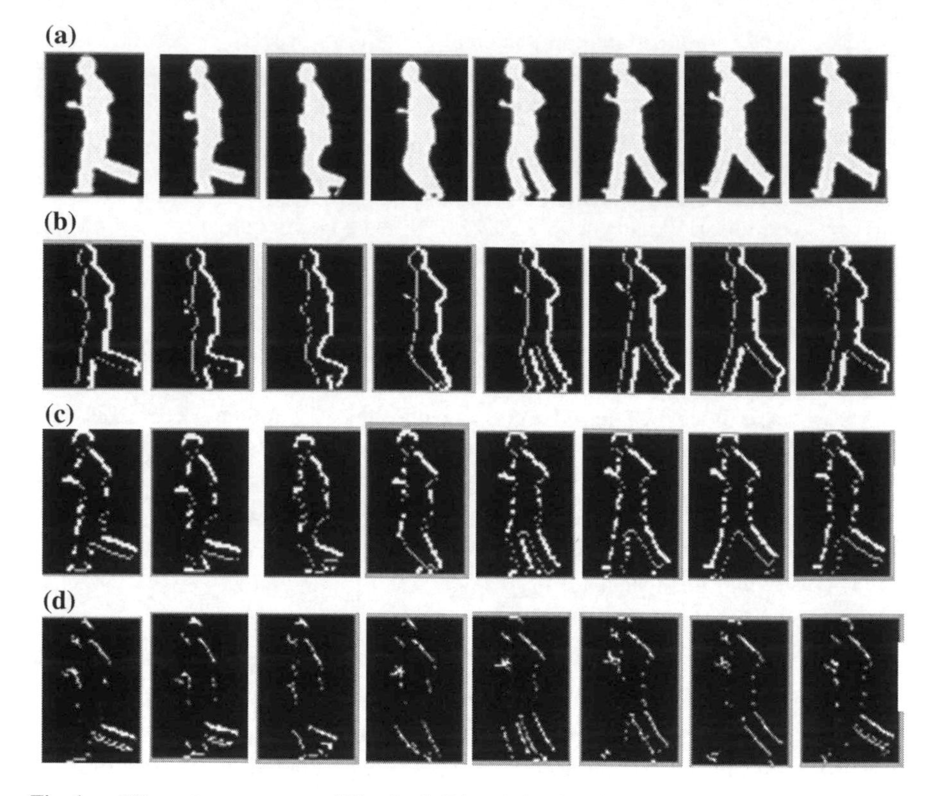

Fig. 1 **a** Silhouette sequences at 10 km/hr. **b** Edge derivative along *X* direction. **c** Edge derivative along *Y* direction, **d** Gradient of silhouette frames

3.1 Naïve Bayseian Technique

Theory: In naïve Bayesian classification technique, the probability density function at the feature point

$x = [x(1)\ldots\ldots\ldots\ldots x(m)]^T \in R^m$ is required to be estimated by the given rule:

$$p(x) = \prod_{i=1}^{m} p(x(i))$$

It has been assumed that the features (attributes) of the constructed feature vector are statistically independent. This assumption has been made for the high-dimensional feature space to deal with the curse of dimensionality problem. It produces a great impact on a large number of training data points to compute a good estimation of multidimensional probability density function. Although the feature's independence assumption is not true indeed, the performance of the naïve Bayesian classifier will still be satisfactory with fewer number sample points to obtain a reliable approximation of one-dimensional probability density function.

Naïve Bayesian classification technique based on the Bayes' conditional probability theorem gives the probability of a hypothesis being true supported by the set of evidences. In the working formula of the Bayes' theorem given underneath $p(Hyi|Ev)$ signifies the probability of the ith hypothesis being true supported by the evidences. Here each hypothesis is defined as the subject under test belongs to the ith class. In our analysis we have a set of nine evidences each corresponding to nine distinct speeds with which the subjects were walking. $p(Evk|Hj)$ symbolizes that the evidence k supports ith hypothesis, whereas $p(Hyj)$ is the prior probability of jth hypothesis.

$$P\left(Hy_i\,|Ev\right) = \frac{\left(\prod_{k=1}^{\text{No of Evidence}} P(Evk|Hyi) * P(Hyi)\right)}{\sum_{j=1}^{\text{No of Subjects}} \left(\prod_{k=1}^{\text{No of evidence}} P(Evk|Hj)\right) * P(Hyj)}$$

$$\text{MaxLikelihood} = \text{Max}\left(P\left(Hyi|Ev\right)\right) \quad \forall\, i = 1 \text{ to No of subjects}$$

We have taken the number of subjects to be ten where each of them was walking at a speed range of 2 km/hr to 10 km/hr with an interval of 1 km/hr. The hypothesis having the maximum likelihood is concluded as true.

4 Result Analysis and Discussion

We have applied Naïve Bayesian technique on ten subjects each walking on nine distinct and constant speeds. Separate datasets were used for training and testing purposes. The system was trained with features extracted from the silhouette images for each distinct speed. Once trained the classifier was tested for ten subjects each walking at nine different speeds. The classifier being a probabilistic one gave the maximum chance of resemblance of the pattern of the individual under test with each of the individuals present in the training database. The subject under test is classified to the corresponding subject in the training with which the likelihood of matching is maximum. In Fig. 2 is shown the graph for ten persons walking at nine different speeds and the percentage of maximum likelihood values in three different axes where one axis denotes the number of subjects, the second axis designates the speeds of walking, and the third axis symbolizes the percentage of maximum likelihood. For instance, the greater likelihood values, in case of person 2 or person 3 suggests the classifier classifying the test data set with a high confidence rate. It implies better probability of resemblance with the corresponding data in the training set. Figure 3 shows the polar plot of dynamic edge.

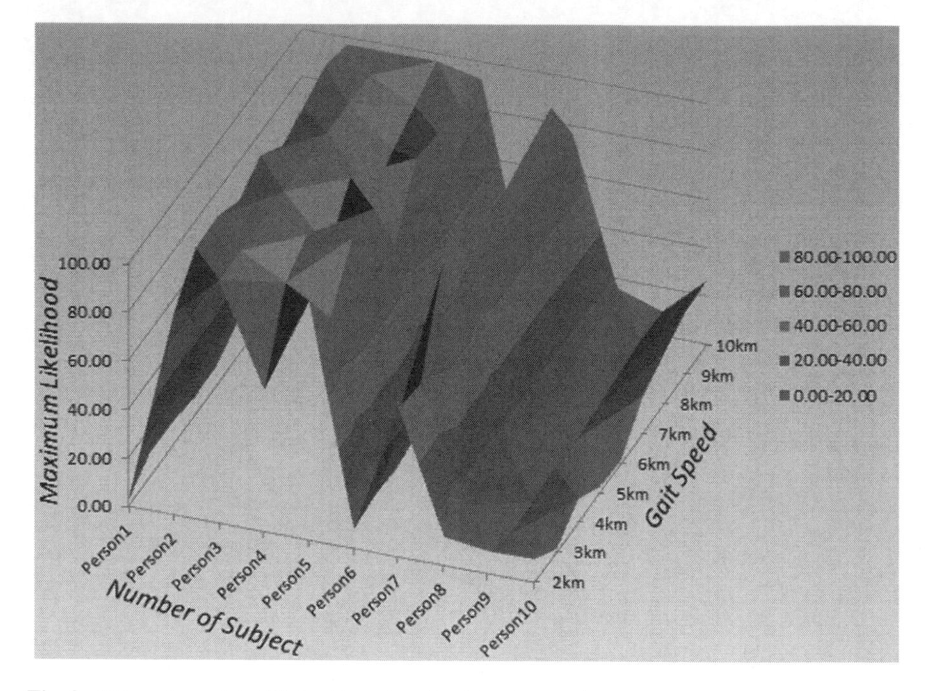

Fig. 2 3-D surface plot of Naïve Bayseian classification result

Orientation histogram at gait speed 10 and 2 km/hr. This plot implies the distribution of edge direction values clubbed together with respect to their magnitude range. Each group is collectively known as one histogram bin.

5 Conclusion and Future Work

It has been investigated and can also be concluded that data procuration is the most simple, as well as flawless, as the subject is freely walking on a treadmill at a constant speed. Moreover, as vision based has been taken as the subjects' Gait signature, it was free from any abnormalities that could have crept in if sensor-based data acquisition methods were applied. However the feature extraction process which needed to be applied for discriminating the individual subjects was quite a challenging job. The feature which was taken for the purpose of recognition should not only reduce the dimension of the dataset but also suffered from the risks of losing valuable information. Hence the performance of the classifiers can be significantly affected. We have after applying the analysis on a population of ten individuals come to the conclusion that the system could identify a person irrespective of its walking speed. We have used Naïve Bayesian technique for recognition purpose which provided the

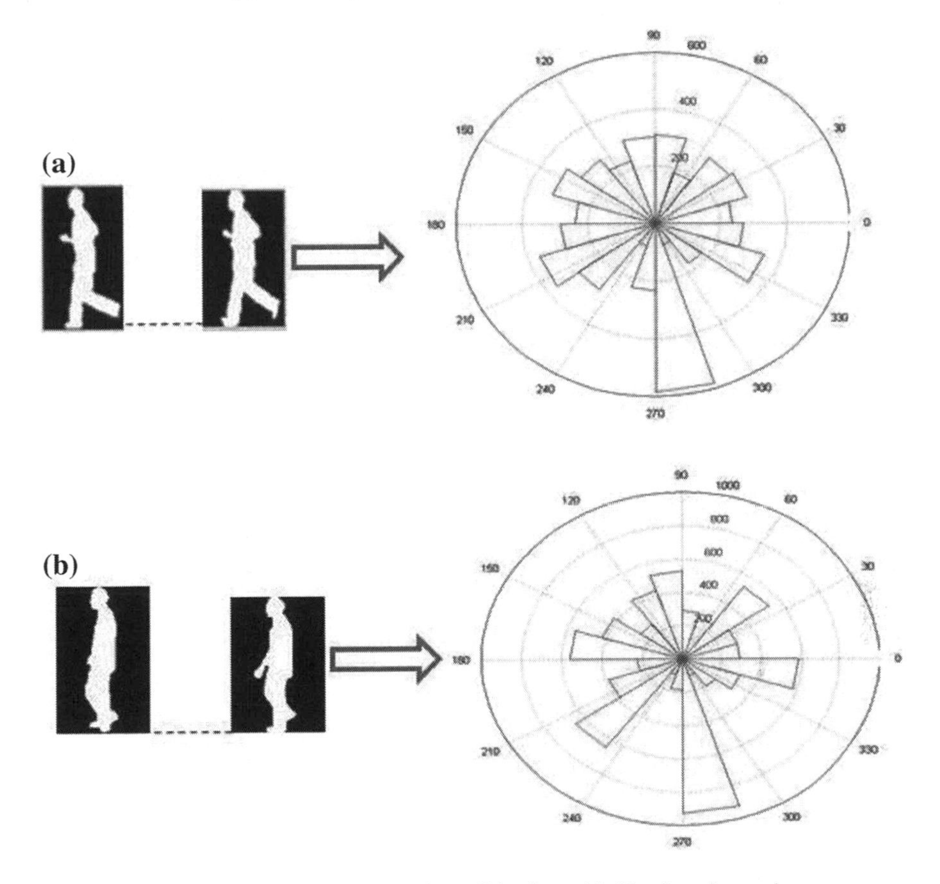

Fig. 3 Polar plot of orientation histogram for **a** 10 km/hr and **b** 2 km/hr gait speed

maximum likelihood for the resemblance of a person with the ones present in the training database.

Population size is a matter of concern as with the increase in the population size drastic change in the performance has been noted. The classifier starts misclassifying and a substantial drop of the likelihood has been observed. It would have been nice if rather selecting all the features the most significant ones could be identified which in course result in significant amendment of the classifier's performance. Apart from it in lieu of incorporating one feature extraction technique ensemble of feature extraction techniques could be taken which may possibly embellish the classifier performance and better likelihood results could have been accomplished. Finally as the human gait is a purely nonlinear oscillatory pattern, Central Pattern Generator-based approach could be taken where from a mathematical model of the biped locomotion could be achieved which would subsequently ease the person identification process.

Acknowledgments We would like to express our warm gratitude to Prof Yasushi Yagi and his entire research team of Osaka University, Japan for providing us OU-ISIR Gait database to accomplish our research work.

References

1. Taga, G., Yamaguchi, Y., Shimizu, H.: Self-organized control of bipedal locomotion by neural oscillators in unpredictable environment. In. Biological Cybernetics **65**(3), 147–159 (1991)
2. Johansson, G.: Visual perception of biological motion and a model for its analysis. In. Perception and Psychophysics, 14(2), 1973.
3. Boulgouris, N.V., Hatzinakos, D., Plataniotis, K.N.: Gait recognition: a challenging signal processing technology for biometric identification. In. IEEE Signal Processing Magazine **22**, 78–90 (2005)
4. Nixon, M.S., Carter, J.N.: Automatic Recognition by Gait.In. Proceedings of the IEEE **94**, 2013–2024 (2006)
5. Morris, S.J.: A shoe-integrated sensor system for wireless gait analysis and real time therapeutic feedback. PhD thesis, Harvard University-MIT Division of Health Sciences and Technology, 2004. http://hdl.handle.net/1721.1/28601
6. Ailisto, H.J., Lindholm, M., M antyj arvi, J., Vildjiounaite, E., M akel a, S.: Identifying people from gait pattern with accelerometers. In: Proceedings of SPIE Volume: 5779; Biometric Technology for Human Identification II, pages 7–14, 2005.
7. Gafurov, D., Snekkenes, E., Buvarp, T.E.: Robustness of biometric gait authentication against impersonation attack. In: First International Workshop on Information Security (IS'06), OnThe-Move Federated Conferences (OTM'06), pages 479–488, Montpellier, France, Oct 30 - Nov 1 2006. Springer LNCS 4277.
8. Orr, R J., Abowd, G.D.:The smart floor: A mechanism for natural user identification and tracking. In: Proceedings of the Conference on Human Factors in, Computing Systems, 2000.
9. Suutala, J., Rning, J.:Towards the adaptive identification of walkers: Automated feature selection of footsteps using distinction sensitive LVQ. In: Int. Workshop on Processing Sensory Information for Proactive Systems (PSIPS 2004), June 14–15 2004.
10. BenAbdelkader, C., Cutler, R., Davis, L.: Stride and cadence as a biometric in automatic person identification and verification. In: Fifth IEEE International Conference on Automatic Face and Gesture Recognition, pages 357–362, May 2002.
11. Johnson, A.Y., Bobick, A.F.: A multi-view method for gait recognition using static body parameters. In: Third International Conference on Audio- and Video-Based Biometric Person Authentication, pages 301–311, June 2001.
12. Liu, Z., Sarkar, S.: Simplest representation yet for gait recognition: Averaged silhouette. In: International Conference on, Pattern Recognition, pp. 211–214, 2004.
13. Liu, Z., Malave, L., Sarkar, S.: Studies on silhouette quality and gait recognition. In: Computer Vision and, Pattern Recognition, pp. 704–711, 2004.
14. Lee, L., Grimson, W.E.L.: Gait analysis for recognition and classification. In: Proc. IEEE Int. Conf. Automatic Face and Gesture Recognition, Washington, DC, May 2002, pp. 148–155.
15. Cunado, D., Nixon, M.S., Carter, J.N.: Automatic extraction and description of human gait models for recognition purposes. In: Comput. Vis. Image Understand **90**(1), 1–14 (2003)
16. Wagg, D.K., Nixon, M.S.: On automated model-based extraction and analysis of gait. In: Proc. IEEE Int. Conf. Automatic Face and Gesture Recognition, Seoul, Korea, May 2004, pp. 11–16.
17. Tafazzoli, F., Safabakhsh, R.: Model-based human gait recognition using leg and arm movements. In. Engineering Applications of Artificial Intelligence **23**(8), 1237–1246 (Dec. 2010)
18. Wang, J., She, M., Nahavandi, S., Kouzani, A.: A Review of Vision-Based Gait Recognition Methods for Human Identification. In: 2010 International Conference on Digital Image Computing: Techniques and Applications, pp. 320–327, Dec. 2010.

19. Sarkar, S., Phillips, P., Liu, Z., Vega, I.R., Grother, P.: J., Bowyer, K.W.: The human ID gait challenge problem: data sets, performance, and analysis. In. IEEE Transactions on Pattern Analysis and Machine Intelligence, **27**, 162–177 (2005)
20. Boulgouris, N. V., Chi, Z. X.: Gait Recognition Using Radon Transform and Linear Discriminant Analysis. In: IEEE Transactions on, Image Processing, vol. 16, pp. 731–740, 2007.
21. Nandy, A., Prasad, J.S., Mondal, S., Chakraborty, P., Nandi, G.C.: Recognition of Isolated Indian Sign Language gesture in Real Time. In: proceeding of BAIP 2010, Springer LNCS-CCIS, Vol. 70, pp. 102–107, March 2010.
22. Hninn, T., Maung, H.: Real-Time Hand Tracking and Gesture Recognition System Using Neural Networks. In: WASET 50, 466–470 (2009).
23. OU-ISIR Gait Database http://www.am.sanken.osaka-u.ac.jp/GaitDB/index.html

XML-Based Authentication to Handle SQL Injection

Nitin Mishra, Saumya Chaturvedi, Anil Kumar Sharma and Shantanu Choudhary

Abstract Structured Query Language (SQL) injection is one of the most devastating vulnerabilities to impact a business, as it can lead to the exposure of sensitive information stored in an application's database. SQL injection can compromise usernames, passwords, addresses, phone numbers, and credit card details. It is the vulnerability that results when an attacker achieves the ability to influence SQL queries that an application passes to a back-end database. The attacker can often leverage the syntax and capabilities of SQL, as well as the power and flexibility of supporting database functionality and operating system functionality available to the database to compromise the web application. In this article we demonstrate two non-web-based SQL injection attacks one of which can be carried out by executing a stored procedure with escalating privileges. We present XML-based authentication approach which can handle this problem in some way.

Keywords Web architecture · SQLIA · HTTP · XML · Web application · Web security · Authentication · Attacker

N. Mishra (✉) S. Chaturvedi · S. Choudhary
Sangam University, Bhilwara, India
e-mail: nitinmishra10@gmail.com

S. Chaturvedi
e-mail: saumyachaturvedi5@gmail.com

S. Choudhary
e-mail: shantunu.chintu@gmail.com

A. K. Sharma
ITM Bhilwara, Bhilwara, India
e-mail: anilsharma8423@gmail.com

B. V. Babu et al. (eds.), *Proceedings of the Second International Conference on Soft Computing for Problem Solving (SocProS 2012), December 28–30, 2012*, Advances in Intelligent Systems and Computing 236, DOI: 10.1007/978-81-322-1602-5_79,

1 Introduction

SQL Injection Attack (SQLIA) is considered one of the top five web application vulnerabilities by the Open Web Application Security Project (OWASP) in the year 2010. A database is an essential component that is necessary in modern web applications. Every web-based application that is developed and deployed over the Internet, requires the interaction of a database, thereby the application becomes fully database driven. It has been noted that, at an average, applications experience, 71 attempts an hour. Some applications experience aggressive attacks and at a peak, were attacked 800–1,300 times per hour.

An SQLIA involves the insertion or "injection" of a SQL query by an attacker via the input data from the client to the application. This injection in the SQL query involves inserting malicious input statements by an attacker. The execution of these malicious input statements by the web server at the database end results in unexpected behavior thus compromising the security of the database. The database just executes the input data provided by a client/attacker as it is. It does not have the ability to differentiate between a valid input string and/or a malicious/injected input string.

A successful SQL injection exploit can

- read sensitive data from the database
- alter database data (Insert/Update/Delete).

1.1 Modern Web Architecture

The diagram above shows the general web architecture. Any web-based architecture typically follows the Client-Server architecture.The client sends a HTTP request to the Web Server. This request will have the user input data. This input data will be sent to the database layer for processing the web server. At the database end, the SQL queries will be processed and the results will be sent to the web server. Hence, the entire web application is database-driven. The database server usually contains many databases, and in turn, each database contains many tables. The database is under huge threat to the attackers Fig. 1.

1.2 Intent

The attacker wants to access the resources of the system so he uses SQL injection for the same. When he gets access to the information of other authentic users of the system he can use the information as he likes. This also fails confidentiality and integrity of the system.

There are lot of users who use the web-based application and store lot of their personal data on it. Facebook is the very common example of web-based application

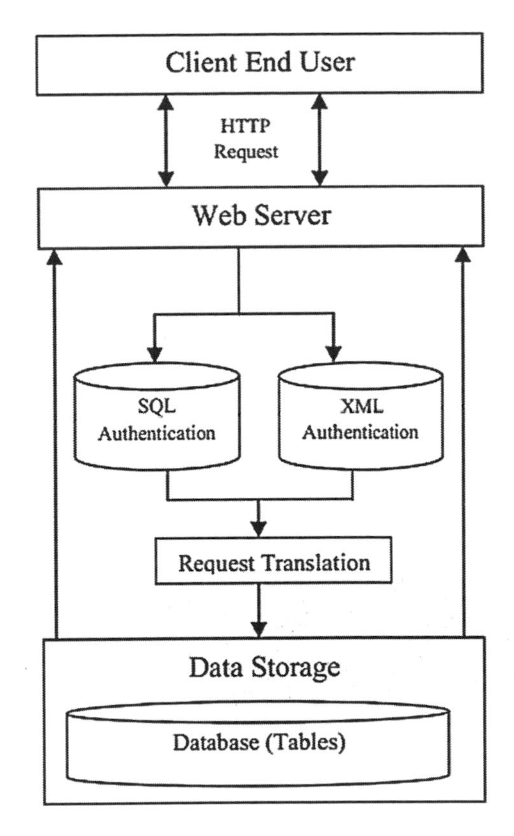

Fig. 1 Web architecture

which has large numbers of users. Most of the web-based applications, social websites, banking websites, and online shopping websites require a user to sign up for the system. Users fill their information and get username and password. A user is identified by the system based on his identity. This process of validating an individual based on a username and password is referred as authentication. The system identifies the user, and provides individual access to the system objects based on his/her identity. This process is referred as authorization. This user credential and other information is stored in the database, and accessing this database is the main goal of the attacker.

Attacker uses SQL injection with following aim in his mind

- Bypass authentication procedure (authentication is lost)
- Extract the existing data from the system (Confidentiality lost)
- Alter the existing data (Integrity lost).

Bypassing an authentication is a serious threat, as it allows an attacker to forge another authorized user identity, perform certain actions on behalf of the other user, and importantly access/modify confidential information that belongs to the user.

Our paper mainly focuses on how the attacker uses various injection techniques to bypass the authentication procedure in a system and presents method for prevention of such attacks.

2 Recent Similar Works

Below I am mentioning some of the recent works which focus on SQLIA. A lot of research has been done in detecting and preventing injection attacks and few approaches are discussed below.

The system proposed by Kiani, Clark, and Mohay [1] uses an anomaly based approach which utilizes the character distribution of certain sections of HTTP requests to detect previously unseen SQLIA. The advantage of the system proposed by Mehdi Kiani et al. is that it does not require any user interaction, or no modification of, or access to the backend database or the source code of the web application. The problem faced is the high rate of false alerts which had to be taken care while implementing the system in real-time environment. This is because of less information available on attacks to the administrator, thus making it difficult to differentiate between false alerts and the real attacks. Shanmughaneethi, Emilin Shyni, and Swamynathan [2] uses a methodology which make use of an independent web service which is intended to generalize the syntactic structure of the SQL query and validate user inputs. The SQL query inputs submitted by the user are parsed through an independent service and the correctness of the syntactic structure of the query is checked. The main advantage of this paper is that the error message generated doesn't contain any Metadata information about the database which could help the attacker. Since the web service is not integrated with the web application, any modification that should be done to the system should be done in such a way that it should be supported by the web service. Ezumalai, Aghila [3], proposed a combinatorial approach for shielding web applications against SQLIA. This combinatorial approach incorporates signature-based method, used to address security problems related to input validation and auditing based method which analyze the transactions to find out the malicious access. This approach requires no modification of the runtime system, and imposes a low execution overhead. It can be inferred from this approach that the public interface exposed by an application becomes the only source of attack. Kosuga, Kono, Hanaoka et al. [4] proposed a technique called Sania for detecting SQL injection vulnerabilities during the development and debugging phases of a web application. It identifies the vulnerable spots by analyzing the SQL queries issued in response to the HTTP requests in which an attacker can insert arbitrary strings. The main feature of Sania is the generation of attacks using the knowledge by this model, thus checking if the SQL injection vulnerabilities lie in the web application. Wei, Muthuprasanna, Kothari [5] proposed a technique to defend attacks against the stored procedures. This technique combines a static application code analysis with a runtime validation to eliminate injection attacks. In the static part, a stored procedure parser is designed, and for any SQL statement that depends on user inputs, and use this parser to instrument the necessary statements in order to compare the original SQL statement structure to that including user inputs. The underlying idea of this technique is that any SQLIA will alter the structure of the original SQL statement and by detecting the difference in the structures, a SQLIA can be identified. Kai-Xiang Zhang, Chia-Jun Lin et al. [6] proposed a translation and validation (TransSQL)-based approach for detecting and preventing SQLIA. The basic idea of this approach relies on how different databases interpret SQL queries and those SQL queries with injection. After detailed analysis on how different databases interpret SQL queries, Kai-Xiang Zhang, et al. proposed an effective solution TransSQL, using which the SQL requests are executed in two different databases to evaluate the responses generated.

3 Overview of the System

SQL injection is a technique that exploits a security vulnerability occurring in the database layer of an application. The key behind this attack is that it alters the structure of the original SQL statement when malicious input statements are added along with the original query.

In this scenario, on bypassing authentication, the injection technique is carried out on login forms where a user has to provide a username and a password, and any other places that has to be provided with a user input. In this paper, we focus the attacker's concentration on a user login form in any web page. A typical login form will contain a username and a password field, and this is where the attacker keeps trying different injection techniques until he compromises the security of the database.

3.1 Consequences of SQLIA

With SQL injections, attackers can take complete remote control of the database, and some of the impacts are:

- Insert a command to obtain access to all account details in the system, including usernames and passwords.
- With the username and password in attacker's hand, he can alter the password; change the privilege of the account.
- Forge an user identity.
- Shutdown a database.
- Upload files.
- Delete a database and its entire contents.

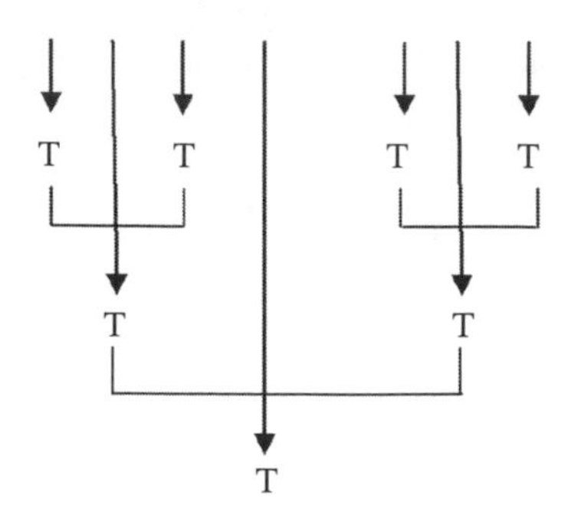

Fig. 2 Tautology

4 Types of SQLIA Techniques

4.1 Tautology

The general goal of a tautology based attack is to inject code in one or more conditional statements so that they always evaluate to true. The most common usages are to bypass the authentication pages and to extract the data. In this type of injection, an attacker exploits an injectable field that is used in query's 'where' conditional. Typically, the attack is successful when the code either displays all of the returned records or performs some action if at least one record is returned.

A typical user authentication SQL statement at the database end will take the following form.

Example. Select name from users where name='$name' and pass='$pass';

In the case of a legitimate user, with username as 'user1' and password as 'pass1', the query will take the form,

Select name from users where namc='user1' and pass='pass1'; [No Injection]

And when these user credentials are validated by the database, the user is authenticated Fig. 2.

Now as an example of a tautology attack, the attacker submits the malicious code [' OR '1'='1] as input for the username and password field, and the query takes the form,

select name from user where name= ' 'OR '1'='1' and pass= ' 'OR '1'='1 ';

select name from users where

name= ' 'OR '1'= '1' and pass=' 'OR '1'='1';

The code injected in the condition [' OR '1'='1] transforms the entire 'where' clause into a tautology. Since the conditional is a tautology, the query evaluates to a true for each row in the table and returns all of them, and finally the attacker will be authenticated into the system with the identity of the first record returned by the SQL query.

4.2 Logically Incorrect Query

This attack lets an attacker gather important information about the type and structure of the back-end database of a web application. The attack is considered to be an information gathering step for other types of attacks. The vulnerability leveraged by this attack is that the default error page returned by the application servers is often overly descriptive. Such error messages generated can often reveal vulnerable/injectable parameters to an attacker. When performing this attack, an attacker tries to inject statements that cause a syntax, or logical error into the database.

4.3 Piggy Backed Query

This kind of attack appends additional queries to the original query string. If the attack becomes successful, the database receives and executes a query string that contains multiple distinct queries. The first query is usually the original legitimate query, whereas the subsequent queries are the injected/malicious queries. This type of attack can be harmful because attackers can use it to inject virtually any type of SQL command.

Example. By using the other injection techniques discussed above, the attacker will have the name(s) of authorized user(s). For subsequent trials and in the case of Piggy Backed queries, he uses the authorized username as input for the user field, and uses the following malicious code for the password field,

pass = 'OR (SELECT COUNT(*) FROM user)=10 AND "='

The entire SQL statement will take the form,

Select name from users where name='user1' and pass= '=' OR (SELECT COUNT(*) FROM user)=10 AND ' '=' ' ;

If this query evaluates to true, then the attacker gains an insight that, there are exactly 15 users in the system. If it is evaluated to be false, then the condition is found to be incorrect and tries different possible techniques. Here, if the 'INSERT into' clause is used, and if the condition evaluates to be true, then the attacker can successfully insert data into the database.

5 Our Method

5.1 Existing Technique

In the existing web applications, authentication process takes place as follows. The user enters his assigned user name and password. The database checks if the particular user name, password combination exists in the database, and if it exists, authenticates the user. If we look at the tautology based attack, an attacker might be able to break into the system even without entering a valid user name in the user field. This is the main issue in few of the existing web applications, that there is no proper authentication procedure. This necessitates the need for a strong user authentication procedure.

This algorithm presents an efficient user authentication procedure, in a way that, an input SQL statement will be processed by the database only if the user is found to be a valid user of the system. This totally isolates the database from such injection attacks. A user is validated against two different databases of the same system.

5.2 Proposed Method

The proposed methodology here is to provide two levels of User authentication at the database level.

1. SQL Authentication
2. XML Authentication

The HTTP request sent by the Client is passed to the Web server. The input user credentials entered in the form are passed to the web server for processing at the database end. Now, the Web server has to pass it to the database. In any form of a SQL-based database, Relational Database management system is used.

The problem faced here is that, the same SQL query no matter in which relational database it is executed, it does not have the ability to differentiate the response or result obtained from the query processed by the database. That is, if a particular SQL request is evaluated to be true in a database, then the same request would evaluate to true on all the other SQL-based databases, which happens because all of these databases work based on the relational database management systems Fig. 3.

Therefore, if the same malicious/injected SQL request is run on hierarchal based database management system, then the response would be different. In a relational database management system like Microsoft Access, SQL Server, MySQL, data is stored in the form of rows and columns in tables, whereas in a hierarchical database, data is stored in the hierarchical tree structures, with the bottom most nodes that store the value. Hence the way of data processing among relational and hierarchical database management will differ, and this is the core concept of this work.

5.2.1 Using XML

Though XML is a widely used language for transportation of data in the web, there have many instances of using the XML language as a means of just storing the data, thus acting as a database. Also XML, stores data in hierarchical structures of trees, that stores data in terminal nodes, with each of the node constituting a root node. The major advantage of using XML is that, it is widely portable, platform independent, and can be integrated very easily into different web technologies. Other

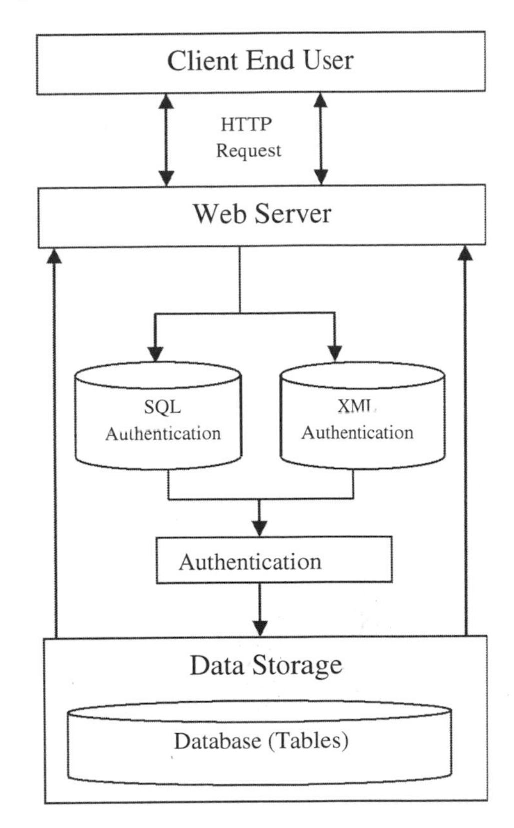

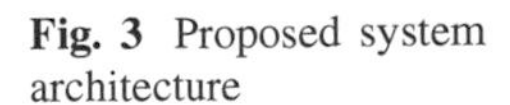
Fig. 3 Proposed system architecture

existing hierarchical database management systems like Microsoft Active Directory, Apache Directory Studio, Open LDAP for Windows are not as flexible as XML, and require a lot of overhead in initial configuration and, not totally reliable in terms of compatibility.

We can convert the database in XML format and then both SQL and XML can be used for authentication. Nowadays databases comes with tools that can convert database table into XML equivalent. XML also has a disadvantage that retrieval time in a hierarchical database is slow as compared to relational databases if the number of users of the system is high. This is because when data is searched in a XML or hierarchical database, all the nodes in present in the database from the beginning are searched and this consumes a lot of time. So we propose, instead of storing the entire user database in a single database file, a single XML file is created for every user of the system, and the corresponding password alone is stored in the XML file. This reduces the search complexity and search time to a large extent and also the size of individual file is of small size.

5.3 How Our System Works?

So, when a user tries to gain access to a system, initially, SQL authentication is done, which might evaluate to true, even in the case of an input by an attacker. But during the XML authentication, initially the corresponding XML database file is searched in the system, and if present the password is checked and only then is the user validated. So, the attacker can hack into a system only in the case of a authorized username. So in the case of an attacker input/injection, even if the SQL authentication evaluates to true, the XML authentication will fail and hence the request will not be processed by the database, thus preventing the direct access to the database. This is the method implemented in this work.

6 Conclusion and Future works

SQL injection is a common technique hackers used to attack these web based applications. SQLIA has also been specified under the top five web security threats by the OWASP in the year 2010. These attacks change the SQL queries, thereby altering the behavior of the program for the benefit of the hacker. In the work carried out, a method is put forward to detect and prevent SQL injection. The technique is based on the intuition that injection codes implicitly perform a different meaning from general queries. An elaborate environment based on XML for distinguishing between legitimate and malicious users has been presented. Here, the main idea is to secure the database from external users/attackers. Also this method helps us to achieve the same by allowing the web server to access the database only if both the levels of authentication have been satisfied. This is the unique functionality of this proposed method. And also there is no necessity to modify/update the legacy application code, as XML can be easily integrated into other languages. There are other various ways of detecting injection attacks and this is just one of them [7–9].

Further, our method can be extended by adding different levels of authentication within the same application.

References

1. Kiani, M., Clark, A., Mohay G.: Evaluation of anomaly based character distribution models in the detection of SQL injection attacks. In: The third International Conference on availability, reliability and security, IEEE. 0–7695-3102-4/08 (2008)
2. Shanmughaneethi, V., Emilin Shyni, C., Swamynathan, S.: SBSQLID: securing web applications with service based SQL injection detection. In: 2009 International Conference on Advances in Computing, Control, and Telecommunication Technologies, IEEE . 978–0-7695-3915-7/09 (2009)

3. Ezumalai, R., Aghila, G.: Combinatorial approach for preventing SQL injection attacks. In: 2009 IEEE International Advance Computing Conference (IACC 2009) Patiala, India, 6–7 March 2009
4. Kosuga, Y., Kono, K., Hanaoka, M., Kohoku-ku, H., Yokohama, F., Hishiyama, M., Takahama, Y., Minato-ku, K.: Sania: syntactic and semantic analysis for automated testing against SQL injection. In: 23rd Annual Computer Security Applications Conference, 2007, 1063–9527/07, 2007 IEEE (2007)
5. Wei, K., Muthuprasanna, M., Kothari, S.: Preventing SQL injection attacks in stored procedures. In: Proceedings of the 2006 Australian, Software Engineering Conference (ASWEC'06) (2006)
6. Kai-Xiang, Z., Chia-Jun, L., Shih-Jen, C., Yanling, H., Hao-Lun, H., Fu-Hau, H.: TransSQL: A translation and validation-based solution for SQL- injection attacks. In: First International Conference on Robot, Vision and Signal Processing, IEEE (2011)
7. Lambert, N., Lin, K.S.: Use of query tokenization to detect and prevent SQL injection attacks. IEEE. 978–1-4244-5540-9/10/2010 (2010)
8. Sushila, M., Supriya, M.: Shielding against SQL injection attacks using ADMIRE model. In: 2009 First International Conference on Computational Intelligence, Communication Systems and Networks, IEEE. 978–0-7695-3743-6/09, 2009
9. Yeole, A.S., Meshram, B.B.: Analysis of different technique for detection of SQL injection. In: International Conference and Workshop on Emerging Trends in Technology (ICWET 2011)—TCET, Mumbai, India, ICWET'11, 25–26 Feb 2011, Mumbai, Maharashtra, India, 2011 ACM (2011)

Observation Probability in Hidden Markov Model for Credit Card Fraudulent Detection System

Ashphak Khan, Tejpal Singh and Amit Sinhal

Abstract The internet has taken its place beside the telephone and the television as on important part of people's lives. Consumers rely on the internet to shop, bank and invest online shoppers use credit card to their purchases. In electronic commerce, credit card has become the most important means of payment due to fast development in information technology around the world. Credit card will be most consentient way to do online shopping, paying bills, online movie ticket booking, fees pay etc., In case of fraud associated with it is also increasing. Credit card fraud come in several ways, Many techniques use for find out the credit card fraud detection. Hidden markov model (HMM) is the statistical tools for Engineering and scientists to solve various problems. In this project, we model the sequence of operations in credit card transaction processing using a HMM and show how it can be used for the detection of frauds.

Keywords Credit card · Hidden markov model · Online transaction · E-commerce · Clustering · Credit card fraud detecting system

1 Introduction

Credit cards are the most popular payment instrument on the internet. The first credit card was introduced decades ago. (Diner's club in 1949, American Express in 1958). These cards have been produced with the magnetic stripes with unencrypted and read-only information. But today many cards are smart cards with the hardware devices offering encryption and far greater storage capacity. The most interesting

A. Khan (✉) · T. Singh · A. Sinhal
Department of Information Technology, Technocrats Institute of Technology, Bhopal, India
e-mail: ashukhan30@gmail.com

T. Singh
e-mail: tejpal1985@gmail.com

A. Sinhal
e-mail: amit_sinhal@rediffmail.com

B. V. Babu et al. (eds.), *Proceedings of the Second International Conference on Soft Computing for Problem Solving (SocProS 2012), December 28–30, 2012*, Advances in Intelligent Systems and Computing 236, DOI: 10.1007/978-81-322-1602-5_80,

event in the whole of this area has been the off again on-again liaison between Master card and Visa to produce what is becoming the de facto Internet standard for secure bankcard payments. Credit cards are by far the most common method of payment for online purchases—60 % of global online consumers used their credit card for a recent online purchase, while one in four online consumers chose PayPal. Of those paying with a credit card, more than half (53 %) used Visa. "Shopping on the Internet with the ease of a credit card is especially appealing to consumers in emerging markets who simply cannot find or buy items they want in their retail trade. Occurrence of credit card fraud is increasing dramatically due to the security weaknesses in contemporary credit card processing systems resulting in loss of billions of dollars every year credit cards can be used for doing shopping either offline or online. In offline transaction, the card must be physically present and is inserted in the payment machine in the merchant's place for making the payment. However, in online transaction, only some of the card details like secure code, expiration date and card number etc., is needed to do the transaction as it is mostly done via phone or internet [1].

Credit card fraudsters employ a large number of techniques to commit fraud. To combat the credit card fraud effectively, it is important to first understand the mechanisms of identifying a credit card fraud. Over the years credit card fraud has stabilized much due to various credit card fraud detection and prevention mechanisms. Those have been suggestion by Benson Edwin Raj, Annie Portia [2]. In day-to-day life, online transactions have increased to purchase goods and services. According to Nielsen study conducted in 2007–2008, 28 % of the world's total population has been using internet [1]. 85 % of total population today have used internet to make online shopping and the rate of making online purchasing has increased by 40 % from 2005 to 2008. In developed countries and in developing countries to some extent, credit card is most acceptable payment mode for online and offline transaction. As usage of credit card increases worldwide, chances of attacker to steal credit card details and then, make fraud transaction are also increasing. There are several ways to steal credit card details such as phishing websites, steal/lost credit cards, counterfeit cards, theft of card details, intercepted cards etc [3]. The total amount of credit card online fraud transaction made in the United States itself was reported to be $1.6 billion in 2005 and estimated to be $1.7 billion in 2006 [4].

In this paper, we show the credit card fraud detection at using statically model for this techniques show hidden markov model (HMM). Show the spending profile of each to be made by the credit card to show each transaction. Hidden markov model, in which the transition probability between hidden state and depend on the observations state.

2 Literature Survey

Credit card fraud detection has received an important attention from researchers in the world. Several techniques have been developed to detect fraud transaction using credit card which are based on neural network, genetic algorithms, data mining, clustering techniques, decision tree, Bayesian networks etc.

Ghosh and Reilly [6] have proposed a neural network method to detect credit card fraud transaction. They have built a detection system, which is trained on a large sample of labeled credit card account transactions. These sample contain example fraud cases due to lost cards, stealing cards, application fraud, stolen card details, counterfeit fraud etc. They tested on a data set of all transactions of credit card account over a subsequent period. Kokkinaki and other have proposed the technics of decision tree. This technics of decision tree are simple and easy to the implementation, decision trees is reduces misclassification of incoming transaction of data, but this is not for use dynamically adaptive of online transaction. A decision tree is defined recursively; it contains nodes and edges that are labeled with attribute names and with values of attributes, respectively [7]. Meas, Suggest of fraud detection technics using the bayesian network, in this technics, improving the fraud detecting by removing highly correlated attribute, ANN was found the credit card fraud predication faster of the testing phase, at using transaction profile. Bayesian algorithm is performed better result of fraud detection only on neural network [8].

Chan and Stalfo, have proposed the a technics of multi-classifier Meta learning issues of credit card transaction, it detecting the fraud detection 46 % improving of overall fraud, to use for each tanning experiment are required to the best distribution determine [9]. Kim, method improving number fraud detection classifier and compare only on the neural network by using the unsupervised algorithm of data mining, this method is only able to find local minima in the error function [10]. Centralize fraudulent transaction from fraud investigation of increasing, accuracy of model a distributed dataset for higher fraud are show chiu and tsai, a web based knowledge sharing scheme using for rule-based algorithms. Since there are millions of transactions processed, every day and their data are highly skewed. The transactions are more legitimate than fraudulent. It requires highly efficient technique to scale down all data and try to identify fraud transaction not legitimate transactions [11]. Syeda has proposed improving the speed of data mining, discovery of knowledge in credit card fraud detection system (FDS) of transaction process using granular neural network. Credit card fraud detecting purpose this system has been implemented [12]. Establish logic rules capable of classifying transactions of credit card into suspicious and non-suspicious classes using Genetic algorithm. This algorithm based on genetic programming this concept suggest by Bentley [13].

Bolton and Hand et al. [14] it has proposed credit card detection using unsupervised method by frequency of transactions and observing abnormal spending behavior. Break point analysis and Group analysis techniques as unsupervised tools, Successful in detection local anomalies and can FDS of behavior in a continuous manner. Those accounts are treating as suspicious ones and fraud analysis is to be done only on these accounts. If break point analysis can identify suspicious behavior such as sudden transaction of high amount and high frequency, then card will be identified as fraudulent. Algorithms do not show differentiate between accounts it show the treats of all accounts equally. We propose in this system of credit card fraud detection using observation probability in HMM. HMM is statically model for best engineering practice. Hidden markov Model is best for using the FDS. Hidden markov process is double embedded random process means it performs transaction of probability if state is “hidden” or state of transaction is “open” of two different levels. In this data mining technics we have divide the three sub categories method. We suggestion of present FDS to alternative sequence to spending profile show online transaction data generate of credit card system Credit card data set is not available to easily its most important part of banking system. Bank should not provide be provide, it is security part of any banking system. We use a dummy data set to credit card FDS; improve fraud-detecting accuracy of system. We propose the observation probabilistic in HMM to detecting “observation” state of cardholder they use the online transaction.

3 Use Hidden Markov Model

We are use application of HMM in credit card fraud detection. Hidden Markov Model is probably the simplest and easiest models, which can be used to model sequential data, i.e. data samples that are dependent from each other. Hidden Markov Model is probably the simplest and easiest models, which can be used to model sequential data, i.e. data samples that are dependent from each other. An HMM is a double embedded random process with two different levels, one is hidden and other is open to all. Hidden Markov Model does not directly use the states, which provide the external observation and gate use external observer find the visible state. Hidden Markov Model technics successfully apply for data mining, speech recognition, bio-information, robotics, artificial intelligence, voice recognition etc.

Hidden Markov Model’s characterizes by the following five traits:-

1. The number of N hidden states within the model. Each state corresponds to a unique state provide by the model. In the model, the states are defines by data set.
2. The amount of M unique observation per state. These symbols are denoted, as this can be trough of as the number of observation that fall in each data set.
3. State transition probability Matrix

$$A = \{a_{ij}\},$$

where

$$a_{ij} = P(q_{t+1} = S_i | q_t = S_i),\ i \leq i,\ j \leq N$$

and

$$\sum_{k=1}^{M} b_j(k) = 1, \quad i \leq i \leq N \tag{1}$$

4. The emission probability Matrix in state j,

$$B = \{b_j(k)\},$$

where

$$b_j = P(V_k, a_t = 1|q_t = S_j),\ 1 \le i \le N, 1 \le j \le M$$

and

$$\sum_{k=1}^{M} b_j(k) = 1, \quad 1 \le j \le N \tag{2}$$

5. The initial provability $\pi = \{\pi_i\}$ of system being in state the observation where

$$\pi_i = P(q_1 = S_i); \quad 1 \le i \le N$$

Such that

$$\sum_{i=1}^{N} \pi_i = 1$$

Then we need the HMM in sequence of observation symbols:

$$\lambda = (A, B, \pi)$$

To denote an HMM with discrete probability distribution, while

$$\lambda = (A,\ C_{jm},\ \mu_{jm},\ \sum{}_{jm},\ \pi)$$

where C_{jm} = Weighting Coefficients, μ_{jm} = Mean Vector,

$$\sum{}_{jm} = \text{Covariance matrices}$$

4 Propose Fraud Detection System and Discussion

In this section, we present credit card FDS based on HMM, which does not require fraud signatures and still is able to detect frauds just by bearing in mind a cardholder's spending habit. The important benefit of the HMM-based approach is an extreme decrease in the number of False Positives transactions recognized as malicious by a FDS even though they are genuine (Fig. 1).

Interstate transition in section. In this FDS, we consider four different spending profiles of the cardholder, which is depending upon range, named Card-I, Card-II, Card-III, Card-IV. In this set of symbols, we define $V = \{C_1, C_2, C_3, C_4\}$ and $M = 4$. The price range of proposed symbols has taken as C_1 (0, \$100], C_2 (€101, €500], C_2(€501, €1000], and C_2 (€1001, Limit of Credit Card], and after finalizing the state and symbol representations, the next step is to determine different components of the HMM, i.e. the probability matrices A, B, and I so that all parameters required

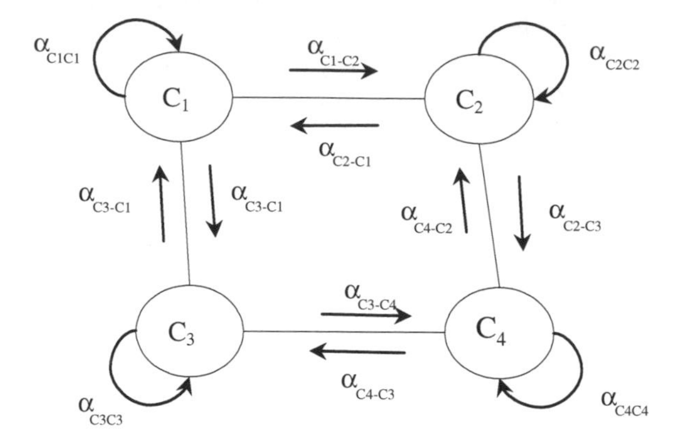

Fig. 1 Different transaction state

for the HMM is known. We show Fig. 2. That every states reach and show each other state, different states denoted and show online purchase, banking, e-cash, pay thought internet. it has been shown that probability of transition from one state to another (for example from C_1 to C_2 and vice versa, show as a_{C1-C2} and a_{C2-C1}, respectively) and also probabilities of transition from a particular state (1, 2, 3 and 4) to different spending C_1, C_2, C_3 or C_4 (for example, a_{C1-C2}, b_{C1-C2}, etc.,). After deciding HMM parameters, we will consider to form an initial sequence of the existing spending behavior of the cardholder. Let O_1, O_2, O_R be consisting of R symbols to form a sequence. This sequence is recorded from cardholder's transaction till time t. We put this sequence in HMM model to compute the probability of acceptance. Let us assume be this probability is α_1, which can be calculated as

$$\delta_1 = P(o_1, o_2, o_3, \ldots o_R|\lambda),$$

Let O_{R+1} be new generated sequence at time $t + 1$, when a transaction is going to process. The total number of sequences is $R + 1$. To consider R sequences only, we will drop O_1 sequence and we will have R sequences from O_2 to O_{R+1}.

$$\delta_2 = P(o_2, o_3, o_4, \ldots o_{R+1}|\lambda),$$

Let the probability of new R sequences be α_2 hence, we will find

$$\Delta\delta = \delta_1 - \delta_2,$$

If $\Delta\alpha > 0$, it means that HMM consider new sequence i.e. O_{R+1} with low probability and therefore, this transaction will be considered as fraud transaction if and only if percentage change in probability is greater than a predefined threshold value.

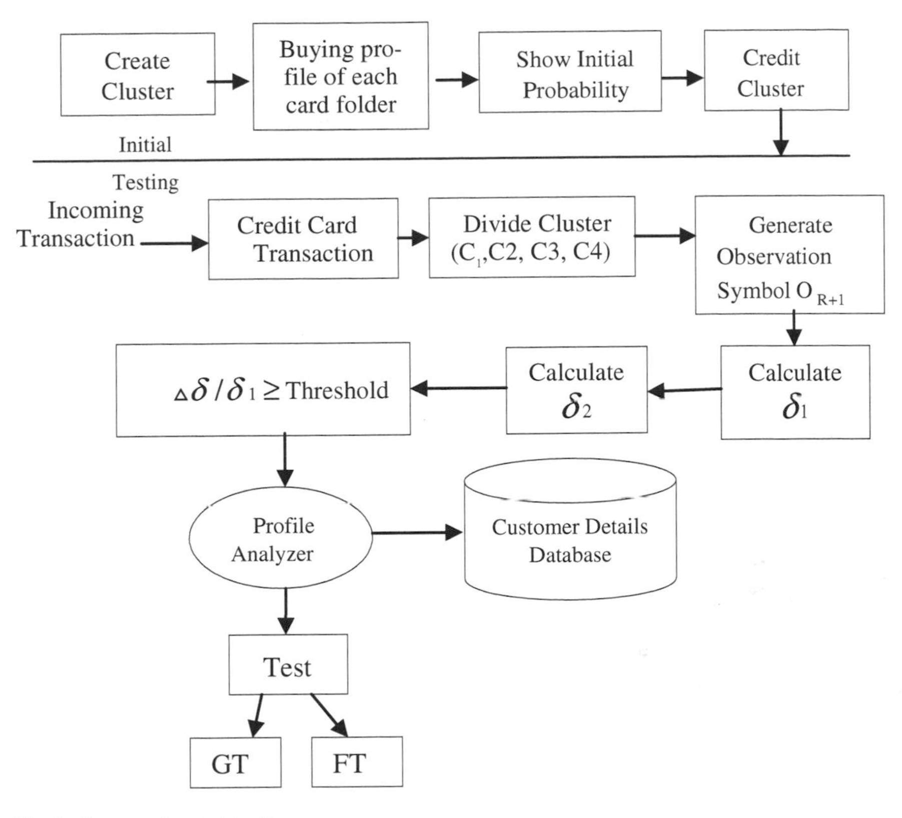

Fig. 2 Propose fraud detection system

$$\Delta\delta/\delta 1 \geq \text{Threshold value}, \tag{3}$$

The threshold value can be calculated empirically. This FDS if finds that the present transaction is a malicious, then credit card issuing bank will regret the transaction and FDS discard to add O_{R+1} symbol to available sequence. If it will be a genuine transaction, FDS will add this symbol in the sequence and will consider in future for fraud detection.

5 Experimental Result and Discussion

It is very difficult to do simulation on real time data set that is not providing from any credit card bank on security reasons. We calculate probability of each spending C_1, C_2, C_3 and C_4 of every category. Fraud detection of incoming transaction will be checked on last 20 transactions (Fig. 3).

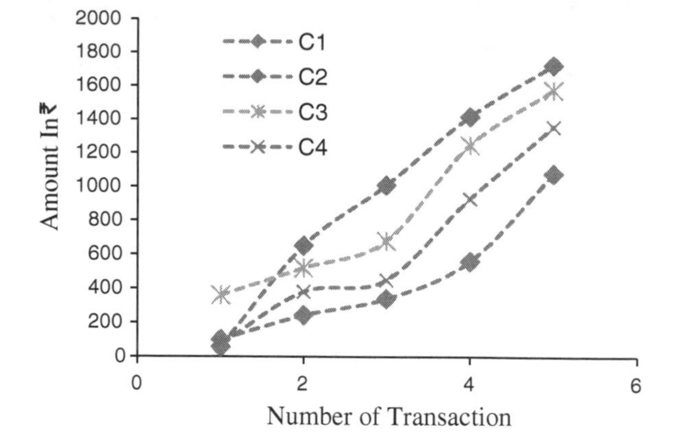

Fig. 3 Amount categories of different transaction

Table 1 All transactions happened until date

Transaction no	Category	Amount in €	Transaction no	Category	Amount in €
1st	C_1	20	11th	C_4	450
2nd	C_3	10	12th	C_3	680
3rd	C_2	40	13th	C_1	560
4th	C_1	75	14th	C_2	1420
5th	C_4	28	15th	C_4	930
6th	C_2	115	16th	C_3	1250
7th	C_4	54	17th	C_1	1080
8th	C_3	110	18th	C_4	1360
9th	C_2	180	19th	C_2	1730
10th	C_1	119	20th	C_3	1580

We have simulated several large data sets; one is shown in Table 1, in our proposed FDS and found out probability mean distribution of false and genuine transactions. When probability of genuine transaction is going down, correspondingly probability of false transaction going up and vice versa. If the percentage change in probability of false transaction will be more than threshold value, then alarm will be generated for fraudulent transaction and credit card bank will decline the same transaction (Fig. 4).

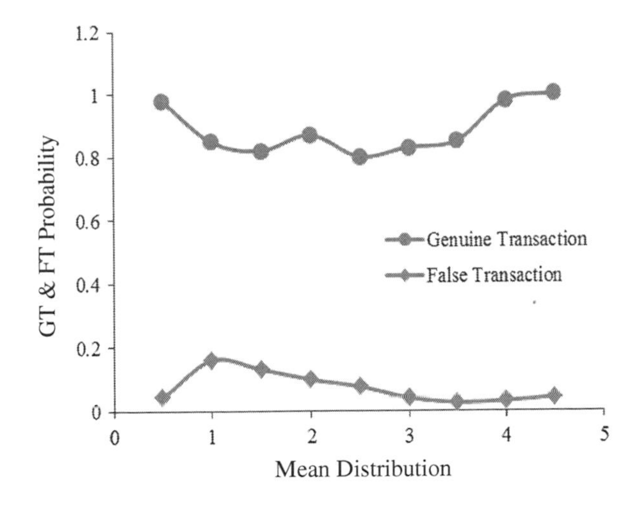

Fig. 4 Mean distribution of GT and FT

6 Conclusion

Credit card FDS is an utmost required for and card issuing bank or all type of online transaction that through using credit card. In this paper, we have implemented of HMM in credit card fraud detection. The very easily detect and remove the complexity for using in this HMM. It has also explained the HMM how can detect whether an incoming transaction is fraudulent or not comparative studies reveal that the accuracy to the system is also 90 % over a wide variation in the input data. We are dividing the transaction amount in three categories that is grouping high, medium and low used on different ranges of transaction amount each group show the aberration symbols. In HMM methods is very low compare techniques using fraud detection rate. It also have been explained low the HMM can detecting whether an incoming transaction is fraudulent or not. The system is also scalable for handling large volumes of transaction.

References

1. Internet usage world statistics. http://www.ternetworldstats.com/stats.htm (2011)
2. Benson Edwin Raj, S., Annie Portia, A.: Analysis on credit card fraud detection method. In: IEEE International Conference on Computer, Communication and, Electrical Technology ICCCET2011 (2011)
3. Trends in online shopping, a global nelson consumer report. http://www.nielsen.com/us/en/insights/reports-downloads/2010/global-trends-in-online-shopping-nielsen-consumer-report.html (2008)

4. European payment cards fraud report, payments, cards and mobiles llp & author. http://www.paymentscardsandmobile.com/payments-cards-mobile-affiliates/fraud-report/pcm_fraud_report_2010.pdf
5. Srivastava, A., Kundu, A., Sural, S., Majumdar, A.K.: Credit card fraud detection using hidden markov model. IEEE Trans. Dependable Secure Comput. **5**(1), 37–48 (2008)
6. Ghosh, S., Reilly, D.l.: Credit card fraud detection with a neural-network. In: Proceedings Of 27th Hawaii International Conference on System Science: Information Systems, Decision Support and Knowledge-Based Systems, vol. 3, pp. 621–630 (2004)
7. Kokkinaki, A.I.: On atypical database transaction: identification of probable using machine learning for using profiling. In: IEEE Knowledge and Data Engineering Exchange Workshop, pp. 107 (1997)
8. Maes, S., Tuyls, K., Vanschoenwinkel, B., Manderick, B.: Credit card fraud detection using bayesian and neural networks. In: Proceedingd of 1st Naiso Congress on Neuro Fuzzy Technologies, Hawana (2002)
9. Stolfo, S.J., Fan, D.W., Lee, W., Prodromidis, A.L., Chan, P.K.: Credit card fraud detection using meta-learing issues and iniitial result. In: Proceedings of AAAI Workshop AI Methods in Fraud and Risk Management, pp. 83–90 (1997)
10. Kim, M.J., Kim, T.S.: A neural classifier with fraud density map for effective credit card fraud detection. In: Proceedings of International Conference on Intelligent Data Engineering and Automated, Learning, pp. 378–383 (2002)
11. Chiu, C., Tsai, C.: A web services-based collaborative scheme for credit card fraud detection. In: Proceedings of 2004 IEEE International Conference on e-Technology, e-Commerce and e-Service (2004)
12. Syeda, M., Zjang, Y.Q., Pan, Y.: Parraller granular networks for fast credit card fraud detection. In: Proceeding of IEEE International Confrence on Fuzzy System, pp. 572–577 (2002)
13. Bentley, P.J., Kim, J., Jung, G.H., Choi, J.U.: Fuzzy darwinian detection of credit card fraud. In: Proceeding of 14th Annual Fall Symposium of Korean Information Processing Society (2000)

Comparative Study of Feature Reduction and Machine Learning Methods for Spam Detection

Basant Agarwal and Namita Mittal

Abstract Nowadays, e-mail is widely used for communication over Internet. A large amount of Internet traffic is of e-mail data. A lot of companies and organizations use e-mail services to promote their products and services. It is very important to filter out spam messages to save users' precious time. Machine learning methods plays vital role in spam detection, but it faces the problem of high dimensionality of feature vector. So feature reduction methods are very important for better results from machine learning approaches. In this paper, Principal Component Analysis (PCA), Singular Value Decomposition (SVD), and Information Gain (IG) methods are used for feature reduction. Further, e-mail messages are classified as spam or ham message using seven different classifiers namely Naïve Baysian, AdaBoost, Random Forest, Support Vector Machine, J48, Bagging, and JRip. Comparative study of these techniques is done on TREC 2007 Spam e-mail Corpus with different feature size.

Keywords Spam classification · E-mail classification · Machine learning algorithms · Feature selection

1 Introduction

With the growth of Internet, e-mail is widely being used for communication over Internet due to fast, efficient, and economical way to communicate. That's why, large number of companies use e-mail facility to promote and advertise their products and services. These unwanted and unsolicited e-mails are known as spam e-mails [1]. Large part of the Internet traffic comprises of these spams. Users have to waste

B. Agarwal (✉) · N. Mittal
Malaviya National Institute of Technology, Jaipur, India
e-mail: thebasant@gmail.com

N. Mittal
e-mail: nmittal@mnit.ac.in

B. V. Babu et al. (eds.), *Proceedings of the Second International Conference on Soft Computing for Problem Solving (SocProS 2012), December 28–30, 2012*, Advances in Intelligent Systems and Computing 236, DOI: 10.1007/978-81-322-1602-5_81,

a lot of time to read spam messages. These spam mail can cause damage to users' system and annoying individual user. That's why it is very important to automatically classify/filter the spam messages from legitimate messages.

Spam filtering is a text classification task. Text classification for spam is challenging due to huge features size and large number of texts [2]. High dimensionality data is a problem for classification algorithm because of the high computational cost and memory usage. So feature reduction methods are essential for better performance of the classification model [3].

Objective of this paper is to do comparative study of machine learning algorithm for building an efficient classifier that can determine if an e-mail message is spam or not. In addition, to analyze the effect of feature reduction methods and feature size on the performance of classifier.

The remainder of the paper is as follows: Sect. 2 discusses related work, Sect. 3 explains spam classification method process, Sect. 4 describes the results and discussions, and Sect. 5 concludes the paper with possible directions for future work.

2 Related Work

It has become necessary to have a filtering system that will classify the e-mails either as spam or ham/legitimate. Spam mails are the unwanted messages which are sent without the consent of the user. Researchers have applied different machine learning methods like Naïve Bayes (NB), Support Vector Machine (SVM), Neural Network (NN), etc., in automatically classifying spam messages from legitimate messages [4, 5]. Principal Component Analysis (PCA) is used for reducing the dimensionality of feature size and then NN is used for spam classification [6]. Latent Semantic Analysis (LSA) is used for anti-spam filtering system, results obtained from LSA is compared with naïve baysian classifier [7].

Several variants of boosting algorithms are performed for e-mail classification and compared the results with naïve baysian and decision tree [8]. Effect of PCA on three different classifier, i.e., C5.0, instance-based learner and naive Bayes are analyzed [9]. Effect of PCA as pre-processing is analyzed on machine learning accuracy with high dimensional dataset [10]. Benefits of dimensionality reduction are explored with the context of latent semantic indexing for e-mail detection [11].

SVM is used for online spam filtering system on large dataset, a Relaxed Online SVM (ROSVM) System is proposed that gives performance comparable to traditional filtering system at reduced computation cost [12]. Performance of the Naïve Bayesian classifier is compared to an alternative memory based learning approach on spam filtering [13]. An anti-spam filtering system is proposed in which a NB classifier is used to detect spam messages considering the effect of attribute size, training dataset size [14].

3 Spam Classification Process

Spam classification is categorized as a problem of text classification. For this, incoming messages will be classified as spam or legitimate message. First, all the documents are pre-processed, after that feature vector will be constructed, then feature reduction techniques can be applied to reduce the feature size. This reduced feature vector is used for spam classification model.

3.1 Pre-processing

Pre-processing tasks are in order to extract unique words which occur in spam e-mails. All the words are lower cased, and symbols like %, $, @ etc. are removed. Further, stop words are removed and stemming is performed.

3.2 Feature Vector Construction

Term Frequency-Inverse Document Frequency (TF-IDF) weighting scheme is used for creating feature vector; the dimensionality of this feature vector is in order of thousands of features [2]. TF-IDF can be calculated by TF-IDF: $\mathrm{w}_{ij} = \mathrm{tf}_{ij} * \mathrm{idf}_i$, where tf_{ij} is the frequency of term i in e-mail j, and idf_i is the inverse document frequency, it measures if a term is common or rare across e-mails.

3.3 Feature Reduction

Feature vector generated using words as a feature is huge in dimension. Therefore, it is required to reduce this feature vector size using dimension reduction techniques like IG, PVA, SVD, etc. Computation complexity of the classifier increases with the size of feature vector. So feature reduction methods are used for removing irrelevant and not important features from the feature vector [15].

IG is one of the important feature selection techniques. It measures the importance of feature globally and top ranked features can be selected based on reduction in the uncertainty after knowing the value of the feature for reducing the feature vector. This reduced feature set is used for better classification results [3]. PCA is a feature reduction technique that transforms high dimensional feature vector into lower dimensional such that maximum variance is extracted from the data. For reducing the size of the feature vector top k principal components are selected. SVD also transforms the high dimensional feature vector into lower dimensional space that is latent semantic space. For reducing the size of feature vector top k singular values are selected [16].

3.4 Classification Methods

Different machine learning algorithms are used for classification of spam messages. Naïve baysian is based on probability and bayes theorem. This classifier assumes independence of feature vectors and tries to predict the probability of new instance.

Boosting and Bagging are two voting based classifiers. In voting classifier, training samples are taken randomly from the dataset multiple times, and different classifiers are learned. To classify a new sample, each classifier gives a different class label; the result of voting classifier is decided by the maximum votes earned for a particular class. Main difference between bagging and boosting is the way; they take the samples for training a classifier. In bagging, training samples are taken with equal weights randomly, and in boosting, more weightage are given to those samples which have been misclassified by previous classifiers.

A single J48 is a classifier based on C4.5 decision tree algorithm [17] and JRip is a java implementation of a propositional rule learner, called Repeated Incremental Pruning to Produce Error Reduction (RIPPER) [18]. Random forest (RF) [19] is an ensemble classifier that comprises of many decision trees and outputs the class that is the mode of the class's output by individual trees. Random forests are generally used when training datasets is huge and input is of the order of thousands of variables. A random forest model is normally composed of tens or hundreds of decision trees. SVM is a supervised learning method that represents input instances as points in space, mapped so that the instances of the separate categories are divided by a gap that is as wide as possible.

4 Results and Discussions

The 2007 TREC Spam e-mail Corpus [20] contains 75,419 e-mail messages at the University of Waterloo. For the experiments, 2,500 sample messages are used for building the classifier out of which 25 % messages are spam messages and others are ham (not spam) messages. After processing of these messages (i.e., stop word removal, stemming,) size of feature vector is 1,400. The classification accuracy is determined using a 5-fold cross-validation.

4.1 Effect of Feature Size and Feature Reduction Methods

IG, PCA, and SVD methods are used for reducing the feature size. It is observed that as features size is reduced, accuracy of the classifier slightly decreases or it remains unchanged. There is not much variation in the performance of all the classifier with reduction of feature size as shown in Tables 1, 2, and 3. For example, using Bagging Technique with all features (without using any feature reduction technique),

Table 1 Accuracy (%) of different classifiers with different feature size when SVD is used for feature reduction

Feature size	NB	Adaboost	Random forest	SVM	J48	Bagging	JRip
400	63	79	80.5	77.3	77.1	83.8	80.5
800	62.2	79.4	80.6	77.7	77.5	83.6	79.7
1100	63.4	79.7	80.4	78.6	76.6	83.2	78.3
1400	63.5	83.6	81.6	79.6	80	86.8	84

Table 2 Accuracy (%) of different classifiers with different feature size when PCA is used for feature reduction

Feature size	Naïve Baysian	Adaboost	Random forest	SVM	J48	Bagging	JRip
400	56.4	80.1	79	77.1	72.8	83.7	77.8
800	58.2	80.2	79.8	77.9	70.2	84.3	80
1100	60.1	80.4	79.3	76.3	74	85.5	81
1400	63.5	83.6	81.6	79.6	80	86.8	84

Table 3 Accuracy (%) of different classifiers with different feature size when IG is used for feature reduction

Feature size	Naïve Baysian	Adaboost	Random forest	SVM	J48	Bagging	JRip
400	56.6	83.7	83.9	82.9	83.3	85.7	80.5
800	57.8	83.7	82.3	81.2	84.5	85.1	79.7
1100	58.8	83.8	82.2	80.1	84.8	86.6	78.3
1400	63.5	83.6	81.6	79.6	80	86.8	84

i.e., 1,400 got accuracy is 86.8 % while with feature size 400 (as shown in Table 3) we got the accuracy 85.7 % (−1.26 %). Similarly accuracy of Adaboost without any feature reduction technique gives the accuracy of 83.6 % and with 400 features it gives accuracy of 83.7 % as shown in Table 3. Therefore without compromising much with accuracy of the classifier, complexity of the model can be decreased and time required for building the model is also decreased.

Accuracies of seven different classifiers are given in Tables 1, 2, and 3 with different feature size when different feature reduction techniques are used, i.e., SVD, PCA, and IG. Experimental results show that information gain gives better results for spam classification when used for feature reduction than PCA and SVD methods as observed from Tables 1, 2, and 3. For example, performance of J48 and random forest slightly increases after reduction of features using information gain (Table 3). Also for all the classifiers information gain gives better accuracy than SVD and PCA as shown in Tables 1, 2, and 3. It is also observed from Tables 2 and 3 that SVD performs better as compared to PCA.

In term of time required for applying the feature reduction technique PCA takes maximum time to reduce the features. And information gain technique take minimum time to create the ranking of important features for reducing the feature size.

4.2 Effect of Machine Learning Algorithm

Different classification methods like Support Vector Machine classifier, Naïve Bayesian classifier, J48, Bagging, Random Forest, JRip are evaluated on TREC 2007 dataset to analyze the efficiency of different machine learning algorithm for spam detection.

Experimental results show that Naïve baysian classifier performs worst among these seven classifiers and Bagging technique outperform all other 6 machine learning methods as shown in Figs. 1, 2, and 3 for all the feature size and feature reduction technique. For example, Bagging gives 86.8 % accuracy when no feature reduction technique is used and after reducing feature size even then it gives good results as compare to others, i.e., 83.8 % as shown in Table 1. It is observed from Figs. 1, 2, and 3 that except naïve baysian classifier all other classifiers perform well. It is also observed that JRip, Adaboost, SVM, J48, and Random forest gives almost equal accuracy for all the feature size as shown in Figs. 1, 2, and 3.

4.3 Time Taken by the Machine Learning Algorithm

As features of spam data are reduced, time taken by each machine learning algorithm is also deceased as shown in Fig. 4. Naïve baysian classifier takes minimum time among all the algorithms, whereas support vector machine and bagging takes maximum time for building the model. Time required for building the model increases very fast for support vector machine and bagging classifiers. Performance (in term of accuracy) and time taken to build the model are minimum for NB classifier. Bagging takes maximum time to build the model also gives maximum accuracy.

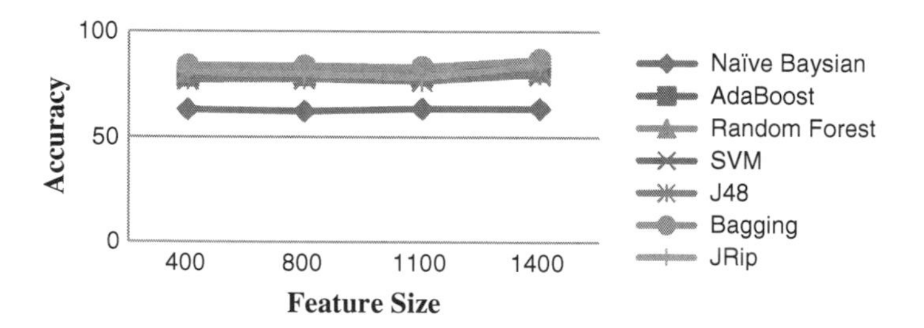

Fig. 1 Accuracy of classifiers with different feature size when SVD is used for feature reduction

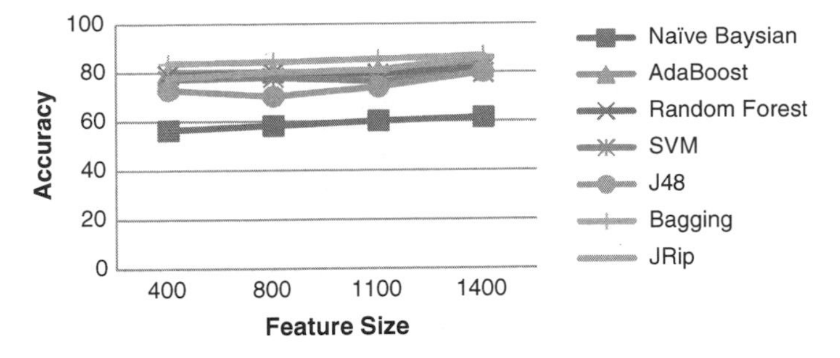

Fig. 2 Accuracy of classifiers with different feature size when PCA is used for feature reduction

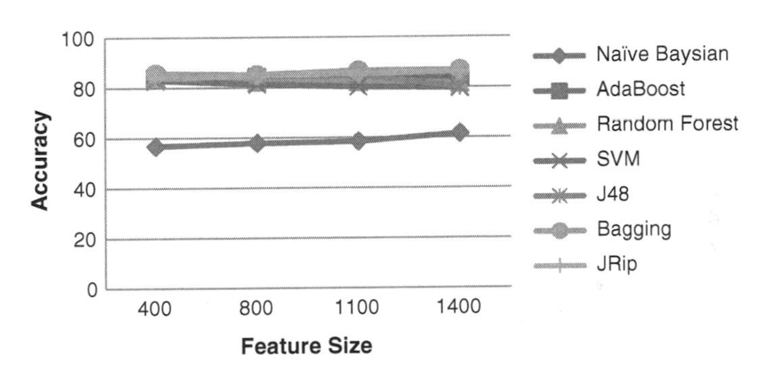

Fig. 3 Accuracy of classifiers with different feature size when information gain is used for feature reduction

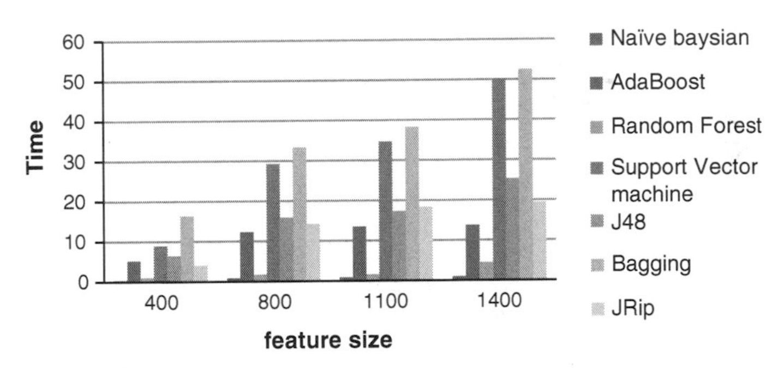

Fig. 4 Time taken by classifier for different feature size when PCA is used for feature reduction

If time is a constraint then random forest should be considered, it takes less time to build the model and performance of spam detection is also comparable with bagging technique, i.e., 81.6 and time taken is 4.4 s for building the model.

5 Conclusions

There are various anti-spam filtering systems which uses different kind of techniques. Therefore, comparative study of these techniques is performed in order to evaluate the performance of classification algorithms. Different algorithms are evaluated with different dataset size and varying feature size. Information gain, PCA, SVD are used for feature reduction methods and seven different machine learning algorithms are used for evaluating the performance for e-mail classification on 2,500 instances using TREC 2007 dataset. It is observed that reducing the feature size does not reduce the performance of classification, whereas information gain technique gives better results for feature reduction in terms of accuracy and time taken for building the model. It is also observed that bagging gives better classification accuracy compare to other machine learning algorithms.

References

1. Cranor, L.F., LaMacchia, B.A.: Spam!. Commun. ACM **41**(8), 74–83 (1998)
2. Sebastiani, F.: Machine learning in automated text categorization. ACM Comput. Surv. **34**(1), 1–47 (2002)
3. Janecek, A., Gansterer, W.N.: On the relationship between feature selection and classi_cation accuracy. In: JMLR:Workshop and Conference Proceedings, vol. 4, pp. 90–105 (2008)
4. Guzella, Thiago S., Caminhas, Walmir M.: A review of machinelearning approaches to Spam filtering. Expert Syst. Appl. **36**(7), 10206–10222 (2009)
5. Fawcett, T.: In vivo spam filtering: a challenge problem for data mining. In: Proceedings of ninth KDD Explorations, vol. 5, No. 2 (2003)
6. Cui, B., Mondal, A., Shen, J., Cong, G., Tan, K.: On effective email classification via neural networks. In: Proceedings of DEXA, pp. 85–94 (2005)
7. Gee, K.R.: Using latent semantic indexing to filter spam In: Proceedings of the 2003 ACM Symposium on Applied Computing, pp. 460–464. ACM press, New York (2003)
8. Carreras, X., Marquez, L.: Boosting trees for anti-spam email filtering. In: Proceedings of RANLP-2001, pp. 58–64, Bulgaria (2001)
9. Popelinsky, L.: Combining the principal components method with different learning algorithms. In: Proceedings of the ECML/PKDD2001 IDDM, Workshop (2001)
10. Howley, Tom, Madden, Michael G., O'Connell, Marie-Louise, Ryder, Alan G.: The effect of principal component analysis on machine learning accuracy with high-dimensional spectral data. Knowl. Based Syst. **19**(5), 363–370 (2006)
11. Gansterer, W.N., Janecek, A.G.K., Neumayer, R.: Spam filtering based on latent semantic indexing. In: Survey of Text Mining II—Clustering, Classification, and Retrieval, vol. 2, pp. 165–185. Springer (2008)
12. Sculley, D., Wachman, G.M.: Relaxed online SVMs for spam filtering. In: Proceedings of the 30th Annual International ACM SIGIR Conference on Research and Development in, Information Retrieval SIGIR '07, pp. 415–422 (2007)
13. Androutsopoulos, I., Koutsias, J., Chandrinos, K.V., Spyropoulos, D.: Learning to filter spam e-mail: A comparison of a naïve bayesian and a memory-based approach. In: 4th PKDD's Workshop on Machine Learning and Textual Information Access (2000)
14. Androutsopoulos, I., Koutsias, J., Chandrinos, K.V., Spyropoulos, D.: An experimental comparison of naive bayesian and keyword based anti-spam filtering with personal e-mail messages. In: Proceedings of the 23rd ACM SIGIR Annual Conference, pp. 160–167 (2000)

15. Parimala, R., Nallaswamy, R.: A study of Spam E-mail classification using feature selection package. Global J. Comput. Sci. Technol. **11**(7) (2011)
16. Kumar, C.A.: Analysis of unsupervised dimensionality reduction techniques. Comput. Sci. Inf. Syst. **6**(2), 217–227 (2009)
17. Quinlan, R.J.: C4.5: Programs for Machine Learning. Morgan Kaufmann, Burlington (1993)
18. Cohen, W.W.: Fast effective rule induction. In: Proceedings of the Twelfth International Conference on Machine Learning, pp. 115–123. Morgan Kaufmann, Burlington (1995)
19. Breiman, Leo: Random forests. Mach. Learn. **45**, 5–32 (2004)
20. TREC 2007 Public Spam Corpus. http://plg.uwaterloo.ca/gvcormac/treccorpus07/

Generation of Key Bit-Streams Using Sparse Matrix-Vector Multiplication for Video Encryption

M. Sivasankar

Abstract The contribution of stream ciphers to cryptography is immense. For fast encryption, stream ciphers are preferred to block ciphers due to their XORing operation, which is easier and faster to implement. In this paper we present a matrix-based stream cipher, in which a $m \times n$ binary matrix single handedly performs the work of m parallel LFSRs. This can be treated as an equivalent way of generating LFSR-based stream ciphers through sparse matrix-vector multiplication (SpMV). Interestingly the output of the matrix multiplication can otherwise be used as a parallel bit/byte generator, useful for encrypting video streams.

Keywords Linear feedback shift registers · Permutations · Primitive polynomials · Sparse matrix-vector multiplication

1 Introduction and Background Study

1.1 Stream Ciphers

Cryptography is basically all about secure communication and broadly revolves around symmetric techniques (communicating parties possess the same key) and asymmetric techniques (communicating parties possess different keys). A key is a data used for encrypting/decrypting purpose. Further symmetric key cryptography is classified into stream ciphers and block ciphers. While block ciphers encrypt data in blocks of fixed size (64, 128 bits, etc.), stream ciphers do encryption in a bit-by-bit manner [1]. Elaborating further, in stream ciphers the data (plain text) is converted

M. Sivasankar (✉)
Department of Mathematics, Amrita School of Engineering,
Amrita Vishwa Vidyapeetham, Coimbatore, India
e-mail: m_sivasankar@cb.amrita.edu

B. V. Babu et al. (eds.), *Proceedings of the Second International Conference on Soft Computing for Problem Solving (SocProS 2012), December 28–30, 2012*, Advances in Intelligent Systems and Computing 236, DOI: 10.1007/978-81-322-1602-5_82,

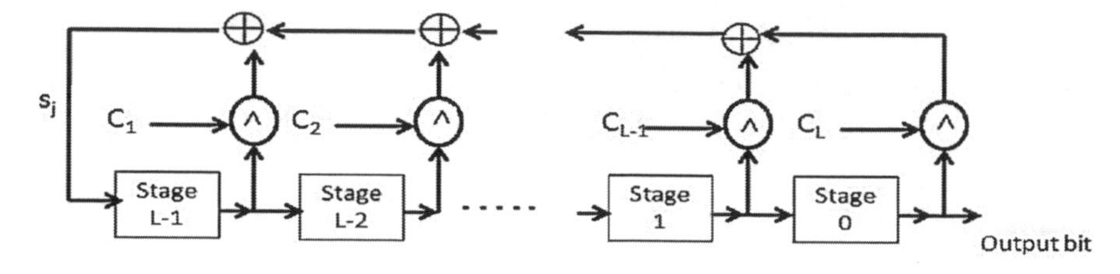

Fig. 1 An LFSR with L stages

into a stream of 0s and 1s and is XORed bit-by-bit with the bits of a bit stream generator. The result is the encrypted plain text called the cipher text.

The bit generator is initialized with a seed and it starts generating bits. The seed bits (called the key) and how the generator is using the present state to generate the next bit are kept secret with the encrypting/decrypting ends. Upon receiving the cipher text, the receiver generates the same bit stream and XORs it with the received cipher text; as XORing a bit with itself nullifies the effect, the plain text is retrieved. Also, as XORing requires less CPU and memory, stream ciphers are preferred when computational and storage overheads are to be minimized.

1.2 Linear Feedback Shift Registers

An LFSR of length L consists of L stages numbered 0, 1, … , $L-1$, each capable of storing one bit and having one input and one output; and a clock which controls the movement of data. During each unit of time the following operations are performed: (i) the content of stage 0 is output and forms part of the output sequence; (ii) the content of stage i is moved to stage $i-1$ for each $1 \leq i \leq L-1$ and (iii) the new content of stage $L-1$ is the feedback bit s which is calculated by adding together mod 2 the previous contents of a fixed subset of stages [1].

The LFSR $< L, C(D)>$, in Fig. 1 has the connection polynomial $C(D) = 1 + c_1 D + c_2 D^2 + ... + C_L D^L \in Z_2(D)$. If the initial content is $[s_{L-1}, .., s_1, s_0] \in \{0,1\}^L$ then the feed back to this LFSR is $s_j = (c_1 s_{j-1} \oplus c_2 s_{j-2} \oplus \ldots \oplus c_L s_{j-L})$ for $j \geq L$. The period of a generated sequence $s_0, s_1, s_2, \ldots\ldots$ is p if $s_i = s_{i+p}$ for all $i \geq 0$.

1.3 Irreducible and Primitive Polynomials

Consider the collection $Z_2[X]$ of all polynomials in variable x of any degree, with coefficients from $\{0,1\} = Z_2$. A polynomial $p(x) \in Z_2[X]$ is said to be irreducible if it cannot be factored into lower degree polynomials of $Z_2[X]$. A primitive polynomial is an irreducible polynomial that generates all elements of an extension field from a base field. For a detailed introduction to primitive polynomials and extension fields the reader can refer [1, 9]. It is interesting to note that if $C(D)$ is a primitive polynomial, then any of the $2^L - 1$ nonzero initial state of the LFSR, produces an output sequence with maximum possible period $2^L - 1$, which is a requirement for secure communications [1].

1.4 Combination Generators and Attacks on Stream Ciphers

Due to the inherent linearity nature of an LFSR, the output of an LFSR is easily predictable. Berlekamp-Massey Algorithm [1, 9], can construct an LFSR for a given sequence. So it is advisable to combine the outputs of many LFSRs to introduce some sort of nonlinearity through a filter generator or through a clock controlled generator [3]. Another way of achieving nonlinearity is by using Boolean functions. A Multi-output Boolean function on n variables is a mapping from $\{0,1\}^n$ into $\{0,1\}^m$ and is denoted by $f(x_1, x_2, \ldots, x_n)$ [4].

Obviously the Boolean function used for inducing nonlinearity should be resistant to various possible attacks mounted on stream ciphers. In fact the Boolean function used should be capable of generating a secure stream cipher possessing the necessary properties of (i) Long period (ii) Large linear complexity (iii) Randomness [10]. A lot of valuable research contributions are available in designing immune Boolean functions [2–4]. These results stress that, Boolean functions should be balanced and possess immunity against fast algebraic attacks and correlation attacks. The linear complexity of the resulting key stream depends on the linear complexity of the constituent streams and it will be maximum if the lengths of individual LFSRs are relatively prime [1].

1.5 Sparse Matrix-Vector Multiplication

In general, matrices used in cryptogrpahy are bigger than available memory. Luckily, many of these matrices are sparse with a special case, binary sparse (entries being 0 or 1). Large binary matrices play an important role in computer science and in information theory. Efficient methods of handling sparse matrices, multiplying them with vectors are in need since their introduction. Due to the crucial requirements of SpMV in (i) Iterative methods for solving large linear systems $Ax = b$ and (ii) Eigenvalue problems $Ax = \lambda x$ it attracted many researchers and more efficient methods are being introduced [8, 13, 15].

2 Proposed Key Bit-Stream Generator

2.1 Binary Matrix Representation of Parallel LFSRs

We observe that the coefficients of a polynomial of degree n, belonging to $Z_2[X]$ can be denoted by a $0-1$ vector of length n. For example, the connection trinomial $C(D) = 1 + D^3 + D^5$ can be represented as (0, 0, 1, 0, 1) $\in \{0,1\}^5$; note that, the feedback coefficient 1 (which is always present in connection polynomials) of $C(D)$, is not used in the vector representation of $C(D)$. Evaluating this polynomial for a state vector $(b_5, b_4, b_3, b_2, b_1)$ is simply the product (0, 0, 1, 0,1) $(b_5, b_4, b_3, b_2, b_1)^T = b_3 + b_1$. This vector–vector multiplication (mod 2) is in fact the output bit of the LFSR for that state $(b_5, b_4, b_3, b_2, b_1)$. Hence shifting the present state vector by one bit and loading the output bit of the just previous vector–vector multiplication as the new entry, we get the next state vector of the LFSR. Continuing this we get the output bit stream of the LFSR. This idea can be extended to represent a set of m parallel LFSRs (i.e., connection polynomials), each having n states, in a binary matrix. The m parallel connection polynomials, $C_i(D) = 1 + c_{i1}D + c_{i2}D^2 + \ldots + c_{in}D^n$, $1 \le i \le m$ can be represented by the binary matrix $C = (c_{ij})$, $i = 1$ to m, $j = 1$ to n.

2.2 Getting the Bit Stream

Using the matrix representation discussed in Sect. 2.1, the ith row of

$$\begin{pmatrix} c_{11} & c_{12} & \ldots & c_{1n} \\ c_{21} & c_{22} & \ldots & c_{2n} \\ & \ldots\ldots & & \\ c_{m1} & c_{m2} & \ldots\ldots & c_{mn} \end{pmatrix} \begin{pmatrix} b_n \\ b_{n-1} \\ \vdots \\ b_1 \end{pmatrix} = \begin{pmatrix} c_{11}b_n + c_{12}b_{n-1} + \ldots\ldots + c_{1n}b_1 \\ c_{21}b_n + c_{22}b_{n-1} + \ldots\ldots + c_{2n}b_1 \\ \ldots\ldots \\ c_{m1}b_n + c_{m2}b_{n-1} + \ldots\ldots + c_{mn}b_1 \end{pmatrix}$$

is the output bit of the ith LFSR, $1 \le i \le m$, for the state $(b_n, b_{n-1}, \ldots., b_2, b_1)$. Let us denote this output vector by $O_k = (o_{k1}, o_{k2}, \ldots., o_{km})^T$, k being the iteration number. Now combining the bits of O_k using a Boolean function f, we get the key bit, $\text{bit}_k = f(o_{k1}, o_{k2}, \ldots., o_{km})$. The generation process is continued as described below:

- The input vector $(b_n, b_{n-1}, \ldots., b_2, b_1)$ is shifted by one bit and the feedback of the first LFSR (i.e., o_{k1}) is loaded as the new entry.
- The Matrix-Vector multiplication is performed.
- The resulting vector O_{k+1} is fed into the Boolean function f to get bit_k.

This way of generation differs from existing schemes, in the randomness happening in loading of the next state of the LFSRs. At any iteration k, the output bit of the

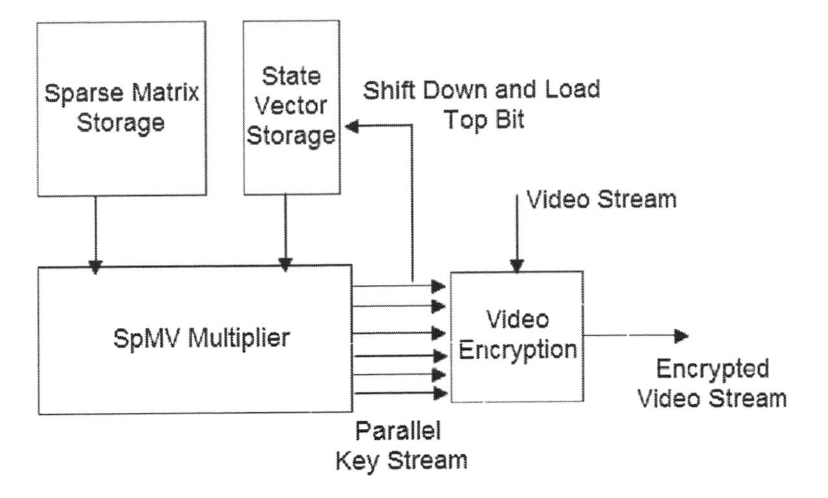

Fig. 2 Video encryption

first LFSR (o_{k1} of O_k) only is used to load the next state. Hence, the first LFSR continues to get its states, according to its connection $C_1(D)$, whereas the other LFSRs are loaded with this state vector, leading to randomness in their states. Since $C_1(D)$ is a primitive connection polynomial, the first LFSR goes through all the possible $2^L - 1$ states and consequently all the other LFSRs also get these different possible states, the main difference (with the existing schemes) being in the order they get the states if they were standing alone (independent). The randomness in the order of acquiring the states has an effect similar to irregular clocking/random deletion and the periodicity of the sequence generated is not affected. Basically one more randomness to the input bits of Boolean function is introduced and it provides more security to the bit-stream generated.

2.3 *Using the Stream for Video Stream Encryption*

The security requirement in digital video streaming while maintaining efficiency and format compliance [14] can also be achieved by eliminating the nonlinear combinator Boolean function f, and instead using the output bits of the parallel LFSRs as such (Fig. 2). If byte encryption is required m can be set to 8 and for more security m can be set to 9 and the 8 bits excluding the first LFSR's output bit can be used.

3 Experimental Works on the Key Stream

Meier and Staffelbach proposed the idea of fast correlation attacks on nonlinear combination generators. In such attacks (i) few bits of the key bit-stream (ii) connection Polynomial of the individual LFSRs (iii) The mth order Boolean function f

Table 1 Partial edit distance matrix for the target LFSR

LFSR2↓ and LFSR3→	10101	11101	10111	11111	10001
10101	2	2	1	1	1
11101	2	1	1	1	2
10111	2	1	2	1	1
11111	2	1	2	1	1
10001	3	2	1	2	1

(iv) The correlation probability of the key stream with the output of the target LFSR are known to the attacker. A detailed study of this type of attack can be found in [7]. In our experimental work we perform another type of attack, the edit distance attack [5], (which is a known plain text attack), on the proposed stream cipher. Basically, stream ciphers generate the key bit-stream from a short key. For a secure stream cipher, it should not be possible to reconstruct the short key from the key stream sequence.The core idea behind edit distance attacks lies in guessing the initial state of the generator and then trying to determine whether the initial guess is consistent.

We proceed with three ($m = 3$) 5 bit LFSRs with connection polynomials $C_1(D) = 1 + D^2 + D^5$, $C_2(D) = 1 + D + D^2 + D^3 + D^5$, $C_3(D) = 1 + D^3 + D^5$, with an initial state (10101) and the Boolean function $f = x_1x_2 + x_3$. The time taken for calculating the edit distance matrix is a crucial step in the attack time. Table 1 shows partial edit distance matrix obtained for various combinations of initial states of LFSR 2 and LFSR 3 with a fixed initial state of (10100) of LFSR 1. From the table it is observed that an increase in the number of known bits leads to a significant increase in the computation time; and comparing with the conventional way of generating the stream, we observe that required properties of the sequence are not compromised.

4 Implementation Issues

Usually, as the number of LFSRs increases the memory/hardware requirement also increases. Boolean function calculation also plays a role in implementation. But here since all the LFSRs are controlled by a Boolean matrix, sparse matrix multiplication implementation is crucial [16]. The number of variables involved in the generation decides the order of the matrix. The space required for LFSR implementation increases linearly with the number of variables and the space required for the Boolean function increases exponentially with the number of variables [11, 12]. Either we can directly implement the Boolean function or we can simplify to some extent so that the hardware requirement is optimized. The possible implementation of the case $m = n$ (number of LFSRs and their number of stages are equal), leading to a square sparse matrix can be analyzed in future.

5 A Variant of SpMV Stream Cipher

5.1 Permutation Matrices

We observe that given a nonzero initial state, an irreducible polynomial makes the LFSR to visit all the possible $2^L - 1$ nonzero states, and hence the LFSR gives out a binary sequence of maximum period possible. The order in which the states are visited is nothing but a permutation of the list of $2^L - 1$ L-bit strings. The possible number of permutations is, as we know $(2^L - 1)!$; and an L-stage LFSR with an irreducible connection polynomial goes through the states in order, which is one order among the $(2^L - 1)!$ ordered lists. This is in fact a combinatorial problem of listing down all the possible permutations of the $2^L - 1$ L-bit strings. It is interesting to note that a permutation of n elements has a matrix representation, the permutation matrix, which is a sparse matrix. A permutation π of n elements is a rearrangement of the n elements. The permutation matrix P, of a permutation is the square matrix of order n whose entries are 0 everywhere except in row i, where the entry $\pi(i)$ is 1.

5.2 Generating Permutations

We can slightly vary the process discussed in Sect. 2.2, by permuting the output bits of the m LFSRs (that is the vector O_k), before feeding it into f. While trying to list down all the possible permutations of a set of elements, we may be interested in whether they have to be listed systematically (with some specific ordering) or randomly. One way to generate all the permutations is using factorial number system and Lehmer codes. It suffers from a drawback of converting back and forth from the permutations to such a number system. Ways of systematically generating all the permutations may become infeasible for large n. An exhaustive list of such algorithms are discussed in [6].

5.3 Including Permutations in Generation

After having access to all possible permutations, we can select one of these permutations and can apply it to the output vector O_k of the SpVM unit. The permutation used is kept secret, adding one more level of diffusion to the key stream generated. Instead of one permutation, two or more permutations can also be applied. As permutation matrices are sparse in nature, we can use the same SpVM unit for performing the permutation. We observed more randomness after applying a permutation over the input before fed to f. This is supported by the partial edit distance matrix (Table 2) for the target LFSR under the same conditions of Sect. 3 except that a permutation

Table 2 Partial edit distance matrix in the SpMV variant case

LFSR2↓ and LFSR3→	10101	11101	10111	11111	10001
10101	3	4	2	3	3
11101	4	2	2	3	4
10111	4	2	3	3	3
11111	3	2	3	3	3
10001	3	4	3	4	3

$C = \begin{pmatrix} 1\,0\,0 \\ 0\,0\,1 \\ 0\,1\,0 \end{pmatrix}$ is introduced. This is so, as the edit distance attack has no information about the permutation introduced in between.

6 Concluding Remarks

In this paper we have proposed a sparse matrix-vector multiplication-based bit/ parallel stream cipher. A variant of the proposed scheme is also discussed in connection with the permutation of the output bits. As in binary matrix multiplication, the possibility of compatibility mismatch with devices never arises, this generator can be implemented in almost all devices. Since research on Boolean functions, Sparse matrix representation, and hardware architecture of multipliers is proceeding at a high pace, we can expect improved , more efficient type of stream cipher generators in the near future. SpMV implementation of the new class of FCSR (feedback carry shift register) can also be explored. The usage of polynomial/multivariate polynomial matriccs can also be analyzed in stream cipher generation that will lead to a new class of stream ciphers for video encryption.

References

1. Menezes, A.J., van Oorschot, P.C., Vanstone, S.A.: Handbook of Applied Cryptography. CRC Press LCC, Boca Raton (1996)
2. Dalai, D.K., Gupta, K.C., Maitra, S.: Results on algebraic immunity for cryptographically significant boolean functions. In: The Proceedings of Progress in Cryptology, INDOCRYPT 2004, vol. 3348, pp 92–106. Springer, LNCS (2004)
3. Ekdhal, P: On LFSR based stream ciphers analysis and design, Ph.D. Thesis, Department of Information Technology, Lund University (2003)
4. Frederik, Armknecht: Matthias, Krause: Constructing Single and Multi-output Boolean Functions with Maximal Algebraic Immunity, In: The Proceedings of ICALP 2006, Venice,Italy vol. 4052. Springer, LNCS (2006)
5. Golic, J.: Edit distances and probabilities for correlation attacks on clock controlled combiners with memory. In: ACISP 96, Wollongong, Australia. IEEE Trans. Inf. Theory **47**(3), (2001)
6. Knuth, D.E.: The Art of Computer Programming, Combinatorial Algorithms, Part 1. Addison-Wesley Professional, New Jersey (2001)

7. Meier, W., Staffelbach, O.: Fast Correlation Attacks on Stream Ciphers. In: The Proceedings of Advances in Cryptography, EUROCRYPT'88, vol. 330, Springer, LNCS, Davos, Switzerland (1988)
8. Bell, N., Garland, M.: Effcient Sparse Matrix-Vector Multiplication on CUDA. NVIDIA Technical, Report, NVR-2008-004 (2008)
9. Lidl, R., Niederreiter, H.: Finite Fields, 2nd edn. Cambridge University Press, Cambridge (1997)
10. Rueppel, RA.: Analysis and Design of Stream Ciphers. Springer, Berlin (1986)
11. Sarkar, P: Efficient implementation of large stream cipher systems. In: The Proceedings of Cryptographic Hardware and Embedded Systems, CHES 2001, vol. 2162, pp. 319–332, Springer, LNCS, Paris, France (2001)
12. Sarkar, P., Maitra, S.: Efficient implementation of cryptographically useful large boolean functions. IEEE Trans. Comput. **52**, 410–417 (2003)
13. Williams, S., Oliker, L., Vuduc, R., Shalf, J., Yelick, K., Demmel, J.: Optimization of sparse matrix-vector multiplication on emerging multicore platforms. In: The Proceedings of ACM/IEEE Conference on Supercomputing (2007)
14. Wong, A., Bishop, W.: An efficient, parallel multikey encryption of compressed video streams. In: The proceedings of International Conference on Signal and Image Processing (2006)
15. Yousef, S.: Iterative Methods for Sparse Linear Systems, 2nd edn. SIAM (2003)
16. Yzelman, A.N.: Fast sparse matrix-vector multiplication by partitioning and reordering. Ph.D. dissertation, Utrecht University (2011)

Steganography-Based Secure Communication

Manjot Bhatia, Sunil Kumar Muttoo and M. P. S. Bhatia

Abstract Security and scalability are important issues for secure group communication in a grid environment. Secure communication involves confidentiality and authenticity of the user and the message between group members. To transmit data in a secure way, a secure and authenticated key transfer protocol should be applied. Key transfer protocol needs an entity responsible for user authentication and secures session keys. Recently, many researchers have proposed secure group communication methods based upon various cryptographic techniques. In this paper, we propose a secure key transfer protocol that uses the concepts of soft dipole representation of the image and steganography to establish secure communication between group of authenticated users. Steganography hides the existence of group keys in images. This protocol is more secure as compare to previously proposed encryption based group communication protocols. This protocol uses a centralized entity key management center (KMC). KMC generates the group key using soft dipoles of the images of the group members and broadcast it securely to all the communicating group members.

Keywords Grid security · Key management protocol · Secure group communication · Soft dipole representation · Communication security

---

M. Bhatia (✉) · S. K. Muttoo
Department of Computer Science, University of Delhi, New Delhi, India
e-mail: manjot_bhatia@hotmail.com

S. K. Muttoo
e-mail: skmuttoo@cs.du.ac.in

M. P. S. Bhatia
Department of Computer Engineering, Netaji Subhas Institute of Technology, New Delhi, India
e-mail: bhatia.mps@gmail.com

B. V. Babu et al. (eds.), *Proceedings of the Second International Conference on Soft Computing for Problem Solving (SocProS 2012), December 28–30, 2012*, Advances in Intelligent Systems and Computing 236, DOI: 10.1007/978-81-322-1602-5_83,

1 Introduction

One of the most important issues in grid environment is security. Since grid environment is based on Internet, various active attackers repeatedly explore security holes existing in hardware, software, processes, or systems to steal information or to distribute the network services [2]. Grid environment is a form of distributed environment in which resources are geographically distributed and owned by different individuals with different technologies. For secure grid environment, grid systems and applications require security functions to (i) provide stronger authentication solutions to user and resources (ii) protect grid applications and data from local applications (iii) protect local policies of different administrative domains (iv) provide secure group communication to users to coordinate and secure their group activities [2]. From these security requirements we focused on secure authenticated group communication environment for authorized users in a secure manner. As we know, secure authenticated group communication is an environment in which members of a group interested in sharing some information in a secure way so that no other person outside the group should be able to perform an attack. The communication among users forming a group must be confidential; this brings the need of protocol for secure group communication [2]. The various applications where secure group communication plays an important role are cyber forensics, multiparty military actions, doctors discussions on serious medical issues, law enforcement practice, and government decisions on critical issues. Apart from these, secure group communications can be used in distributed interactive simulations, interactive games, and real-time information services [2]. Group key management is an important issue for secure group communication. The secure group communication needs to consider the following security functions [2]:

- Group key generation: The group key is generated for carrying out confidential communication among the group members of that specified group.
- Group Key confidentiality: Broadcasting group key to the group members securely, so that it could be recovered by the authorized group members only.
- Message confidentiality: To ensure that the message should be read only by the specified receiver.
- Message verification: Each user verifies the authenticity of the message to ensure that the message was sent by the authorized group member of its group.

Most of the secure group communication protocols presented by the various researchers are either implementing key generation and key confidentiality functions or message confidentiality and message authentication functions. Since group communication involves communication and information sharing among several members, so above listed security functions can be provided only by establishing a secure group key or secure session key among the group members for confidential communication [2]. The group key should be generated by the trusted Key management center and distributed to all the members of the communicating group. To send the group key securely to the group members, it can be encrypted using another secret key. Once the group Key received, members of the group can communicate securely with each other. Several group key management protocols have been proposed by various researchers [2].

(1) Centralized Group Key Management Protocol: A centralized single entity as key management center is responsible for generating group key for the entire group, reducing storage requirements and computational effort of the clients. The threat lies in the failure of the single entity.
(2) Distributed Key Management Protocol: ln distributed key management protocol there is no explicit KMC, the members of the group communicate with each other and can generate the group key themselves. They do not need to depend on thc third party. Each member can contribute some information to generate the group key. So the security level has been raised but this method is suitable for a small group only. For large groups it is very time consuming to collect information from every user due to the scalability factor of the group users.
(3) Decentralized Key Management Protocol: In a decentralized architecture the management of a large group is divided among subgroup managers, trying to minimize the problem of concentrating the work in a single place [7, 26].

This paper is an implementation of our previous work in [2] that explains the proposed protocol for user authentication and secure group communication.

2 Related Work

Many researchers have proposed various protocols for secure group communication between grid entities. Grid security infrastructure (GSI) above all provides confidentiality and security for transferring sensitive information on the Internet [10]. Most of the researchers have used centralized group key management protocols for secure group communication. Sudha et al. [21] have proposed secret keys multiplication protocol based on modular polynomial arithmetic (SKMP), which eliminates the need for the encryption/decryption during the group re-keying. Valli et al. [22] have proposed a new technique (SGKP-l), using hybrid key trees, has certain advantages like secure channel establishment for the distribution of the key material, reducing the storage requirements and burden at each member, minimization of time requirement to become a new member of a group. The computational complexity further reduced using both the combination of public and private key cryptosystems. The dual level key management protocol proposed by Zoua et al. (DLKM) [26] uses access control polynomial (ACP) and one-way functions that provides flexibility, security, and hierarchical access control. Researchers used encryptions to update the group key for forward and backward secrecy, Harney [11] group key management protocol based on encryptions is of $O(n)$, where n is the size of group. Zhenga et al. [25] used identity-based signature (IBS) scheme for grid authentication. Park et al. [18] have proposed an ID-based key distribution scheme which is secure against session state reveal attacks and long-term key reveal attacks, this scheme offers the scalability, non-usage of additional cryptographic algorithms, and efficiency similar to those of the existing schemes. The distributed group key management protocol proposed by Steiner et al. [20], Burmester and Desmedt [3] are based on DH key agreement protocol. Li et al. [15] proposed a scalable service scheme using digital signatures and used Huffman binary tree to distribute and manage keys. Li et al. [16] have proposed an authenticated encryption mechanism for group communication in terms of the basic theory of threshold signature and the basic characteristics of group communication in grid. In this approach, each group member in the grid can verify the identity of the signer and hold the private key. Li et al. [17] have proposed the service infrastructure of middleware for pervasive grid. Ingle and Sivakumar [12] presented an extended grid security infrastructure (EGSI) that includes an authentication and access control scheme at virtual organization (VO) level for group communication in grid environment. EGSI introduced the concept of application class awareness in grid. This paper is organized as follows. We discuss introduction in Sect. 1 and related work in Sect. 2. In Sect. 3, we discuss model of our proposed protocol for secure group communication. Section 4 discusses security analysis of the proposed protocol.

3 Proposed Protocol: Secure Communication

The proposed architecture of secure protocol tries to fulfill the security constraints of the grid environment: single sign-on, secure group communication (Fig. 1).

3.1 Model of the Proposed Protocol

We proposed a secure group communication protocol which consists of total of m users $\mathbf{usr}_i (i = 1, 2, \ldots, m)$. Our proposed protocol consists of the following phases (Table 1):

(i) Registration Process

Each user needs to register with one centralized authority, key management center. During registration, users need to submit their images $I_i (i = 1, 2, \ldots, m)$ to the KMC as the part of the registration process through the secured channel. KMC stores the images $I_1, I_2 \ldots I_n$ of users. KMC assigns a unique identification number (uid) and unique password (pw) to every user. Each user's password generated by the KMC is a soft dipole representation (S_I) of user' image, i.e., $\mathrm{pw}_i = S_{I(i)}$, soft dipole representation of an image is a triple $S_I(d, \alpha, \beta)$ that uniquely represents the image [27]. Password (pw) of each user is embedded into its corresponding image I using any embedding algorithm given by Neil and Stefan [14] and stego image (sp) is transferred to its corresponding user.

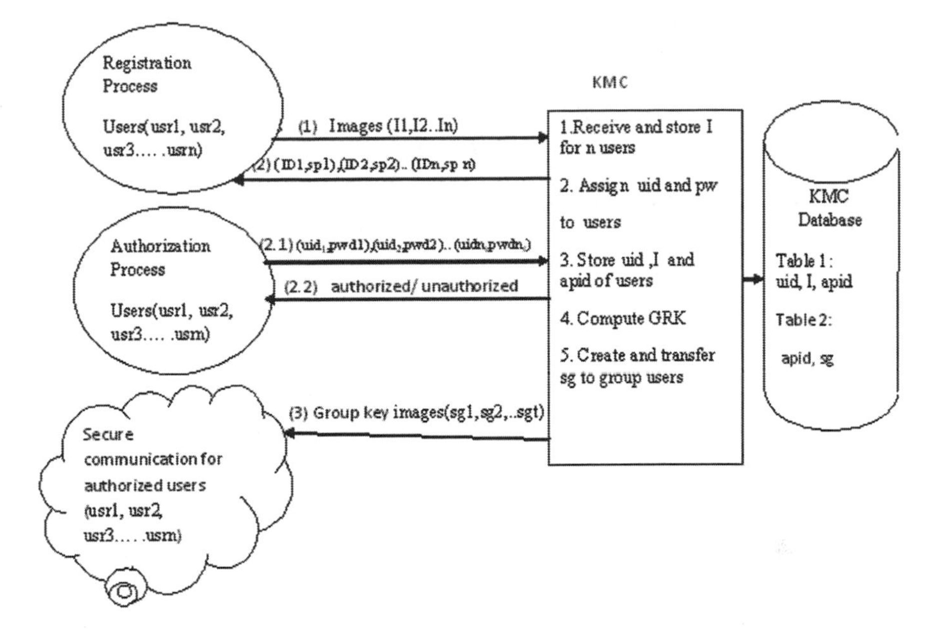

Fig. 1 Architecture of proposed model [2]

Table 1 Notations

usr_i	User
I_i	User's image
uid_i	Every registered user is assigned a unique identification number
pw_i	Every registered user is assigned a password
$Ausr_i$	Authorized user
$S_{I(i)}$	Soft dipole representation for image of ith user
sp_i	Stego-image transferred to each user with embedded password
sg_i	Stego-image transferred to each group member with embedded group key
apid	Unique application ID of applications required by users in groups
m	Total number of registered users
t	Total number of group members
n	nonce (number used only once)
GRK	Group key

Process: registration

```
KMC_Registration(int i)
{
Get I_i
uid_i=generate_uid(int i)//using dicomuid function of matlab
pw_(i)=compute_ S_I (image I_i)
sp_i=embed(S_I(i), I_i) // using any steganography technique
store uid_i, I_i into KMC DB
return uid_i,sp_i
}
```

Process: password computation

```
compute_ S_I (image I_i)
{
  N= total no. of pixels in an image
  Pi= 3.14159, sigma=1, pw=0
  A=sum of color values of 1st pixel at (0,0) position of the image
  B=sum of color values of dth pixel
  For d=0 to m // where m can be any value ranging 0 to N-1
  {
    For r=0 to N-1-d
    {
      C=(sum of color values of rth pixel-A)
      E=(sum of color values of (r+d)th pixel - B)
      Gc=probability density function based on value of C
      Ge= probability density function based on value of E
      pw=∑(Gc*Ge)
    }
  }
    Return pw
}
```

(ii) Authentication Process

For authentication process, KMC stores the uid and its corresponding image (I) for every user. When user enters the uid and pw, KMC computes the S_I values of the user from the already stored I. As S_I value of user's image is used as password, KMC compares the computed S_I with pw entered by the user for authentication.

Process: user login

```
for all users (i=1..m)
{
receive uid_i, sp_i
pw_i=extract(sp_i) // using any extraction technique
}
```

Process: user authentication

```
user_authentication(int i)
{
Get uid_i
Verify uid_i from the KMC DB
if exists(uid_i)
{
get "password" , pw_i
retrieve I_i for uid_i from KMC DB
S_I(i)= compute_ S_I (image I_i)
If pw_i= S_I(i)
result="authorized"
else
rcsult="unauthorized"
}
else
result="uid_i unauthorized"
return result for uid
}
```

(iii) Group Key Generation

After the authentication process is over KMC computes the group key for group communication of the users with common application ID (apid). Groups are identified by the unique application ID (apid). The registered users have to go through the authentication process for secure group communication. At the initial set up, when all the registered users of any particular group authenticates with KMC, group key for that group is computed and distributed to all the members of that group.

$$\mathrm{GRK}_j = \sum(S_{I(i)}) + n, k = (1, 2, \ldots, t_j), \quad i \in k$$

where k denotes the number of group members of the group [2] under consideration and n represents nonce.

(iv) Group Key Confidentiality

The group key is transferred securely and confidentially to the group members by hiding it in such a way that does not allow any message trapper to even detect the presence of GRK. To conceal the existence of GRK while transferring it to group members we use steganography. The security of group key in our protocol is kept by hiding the GRK into user's image (I) and transferring image I as GRK.

(v) Password/Group Key Extraction

Each group member usr_i knows the extraction method and can extract the password and group key from the received stego-image file using the extraction algorithm [14].

Process P3: Group Key generation

```
Group Key generation
{
select unique apid from KMC DB
for all apids
GRK(apid)=0;
for all user usr_i (i=1..m)
{
user_authentication (int i)
if result="authorized"
{
Select apid for uid_i from KMC DB
GRK(apid)=GRK(apid)+S_I(i)
}
unauthorized
}
}
```

Process P3: Group Key generation

```
for all apids from KMC DB
{
sg(apid)=embed (GRK(app_id), I(uid) ) //using any steganography technique
return sg(uid)
}
```

4 Security Analysis

Proposed protocol can handle following security attacks from two enemies: internal and external.

4.1 Insider Attack

Internal attack that a group member can extract the information from sg files of other group members. The sg files transferred to the t group members are protected with user's login password. Any *j*th user inside the group trying to gain access to the image file sg_i of the *i*th user would not be able to do as $sg_l, sg_2, \ldots, sg_m$ files for all the group users are locked with the unique password, i.e., pw.

4.2 Outsider Attack

KMC keeps the record of unique apid of every group ids of its corresponding group members in KMC database. Any other registered user sending the request for the group key needs to send the uid and apid of the application for which it is sending

Table 2 User's image, password computed and embedded, stego image, PSNR of IM and stego image, and password extracted

User's image	Password computed	Stego image (stmp)	PSNR	Password extracted
	4911739		88.6851	4911739
	49695168		87.9885	49695168
	29602972		Infinity	29602972
	48341515		91.1411	48341515
	10548923		Infinity	10548923
	7280785		95.4008	7280785
EXIT	3635416	EXIT	Infinity	3635416
	14759021		88.3967	14759021
	46550076		79.3712	46550076
	11089453		79.1484	11089453

the request. KMC checks the database for uids of the users for the received app_id. If that uid is not in the list, KMC will not respond for that request.

5 Simulation Result

The above-proposed protocol is divided into three phases: (i) user registration, (ii) user authentication, (iii) generating group key.

To simulate the first two phases of our protocol, we took the example of ten users: The simulation is done using java programming and Matlab. We used java for creating client and KMC server environment, submitting images to the KMC by the clients and computing passwords from the submitted images. Matlab programming is used for embedding and extracting password in user's image. After embedding password in original image, stego image is passed to the user as password. To compare the original images and the stego images, peak signal to noise ratio (PSNR) is calculated through Matlab code. The PSNR values in Table 2 shows that there is not any visible difference between the original and the stego images of the users.

6 Conclusion

We have proposed a secure group key generation protocol for secure group communication where every user needs to register with an image with centralized entity KMC. KMC is responsible for generating passwords and group keys for the users. The security of the password and group key is achieved with steganography based embedding algorithm. Only group users can extract the group key. In future we will work on completing its implementation part and try to enhance security features of secure group communication.

Appendix 1

Screen shots of simulation [28].

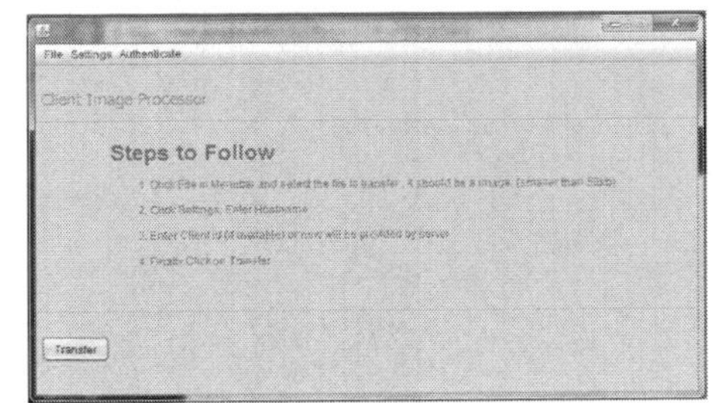

Client side screen

Login Screen

ID	IMAGE	APPID
16	[illegible]	5
17	[illegible]	5
18	[illegible]	5
19	[illegible]	374628
20	[illegible]	374628
21	[illegible]	374628
22	[illegible]	34
23	[illegible]	374628
24	[illegible]	5
25	[illegible]	5
26	[illegible]	5
27	[illegible]	5
28	[illegible]	5
29	[illegible]	5
30	[illegible]	5
31	[illegible]	5
32	[illegible]	5

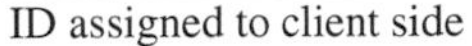

Client table in KMC database

Server generating password

ID assigned to client side

References

1. Aparna, R., Amberker, B.B.: Dynamic authenticated secure group communication. World Acad. Sci. Eng. Technol. **34** (2007)
2. Bhatia, M., Mutto, S.K., Bhatia, M.P.S.: Secure group communication protocol. Int. J. Adv. Eng. Sci. Technol. (2011)
3. Burmester, M., Desmedt, Y.G.: A secure and efficient conference key distribution system. In: Proceedings of Eurocrypt '94 Workshop Advances in Cryptology, pp. 275–286 (1994)
4. Chadramouli, R., Memon, N.: Analysis of LSB based image steganography techniques. IEEE (2001)
5. Chubb, C., Yellott, J.I.: Dipole statistics of discrete finite images: two visually motivated representation theorems. J. Opt. Soc. Am. A (2002)
6. Cody, E., Sharman, R., Rao, R.H., Upadhyaya, S., et al.: Security in grid computing: a review and synthesis. Decis. Supp. Syst. **44**(4), 749–764 (2008)

7. Foster, I., et al.: Grid services for distributed systems integration. IEEE Comput. Soc. **35**(6), 37–46 (2002)
8. Foster, I., et al.: In the Grid: Blueprint for a New Computing Infrastructure. Morgan Kaufmann, San Francisco (1999)
9. Foster, I., et al.: The anatomy of the grid: enabling scalable virtual organizations. Int. J. High Perform. Comput. Appl. ACM Digital Library, pp. 200–222 (2001)
10. Foster, I., Kesselman, C., Tsudik, G., Tuecke, S.: A security architecture for computational grids. In: Proceedings of ACM Conference on Computers and Security (1998)
11. Harney, H., Muckenhirn, C., Rivers, T.: Group Key Management Protocol (GKMP), Architecture RFC2094 (1997)
12. Ingle, R., Sivakumar, G.: EGSI: TGKA based security architecture for group communication in grid. In: Proceedings of the 10th IEEE/ACM International Conference on Cluster, Cloud and Grid Computing, May 17–20, IEEE Computer Society, Australia, pp. 34–42 (2010). doi:10.1109/CCGRID.2010.28
13. Johnson, N. F., Katzenbeisser Stefan, C.: A Survey of Steganographic Techniques. Artech House Books, London (2000)
14. Katzenbeisser, S., Petitcolas, F.A.P.: Information Hiding Techniques for Steganogarphy and Digital Watermarking. Artech House, London (2000)
15. Li, Y., Jin, H., Zou, D., Liu, S., Han, Z.: A scalable service scheme for secure group communication in grid. In: 31st Annual International Computer Software and Applications Conference (COMPSAC 2007), pp. 31–38 (2007)
16. Li, Y., Xu, X., Wan, J., Jin, H., Han, Z.: An authenticated encryption mechanism for secure group communication in grid. In: Proceedings of the International Conference on Internet Computing in Science and Engineering, Jan. 28–29, USA, pp. 298–305 (2008). doi:10.1109/ICICSE.2008.80
17. Li, Y., Jin, H., Han, Z., Liu, S., et al.: A secure mechanism of group communication for pervasive grid. Int. J. Ad Hoc Ubiquitous Comput. **4**, 344–353 (2009). doi:10.1504/IJAHUC.2009.028662
18. Park, H., Yi, W.S., Lee, G.: Simple ID-based key distribution scheme. In: Fifth International Conference on Internet and Web Applications and Services **2010**, 369–373 (2010)
19. Steganographic techniques, a brief survey. Department of computer science, Wellesley College
20. Steiner, M., Tsudik, G., Waidner, M.: Diffie-Hellman key distribution extended to group communication. In: Proceedings of Third ACM Conference on Computer and Communication Security (CCS '96), pp. 31–37 (1996)
21. Sudha, S., Samsudin, A., Alia, M.A.: Group rekeying protocol based on modular polynomial arithmetic over Galois field GF (2n). Am. J. Appl. Sci. **6**, 1714–1717 (2009). doi:10.3844/ajassp.2009.1714.1717
22. Valli, V., Kumari, D., NagaRaju, V., Soumy, K., Raju, K.V.S.V.N.: Secure group key distribution using hybrid cryptosystem. In: Proceedings of the 2nd International Conference on Machine Learning and Computing, IEEE Computer Society, Washington, pp. 188–192 (2010). doi:10.1109/ICMLC.2010.4110
23. Wikipedia, Digital Image processing, available at: http://en.wikipedia.org/wiki/Digital_image_processing, accessed on March 2012
24. Xu, S.: On the Security of Group Communication Schemes
25. Zhenga, Y., Wanga, H.Y., Wang, R., et al.: Grid authentication from identity-based cryptography without random oracles. J. China Univ. Posts Telecommun., Elsevier **15**(4), 55–59 (2008)
26. Zoua X., Dai Y.S., Rana X.: Dual-level key management for secure grid communication in dynamic and hierarchical groups. Future Gen. Comput. Syst., Science Direct, **23**(6), 776–78 (2007)

Printed by Publishers' Graphics LLC